12/15/89

Genetic Recombination

Genetic Recombination

Edited by

Raju Kucherlapati

Department of Genetics
University of Illinois College of Medicine
Chicago, Illinois

Gerald R. Smith

Fred Hutchinson Cancer Research Center
Seattle, Washington

American Society for Microbiology
Washington, D.C.

Library of Congress Cataloging-in-Publication Data

Genetic recombination.

  Includes bibliographies and index.
  1. Genetic recombination. I. Kucherlapati, Raju.
II. Smith, Gerald R. (Gerald Ralph), 1944–
[DNLM: 1. Recombination, Genetic. QH 443 G3281]
QH443.G455   1988   575.1'3   88-24227
ISBN 1-55581-004-7

To our teachers

Contents

Authors

Neca D. Allgood
Department of Molecular Biology, Princeton University, Princeton, NJ 08544

Peter A. Bullock
Memorial Sloan-Kettering Cancer Center, New York, NY 10021

Adelaide T. C. Carpenter
Department of Biology, University of California, San Diego, La Jolla, CA 92093

James J. Champoux
Department of Microbiology, School of Medicine, University of Washington, Seattle, WA 98195

Michael M. Cox
Department of Biochemistry, College of Agricultural and Life Sciences, University of Wisconsin–Madison, Madison, WI 53706

Edward H. Egelman
Department of Molecular Biophysics and Biochemistry, Yale University, New Haven, CT 06511

Peter Engler
Department of Molecular Genetics and Cell Biology, University of Chicago, Chicago, IL 60637

Craig N. Giroux
Cellular and Genetic Toxicology Branch, NIEHS, Research Triangle Park, NC 27709

Nigel D. F. Grindley
Department of Molecular Biophysics and Biochemistry, Yale University, New Haven, CT 06510

P. J. Hastings
Department of Genetics, University of Alberta, Edmonton, Alberta, Canada T6G 2E9

Graham F. Hatfull
Department of Molecular Biophysics and Biochemistry, Yale University, New Haven, CT 06510

R. Scott Hawley
Department of Genetics, Albert Einstein College of Medicine, Bronx, NY 10461

Raju Kucherlapati
Department of Genetics, University of Illinois College of Medicine, Chicago, IL 60612

Sanford A. Lacks
Biology Department, Brookhaven National Laboratory, Upton, NY 11973

Suresh K. Mahajan
Molecular Biology and Agriculture Division, Bhabha Atomic Research Centre, Bombay 400 085, India

Peter D. Moore
Department of Microbiology and Immunology, University of Illinois College of Medicine, Chicago, IL 60612

Gisela Mosig
Department of Molecular Biology, Vanderbilt University, Nashville, TN 37235

Ronald D. Porter
Department of Molecular and Cell Biology, Pennsylvania State University, University Park, PA 16802

Charles M. Radding
Department of Human Genetics, Yale University School of Medicine, New Haven, CT 06510

Miroslav Radman
Institute Jacques Monod, 75251 Paris Cedex 05, France

David Roth
Laboratory of Pathology, National Cancer Institute, Bethesda, MD 20892

Brent L. Seaton
Department of Biology and Center for Molecular Genetics, University of California, San Diego, La Jolla, CA 92093

Thomas J. Silhavy
Department of Molecular Biology, Princeton University, Princeton, NJ 08544

Anna Marie Skalka
Institute for Cancer Research, Fox Chase Cancer Center, Philadelphia, PA 19111

Andrzej Stasiak
Institute for Cell Biology, E.T.H.-Honggerberg, 8093-Zurich, Switzerland

Ursula Storb
Department of Molecular Genetics and Cell Biology, University of Chicago, Chicago, IL 60637

Jeffery N. Strathern
BRI-Basic Research Program, NCI-Frederick Cancer Research Facility, Frederick, MD 21701

Suresh Subramani
Department of Biology and Center for Molecular Genetics, University of California, San Diego, La Jolla, CA 92093

Michael Syvanen
Department of Medical Microbiology and Immunology, School of Medicine, University of California, Davis, CA 95616

Andrew F. Taylor
Hutchinson Cancer Center, Seattle, WA 98104

Larry Thompson
Biomedical Sciences Division, Lawrence Livermore National Laboratory, Livermore, CA 94550

John Wilson
Department of Biochemistry, Baylor College of Medicine, Houston, TX 77030

Preface

In the early 20th century genetic recombination was recognized as a key feature in generating species diversity in eucaryotic organisms. In this process, homologous chromosomes align, and the exchange of genetic information at one or more regions along the chromosomes produces new combinations of genes. In the mid-20th century it became clear that a similar exchange of genetic information (and material) occurs in procaryotes as well. Recombination requiring extensive nucleotide sequence homology in the DNA of the chromosomes is referred to as homologous recombination. It is important not only in generating species diversity, but also in the repair of damaged chromosomes and, in eucaryotes, for the faithful segregation of homologous chromosomes during meiosis.

Numerous organisms have developed specialized mechanisms by which recombination occurs at specific chromosomal sites that share little or no sequence homology. These site-specific recombination events include insertion of certain viruses into procaryotic and eucaryotic chromosomes, the movement of transposable elements, and the rearrangement of vertebrate immunoglobulin genes. Site-specific recombination is important for the generation of new arrangements of genes in chromosomes and for the proper timing of gene expression in the development of numerous organisms.

In a third class of events, illegitimate recombination, DNA recombines without apparent regard to special sites or extensive sequence homology. Such recombination generates deletions, insertions, and other rearrangements. Although its biological role is unclear, illegitimate recombination may reflect errors of replication or "fail-safe" measures that cells use to rescue otherwise dead chromosomes.

Studies of each of these classes of recombination have been intensive and have revealed a fundamental fluidity of genomes. These studies have also revealed a remarkable similarity in the mechanisms used for recombination in diverse organisms, many of which have proven particularly useful for the study of different aspects of recombination. For example, in certain fungi all of the products of a single meiosis can be recovered; the frequencies and the types of recombinants observed have provided much information about the genetics of recombination. The identification and study of recombination mutants and the purification of the proteins required for recombination in certain bacteria, especially *Escherichia coli,* have provided methods to study the biochemical mechanisms of recombination. The studies of viral integration in bacteria and immunoglobulin gene rearrangements in mammals are providing new insights into site-specific recombination. Studies of homologous recombination in mammalian cells promise to facilitate the genetic manipulation of mammals.

This book is intended to illustrate the rich diversity of approaches that are used to study recombination and what we have learned from these approaches. It is not intended to be comprehensive or exhaustive, and we realize that certain areas of recombination are not represented. The extensive references at the end of each chapter are intended to lead the reader to the pertinent literature on the areas not explicitly covered. We have included chapters dealing with biological, genetic, and biochemical aspects of recombination in both procaryotes and eucaryotes. We have attempted to provide adequate background on each topic so that students new to the field can read this book after a course in genetics or molecular biology. Each chapter also provides an up-to-date account of the subject designed to be useful for researchers in the field.

We are indebted to many people for helping us produce this book. Above all, we thank the authors for their timely contributions and patient revisions. At the American Society for Microbiology publications department, Kirk Jensen initiated the idea, Dennis Burke helped us with the organization, and Susan Birch was an able production editor. Many of the ideas in this book were generated at recent recombination meetings sponsored by the Cold Spring Harbor Laboratory, the European Molecular Biology Organization, and the Federation of American Societies for Experimental Biology. The recombination community owes them a great deal for facilitating the gathering of researchers for the free exchange of ideas. We hope that this book also helps in this exchange.

Raju Kucherlapati
Gerald R. Smith

Chapter 1

Modes of Gene Transfer in Bacteria

Ronald D. Porter

I. INTRODUCTION

The transfer of genetic information between individual microorganisms occurs frequently in nature. The ability of microorganisms to exchange genetic information is also frequently exploited for a variety of purposes in the laboratory. Much of classical bacterial genetics involves the introduction of DNA that must recombine with a replicon (typically the chromosome) in the recipient cell in order to be expressed and stably inherited. In other situations the introduced donor DNA functions as an independent replicon in the absence of recombination. Whereas gene transfer events that depend on recombination between homologous sequences typically require that the donor and recipient cells be closely related, those events that involve recombination-independent gene transfer can sometimes occur between individual cells of different species, or even different genera.

Although recombination usually includes all facets of the introduction and stabilization of new genetic information in a recipient cell (Lederberg, 1955), gene

1

transfer and recombination will be treated separately here. As in a previous review (Low and Porter, 1978), genetic recombination will be defined as the reassortment of a series of nucleotides along nucleic acid molecules (Clark, 1971). Reassortments of nucleotides can occur within a single DNA molecule or between two or more DNA molecules. This article focuses on the types of gene transfer that occur between bacterial cells and on the mechanisms involved. Recombination is also discussed, since the mode by which the donor DNA is introduced into the recipient cell often has a major effect on recombination frequency and/or the types of recombination events that are either required or permissible. In particular, the gene transfer-dependent production of single strandedness in donor DNA often has a major impact on recombination frequency.

II. MODES OF RECOMBINATION

A large number of terms have been used to describe different types of recombination events. For purposes of consistency, a previously proposed scheme involving three basic types of recombination (Low and Porter, 1978) will be used; these are (i) general, (ii) site-specific (single or double), and (iii) illegitimate recombination. Although in some situations these distinctions are not totally clear, this classification scheme is generally useful.

A. General Recombination

General recombination typically involves extended stretches of nucleotide homology. Such recombination is called "homologous" or, less frequently, "generalized" (Dove, 1968; Gottesman and Yarmolinsky, 1968), "normal" (Echols et al., 1968; Franklin, 1971), "equal" (Campbell, 1964), "genetic" (Hotchkiss, 1974), "*rec* dependent" (Franklin, 1971), "chromosomal" (Gottesman and Yarmolinsky, 1968), "legitimate" (Falkow, 1975), "nonspecific" (Gottesman and Yarmolinsky, 1968; Signer and Beckwith, 1966), "equational" (Campbell, 1960), "mating" (Gottesman and Yarmolinsky, 1968), or "vegetative" (Hotchkiss, 1974; Gingery and Echols, 1967) (the last term referring to phage recombination). With the exception of phage recombination, general recombination in *Escherichia coli* is usually dependent on the RecA protein and other Rec functions active in the various *E. coli* recombination pathways (see Mahajan, this volume; Radding, this volume). Genes analogous to *recA* occur in numerous gram-negative and gram-positive bacteria (Better and Helinski, 1983; Eitner et al., 1982; Finch et al., 1986; Genthner and Wall, 1984; Goldberg and Mekalanos, 1986; Hamood et al., 1986; Keener et al., 1984; Kokjohn and Miller, 1985; Koomey and Falkow, 1987; Miles et al., 1986; Resnick and Nelson, 1988; Yagi and Clewell, 1980; Marrero and Yasbin, 1988) and in cyanobacteria (Owttrim and Coleman, 1987; Murphy et al., 1987).

Since recombination can occur between very closely spaced mutations (Benzer, 1957; Yanofsky et al., 1964), the "crossover points" in general recombination are often thought to be randomly distributed. Although the precise location at which recombination junctions are resolved may well be essentially random, it seems quite unlikely that the initiation of recombination or the regions in which resolution occurs, or both, are totally random (for reviews, see Stahl, 1979; G. R. Smith, *in* K. B. Low, ed., *The Recombination of Genetic Material*, in press; Mahajan, this volume; Mosig, this volume; Taylor, this volume). Nonrandom breakage of DNA yields nonrandom distribution of exchanges. Such breakage may occur prior to or during transfer of donor DNA into a recipient cell; as discussed later, Hfr donor DNA is broken at a special *oriT* site in preparation for conjugal transfer in *E. coli*, and packaging of generalized transducing DNA by bacteriophages P1 in *E. coli* and P22 in *Salmonella typhimurium* appears to be initiated at sites resembling the phages' *pac* sites for packaging viral DNA. In addition, "hot spots" of recombination enhance recombination near the sites by inducing breaks in the DNA for the initiation or resolution of genetic exchange. Such nonrandom breakage and exchange produce a nonconstant relation between genetic distance (recombination frequency) and physical distance (for further discussion, see Mosig, this volume). It has been proposed that *fre* sites are required for the initiation of RecF pathway-promoted recombination following Hfr conjugation in *E. coli* (Bresler et al., 1978, 1981a, b), but the location and mechanism of these sites are uncertain. Chi sites, first observed in vegetative crosses of bacteriophage λ whose own recombination functions had been disabled, locally enhance RecBCD pathway-promoted recombination in *E. coli* cells. Chi sites may stimulate either the formation (Smith et al., 1984; Taylor et al., 1985) or the resolution (Stahl, 1979; Rosenberg, 1987) of recombination intermediates, or both (see Taylor, this volume; Mosig, this volume).

General recombination is most efficient when it operates on long stretches of highly homologous DNA, but *recA*-dependent recombination occurs at low frequency between DNA molecules with homologies as short as about 20 base pairs (Watt et al., 1985; Shen and Huang, 1986) or with a significant amount of heterology. Although intergeneric general recombination is extremely rare, the higher efficiencies of interspecific recombination are often used as measures of homology, and hence relatedness, between different species.

B. Site-Specific Recombination

Recombination can also occur at significant frequencies between DNA molecules sharing little or no DNA sequence homology. Such recombination typically involves one or more special proteins not involved in general recombination and does not require the product of the *E. coli recA* or other genes (or analogous proteins in other organisms) needed for general recombination. In contrast to general recombination, recombination events of this class normally occur at highly preferred sites (on one or both molecules) and are therefore termed site specific.

The classic example of site-specific recombination is the integration of the circularized bacteriophage λ genome into the bacterial chromosome and its subsequent excision (Weil and Signer, 1968; Gottesman and Yarmolinsky, 1968; Echols et al., 1968), which normally requires highly preferred sites on both of the

participating DNA molecules. Such double-site-specific recombination also occurs in the resolution of certain classes of transpositional cointegrates (see
Grindley and Reed, 1985) and in the partitioning of plasmid DNA into daughter
cells (see Craig and Kleckner, 1987; Syvanen, this volume; Hatfull and Grindley,
this volume). In other cases, a special site is required on only one of the
participating DNA molecules; examples of such single-site-specific recombination
include the insertion of transposable elements and the integration of bacteriophage Mu. Although integration of these elements can occur at many sites in the
target DNA molecule, it is not totally random and the degree of randomness
depends on the particular element (see Kleckner, 1981).

C. Illegitimate Recombination

Illegitimate recombination includes those events not classified as general or
site specific. These events require little or no DNA sequence homology and yet do
not require special sites. They do not appear to require the proteins promoting
general or site-specific recombination. Illegitimate recombination also normally
occurs at a considerably lower frequency than either general or site-specific
recombination (see Allgood and Silhavy, this volume).

D. Other Terminology

The term "addition" recombination is used when an entire circular DNA
molecule is incorporated into another DNA molecule by general recombination.
For example, in specialized transduction the transducing phage DNA molecule
may be inserted into homologous chromosomal or plasmid DNA. Although the
end products of addition recombination and site-specific recombination at the
phage attachment site ("integrative" recombination) may be physically indistinguishable if the *att* site is adjacent to the transduced DNA, addition recombination
events can be intragenic while integrative recombination events cannot. In
"substitution" recombination, DNA from the transducing phage DNA replaces a
homologous region of the recipient chromosome or plasmid. The replaced
recipient DNA and the remainder of the transducing phage DNA are lost.
Substitution recombinants remain haploid, while addition or site-specific recombinants become merodiploid for the chromosomal or plasmid sequence carried by
the specialized transducing phage. Generalized transduction, Hfr conjugation,
and transformation with chromosomal DNA fragments essentially always produce
substitution recombinants, as the DNA fragments introduced into the recipient
cell in those situations do not normally circularize, as required for addition
recombination.

III. MODES OF GENE TRANSFER

A. Transformation

In transformation, naked DNA molecules are taken up by the bacterial cell.
The internalized donor DNA may be incorporated into the chromosome or a

plasmid in the recipient cell by recombination, or it may become established as a replicon if it contains an origin of replication and can circularize. For the cells to take up DNA, they must achieve a state called competence. This state occurs naturally in some organisms (physiological competence) and can be artificially produced in others (artificial competence) (for general reviews on transformation, see Lacks, 1977; Fox, 1978; Low and Porter, 1978; Smith et al., 1981; Goodgal, 1982; Lacks, this volume).

Depending on the nature of the donor DNA, there are three basic types of transformation: replacement, plasmid, and facilitated plasmid transformation. In replacement transformation the donor DNA is substituted for homologous sequences in the chromosome or a plasmid in the recipient cell. The latter case is sometimes called plasmid marker rescue. Transformation with a nonreplicating plasmid that contains homology to DNA within the recipient cell may result in replacement transformation, or the entire donor DNA molecule may be incorporated by addition recombination. As this latter situation involves a somewhat different set of constraints, it will be discussed separately.

Plasmid transformation occurs when a new plasmid replicon is established in a recipient cell without recombination between donor and recipient DNA. If the donor plasmid DNA contains homology to DNA (chromosomal or plasmid) in the recipient cell, recombination may participate in the establishment of the plasmid replicon. In this case, the process is called facilitated plasmid transformation. A replicating plasmid containing homology with DNA in the recipient cell may give rise to both plasmid transformation and facilitated plasmid transformation in the same culture. Plasmid transformation appears to be mechanistically distinct from replacement transformation, while facilitated plasmid transformation has features in common with both of the other types of transformation.

1. Replacement transformation

Physiological competence. Transformation does not appear to be common in bacteria, and most studies have used a few strains of *Streptococcus* spp. (primarily *S. pneumoniae*, or pneumococci, and *S. sanguis*), *Bacillus* spp. (primarily *B. subtilis*), *Haemophilus* spp. (primarily *H. influenzae* and *H. parainfluenzae*), *Neisseria gonorrhoeae*, and some of the cyanobacteria (see Porter, 1986).

Development of competence. Competence for transformation usually requires complex physiological changes that occur at particular stages of growth in special media for each organism. Exponentially growing streptococci dramatically shift their pattern of protein synthesis when they become competent (Morrison and Baker, 1979; Raina and Ravin, 1980), and competence of essentially all of the cells in a culture occurs after an initial release of an autocatalytic competence-inducing factor (Tomasz, 1966). This phenomenon appears to be similar to a hormonal response, as competence can be induced in a noncompetent culture by the addition of competence factor to the medium. A pneumococcus mutant that fails to become competent because of an inability to release competence factor can be made fully competent by the addition of sufficient exogenous competence factor (Morrison et al., 1984). An autolysin activity released from cells treated with competence factor appears to be involved in the unmasking of DNA binding or uptake sites, or both, as autolysin-defective strains of *S. pneumoniae* are incapable of achieving competence (Seto and Tomasz, 1975).

Nucleic acid synthesis is reduced at the onset of competence in *B. subtilis* (Dooley et al., 1971), and competent cells are therefore less dense than noncompetent cells. This density difference has allowed the isolation on Renografin gradients of only that fraction of the cells that are competent in a culture (Cahn and Fox, 1968; Hadden and Nester, 1968). Although the cells remain relatively quiescent during competence, a number of SOS-like phenomena are induced (Yasbin et al., 1975; Love and Yasbin, 1985). Like *S. pneumoniae*, *B. subtilis* synthesizes competence factors that may involve an autolysin activity (Akrigg and Ayad, 1970).

In *Haemophilus* spp. competence also involves major metabolic shifts and is generally associated with nondividing cells (Spencer and Herriott, 1965; Herriott et al., 1970). Piliation and competence appear to be intimately related in *N. gonorrhoeae*, as piliated cells are constitutively competent and nonpiliated cells fail to achieve competence (Sparling, 1966; Biswas et al., 1977). Competence appears to be constitutive in exponentially growing cells for most of the cyanobacteria examined (see Porter, 1986).

DNA binding and uptake. Most physiologically competent bacteria take up primarily double-stranded donor DNA, but the uptake of limited amounts of single-stranded DNA may also occur. Upon binding to competent pneumococci the double-stranded DNA is nicked (Lacks and Greenberg, 1976), but the binding of DNA appears to be independent of DNA uptake. Firstly, in the presence of EDTA, cells bind DNA but cannot internalize it (Seto and Tomasz, 1974). Secondly, pneumococcus *end* mutants bind DNA normally but internalize it poorly or not at all (Lacks et al., 1974, 1975). Thirdly, in pneumococcal transfection (transformation with phage DNA) DNA uptake occurs in the absence of surface binding at very high DNA concentrations (Porter and Guild, 1978). DNA binding may increase the effective DNA concentration near the uptake sites. There are about 50 receptor sites on the surface of a competent pneumococcal cell (see Lacks, 1977), and these receptor sites presumably involve the appropriate juxtaposition of DNA-binding and DNA-uptake sites.

The specificity of DNA uptake appears to differ greatly for different bacteria. The well-characterized gram-positive transformable bacteria (*Streptococcus* and *Bacillus* spp.) take up almost any double-stranded DNA present in the culture medium, while the well-characterized gram-negative transformable bacteria (*Haemophilus* and *Neisseria* spp.) largely take up only DNA from the same or closely related species (Scocca et al., 1974; Dougherty et al., 1979). The cyanobacteria, although gram negative, do not demonstrate a significant uptake specificity (see Porter, 1986).

During uptake by *S. pneumoniae* and *B. subtilis*, DNA undergoes double-stranded breakage and is considerably reduced in size even before it enters the cell (Dubnau and Cirigliano, 1972; Morrison and Guild, 1973a, b). The double-stranded break may result from the introduction of a second nick roughly opposite the nick introduced by binding. The major endonuclease activity of *S. pneumoniae* is required for DNA uptake (Lacks et al., 1974, 1975), and this membrane-bound protein appears to pull one strand into the cell while the complementary strand is degraded to oligonucleotides which are released into the growth medium (Morrison and Guild, 1973a; Lacks and Greenberg, 1973). *B. subtilis* contains an envelope-associated nuclease that may play a similar role in DNA uptake (Mulder and Venema, 1982). In both organisms, the internalized donor DNA is found as single-stranded DNA (Lacks, 1962; Piechowska and Fox, 1971). At this stage, called DNA eclipse (Fox, 1960; Venema et al., 1965; Ghei and Lacks, 1967), the internalized donor DNA has temporarily lost its ability to serve as donor DNA because single-stranded DNA is such a poor substrate for DNA binding and uptake. Once the genetic information in the internalized donor DNA has been restored to double strandedness by recombination with the genome, it regains donor transforming activity.

The species-specific uptake of DNA by *Haemophilus* and *Neisseria* cells proceeds by a different mechanism. A specific nucleotide sequence within the DNA is required for internalization in *Haemophilus* cells. This 11-base-pair sequence (5′-AAGTGCGGTCA-3′) (Danner et al., 1980) occurs about 600 times in the *Haemophilus* genome (Sisco and Smith, 1979). This frequency is much higher than would be expected on a purely random basis and appears to account for the poor ability of DNA from other organisms to be taken up by *Haemophilus* cells. Synthetic DNA with this sequence introduced at different sites in pBR322 led to a highly variable increase in uptake ability, suggesting that flanking sequences also play a role in determining uptake (Danner et al., 1982).

The absence of DNA eclipse in *Haemophilus* transformation (Voll and Goodgal, 1961; Stuy, 1965) has been explained by the role of special structures in DNA uptake. Double-stranded donor DNA is first taken into membrane-bound vesicles called transformasomes, where it becomes resistant to externally added DNase (Kahn et al., 1982, 1983). It appears, however, that passage from the transformasome to the cytoplasm involves linear transport starting at a double-stranded end; one strand is degraded while the other strand enters, as in pneumococcal uptake (Barany et al., 1983). Although no enzyme involved in this passage has been identified in *Haemophilus* spp., the DNA substrate requirements for such an enzyme appear to be those of the ATP-dependent DNase, or exonuclease V, of *Haemophilus* spp. (Barany et al., 1983). Although *Haemophilus add* mutants that lack the nuclease activity of that enzyme are still transformable (LeClerc and Setlow, 1974), another activity of the enzyme, such as DNA unwinding, may still be present in that mutant (for further discussion, see below and Taylor, this volume).

Recombination with recipient cell sequences. Donor and recipient DNAs interact to yield a product that is solely heteroduplex DNA in *S. pneumoniae* (Guild and Robison, 1963; Fox and Allen, 1964), *B. subtilis* (Bodmer and Ganesan, 1964), and *H. influenzae* (Notani and Goodgal, 1966). The internalized single-stranded donor DNA in both *S. pneumoniae* and *S. sanguis* associates with one of the proteins made in large amounts during the development of competence (Morrison, 1977; Raina and Ravin, 1978), and a comparable eclipse complex may also be formed in *B. subtilis* (Eisenstadt et al., 1975). With both *S. pneumoniae* and *B. subtilis*, the final heteroduplex product is approximately as long as the internalized single-stranded DNA fragment found in the eclipse complex. No single-stranded intermediate has been described for *H. influenzae*; the internalized donor DNA may be incorporated into recipient DNA to form a heteroduplex DNA product as it enters the cytoplasm from the transformasome (Barany et al., 1983).

In the pneumococcal eclipse complex the protein confers partial resistance to a variety of nucleases (Morrison and Mannarelli, 1979); one of the roles of this complex may be to protect the donor DNA from rapid degradation. The pneumococcal eclipse complex appears to be an integration intermediate (Vijayakumar and Morrison, 1983), and its protein component may also participate in DNA-DNA interactions such as synapsis and strand transfer, as promoted by the *E. coli* RecA protein (see Radding, this volume). This speculation is consistent with the similar kinetics of donor DNA integration in *S. pneumoniae* and D-loop formation in vitro (Shoemaker and Guild, 1972).

In both *B. subtilis* (Harris and Barr, 1971) and *H. influenzae* (LeClerc and Setlow, 1975; McCarthy and Kupfer, 1987), single-stranded gaps in the recipient DNA of competent cells have been described and proposed to be sites for the integration of single-stranded donor DNA. A connection between gap formation and transformability had been inferred from the inability of *rec-2* mutants of *H. influenzae* to make gaps or be transformed (LeClerc and Setlow, 1975), but it has subsequently been shown that DNA fails to exit from the transformasomes in these mutants (Barouki and Smith, 1985). The role, if any, that the single-stranded gaps in recipient DNA actually play in transformation remains to be elucidated.

The product of the *recA* gene of *E. coli* can promote both D-loop formation and the renaturation of complementary single-stranded DNA (see Radding, this volume), and either of these activities could be involved in generating the heteroduplex product of transformation. The *recE* gene of *B. subtilis* (Lovett and Roberts, 1985; Love and Yasbin, 1986; Marrero and Yasbin, 1988) and the *rec-1* gene of *H. influenzae* (Setlow et al., 1972) appear by a variety of criteria to be analogous to the *recA* gene of *E. coli*, and mutations in these genes abolish or drastically reduce replacement transformation in their respective organisms (Setlow et al., 1972; Dubnau et al., 1973). Although the pneumococcal Rec mutants have not been characterized in detail (Morrison et al., 1983), the kinetics of integration in that organism suggest recombination by D-loop formation, as mentioned above. It is possible, however, that RecA-like proteins generate heteroduplex product by a renaturation mechanism in the *Bacillus* and *Haemophilus* systems.

Artificial competence. A variety of regimens for producing artificial competence in numerous bacterial species have been described. CaCl$_2$ treatment of *E. coli* first allowed transfection (Mandel and Higa, 1970) and later allowed replacement (Oishi and Cosloy, 1972; Cosloy and Oishi, 1973; Wackernagel, 1973) and plasmid transformation (Cohen et al., 1972). Although CaCl$_2$ treatment and other artificial competence protocols are widely used for plasmid transformation, they are not widely used for replacement transformation because of their very low efficiencies.

Very few replacement transformants occur in Rec$^+$ strains of *E. coli* (RecBCD pathway strains), but more occur with strains utilizing the RecE or RecF pathway (Oishi and Cosloy, 1972; Cosloy and Oishi, 1973; Wackernagel, 1973; for a description of these pathways, see Mahajan, this volume). Transformation frequencies are greater with specialized transducing phage DNA than with chromosomal DNA as donor, as a consequence of increased gene dosage (Oishi and Cosloy, 1974). *recBC* mutants yield more transformants than Rec$^+$ strains, presumably because the RecBCD enzyme degrades linear DNA in the Rec$^+$ strains (for a description of the RecBCD enzyme, see Taylor, this volume). Transformation frequencies are even greater in *recBC sbcA* and *recBC sbcB* mutants in which the RecE or RecF recombination pathway is active. Because the primary problem in achieving transformants in a RecBCD pathway strain may be the rapid disappearance of the donor DNA, this observation does not necessarily imply that the RecE and RecF pathways are actually that much more efficient than the RecBCD pathway in carrying out the necessary recombination events.

The extremely low transformation efficiencies with CaCl$_2$-treated *E. coli* cells have precluded detailed analysis of replacement transformation mechanism. Although linkage following transformation has been described (Hoekstra et al., 1976), physical studies have not been possible. The cell apparently cannot efficiently utilize the double-stranded donor DNA. This sharply contrasts with replacement transformation in physiologically competent organisms, in which the donor DNA is converted to single-stranded DNA before recombination. As discussed below, transduction also presents the cell with a double-stranded donor DNA and produces lower recombination frequencies than occur in those situations in which single-stranded donor DNA is produced. Although transformation of *E. coli* with specialized transducing phage DNA yields fewer recombinants than a parallel transduction experiment (Wackernagel and Radding, 1974), the mode of donor DNA introduction may not play a major role in the observed difference in frequency. Degradation of the donor DNA by the RecBCD enzyme presumably occurs in transformation, but is prevented by the circularization of the phage DNA molecule in specialized transduction. Since the amount of DNA surviving in the two situations is undetermined, it is unclear whether the mechanism and frequency of recombination per unit of surviving donor DNA actually differ.

2. Plasmid transformation

Physiological competence. The binding and uptake of circular plasmid DNA appears to occur by the same mechanisms as for linear chromosomal fragments (Lacks, 1979), as described above. Plasmid transformation may therefore occur by the reassembly of internalized single-stranded DNA fragments, as envisioned for transfection in both *B. subtilis* (Loveday and Fox, 1978) and *S. pneumoniae* (Porter and Guild, 1978). The relatively low frequency of plasmid transformation may, however, reflect low-frequency internalization of double-stranded DNA. In *N. gonorrhoeae*, most of the donor plasmid DNA is cut at nonspecific sites into linear double-stranded molecules during uptake (Biswas et al., 1986), but some intact circles may also enter. Either repair of linearized molecules or the occasional entry of intact circles may produce the plasmid transformants. In *H. influenzae*, plasmid transformation is thought to occur only when a circular molecule is occasionally internalized (Pifer, 1986). Whether these molecules are converted to single strands during a later stage of the uptake process is unknown. While monomeric plasmids do transform *S. pneumoniae* (Saunders and Guild, 1981a, b), the dependence of the transformation frequency on the square of the DNA concentration indicates that two monomeric plasmid DNA molecules are required. On the other hand, only multimeric plasmids transform *B. subtilis* (Canosi et al., 1978; Mottes et al., 1979). With both *S. pneumoniae* and *B. subtilis*, the multimeric plasmid transformation frequency depends approximately on the square of the time of DNA exposure. This observation demonstrates that two separate DNA uptake events are required (as with double transformants for unlinked chromosomal markers) and supports the argument that the reassembly of complementary single strands of donor DNA is required for plasmid transformation in these organisms. The low frequency of plasmid transformation in physiologically competent cells precludes firm conclusions regarding mechanism.

Artificial competence. In numerous species, $CaCl_2$ treatment and various other regimens allow plasmid DNA molecules to enter the cell. They apparently enter as double-stranded DNA, as even partially three-stranded (D-loop) structures can enter these treated cells (Holloman and Radding, 1976). The prototype *E. coli* system is reviewed elsewhere (Hanahan, 1987). The polyethylene glycol treatment for *B. subtilis* protoplasts (Chang and Cohen, 1979) may be the preferred method for introducing plasmids into that organism since the efficiency is high and monomeric plasmids can be used. Electroporation, recently used to achieve plasmid transformation in *Campylobacter jejuni* (Miller et al., 1988), may also allow plasmid transformation in other bacteria.

3. Facilitated plasmid transformation

Transformation of physiologically competent bacteria occurs at a much higher frequency with plasmids carrying homology to recipient cell DNA (chromosomal or plasmid) than with plasmids without homology. Recombination between donor DNA and recipient DNA is apparently responsible for the increase in frequency seen in the former case (see Fig. 1 for possible mechanism). Facilitated plasmid transformation has been demonstrated in a number of organisms, including *B. subtilis* (Contente and Dubnau, 1979a, b; Canosi et al., 1981; Weinrauch and Dubnau, 1987), *S. pneumoniae* (Lopez et al., 1982), *H. influenzae* (Albritton et al., 1981; Balganesh and Setlow, 1985), and *N. gonorrhoeae* (Biswas et al., 1982). When there is homology between the donor plasmid DNA and the recipient chromosome, either replacement transformation or facilitated plasmid transformation (establishment) can occur. In one such study with *H. influenzae*, replacement transformation was 100-fold more frequent than facilitated plasmid transformation (Setlow et al., 1981). In another study with the cyanobacterium *Anacystis nidulans*, replacement transformation occurred at high frequency,

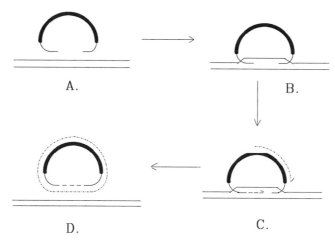

Figure 1. Possible mechanism for facilitated plasmid transformation. (A) An incomplete single strand from the donor plasmid locates homologous double-stranded DNA (chromosome or plasmid) in the recipient cell; the heavy line represents a heterologous plasmid vector. (B) D-loop-type pairing occurs. (C) Replication fills the gap in the donor single strand. Complementary strand synthesis may be primed either randomly or by pairing with an internalized complementary strand fragment. (D) Resolution of the Holliday-like junctions can release the restored plasmid; exchanges of alleles between donor and recipient homologies may occur. Although this figure depicts a single-stranded intermediate, a similar mechanism could operate with an incomplete double-stranded donor DNA molecule possessing single-stranded tails.

while plasmid establishment was never observed (Tandeau de Marsac et al., 1982).

The mechanism of facilitated plasmid transformation can be compared with that of transformation of a physiologically competent cell with a nonreplicating plasmid. In one study, *A. nidulans* was transformed with *E. coli* plasmids carrying inserts of *A. nidulans* chromosomal DNA (Williams and Szalay, 1983). The donor plasmids were not capable of independent replication in the recipient organism, but addition transformants were readily obtained by selecting for a plasmid-borne drug resistance marker. These addition transformants were found at a considerably lower frequency, however, than replacement transformants (Williams and Szalay, 1983). Addition recombinants were also found in which both copies of the tandemly repeated chromosomal DNA fragment carried the genetic information from the recipient cell (see Fig. 2).

Similarly, in the cyanobacterium *Agmenellum quadruplicatum* PR-6 (*Synechococcus* sp. strain PCC 7002), addition recombinants obtained with nonreplicating plasmids appear to exist in equilibrium with a free plasmid form (A. Pilon and R. D. Porter, manuscript in preparation). The original plasmid can easily be recovered by transforming *E. coli* with DNA extracted from stabilized addition recombinants of the cyanobacterium. If the donor plasmid can replicate in the recipient cell, facilitated plasmid transformation may occur by integration of the plasmid to form an addition recombinant and then release of the plasmid from the chromosome by recombination between the flanking direct repeats.

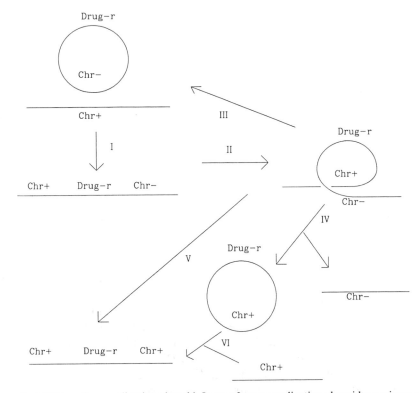

FIGURE 2. Addition of nonreplicating plasmid. In step I, a nonreplicating plasmid carrying a selectable drug resistance marker (Drug-r) and a distinguishable copy of a chromosomal gene (Chr–) recombines with a recipient chromosome (Chr+) to give a heterozygous addition recombinant. In step II, the homologous tandem repeats align to allow one of three possible types of recombination event. The first possible type of event (step III) simply reverses the original integration event. The second possible type of event (step V) involves nonreciprocal recombination between the tandem repeats without the excision of the plasmid from the chromosome. The third possible type of event (step IV) exchanges the chromosomal alleles while excising the plasmid from the chromosome. The plasmid resulting from a step IV event can recombine with another copy of the chromosome (step VI) to yield an addition recombinant where both of the flanking repeats possess the original chromosomal allele. Other variations on these events are also possible.

In *S. pneumoniae* (Lopez et al., 1982) and *H. influenzae* (Setlow et al., 1984), when a piece of the homologous DNA is removed from the donor plasmid with restriction enzymes prior to transformation (generating a double-strand gap), the plasmid is restored to full length in the transformants. Such restoration of double-strand gaps has also been observed in yeast cells (Orr-Weaver et al., 1981). A mechanism for repair of double-strand breaks and its application in yeast cells has been proposed (see Szostak et al., 1983; Hastings, this volume). This mechanism can account for double-strand gap repair in procaryotes, but the multiple copies of the chromosome in procaryotes do not rule out reciprocal recombination, without DNA synthesis to fill the gap, since the loss of one copy of the chromosome might not be detectable.

The double-strand gap repair model can also explain the results of *A. nidulans* transformation with nonreplicating plasmids containing gaps in the homologous DNA (Kolowsky and Szalay, 1986). The repair of gaps up to 20 kilobases (kb) long can be observed by the recovery of repaired plasmids in *E. coli*. In addition, Southern blots of DNA from these transformants show that the integrated plasmid is flanked by two intact copies of the homologous fragment carried by the plasmid. These addition recombinants may also result from the gapped plasmid recombining reciprocally with one copy of the chromosome (which is gapped and thus lost) and then integrating into another copy of the chromosome.

4. Ectopic transformation

In ectopic integration chromosomal genes are inserted into the chromosome at places other than their normal location (Mannarelli and Lacks, 1984) or foreign DNA is incorporated into the chromosomal or plasmid DNA of the recipient cell.

In interposon mutagenesis (see Fig. 3), a piece of foreign DNA—most often a drug resistance determinant—is inserted in vitro within a cloned DNA fragment from the desired recipient or substituted for a piece of the cloned DNA fragment. When the recipient is transformed with a linear DNA fragment containing the interposon and flanking recipient DNA, the interposed piece of foreign DNA is efficiently incorporated into the homologous region of the recipient DNA by recombination within the flanking homologous DNA. Such an event is a minor variation on replacement transformation. If the vector is not removed before transformation, addition transformants can also be observed. Interposons up to 20 kb long have been incorporated into the chromosome of *A. nidulans* (Kolowsky et al., 1984).

By other methods using "random" ligation, foreign DNA can be incorporated into the recipient DNA of a physiologically competent cell. For example, when random recipient DNA fragments are ligated to a selectable foreign DNA fragment, subsequent transformation can insert the foreign DNA into numerous sites in the genome of the recipient cell. At the same time, this procedure can result in homologous DNA being integrated at sites where it is not normally found. In some cases, it appears that a ligation of the foreign and recipient DNA (or two normally unlinked fragments of recipient DNA) produces a circle which integrates into the recipient genome to produce an unstable addition recombinant. This procedure has been characterized in detail in *S. pneumoniae* (Mannarelli and Lacks, 1984). Variations on this theme have been used to mutagenize the genome of both *S. pneumoniae* (Morrison et al., 1984) and *B. subtilis* (Ferrari et al., 1983).

Some ectopic transformants produced by this method are extremely stable, apparently as a result of alterations of the recipient DNA. For example, when *S. pneumoniae* was transformed with DNA from λ cloning phages containing *S. pneumoniae* DNA, λ DNA was incorporated into the recipient genome (Claverys et al., 1980). This apparently involved one homologous recombination event and one illegitimate recombination event, as deletions in the recipient DNA were seen where it had undergone recombination with λ DNA (Claverys et al., 1980). Stable ectopic mutants of *A. quadruplicatum* PR-6 also resulted from transformation

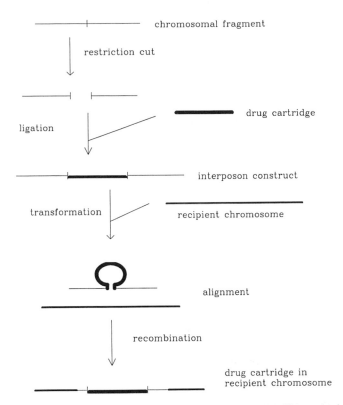

FIGURE 3. Interposon mutagenesis. A fragment of chromosomal DNA (still in a cloning vector which is not shown) is cut with one or more restriction endonucleases and ligated to a properly prepared drug resistance cartridge (or any DNA fragment carrying a suitable selectable marker). The interposon may simply be inserted or may replace part of the chromosomal DNA fragment. A recipient cell is transformed with the interposon construct fragment, generally after it is removed from the cloning vector. The homologous regions flanking the drug cartridge align with a recipient chromosome, and the resulting recombination event inserts the cartridge into the recipient chromosome. If the cartridge is near the gene of interest, a selectable linked marker is now available. If the cartridge has been inserted within a gene, that gene has been mutagenized in the transformant.

with recipient DNA ligated to a DNA fragment containing a drug resistance determinant under conditions blocking circularization (Buzby et al., 1985). Although the DNA may have circularized inside the recipient cell before recombination with the genome, transformation experiments with plasmids linearized in nonessential regions fail to demonstrate such circularization at detectable levels (Pilon and Porter, in preparation). Properly prepared hybrid DNA fragments can be targeted to the desired region of the recipient genome by homologous recombination, while mutations in that region may be produced by illegitimate recombination involving the opposite end of the foreign DNA (see Fig. 4).

B. Transduction

In transduction bacterial DNA is moved between cells by bacteriophage. In specialized transduction, the DNA involved is a hybrid of phage and bacterial

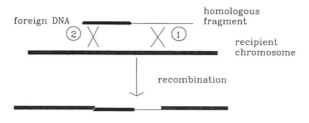

FIGURE 4. Deletion generation by ectopic transformation. Random or specific fragments of DNA from a transformable organism are ligated to a foreign DNA fragment containing a selectable marker by a protocol tht precludes circularization. Upon transformation, the homologous portion of the hybrid construct undergoes general recombination (crossover 1) with the recipient chromosome. Transformants for the marker carried on the foreign DNA fragment will have undergone an illegitimate recombination event (crossover 2) to incorporate the marker and restore the continuity of the recipient chromosome. The hybrid DNA junction is preserved in the ectopic transformant, while recipient sequences adjacent to that junction point have been altered.

DNA, and any particular line of specialized transducing phage contains a specific fragment of the bacterial genome. In generalized transduction the DNA is bacterial DNA erroneously packaged in a virion; generalized transducing phages may contain any piece of the bacterial genome.

1. Specialized transduction

Specialized transduction was first demonstrated with phage λ, which transduced only those bacterial markers (*gal* and *bio*) closely linked to the phage attachment site on the *E. coli* chromosome (Morse et al., 1956). Integration of phage λ at secondary attachment sites (Shimada et al., 1972) led to specialized transducing phages for other bacterial markers. More recently, insertion of DNA fragments into λ cloning phages by recombinant DNA techniques has produced an almost unlimited variety of specialized transducing phages for *E. coli* (for a partial listing, see Weisberg, 1987).

The formation of a specialized transducing phage requires illegitimate recombination between a point within the integrated phage DNA and a point within flanking bacterial DNA (see Fig. 5). These two points can vary, as long as the hybrid DNA is within the packageable size range. Hybrid DNAs lacking functions required for virion formation generate defective specialized transducing phage, such as λ d*gal*, which require the assistance of a nondefective helper phage for growth. Hybrid DNAs retaining the functions required for virion formation produce plaque-forming specialized transducing phage, such as λ p*lac*. Phages, such as P1, that lysogenize by plasmid formation apparently generate specialized transducing phages via events involving insertion sequences in either the phage or bacterial DNA.

Specialized transducing phages arise very infrequently, and the phage lysates containing them are called LFT, or low frequency transduction, lysates. HFT, or

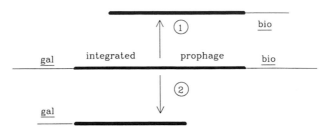

FIGURE 5. Production of specialized transducing phage DNA. The integrated prophage is shown between the chromosomal *gal* and *bio* operons, as in the case of phage λ. An illegitimate recombination event between a point within the prophage and a point within the flanking chromosomal DNA (for example, event 1 or event 2) can give rise to a transducing phage DNA. Further considerations are discussed in the text.

high frequency transduction, lysates can be obtained from a transductant produced with an LFT lysate. Such transductants often contain both the transducing phage and a helper phage, which may be required for the integration, excision, and/or growth of the transducing phage if it is defective. Induction of such a double lysogen produces an HFT lysate containing both phages. The two types of phages can often be separated on CsCl gradients as a result of differences in size between the transducing phage DNA and the helper phage DNA.

Upon injection into a recipient cell, the transducing phage DNA becomes circular and supercoiled as in a typical phage infection. Thereafter, the phage may propagate vegetatively and lyse the cell, it may remain inactive and be lost by segregation, or it may undergo recombination. If the recipient cell is not killed, three types of transduction events can occur. Firstly, a transducing phage that carries the phage site for integration and has the requisite gene products available, possibly provided by a helper phage, can undergo double-site-specific recombination with the phage attachment site on the bacterial chromosome, producing an integrative transductant. A merodiploid for the bacterial DNA carried on the phage generally results. Intragenic recombination does not occur in this case, and the selection of transductants requires that the transducing phage carry a wild-type version of the gene on which the selection is based.

The other two types of transduction events involve general recombination. If recombination between the two copies of the bacterial DNA results in the incorporation of the entire transducing phage DNA molecule, a merodiploid addition transductant is produced. The product is physically indistinguishable from the integrative transductant if the bacterial DNA is adjacent to the phage attachment site. Intragenic recombination can occur in this situation, however, and the infection of a *lacZ1* cell with a *lacZ2* phage, for example, can yield a Lac[+] transductant. Addition transductants are normally somewhat unstable, as subsequent recombination between the flanking homologies can remove the transducing phage DNA and leave one copy of the bacterial DNA. If the exchange points of the two recombination events are in the same interval, the single copy remains as it was before transduction. If the exchanges are in different intervals, a recombinant copy remains (see Fig. 6). General recombination can also give rise to

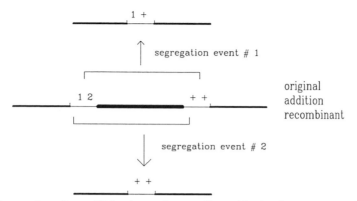

FIGURE 6. Segregation of an additional transductant. Recombination between a specialized trans-ducing phage carrying mutant allele 2 and a recipient chromosome carrying mutant allele 1 has given rise to the phenotypically positive addition transductant shown in the middle of the figure. Subsequent recombination between the flanking homologies can result in the loss of the transducing phage DNA. Segregation event #1 involves a crossover between the mutant alleles and results in the restoration of the original chromosome (phenotypically negative because of mutant allele 1) configuration. Segregation event #2 involves a crossover to the left of both mutant alleles and yields a chromosome with a wild-type genotype. A crossover to the right of both mutant alleles would result in a segregant whose chromosome carries both mutant alleles.

substitution transduction, in which sequence information from the transducing phage DNA is transferred to the bacterial DNA copy without the incorporation of the transducing phage DNA. This type of transductant is stable and remains haploid.

Addition transductants may be most simply generated by the formation of a single Holliday junction that is resolved to integrate the transducing phage DNA, analogous to integrative transduction by site-specific recombination. Migration of the Holliday junction past one or both markers before resolution yields a heteroduplex product which is converted to homoduplex by mismatch correction or replication (see Radman, this volume). Alternatively, resolution of the Holliday junction after branch migration past the mutant allele in the recipient cell copy would give rise to a substitution transductant. The alternate resolutions of a single Holliday junction are shown in Fig. 7. A substitution transductant could also be generated by two crossovers between the two copies of bacterial DNA. The single-crossover scenario would yield exclusively heteroduplex recombination product, while the double crossover could result in the direct incorporation of double-stranded donor DNA. Experiments to distinguish these two possibilities have not been conducted with any specialized transduction system.

Addition transduction and substitution transduction normally occur at very low frequencies in *E. coli recA* mutant recipients (Porter et al., 1978, 1981; but also see Weisberg and Sternberg, 1974; Wackernagel and Radding, 1974). Although *recB* mutants yield few recombinants in Hfr crosses (see below), they produce addition and substitution transductants at about the same frequency as *recB*[+] cells (Weisberg and Sternberg, 1974; Porter et al., 1982). Strains using either the RecE or RecF recombination pathway undergo specialized transduction

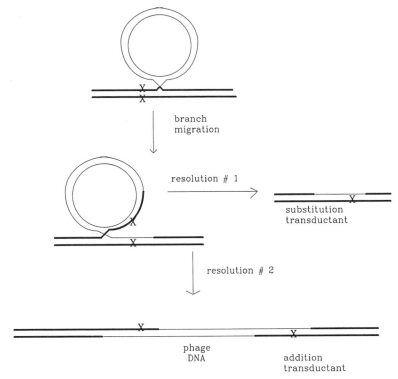

FIGURE 7. Addition or substitution recombination can result from a single crossover. A single Holliday junction is formed and undergoes branch migration past the mutant allele (×) originally present in the recipient chromosome. Depending on whether the "crossed" or "uncrossed" strand of the Holliday junction undergoes resolution, either a substitution (resolution #1) or an addition (resolution #2) transductant can result. In both cases, the recombination product contains heteroduplex DNA; homozygosity can result from either mismatch correction or replication.

at somewhat higher frequency than wild-type strains, in which the RecBCD pathway predominates (Wackernagel and Radding, 1974; Porter et al., 1978; Porter, 1983).

The question of addition versus substitution transduction has received little attention. One report indicated that the proportion of the two types differed little between a RecBCD pathway strain and a *recB* mutant, while the proportion of substitution transductants was higher in a RecF pathway strain for λ d*gal* transduction (Wackernagel and Radding, 1974). Two later reports compared addition and substitution transduction of *lacZ* mutant recipients with λ p*lacZ*$^+$ or λ p*lacZ* (mutant *lacZ* gene) phage (Porter et al., 1982; Porter, 1983). Both the recombination frequency and the proportion of addition and substitution trans-ductants were about the same (65% addition for λ p*lacZ*$^+$ and 30% addition for λ p*lacZ*) in either type of cross for wild-type (RecBCD pathway) strains and *recB* mutant strains. In RecF pathway strains, the increase in transduction frequency was accompanied by an increase in the proportion of substitution transductants for both types of crosses. In RecE pathway strains the proportion of addition and substitution transductants was comparable to that in the RecBCD pathway strain for λ p*lacZ*, while the proportion of substitution transductants increased from 35 to 60% with λ p*lacZ*$^+$. Although the recombination mechanism is not clear from these studies, different recombination pathways yield different recombination end products, presumably as a result of different mechanisms of the pathways.

The frequency of specialized transduction is lower than one might expect, since essentially every recipient cell receives a transducing phage under appropriate conditions. The low frequency is probably due to the fact that the parental DNA is double stranded without free ends. UV irradiation of the transducing phage prior to infection, which may lead to ends and single-stranded DNA during repair of the UV-induced lesions, enhances the frequency of transduction. The exact mechanism of UV-mediated enhancement is unclear, however, as the damage to the DNA can be located very far from the actual recombining region (Golub and Low, 1983). The transductants which are observed after UV irradiation of the phage are primarily substitution transductants (Porter et al., 1982).

Specialized transduction is also enhanced when the recipient DNA is on a fertile F′ factor, such as F42*lac* (Porter et al., 1978). The enhancement depends on a functional RecBCD enzyme in the recipient (Porter et al., 1978, 1982) and on the constitutive expression of the *tra* regulon of the F′ (Porter, 1981; Seifert and Porter, 1984a). The *oriT* site of the F factor, from which DNA transfer is initiated during conjugation, is the only part of the *tra* regulon that must be located *cis* to the recipient DNA, while the required *tra* gene products can be provided in *trans* (Seifert and Porter, 1984b). The action of those *tra* gene products at *oriT* presumably allows the RecBCD enzyme to gain access to the F′ molecule, but the mechanism by which this leads to enhancement is not established. Preliminary evidence suggests that RecBCD enzyme acting at Chi sites in the F′ may be responsible for the enhancement (R. D. Porter, unpublished data). Although enhancement appears to occur primarily as a result of higher levels of recombination initiation (Porter et al., 1978), enhancement by the F′ also increases the proportion of substitution transductants with λ p*lacZ*, presumably reflecting a mechanistic difference between enhanced and nonenhanced specialized transduction (Porter et al., 1982).

2. Generalized transduction

Generalized transducing particles arise from the accidental packaging of a piece of bacterial DNA in an otherwise normal phage particle (Ozeki and Ikeda, 1968; also see Low and Porter, 1978). Upon injection into a suitable recipient cell, the transducing DNA may recombine with homologous DNA in the recipient cell. Most studies of generalized transduction have utilized phage P22 of *S. typhimurium* or phage P1 of *E. coli* (for reviews, see Masters, 1985; Margolin, 1987).

Linkage studies in generalized transduction led early workers to postulate specific breakage sites on the bacterial genome from which packaging is initiated (Ozeki, 1959). Both P1 and P22 initiate packaging of their DNA from linear concatemers at special sites (*pac*) and continue one "headful" after another. Although *pac* sites are not found in the bacterial genome, related sequences presumably are occasionally used by the phage DNA packaging machinery to initiate the packaging of bacterial DNA. Once initiated, packaging of bacterial DNA continues by the headful mechanism. There seem to be about 10 to 15 degenerate P22 *pac* sites in the genome of *S. typhimurium* (Chelala and Margolin, 1974) and substantially more degenerate P1 *pac* sites in the *E. coli* genome. In agreement with this hypothesis, certain phage mutants have altered ability to form generalized transducing particles (Schmieger, 1972; Wall and Harriman, 1974). The possibility that P1 packaging of bacterial DNA occurs only after P1 DNA containing a normal *pac* site has integrated into degenerate *loxP* sites in the *E. coli* genome has largely been discounted (Sternberg and Hoess, 1983). A protein attached to the bacterial DNA in P1 transducing particles (Ikeda and Tomizawa, 1965) may be a component of the packaging machinery of P1 phage DNA, but this has not been established.

There is considerable variation in the production of transducing particles among cells infected by P1; some cells produce none, while in others about 20% of the bacterial chromosome is encapsidated (Harriman, 1972). In mass lysates of both P1 and P22, however, all segments of the bacterial chromosome are represented even though not all loci can be transduced at the same frequency. With P22, this frequency differential appears to be due to differences in packaging frequency for different regions of the bacterial chromosome (Schmieger, 1972, 1982). With P1 phage, on the other hand, the differences in transduction frequency for various markers are more a consequence of differential recombination than packaging (Newman and Masters, 1980).

After the injection of bacterial DNA into a recipient cell, stable transductants arise with low efficiency. Approximately 90% of the transductants are abortive: the new genetic information does not replicate with the genome of its host cell (Ozeki, 1956; Ozeki and Ikeda, 1968; Schmieger, 1982). Similarly, physical measurements show that only a small percentage of the donor DNA becomes associated with the recipient DNA (Ebel-Tsipis et al., 1972; Sandri and Berger, 1980a). The abortive transductants form microcolonies in which the donor DNA is unilinearly inherited. The stability of the donor DNA in the abortive transductants is apparently due to circularization of the donor DNA fragment (Sandri and Berger, 1980b) which is refractory to degradation by the RecBCD enzyme (Karu et al., 1973). The linearization of the extracted circular DNA by treatments that degrade proteins (Sandri and Berger, 1980b) implies that a protein, such as that on the ends of the bacterial DNA inside P1 transducing particles (Ikeda and Tomizawa, 1965), maintains the circles. Stable transductants are thought to result primarily from recombination events involving linear donor DNA fragments, as abortive transductants are rarely converted to stable transductants (Ozeki and Ikeda, 1968).

Recombination involving linear fragments of donor DNA is expected to yield exclusively substitution transductants. Density labeling experiments have shown that small, double-stranded donor DNA fragments can be directly incorporated into the recipient DNA (Ebel-Tsipis et al., 1972; Sandri and Berger, 1980a). However, there may be concomitant or independent incorporation of single-stranded donor DNA as well. Marker effects (perturbations of recombination frequency between closely linked markers arising, for example, from the correction of mismatches in heteroduplex DNA) in P1 transduction (Crawford and Preiss, 1972) suggest extensive single-strand incorporation, since heteroduplex DNA is presumably the basis of such marker effects (see Radman, this volume).

The genetic linkages observed in generalized transduction also suggest the incorporation of larger donor DNA fragments than indicated by the physical studies (Ebel-Tsipis et al., 1972; Sandri and Berger, 1980a), and this apparent paradox could be resolved if long heteroduplex DNA typically flanks the incorporated double-stranded fragment. Although heteroduplex DNA could be formed by branch migration of a single Holliday junction between donor and recipient DNA, it might also be formed by the divergent migration of two Holliday junctions that were initially at the ends of an incorporated double-stranded fragment. In either case, the multiple exchanges that are occasionally seen in transductants could arise by the correction of mismatches within the heteroduplex DNA (see Radman, this volume) without the need for forming additional Holliday junctions.

As expected for general recombination, P1 transduction is reduced 100- to 1,000-fold in *recA* mutants (Hertman and Luria, 1967), which means that it very rarely, if ever, occurs (Sandri and Berger, 1980a). P1 transduction is also reduced between 10- and 100-fold in *recB* or *recC* mutants (Willetts and Mount, 1969). Chi sites can function in P1 transduction of *recBCD*$^+$ strains (Dower and Stahl, 1981), but strains containing altered RecBCD enzyme without Chi activity show normal levels of P1 transduction (Schultz et al., 1983; Chaudhury and Smith, 1984). RecF pathway strains (*recBCD sbcB* mutants) yield essentially wild-type levels of P1 transduction with less recombination-dependent marker discrimination than is seen in RecBCD pathway strains (Masters et al., 1984). The RecF pathway may integrate longer pieces of donor DNA and, in contrast to Hfr conjugation, appears to operate in cells where the RecBCD pathway is also functional (Cosloy, 1982).

Generalized transduction can be achieved not only by temperate or pseudo-temperate phages (Ozeki and Ikeda, 1968), but also by virulent phages if the recipient cell survives the attachment of the transducing phage particle. For example, a T4 mutant can transduce *E. coli* (Wilson et al., 1979) even though T4 particle adsorption usually kills the cell (Duckworth, 1970). Since virulent phage particles greatly outnumber transducing phage particles, potential transductants must be protected from the virulent particles. Low multiplicities of infection and the addition of phage antiserum or agents chelating adsorption cofactors after the initial infection can help in this regard.

Since generalized transduction requires the packaging of intact bacterial DNA, the bacterial DNA must not be extensively degraded before encapsidation begins. Phage T1 extensively degrades the DNA of its host, and as expected, transducing particles are produced early in the infection (Kylberg et al., 1975). In contrast, P1 does not extensively degrade the DNA of its host, and transducing particles are formed late (Harriman, 1971). Although "headful" packaging is usually required for formation of transducing particles, phage λ, which normally packages from *cos* to *cos*, can form generalized transducing particles if DNA protruding from the phage head is nucleolytically removed to permit tails to attach to filled heads (Sternberg and Weisberg, 1975).

Extrachromosomal elements can also be transduced in a number of systems (Arber, 1960; Fredericq, 1965; Dubnau and Stocker, 1967; Watanabe et al., 1968; Grubb and O'Reilly, 1971; Ruby and Novick, 1975; Bramucci and Lovett, 1976). The production of deletions, called transductional shortening, occurs with plasmids whose DNA is larger than can be accommodated by the virus particle (Watanabe et al., 1968; Yoshikawa and Hirota, 1971; see Low, 1972). Two or more plasmids may be simultaneously transferred, presumably by reversible recombination between the multiple plasmid species involved (Dubnau and Stocker, 1967; Stiffler et al., 1974; Iordanescu, 1977). The transfer of small plasmids is thought to require linear DNA concatemers, and their packaging presumably depends on *pac*-like sites (Schmidt and Schmieger, 1984).

C. Conjugation

Conjugation involves the transfer of DNA between two cells in direct contact. Contact initially occurs between the tip of the sex pilus of the donor cell and the exterior envelope of the recipient cell; direct contact, presumably achieved by disassembly of the pilus, produces an unstable mating pair. Multiple cell interactions frequently give rise to mating cell aggregates. Although some DNA transfer between the cells may occur at these early stages, most DNA transfer occurs between pairs of cells specifically stabilized within the mating aggregate. Cells are called exconjugants after mating pair dissociation, and recipient cells that have received DNA from donor cells are called merozygotes. These merozygotes become transconjugants after the donor DNA has become stabilized in the recipient cell, either by recombination with recipient DNA or, in the case of transferred plasmid DNA, by establishment as an independent replicon in the recipient cell. This outline is discussed in detail elsewhere (Clark and Warren, 1979).

Conjugation is used extensively in bacterial genetics and often involves the simple transfer of plasmids between cells without any interaction with the DNA of the recipient. Some plasmids isolated from natural sources or derived by mutation in the laboratory can carry out the specific cell contact cycle but cannot transfer their DNA; such plasmids are designated conjugative but nonmobilizable. Other plasmids have the converse properties and are designated mobilizable but nonconjugative. Plasmids with both properties (conjugative and mobilizable) are designated self-transmissible (Clark and Warren, 1979). Plasmids that are mobilizable but nonconjugative are often efficiently transferred to recipient cells when other plasmids present in the donor cell provide the necessary cell contact.

1. *E. coli* F-factor-based conjugation

Conjugation promoted by the F factor of *E. coli* has played a seminal role in the development of bacterial genetics. Although extensively studied, F-factor-based conjugation is still incompletely understood (for reviews, see Willetts and Skurray, 1980, 1987; Willetts and Wilkins, 1984; Ippen-Ihler and Minkley, 1986).

Regulation of F-factor fertility. The F factor has 100 kb, of which 35 kb are involved in making it a self-transmissible plasmid (Willetts and Skurray, 1980). This 35-kb region contains the *oriT* site, at which DNA transfer is initiated, and at least 28 genes, of which most are designated *tra* and some are designated *trb*. Two genes (*traM* and *traJ*) produce separate transcripts, but all of the other genes form a single operon. Although this huge operon has secondary promoters, one promoter is dominant under conjugative conditions. The overall structure of the *tra* regulon is shown in Fig. 8.

The main *tra* operon and perhaps *traM* are positively regulated by the *traJ* gene product (Willetts, 1977; Gaffney et al., 1983), although the mechanism of this regulation remains obscure. The *traJ* gene is negatively regulated by the *finO* and *finP* genes (*fin* = fertility inhibition). A virtue of the F factor, however, is its lack of a functional *finO* gene; the *tra* genes and, hence, conjugal ability are constitutively expressed unless *finO* is provided in *trans* by a *fin⁺* F-like plasmid. The mysterious lack of fertility regulation in F is due to an IS*3* insertion, which traditionally marks one end of the *tra* regulon, in the *finO* gene (Yoshioka et al., 1987). The *finP* gene is transcribed from the antisense strand in the mRNA leader region of the *traJ* gene, but the *finP* transcript apparently does not code for a protein (Johnson et al., 1981). Mullineaux and Willetts (1985) speculate that a complex of the *finP* RNA and the *finO* protein interacts with the leader portion of the *traJ* transcript to prevent its translation.

Cell contact and DNA transfer. The F sex pilus consists of a single protein, pilin, encoded by the *traA* gene (Frost et al., 1984). The *traQ* gene product apparently converts the initial 121-amino-acid *traA* polypeptide into the functional 70-amino-acid polypeptide, perhaps in several steps (Laine et al., 1985). The N-terminal amino acid of the mature pilin is acetylated (Frost et al., 1984); phosphorylation and glycosylation of pilin have also been proposed (Date et al., 1977; Armstrong et al., 1981). At least 11 additional genes are required for the assembly of a functional pilus, but their roles are not known.

Effective mating pair formation between two donor cells is prevented by surface exclusion, which requires the *traS* and *traT* gene products located in the inner and outer membranes, respectively (Achtman et al., 1977, 1979; Minkley and Willetts, 1984; Cheah et al., 1986). The *traT* gene product also plays a role in serum resistance and in reducing the susceptibility of cells to phagocytosis (Moll et al., 1980; Aguero et al., 1984). Although pilus-to-envelope contacts between donor cells do occur, surface exclusion prevents mating pair stabilization between two donor cells. DNA transfer can occur through an extended F pilus (Ou and Anderson, 1970), but mating pair stabilization involving the *traG* and *traN* gene products occurs (Manning et al., 1981) prior to extensive DNA transfer. The exact nature of the final surface-to-surface interaction between two mating cells is not well understood, but the *traD* gene product may be involved in the formation of a pore between the two inner membranes (Panicker and Minkley, 1985). The destabilization of the mating pair and its separation remain a total mystery. Mechanical disruption of the mating pairs leaves little apparent lasting damage

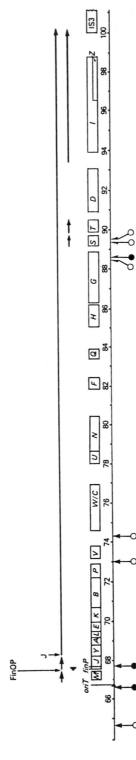

FIGURE 8. Map of the F transfer region. The numbers show kilobase coordinates, and the horizontal lines above the genes represent transcripts. The vertical arrows with open and solid circles indicate the positions of EcoRI and BglII cleavage sites, respectively. The sites of the genes are derived from sequence data where available or from the apparent molecular weights of the protein products. The latter sizes, and the gene locations, are approximations only. (From Willetts and Skurray, 1987.)

(Low and Wood, 1965), and it is therefore possible that mating pair disruption is sometimes a spontaneous and random process. The expression of F-factor genes in the recipient cell may also play an active role in mating pair disaggregation (Achtman et al., 1978).

Four or five *tra* gene products are concerned with donor conjugal DNA synthesis. Several events must occur before this DNA synthesis begins. First, one strand of the DNA is nicked at one of three closely spaced, major sites within the *oriT* region (Thompson et al., 1984), and a constitutive nicking-closing reaction at *oriT* occurs even in the absence of recipient cells (Everett and Willetts, 1980). The *traY* gene product and the amino-terminal domain of the *traI* gene product (formerly *traZ*) apparently constitute the *oriT*-specific endonuclease (Everett and Willetts, 1980; Traxler and Minkley, 1987). The *traI* gene product is a DNA helicase (Abdel-Monem et al., 1983) responsible for separating the two DNA strands during transfer.

The *traM* gene product seems to trigger the start of donor conjugal DNA synthesis at the *oriT* site in response to a signal arising after the tip of the F pilus contacts a suitable recipient cell (Everett and Willetts, 1980). Although the exact function of the *traM* gene product in triggering donor conjugal DNA synthesis is not known, it may expose sufficient single-stranded DNA at the nicked *oriT* site to allow binding of the *traI* helicase (Willetts and Skurray, 1987). The transfer of the 5' end of the displaced strand is normally accompanied by replacement synthesis in the donor cell, but transfer can occur in the absence of this synthesis (Sarathy and Siddiqi, 1973; Kingsman and Willetts, 1978). This replacement DNA synthesis requires priming by RNA polymerase (Kingsman and Willetts, 1978), and this requirement may reflect blockage of the 3' end by one of the proteins involved in donor conjugal DNA synthesis.

A single strand of F-factor DNA is transferred to the recipient cell (Rupp and Ihler, 1968; Ohki and Tomizawa, 1968), and a new complementary strand is synthesized by the normal DNA synthesis machinery of the recipient cell. The stabilization of the F factor in the recipient cell is a *recA*-independent process (Clark, 1967) that typically requires the transfer of both ends of *oriT* (Everett and Willetts, 1982). Despite earlier evidence for the transfer of single-stranded concatemers of F-factor DNA (Ohki and Tomizawa, 1968; Matsubara, 1968), the observed requirements for recircularization favor the transfer of unit-length DNA strands (Willetts and Skurray, 1987). Detailed models for F-factor transfer are discussed in several recent review articles (Willetts and Wilkins, 1984; Ippen-Ihler and Minkley, 1986; Willetts and Skurray, 1987).

Hfr conjugation and chromosomal transfer. Four transposable elements on the F factor (two copies of IS*3*, one copy of IS*2*, and one copy of Tn*1000*) (Davidson et al., 1975) allow it to integrate into the host cell chromosome to form an Hfr strain. This process most often involves *recA*-dependent recombination between two copies of a transposable element, and certain Hfr strains therefore repeatedly arise (Low, 1972). Hfr strains can arise in *recA* mutants (Deonier and Mirels, 1977; Cullum and Broda, 1979), presumably as a result of transposition (see Syvanen, this volume). The limited ability of an autonomous F factor (F$^+$) to transfer the donor chromosome is thought to involve both Hfr formation and conduction of the chromosome by transient replicon fusions formed when Tn*1000* transposes from the F factor to the chromosome (Guyer, 1978). Other transient associations between the F factor and the chromosome may also promote chromosome transfer by F^+ donor cells (Goto et al., 1984).

The uniform transfer of DNA from each Hfr makes conjugation a powerful tool for genetic mapping over very long distances. Transfer is initiated at *oriT* and proceeds in the direction dictated by the orientation of its integration. Transfer initiates rapidly and proceeds at a reasonably uniform rate (Wood, 1968), making time of entry a good criterion for determining the distance of a marker from the Hfr origin (Low, 1987). Since transfer is initiated in the middle of the integrated F factor, the recipient cells remain F^- (Hayes, 1953) unless the entire donor chromosome is transferred. Complete transfer is rare, but can be detected by selecting recombinants for a marker transferred late. Chromosomal DNA transferred by an Hfr can also recombine with homologous plasmid-borne DNA in the recipient (Porter, 1982).

Recombination after Hfr conjugation. Very long linkage groups are typically observed by genetic criteria in Hfr conjugation. One study of marker linkage in Hfr conjugation estimated a 20% probability of a crossover per "minute" (1 min is 1% of the *E. coli* chromosome—about 47 kb) of transferred DNA (Low, 1965), while another study estimated an even lower frequency of crossovers (Pittard and Walker, 1967). Among recombinants selected for a distal marker, roughly 50% inherit a proximal marker from the donor, as expected if these are frequent, multiple exchanges between donor and recipient DNA. These selected recombinants only rarely inherit an even more distal marker, however, probably as a result of subsequent interruption of DNA transfer.

Markers very near the Hfr origin, however, are not frequently inherited. The rare inheritance of very early markers (1 to 5 min from the origin) from the donor was proposed to be due to a length exclusion effect (Low, 1965). However, crossovers do occur frequently in this very early region (Pittard and Walker, 1967), suggesting that increased crossover frequency leads to the reduced recovery of these markers. An antipairing effect of the leading F-factor DNA has also been suggested (Pittard and Walker, 1967). In contrast, the probability of crossover per minute of transferred DNA is less for very late markers; this effect leads to physically larger linkage groups (Verhoff and DeHaan, 1966), but the basis for this phenomenon is unknown.

The long linkage groups observed genetically are at variance with the results of physical studies of recombination following conjugation. Differentially labeled donor DNA and recipient DNA become covalently associated, but only short pieces (mostly about 0.4 kb) of single-stranded donor DNA appeared to be integrated (Siddiqi and Fox, 1973). Incorporation of double-stranded donor DNA was not detected (Siddiqi and Fox, 1973). The dramatic marker effects in Hfr conjugation (Norkin, 1970) also argue for the generation of heteroduplex DNA during recombination, since such marker effects are thought to be due to the differential susceptibility of mismatches in heteroduplex DNA to correction (see Radman, this volume). Neither result rules out the possibility that double-stranded incorporation events also occur, and the apparent discrepancy in linkage group size between the genetic and the physical results remains to be resolved.

The observations discussed above involved the use of Rec$^+$ (RecBCD recombination pathway) strains of *E. coli*. *recB* and *recC* mutants, lacking the RecBCD enzyme, yield 100- to 1,000-fold fewer viable recombinants after Hfr conjugation (Clark, 1967; Low, 1968; Birge and Low, 1974). Suppressor mutations (*sbcA* or *sbcB*) that activate the RecE or RecF recombination pathway restore recombination proficiency (Clark, 1973; see Mahajan, this volume). In strains with an active RecF pathway, linkage is considerably reduced (Mahajan and Datta, 1977), recombination is slower (Mahajan and Datta, 1979; Lloyd and Thomas, 1983), and efficient recombination requires higher levels of RecA protein than are required by the RecBCD pathway (Lloyd and Thomas, 1983). It has been suggested that the RecF pathway integrates large pieces of single-stranded donor DNA, while the RecBCD pathway integrates double-stranded DNA (Mahajan and Datta, 1979; Lloyd and Thomas, 1983) and perhaps also small segments of single-stranded DNA (Lloyd and Thomas, 1983). For further discussion of recombination following Hfr conjugation, see Mahajan and Datta (1979), Lloyd and Thomas (1983, 1984), Clark et al. (1984), and Mahajan (this volume).

Although homologous recombination occurs after Hfr conjugation, site-specific recombination may also be involved. It has been proposed that a site-specific conjugal recombination (or *ssr*) system recognizes the transferred portion of the F-factor DNA and catalyzes an exchange within the leading chromosomal DNA independently of both *recA* and *recBCD* (Birge, 1983). Recombinants are not actually observed, however, unless a subsequent *recA*-dependent event completes the genetic exchange. Frequent recombination exchange (or *fre*) regions where genetic exchanges by the RecF pathway are clustered on the *E. coli* genome have also been suggested (Bresler et al., 1978, 1981a). Chi sites within integrated λ prophages alter the distribution of genetic exchanges in their immediate vicinity when the RecBCD pathway acts after conjugation (Dower and Stahl, 1981). Genetic exchanges are therefore not entirely random following conjugation, as in other situations (see Mosig, this volume).

F′ conjugation and merodiploids. F′ factors (for reviews, see Low, 1972; Holloway and Low, 1987) arise from Hfr strains by illegitimate recombination or abortive intramolecular transposition (see Fig. 9). Type I F′s incorporate host DNA from only one side of the integrated F factor and leave behind part of the F factor. Type II F′s incorporate host DNA from both sides of the integrated F factor and retain the complete F factor. The primary F′ strain, in which the F′ arises, remains haploid and may require the F′ for viability. These strains cannot mobilize the chromosome because the F′ cannot readily recombine with it (Pittard and Ramakrishnan, 1964).

Transfer of an F′ to a recipient produces a secondary F′ strain, which is merodiploid, or partially diploid, for the host DNA carried by the F′. Such merodiploids are frequently used for genetic complementation experiments, but *recA* mutants should be used to preclude recombination, which can lead to confusing results. Recombination between the chromosomal and F′ copies of host DNA, and subsequent segregation (which may not be necessary if the recombi-

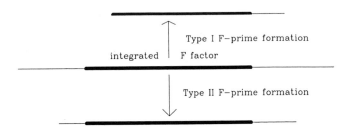

FIGURE 9. F′ formation. An Hfr strain with the F factor integrated in its chromosome is shown in the central portion of the figure. A recombination event between a point within the integrated F factor and flanking chromosomal DNA will yield a type I F′. A recombination event between points in the flanking DNA on either side of the integrated F factor will give rise to a type II F′ factor. Although depicted as linear DNA molecules in this figure, both types of F′ formation events give rise to circular molecules.

nation event is nonreciprocal), can convert an initial heterogenote into a homogenote. Such "homogenotization" is commonly used to move alleles between strains. The enhanced recombination properties of F′ plasmids cause them to demonstrate homogenotization more frequently than most other plasmids (Yancey and Porter, 1985).

As a result of general recombination in secondary F′ strains, there is an equilibrium between the integrated and autonomous states of the F′. Because of their ability to reintegrate by homologous recombination. F′'s were initially described as F factors that could "remember" where they had been integrated (Richter, 1957). In its autonomous state, the F′, like the F^+ factor, promotes conjugation and self-transfer. Such conjugation establishes the F′ in the recipient cell as an independent replicon (replication), but the host DNA carried by the F′ can recombine with recipient DNA even when complete replication does not occur (for example, if part of the F DNA is not transferred, as frequently happens with large F′'s). The integrated form of the F′ behaves like an Hfr, and F′ mobilization of the chromosome can be used to move specific markers by conjugation when a suitable Hfr is not available.

2. Other conjugation systems in gram-negative bacteria

Gram-negative bacteria commonly contain plasmids, many of which have conjugative or mobilizable properties, or both. Some of these plasmids possess a transfer system that is highly analogous to that of the F factor (for example, overlapping *fin* specificity or specific *tra* gene complementation), but transfer systems quite distinct from F are also found. Although most such plasmids rarely mobilize or conduct (mobilization due to very temporary interactions such as transposition cointegrates) donor chromosomal DNA to the recipient, some move donor DNA to a recipient by either an Hfr-like (very stable integration of the plasmid) or an F′-like (recombinational equilibrium) mechanism (for examples of genome mobilization by non-F plasmids, see Holloway, 1979).

Although low-level chromosome mobilization occurs without firm evidence for plasmid integration, various plasmids integrate into the chromosome at apparently random locations, forming Hfr-like strains in which the DNA replication defect of a *dnaA*(Ts) mutation is suppressed at high temperature (Datta and Barth, 1976). Most chromosome mobilization appears more akin to F′ mobilization (a recombinational equilibrium between the autonomous and integrated states) than to stable Hfr-like mobilization. The same plasmid often gives quite different results in different bacterial species. This may often be due to differing degrees and amounts of homology; for example, with plasmid RP4 in *E. coli* the amount of homology between the plasmid and the chromosome was correlated with the ability to mobilize the chromosome by an apparently *recA*⁺-dependent mechanism (Grinter, 1981).

Nonconjugative but mobilizable plasmids can be transferred to a recipient cell if the necessary conjugative functions are provided by a suitable plasmid in the donor. The nonconjugative plasmids must have an *oriT*-like site (often called *nic* or *bom*, for basis of mobilization) and *mob* (mobilization) genes whose products act at the *oriT*-like site. The conjugal transfer of such plasmids is reviewed elsewhere (Clark and Warren, 1979; Willetts and Wilkins, 1984). Nonconjugative, nonmobilizable plasmids are transferred by conjugation at very low frequency via transposition cointegrates formed with a self-transmissible plasmid in the donor. For example, Tn*1000* forms cointegrates between the F factor and pBR322 (a nonconjugative, nonmobilizable plasmid) and the two are transferred together (Guyer, 1978); other self-transmissible plasmids containing a suitable transposable element may be able to act similarly. The plasmids transferred by this mechanism contain one copy of the transposable element left by resolution of the cointegrate in the recipient (see Syvanen, this volume; Hatfull and Grindley, this volume). Mutations in genes on nonmobilizable plasmids can thus be readily obtained (Sancar et al., 1981).

3. Conjugation in gram-positive bacteria

Conjugation in gram-positive bacteria has been less extensively characterized than that in gram-negative bacteria (see Clewell, 1981; Clewell et al., 1987; Lacks, this volume).

Sex pili have not been observed in conjugative gram-positive bacteria, and conjugal transfer frequently depends on the coprecipitation of donor and recipient cells onto a solid surface. High-efficiency mating in liquid culture can be obtained, however, with a few plasmid-containing *Streptococcus faecalis* strains, whose mating is controlled by small peptide sex pheromones (Mori, 1987). The recipient cells produce a pheromone that induces the donor cell to produce a surface protein that promotes cell clumping (Dunny et al., 1978). After receiving the plasmid, the recipient cell ceases production of its particular pheromone, but continues to produce other pheromones (Dunny et al., 1979; Ike et al., 1983). The nature of the cell aggregation and the mechanism of DNA transfer remain unknown. Some of these plasmids can mobilize nonconjugative plasmids (Dunny and Clewell, 1975; Smith et al., 1980), and others can mobilize chromosomal DNA (Franke et al., 1978), but the basis of mobilization is not yet characterized.

Even less is known about the mechanism of transfer for plasmids that do not utilize the pheromone system. These plasmids have a broader host range and generally transfer at considerably lower frequencies. They are frequently grouped with numerous chromosomal elements possessing both conjugative and transposition properties (for reviews, see Lacks, this volume; Clewell and Gawron-Burke, 1986).

IV. SUMMARY

The ability of bacterial cells to exchange genetic information is an important component of their genetic variability. Transformation, transduction, and conjugation are the three major modes by which such exchange is achieved. Although gene transfer may sometimes occur without the involvement of recombination, recombination is often required to complete the process effectively. Recombination, in turn, is frequently dependent on the gene transfer process for providing the donor DNA in a form in which it can readily be acted upon. Although the interrelationships between gene transfer and recombination are not always well understood, an understanding of gene transfer mechanisms promises to be valuable in ultimately understanding the mechanistic basis of recombination.

LITERATURE CITED

Abdel-Monem, M., G. Taucher-Scholz, and M.-Q. Klinkert. 1983. Identification of *Escherichia coli* DNA helicase I as the *traI* gene product of the F sex factor. *Proc. Natl. Acad. Sci. USA* **80:**4659–4663.

Achtman, M., N. Kennedy, and R. Skurray. 1977. Cell-cell interactions in conjugating *Escherichia coli*: role of *traT* protein in surface exclusion. *Proc. Natl. Acad. Sci. USA* **74:**5104–5108.

Achtman, M., P. A. Manning, C. Edelbluth, and P. Herrlich. 1979. Export without proteolytic processing of inner and outer membrane proteins encoded by F sex factor *tra* cistrons in *Escherichia coli* minicells. *Proc. Natl. Acad. Sci. USA* **76:**4837–4841.

Achtman, M., G. Morelli, and S. Schwuchow. 1978. Cell-cell interactions in conjugating *Escherichia coli*: role of F pili and fate of mating aggregates. *J. Bacteriol.* **135:**1053–1061.

Aguero, C. W., L. Aron, A. G. DeLuca, K. N. Timmis, and F. C. Cabello. 1984. A plasmid-encoded outer membrane protein, TraT, enhances resistance of *Escherichia coli* to phagocytosis. *Infect. Immun.* **46:**740–746.

Akrigg, A., and S. R. Ayad. 1970. Studies on competence inducing factor of *Bacillus subtilis. Biochem. J.* **117:**397–403.

Albritton, W. L., J. W. Bendler, and J. K. Setlow. 1981. Plasmid transformation in *Haemophilus influenzae. J. Bacteriol.* **145:**1099–1101.

Arber, W. 1960. Transduction of chromosomal genes and episomes in *Escherichia coli. Virology* **11:**273–288.

Armstrong, G. D., L. S. Frost, H. J. Vogel, and W. Paranchych. 1981. Nature of the carbohydrate and phosphate associated with ColB2 and EDP208 pilin. *J. Bacteriol.* **145:**1167–1176.

Balganesh, M., and J. K. Setlow. 1985. Differential behavior of plasmids containing chromosomal DNA insertions of various sizes during transformation and conjugation in *Haemophilus influenzae. J. Bacteriol.* **161:**141–146.

Barany, F., M. E. Kahn, and H. O. Smith. 1983. Directional transport and integration of donor DNA in *Haemophilus influenzae* transformation. *Proc. Natl. Acad. Sci. USA* **80:**7274–7278.

Barouki, R., and H. O. Smith. 1985. Reexamination of phenotypic defects in *rec-1* and *rec-2* mutants of *Haemophilus influenzae* Rd. *J. Bacteriol.* **163**:629–634.

Benzer, S. 1957. The elementary units of heredity, p. 70–93. *In* W. D. McElroy and B. Glass (ed.), *The Chemical Basis of Heredity*. Johns Hopkins Press, Baltimore.

Better, M., and D. R. Helinski. 1983. Isolation and characterization of the *recA* gene of *Rhizobium meliloti*. *J. Bacteriol.* **155**:311–316.

Birge, E. A. 1983. Site-specific recombination following conjugation in *Escherichia coli* K-12. *Mol. Gen. Genet.* **192**:366–372.

Birge, E. A., and K. B. Low. 1974. Detection of transcribable recombination products following conjugation in Rec$^+$, RecB$^-$ and RecC$^-$ strains of *Escherichia coli* K-12. *J. Mol. Biol.* **83**:447–457.

Biswas, G. D., K. L. Burnstein, and P. F. Sparling. 1986. Linearization of donor DNA during plasmid transformation in *Neisseria gonorrhoeae*. *J. Bacteriol.* **168**:756–761.

Biswas, G. D., J. Graves, T. E. Sox, F. C. Tenover, and P. F. Sparling. 1982. Marker rescue by a homologous recipient plasmid during transformation of gonococci by a hybrid Pc^r plasmid. *J. Bacteriol.* **151**:77–82.

Biswas, G. D., T. Sox, E. Blackman, and P. F. Sparling. 1977. Factors affecting genetic transformation of *Neisseria gonorrhoeae*. *J. Bacteriol.* **129**:983–992.

Bodmer, W. F., and A. T. Ganesan. 1964. Biochemical and genetic studies of integration and recombination in *Bacillus subtilis* transformation. *Genetics* **50**:717–738.

Bramucci, M. G., and P. S. Lovett. 1976. Low-frequency, PBS-1 mediated plasmid transduction in *Bacillus pumilus*. *J. Bacteriol.* **127**:829–831.

Bresler, S. E., S. V. Krivonogov, and V. A. Lanzov. 1978. Scale of the genetic map and genetic control of recombination after conjugation in *Escherichia coli* K-12. *Mol. Gen. Genet.* **166**:337–346.

Bresler, S. E., I. Yu. Goryshin, and V. A. Lanzov. 1981a. The process of general recombination in *Escherichia coli* K-12: structure of intermediate products. *Mol. Gen. Genet.* **183**:139–143.

Bresler, S. E., S. V. Krivonogov, and V. A. Lanzov. 1981b. Recombinational instability of F′ plasmids in *Escherichia coli* K-12: location of *fre*-sites. *Mol. Gen. Genet.* **183**:192–196.

Buzby, J. S., R. D. Porter, and S. E. Stevens, Jr. 1985. Expression of the *Escherichia coli lacZ* gene on a plasmid vector in a cyanobacterium. *Science* **230**:805–807.

Cahn, F. H., and M. S. Fox. 1968. Fractionation of transformable bacteria from competent cultures of *Bacillus subtilis* on Renografin gradients. *J. Bacteriol.* **95**:867–875.

Campbell, A. 1960. On the mechanism of the recombinational event in the formation of transducing phages. *Virology* **11**:339–348.

Campbell, A. 1964. Transduction, p. 49–85. *In* I. C. Gunsalus and R. Y. Stanier (ed.), *The Bacteria*, vol. 5. Academic Press, Inc., New York.

Canosi, U., A. Iglesias, and T. A. Trautner. 1981. Plasmid transformation in *Bacillus subtilis*: effects of insertion of *Bacillus subtilis* DNA in plasmid pC194. *Mol. Gen. Genet.* **181**:434–440.

Canosi, U., G. Morelli, and T. A. Trautner. 1978. The relationship between molecular structure and transformation efficiency of some *S. aureus* plasmids isolated from *B. subtilis*. *Mol. Gen. Genet.* **166**:259–267.

Chang, S., and S. N. Cohen. 1979. High frequency transformation of *Bacillus subtilis* protoplasts by plasmid DNA. *Mol. Gen. Genet.* **168**:111–115.

Chaudhury, A. M., and G. R. Smith. 1984. A new class of *Escherichia coli recBC* mutants: implications for the role of RecBC enzyme in homologous recombination. *Proc. Natl. Acad. Sci. USA* **81**:7850–7854.

Cheah, K.-C., A. Ray, and R. Skurray. 1986. Expression of F plasmid *traT*: independence of *traY* → *Z* promoter and *traJ* control. *Plasmid* **16**:101–107.

Chelala, C. A., and P. Margolin. 1974. Effects of deletions on contransduction linkage in *Salmonella typhimurium*: evidence that bacterial chromosome deletions affect the formation of transducing DNA fragments. *Mol. Gen. Genet.* **131**:97–112.

Clark, A. J. 1967. The beginning of a genetic analysis of recombination-proficiency. *J. Cell. Physiol.* **70**(Suppl. 1):165–180.

Clark, A. J. 1971. Toward a metabolic interpretation of genetic recombination of *E. coli* and its phages. *Annu. Rev. Microbiol.* **25**:437–464.

Clark, A. J. 1973. Recombination deficient mutants of *E. coli* and other bacteria. *Annu. Rev. Genet.* 7:67–86.

Clark, A. J., S. J. Sandler, D. K. Willis, C. C. Chu, M. A. Blanar, and S. T. Lovett. 1984. Genes of the RecE and RecF pathways of conjugational recombination in *Escherichia coli*. Cold Spring Harbor Symp. Quant. Biol. 49:453–462.

Clark, A. J., and G. J. Warren. 1979. Conjugal transmission of plasmids. *Annu. Rev. Genet.* 13:99–125.

Claverys, J. P., J. C. Lefevre, and A. M. Sicard. 1980. Transformation of *Streptococcus pneumoniae* with *S. pneumoniae*-lambda hybrid DNA: induction of deletions. *Proc. Natl. Acad. Sci. USA* 77:3534–3538.

Clewell, D. B. 1981. Plasmids, drug resistance, and gene transfer in the genus *Streptococcus*. *Microbiol. Rev.* 45:409–436.

Clewell, D. B., E. E. Ehrenfeld, F. An, R. E. Kessler, R. Wirth, M. Mori, C. Kitada, M. Fujino, Y. Ike, and A. Suzuki. 1987. Sex pheromones and plasmid-related conjugation phenomena in *Streptococcus faecalis*, p. 2–7. *In* J. J. Ferretti and R. Curtiss III (ed.), *Streptococcal Genetics*. American Society for Microbiology, Washington, D.C.

Clewell, D. B., and C. Gawron-Burke. 1986. Conjugative transposons and the dissemination of antibiotic resistance in streptococci. *Annu. Rev. Microbiol.* 40:635–659.

Cohen, S. N., A. C. Y. Chang, and L. Hsu. 1972. Nonchromosomal antibiotic resistance in bacteria: genetic transformation of *Escherichia coli* by R-factor DNA. *Proc. Natl. Acad. Sci. USA* 69:2110–2114.

Contente, S., and D. Dubnau. 1979a. Characterization of plasmid transformation in *Bacillus subtilis*: kinetic properties and the effect of DNA concentration. *Mol. Gen. Genet.* 167:251–258.

Contente, S., and D. Dubnau. 1979b. Marker rescue transformation by linear plasmid DNA in *Bacillus subtilis*. *Plasmid* 2:555–571.

Cosloy, S. D. 1982. Analysis of genetic recombination by the RecBC and RecF pathways of *Escherichia coli* K12, p. 261–273. *In* U. N. Streips, S. H. Goodgal, W. R. Guild, and G. A. Wilson (ed.), *Genetic Exchange: a Celebration and a New Generation*. Marcel Dekker, Inc., New York.

Cosloy, S. D., and M. Oishi. 1973. Genetic transformation in *Escherichia coli* K-12. *Proc. Natl. Acad. Sci. USA* 70:84–87.

Craig, N. L., and N. Kleckner. 1987. Transposition and site-specific recombination, p. 1054–1070. *In* F. C. Neidhardt, J. L. Ingraham, K. B. Low, B. Magasanik, M. Schaechter, and H. E. Umbarger (ed.), *Escherichia coli and Salmonella typhimurium: Cellular and Molecular Biology*, vol. 2. American Society for Microbiology, Washington, D.C.

Crawford, I. P., and J. Preiss. 1972. Distribution of closely linked markers following intragenic recombination in *Escherichia coli*. *J. Mol. Biol.* 71:717–733.

Cullum, J., and P. Broda. 1979. Chromosome transfer and Hfr formation by F in *rec*⁺ and *recA* strains of *Escherichia coli* K-12. *Plasmid* 2:358–365.

Danner, D. B, R. A. Deich, K. L. Sisco, and H. O. Smith. 1980. An 11-base-pair sequence determines the specificity of DNA uptake in *Haemophilus* transformation. *Gene* 11:311–318.

Danner, D. B., H. O. Smith, and S. A. Narang. 1982. Construction of DNA recognition sites active in *Haemophilus* transformation. *Proc. Natl. Acad. Sci. USA* 79:2393–2397.

Date, T., M. Inuzuka, and M. Tomoeda. 1977. Purification and characterization of F pili from *Escherichia coli*. *Biochemistry* 16:5579–5585.

Datta, N., and P. T. Barth. 1976. Hfr formation by I pilus-determining plasmids in *Escherichia coli* K-12. *J. Bacteriol.* 125:811–817.

Davidson, N., R. C. Deonier, S. Hu, and E. Ohtsubo. 1975. Electron microscope heteroduplex studies of sequence relations among plasmids of *Escherichia coli*. X. Deoxyribonucleic acid sequence organization of F and of F-primes, and the sequences involved in Hfr formation, p. 56–65. *In* D. Schlessinger (ed.), *Microbiology—1974*. American Society for Microbiology, Washington, D.C.

Deonier, R. C., and L. Mirels. 1977. Excision of F plasmid sequences by recombination at directly repeated insertion sequence 2 elements: involvement of *recA*. *Proc. Natl. Acad. Sci. USA* 74:3965–3969.

Dooley, D. C., C. T. Hadden, and E. W. Nester. 1971. Macromolecular synthesis by *Bacillus subtilis* during development of the competent state. *J. Bacteriol.* 108:668–679.

Dougherty, T. J., A. Asmus, and A. Tomasz. 1979. Specificity of DNA uptake in genetic transformation of gonococci. *Biochem. Biophys. Res. Commun.* **86**:97–104.

Dove, W. F. 1968. The genetics of the lambdoid phages. *Annu. Rev. Genet.* **2**:305–340.

Dower, N. A., and F. W. Stahl. 1981. χ activity during transduction associated recombination. *Proc. Natl. Acad. Sci. USA* **78**:7033–7037.

Dubnau, D., and C. Cirigliano. 1972. Fate of transforming DNA following uptake by competent *Bacillus subtilis*. III. Formation and properties of products isolated from transformed cells which are derived entirely from donor DNA. *J. Mol. Biol.* **64**:9–29.

Dubnau, D., R. Davidoff-Abelson, B. Scher, and C. Cirigliano. 1973. Fate of transforming deoxyribonucleic acid after uptake by competent *Bacillus subtilis*: phenotypic characterization of radiation-sensitive recombination-deficient mutants. *J. Bacteriol.* **114**:273–286.

Dubnau, E., and B. A. D. Stocker. 1967. Behavior of three colicin E factors and an R (drug-resistance) factor in Hfr crosses in *Salmonella typhimurium*. *Genet. Res.* **9**:283–297.

Duckworth, D. H. 1970. Biological activity of bacteriophage ghosts and "take-over" of host functions by bacteriophage. *Bacteriol. Rev.* **34**:344–363.

Dunny, G., B. Brown, and D. B. Clewell. 1978. Induced cell aggregation and mating in *Streptococcus faecalis*. Evidence for a bacterial sex pheromone. *Proc. Natl. Acad. Sci. USA* **75**:3479–3483.

Dunny, G. M., and D. B. Clewell. 1975. Transmissible toxin (hemolysin) plasmid in *Streptococcus faecalis* and its mobilization of a noninfectious drug resistance plasmid. *J. Bacteriol.* **124**:784–790.

Dunny, G., R. Craig, R. Carron, and D. Clewell. 1979. Plasmid transfer in *Streptococcus faecalis*. Production of multiple sex pheromones by recipients. *Plasmid* **2**:454–465.

Ebel-Tsipis, J., M. S. Fox, and D. Botstein. 1972. Generalized transduction by bacteriophage P22 in *Salmonella typhimurium*. II. Mechanism of integration of transducing DNA. *J. Mol. Biol.* **71**:449–469.

Echols, H., R. Gingery, and L. Moore. 1968. Integrative recombination function of bacteriophage λ; evidence for a site-specific recombination enzyme. *J. Mol. Biol.* **34**:251–260.

Eisenstadt, E., R. Lange, and K. Willecke. 1975. Competent *Bacillus subtilis* cultures synthesize a denatured DNA binding activity. *Proc. Natl. Acad. Sci. USA* **72**:323–327.

Eitner, G., B. Adler, V. A. Lanzov, and J. Hofmeister. 1982. Interspecies *recA* protein substitution in *Escherichia coli* and *Proteus mirabilis*. *Mol. Gen. Genet.* **185**:481–486.

Everett, R., and N. Willetts. 1980. Characterization of an in vivo system for nicking at the origin of conjugal DNA transfer of the sex factor F. *J. Mol. Biol.* **136**:129–150.

Everett, R., and N. Willetts. 1982. Cloning, mutation, and location of the F origin of conjugal transfer. *EMBO J.* **1**:747–753.

Falkow, S. 1975. *Infectious Multiple Drug Resistance*. Pion Press, London.

Ferrari, F. A., A. Nguyen, D. Lang, and J. A. Hoch. 1983. Construction and properties of an integrable plasmid for *Bacillus subtilis*. *J. Bacteriol.* **154**:1513–1515.

Finch, P. W., C. L. Brough, and P. T. Emmerson. 1986. Molecular cloning of a *recA*-like gene of *Methylophilus methylotrophus* and identification of its product. *Gene* **44**:47–53.

Fox, M. S. 1960. Fate of transforming deoxyribonucleate following fixation by transforming bacteria. II. *Nature* (London) **187**:1004–1006.

Fox, M. S. 1978. Some features of genetic recombination in prokaryotes. *Annu. Rev. Genet.* **12**:47–68.

Fox, M. S., and M. K. Allen. 1964. On the mechanism of deoxyribonucleate integration in pneumococcal transformation. *Proc. Natl. Acad. Sci. USA* **52**:412–419.

Franke, A. E., G. M. Dunny, B. L. Brown, F. An, D. R. Oliver, S. P. Damle, and D. B. Clewell. 1978. Gene transfer in *Streptococcus faecalis*: evidence for the mobilization of chromosomal determinants by transmissible plasmids, p. 45–47. *In* D. Schlessinger (ed.), *Microbiology—1978*. American Society for Microbiology, Washington, D.C.

Franklin, N. C. 1971. Illegitimate recombination, p. 175–194. *In* A. D. Hershey (ed.), *The Bacteriophage Lambda*. Cold Spring Harbor Laboratory Press, Cold Spring Harbor, N.Y.

Fredericq, P. 1965. Genetics of colicinogenic factors. *Zentralbl. Bakteriol. Parasintenkd. Infektionskr. Hyg.* Abt. 1 Orig. Reihe A **196**:142–151.

Frost, L. S., W. Paranchych, and N. S. Willetts. 1984. DNA sequence of the F *traALE* region that includes the gene for F pilin. *J. Bacteriol.* **160**:395–401.

Gaffney, D., R. Skurray, and N. Willetts. 1983. Regulation of the F conjugation genes studied by hybridization and *tra-lacZ* fusion. *J. Mol. Biol.* **168:**103–122.

Genthner, F. J., and J. D. Wall. 1984. Isolation of a recombination-deficient mutant of *Rhodopseudomonas capsulata. J. Bacteriol.* **160:**971–975.

Ghei, O. K., and S. Lacks. 1967. Recovery of donor deoxyribonucleic acid marker activity from eclipse in pneumococcal transformation. *J. Bacteriol.* **93:**816–829.

Gingery, R., and H. Echols. 1967. Mutants of bacteriophage λ unable to integrate into the host chromosome. *Proc. Natl. Acad. Sci. USA* **58:**1507–1514.

Goldberg, I., and J. J. Mekalanos. 1986. Cloning of the *Vibrio cholerae recA* gene and construction of a *Vibrio cholerae recA* mutant. *J. Bacteriol.* **165:**715–722.

Golub, E. I., and K. B. Low. 1983. Indirect stimulation of genetic recombination. *Proc. Natl. Acad. Sci. USA* **80:**1401–1405.

Goodgal, S. H. 1982. DNA uptake in *Haemophilus* transformation. *Annu. Rev. Genet.* **16:**169–192.

Goto, N., A. Shoji, S. Horiuchi, and R. Nakaya. 1984. Conduction of nonconjugative plasmids by F′*lac* is not necessarily associated with transposition of the γ-δ sequence. *J. Bacteriol.* **159:**590–596.

Gottesman, M. E., and M. B. Yarmolinsky. 1968. Integration-negative mutants of bacteriophage lambda. *J. Mol. Biol.* **31:**487–505.

Grindley, N. D. F., and R. R. Reed. 1985. Transpositional recombination in prokaryotes. *Annu. Rev. Genet.* **54:**863–896.

Grinter, N. J. 1981. Analysis of chromosome mobilization using hybrids between plasmid RP4 and a fragment of bacteriophage λ carrying IS1. *Plasmid* **5:**267–276.

Grubb, W. B., and R. J. O'Reilly. 1971. Joint transduction of separate extrachromosomal drug resistance determinants in *Staphylococcus aureus* E169. *Biochem. Biophys. Res. Commun.* **42:**377–383.

Guild, W. R., and M. Robison. 1963. Evidence for message reading from a unique strand of pneumococcal DNA. *Proc. Natl. Acad. Sci. USA* **50:**106–112.

Guyer, M. S. 1978. The γ-δ sequence of F is an insertion sequence. *J. Mol. Biol.* **125:**233–247.

Hadden, C., and E. W. Nester. 1968. Purification of competent cells in the *Bacillus subtilis* transformation system. *J. Bacteriol.* **95:**876–885.

Hamood, A. N., G. S. Pettis, C. D. Parker, and M. A. McIntosh. 1986. Isolation and characterization of a *Vibrio cholerae recA* gene. *J. Bacteriol.* **167:**375–378.

Hanahan, D. 1987. Mechanisms of DNA transformation, p. 1177–1183. In F. C. Neidhardt, J. L. Ingraham, K. B. Low, B. Magasanik, M. Schaechter, and H. E. Umbarger (ed.), *Escherichia coli and Salmonella typhimurium: Cellular and Molecular Biology*, vol. 2. American Society for Microbiology, Washington, D.C.

Harriman, P. 1971. Appearance of transducing activity in P1-infected *Escherichia coli. Virology* **45:**324–325.

Harriman, P. D. 1972. A single-burst analysis of the production of P1 infectious and transducing particles. *Virology* **48:**595–600.

Harris, W. J., and G. C. Barr. 1971. Structural features of DNA in competent *Bacillus subtilis. Mol. Gen. Genet.* **113:**316–330.

Hayes, W. 1953. The mechanism of genetic recombination in *Escherichia coli. Cold Spring Harbor Symp. Quant. Biol.* **18:**75–93.

Herriott, R. M., E. M. Meyer, and M. Vogt. 1970. Refined nongrowth media for stage II development of competence in *Haemophilus influenzae. J. Bacteriol.* **101:**517–524.

Hertman, I., and S. E. Luria. 1967. Transduction studies on the role of a *rec*+ gene in the ultraviolet induction of prophage lambda. *J. Mol. Biol.* **23:**117–133.

Hoekstra, W. P. M., P. G. de Haan, J. E. N. Bergmans, and E. M. Zuidweg. 1976. Transformation in *E. coli* K12: relation of linkage to distance between markers. *Mol. Gen. Genet.* **145:**109–110.

Holloman, W. K., and C. M. Radding. 1976. Recombination promoted by superhelical DNA and the *recA* gene of *Escherichia coli. Proc. Natl. Acad. Sci. USA* **73:**3910–3914.

Holloway, B. W. 1979. Plasmids that mobilize the bacterial chromosome. *Plasmid* **2:**1–19.

Holloway, B., and K. B. Low. 1987. F-prime and R-prime factors, p. 1145–1153. *In* F. C. Neidhardt, J. L. Ingraham, K. B. Low, B. Magasanik, M. Schaechter, and H. E. Umbargar (ed.), *Escherichia* *coli and Salmonella typhimurium: Cellular and Molecular Biology*, vol. 2. American Society for Microbiology, Washington, D.C.

Hotchkiss, R. D. 1974. Models of genetic recombination. *Annu. Rev. Genet.* **8**:445–468.

Ike, Y., R. Craig, B. White. Y. Yagi, and D. Clewell. 1983. Modification of *Streptococcus faecalis* sex pheromones after acquisition of plasmid DNA. *Proc. Natl. Acad. Sci. USA* **80**:5369–5373.

Ikeda, H., and J. Tomizawa. 1965. Transducing fragments in generalized transduction by phage P1. II. Association of DNA and protein in the fragments. *J. Mol. Biol.* **14**:110–119.

Iordanescu, S. 1977. Relationships between co-transducible plasmids in *Staphylococcus aureus*. *J. Bacteriol.* **129**:71–75.

Ippen-Ihler, K., and E. G. Minkley, Jr. 1986. The conjugation system of F, the fertility factor of *Escherichia coli*. *Annu. Rev. Genet.* **20**:593–624.

Johnson, D., R. Everett, and N. Willetts. 1981. Cloning of F DNA fragments carrying the origin of transfer *oriT* and the fertility inhibition gene *finP*. *J. Mol. Biol.* **153**:187–202.

Kahn, M. E., F. Barany, and H. O. Smith. 1983. Transformasomes: specialized membranous structures that protect DNA during *Haemophilus* transformation. *Proc. Natl. Acad. Sci. USA* **80**: 6927–6931.

Kahn, M. E., G. Maul, and S. H. Goodgal. 1982. Possible mechanism for donor DNA binding and transport in *Haemophilus*. *Proc. Natl. Acad. Sci. USA* **79**:6370–6374.

Karu, A. E., V. MacKay, P. J. Goldmark, and S. Linn. 1973. The *recBC* deoxyribonuclease of *Escherichia coli* K-12: substrate specificity and reaction intermediates. *J. Biol. Chem.* **248**:4874–4884.

Keener, S. L., K. P. McNamee, and K. McEntee. 1984. Cloning and characterization of *recA* genes from *Proteus vulgaris*, *Erwinia carotovora*, *Shigella flexneri*, and *Escherichia coli* B/r. *J. Bacteriol.* **160**: 153–160.

Kingsman, A., and N. Willetts. 1978. The requirements for conjugal DNA synthesis in the donor strain during F*lac* transfer. *J. Mol. Biol.* **122**:287–300.

Kleckner, N. 1981. Transposable elements in prokaryotes. *Annu. Rev. Genet.* **15**:341–404.

Kokjohn, T. A., and R. V. Miller. 1985. Molecular cloning and characterization of the *recA* gene of *Pseudomonas aeruginosa* PAO. *J. Bacteriol.* **163**:568–572.

Kolowsky, K. S., and Λ. A. Szalay. 1986. Double-stranded gap repair in the photosynthetic prokaryote *Synechococcus* R2. *Proc. Natl. Acad. Sci. USA* **83**:5578–5582.

Kolowsky, K. S., J. G. K. Williams, and A. A. Szalay. 1984. Length of foreign DNA in chimeric plasmids determines the efficiency of its integration into the chromosome of the cyanobacterium *Synechococcus* R2. *Gene* **27**:289–299.

Koomey, J. M., and S. Falkow. 1987. Cloning of the *recA* gene of *Neisseria gonorrhoeae* and construction of gonococcal *recA* mutants. *J. Bacteriol.* **169**:790–795.

Kylberg, K. J., M. M. Bending, and H. Drexler. 1975. Characterization of transduction by bacteriophage T1: time of production and density of transducing particles. *J. Virol.* **16**:854–858.

Lacks, S. 1962. Molecular fate of DNA in genetic transformation of pneumococcus. *J. Mol. Biol.* **5**: 119–131.

Lacks, S. 1977. Binding and entry of DNA in bacterial transformation, p. 179–232. *In* J. Reissig (ed.), *Microbial Interactions*. Chapman & Hall, Ltd., London.

Lacks, S. 1979. Uptake of circular deoxyribonucleic acid and mechanism of deoxyribonucleic acid transport in genetic transformation of *Streptococcus pneumoniae*. *J. Bacteriol.* **138**:404–409.

Lacks, S., and B. Greenberg. 1973. Competence for deoxyribonucleic acid uptake and deoxyribonuclease action external to cells in the genetic transformation of *Diplococcus pneumoniae*. *J. Bacteriol.* **114**:152–163.

Lacks, S., and B. Greenberg. 1976. Single-strand breakage on binding of DNA to cells in the genetic transformation of *Diplococcus pneumoniae*. *J. Mol. Biol.* **101**:255–275.

Lacks, S., B. Greenberg, and M. Neuberger. 1974. Role of a deoxyribonuclease in the genetic transformation of *Diplococcus pneumoniae*. *Proc. Natl. Acad. Sci. USA* **71**:2305–2309.

Lacks, S., B. Greenberg, and M. Neuberger. 1975. Identification of a deoxyribonuclease implicated in genetic transformation of *Diplococcus pneumoniae*. *J. Bacteriol.* **123**:222–232.

Laine, S., D. Moore, P. Kathir, and K. Ippen-Ihler. 1985. Genes and gene products involved in the synthesis of F pili, p. 535–553. *In* D. R. Helinski, S. N. Cohen, D. B. Clewell, D. A. Jackson, and A. Hollaender (ed.), *Plasmids in Bacteria*. Plenum Publishing Corp., New York.

LeClerc, J. E., and J. K. Setlow. 1974. Transformation in *Haemophilus influenzae*, p. 187–207. *In* R. F. Grell (ed.) *Mechanisms in Recombination*. Plenum Publishing Corp., New York.

LeClerc, J. E., and J. K. Setlow. 1975. Single-strand regions in the deoxyribonucleic acid of competent *Haemophilus influenzae*. *J. Bacteriol.* **122**:1091–1102.

Lederberg, J. 1955. Recombination mechanisms in bacteria. *J. Cell. Comp. Physiol.* **45**(Suppl. 2):75–107.

Lloyd, R. G., and A. Thomas. 1983. On the nature of the RecBC and RecF pathways of conjugal recombination in *Escherichia coli*. *Mol. Gen. Genet.* **190**:156–161.

Lloyd, R. G., and A. Thomas. 1984. A molecular model for conjugational recombination in *Escherichia coli* K12. *Mol. Gen. Genet.* **197**:328–336.

Lopez, P., M. Espinosa, D. L. Stassi, and S. A. Lacks. 1982. Facilitation of plasmid transfer in *Streptococcus pneumoniae* by chromosomal homology. *J. Bacteriol.* **150**:692–701.

Love, P. E., and R. E. Yasbin. 1985. DNA-damage-inducible (*din*) loci are transcriptionally activated in competent *Bacillus subtilis*. *Proc. Natl. Acad. Sci. USA* **82**:6201–6205.

Love, P. E., and R. E. Yasbin. 1986. Induction of the *Bacillus subtilis* SOS-response by *Escherichia coli* RecA protein. *Proc. Natl. Acad. Sci. USA* **83**:5204–5208.

Loveday, K. S., and M. S. Fox. 1978. The fate of bacteriophage φe transfecting DNA. *Virology* **85**:387–403.

Lovett, P. S., and J. R. Roberts. 1985. Purification of a RecA protein analogue from *Bacillus subtilis*. *J. Biol. Chem.* **260**:3305–3313.

Low, B. 1965. Low recombination frequency for markers very near the origin in conjugation in *E. coli*. *Genet. Res.* **6**:469–473.

Low, B. 1968. Formation of merodiploids in matings with a class of Rec⁻ recipient strains of *Escherichia coli* K12. *Proc. Natl. Acad. Sci. USA* **60**:160–167.

Low, B., and T. H. Wood. 1965. A quick and efficient method for interruption of bacterial conjugation. *Genet. Res.* **6**:300–303.

Low, K. B. 1972. *Escherichia coli* K-12 F-prime factors, old and new. *Bacteriol. Rev.* **36**:587–607.

Low, K. B. 1987. Mapping techniques and determination of chromosome size, p. 1184–1189. *In* F. C. Neidhardt, J. L. Ingraham, K. B. Low, B. Magasanik, M. Schaechter, and H. E. Umbargar (ed.), *Escherichia coli and Salmonella typhimurium: Cellular and Molecular Biology*, vol. 2. American Society for Microbiology, Washington, D.C.

Low, K. B., and R. D. Porter. 1978. Modes of gene transfer and recombination in bacteria. *Annu. Rev. Genet.* **12**:249–287.

Mahajan, S. K., and A. R. Datta. 1977. Nature of recombinants produced by the RecBC and the RecF pathways in *Escherichia coli*. *Nature* (London) **266**:652–653.

Mahajan, S. K., and A. R. Datta. 1979. Mechanisms of recombination by the RecBC and the RecF pathways following conjugation in *Escherichia coli*. *Mol. Gen. Genet.* **169**:67–78.

Mandel, M., and A. Higa. 1970. Calcium-dependent bacteriophage DNA infection. *J. Mol. Biol.* **53**:159–162.

Mannarelli, B. M., and S. A. Lacks. 1984. Ectopic integration of chromosomal genes in *Streptococcus pneumoniae*. *J. Bacteriol.* **160**:867–873.

Manning, P. A., G. Morelli, and M. Achtman. 1981. *traG* protein of the F sex factor of *Escherichia coli* and its role in conjugation. *Proc. Natl. Acad. Sci. USA* **78**:7487–7491.

Margolin, P. 1987. Generalized transduction, p. 1154–1168. *In* F. C. Neidhardt, J. L. Ingraham, K. B. Low, B. Magasanik, M. Schaechter, and H. E. Umbarger (ed.), *Escherichia coli and Salmonella typhimurium: Cellular and Molecular Biology*, vol. 2. American Society for Microbiology, Washington, D.C.

Marrero, R., and R. E. Yasbin. 1988. Cloning of the *Bacillus subtilis recE⁺* gene and functional expression of *recE⁺* in *Bacillus subtilis*. *J. Bacteriol.* **170**:335–344.

Masters, M. 1985. Generalized transduction, p. 197–215. *In* J. Scaife, D. Leach, and A. Galizzi (ed.), *Genetics of Bacteria*. Academic Press, Inc., New York.

Masters, M., B. J. Newman, and C. M. Henry. 1984. Reduction of marker discrimination in transductional recombination. *Mol. Gen. Genet.* **196**:85–90.

Matsubara, K. 1968. Properties of sex factor and related episomes isolated from purified *Escherichia coli* zygote cells. *J. Mol. Biol.* **38**:89–108.

McCarthy, D., and D. M. Kupfer. 1987. Electron microscopy of single-stranded structures in the DNA of competent *Haemophilus influenzae* cells. *J. Bacteriol.* **169**:565–571.

Miles, C. A., A. Mountain, and G. R. K. Sastry. 1986. Cloning and characterization of the *recA* gene of *Agrobacterium tumefaciens* C58. *Mol. Gen. Genet.* **204**:161–165.

Miller, J. F., W. J. Dower, and L. S. Tompkins. 1988. High-voltage electroporation of bacteria: genetic transformation of *Campylobacter jejuni* with plasmid DNA. *Proc. Natl. Acad. Sci. USA* **85**:856–860.

Minkley, E. G., Jr., and N. S. Willetts. 1984. Overproduction, purification, and characterization of the F *traT* protein. *Mol. Gen. Genet.* **196**:225–235.

Moll, A., P. A. Manning, and K. N. Timmis. 1980. Plasmid-determined resistance to serum bactericidal activity: a major outer membrane protein, the *traT* gene product, is responsible for plasmid-specified serum resistance in *Escherichia coli. Infect. Immun.* **28**:359–367.

Mori, M. 1987. Isolation and structure of *Streptococcus faecalis* sex pheromone cAM373, also produced by *Staphylococcus aureus*, p. 8–10. In J. J. Ferretti and R. Curtiss III (ed.), *Streptococcal Genetics*. American Society for Microbiology, Washington, D.C.

Morrison, D. A. 1977. Transformation in pneumococcus: existence and properties of a complex involving donor deoxyribonucleate single strands in eclipse. *J. Bacteriol.* **132**:576–583.

Morrison, D. A., and M. Baker. 1979. Competence for genetic transformation in pneumococcus depends on the synthesis of a small set of proteins. *Nature* (London) **282**:215–217.

Morrison, D. A., and W. R. Guild. 1973a. Breakage prior to entry of donor DNA in pneumococcus transformation. *Biochim. Biophys. Acta* **299**:545–556.

Morrison, D. A., and W. R. Guild. 1973b. Structure of deoxyribonucleic acid on the cell surface during uptake by pneumococcus. *J. Bacteriol.* **115**:1055–1062.

Morrison, D. A., S. A. Lacks, W. R. Guild, and J. M. Hageman. 1983. Isolation and characterization of three new classes of transformation-deficient mutants of *Streptococcus pneumoniae* that are defective in DNA transport and genetic recombination. *J. Bacteriol.* **156**:281–290.

Morrison, D. A., and B. Mannarelli. 1979. Transformation in pneumococcus: nuclease resistance of deoxyribonucleic acid in the eclipse complex. *J. Bacteriol.* **140**:655–665.

Morrison, D. A., M.-C. Trombe, M. K. Hayden, G. A. Waszak, and J.-D. Chen. 1984. Isolation of transformation-deficient *Streptococcus pneumoniae* mutants defective in control of competence, using insertion-duplication mutagenesis with the erythromycin resistance determinant of pAMβ1. *J. Bacteriol.* **159**:870–876.

Morse, M. L., E. M. Lederberg, and J. Lederberg. 1956. Transduction in *Escherichia coli* K12. *Genetics* **41**:142–156.

Mottes, M., G. Grandi, V. Sgaramella, U. Canosi, G. Morelli, and T. A. Trautner. 1979. Different specific activities of the monomeric and oligomeric forms of plasmid DNA in the transformation of *B. subtilis* and *E. coli. Mol. Gen. Genet.* **174**:281–286.

Mulder, J. A., and G. Venema. 1982. Transformation-deficient mutants of *Bacillus subtilis* impaired in competence-specific nuclease activities. *J. Bacteriol.* **152**:166–174.

Mullineaux, P., and N. Willetts. 1985. Promoters in the transfer region of plasmid F, p. 605–614. *In* D. R. Helinski, S. N. Cohen, D. B. Clewell, D. A. Jackson, and A. Hollaender (ed.), *Plasmids in Bacteria*. Plenum Publishing Corp., New York.

Murphy, R. C., D. A. Bryant, R. D. Porter, and N. Tandeau de Marsac. 1987. Molecular cloning and characterization of the *recA* gene from the cyanobacterium *Synechococcus* sp. strain PCC 7002. *J. Bacteriol.* **169**:2739–2747.

Newman, B. J., and M. Masters. 1980. The variation in frequency with which markers are transduced by phage P1 is primarily a result of discrimination during recombination. *Mol. Gen. Genet.* **180**:585–589.

Norkin, L. C. 1970. Marker-specific effects in genetic recombination. *J. Mol. Biol.* **51**:633–655.

Notani, N. K., and S. H. Goodgal. 1966. On the nature of recombinants formed during transformation of *Haemophilus influenzae. J. Gen. Physiol.* **49**(Part 2):197–209.

Ohki, M., and J. Tomizawa. 1968. Asymmetric transfer of DNA strands in bacterial conjugation. *Cold Spring Harbor Symp. Quant. Biol.* **33**:651–657.

Oishi, M., and S. D. Cosloy. 1972. The genetic and biochemical basis of the transformability of *Escherichia coli* K-12. *Biochem. Biophys. Res. Commun.* **49:**1568–1572.

Oishi, M., and S. D. Cosloy. 1974. Specialized transformation in *Escherichia coli* K-12. *Nature* (London) **248:**112–116.

Orr-Weaver, T. L., J. W. Szostak, and R. J. Rothstein. 1981. Yeast transformation: a model system for the study of recombination. *Proc. Natl. Acad. Sci. USA* **78:**6354–6358.

Ou, J. T., and T. F. Anderson. 1970. Role of pili in bacterial conjugation. *J. Bacteriol.* **102:**648–654.

Owttrim, G. W., and J. R. Coleman. 1987. Molecular cloning of a *recA*-like gene from the cyanobacterium *Anabaena variabilis. J. Bacteriol.* **169:**1824–1829.

Ozeki, H. 1956. Abortive transduction in purine-requiring mutants of *Salmonella typhimurium. Carnegie Inst. Wash. Publ.* **692:**97–106.

Ozeki, H. 1959. Chromosome fragments participating in transduction in *Salmonella typhimurium. Genetics* **44:**457–470.

Ozeki, H., and H. Ikeda. 1968. Transduction mechanisms. *Annu. Rev. Genet.* **2:**245–278.

Panicker, M. M., and E. G. Minkley, Jr. 1985. DNA transfer occurs during a cell surface contact stage of F sex factor-mediated bacterial conjugation. *J. Bacteriol.* **162:**584–590.

Piechowska, M., and M. S. Fox. 1971. Fate of transforming deoxyribonucleate in *Bacillus subtilis. J. Bacteriol.* **108:**680–689.

Pifer, M. L. 1986. Plasmid establishment in competent *Haemophilus influenzae* occurs by illegitimate transformation. *J. Bacteriol.* **168:**683–687.

Pittard, J., and T. Ramakrishnan. 1964. Gene transfer by F′ strains of *Escherichia coli.* IV. Effect of a chromosomal deletion on chromosome transfer. *J. Bacteriol.* **88:**367–373.

Pittard, J., and E. M. Walker. 1967. Conjugation in *Escherichia coli*: recombination events in terminal regions of transferred deoxyribonucleic acid. *J. Bacteriol.* **94:**1656–1663.

Porter, R. D. 1981. Enhanced recombination between F42*lac* and λ*plac*5: dependence on F42*lac* fertility functions. *Mol. Gen. Genet.* **184:**355–358.

Porter, R. D. 1982. Recombination properties of P1 d*lac. J. Bacteriol.* **152:**345–350.

Porter, R. D. 1983. Specialized transduction with λ*plac*5: involvement of the RecE and RecF recombination pathways. *Genetics* **105:**247–257.

Porter, R. D. 1986. Transformation in cyanobacteria. *Crit. Rev. Microbiol.* **13:**111–132.

Porter, R. D., and W. R. Guild. 1978. Transfection in pneumococcus: single-strand intermediates in the formation of infective centers. *J. Virol.* **25:**60–72.

Porter, R. D., M. W. Lark, and K. B. Low. 1981. Specialized transduction with λ*plac*5: dependence on *recA* and on configuration of *lac* and *att*λ. *J. Virol.* **38:**487–503.

Porter, R. D., T. McLaughlin, and K. B. Low. 1978. Transduction versus "conjuction": evidence for multiple roles for exonuclease V in genetic recombination in *Escherichia coli. Cold Spring Harbor Symp. Quant. Biol.* **43:**1043–1047.

Porter, R. D., R. A. Welliver, and T. A. Witkowski. 1982. Specialized transduction with λ*plac*5: depencence on *recB. J. Bacteriol.* **150:**1485–1488.

Raina, J. L., and A. W. Ravin. 1978. Fate of homospecific transforming DNA bound to *Streptococcus sanguis. J. Bacteriol.* **133:**1212–1223.

Raina, J. T., and A. W. Ravin. 1980. Switches in macromolecular synthesis during induction of competence for transformation of *Streptococcus sanguis. Proc. Natl. Acad. Sci. USA* **77:**6062–6066.

Resnick, D., and D. R. Nelson. 1988. Cloning and characterization of the *Aeromonas caviae recA* gene and construction of an *A. caviae recA* mutant. *J. Bacteriol.* **170:**48–55.

Richter, A. 1957. Complementary determinants on an Hfr phenotype in *E. coli* K12. *Genetics* **42:**391.

Rosenberg, S. M. 1987. Chi-stimulated patches are heteroduplex, with recombinant information of the phage λ *r* chain. *Cell* **48:**855–865.

Ruby, C., and R. P. Novick. 1975. Plasmid interactions in *Staphylococcus aureus.* Nonadditivity of compatible plasmid DNA pools. *Proc. Natl. Acad. Sci. USA* **72:**5031–5035.

Rupp, W. D., and G. Ihler. 1968. Strand selection during bacterial mating. *Cold Spring Harbor Symp. Quant. Biol.* **33:**647–650.

Sancar, A., R. P. Wharton, S. Seltzer, B. M. Kacinski, N. D. Clark, and W. D. Rupp. 1981. Identification of the *uvrA* gene product. *J. Mol. Biol.* **148:**45–62.

Sandri, R. M. and H. Berger. 1980a. Bacteriophage P1-mediated generalized transduction in *Escherichia coli*: fate of transduced DNA in Rec⁺ and RecA⁻ recipients. *Virology* **106**:14–29.

Sandri, R. M., and H. Berger. 1980b. Bacteriophage P1-mediated generalized transduction in *Escherichia coli*: structure of abortively transduced DNA. *Virology* **106**:30–40.

Sarathy, P. V., and O. Siddiqi. 1973. DNA synthesis during bacterial conjugation. II. Is DNA replication in the Hfr obligatory for chromosome transfer? *J. Mol. Biol.* **78**:443–451.

Saunders, C. W., and W. R. Guild. 1981a. Monomer plasmid DNA transforms *S. pneumoniae*. *Mol. Gen. Genet.* **181**:57–62.

Saunders, C. W., and W. R. Guild. 1981b. Pathway of plasmid transformation in pneumococcus: open circular and linear molecules are active. *J. Bacteriol.* **146**:517–526.

Schmidt, C., and H. Schmieger. 1984. Selective transduction of recombinant plasmids with cloned *pac* sites by *Salmonella* phage P22. *Mol. Gen. Genet.* **196**:123–128.

Schmieger, H. 1972. Phage P22-mutants with increased or decreased transduction abilities. *Mol. Gen. Genet.* **119**:75–78.

Schmieger, H. 1982. Packaging signals for phage P22 on the chromosome of *Salmonella typhimurium*. *Mol. Gen. Genet.* **187**:516–518.

Schultz, D. W., A. F. Taylor, and G. R. Smith. 1983. *Escherichia coli* RecBC pseudorevertants lacking Chi recombinational hotspot activity. *J. Bacteriol.* **155**:664–680.

Scocca, J. J., R. L. Poland, and K. C. Zoon. 1974. Specificity in deoxyribonucleic acid uptake by transformable *Haemophilus influenzae*. *J. Bacteriol.* **118**:369–373.

Seifert, H. S., and R. D. Porter. 1984a. Enhanced recombination between λ*plac5* and mini-F-*lac*: the *tra* regulon is required for recombination enhancement. *Mol. Gen. Genet.* **193**:269–274.

Seifert, H. S., and R. D. Porter. 1984b. Enhanced recombination between λ*plac5* and F42*lac*: identification of *cis-* and *trans*-acting factors. *Proc. Natl. Acad. Sci. USA* **81**:7500–7504.

Setlow, J. K., M. E. Boling, K. L. Beattie, and R. F. Kimball. 1972. A complex of recombination and repair genes in *Haemophilus influenzae*. *J. Mol. Biol.* **68**:361–378.

Setlow, J. K., N. K. Notani, D. McCarthy, and N.-L. Clayton. 1981. Transformation of *Haemophilus influenzae* by plasmid RSF0885 containing a cloned segment of chromosomal deoxyribonucleic acid. *J. Bacteriol.* **148**:804–811.

Setlow, J. K., E. Cabrera-Juarez, and K. Griffin. 1984. Mechanism of acquisition of chromosomal markers by plasmids in *Haemophilus influenzae*. *J. Bacteriol.* **160**:662–667.

Seto, H., and A. Tomasz. 1974. Early stages in DNA binding and uptake during genetic transformation of pneumococci. *Proc. Natl. Acad. Sci. USA* **71**:1493–1498.

Seto, H., and A. Tomasz. 1975. Protoplast formation and leakage of intramembrane cell components: induction by the competence activator substance of pneumococci. *J. Bacteriol.* **121**:344–353.

Shen, P., and H. V. Huang. 1986. Homologous recombination in *Escherichia coli*: dependence on substrate length and homology. *Genetics* **112**:441–457.

Shimada, K., R. A. Weisberg, and M. E. Gottesman. 1972. Prophage lambda at unusual chromosomal locations. I. Location of the secondary attachment sites and the properties of the lysogens. *J. Mol. Biol.* **63**:483–503.

Shoemaker, N. B., and W. R. Guild. 1972. Kinetics of integration of transforming DNA in pneumococcus. *Proc. Natl. Acad. Sci. USA* **69**:3331–3335.

Siddiqi, O., and M. S. Fox. 1973. Integration of donor DNA in bacterial conjugation. *J. Mol. Biol.* **77**:101–123.

Signer, E., and J. R. Beckwith. 1966. Transposition of the *lac* region of *Escherichia coli*. III. The mechanism of attachment of bacteriophage φ80 to the bacterial chromosome. *J. Mol. Biol.* **22**:33–51.

Sisco, K. L., and H. O. Smith. 1979. Sequence-specific DNA uptake in *Haemophilus* transformation. *Proc. Natl. Acad. Sci. USA* **76**:972–976.

Smith, G. R., S. K. Amundsen, A. M. Chaudhury, K. C. Cheng, A. S. Ponticelli, C. M. Roberts, D. W. Schultz, and A. F. Taylor. 1984. Roles of RecBC enzyme and Chi sites in homologous recombination. *Cold Spring Harbor Symp. Quant. Biol.* **49**:485–495.

Smith, H. O., D. B. Danner, and R. A. Deich. 1981. Genetic transformation. *Annu. Rev. Biochem.* **50**:41–68.

Smith, M. D., N. B. Shoemaker, V. Burdett, and W. R. Guild. 1980. Transfer of plasmids by conjugation in *Streptococcus pneumoniae*. *Plasmid* **3**:70–79.

Sparling, P. F. 1966. Genetic transformation of *Neisseria gonorrhoeae* to streptomycin resistance. *J. Bacteriol.* **92**:1364–1371.

Spencer, H. T., and R. M. Herriott. 1965. Development of competence of *Haemophilus influenzae*. *J. Bacteriol.* **90**:911–920.

Stahl, F. W. 1979. Special sites in generalized recombination. *Annu. Rev. Genet.* **13**:7–24.

Sternberg, N., and R. Hoess. 1983. The molecular genetics of bacteriophage P1. *Annu. Rev. Genet.* **17**:123–154.

Sternberg, N., and R. Weisberg. 1975. Packaging of prophage and host DNA by coliphage λ. *Nature* (London) **256**:97–103.

Stiffler, P. W., H. M. Sweeney, and S. Cohen. 1974. Co-transduction of plasmids mediating resistance to tetracycline and chloramphenicol in *Staphylococcus aureus*. *J. Bacteriol.* **120**:934–944.

Stuy, J. H. 1965. Fate of transforming DNA in the *Haemophilus influenzae* transformation system. *J. Mol. Biol.* **13**:554–570.

Szostak, J. W., T. L. Orr-Weaver, R. J. Rothstein, and F. W. Stahl. 1983. The double-strand-break repair model for recombination. *Cell* **33**:25–35.

Tandeau de Marsac, N., W. E. Borrias, C. J. Kuhlemeier, A. M. Castets, G. A. van Arkel, and C. A. M. J. J. van den Hondel. 1982. A new approach for molecular cloning in cyanobacteria: cloning of an *Anacystis nidulans met* gene using a Tn901-induced mutant. *Gene* **20**:111–119.

Taylor, A. F., D. W. Schultz, A. S. Ponticelli, and G. R. Smith. 1985. RecBC enzyme nicking at Chi sites during DNA unwinding: location and orientation-dependence of the cutting. *Cell* **41**:153–163.

Thompson, R., L. Taylor, K. Kelly, R. Everett, and N. Willetts. 1984. The F plasmid origin of transfer: DNA sequence of wild-type and mutant origins and location of origin-specific nicks. *EMBO J.* **3**:1175–1180.

Tomasz, A. 1966. Model for the mechanism controlling the expression of the competent state in pneumococcus cultures. *J. Bacteriol.* **91**:1050–1061.

Traxler, B. A., and E. G. Minkley, Jr. 1987. Revised genetic map of the distal end of the F transfer operon: implications for DNA helicase I, nicking at *oriT*, and conjugal DNA transport. *J. Bacteriol.* **169**:3251–3259.

Venema, G., R. H. Pritchard, and T. Venema-Schroder. 1965. Fate of transforming deoxyribonucleic acid in *Bacillus subtilis*. *J. Bacteriol.* **89**:1250–1255.

Verhoff, C., and P. G. DeHaan. 1966. Genetic recombination in *Escherichia coli*. I. Relation between linkage of unselected markers and map distance. *Mutat. Res.* **3**:101–110.

Vijayakumar, M. N., and D. A. Morrison. 1983. Fate of DNA in eclipse complex during genetic transformation in *Streptococcus pneumoniae*. *J. Bacteriol.* **156**:644–648.

Voll, M. J., and S. H. Goodgal. 1961. Recombination during transformation in *Haemophilus influenzae*. *Proc. Natl. Acad. Sci. USA* **47**:505–512.

Wackernagel, W. 1973. Genetic transformation of *E. coli*: the inhibitory role of the *recBC* DNase. *Biochem. Biophys. Res. Commun.* **51**:306–311.

Wackernagel, W., and C. M. Radding. 1974. Transformation and transduction of *Escherichia coli*: the nature of recombinants formed by Rec, RecF, and λRed, p. 111–122. *In* R. F. Grell (ed.), *Mechanisms in Recombination*. Plenum Publishing Corp., New York.

Wall, J. D., and P. D. Harriman. 1974. Phage P1 mutants with altered transducing abilities for *Escherichia coli*. *Virology* **59**:532–544.

Watanabe, T., C. Furuse, and S. Sakaizumi. 1968. Transduction of various R factors by phage P1 in *Escherichia coli* and by phage P22 in *Salmonella typhimurium*. *J. Bacteriol.* **96**:1791–1795.

Watt, V. M., C. J. Ingles, M. S. Urdea, and W. J. Rutter. 1985. Homology requirements for recombination in *Escherichia coli*. *Proc. Natl. Acad. Sci. USA* **82**:4768–4772.

Weil, J., and E. R. Signer. 1968. Recombination in bacteriophage λ. II. Site-specific recombination promoted by the integration system. *J. Mol. Biol.* **34**:273–279.

Weinrauch, Y., and D. Dubnau. 1987. Plasmid marker rescue transformation proceeds by breakage-reunion in *Bacillus subtilis*. *J. Bacteriol.* **169**:1205–1211.

Weisberg, R. A. 1987. Specialized transduction, p. 1169–1176. *In* F. C. Neiderhardt, J. L. Ingraham, K. B. Low, B. Magasanik, M. Schaechter, and H. E. Umbarger (ed.), *Escherichia coli and Salmonella typhimurium: Cellular and Molecular Biology*, vol. 2. American Society for Microbiology, Washington, D.C.

Weisberg, R. A., and N. Sternberg. 1974. Transduction of *recB⁻* hosts is promoted by λ*red⁺* function, p. 107–109. *In* R. F. Grell (ed.), *Mechanisms in Recombination.* Plenum Publishing Corp., New York.

Willetts, N. 1977. The transcriptional control of fertility in F-like plasmids. *J. Mol. Biol.* **112:**141–148.

Willetts, N. S., and D. W. Mount. 1969. Genetic analysis of recombination-deficient mutants of *Escherichia coli* K-12 carrying *rec* mutations cotransducible with *thyA. J. Bacteriol.* **100:**923–934.

Willetts, N. S., and R. Skurray. 1980. The conjugative system of F-like plasmids. *Annu. Rev. Genet.* **14:**41–76.

Willetts, N., and R. Skurray. 1987. Structure and function of the F factor and mechanism of conjugation, p. 1110–1133. *In* F. C. Neidhardt, J. L. Ingraham, K. B. Low, B. Magasanik, M. Schaechter, and H. E. Umbargar (ed.), *Escherichia coli and Salmonella typhimurium: Cellular and Molecular Biology,* vol. 2. American Society for Microbiology, Washington, D.C.

Willetts, N. S., and B. Wilkins. 1984. Processing of plasmid DNA during bacterial conjugation. *Microbiol. Rev.* **48:**24–41.

Williams, J. G. K., and A. A. Szalay. 1983. Stable integration of foreign DNA into the chromosome of the cyanobacterium *Synechococcus* R2. *Gene* **24:**37–51.

Wilson, G. G., K. K. Y. Young, G. J. Edlin, and W. Konigsberg. 1979. High frequency generalized transduction by bacteriophage T4. *Nature* (London) **280:**80–82.

Wood, T. H. 1968. Effects of temperature, agitation, and donor strain on chromosome transfer in *Escherichia coli* K-12. *J. Bacteriol.* **96:**2077–2084.

Yagi, Y., and D. B. Clewell. 1980. Recombination-deficient mutant of *Streptococcus faecalis. J. Bacteriol.* **143:**966–970.

Yancey, S. D., and R. D. Porter. 1985. General recombination in *Escherichia coli* K-12: in vivo role of RecBC enzyme. *J. Bacteriol.* **162:**29–34.

Yanofsky, C., B. C. Carlton, J. R. Guest, D. R. Helinski, and U. Henning. 1964. On the colinearity of gene structure and protein structure. *Proc. Natl. Acad. Sci. USA* **51:**266–272.

Yasbin, R. E., G. A. Wilson, and F. E. Young. 1975. Transforamtion and transfection in lysogenic strains of *Bacillus subtilis*: evidence for selection induction of prophage in competent cells. *J. Bacteriol.* **121:**296–304.

Yoshikawa, M., and Y. Hirota. 1971. Impaired transduction of R213 and its recovery by a homologous resident R factor. *J. Bacteriol.* **106:**523–528.

Yoshioka, Y., H. Ohtsubo, and E. Ohtsubo. 1987. Repressor gene *finO* in plasmids R100 and F: constitutive transfer of plasmid F is caused by insertion of IS*3* into F *finO. J. Bacteriol.* **169:**619–623.

Chapter 2

Mechanisms of Genetic Recombination in Gram-Positive Bacteria

Sanford A. Lacks

INTRODUCTION

Genetic recombination in gram-positive bacteria was first observed by Griffith (1928), when he found that heat-killed pneumococci could pass on the ability to make capsular polysaccharide to live noncapsulated pneumococci simultaneously injected into a mouse. At the time of this discovery of bacterial transformation, however, the genetic nature of the transformation was not immediately obvious. Only after Avery et al. (1944) showed that the transforming principle consisted of DNA did it become clear that transformation was a form of genetic recombination.

During the past 35 years, bacterial transformation has been intensively studied both as a mode of gene transfer and with respect to the genetic recombination it entails. Other forms of gene transfer first observed in gram-negative bacteria, such as conjugation and transduction, have been found also in gram-positive microorganisms. During this period, plasmids were discovered in most bacterial species, and investigation of plasmid transfer and recombination has proceeded rapidly. In addition to these mechanisms of partial genetic exchange, fusion of protoplasts, which has been studied particularly with gram-positive cells, allows the admixture of entire genomes.

The aforementioned processes of gene transfer appear to occur naturally and to provide bacteria with the means of genetic exchange and recombination afforded by sexual processes in higher organisms. They can also be useful in the genetic engineering of living cells for human ends. Unlike sexual exchange in plants and animals, which generally involves the interaction and reassortment of their entire haploid genomes, the bacterial modes generally entail transfer of only fragments of the genome. Thus, if genetic recombination is viewed as the separation of genes from one cell and their integration with genes from another cell, the mode of gene transfer itself contributes to the first part of the recombination process. Several of the modes by which DNA is exchanged between cells will be examined in detail. However, the emphasis of this chapter will be on transformation and conjugation—those modes that introduce single-stranded DNA into the cell—and on the recombination mechanisms in which the introduced strands of DNA participate.

Recent studies on the transfer of plasmids, particularly plasmids containing DNA homologous to the chromosome (or otherwise circularly ligated DNA, as discussed in section III.C. on ectopic integration), into gram-positive bacteria by the mode of transformation have elucidated novel forms of genetic recombination. The single-stranded DNA that enters the cell can interact with the chromosome by either linear or circular synapsis (Lacks, 1984). The former gives rise to classical chromosomal or plasmid transformation; the latter, to recombinative processes called chromosomal facilitation of plasmid establishment, additive (or duplicative) insertion, and ectopic integration. It should be noted that the mechanisms to be discussed and the experimental findings on which they are based come from the work of many laboratories. I beg pardon from colleagues in the field for any failure to directly cite pertinent contributions. Although much progress has been made in the investigation of recombination in various gram-positive bacteria, the discussion will focus mainly on the genera *Bacillus* and *Streptococcus*, in which the recombinatory behavior of single stranded DNA has been most clearly delineated.

II. MODES OF GENE TRANSFER

At least four modes of genetic exchange have been observed in gram-positive bacteria. (i) Transformation is mediated by free DNA, which leaks out or is extracted from donor cells and is taken up by competent recipient cells. (ii)

Conjugation requires cell contact; it is mediated by specific cellular functions that mobilize the transfer of a segment of DNA from donor to recipient. (iii) Transduction makes use of the infective ability of a bacterial virus to transfer nonviral genes by incorporating them into a viral particle. Entry of DNA in transduction depends on mechanisms of viral infection and falls outside the scope of this chapter. These first three modes all transfer genes from a donor cell to a recipient cell. (iv) In protoplast fusion, the fourth mode, the contents of two cells are merged. Genetic recombination in *Bacillus* species after protoplast fusion was reviewed recently (Hotchkiss and Gabor, 1985).

Plasmids can be transferred from cell to cell by either transformation or conjugation; therefore, their modes of transmission and establishment will also be examined. Transfection, which is the term for viral infection by free viral DNA entering by the transformation pathway, is similar in many ways to the establishment of heterologous plasmid replicons (for a review of transfection, see Trautner and Spatz, 1973)

A. Transformation

Genetic transformation appears to be a natural process of gene transfer in several, but by no means all, species of gram-positive bacteria. It has been studied intensively in *Streptococcus pneumoniae*, *Streptococcus sanguis*, and *Bacillus subtilis*. Cells of these species are able to take up large amounts of DNA, up to 10% of their genomic content, as judged by the frequency of transformation of competent cells (Hotchkiss, 1954; Piechowska and Shugar, 1967; Singh and Pitale, 1968). Mutations in a large number of host cell genes can interfere with DNA uptake (see below), so it appears that the responsible systems are highly evolved. In some cases, transformation has been demonstrated in the natural bacterial environment, for example, with *S. pneumoniae* injected into animals (Griffith, 1928; Ottolenghi and MacLeod, 1963). Cells of these highly transformable species are not always competent to take up DNA. Rather, they become competent at late stages of the culture growth cycle, when the cells have reached a high density. This increases the likelihood of interaction between DNA leaking out of moribund cells and viable competent cells in the culture, and it provides the possibility for genetic recombination when selective pressures would be strongest.

In cultures of *S. pneumoniae* and *S. sanguis*, competence develops during late logarithmic growth. It is controlled by a competence factor or activator (Pakula et al., 1962; Tomasz and Hotchkiss, 1964), a polypeptide of 5 to 10 kilodaltons (kDa) (Tomasz and Mosser, 1966; Leonard and Cole, 1972), which is excreted by the cells into the medium where it accumulates to a sufficient concentration to act back on the cells to elicit competence. This hormonelike mechanism assures the development of competence in all cells of the culture, but only when the culture reaches a high cell density. *B. subtilis*, in contrast, develops competence at the onset of stationary phase, in only a subset of the population (Singh and Pitale, 1968) that is particularly retarded in its synthetic abilities (Nester and Stocker, 1963). No convincing evidence for an extracellular competence factor in this species has been obtained.

TABLE 1

Mutants of *S. pneumoniae* affected in competence for transformation and DNA uptake

Genotype	Phenotype	Reference
xfo	Fails to transform	Morrison et al. (1983)
ntr	Fails to bind DNA	Lacks and Greenberg (1973)
com	Requires added external competence factor	Chandler and Morrison (1987)
trt	Transforms in presence of trypsin; no external competence factor needed	Lacks and Greenberg (1973)
end, *noz*	Lacks membrane DNase; binds DNA; no DNA entry	Lacks et al. (1974, 1975)
ent	Degrades bound DNA; no DNA entry	Morrison et al. (1983)

With the advent of competence, gene expression is altered in the competent cell population. As first shown in *S. pneumoniae* (Morrison and Baker, 1979) and later in *S. sanguis* (Raina and Ravin, 1980) and *B. subtilis* (Green and Coughlin, 1982), normal protein synthesis is reduced and a set of up to 20 competence-specific proteins is made. The function of most of these proteins is not yet known, but a 19.5-kDa protein of *S. pneumoniae* has been shown to be a single-stranded DNA-binding protein (Morrison and Mannarelli, 1979), and a 51-kDa protein of *S. sanguis* functions in genetic recombination (Raina and Macrina, 1982). Several kinds of mutations have been found to prevent or otherwise affect competence (Tables 1 and 2). Early studies in *S. pneumoniae* revealed *ntr* (*n*ontransformable) mutations, which prevented any manifestation of competence, and a *trt* (*t*rypsin *t*ransformable) mutation that circumvented the need for external competence factor (Lacks and Greenberg, 1973). A systematic search for transformation-negative (*xfo*; [*x* = trans]*fo*rmation) mutations in *S. pneumoniae* has revealed a number of loci affecting transformation (Morrison et al., 1983), including a gene for production of competence factor (Chandler and Morrison, 1987). Similar

TABLE 2

Mutants of *B. subtilis* affected in competence for transformation and DNA uptake

Mutation	Map[a]	Class[b]	Phenotype	Reference
com-38	245	I	No binding of DNA	Hahn et al. (1987)
com-524	300	II	No binding of DNA	Hahn et al. (1987)
com-540, *com-9*	280	III	No competence-related density shift	Hahn et al. (1987), Fani et al. (1984)
com-43	225	IV	Binds DNA; no DNA entry	Hahn et al. (1987)
com-12, *com-71*	210	V	No competence-related density shift	Hahn et al. (1987), Fani et al. (1984)
com-22	220	VI	No binding of DNA	Hahn et al. (1987)
com-162	230	VII	Not well characterized	Hahn et al. (1987)
com-30	80		No binding of DNA	Fani et al. (1984)
com-104	340		Binds DNA; no DNA entry	Fani et al. (1984)
end	25		Binds DNA; no DNA entry	Vosman et al. (1987)

[a] Angular degrees measured on circular chromosome from *guaA* in direction of *cysA* (Henner and Hoch, 1982).
[b] Hahn et al. (1987).

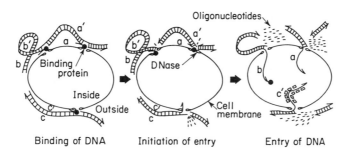

FIGURE 1. DNA uptake in transformation of gram-positive bacteria. Double-stranded DNA is irreversibly bound to the cell surface and undergoes single-strand cleavage at random sites, presumably by the action of a hypothetical binding protein. A membrane-located DNase initiates entry of the bound strand by endonucleolytic cleavage of the complementary strand to give a double-strand break. Processive action of the DNase degrades the complementary strand to oligonucleotides, which remain outside the cell, allowing donor strands to enter. It is not known whether the strand enters without (a) or with (b) the surface binding protein. The entering DNA is rapidly covered with a single-strand binding protein (c'). Segments from either strand may enter.

searches in *B. subtilis* have revealed nine competence-related loci (Hahn et al., 1987; Fani et al., 1984). For the most part the functions of these competence genes remain to be determined.

Despite differences in the control of competence and in detailed properties of the systems, the general mechanism of DNA binding and uptake appears to be the same in the various naturally transformable gram-positive bacteria (Fig. 1). In particular, double-stranded donor DNA is bound to the cell surface where it undergoes, first, single-strand breaks and, then, double-strand breaks. Following this breakage, a single-strand segment of donor DNA is introduced into the cell. Critical for subsequent genetic recombination are (i) strand breakage during binding, which can separate previously linked genes and disrupt replicons, and (ii) conversion to single strands on entry, which can facilitate interaction with either donor or host DNA.

When DNA is bound to competent cells of *S. pneumoniae*, it undergoes single-strand breaks, or nicks, apparently randomly in either strand, that are spaced on the average about 6,000 nucleotides apart (Lacks and Greenberg, 1976). The molecular state of such bound DNA can be investigated by blocking the subsequent entry step either by addition of a chelating agent, such as EDTA, which binds divalent cations necessary for entry (Lacks et al., 1974; Seto and Tomasz, 1974), or by using a mutant recipient, either *end* or *noz* (see Table 1), which is blocked in entry (Lacks et al., 1974). It is likely that the binding reaction itself produces the nick, because any covalently closed circular DNA that is bound has suffered at least one such nick (Lacks, 1979b). It is obvious, therefore, that a circular replicon will be disrupted during uptake. The binding-nicking reaction may require energy inasmuch as competent cells deprived of sugar will not bind DNA (Lacks et al., 1974). It also requires potassium ions, at least in cells previously treated with EDTA (Lacks, 1979a). Extraction of competent cells of *S. pneumoniae* with 1 M LiCl releases a DNA nicking activity that appears to be the one implicated in DNA binding (Fujii et al., 1987). At this stage the donor DNA is external to the cell since it can be removed by enzymatic treatment or shear forces.

Initiation of entry, which can be observed either by adding back divalent cations or by use of a leaky *noz* mutant recipient, produces double-strand breaks in DNA bound at the surface, presumably by action of the entry nuclease (see below) on the strand opposite the original nick (Lacks and Greenberg, 1976; Lacks, 1979b). Precisely how close the initial scisson by the entry nuclease is to the original nick is not known. These double-strand breaks occur normally in *S. pneumoniae* at a number-average spacing of about 3,000 nucleotides (Lacks and Greenberg, 1976; Lacks, 1977). Double-strand breaks were observed originally with *B. subtilis* in externally bound DNA, at a number-average spacing of 9,000 nucleotides (Dubnau and Cirigliano, 1972). The double-strand breaks appear to reflect the initiation of action of a membrane-bound nuclease on one strand of DNA concomitant with initiation of entry of the complementary strand.

Analysis of the molecular state of donor DNA immediately after entry into *S. pneumoniae* (Lacks, 1962), *S. sanguis* (Raina and Ravin, 1978), or *B. subtilis* (Piechowska and Fox, 1971; Davidoff-Abelson and Dubnau, 1973) shows it to be in the form of single strands. The number-average size of these strand segments is 3,000 nucleotides, or less if the donor DNA length is shorter (Morrison and Guild, 1972). Concomitant with entry of the single-strand segments an equivalent amount of donor DNA, presumably corresponding to their complement, is degraded to oligonucleotides, which are released into the medium (Lacks and Greenberg, 1973; Morrison and Guild, 1973). A membrane-located nuclease (Lacks and Neuberger, 1975), the major DNase of *S. pneumoniae* (Lacks and Greenberg, 1967), is responsible for this process. This enzyme corresponds to a polypeptide of 25 kDa (Rosenthal and Lacks, 1977), but appears to be part of a much larger structure in its native form (Rosenthal and Lacks, 1980). Mutants that lack the nuclease (more precisely, those that show less than 1% of its normal activity in vitro) are defective in DNA entry (Lacks et al., 1974, 1975). One of several DNases that are found in *B. subtilis* has also been implicated in DNA uptake (Mulder and Venema, 1982). This 17-kDa *B. subtilis* enzyme has been isolated in a 70-kDa complex together with a DNA-binding protein (Smith et al., 1984), and the gene encoding the nuclease has been cloned (Vosman et al., 1987). It is not known, however, whether the associated binding protein is necessary for DNA uptake. Both the *S. pneumoniae* and *B. subtilis* enzymes require Mg^{2+} for activity. The entry process in *S. pneumoniae* also requires Ca^{2+} (Seto and Thomasz, 1976; Lacks, 1977), but the role of this cation has not been determined.

The nuclease-dependent pathway appears to account for over 99% of the DNA uptake in transformation of *S. pneumoniae*. However, a residual level of transformation (0.2% of normal) is found in *end* mutants with undetectable enzyme activity in vitro, including a nonsense mutant, *end-14* (Weinrauch and Lacks, 1981). Furthermore, single-stranded DNA can donate markers in transformation of *S. pneumoniae* to a similar weak extent, that is, less than 1% of double-stranded DNA (Miao and Guild, 1970; Barany and Boeke, 1983). In *B.*

subtilis, transformation by single-stranded donor DNA is not inhibited by EDTA (Chilton, 1967).

Because single-stranded DNA is such a poor donor, immediately after uptake of single strands from double-stranded donor DNA the genetic markers that they contain are in a state of eclipse in that their presence cannot be demonstrated in a recipient cell lysate by transformation of a tester culture (Fox, 1960; Lacks, 1962; Venema et al., 1965). Marker activity is recovered as the donor DNA is integrated into a double-stranded form. Within the cell, single-stranded DNA is sequestered in an "eclipse complex" with a single-stranded DNA-binding protein, which is produced during the development of competence (Morrison, 1977; Raina and Ravin, 1978; Eisenstadt et al., 1975). Formation of this complex may facilitate entry of single strands formed from double-stranded donor DNA and perhaps enable the limited entry of single-stranded donor DNA. The complex may also protect donor strands from nucleases within the cell and assist in subsequent recombinative processes. It is also possible that proteins which nick donor DNA at the surface during binding remain fixed to the ends of the incoming strand segment. Such proteins could conceivably mark the ends of the segment and play a role in its genetic integration.

Uptake of a large molecule of DNA into a cell requires the expenditure of energy. Conversion of native DNA to a single strand presumably facilitates its passage through the cell membrane. Hydrolysis of the complementary strand and binding to the eclipse protein could provide energy for uptake. Binding of DNA to the cell surface itself requires energy (Lacks et al., 1974), and it has been reported that bound DNA will enter only in the presence of an energy source (Seto and Tomasz, 1974). Recent investigation of energy-coupled membrane processes in *S. pneumoniae* indicates that neither a membrane potential nor a proton gradient is essential for DNA uptake; this result implicates ATP itself in the process (M. C. Trombe, personal communication). A possible role for ATP is activation of the binding-nicking protein at the initial stage of DNA uptake.

Although a natural mechanism for DNA-mediated transformation is found in only a minority of bacterial species, by artificial treatment cells of almost any species can be made to take up some DNA and become genetically transformed. For example, intact cells of *Staphylococcus aureus* can be transformed after treatment with unphysiologically high concentrations (50 mM) of Ca^{2+} (Lindberg et al., 1972), a procedure developed for the transfection of *Escherichia coli* (Mandel and Higa, 1970). A more general method of artificial bacterial transformation is conversion of cells to protoplasts and treatment of the protoplasts with DNA in the presence of polyethylene glycol (Chang and Cohen, 1979). The protoplasts are then allowed to regenerate cell walls prior to selection for a genetic marker. These procedures allow only a very small amount of DNA uptake compared with natural transformation, so that they are more suited to transformation with homogeneous donor DNA, such as plasmids, rather than heterogeneous chromosomal DNA. It is not known how DNA gets into the cell as a result of these treatments, but it is clear that the molecular structure of the DNA is not altered on entry. The donor DNA is found within the cell in its original form, at least in *E. coli* (Lacks, 1977) and *B. subtilis* (de Vos and Venema, 1981). Because the DNA is not altered upon entry, as it is in natural transformation, small amounts of introduced DNA can be particularly efficient in plasmid establishment. This mode of entry also allows the use of certain techniques for recombinant plasmid recovery, such as phosphatase treatment of vectors, which cannot be used with the natural systems (Balganesh and Lacks, 1984). For other reviews of transformation, see Lacks (1977), Venema (1979), Smith et al. (1981), and Dubnau (1982).

B. Plasmids in Gram-Positive Bacteria

The first bacterial plasmid to be identified, the fertility factor F1 of *E. coli* (Lederberg, 1952), and various drug resistance and colicinogenic factors discovered soon after (Watanabe and Fukusawa, 1960; Clowes, 1961) were found in gram-negative bacteria. However, plasmids were also shown to be responsible for penicillin resistance in *S. aureus* (Novick and Richmond, 1965). By now, plasmids have been found in almost every gram-positive species examined. The original isolate of *S. pneumoniae* used by Avery et al. (1944) as a recipient in transformation, R36A, carried a cryptic, 3-kilobase (kb) plasmid, as shown by its retention in a sibling strain, R36NC (Smith and Guild, 1979). Antibiotic resistance plasmids and cryptic plasmids were found in various *Streptococcus* species, including *S. faecalis*, *S. mutans*, *S. pyogenes*, and *S. agalactiae* (for review, see Clewell, 1981), and in the dairy streptococci (for review, see Kondo and McKay, 1985). Some of these plasmids have been transferred to *S. pneumoniae* (Smith et al., 1980) and *S. sanguis* (LeBlanc and Hassel, 1976); the natural transformability of these last two species makes them useful cloning systems (Stassi et al., 1981; Behnke and Feretti, 1980; Macrina et al., 1980). Several antibiotic resistance plasmids from *S. aureus* were transferred to *B. subtilis* (Ehrlich, 1977; Gryczan et al., 1978), and they have formed the basis for cloning systems in this species (Keggins et al., 1978). In several cases, namely, pSA5700 (Barany et al., 1982), pMV158 and pLS1 (Lacks et al., 1986a), and derivatives of pWV01 (Kok et al., 1984), gram-positive plasmids replicate also in gram-negative hosts. Such broad-host-range plasmids can serve as shuttle vectors without further modification.

Several gram-positive plasmids replicate by a mechanism distinct from those found in gram-negative plasmids. This mechanism involves nicking of the plasmid DNA (+) strand at its replication origin by a plasmid-encoded replication protein (Koepsel et al., 1985), and rolling circle replication primed by the 3′ end peels off the parental (+) strand, which then circularizes (te Riele et al., 1986). Subsequently, the (−) strand is synthesized on the circular single-strand template from a (−)-strand origin. This mechanism of replication has been observed for plasmids pT181 in *S. aureus* (Koepsel et al., 1985), pC194 in *S. aureus* and *B. subtilis* (te Riele et al., 1986), and pLS1 in *S. pneumoniae*, *B. subtilis*, and *E. coli*, as well (del Solar et al., 1987).

The gram-positive plasmids can be transferred to naturally transformable strains of *S. pneumoniae*, *S. sanguis*, or *B. subtilis* by the same mechanism of DNA uptake used in chromosomal transformation. This process is best referred to as DNA-mediated plasmid transfer or establishment (Stassi et al., 1981), rather than plasmid transformation, inasmuch as plasmid DNA can also transform a resident plasmid by a process analogous to chromosomal transformation, which for clarity is best referred to as plasmid marker-rescue transformation (Weinrauch and Dubnau, 1983). The mechanisms of plasmid transfer and plasmid transformation are considered in detail below. Some gram-positive plasmids encode transfer functions that enable their transmission by conjugative processes, as well.

C. Restriction Effects on Gene Transfer

Restriction systems that prevent viral infection are ubiquitous in bacteria. An interesting example is provided by the *Dpn*I and *Dpn*II systems present in different strains of *S. pneumoniae*. These restriction systems are encoded by alternative genetic cassettes located at the same position in the chromosome (Lacks et al., 1986b). *Dpn*I is an unusual restriction endonuclease in that it acts only on methylated sites in DNA (Lacks and Greenberg, 1975). The two systems are complementary in that DNA from a *Dpn*I strain, which is not methylated at GATC sites, is susceptible to *Dpn*II, and DNA from a *Dpn*II strain, which contains GmeATC, is susceptible to *Dpn*I. Thus, when grown on the complementary strain, phage is reduced in infectivity to a level of $<10^{-5}$ (Muckerman et al., 1982). Existence of the complementary systems can enhance survival of the species in the face of viral (phage) epidemics (Lacks et al., 1987). The susceptibility of phage results from its introduction into the cell as double-stranded DNA. Both *Dpn*I and *Dpn*II act on double-stranded DNA only when both strands are methylated or unmethylated, respectively, at GATC sites; neither enzyme acts on single-stranded DNA or hemimethylated double-stranded DNA, in which one strand is methylated and the other strand is not (Vovis and Lacks, 1977).

Chromosomal transformation is not affected by restriction enzymes in the recipient cell. Markers are transformed at the same frequency whether or not the donor DNA is methylated in both *Dpn*I- and *Dpn*II-containing recipients (Lacks and Greenberg, 1975; Lacks and Springhorn, 1984). Similarly, *B. subtilis* transformation is not affected by the presence of the restriction enzyme *Bsu*R (Trautner et al., 1974). In contrast, phage-mediated transduction of chromosomal markers is strongly affected by restriction in *B. subtilis* (Prozorov et al., 1980). The lack of a restriction effect in transformation reflects the molecular fate of DNA: neither the single strands that enter the cell nor the hemimethylated heteroduplex DNA that results from integration into the chromosome is susceptible to the restriction enzymes. Treatment of donor DNA with restriction enzymes in vitro, of course, destroys most of its transforming activity.

Plasmid transfer in *S. pneumoniae* via the transformation system for DNA uptake is mildly restricted by the *Dpn*I and *Dpn*II systems. Transfer of pMV158, which contains eight susceptible GATC sites, is reduced to 40% in the cross-transformation (Lacks and Springhorn, 1984). Greater susceptibility for plasmid transfer than for chromosomal transformation is expected from the need to reconstitute double-strand plasmids from the single-strand fragments that enter (see section III.B.). However, such reconstitution presumably requires considerable new synthesis in repair of gaps in the reconstituted structure (Saunders and

Guild, 1981), which in the case of *Dpn*I-containing recipients could allow the plasmid to escape restriction (Lacks and Springhorn, 1984). In the reciprocal situation, transfer of an unmethylated plasmid to a *Dpn*II-containing recipient, methylation of single-stranded DNA prior to plasmid reconstitution could allow escape from restriction (W. Guild, personal communication). Whether the *Dpn*II system methylases (de la Campa et al., 1987) can act on single-stranded DNA has not yet been determined. The effects of restriction on conjugative plasmid transfer will be considered in the next section.

D. Conjugation

Conjugative plasmids are able to mediate their transfer by cell-to-cell contact. Such self-transmissible plasmids, for example, the sex factor F, the colicinogenic factor Col I, and the resistance factor R1, were originally observed in the gram-negative enterobacteria (for review, see Falkow, 1975). Subsequently, conjugative plasmids have been found in various gram-positive species including *S. faecalis*, *S. agalactiae*, and *S. pyogenes* (for review, see Clewell, 1981). The streptococcal conjugative plasmids fall into two classes. One class, which includes pAD1, pPD1, and pJH2, transfers at high frequencies (~1% of the donors) to cells of *S. faecalis* in broth. The other class, which includes pAMβ1, pIP501, and pSM15346, transfers poorly in broth but allows mating at reasonably high frequencies (~0.1% of donors) when cells are concentrated on filter membranes. Plasmids from the latter class can be conjugatively transferred among various gram-positive species. As well as effecting their own transfer, plasmids from both classes can mobilize otherwise nontransmissible plasmids. For example, pIP501, which can be transmitted between strains of *S. pneumoniae* by filter mating at a frequency of 10^{-3}, will mobilize pMV158 at a frequency of 10^{-4} (Guild et al., 1982). However, there is little evidence for conjugative transfer of chromosomal markers by these plasmid systems.

Sex pheromones secreted by cells of *S. faecalis* appear to be responsible for the high transmission frequency in liquid medium of plasmids of the first class (Dunny et al., 1978). These pheromones act on donor cells containing the plasmids to produce surface substances that cause aggregation of donor and recipient cells, thereby facilitating conjugation. The pheromones are short polypeptides containing seven or eight amino acid residues (Clewell et al., 1987). Various strains of *S. faecalis* each produce at least three different pheromones. A particular donor plasmid reacts to a specific one of these, the synthesis of which is turned off by the presence of the plasmid. So far, this interesting class of pheromone-induced conjugative plasmids appears to be confined to *S. faecalis*.

In the absence of direct knowledge of the mechanism of conjugative transfer in gram-positive bacteria, a useful model is the F factor of *E. coli*, which has been studied in considerable detail (for reviews, see Willetts and Wilkins, 1984; Ippen-Ihler and Minkley, 1986). Not long after it was observed that single strands of donor DNA entered recipient cells in pneumococcal transformation, it was found that in conjugation, also, a single strand of the donor F plasmid was transmitted to the recipient (Cohen et al., 1968). Transmission begins at a site called *oriT* by nicking of one strand, and that strand proceeds 5' to 3' into the recipient. De novo synthesis restores the plasmid in the donor cell. Soon after transfer into the recipient cell, the strand complementary to the one introduced is synthesized. Although this requires host replicative functions, it makes use of priming enzymes transmitted along with the plasmid strand, which apparently ensure its rapid conversion to a duplex structure. The incoming strand quickly circularizes, apparently aided by the attachment of proteins to its ends during transfer. Approximately 20 genes in F govern the transfer (*tra*) functions. In addition to the proteins already mentioned, they encode various surface structures necessary for cell contact and transfer of DNA. A plasmid lacking transfer functions, such as ColE1, can be mobilized by a conjugative plasmid, such as F, if it contains an *oriT* site (Warren et al., 1978), while the other *tra* functions act "in *trans*."

Restriction effects on conjugative transfer of gram-positive plasmids are consistent with the F conjugation model. On transfer from a *Dpn*II strain of *S. pneumoniae*, which methylates DNA, to a *Dpn*I strain, pIP501 was not restricted (W. Guild, personal communication). This lack of susceptibility to *Dpn*I indicates that only a single strand was transferred. Replication of the complementary strand would give a hemimethylated duplex, which is also not a substrate for *Dpn*I. In contrast, pIP501 grown in a strain that does not methylate its DNA was restricted to a level of 10^{-4} when transferred to a *Dpn*II strain (Guild et al., 1982). Even if a single strand were transferred, rapid synthesis of the complementary strand would produce DNA fully unmethylated at GATC sites, rendering it susceptible to *Dpn*II. Even the smaller pMV158, when mobilized by pIP501, was restricted to 2% under these conditions. This contrasts with its restriction to only 40% when introduced by the transformation pathway (see section II.C.). In the case of conjugative transfer rapid synthesis of the complementary strand may not allow as much methylation of the single-stranded DNA as does the transformation pathway before action of the endonuclease.

1. Conjugative transposons

Certain strains of streptococci contain sets of genes responsible for multiple drug resistances that are not located on plasmids but rather on their chromosomes (Jacobs et al., 1978; Shoemaker et al., 1979; Horodniceanu et al., 1981). These genes are found in chromosomal elements that can mediate their own conjugative transmission in filter matings (Shoemaker et al., 1980; Buu-Hoi and Horodniceanu, 1980). Such resistance elements have important implications in the antibiotic treatment of diseases such as pneumonia.

Guild and co-workers (Vijayakumar et al., 1986a, b) have intensively investigated several conjugative resistance elements, which they call Omega elements. These elements can be transferred intact by transformation with isolated chromosomal DNA as well as by conjugation. Parts of an element carrying a single antibiotic resistance determinant can be transferred by transformation, but the intact element is required for conjugative transmissibility. One element, BM6001, which contains *cat* and *tet* genes conferring resistance to chloramphenicol and tetracycline, respectively, was analyzed by cloning of fragments, restriction site mapping, and hybridization to host chromosomal DNA (Vijayakumar et al., 1986a, b). The element is 65.5 kb in size. It is transferred conjugatively always to the same target site in the *S. pneumoniae* chromosome. No extensive repetition of DNA sequence was evident at its termini in the chromosome. Another element, B109, carrying *cat*, *tet*, and *erm* genes, was analyzed by Inamine and Burdett (1985). This 67-kb element is remarkably similar in its restriction map to BM6001, and it may have been formed by addition of a 2.5-kb *erm* gene segment, conferring resistance to erythromycin, to the latter element (Vijayakumar et al., 1986a).

Smaller conjugative elements carrying tetracycline resistance genes were identified in the chromosome of *S. faecalis* (Franke and Clewell, 1981) and *S. sanguis* (Fitzgerald and Clewell, 1985). These elements behave like transposons in that they insert in various positions in the chromosome, can transfer from a chromosome to a plasmid, and can be removed from the site of integration within a plasmid gene by precise excision. They were therefore called conjugative transposons and named Tn*916* and Tn*919*, respectively. Both elements are approximately 16 kb in size and similar, but not identical, in their properties. A similarly conjugative and transposable element, Tn*1545*, approximately 25 kb in size and containing resistance genes for kanamycin, erythromycin, and tetracycline, was isolated from a strain of *S. pneumoniae* (Courvalin et al., 1987). The large (67-kb) B109 element mentioned above, when transferred to the *S. faecalis* chromosome, was also shown to transpose to plasmid pAD1 (Smith and Guild, 1982). Thus, it appears that all of the conjugative chromosomal elements may be transposable. It is reasonable, therefore, to call them conjugative transposons, but their mechanism of transposition may be different from that of the smaller, nonconjugative transposons typified by the transposons of the gram-negative enterobacteria.

It has been suggested that in the transfer of conjugative transposons (Clewell, 1981) and Omega elements (Imamine and Burdett, 1985), the elements are excised from the chromosome and circularized prior to transfer. Their subsequent conjugative transfer was then supposed to be similar to conjugative plasmid transfer. However, unlike plasmid conjugation, Omega element transfer is not subject to restriction enzyme destruction in the host cell (Guild et al., 1982). This is consistent with transfer of a single DNA strand, but without immediate synthesis of its complement in the recipient cell. Thus, the transferred conjugative element strand may be directly inserted into the recipient chromosome. Host *rec* functions needed for chromosomal transformation in *S. pneumoniae* are not needed for this integration (Morrison et al., 1983). Insertion mutagenesis of the Tn*916* conjugative transposon showed that over half of the 16.4-kb element contained genes necessary for conjugative transfer (Jones et al., 1987). Integration into the recipient chromosome may depend on specific *tra* gene products.

III. RECOMBINATION MECHANISMS

This section deals mainly with the molecular mechanisms by which DNA undergoes recombination after introduction into the cells as single strands by the

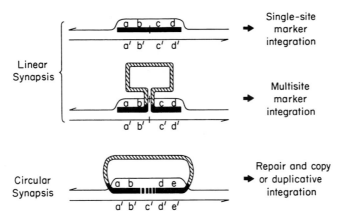

FIGURE 2. Linear and circular synapsis. Depending on the manner of attachment of homologous (solid bar) and heterologous (hatched bar) donor DNA, entering donor strands can interact with recipient double-stranded DNA (thin lines) by either linear or circular synapsis. Their subsequent fate is discussed later in the text. Vertical marks indicate mutated sites. Broken bar indicates repair synthesis.

transformation pathway. Its conclusions are based on the physical and genetic fate of isotopic and genetic markers in the donor DNA. The interactions depend for the most part on homology between DNA structures mediated by complementary base pairing. It is useful, here, to distinguish two types of interaction: *linear* synapsis, in which the ends of the homologous, pairing portions of the donor strand are not attached to each other, and *circular* synapsis, in which the ends are attached (Fig. 2). In addition to these single-strand interactions, some instances of duplex DNA recombination in gram-positive bacteria are also considered.

A. Chromosomal Transformation

In the absence of homology with recipient DNA, for example, when *E. coli* DNA is introduced into *S. pneumoniae*, donor single strands persist in the cell with a half-life of 30 min at 30°C, but they are eventually degraded, presumably by cellular nucleases (Lacks et al., 1967). In contrast, homologous donor strands are rapidly integrated physically into the chromosome, with a half-life of 5 min at 30°C, as judged by recovery of donor markers from eclipse (Fox, 1960; Ghei and Lacks, 1967; for an example, see Fig. 6, below) and incorporation of isotopic label into chromosomal DNA (Lacks, 1962; Fox and Allen, 1964; Lacks et al., 1967). This integration is very efficient; at least half of the introduced donor DNA markers are genetically integrated (Fox, 1957; Lacks et al., 1967). Although single base differences and small deletions do not affect marker integration efficiency (Lacks, 1966), greater nonhomology, for example, in heterospecific transformation between *S. pneumoniae* and *S. sanguis* (Bracco et al., 1957) or among *Bacillus* species (Dubnau et al., 1965), can reduce transformation to 10% or less of the homospecific transformation, depending on the relatedness of the strains and the gene containing the marker. The high efficiency of chromosomal integra-

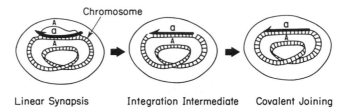

Linear Synapsis   Integration Intermediate   Covalent Joining

FIGURE 3. Chromosomal transformation mediated by linear synapsis. Heavy line, donor DNA strand segment; thin line, chromosomal DNA. Cross hatches indicate hydrogen bonding. A and a, marker difference between donor and recipient. For plasmid marker-rescue transformation substitute resident plasmid for chromosomal DNA.

tion of homologous DNA is particularly evident in transformation with cloned chromosomal DNA segments as donor DNA. Because the DNA is homogeneous and all the cells in a population of *S. pneumoniae* are competent, almost every cell takes up the genetic marker. Thus, transformation frequencies exceeding 50% of the recipient cell genomes are obtained (Claverys et al., 1981a; Stassi et al., 1981).

Physical integration of donor DNA strand segments gives rise to heteroduplex regions in the recipient chromosomal DNA (Fox and Allen, 1964; Dubnau and Cirigliano, 1972). As indicated in Fig. 3, this presumably occurs by linear synapsis and displacement of the corresponding recipient strand. Formation of a D-loop structure is presumably dependent on the superhelical state of bacterial chromosomal DNA (Worcel and Burgi, 1972) and catalysis by a RecA-like protein, as observed in vitro (Shibata et al., 1979; see Radding, this volume). Inhibition of transformation by coumermycin may reflect a decrease in gyrase-mediated chromosomal supercoiling (Raina and Ravin, 1979). The displaced recipient strand is eliminated by an unknown mechanism. For a brief interval, the donor strand segment remains unjoined to the corresponding recipient strand ends adjacent to it. These interruptions apparently allow donor strand targeting in mismatch repair, as discussed in section III.A. Physical studies distinguish this state of the heteroduplex donor-recipient complex from the covalently joined form that follows ligation (Dubnau and Cirigliano, 1973; Arwert and Venema, 1973). Integration from this point on is irreversible, although restoration of the donor marker to a homoduplex form requires completion of a round of chromosomal replication. The process of integration is schematized in Fig. 3.

The minimum length of DNA homology that can give chromosomal transformation is of the order of 100 nucleotides. However, the efficiency of such transformation is much reduced relative to that of larger segments (Cato and Guild, 1968). Integration efficiency increases linearly with the size of donor DNA within the cell above 100 nucleotides, so that it is half maximum at about 800 nucleotides (Morrison and Guild, 1972). Because donor DNA must be cleaved during uptake, the number-average length of donor single strands in the cell is approximately 3,000 nucleotides (Lacks and Greenberg, 1976), but the randomness of the breaks allows DNA as long as 60 kb (e.g., the conjugative element BM6001) to enter intact, albeit at a low frequency (Shoemaker et al., 1979). Transformation frequencies of chromosomal deletion mutants of various lengths are inversely proportional to the size of the insert corresponding to the deletion (Lacks, 1966). Again, this appears to be due to breakage during entry because inserts corresponding to small deletions are integrated very well. The presence of heterologous DNA within a linearly synapsed region (Fig. 2) does not impede transformation. However, complete restoration to the duplex wild-type form requires chromosomal replication, even for very small deletions less than 100 nucleotides in length (Ghei and Lacks, 1967).

Genetic recombination in transformation results from several factors. In the first instance separation of markers occurs during the release of DNA from donor cells. Shear forces alone break up the donor chromosome, and nucleases in the external milieu may further fragment the DNA. Breaks that are part of the DNA uptake process also separate markers. However, several different segments can be introduced from a single adsorbed DNA particle, as demonstrated by hetero-duplex analysis of integrated bromouracil-labeled donor DNA in *B. subtilis* (Fornili and Fox, 1977). Some trimming of introduced donor DNA strands by the action of cellular exonucleases may occur after entry. It has been proposed that the resulting strand segments in recipient cells are integrated in their entirety (Morrison and Guild, 1972). This would exclude additional recombination events after synapsis. However, an appreciable frequency of insertion of small segments in three-factor crosses (unpublished data) suggests that multiple changes can occur within the linearly synapsed fragment. Also, the presence of heterologous DNA external to the homologous fragment in transformation with cloned DNA does not dramatically interfere with the homologous transformation, although in rare cases illegitimate recombination (i.e., that not requiring appreciable homol-ogy) between vector and host DNA in such instances does give a low frequency of deletion mutants (Claverys et al., 1980). With homologous DNA, high effi-ciency of integration is the rule for gram-positive transformation, but a notable exception results from the action of a mismatch correction system in *S. pneumo-niae*.

1. Heteroduplex mismatch repair

Wild-type strains of *S. pneumoniae* contain the Hex mismatch repair system. This system recognizes certain base mismatches in the heteroduplex DNA integration product of transformation and eliminates the corresponding donor marker (for review, see Claverys and Lacks, 1986). Affected markers are reduced in transformation efficiency by as much as 20-fold relative to unaffected markers (Lacks and Hotchkiss, 1960; Ephrussi-Taylor et al., 1965; Lacks, 1966). Mis-matches of markers corresponding to transition mutations (A/C and G/T) are very well recognized; those due to transversion mutations are less well recognized, some not at all, depending on the adjacent sequence (Claverys et al., 1981b, 1983; Lacks et al., 1982). Deletion (or insertion) mismatches of one or two nucleotides are also recognized, but deletion mismatches greater than four nucleotides are not affected by the repair system (Gasc et al., 1987; Lopez et al., 1987). Ephrussi-Taylor and Gray (1966) proposed that mismatches are recognized by a repair system similar to those that recognize DNA damage; they suggested that either strand may be excised.

Whenever a mismatch is recognized, it appears that it is always the donor contribution to the mismatch that is corrected (Lacks, 1966). A linked marker that is normally not recognized is excluded together with a mismatched marker, and the dependence of such exclusion on distance between the markers is similar to the recombination dependence on distance between the marker sites (Lacks, 1966). Thus, the average excluded segment, like the average segment normally integrated, is several kilobases in length. Kinetic studies of the recovery of excluded markers from eclipse indicate that the mismatch correction occurs early in the integration process; loss of a donor marker is complete within 10 min at 30°C (Ghei and Lacks, 1967; Shoemaker and Guild, 1974). That the donor DNA is physically degraded during this process was shown by transformation with isotopically labeled cloned marker DNA (Mejean and Claverys, 1984). These findings are all consistent with a model in which targeting of the strand to be corrected (i.e., the donor strand in the transformation heteroduplex) depends on single-strand breaks in the initial integration product prior to covalent joining, with the entire donor strand segment eliminated by the repair process.

Chromosomal mutations called *hex* eliminate differences in transformation efficiency for single-site markers (Lacks, 1970). Hex⁻ cells also show high spontaneous mutation rates (Tiraby and Fox, 1973), which indicates that the Hex repair system normally prevents mutations resulting from chromosomal replication errors. Two loci, *hexA* and *hexB*, have been implicated (Claverys et al., 1984; Balganesh and Lacks, 1985). These genes have been cloned, and their products have been identified as polypeptides approximately 90 and 80 kDa in size, respectively (Balganesh and Lacks, 1985; Martin et al., 1985; Prats et al., 1985). Although the biochemical functions of the HexA and HexB proteins remain unknown, sequence analysis of the *hexA* gene (Priebe et al., 1988) revealed a potential ATP- or GTP-binding site in the HexA protein. Interestingly, the protein proved to be homologous to MutS, a component of the Mut mismatch repair system of the gram-negative enterobacteria (Priebe et al., 1988, Haber et al., 1988).

Despite uncertainty concerning the number of components in the system and their function, a tentative model for Hex system action is shown in Fig. 4. It is proposed that the two gene products carry out the two functions implicated in repair, that is, mismatch recognition and donor strand elimination. HexA may be responsible for mismatch recognition, since that is the function attributed to MutS (Su and Modrich, 1986), and HexB may be an exonuclease or helicase that could remove the donor strand. Inasmuch as the Hex system must recognize both the mismatch and the breaks in the donor strand of the integration intermediate, one or more of its components may enter at a break and travel to the mismatch and to the following break, possibly retaining hold of all three points before eliminating the donor segment (for a more complete discussion of models for mismatch repair, see Claverys and Lacks, 1986).

In addition to the Hex system, a sequence-specific, short-patch repair of A/G to C/G in 5′-ATTAAT/3′-TAAGTA mismatches is found in *S. pneumoniae* (Sicard

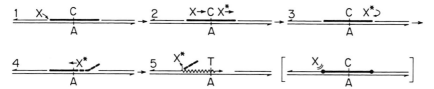

FIGURE 4. Heteroduplex DNA mismatch repair during transformation of *S. pneumoniae*. (1) The Hex system monitor (X) enters DNA at a nick and scans for a mismatch (exemplified here by a C/A mismatch). (2) Recognition of mismatched base pair triggers monitor. (3) Activated monitor (X*), on reaching end of segment, reverses course. (4) Donor segment is degraded. (5) Repair synthesis (wavy line) restores recipient chromosome. Shown in brackets is the inability of the Hex system to act on donor DNA already ligated into the chromosome. Variations of this hypothetical mechanism, also consistent with the data, are discussed in the text.

et al., 1985). Unlike the Hex system, which acts symmetrically and, therefore, does not affect genetic mapping (Lacks, 1966), this sequence-specific system gives excess recombinants when a cross is carried out in one direction. A mismatch repair system similar to Hex affects transformation in the gram-negative bacterium *Haemophilus influenzae* (Bagci and Stuy, 1979). No Hex-like system has yet been reported for *B. subtilis* or other transformable gram-positive species. However, a mismatch repair system found in *Saccharomyces cerevisiae* (Williamson et al., 1985) appears to be similar to Hex in several respects (Claverys and Lacks, 1986).

2. Plasmid marker-rescue transformation

Resident plasmids in cells can be transformed by intact or fragmented plasmid donor DNA containing genetic markers. Such plasmid marker-rescue transformation was demonstrated in *B. subtilis* for multicopy plasmids of staphylococcal origin (Contente and Dubnau, 1979b). It was also observed in *S. pneumoniae* with staphylococcal plasmids (Barany and Tomasz, 1980) and multicopy streptococcal plasmids (Stassi et al., 1981).

The mechanism of such transformation has been examined in detail for a single-copy plasmid in *B. subtilis* (Weinrauch and Dubnau, 1983, 1987). Isotopically labeled donor DNA was used to show that single-strand segments were physically integrated into the resident plasmid similarly to chromosomal transformation. Rec⁻ mutants that were deficient in chromosomal transformation were similarly deficient in plasmid marker-rescue transformation. Very little DNA synthesis was associated with the recombination event. This result is consistent with earlier findings that DNA replication is not necessary for chromosomal transformation (Lacks et al., 1967). Thus, it appears that the mechanism of linear synapsis, D-loop formation, and donor strand integration operates also in plasmid marker-rescue transformation.

It might be argued that transformation of a single-copy resident plasmid, which in effect is like an accessory chromosome, is not typical of multicopy plasmid transformation. However, the latter also is dependent on Rec functions

(Prozorov et al., 1982). Further complications may arise with multicopy plasmids. For example, depending on selective pressures, expression of a transformed marker might follow its integration into a single plasmid molecule, or conversion of the entire pool might be required. If resident plasmids share homology with the chromosome, regions adjacent to the marker may determine whether chromosomal or plasmid transformation takes place. These questions remain to be explored. Initial difficulties in cloning genes in *B. subtilis* by direct introduction of recombinant plasmids led to the suggestion that plasmid marker-rescue transformation may be a more effective route for cloning (Gryczan et al., 1980). However, as will be seen below, despite the degradative effects of uptake in the transformation pathway, recombinant plasmids can readily be established de novo in gram-positive bacteria.

B. Plasmid Transfer via the Transformation Pathway

Transfer of plasmids by treatment of competent cells of gram-positive bacteria with free plasmid DNA was first demonstrated by the introduction of staphylococcal drug resistance plasmids into *B. subtilis* (Ehrlich, 1977). It was found that covalently closed circular duplex DNA suffered the same degradative fate as linear chromosomal DNA on entry into cells of *S. pneumoniae* (Lacks, 1979b) or *B. subtilis* (de Vos et al., 1981). Thus, entry of a monomeric plasmid would give rise at most to a linear single strand, which could not circularize to allow replication of the plasmid genome. How, then, could a plasmid be established?

1. Independent reconstitution of plasmids

Saunders and Guild (1980) showed that monomeric forms of plasmid pMV158 could give rise to plasmid establishment in the transformation of *S. pneumoniae*. A key finding of this work was that unlike chromosomal transformation, which is linearly dependent on DNA concentration, the frequency of plasmid transfer is dependent on the square of the monomer plasmid DNA concentration. This indicates that two plasmid molecules must enter the cell for establishment to occur. If the two entering strand segments come from complementary strands, they can anneal and, following repair synthesis of the gaps, give rise to a circular replicon as shown in Fig. 5A. A quadratic dependence on monomer plasmid concentration is observed also for plasmid transfer in *S. sanguis* (Macrina et al., 1981).

Multimeric forms of plasmids, which constitute a variable but significant proportion of the plasmid content of bacteria, also give rise to plasmid establishment in *S. pneumoniae*. Such transfer is linearly dependent on plasmid concentration (Saunders and Guild, 1980), so that a single entering particle is sufficient. In this case, an entering strand segment could contain an entire plasmid genome flanked by a terminal repetition. A possible scheme by which such a segment could circularize is shown in Fig. 5B. If the entering segment is from the (+) strand of the multimeric plasmid, synthesis of the complementary strand could

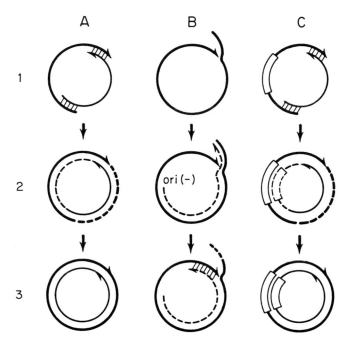

FIGURE 5. Plasmid establishment after entry by the transformation pathway. A, B, and C occur in *Streptococcus* species; only B occurs in *B. subtilis*. (A) Reconstitution from separately entering monomeric plasmid strand fragments: 1, annealing of complementary fragments; 2, repair synthesis; 3, completed replicon. (B) Reconstitution from an entering multimeric plasmid fragment: 1, entering strand segment with terminal repeats; 2, partial replication of the complementary strand from a plasmid origin [ori(−)], 3, switchback of newly synthesized strand end to circularize plasmid and subsequent degradation of terminal repeat. The replicon is completed by repair synthesis as in A. (C) Reconstitution of recombinant plasmid from separately entering recombinant (rare) and vector (common) plasmid fragments: 1, annealing of complementary fragments; 2, repair synthesis of vector and recombinant portions; 3, completed recombinant plasmid replicon. Heavy line, plasmid (+) strand; thin line, plasmid (−) strand; open bar, cloned chromosomal DNA insert; broken line, newly synthesized DNA (except in B3, where heavy broken line indicates strand degradation); hatch marks, hydrogen-bonded regions.

proceed from the (−)-strand origin along the segment and through the terminal repetition. Displacement of the newly synthesized end to the other complementary terminus would restore the circular form, and trimming of the repeated portion would give rise to a complete monomeric plasmid.

In *B. subtilis* it had previously been shown (Canosi et al., 1978) that multimeric plasmids were 1,000-fold more effective than monomers in transfer to competent cells. Here, too, the frequency of transfer was linearly dependent on DNA concentration. It appears, however, that monomers are much less effective in plasmid transfer to *B. subtilis* than to *S. pneumoniae*. One possible explanation is that fragments entering at separate points on the cell surface are not so freely diffusible and able to associate in *B. subtilis*. When donor plasmids contain a directly repeated sequence, monomeric forms can be transferred to *B. subtilis*

(Michel et al., 1982). Although most of the plasmids established consisted of only the portion of the original plasmid located between the repeats (plus one copy of the repeat), a significant proportion of transformants (~10%) contained the entire donor plasmid. In both cases the frequency of establishment was linearly dependent on DNA concentration, which indicates the effectiveness of a single donor molecule. Establishment of the truncated plasmids can be attributed to the mechanism shown in Fig. 5B, with the direct repeats taking the place of terminal repetitions. It is more difficult to explain establishment of the entire donor plasmid, but one possibility is an interaction of complementary strand segments situated between the repeats, which can be introduced from a single donor molecule (see the model presented by Stassi et al. [1981]).

The requirement for a region of overlap in the introduced plasmid DNA strands (to establish a circular replicon) makes plasmid transfer sensitive to treatment of donor plasmids in vitro with restriction endonucleases. A double-strand break at a particular site in every genome will eliminate either the complementary overlap required by donor monomers or the sequence repetition required by donor multimers. Such restriction enzyme sensitivity was found for plasmid transfer in both *B. subtilis* (Contente and Dubnau, 1979a) and *S. pneumoniae* (Barany and Tomasz, 1980). In *S. pneumoniae* but not *B. subtilis*, mixing of plasmids cut separately with different restriction enzymes restored activity. Random double-strand breaks, such as those introduced by nuclease S1 treatment of a plasmid preparation, also did not prevent plasmid transfer in *S. pneumoniae*, as expected from the ability of differently cut fragments to complement each other (Saunders and Guild, 1981).

The degradative fate of DNA on entry in the gram-positive transforming systems raised questions, at first, of the usefulness of these systems for cloning chromosomal genes. For example, in a shotgun cloning experiment with total chromosomal DNA ligated to a vector plasmid, it is unlikely that two identical recombinant plasmids would enter the same cell and be reconstituted. However, further consideration leads to the realization that only one recombinant segment need enter. It could be complemented simply by a vector segment from any of the many plasmids in a ligation mixture, as shown in Fig. 5C. Furthermore, ligation of a chromosomal DNA fragment to two vector molecules would give rise to terminal repeats and allow recombinant plasmid establishment by the mechanism of Fig. 5B. Since the experimental demonstration of such cloning of chromosomal genes in *B. subtilis* (Keggins et al., 1978) and *S. pneumoniae* (Stassi et al., 1981), numerous genes have been cloned in these species. Frequently, genes that cannot be cloned in *E. coli* host/vector systems because of their excessive transcription or the toxicity of their products can be successfully cloned in a gram-positive system (Stassi and Lacks, 1982). Inasmuch as the mechanism of DNA uptake in these systems is different from *E. coli* transformation, caution must be exercised in adapting procedures useful in the latter system. For example, removal of vector terminal phosphate groups to prevent simple recircularization of vector plasmids in a ligation cannot be employed for chromosomal gene cloning in the gram-positive systems (Balganesh and Lacks, 1984). Because the cloned DNA is introduced on a single strand, it must be ligated at both ends to the vector.

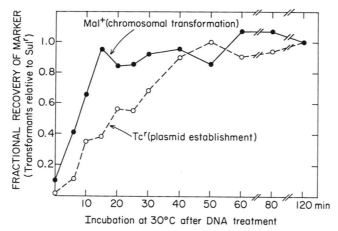

FIGURE 6. Kinetics of plasmid establishment by chromosomal facilitation. Recovery from eclipse of Mal$^+$ marker in chromosomal transformation and recovery of Tc^r marker in plasmid establishment and filling of pool. A culture of *S. pneumoniae* 772 (*malM594 sul-d*) at 1.1×10^8 CFU/ml was treated with recombinant plasmid pLS70 (*malM$^+$ tet*) for 9 min at 30°C. DNA uptake was terminated by the addition of pancreatic DNase. Samples were incubated for various times as indicated; they were then chilled, and crude DNA extracts were prepared. These extracts were used to transform tester strain 708 (*malM594*), and the frequencies of Mal$^+$, Tc^r, and Sul^r transformants were determined. Frequencies relative to Sul^r, and normalized to 1.0 at 120 min, are plotted. The absolute frequencies, per milliliter of tester culture, of transformants obtained with the 120-min sample extract were Mal$^+$, 525,000; Tc^r, 4,330; and Sul^r, 1,840,000. For details of the procedure see Ghei and Lacks (1967).

2. Chromosomal facilitation of plasmid establishment

Introduction of a segment of chromosomal DNA into a plasmid so that the plasmid shares homology with the recipient cell chromosome greatly enhances the ability of monomer plasmid DNA to establish itself in *B. subtilis* transformation (Canosi et al., 1981). The frequency of such establishment can be increased over 1,000-fold, and it is strongly dependent on the length of the homologous segment, showing a cubic dependence on length over the range from 0.6 to 6.0 kb (Bensi et al., 1981). Facilitation of plasmid transfer was independently observed in *S. pneumoniae* (Lopez et al., 1982). In this species the frequency of plasmid establishment was also dependent on the length of homology with the chromosome, but inasmuch as monomeric donor plasmids without homology can themselves be established at an appropriate frequency, the maximum facilitation did not exceed 100-fold. Most significantly, unlike the quadratic dependence on DNA concentration shown by donor monomers in the absence of homology, the homology-facilitated establishment of plasmids in *S. pneumoniae* was linearly dependent on concentration (Lopez et al., 1982). This indicated that a single plasmid entry event was sufficient and that an interaction of the introduced plasmid strand fragment with the chromosome took the place of interaction with a complementary strand fragment in reconstitution of the replicon.

To clarify the mechanism of chromosomal facilitation, two experimental approaches were taken. In one, illustrated in Fig. 6, the kinetics of facilitated plasmid establishment were examined. A recombinant plasmid, pLS70, in which the chromosomal $malM^+$ gene was inserted into the vector pMV158, was used to transform a strain of *S. pneumoniae* containing a *malM* mutation in its chromosome. The recovery of maltose utilization (Mal^+) and tetracycline resistance (Tc^r) transforming activity from eclipse was determined at different times after DNA treatment by transformation of a tester strain. Inasmuch as the frequency of chromosomal transformation is much higher than even facilitated plasmid transfer, Mal^+ represents recovery of the chromosomal marker from eclipse, and the plasmid marker Tc^r represents plasmid establishment and filling of the plasmid pool. Recovery of Mal^+, which corresponds to the chromosomal integration of single-stranded DNA, is complete by 15 min at 30°C. Interestingly, the recovery of Tc^r, although slower, is complete by 50 min, which is less than the generation time of the bacteria at this temperature. During this period the plasmid replicon had to be not only established but also replicated to fill the plasmid pool of approximately 25 copies. Thus, facilitated plasmid establishment must be a rapid process which occurs with a time course similar to chromosomal integration.

A second experimental approach to the mechanism was treatment in vitro of a recombinant plasmid—in this case pLS80, which contains a *sul-d* mutation that confers sulfonamide resistance (Sul^r) to *S. pneumoniae*—with various restriction endonucleases (Lopez et al., 1982, 1984b). Cleavage in the heterologous, vector portion of the plasmid completely prevented plasmid establishment. Cleavage within the homologous region, even removal of a segment of that region, did not prevent plasmid transfer, and in every case the plasmid established was the size of the original donor. An incoming plasmid strand fragment broken within the region of homology can apparently be reconstituted by interaction with the chromosome.

A mechanism for facilitation of plasmid establishment, which is consistent with the above findings, is depicted in Fig. 7A. The incoming plasmid strand with the vector portion intact and regions of chromosomal homology adjacent on both sides interacts by circular synapsis with the homologous portion of the chromosome. Any gap is filled in by DNA repair synthesis using the complementary chromosomal strand as template. Synthesis of the complementary plasmid strand then begins at a plasmid origin. Synthesis through the homologous region releases the plasmid from the chromosome, and its completion produces a complete plasmid, which then replicates to fill the pool.

Chromosomal facilitation of plasmid establishment can introduce chromosomal markers into the plasmid in both *S. pneumoniae* (Lopez et al., 1982) and *B. subtilis* (Iglesias et al., 1981). This is expected from the chromosomal copy mechanism of facilitation, as illustrated in Fig. 7B. In general (but with exceptions noted below), the resultant plasmid pool is homogeneous with respect to the marker. The probability of introduction of a chromosomal marker into the plasmid varies between 15 and 85% of the transfer events, depending on whether the marker is located terminally or centrally, respectively, within the homologous segment (Lopez et al., 1982).

In some cases, perhaps 20% of the total, it appears that heteroduplex plasmids are established, which give rise to heterogeneous pools of plasmids

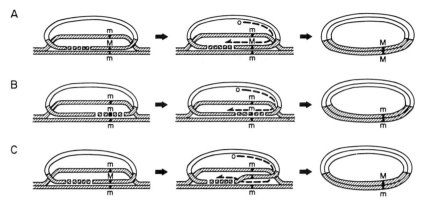

FIGURE 7. Mechanism of chromosomal facilitation of plasmid establishment. (A) Formation of plasmid with homoduplex plasmid marker. Circular synapsis with the donor strand break distant from the marker is followed by repair synthesis and ligation to close the plasmid strand. Synthesis of the comlementary strand from the plasmid origin replicates the plasmid marker. The plasmid is established with the plasmid marker present in both strands. (B) Formation of plasmid with homoduplex chromosomal marker. Circular synapsis with the donor strand gap covering the marker is followed by repair synthesis that replicates the chromosomal marker. Complementary strand synthesis again replicates the chromosomal marker. The chromosomal marker is present in both strands of established plasmid. (C) Formation of plasmid with heteroduplex marker configuration. Circular synapsis occurs with the break distant from the marker. Complementary strand synthesis shifts to the chromosomal template in the region of the marker before it returns to the plasmid strand. A heteroduplex plasmid is established; upon replication, it gives rise to a mixed plasmid pool in the host cell. In each case the chromosome is released unchanged (not shown) after establishment of the plasmid. Linear bar, segment of double-strand chromosomal DNA; curvilinear bar, introduced fragment of plasmid DNA; hatched portion, homologous DNA; open portion, heterologous (vector) DNA; broken bar, repair synthesis of introduced plasmid strand; broken line, synthesis of complementary plasmid strand; M and m, marker alleles; o, origin of plasmid replication.

containing either the original plasmid or the chromosomal allele of the marker gene. Evidence for this came from transfer of pLS83 to cells of *S. pneumoniae* (Lopez et al., 1984b). This plasmid is a deleted derivative of the original *sul-d* recombinant plasmid, pLS80. It contains a Tc^r marker and the *sul-d* mutation, but it cannot elicit high levels of sulfonamide resistance unless a chromosomal segment corresponding to a deletion in the plasmid (1 kb downstream from *sul-d*) is restored to it. Approximately 20% of the initial transformants appeared to contain mixtures of plasmids with and without the additional chromosomal segment. Depending on subsequent selection, either one or the other type was lost. A possible mechanism for producing mixed plasmid pools by chromosomal facilitation is shown in Fig. 7C. In this case the original plasmid marker persists in the introduced strand segment, but synthesis of the complementary strand is diverted within the region of homology to the chromosomal template strand, thereby producing a heteroduplex product.

As well as providing insight into the phenomenon of circular synapsis, chromosomal facilitation of plasmid transfer can be of considerable practical use. It provides a convenient means for introducing chromosomal mutations into plasmids, thereby forgoing the need for separate cloning of mutant alleles. Although plasmid marker-rescue transformation is another feasible approach to introducing a mutant allele, it requires an additional transformation step and a selective procedure to isolate the transformed plasmid from its untransformed sisters in the plasmid pool. Transfer of mutant alleles to plasmids can be useful both for determining DNA sequence changes corresponding to mutations (Lacks et al., 1982) and for analyzing the ability of different alleles to complement each other (Balganesh and Lacks, 1985). Chromosomal facilitation of plasmid transfer can also be used in cloning experiments to enrich for recombinant plasmids containing DNA homologous to the recipient chromosome (Balganesh and Lacks, 1984).

C. Ectopic Integration

Ectopic integration can be defined as the introduction of a gene, or any segment of DNA, into a genome at a place in which it is not normally found. The term has been used particularly to refer to the mechanism by which a strand deriving from a circular donor structure, which contains adjacent segments of DNA homologous and heterologous with the chromosome of the recipient cell, is introduced into that chromosome via circular synapsis to give an integrated product containing a duplication of the homologous segment flanking the heterologous portion (Stassi et al., 1981; Mannarelli and Lacks, 1984). However, there are several mechanisms that can integrate a heterologous segment of DNA into the chromosome.

Heterologous DNA can be integrated into a bacterial chromosome by the transformation pathway if it is inserted within, rather than adjacent to, a stretch of homology. The integration occurs by linear synapsis and is essentially the same as transformation of a chromosomal deletion mutant by wild-type DNA (see Fig. 2, multisite marker). An example of this is the insertion of pHV32 (an *E. coli* plasmid that does not replicate but expresses its *cat* gene in *B. subtilis*) into the *ilvA* gene of *B. subtilis* to give strain c26 (Niaudet et al., 1982).

Another way that heterologous DNA can be integrated is by attaching it at both ends to identical pieces of DNA homologous to the chromosome so that the flanking DNA is directly repeated. Such donor DNA can be produced either by ligation of appropriate fragments or by fragmentation of multimeric forms of plasmids containing cloned chromosomal genes. In this case DNA is also integrated into the chromosome by linear synapsis; it differs from the foregoing case in that a duplication of the homologous piece of DNA is present in the chromosomal product. An example is the mutagenic insertion into the *amiA* locus of *S. pneumoniae* of the *E. coli* plasmid pBR325 by treatment of the cells with dimeric forms of a recombinant plasmid containing a fragment of the *amiA* locus cloned in pBR325 (Vasseghi et al., 1981).

A more unusual mechanism for integration of nonhomologous DNA depends on circular synapsis. When the donor DNA consists of adjacent homologous and heterologous segments in a circular structure, as in the product of a simple ligation or in the monomeric form of a recombinant plasmid, this mechanism appears to be the only way that the heterologous portion can be integrated. The linear single-strand fragment that is introduced into the cell undergoes circular synapsis just as in chromosomal facilitation of plasmid establishment. However, inasmuch as the circular structure has no functional origin of replication, it does not replicate but remains associated with the chromosome until either it is degraded or a single-strand recombination event with its chromosomal homolog integrates it into the chromosome. Subsequent replication of the chromosome converts the integrated material to a double-stranded form. The product thus contains a direct repeat of the homologous segment flanking the heterologous DNA. Experimental evidence for the novel form of ectopic integration came initially from the observation that ligated restriction fragments of wild-type DNA could transform a *malM* deletion mutant of *S. pneumoniae* to maltose utilization (Mal⁺), whereas the unligated fragments could not (Stassi et al., 1981). Because the deletion in the recipient removed the locus of the *malM* gene, the wild-type gene was apparently integrated elsewhere in the chromosome, hence the term ectopic integration. Subsequent studies showed that the mechanism of this phenomenon was ligation of the *malM* restriction fragment with some other fragment to give a circular structure so that the strand introduced into the cell could interact by circular synapsis with the chromosomal region homologous to the ligated partner to insert the *malM* gene (Mannarelli and Lacks, 1984). In particular, a circular construction containing the cloned *mal* and *sul* segments integrated the *malM* gene at the *sul* locus. Figure 8A illustrates this mechanism. An independent demonstration of the same type of ectopic integration is the mutagenic insertion into the *S. pneumoniae amiA* locus of the *E. coli* plasmid pBR325 after treatment of the cells with monomeric forms of the recombinant plasmid containing a fragment of the *amiA* locus cloned in pBR325 (Vasseghi et al., 1981). The interaction envisaged for this case is shown in Fig. 8B. The probability of ectopic integration appears to be rather low compared with normal chromosomal integration, being only of the order of 1% of the latter. The type of recombination event needed here, a single crossover within a circular structure, may be less frequent than recombination events at the ends of a linear segment.

In general, when plasmid preparations or mixtures of ligated DNA fragments are used to insert heterologous DNA at a particular position in the chromosome, the mechanism of integration can be by either linear synapsis (as with a dimeric plasmid) or circular synapsis (as with a monomeric plasmid). In either case, however, the resulting chromosomal product contains heterologous DNA flanked by repeats of the homologous DNA segment used to direct the heterologous material. Early demonstrations of such integration in *B. subtilis*, without addressing the mechanism of the process, were reported by Duncan et al. (1978) and Haldenwang et al. (1980). Because the final products of these interactions are similar to what is obtained in a single crossover of a double-stranded circle with the chromosome—an interaction proposed by Campbell (1962) to explain integration of temperate phage DNA—the mechanism of ectopic integration in gram-positive bacterial transformation is often referred to as Campbell type. However, it should be emphasized that in ectopic integration the interaction and crossover occur at the level of single-stranded DNA.

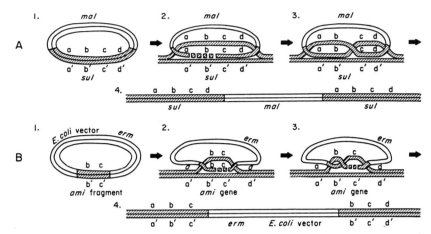

FIGURE 8. Mechanisms of ectopic integration. (A) Ectopic integration of the *mal* marker in the vicinity of the *sul* locus. (1) Restriction fragments containing cloned *mal* and *sul* genes ligated together in a circular construct were used as donor DNA. (2) Circular synapsis of the donor strand fragment occurs at the *sul* locus of the chromosome in a *mal* deletion mutant recipient; the gap in the homologous portion of the donor fragment is filled by repair synthesis. (3) A single-strand crossover integrates the donor strand into the chromosome. (4) Replication of the chromosome converts the integrated single-strand segment to a duplex form so that the heterologous *mal* segment is inserted between duplicated homologous *sul* segments. The other product of the replication is an unchanged recipient chromosome (not shown). (B) Mutagenesis of the *ami* gene by additive insertion of a nonreplicating plasmid. (1) A chromosomal *ami* gene fragment and an *E. coli* vector plasmid containing an *erm* gene expressible in *S. pneumoniae* are ligated together in a circular construct and used as donor DNA. (2) Circular synapsis of the donor strand fragment with the chromosome of the *S. pneumoniae* recipient occurs at the *ami* locus; the gap in the homologous portion of the donor fragment is filled by repair synthesis. (3) A single-strand crossover integrates the donor strand into the chromosome. (4) Replication of the chromosome converts the integrated single-strand segment to a duplex form so that the heterologous *E. coli* vector segment is inserted between duplicated homologous *ami* segments, thereby giving rise to an Ami[r] mutation. Such ectopic integrants can be selected on the basis of aminopterin (*ami*) and erythromycin (*erm*) resistance. The other product of the replication is an unchanged recipient chromosome (not shown).

From the practical point of view, the mechanisms of ectopic integration in gram-positive transformation can be very useful in genetic manipulations. If the homologous segment is entirely within a gene, insertion of the heterologous segment will mutate the gene, so that ectopic integration is a useful mutagenic tool. Because the insert can be removed by recombination between the direct repeats at its flanks, these mutations are characterized by reversion at an appreciable frequency (Vasseghi et al., 1981). For example, the rate of loss of a *mal(sul)* ectopic insertion with 10-kb *sul* repeats was 5×10^{-4} per generation in the absence of selection (Mannarelli and Lacks, 1984). This instability would not interfere with dominance and complementation analyses, especially if the ectopic allele is joined to a selectable marker. If a detectable marker gene is present within the heterologous segment, the locus of the ectopic integration can be mapped genetically (Haldenwang et al., 1980). If the heterologous segment corresponds to a plasmid that cannot replicate in the recipient cell but can replicate in another species, then it can be used to rapidly clone the gene carrying the novel mutation (Mejean et al., 1981; Niaudet et al., 1982).

D. Recombination of Double-Stranded DNA

Mechanisms for double-stranded DNA recombination, which would integrate DNA introduced by transduction, have not been studied as intensively in gram-positive bacteria as in gram-negative species (see Mahajan, this volume; Porter, this volume). However, since the suggestion by Hotchkiss (1974) that single-strand invasion initiates double-strand recombination, most proposed mechanisms for the latter have incorporated a role for a single-strand interaction in the process. Thus, some steps may be common and others specific in single- and double-strand recombination pathways.

Investigations of plasmid transfer and maintenance have revealed several novel modes of double-strand recombinatory behavior. One is the integration and excision of plasmids, which can lead to the equilibration of genetic markers between the chromosome and the plasmid pool. The others relate to mechanisms for generation of deletions in plasmids.

1. Plasmid integration and excision

The outcome of interactions between plasmid and chromosome depends on the ability of the plasmid to replicate in the host cell and on the extent of homology between plasmid and chromosome. When plasmids can replicate well, as in the case of pMV158 and its derivatives in *S. pneumoniae*, the plasmid will be transiently inserted into the chromosome and excised so as to equilibrate markers between the chromosome and the plasmid pool. For example, after establishment of *malM* mutant derivatives of pLS70 (copy number, ~30) in a wild-type host, such equilibration will produce a plasmid pool containing one wild-type *malM* allele per 30 *malM* mutant alleles (Lopez et al., 1982). When plasmids cannot replicate so well, derivatives that lack free plasmids but have the plasmid inserted into the chromosome will arise, provided that the plasmid shares homology with the chromosome. This appears to be the case for pPV21 in *B. subtilis* (Galizzi et al., 1981). In both of these cases insertion apparently occurs by a single reciprocal recombination event between the plasmid and the chromosome in the region of homology, according to the model of Campbell (1962), and excision occurs by reversal of this reaction. When a plasmid lacks homology with the chromosome and its replication is inhibited, by high temperature, for example, illegitimate recombination events can insert the plasmid into the chromosome at low frequency (Hofemeister et al., 1983).

2. Deletion mutagenesis

Deletions occur frequently enough in hybrid plasmids or those that contain cloned DNA so that if the deleted form confers a selective advantage to either the plasmid replicon or the cell, it will accumulate in the plasmid pool. One mechanism for producing deletions appears to be a double-strand recombination event between directly repeated sequences, which excises the intervening segment (Lopez et al., 1984a). Deletion could also occur by slippage during DNA replication (Albertini et al., 1982). Another hypothesized mechanism is removal of a segment between topoisomerase recognition sites by faulty closure (Lopez et al., 1984a; see also Champoux and Bullock, this volume). An additional mechanism, which can affect hybrid plasmids that replicate by a rolling circle mechanism (see section II.B.), has recently come to light (Michel and Ehrlich, 1986). In a hybrid plasmid the presence of an additional nicking site can lead to deletions between that site and a distal point in the plasmid as a consequence of aberrant rolling circle synthesis.

IV. GENETICS AND BIOCHEMISTRY OF RECOMBINATION

Most of the information on recombination described in section III came from analysis of the molecular and genetic fate of isotopically and genetically marked donor DNA. Another approach to understanding recombination mechanisms is the isolation of mutants blocked in the process, identification of the mutant genes, and determination of the normal biochemical functions of those genes. This approach has proved fruitful for the analysis of recombination in *E. coli* (Mahajan, this volume), and it has been adopted for the analysis of recombination in gram-positive systems.

A. *B. subtilis*

Progress in identifying genes in which mutations lead to recombination deficiency (*rec*) has been greatest in *B. subtilis*. As many as 10 genes have been implicated in one type of recombination or another. However, there are three problems relating to this body of work that require consideration. The first concerns nomenclature. Because various mutant loci were reported by different laboratories at approximately the same time, there has been duplication of gene letter designations. Mazza et al. (1975) attempted to resolve this problem by renaming loci reported by other laboratories. In Table 3 the *rec* genes are listed according to this nomenclature; prior designations are also indicated.

Another problem with the *rec* mutants of *B. subtilis* is that almost all of them were isolated as mutants sensitive to radiation or chemicals that damage DNA. Genes affecting recombination but not DNA repair would not be identified by this approach. Furthermore, some of the mutations may affect DNA repair but not directly affect recombination. An added complication in this regard is the presence of defective prophages in the *B. subtilis* strains used to isolate the mutants. Examination of the putative *rec* mutations in strains cured of the prophages is necessary to ensure that they actually block DNA repair and do not just induce a prophage (Yasbin, 1985). This problem is particularly acute in gauging the effect of a *rec* mutation on recombination during transformation. Induction of a prophage in potential transformants would reduce the yield of

TABLE 3
Recombination-defective mutants of *B. subtilis*

Gene designation		Map position[a]	Relative recombination ability[b]		Reference
Present	Prior		Transfor-mation[c]	Transduc-tion[d]	
recE	*recK,*[e] *recN,*[e] *rec-45*[f]	150	<0.001	<0.001	Dubnau and Cirigliano (1974), de Vos and Venema (1983)
recA	*rec-73*[g]	150	0.1	0.5[h]	De Vos and Venema (1983), Hoch and Anagnostopoulos (1970)
recB	*recD3*[i]	245	0.1	0.1	Dubnau and Cirigliano (1974), Hoch and Anagnostopoulos (1970)
recG		205	0.1	1.0[h]	Mazza et al. (1975)
recJ	*recC*[i]	11	0.01	0.5	De Vos and Venema (1983)
recC		?	0.05	?	Sinha and Iyer (1972), Devanathan and Iyer (1978)
recD	*rec-43*[f]	5	0.5	0.1	Mazza et al. (1975)
recM	*recG*[i]	5	0.5	0.02	Dubnau and Cirigliano (1974)
recF		0	0.5	0.05	Mazza et al. (1975), Perego et al. (1987)
recL	*recF*[i]	?	0.8	0.1	Dubnau and Cirigliano (1974)
recH	*rec-342*[j]	70	0.02		Prozorov et al. (1982)
recI	*rec-59*[j]	70	0.001		Prozorov et al. (1982)
recO	*recE*[k]	?			

[a] As recipient, wild type, 1.0.
[b] Angular degrees measured on circular chromosome from *guaA* in direction of *cysA* as described by Henner and Hoch (1982), except as indicated in the text.
[c] Relative to DNA uptake, which was in the normal range.
[d] Both homologous (within strain 168) and heterologous (between strains 168 and W23), except as indicated.
[e] Mazza et al. (1975).
[f] Sadaie and Kada (1976).
[g] De Vos and Venema (1983).
[h] Value for heterologous transduction, 0.02.
[i] Dubnau and Cirigliano (1974).
[j] Prozorov et al. (1982).
[k] Harford (1974).

viable transformed cells without directly affecting the recombination process. Even in the absence of prophages, some *rec* mutations may be deleterious enough by themselves to reduce viability of potential transformants, especially in *B. subtilis*, where competent cells constitute a metabolically retarded subset of the population (Nester and Stocker, 1963). Thus, marginal effects, which reduce recombination yields by a factor of only 2 or 3, may not indicate a real role for the *rec* gene product in the recombination process.

Table 3 lists the putative *rec* genes of *B. subtilis*. Genes *C, H, I, L, M,* and *O* are each represented by a single mutation; the others, by two or more. All mutants except *recC* were identified by sensitivity to DNA-damaging agents, and relative to the wild type they show greatly reduced resistance to UV light, methyl methanesulfonate, and mitomycin C. The *recC* mutation was obtained accidentally in a strain rendered resistant to mitomycin C by a separate mutation (Sinha and Iyer, 1972). Few properties of this mutation have been examined, but it reduces chromosomal transformation significantly without interfering with DNA uptake or donor-recipient complex formation (Devanathan and Iyer, 1978), and it does not affect DNA repair.

Genetic map locations of most of the *rec* genes have been determined. Map positions are taken from Henner and Hoch (1982) but have been modified to show that *recA* and *recE* are closely linked by transformation mapping (de Vos and Venema, 1983) and that cloning of *recF* locates it at the map origin (Perego et al., 1987).

Values listed in Table 3 indicate the level of chromosomal transformation in a typical mutant of each class relative to the wild-type strain. Allowance is made for variations in competence by measuring DNA uptake; none of these mutants are significantly impaired in uptake. Also indicated are the residual levels of transduction by phage PBS-1. Thus, each *rec* gene is examined with respect to a single-strand (transformation) and a double-strand (transduction) recombination pathway. The values under transduction are for donor phage grown on *B. subtilis* 168 derivatives, that is, homologous transduction. In two instances, *recA* and *recG*, heterologous transduction (with phage grown on strain W23) gave much less recombination. It is not clear what differences between mechanisms of recombination in homologous and heterologous transduction could give rise to these different responses. None of the mutant *rec* genes, with the exception of *recA*, interfere with plasmid establishment via transformation, which in the case of *B. subtilis* is the pathway shown in Fig. 5B. The mutation *rec-73* (which maps in *recA* [de Vos and Venema, 1983]) reduces plasmid transfer more than chromosomal transformation (de Vos and Venema, 1981).

The best-characterized recombination gene in *B. subtilis* is *recE*. Its *recE4* allele reduces both transformation and transduction to less than 0.1% of their normal level. A protein containing a 45-kDa polypeptide is absent in a *recE4* strain (de Vos and Venema, 1982). The *rec-45* mutation (Sadaie and Kada, 1976), which maps close to *recE4* (only 3% recombination in transformation), gives rise to an altered form of this protein that differs in isoelectric point (de Vos and Venema, 1983). The *recE45* mutant is defective in transformation and transduction, but less so than *recE4*. The phenotypes of both *recE* mutants are reversed by introducing into the cells a plasmid that produces the *E. coli recA* gene product, RecA (de Vos et al., 1983). Thus, the *recE* gene product is the *B. subtilis* counterpart of RecA. This *E. coli* protein can facilitate pairing of complementary homologous strands (Weinstock et al., 1979; see Radding, this volume); the purified RecE protein of *B. subtilis* also has this ability (Lovett and Roberts, 1985).

Two independent mutations of *B. subtilis*, *recA1* and *rec-73*, were mapped by transformation with respect to each other and to the two above-mentioned *recE* mutations (de Vos and Venema, 1983). Inasmuch as *rec-73* showed only 0.2% recombination with *recA1*, both mutations appear to be in the *recA* gene. They both recombined at frequencies of 3% with *recE45* and 7% with *recE4*. These data indicate that the *recA* mutations fall into either the *recE* gene or a gene adjacent to it. Because *recA* mutants differ phenotypically from *recE* mutants, in that the former affect plasmid transfer more and homologous transduction less, they may fall into two different genes. However, the *recA* mutants do not contain detectable RecE protein, and their ability to repair DNA damage, at least, is restored by *E. coli* RecA (de Vos and Venema, 1983; de Vos et al., 1983). These data are compatible with the hypothesis that the *B. subtilis recA* product acts as a repressor of *recE* expression, similarly to the *lexA* product in *E. coli* (Little and Mount, 1982), and that the two *recA* mutants studied make superrepressors, which prevent detectable RecE expression (de Vos and Venema, 1983). The *recA* mutants may make low levels of RecE, adequate for transduction and a low level of transformation; their specific defect in plasmid transfer may reflect a different function of the *B. subtilis* RecA protein. Alternatively, *recA* and *recE* may affect different functions of a multifunctional RecE protein.

Limitation of RecE, in either *recE* or *recA* mutants, would prevent homologous strand interaction. With minimal impairment, as in *recA* mutants, homologous transduction may be less sensitive than transformation or heterologous transduction because longer stretches of pairing are needed in the latter processes. Inasmuch as plasmid transfer by the mechanism assumed for *B. subtilis* (Fig. 5B) does not require interaction between two strands, absence of RecE may not affect it. The switchback needed to circularize the plasmid replicon may conceivably be catalyzed by a protein encoded by a gene that is separately and more fully repressed by a *recA* superrepressor mutant than is *recE*. By blocking homologous strand interaction at an early stage of transformation, the *recE* and *recA* defects prevent the association of donor and recipient DNA, that is, formation of a donor-recipient complex (Dubnau et al., 1973). The *recE4* mutation eliminates chromosomal facilitation of monomer plasmid transfer (Canosi et al., 1981) and transfer of recipient chromosomal markers to plasmids during establishment (Iglesias et al., 1981). Thus, RecE is required for circular synapsis as well as linear synapsis.

Mutations in *recB* decrease both transformation and transduction, but neither one strikingly. Not much is known about *recB* function. Genes *recG* and *recJ* both affect transformation more than they do transduction. In *recJ* mutants, a donor-recipient complex is formed, but it is subsequently degraded. It is possible that *recG* and *recJ* function at a later stage in single-stranded DNA integration. Mutants in *recG* are blocked in heterologous transduction also, perhaps because it entails more single-strand integration.

In contrast to the *rec* genes discussed above, which function in transformation, another set of genes—*recD*, *recF*, *recL*, and *recM*—do not appreciably affect transformation but are required for transduction. Presumably, they encode proteins necessary only for double-stranded DNA integration. The nature of these proteins and their function are not known. There may be only three distinct genes in this group inasmuch as *recD* and *recM* mutations exhibit similar phenotypes and map at the same position. The location of *recF*, which has been cloned (Perego et al., 1987), is distinct, and although *recL* has not been located, mapping attempts show that it is not present in the 0–5 portion of the *B. subtilis*

TABLE 4
Recombination-defective mutants of *S. pneumoniae* and *S. sanguis*

Species	Gene	Relative transformability[a]		Reference
		Chromosomal transformation	Plasmid transfer	
S. pneumoniae	*recP*	0.05	0.8[b]	Morrison et al. (1983)
	recQ	0.01	0.01[b]	Morrison et al. (1983)
S. sanguis	*cipA*	0.000001	0.001[c]	Raina and Macrina (1982)

[a] As recipient, wild type, 1.0. Relative to DNA uptake, which was in the normal range.
[b] With plasmid pLS1.
[c] With plasmid pVA736.

chromosome, which contains *rec* genes conferring similar phenotypes. The *rec-43* mutation, which corresponds phenotypically to *recD* (Sadaie and Kada, 1976), does map to this region.

The remaining *rec* loci—*recH*, *recI*, and *recO*—have not been sufficiently characterized to ascertain their function in transformation and transduction. The original isolate containing *recH342* was reportedly defective in an ATP-dependent DNase (Chestukhin et al., 1972), but that defect was not confirmed by subsequent analysis of the strain (Dubnau, 1982). Although the ATP-dependent DNase encoded by *recBCD* of *E. coli* functions in recombination in that bacterium (see Taylor, this volume; Mahajan, this volume), there has been no conclusive evidence of a role in recombination for such enzymes in gram-positive bacteria. Two independently isolated mutants of *S. pneumoniae* defective in ATP-dependent DNase did not significantly affect transformation (Vovis, 1973; Lacks, unpublished data), although it is conceivable that the mutant enzymes retain their recombination function, as in *recD* mutants of *E. coli* (Amundsen et al., 1986).

B. *S. pneumoniae*

Two sets of *rec* mutants with different phenotypes were obtained in *S. pneumoniae* by screening for transformation-defective clones that showed normal uptake of DNA (Morrison et al., 1983). Although none of the mutations have been mapped, they can tentatively be assigned to two loci, *recP* and *recQ* (Table 4). Two independent *recP* isolates reduced chromosomal transformation to 5% of the wild-type level but did not affect plasmid transfer at all. Seven independent *recQ* mutants reduced chromosomal transformation and plasmid transfer equally to levels of 0.1 to 3%, depending on the mutation. The mutants were not examined for defects in DNA repair abilities or for effects on recombinatory processes responsible for chromosomal facilitation of plasmid transfer and plasmid equilibration with the chromosome.

The *recP* gene was cloned (Rhee and Morrison, 1988), and its extent was defined by insertional mutagenesis to be 2.1 kb. In vitro it gave rise to a 72-kDa protein. Inasmuch as this size is greater than any competence-specific protein (see section II.A.), RecP appears to be constitutively made. Its biochemical function is unknown, but because it is required for chromosomal integration and not for plasmid establishment, it may be required for elimination of the recipient DNA strand segment corresponding to the integrated donor strand. Thus, it could be an endonuclease, exonuclease, or helicase.

A protein that cross-reacts with antibody to *E. coli* RecA protein can be identified in extracts of *S. pneumoniae* (Rhee and Morrison, 1988). This protein is present in *recP* mutants, but it may be absent in *recQ* strains. If so, RecQ might function like the *E. coli* RecA protein, and it could be responsible for the annealing of complementary homologous strands in both chromosomal transformation and plasmid establishment (see Fig. 3 and 5A; also see Radding, this volume). Plasmid transfer by conjugation and conjugative transfer of chromosomal elements do not require either *recP* or *recQ* functions (Morrison et al., 1983).

C. *S. sanguis*

A *rec* mutant was obtained in *S. sanguis* by screening for resistance to DNA-damaging agents (Raina and Macrina, 1982). This mutation was called *cipA9* because it appeared to modify a competence-inducible protein, 51 kDa in polypeptide size, so that it migrated slightly more slowly in sodium dodecyl sulfate-gel electrophoresis. Chromosomal transformation is reduced in *cipA9* mutant strains to 0.0001% of the wild-type level. Transfer of plasmid pVA736 is reduced to 0.1% of wild type. Other plasmids, which transform the wild-type strain at lower frequencies than pVA736, were less affected by the mutation. Except for its somewhat lesser importance for plasmid transfer, the *cipA* gene appears to correspond to the *S. pneumoniae recQ* gene in its importance for both chromosomal transformation and plasmid transfer. It, too, is not necessary for conjugative plasmid transfer (Raina and Macrina, 1982).

V. SINGLE-STRAND RECOMBINATION IN GRAM-NEGATIVE BACTERIA

Although it is generally acknowledged that the natural transformation of gram-positive bacteria proceeds through single-strand intermediates, the situation in the natural transformation of gram-negative bacteria is less clear. The transformation systems in question are those of *Haemophilus* and cyanobacterial species.

Unlike competent gram-positive bacteria, in which bound double-stranded DNA is accessible to external agents, *Haemophilus influenzae* and *Haemophilus parainfluenzae* sequester such bound DNA in vesicles (Kahn et al., 1982, 1983). However, the donor DNA seems to enter the cell proper in the form of single strands (Barany et al., 1983), and single-strand segments are integrated into the recipient chromosome (Notani and Goodgal, 1966). Yet donor single-stranded DNA does not accumulate in the cell (Stuy, 1965). This paradox has been resolved by postulating that to enter the cell the donor strands must interact with homologous portions of the chromosome (Sedgwick and Setlow, 1976; Barany et al., 1983). Donor strands could interact with the recipient chromosome by either linear or circular synapsis, so it is not surprising to find evidence of both types of interaction. In addition to chromosomal transformation, which is not sensitive to restriction enzyme effects (Gromkova and Goodgal, 1974), *H. influenzae* shows chromosomal facilitation of plasmid transfer (Balganesh and Setlow, 1985; Stuy and Walter, 1986) and introduction of recipient markers during such transfer (Setlow et al., 1981).

Transformation of the cyanobacterial species *Anacystis nidulans* (also known as *Synechococcus*) and *Agmenellum quadruplicatum* occurs under physiological conditions (Shestakov and Khyen, 1970; Stevens and Porter, 1980; Porter, this volume). The molecular mechanisms of DNA uptake and recombination have not been elucidated in these cases, but the behavior of various donor molecules containing homologous and heterologous DNA (Williams and Szalay, 1983) is consistent with single-strand entry and linear or circular synapsis, as indicated in sections III.A. and III.C. It should be pointed out, though, that the outcome of linear and circular synapsis in single-stranded DNA interactions can be obtained as well by double-stranded DNA interactions. An example of this is the artificial transformation of yeast protoplasts, in which introduced double-stranded DNA can undergo chromosomal facilitation of plasmid transfer and ectopic integration as well as linear chromosomal integration (Orr-Weaver and Szostak, 1983).

VI. SUMMARY

Genetic recombination in gram-positive bacteria was reviewed with special emphasis on those processes that introduce single-stranded DNA into the cell—transformation and conjugation. Depending on the mode of transfer, introduced DNA is fragmented to various degrees. Several different mechanisms were described for establishment of plasmids after their introduction in altered form. Single-stranded DNA that is homologous to the chromosome can interact with it by either linear or circular synapsis. Examples of the former are "classical" chromosomal transformation and plasmid marker-rescue transformation. Examples of the latter are chromosomal facilitation of plasmid establishment and ectopic integration of heterologous DNA. During chromosomal transformation, donor DNA is subject to mismatch correction. The broad outlines of these recombination mechanisms have been determined mainly by tracing the fate of genetically and isotopically labeled chromosomal and plasmid DNA. It is hoped that the investigations of genes and their products that affect various recombination functions, which are currently under way, will clarify the precise biochemical steps underlying the various mechanisms.

ACKNOWLEDGMENTS. I am grateful to my associates past and present for their contribution of ideas and experimental work to the substance of this chapter. I thank J. C. Alonso, M. Espinosa, W. Guild, D. Morrison, and M. C. Trombe for communicating results to me prior to publication.
This chapter was written at Brookhaven National Laboratory under the

auspices of the U.S. Department of Energy Office of Health and Environmental Research with support from Public Health Service grants AI14885 and GMAI29721 from the National Institutes of Health.

LITERATURE CITED

Albertini, A. M., M. Hofer, M. P. Calos, and J. H. Miller. 1982. On the formation of spontaneous deletions: the importance of short sequence homologies in the generation of large deletions. *Cell* **29:** 319–328.

Amundsen, S. K., A. F. Taylor, A. M. Chaudhury, and G. R. Smith. 1986. *recD*: the gene for an essential third subunit of exonuclease V. *Proc. Natl. Acad. Sci. USA* **83:**5558–5562.

Arwert, F., and G. Venema. 1973. Transformation in *Bacillus subtilis*. Fate of newly introduced transforming DNA. *Mol. Gen. Genet.* **123:**185–198.

Avery, O. T., C. M. MacLeod, and M. McCarty. 1944. Studies on the chemical nature of the substance inducing transformation of pneumococcal types. Induction of transformation by a desoxyribonucleic acid fraction isolated from pneumococcus type III. *J. Exp. Med.* **89:**137–158.

Bagci, H., and J. H. Stuy. 1979. A *hex* mutant of *Haemophilus influenzae. Mol. Gen. Genet.* **175:**175–179.

Balganesh, M., and J. K. Setlow. 1985. Differential behavior of plasmids containing chromosomal DNA insertions of various sizes during transformation and conjugation in *Haemophilus influenzae. J. Bacteriol.* **161:**141–146.

Balganesh, T. S., and S. A. Lacks. 1984. Plasmid vector for cloning in *Streptococcus pneumoniae* and strategies for enrichment for recombinant plasmids. *Gene* **29:**221–230.

Balganesh, T. S., and S. A. Lacks. 1985. Heteroduplex DNA mismatch repair system of *Streptococcus pneumoniae*: cloning and expression of the *hexA* gene. *J. Bacteriol.* **162:**979–984.

Barany, F., and J. D. Boeke. 1983. Genetic transformation of *Streptococcus pneumoniae* by DNA cloned into the single-stranded bacteriophage f1. *J. Bacteriol.* **153:**200–210.

Barany, F., J. D. Boeke, and A. Tomasz. 1982. Staphylococcal plasmids that replicate and express erythromycin resistance in both *Streptococcus pneumoniae* and *Escherichia coli. Proc. Natl. Acad. Sci. USA* **79:**2991–2995.

Barany, F., M. E. Kahn, and H. O. Smith. 1983. Directional transport and integration of donor DNA in *Haemophilus influenzae* transformation. *Proc. Natl. Acad. Sci. USA* **80:**7274–7278.

Barany, F., and A. Tomasz. 1980. Genetic transformation of *Streptococcus pneumoniae* by heterologous plasmid deoxyribonucleic acid. *J. Bacteriol.* **144:**698–709.

Behnke, D., and J. J. Ferretti. 1980. Molecular cloning of an erythromycin resistance determinant in streptococci. *J. Bacteriol.* **144:**806–813.

Bensi, G., A. Iglesias, U. Canosi, and T. A. Trautner. 1981. Plasmid transformation in *Bacillus subtilis*. The significance of partial homology between plasmid and recipient cell DNA. *Mol. Gen. Genet.* **184:**400–404.

Bernheimer, H. P. 1979. Lysogenic pneumococci and their bacteriophages. *J. Bacteriol.* **138:**618–624.

Bracco, R. M., M. R. Krauss, A. S. Roe, and C. M. MacLeod. 1957. Transformation reactions between pneumococcus and three strains of streptococci. *J. Exp. Med.* **106:**247–259.

Buu-Hoi, A., and T. Horodniceanu. 1980. Conjugative transfer of multiple antibiotic resistance markers in *Streptococcus pneumoniae. J. Bacteriol.* **143:**313–320.

Campbell, A. 1962. Episomes. *Adv. Genet.* **11:**101–145.

Canosi, U., A. Iglesias, and T. A. Trautner. 1981. Plasmid transformation in *Bacillus subtilis*: effects of insertion of *Bacillus subtilis* DNA in plasmid pC194. *Mol. Gen. Genet.* **181:**434–440.

Canosi, U., G. Morelli, and T. A. Trautner. 1978. The relationship between molecular structure and transforming efficiency of some *S. aureus* plasmids isolated from *B. subtilis. Mol. Gen. Genet.* **166:** 259–267.

Cato, A., and W. R. Guild. 1968. Transformation and DNA size. I. Activity of fragments of defined size and a fit to a random double cross-over model. *J. Mol. Biol.* **37:**157–180.

Chandler, M., and D. A. Morrison. 1987. Competence for genetic transformation in *Streptococcus pneumoniae*: molecular cloning of *com*, a competence control locus. *J. Bacteriol.* **169**:2005–2011.

Chang, S., and S. N. Cohen. 1979. High frequency transformation of *Bacillus subtilis* protoplasts by plasmid DNA. *Mol. Gen. Genet.* **168**:111–115.

Chestukhin, A. V., M. F. Shemyakin, N. A. Kalinina, and A. A. Prozorov. 1972. Some properties of ATP dependent deoxyribonucleases from normal and rec-mutant strains of *Bacillus subtilis*. *FEBS Lett.* **24**:121–125.

Chilton, M. D. 1967. Transforming activity in both complementary strands of *Bacillus subtilis* DNA. *Science* **157**:817–819.

Claverys, J. P., and S. A. Lacks. 1986. Heteroduplex DNA base mismatch repair in bacteria. *Microbiol. Rev.* **50**:133–165.

Claverys, J. P., J. C. Lefevre, and A. M. Sicard. 1980. Transformation of *Streptococcus pneumoniae* with *S. pneumoniae*-lambda phage hybrid DNA: induction of deletions. *Proc. Natl. Acad. Sci. USA* **77**:3534–3538.

Claverys, J. P., J. M. Louarn, and A. M. Sicard. 1981a. Cloning of *Streptococcus pneumoniae* DNA: its use in pneumococcal transformation and in studies of mismatch repair. *Gene* **13**:65–73.

Claverys, J. P., V. Mejean, A. M. Gasc, F. Galibert, and A. M. Sicard. 1981b. Base specificity of mismatch repair in *Streptococcus pneumoniae*. *Nucleic Acids Res.* **9**:2267–2280.

Claverys, J. P., V. Mejean, A. M. Gasc, and A. M. Sicard. 1983. Mismatch rpeair in *Streptococcus pneumoniae*: relationship between base mismatches and transformation efficiencies. *Proc. Natl. Acad. Sci. USA* **80**:5956–5960.

Claverys, J. P., H. Prats, H. Vasseghi, and M. Gherardi. 1984. Identification of *Streptococcus pneumoniae* mismatch repair genes by an additive transformation approach. *Mol. Gen. Genet.* **196**:91–96.

Clewell, D. B. 1981. Plasmids, drug resistance, and gene transfer in the genus *Streptococcus*. *Microbiol. Rev.* **45**:409–436.

Clewell, D. B., E. E. Ehrenfeld, F. An, R. E. Kessler, R. Wirth, M. Mori, C. Kitada, M. Fujino, Y. Ike, and A. Suzuki. 1987. Sex pheromones and plasmid-related conjugation phenomena in *Streptococcus faecalis*, p. 2–7. *In* J. J. Ferretti and R. Curtiss III (ed.), *Streptococcal Genetics*. American Society for Microbiology, Washington, D.C.

Clowes, R. C. 1961. Colicin factors as fertility factors in bacteria: *Escherichia coli*. *Nature* (London) **190**:988–989.

Cohen, A., W. D. Fisher, R. Curtiss III, and H. I. Adler. 1968. DNA isolated from *Escherichia coli* minicells mated with F+ cells. *Proc. Natl. Acad. Sci. USA* **61**:61–68.

Contente, S., and D. Dubnau. 1979a. Characterization of plasmid transformation in *Bacillus subtilis*: kinetic properties and the effect of DNA concentration. *Mol. Gen. Genet.* **167**:251–258.

Contente, S., and D. Dubnau. 1979b. Marker rescue transformation by linear plasmid DNA in *Bacillus subtilis*. *Plasmid* **2**:555–571.

Courvalin, P., C. Carlier, and F. Caillaud. 1987. Functional anatomy of the conjugative shuttle transposon Tn*1545*, p. 61–64. *In* J. J. Ferretti and R. Curtiss III (ed.), *Streptococcal Genetics*. American Society for Microbiology, Washington, D.C.

Davidoff-Abelson, R., and D. Dubnau. 1973. Kinetic analysis of the products of donor deoxyribonucleate in transformed cells of *Bacillus subtilis*. *J. Bacteriol.* **116**:154–162.

de la Campa, A. G., P. Kale, S. S. Springhorn, and S. A. Lacks. 1987. Proteins encoded by the *Dpn*II restriction gene cassette: two methylases and an endonuclease. *J. Mol. Biol.* **196**:457–469.

del Solar, G. H., A. Puyet, and M. Espinosa. 1987. Initiation signals for the conversion of single stranded to double stranded DNA forms in the streptococcal plasmid pLS1. *Nucleic Acids Res.* **15**:5561–5580.

Devanathan, T., and V. N. Iyer. 1978. Genetic fate of DNA in a strain of *Bacillus subtilis* which is impaired in genetic transformation. *Can. J. Microbiol.* **24**:282–288.

de Vos, W. M., S. C. de Vries, and G. Venema. 1983. Cloning and expression of the *Escherichia coli recA* gene in *Bacillus subtilis*. *Gene* **25**:301–308.

de Vos, W. M., and G. Venema. 1981. Fate of plasmid DNA in transformation of *Bacillus subtilis* protoplasts. *Mol. Gen. Genet.* **182**:39–43.

de Vos, W. M., and G. Venema. 1982. Transformation of *Bacillus subtilis* competent cells: identification of a protein involved in recombination. *Mol. Gen. Genet.* **187:**439–445.

de Vos, W. M., and G. Venema. 1983. Transformation of *Bacillus subtilis* competent cells: identification and regulation of the *recE* gene product. *Mol. Gen. Genet.* **190:**56–64.

de Vos, W. M., G. Venema, U. Canosi, and T. A. Trautner. 1981. Plasmid transformation in *Bacillus subtilis*: fate of plasmid DNA. *Mol. Gen. Genet.* **181:**424–433.

Dubnau, D. 1982. Genetic transformation in *Bacillus subtilis*, p. 147–178. *In* D. A. Dubnau (ed.), *The Molecular Biology of the Bacilli*, vol. 1. Academic Press, Inc., New York.

Dubnau, D., and C. Cirigliano. 1972. Fate of transforming DNA following uptake by competent *Bacillus subtilis*. III. Formation and properties of products isolated from transformed cells which are derived entirely from donor DNA. *J. Mol. Biol.* **64:**9–29.

Dubnau, D., and C. Cirigliano. 1973. Fate of transforming DNA following uptake by competent *Bacillus subtilis*. VI. Non-covalent association of donor and recipient DNA. *Mol. Gen. Genet.* **120:**101–106.

Dubnau, D., and C. Cirigliano. 1974. Genetic characterization of recombination-deficient mutants of *Bacillus subtilis*. *J. Bacteriol.* **117:**488–493.

Dubnau, D., R. Davidoff-Abelson, B. Scher, and C. Cirigliano. 1973. Fate of transforming deoxyribonucleic acid after uptake by competent *Bacillus subtilis*: phenotypic characterization of radiation-sensitive recombination-deficient mutants. *J. Bacteriol.* **114:**273–286.

Dubnau, D., I. Smith, P. Morell, and J. Marmur. 1965. Gene conservation in *Bacillus* species. I. Conserved genetic and nucleic acid base sequence homologies. *Proc. Natl. Acad. Sci. USA* **54:**491–498.

Duncan, C. H., G. A. Wilson, and F. E. Young. 1978. Mechanism of integrating foreign DNA during transformation of *Bacillus subtilis*. *Proc. Natl. Acad. Sci. USA* **75:**3664–3668.

Dunny, G. M., B. L. Brown, and D. B. Clewell. 1978. Induced cell aggregation and mating in *Streptococcus faecalis*: evidence for a bacterial sex pheromone. *Proc. Natl. Acad. Sci. USA* **75:**3475–3483.

Ehrlich, S. D. 1977. Replication and expression of plasmids from *Staphylococcus aureus* in *Bacillus subtilis*. *Proc. Natl. Acad. Sci. USA* **74:**1680–1682.

Eisenstadt, E., R. Lange, and K. Willecke. 1975. Competent *Bacillus subtilis* cultures synthesize a denatured DNA binding activity. *Proc. Natl. Acad. Sci. USA* **72:**323–327.

Ephrussi-Taylor, H., and T. C. Gray. 1966. Genetic studies of recombining DNA in pneumococccal transformation. *J. Gen. Physiol.* **49**(Part 2):211–231.

Ephrussi-Taylor, H., A. M. Sicard, and R. Kamen. 1965. Genetic recombination in DNA-induced transformation in pneumococcus. I. The problem of relative efficiency of transforming factors. *Genetics* **51:**455–475.

Falkow, S. 1975. *Infectious Multiple Drug Resistance*. Pion Ltd., London.

Fani, R., G. Mastromei, M. Polsinelli, and G. Venema. 1984. Isolation and characterization of *Bacillus subtilis* mutants altered in competence. *J. Bacteriol.* **157:**152–157.

Fitzgerald, G. F., and D. B. Clewell. 1985. A conjugative transposon (Tn*919*) in *Streptococcus sanguis*. *Infect. Immun.* **47:**415–420.

Fornili, S. L., and M. S. Fox. 1977. Electron microscope visualization of the products of *Bacillus subtilis* transformation. *J. Mol. Biol.* **113:**181–191.

Fox, M. S. 1957. Deoxyribonucleic acid incorporation by transformed bacteria. *Biochim. Biophys. Acta* **26:**83–85.

Fox, M. S. 1960. Fate of transforming deoxyribonucleate following fixation by transforming bacteria. II. *Nature* (London) **187:**1004–1006.

Fox, M. S., and M. K. Allen. 1964. On the mechanism of deoxyribonucleate integration in pneumococcal transformation. *Proc. Natl. Acad. Sci. USA* **52:**412–419.

Franke, A. E., and D. B. Clewell. 1981. Evidence for a chromosome-borne resistance transposon (Tn*916*) in *Streptococcus faecalis* that is capable of "conjugal" transfer in the absence of a conjugative plasmid. *J. Bacteriol.* **145:**494–502.

Fujii, T., D. Naka, N. Toyoda, and H. Seto. 1987. LiCl treatment releases a nickase implicated in genetic transformation of *Streptococcus pneumoniae*. *J. Bacteriol.* **169:**4901–4906.

Galizzi, A., F. Scoffone, G. Milanesi, and A. M. Albertini. 1981. Integration and excision of a plasmid in *Bacillus subtilis. Mol. Gen. Genet.* **182:**99–105.

Gasc, A. M., P. Garcia, D. Batz, and A. M. Sicard. 1987. Mismatch repair during pneumococcal transformation of small deletions produced by site-directed mutagenesis. *Mol. Gen. Genet.* **210:**369–372.

Ghei, O. K., and S. A. Lacks. 1967. Recovery of donor deoxyribonucleic acid marker activity from eclipse in pneumococcal transformation. *J. Bacteriol.* **93:**816–829.

Green, D. M., and S. A. Coughlin. 1982. Coordinated changes in competence associated polypeptides among transformation defective mutants of *Bacillus subtilis*, p. 227–236. *In* A. T. Ganesan, S. Chang, and J. A. Hoch (ed.), *Molecular Cloning and Gene Regulation in Bacilli*. Academic Press, Inc., New York.

Griffith, R. 1928. The significance of pneumococcal types. *J. Hyg.* **27:**113–159.

Gromkova, R., and S. H. Goodgal. 1974. On the role of restriction enzymes of *Haemophilus* in transformation and transfection, p. 209–215. *In* R. Grell (ed.), *Mechanisms in Recombination*. Plenum Publishing Corp., New York.

Gryczan, T. J., S. Contente, and D. Dubnau. 1978. Characterization of *Staphylococcus aureus* plasmids introduced by transformation into *Bacillus subtilis. J. Bacteriol.* **134:**318–329.

Gryczan, T. J., S. Contente, and D. Dubnau. 1980. Molecular cloning of heterologous chromosomal DNA by recombination between a plasmid vector and a homologous resident plasmid in *Bacillus subtilis. Mol. Gen. Genet.* **177:**459–467.

Guild, W. R., M. D. Smith, and N. B. Shoemaker. 1982. Conjugative transfer of chromosomal R determinants in *Streptococcus pneumoniae*, p. 88–92. *In* D. Schlessinger (ed.), *Microbiology—1982*. American Society for Microbiology, Washington, D.C.

Haber, L. T., P. P. Pang, D. I. Sobell, J. A. Mankovich, and G. C. Walker. 1988. Nucleotide sequence of the *Salmonella typhimurium mutS* gene required for mismatch repair: homology of MutS and HexA of *Streptococcus pneumoniae. J. Bacteriol.* **170:**197–202.

Hahn, J., M. Albano, and D. Dubnau. 1987. Isolation and characterization of Tn*917lac*-generated competence mutants of *Bacillus subtilis. J. Bacteriol.* **169:**3104–3109.

Haldenwang, W. G., C. D. B. Banner, J. F. Ollington, R. Losick, J. A. Hoch, M. B. O'Connor, and A. L. Sonenshein. 1980. Mapping a cloned gene under sporulation control by insertion of a drug resistance marker into the *Bacillus subtilis* chromosome. *J. Bacteriol.* **142:**90–98.

Harford, N. 1974. Genetic analysis of *rec* mutants of *Bacillus subtilis*. Evidence for at least six linkage groups. *Mol. Gen. Genet.* **129:**269–274.

Henner, D. J., and J. A. Hoch. 1982. The genetic map of *Bacillus subtilis*, p. 1–33. *In* D. A. Dubnau (ed.), *The Molecular Biology of the Bacilli*, vol. 1. Academic Press, Inc., New York.

Hoch, J. A., and C. Anagnostopoulos. 1970. Chromosomal location and properties of radiation sensitivity mutations in *Bacillus subtilis. J. Bacteriol.* **103:**295–301.

Hofemeister, J., M. Israeli-Reches, and D. Dubnau. 1983. Integration of plasmid pE194 at multiple sites on the *Bacillus subtilis* chromosome. *Mol. Gen. Genet.* **189:**58–68.

Horodniceanu, T., L. Bougueleret, and G. Bieth. 1981. Conjugative transfer of multiple-antibiotic resistance markers in beta-hemolytic group A, B, F, and G streptococci in the absence of extrachromosomal deoxyribonucleic acid. *Plasmid* **5:**127–137.

Hotchkiss, R. D. 1954. Cyclical behavior in pneumococcal growth and transformability occasioned by environmental changes. *Proc. Natl. Acad. Sci. USA* **40:**49–55.

Hotchkiss, R. D. 1974. Molecular basis for genetic recombination. *Genetics* **78:**247–257.

Hotchkiss, R. D., and M. H. Gabor. 1985. Protoplast fusion in *Bacillus* and its consequences, p. 109–149. *In* D. A. Dubnau (ed.), *The Molecular Biology of the Bacilli*, vol. 2. Academic Press, Inc., New York.

Iglesias, A., G. Bensi, U. Canosi, and T. A. Trautner. 1981. Plasmid transformation in *Bacillus subtilis*. Alterations introduced into the recipient-homologous DNA of hybrid plasmids can be corrected in transformation. *Mol. Gen. Genet.* **184:**405–409.

Inamine, J. M., and V. Burdett. 1985. Structural organization of a 67-kilobase streptococcal conjugative element mediating antibiotic resistance. *J. Bacteriol.* **161:**620–626.

Ippen-Ihler, K. A., and E. G. Minkley, Jr. 1986. The conjugation system of F, the fertility factor of *Escherichia coli. Annu. Rev. Genet.* **20:**593–624.

Jacobs, M. R., H. J. Koornhof, R. M. Robins-Browne, C. M. Stevenson, Z. A. Vermaak, I. Freiman, M. A. Miller, M. A. Witcomb, M. Isaacson, J. I. Ward, and R. Austrian. 1978. Emergence of multiple resistant pneumococci. *N. Engl. J. Med.* **299**:735–740.

Jones, J. M., C. Gawron-Burke, S. E. Flannagan, M. Yamamoto, E. Senghas, and D. B. Clewell. 1987. Structural and genetic studies of the conjugative transposon Tn*916*, p. 54–60. *In* J. J. Ferretti and R. Curtiss III (ed.), *Streptococcal Genetics*. American Society for Microbiology, Washington, D.C.

Kahn, M. E., F. Barany, and H. O. Smith. 1983. Transformasomes: specialized membranous structures that protect DNA during *Haemophilus* transformation. *Proc. Natl. Acad. Sci. USA* **89**:6927–6931.

Kahn, M. E., G. Maul, and S. H. Goodgal. 1982. Possible mechanism for donor DNA binding and transport in *Haemophilus*. *Proc. Natl. Acad. Sci. USA* **79**:6370–6374.

Keggins, K. M., P. S. Lovett, and E. J. Duvall. 1978. Molecular cloning of genetically active fragments of *Bacillus* DNA in *Bacillus subtilis* and properties of the vector pUB110. *Proc. Natl. Acad. Sci. USA* **75**:1423–1427.

Koepsel, R. R., R. W. Murray, W. D. Rosenblum, and S. A. Khan. 1985. The replication initiator protein of plasmid pT181 has sequence-specific endonuclease and topoisomerase-like activities. *Proc. Natl. Acad. Sci. USA* **82**:6845–6849.

Kok, J., J. M. B. M. van der Vossen, and G. Venema. 1984. Construction of plasmid cloning vectors for lactic streptococci which also replicate in *Bacillus subtilis* and *Escherichia coli*. *Appl. Environ. Microbiol.* **48**:726–731.

Kondo, J. K., and L. L. McKay. 1985. Gene transfer systems and molecular cloning in group N streptococci: a review. *J. Dairy Sci.* **68**:2143–2159.

Lacks, S. 1962. Molecular fate of DNA in genetic transformation of pneumococcus. *J. Mol. Biol.* **5**:119–131.

Lacks, S. 1966. Integration efficiency and genetic recombination in pneumococcal transformation. *Genetics* **53**:207–235.

Lacks, S. 1970. Mutants of *Diplococcus pneumoniae* that lack deoxyribonucleases and other activities possibly pertinent to genetic transformation. *J. Bacteriol.* **101**:373–383.

Lacks, S. A. 1977. Binding and entry of DNA in bacterial transformation, p. 179–232. *In* J. Reissig (ed.), *Microbial Interactions*. Chapman & Hall, Ltd., London.

Lacks, S. 1979a. Steps in the process of DNA binding and entry in transformation, p. 27–41. *In* S. W. Glover and L. O. Butler (ed.), *Transformation 1978*. Cotswold Press, Oxford.

Lacks, S. 1979b. Uptake of circular deoxyribonucleic acid and mechanism of deoxyribonucleic acid transport in genetic transformation of *Streptococcus pneumoniae*. *J. Bacteriol.* **138**:404–409.

Lacks, S. A. 1984. Modes of DNA interaction in bacterial transformation, p. 149–158. *In* V. L. Chopra, B. C. Joshi, R. P. Sharma, and H. C. Bansal (ed.), *Genetics: New Frontiers*, vol. 1. Oxford and IBH, New Delhi, India.

Lacks, S. A., J. J. Dunn, and B. Greenberg. 1982. Identification of base mismatches recognized by the heteroduplex-DNA-repair system of *Streptococcus pneumoniae*. *Cell* **31**:327–336.

Lacks, S., and B. Greenberg. 1967. Deoxyribonucleases of pneumococcus. *J. Biol. Chem.* **242**:3108–3120.

Lacks, S., and B. Greenberg. 1973. Competence for deoxyribonucleic acid uptake and deoxyribonuclease action external to cells in the genetic transformation of *Diplococcus pneumoniae*. *J. Bacteriol.* **114**:152–163.

Lacks, S., and B. Greenberg. 1975. A deoxyribonuclease of *Diplococcus pneumoniae* specific for methylated DNA. *J. Biol. Chem.* **250**:4060–4066.

Lacks, S., and B. Greenberg. 1976. Single-strand breakage on binding of DNA to cells in the genetic transformation of *Diplococcus pneumoniae*. *J. Mol. Biol.* **101**:255–275.

Lacks, S., and B. Greenberg. 1977. Complementary specificity of restriction endonucleases of *Diplococcus pneumoniae* with respect to DNA methylation. *J. Mol. Biol.* **114**:153–168.

Lacks, S., B. Greenberg, and K. Carlson. 1967. Fate of donor DNA in pneumococcal transformation. *J. Mol. Biol.* **29**:327–347.

Lacks, S., B. Greenberg, and M. Neuberger. 1974. Role of a deoxyribonuclease in the genetic transformation of *Diplococcus pneumoniae*. *Proc. Natl. Acad. Sci. USA* **71**:2305–2309.

Lacks, S., B. Greenberg, and M. Neuberger. 1975. Identification of a deoxyribonuclease implicated in genetic transformation of *Diplococcus pneumoniae*. *J. Bacteriol.* **123**:222–232.

Lacks, S., and R. D. Hotchkiss. 1960. A study of the genetic material determining an enzyme activity in pneumococcus. *Biochim. Biophys. Acta* **39**:508–518.

Lacks, S. A., P. Lopez, B. Greenberg, and M. Espinosa. 1986a. Identification and analysis of genes for tetracycline resistance and replication functions in the broad-host-range plasmid pLS1. *J. Mol. Biol.* **192**:753–765.

Lacks, S. A., B. M. Mannarelli, S. S. Springhorn, and B. Greenberg. 1986b. Genetic basis of the complementary DpnI and DpnII restriction systems of *S. pneumoniae*: an intercellular cassette mechanism. *Cell* **46**:993–1000.

Lacks, S. A., B. M. Mannarelli, S. S. Springhorn, B. Greenberg, and A. G. de la Campa. 1987. Genetics of the complementary restriction systems *Dpn*I and *Dpn*II revealed by cloning and recombination in *Streptococcus pneumoniae*, p. 31–41. *In* J. J. Ferretti and R. Curtiss III (ed.), *Streptococcal Genetics*. American Society for Microbiology, Washington, D.C.

Lacks, S., and M. Neuberger. 1975. Membrane location of a deoxyribonuclease implicated in the genetic transformation of *Diplococcus pneumoniae*. *J. Bacteriol.* **124**:1321–1329.

Lacks, S. A., and S. S. Springhorn. 1984. Transfer of recombinant plasmids containing the gene for *Dpn*II DNA methylase into strains of *Streptococcus pneumoniae* that produce *Dpn*I or *Dpn*II restriction endonucleases. *J. Bacteriol.* **158**:905–909.

LeBlanc, D. J., and F. P. Hassel. 1976. Transformation of *Streptococcus sanguis* Challis by plasmid deoxyribonucleic acid from *Streptococcus faecalis*. *J. Bacteriol.* **128**:347–355.

Lederberg, J. 1952. Cell genetics and hereditary symbiosis. *Physiol. Rev.* **32**:403–430.

Leonard, C. G., and R. M. Cole. 1972. Purification and properties of streptococcal competence factor isolated from chemically defined medium. *J. Bacteriol.* **110**:273–280.

Lindberg, M., J.-E. Sjostrom, and T. Johansson. 1972. Transformation of chromosomal and plasmid characters in *Staphylococcus aureus*. *J. Bacteriol.* **109**:844–847.

Little, J. W., and D. W. Mount. 1982. The SOS regulatory system of *Escherichia coli*. *Cell* **29**:11–22.

Lopez, P., M. Espinosa, B. Greenberg, and S. A. Lacks. 1984a. Generation of deletions in pneumococcal *mal* genes cloned in *Bacillus subtilis*. *Proc. Natl. Acad. Sci. USA* **81**:5189–5193.

Lopez, P., M. Espinosa, B. Greenberg, and S. A. Lacks. 1987. Sulfonamide resistance in *Streptococcus pneumoniae*: DNA sequence of the gene encoding dihydropteroate synthase and characterization of the enzyme. *J. Bacteriol.* **169**:4320–4326.

Lopez, P., M. Espinosa, and S. A. Lacks. 1984b. Physical structure and genetic expression of the sulfonamide-resistance plasmid pLS80 and its derivatives in *Streptococcus pneumoniae*. *Mol. Gen. Genet.* **195**:402–410.

Lopez, P., M. Espinosa, D. L. Stassi, and S. A. Lacks. 1982. Facilitation of plasmid transfer in *Streptococcus pneumoniae* by chromosomal homology. *J. Bacteriol.* **150**:692–701.

Lovett, C. M., Jr., and J. W. Roberts. 1985. Purification of a RecA protein analogue from *Bacillus subtilis*. *J. Biol. Chem.* **260**:3305–3313.

Macrina, F. L., K. R. Jones, and R. A. Welch. 1981. Transformation of *Streptococcus sanguis* with monomeric pVA736 plasmid deoxyribonucleic acid. *J. Bacteriol.* **146**:826–830.

Macrina, F. L., K. R. Jones, and P. H. Wood. 1980. Chimeric streptococcal plasmids and their use as molecular cloning vehicles in *Streptococcus sanguis* (Challis). *J. Bacteriol.* **143**:1425–1435.

Mandel, M., and A. Higa. 1970. Calcium-dependent bacteriophage DNA infection. *J. Mol. Biol.* **53**:159–162.

Mannarelli, B. M., and S. A. Lacks. 1984. Ectopic integration of chromosomal genes in *Streptococcus pneumoniae*. *J. Bacteriol.* **160**:867–873.

Martin, B., H. Prats, and J. P. Claverys. 1985. Cloning of the *hexA* mismatch repair gene of *Streptococcus pneumoniae* and identification of the product. *Gene* **34**:293–303.

Mazza, G., A. Fortunato, E. Ferrari, U. Canosi, A. Falaschi, and M. Polsinelli. 1975. Genetic and enzymic studies on the recombination process in *Bacillus subtilis*. *Mol. Gen. Genet.* **136**:9–30.

Mejean, V., and J. P. Claverys. 1984. Effect of mismatched bases on the fate of donor DNA in transformation of *Streptococcus pneumoniae*. *Mol. Gen. Genet.* **197**:467–471.

Mejean, V., J. P. Claverys, H. Vasseghi, and A. M. Sicard. 1981. Rapid cloning of specific DNA fragments of *Streptococcus pneumoniae* by vector integration into the chromosome followed by endonucleolytic excision. *Gene* **15**:289–293.

Miao, R., and W. R. Guild. 1970. Competent *Diplococcus pneumoniae* accept both single- and double-stranded deoxyribonucleic acid. *J. Bacteriol.* **101:**361–364.

Michel, B., and S. D. Ehrlich. 1986. Illegitimate recombination occurs between the replication origin of the plasmid pC194 and a progressing replication fork. *EMBO J.* **5:**3691–3696.

Michel, B., B. Niaudet, and S. D. Ehrlich. 1982. Intramolecular recombination during plasmid transformation of *Bacillus subtilis* competent cells. *EMBO J.* **1:**1565–1571.

Morrison, D. A. 1977. Transformation in pneumococcus: existence and properties of a complex involving donor deoxyribonucleate single strands in eclipse. *J. Bacteriol.* **132:**576–583.

Morrison, D. A., and M. F. Baker. 1979. Association of competence for genetic transformation in pneumococcus with synthesis of a small set of proteins. *Nature* (London) **282:**215–217.

Morrison, D. A., and W. R. Guild. 1972. Transformation and deoxyribonucleic acid size: extent of degradation on entry varies with size of donor. *J. Bacteriol.* **112:**1157–1168.

Morrison, D. A., and W. R. Guild. 1973. Breakage prior to entry of donor DNA in pneumococcus transformation. *Biochim. Biophys. Acta* **299:**545–556.

Morrison, D. A., S. A. Lacks, W. G. Guild, and J. M. Hageman. 1983. Isolation and characterization of three new classes of transformation-deficient mutants of *Streptococcus pneumoniae* that are defective in DNA transport and genetic recombination. *J. Bacteriol.* **156:**281–290.

Morrison, D. A., and B. Mannarelli. 1979. Transformation in pneumococcus: nuclease resistance of deoxyribonucleic acid in the eclipse complex. *J. Bacteriol.* **140:**655–665.

Muckerman, C. C., S. S. Springhorn, B. Greenberg, and S. A. Lacks. 1982. Transformation of restriction endonuclease phenotype in *Streptococcus pneumoniae*. *J. Bacteriol.* **152:**183–190.

Mulder, J. A., and G. Venema. 1982. Transformation-deficient mutants of *Bacillus subtilis* impaired in competence-specific nuclease activities. *J. Bacteriol.* **152:**166–174.

Nester, E. W., and B. A. D. Stocker. 1963. Biosynthetic latency in early stages of deoxyribonucleic acid transformation in *Bacillus subtilis*. *J. Bacteriol.* **86:**785–796.

Niaudet, B., A. Goze, and S. D. Ehrlich. 1982. Insertional mutagenesis in *Bacillus subtilis*: mechanism and use in gene cloning. *Gene* **19:**277–284.

Notani, N., and S. H. Goodgal. 1966. On the nature of recombinants formed during transformation in *Hemophilus influenzae*. *J. Gen. Physiol.* **49**(Part 2):197–209.

Novick, R. P., and M. H. Richmond. 1965. Nature and interactions of the genetic elements governing penicillinase synthesis in *Staphylococcus aureus*. *J. Bacteriol.* **90:**467–480.

Orr-Weaver, T. L., and J. W. Szostak. 1983. Yeast recombination: the association between double-strand gap repair and crossing-over. *Proc. Natl. Acad. Sci. USA* **80:**4417–4421.

Ottolenghi, E., and C. M. MacLeod. 1963. Genetic transformation among living pneumococci in the mouse. *Proc. Natl. Acad. Sci. USA* **50:**417–419.

Pakula, R., M. Piechowska, E. Bankowska, and W. Walczak. 1962. A characteristic of DNA mediated transformation systems of two streptococcal strains. *Acta Microbiol. Pol.* **11:**205–222.

Perego, M., E. Ferrari, M. T. Bassi, A. Galizzi, and P. Mazza. 1987. Molecular cloning of *Bacillus subtilis* genes involved in DNA metabolism. *Mol. Gen. Genet.* **209:**8–14.

Piechowska, M., and M. S. Fox. 1971. Fate of transforming deoxyribonucleate in *Bacillus subtilis*. *J. Bacteriol.* **108:**680–689.

Piechowska, M., and D. Shugar. 1967. Streptococcal group H transforming system with reproducible high transformation yields. *Acta Biochim. Pol.* **14:**349–360.

Prats, H., B. Martin, and J. P. Claverys. 1985. The *hexB* mismatch repair gene of *Streptococcus pneumoniae*: characterization, cloning and identification of the product. *Mol. Gen. Genet.* **200:**482–489.

Priebe, S. D., S. M. Hadi, B. Greenberg, and S. A. Lacks. 1988. Nucleotide sequence of the *hexA* gene for DNA mismatch repair in *Streptococcus pneumoniae* and homology of *hexA* to *mutS* of *Escherichia coli* and *Salmonella typhimurium*. *J. Bacteriol.* **170:**190–196.

Prozorov, A. A., V. I. Bashkirov, N. M. Lakomova, and N. N. Surikov. 1982. Study of the phenomenon of marker rescue in plasmid transformation and transduction of intact cells, protoplasts and different *rec* mutants of *Bacillus subtilis*. *Mol. Gen. Genet.* **185:**363–364.

Prozorov, A. A., T. S. Belova, and N. N. Surikov. 1980. Transformation and transduction of *Bacillus subtilis* strains with the *Bsu*R restriction-modification system by means of modified and unmodified DNA of pUB110 plasmid. *Mol. Gen. Genet.* **180:**135–138.

Raina, J. L., and F. L. Macrina. 1982. A competence specific inducible protein promotes in vivo recombination in *Streptococcus sanguis*. *Mol. Gen. Genet.* **185**:21–29.

Raina, J. L., and A. W. Ravin. 1978. Fate of homospecific transforming DNA bound to *Streptococcus sanguis*. *J. Bacteriol.* **133**:1212–1223.

Raina, J. L., and A. W. Ravin. 1979. Superhelical DNA in *Streptococcus sanguis*: role in recombination in vivo. *Mol. Gen. Genet.* **176**:171–181.

Raina, J. L., and A. W. Ravin. 1980. Switches in macromolecular synthesis during induction of competence for transformation of *Streptococcus sanguis*. *Proc. Natl. Acad. Sci. USA* **77**:6062–6066.

Rhee, D.-K., and D. A. Morrison. 1988. Genetic transformation in *Streptococcus pneumoniae*: molecular cloning and characterization of *recP*, a gene required for genetic recombination. *J. Bacteriol.* **170**:630–637.

Rosenthal, A. L., and S. A. Lacks. 1977. Nuclease detection in SDS-polyacrylamide gel electrophoresis. *Anal. Biochem.* **80**:76–90.

Rosenthal, A. L., and S. A. Lacks. 1980. Complex structure of the membrane nuclease of *Streptococcus pneumoniae* revealed by two-dimensional electrophoresis. *J. Mol. Biol.* **141**:133–146.

Sadaie, Y., and T. Kada. 1976. Recombination-deficient mutants of *Bacillus subtilis*. *J. Bacteriol.* **125**:489–500.

Saunders, C. W., and W. R. Guild. 1980. Monomer plasmid DNA transforms *Streptococcus pneumoniae*. *Mol. Gen. Genet.* **181**:57–62.

Saunders, C. W., and W. R. Guild. 1981. Pathway of plasmid transformation in pneumococcus: open circular and linear molecules are active. *J. Bacteriol.* **146**:517–526.

Sedgwick, B., and J. K. Setlow. 1976. Single-stranded regions in transforming deoxyribonucleic acid after uptake by competent *Haemophilus influenzae*. *J. Bacteriol.* **125**:588–594.

Setlow, J. K., N. K. Notani, D. McCarthy, and N.-L. Clayton. 1981. Transformation of *Haemophilus influenzae* by plasmid RSF0885 containing a cloned segment of chromosomal deoxyribonucleic acid. *J. Bacteriol.* **148**:804–811.

Seto, H., and A. Tomasz. 1974. Early stages in DNA binding and uptake during genetic transformation of pneumococci. *Proc. Natl. Acad. Sci. USA* **71**:1493–1498.

Seto, H., and A. Tomasz. 1976. Calcium-requiring step in the uptake of deoxyribonucleic acid molecules through the surface of competent pneumococci. *J. Bacteriol.* **126**:1113–1118.

Shestakov, S. V., and N. T. Khyen. 1970. Evidence for genetic transformation in blue-green alga *Anacystis nidulans*. *Mol. Gen. Genet.* **107**:372–375.

Shibata, T., C. DasGupta, R. P. Cunningham, and C. M. Radding. 1979. Purified *Escherichia coli recA* protein catalyzes homologous pairing of superhelical DNA and single-stranded fragments. *Proc. Natl. Acad. Sci. USA* **76**:1638–1642.

Shoemaker, N. B., and W. R. Guild. 1974. Destruction of low efficiency markers is a slow process occurring at a heteroduplex stage of transformation. *Mol. Gen. Genet.* **128**:283–290.

Shoemaker, N. B., M. D. Smith, and W. R. Guild. 1979. Organization and transfer of heterologous chloramphenicol and tetracycline resistance genes in pneumococcus. *J. Bacteriol.* **139**:432–441.

Shoemaker, N. B., M. D. Smith, and W. R. Guild. 1980. DNase-resistance transfer of chromosomal *cat* and *tet* insertions by filter mating in pneumococcus. *Plasmid* **3**:80–87.

Sicard, M., J.-C. Lefevre, P. Mostachfi, A.-M. Gasc, and C. Sarda. 1985. Localized conversion in *Streptococcus pneumoniae* recombination: heteroduplex preference. *Genetics* **110**:557–568.

Singh, R. N., and M. P. Pitale. 1968. Competence and deoxyribonucleic acid uptake in *Bacillus subtilis*. *J. Bacteriol.* **95**:864–866.

Sinha, R. P., and V. N. Iyer. 1972. Isolation and some distinctive properties of a new type of recombination-deficient mutant of *Bacillus subtilis*. *J. Mol. Biol.* **72**:711–724.

Smith, H., K. Wiersma, S. Bron, and G. Venema. 1984. Transformation in *Bacillus subtilis*: a 75,000-dalton protein complex is involved in binding and entry of donor DNA. *J. Bacteriol.* **157**:733–738.

Smith, H. O., D. B. Danner, and R. A. Deich. 1981. Genetic transformation. *Annu. Rev. Biochem.* **50**:41–68.

Smith, M. D., and W. R. Guild. 1979. A plasmid in *Streptococcus pneumoniae*. *J. Bacteriol.* **137**:735–739.

Smith, M. D., and W. R. Guild. 1982. Evidence for transposition of the conjugative R determinants of *Streptococcus agalactiae* B109, p. 109–111. *In* D. Schlessinger (ed.), *Microbiology—1982*. American Society for Microbiology, Washington, D.C.

Smith, M. D., N. B. Shoemaker, V. Burdett, and W. R. Guild. 1980. Transfer of plasmids by conjugation in *Streptococcus pneumoniae*. *Plasmid* **3:**70–79.

Stassi, D. L., and S. A. Lacks. 1982. Effect of strong promoters on the cloning in *Escherichia coli* of DNA fragments from *Streptococcus pneumoniae*. *Gene* **18:**319–328.

Stassi, D. L., P. Lopez, M. Espinosa, and S. A. Lacks. 1981. Cloning of chromosomal genes in *Streptococcus pneumoniae*. *Proc. Natl. Acad. Sci. USA* **78:**7028–7032.

Stevens, S. E., Jr., and R. D. Porter. 1980. Transformation in *Agmenellum quadruplicatum*. *Proc. Natl. Acad. Sci. USA* **77:**6052–6056.

Stuy, J. H. 1965. Fate of transforming DNA in the *Hemophilus influenzae* transformation system. *J. Mol. Biol.* **13:**554–570.

Stuy, J. H., and R. B. Walter. 1986. Homology-facilitated plasmid transfer in *Hemophilus influenzae*. *Mol. Gen. Genet.* **203:**288–295.

Su, S.-S., and P. Modrich. 1986. *Escherichia coli mutS*-encoded protein binds to mismatched DNA base pairs. *Proc. Natl. Acad. Sci. USA* **83:**5057–5061.

te Riele, H., B. Michel, and S. D. Ehrlich. 1986. Single-stranded plasmid DNA in *Bacillus subtilis* and *Staphylococcus aureus*. *Proc. Natl. Acad. Sci. USA* **83:**2541–2545.

Tiraby, G., and M. S. Fox. 1973. Marker discrimination in transformation and mutation of pneumococcus. *Proc. Natl. Acad. Sci. USA* **70:**3541–3545.

Tomasz, A., and R. D. Hotchkiss. 1964. Regulation of the transformability of pneumococcal cultures by macromolecular cell products. *Proc. Natl. Acad. Sci. USA* **51:**480–487.

Tomasz, A., and J. L. Mosser. 1966. On the nature of the pneumococcal activator substance. *Proc. Natl. Acad. Sci. USA* **55:**58–66.

Trautner, T., B. Pawlek, S. Bron, and C. Anagnostopoulos. 1974. Restriction and modification in *Bacillus subtilis*. Biological aspects. *Mol. Gen. Genet.* **131:**181–191.

Trautner, T. A., and H. C. Spatz. 1973. Transfection in *B. subtilis*. *Curr. Top. Microbiol. Immunol.* **62:**61–88.

Vasseghi, H., J. P. Claverys, and A. M. Sicard. 1981. Mechanism of integrating foreign DNA during transformation in *Streptococcus pneumoniae*, p. 137–154. *In* M. Polsinelli and G. Mazza (ed.), *Transformation 1980*. Cotswold Press, Oxford.

Venema, G. 1979. Bacterial transformation. *Adv. Microb. Physiol.* **19:**245–331.

Venema, G., R. H. Pritchard, and T. Venema-Schroder. 1965. Fate of transforming deoxyribonucleic acid in *Bacillus subtilis*. *J. Bacteriol.* **89:**1250–1255.

Vijayakumar, M. N., S. D. Priebe, and W. R. Guild. 1986a. Structure of a conjugative element in *Streptococcus pneumoniae*. *J. Bacteriol.* **166:**978–984.

Vijayakumar, M. N., S. D. Priebe, G. Pozzi, J. M. Hageman, and W. R. Guild. 1986b. Cloning and physical characterization of chromosomal conjugative elements in streptococci. *J. Bacteriol.* **166:**972–977.

Vosman, B., J. Kooistra, J. Olijve, and G. Venema. 1987. Cloning in *Escherichia coli* of the gene specifying the DNA-entry nuclease of *Bacillus subtilis*. *Gene* **52:**175–183.

Vovis, G. 1973. Adenosine triphosphate-dependent deoxyribonuclease from *Diplococcus pneumoniae*: fate of transforming deoxyribonucleic acid. *J. Bacteriol.* **113:**718–723.

Vovis, G. F., and S. Lacks. 1977. Complementary action of restriction enzymes Endo R.*Dpn*I and Endo R.*Dpn*II on bacteriophage f1 DNA. *J. Mol. Biol.* **115:**525–538.

Warren, G. J., A. J. Twigg, and D. J. Sherratt. 1978. ColE1 plasmid mobility and relaxation complex. *Nature* (London) **274:**259–261.

Watanabe, T., and T. Fukasawa. 1960. Resistance transfer factor, an episome in Enterobacteriaceae. *Biochem. Biophys. Res. Commun.* **3:**660–665.

Weinrauch, Y., and D. Dubnau. 1983. Plasmid marker rescue transformation in *Bacillus subtilis*. *J. Bacteriol.* **154:**1077–1087.

Weinrauch, Y., and D. Dubnau. 1987. Plasmid marker rescue transformation proceeds by breakage-reunion in *Bacillus subtilis*. *J. Bacteriol.* **169:**1205–1211.

Weinrauch, Y., and S. A. Lacks. 1981. Nonsense mutations in the amylomaltase gene and other loci of *Streptococcus pneumoniae*. *Mol. Gen. Genet.* **183:**7–12.

Weinstock, G. M., K. McEntee, and I. R. Lehman. 1979. ATP-dependent renaturation of DNA catalyzed by the *recA* protein of *Escherichia coli. Proc. Natl. Acad. Sci. USA* **76:**126–130.

Willetts, N., and B. Wilkins. 1984. Processing of plasmid DNA during bacterial conjugation. *Microbiol. Rev.* **48:**24–41.

Williams, J. G. K., and A. A. Szalay. 1983. Stable integration of foreign DNA into the chromosome of the cyanobacterium *Synechococcus* R2. *Gene* **24:**37–51.

Williamson, M. S., J. C. Game, and S. Fogel. 1985. Meiotic gene conversion in *Saccharomyces cerevisiae*. I. Isolation and characterization of pms1-1 and pms1-2. *Genetics* **110:**609–646.

Worcel, A., and E. Burgi. 1972. On the structure of folded chromosome of *Escherichia coli. J. Mol. Biol.* **71:**127–147.

Yasbin, R. E. 1985. DNA repair in *Bacillus subtilis*, p. 33–52. *In* D. A. Dubnau (ed.), *The Molecular Biology of the Bacilli*, vol. 2. Academic Press, Inc., New York.

Chapter 3

Pathways of Homologous Recombination in *Escherichia coli*

Suresh K. Mahajan

I. INTRODUCTION

Homologous genetic recombination can be defined as a series of interactions between two largely homologous DNA sequences, present on two different molecules or on a single molecule, to produce a mixed sequence (or two mixed sequences) derived partly from one parental sequence and partly from the other.

The most remarkable feature of homologous recombination is its precision. The recombinational exchange between two parental molecules almost never leads to any net gain or loss of the parental information near the exchange point. This precision is presumably accomplished by base pairing between complementary single strands derived from the two parents to generate a heteroduplex region near the point of exchange (Whitehouse, 1963; Holliday, 1964).

The second outstanding feature of homologous recombination is its efficiency: whenever sufficiently long homologous sequences are brought together in a single cell under appropriate conditions, the production of recombinant sequences is more of a rule than an exception (Fox, 1978). Since the recombinant DNA molecules normally inherit physical material from both parental DNAs (Meselson and Weigle, 1961; Kellenberger et al., 1961; Siddiqi, 1963; Oppenheim and Riley, 1966; Tomizawa, 1967; Kellenberger-Gujer and Weisberg, 1971; Siddiqi and Fox, 1973), recombination must involve breaking and rejoining of covalent phosphodiester bonds along the DNA. Enzymes and other proteins must be involved in these and other steps, e.g., in homology search and heteroduplex formation.

An understanding of general recombination should, therefore, involve two aspects: (i) the stepwise structural changes of the two parental DNAs and the nucleoprotein complexes generated by their interaction to produce recombinant molecules capable of replication and expression of the recombinant phenotype, and (ii) the enzymes and other molecules that catalyze or facilitate these structural changes and regulate their stability, duration, and extent at each step.

Such a complete description of recombination would constitute a pathway of recombination. There is considerable evidence, at least in *Escherichia coli*, for several distinguishable recombination pathways. This chapter reviews this evidence.

Several systems have been used for investigating recombination in *E. coli*. These differ from one another in the materials used (e.g., male and female cells, transducing phages, plasmids), the manner in which the genetic cross is made (i.e., how the two interacting DNAs are brought together in a single organism), and the recombinant product monitored (e.g., recombinant organisms, recombinant DNA molecules, or the functional proteins produced by recombinant genes (Clark and Low, 1988). The Hfr $\times$ F$^-$ conjugation system was the first to provide evidence for multiple pathways of recombination in *E. coli*. This evidence and the evolution of the hypothesis of multiple pathways as it applies to this system have been frequently reviewed (Clark, 1971, 1973, 1974, 1980; Clark et al., 1984; Smith, 1987; Clark and Low, 1988). Here I have briefly summarized and updated this evidence (section III.).

Some modifications in the original hypothesis of multiple pathways of recombination in *E. coli* conjugation have been suggested to accommodate recent findings (section III.). These modifications deal primarily with the number of pathways and their names. The considerable evidence for multiple pathways in other *E. coli* systems is summarized in section IV. An attempt to reconcile the apparently disparate results given by different systems and some ideas about the DNA structures and interactions involved in different pathways are presented in section V. A summary of the properties of different recombination-associated genes and their products is also included (section II.).

II. GENES AND ENZYMES OF RECOMBINATION IN *E. COLI*

A. General Approaches

Four approaches have been useful in investigating recombination: (i) *phenotypic analysis* of the recombinant organisms; (ii) *physical analysis* of the recombinant DNA molecules; (iii) *genetic and biochemical analysis* of the cellular catalysts of recombination; and (iv) *intellectual analysis* or model building. All these approaches have contributed to our understanding of recombination in *E. coli*, but most of the evidence for multiple pathways has come from genetic and biochemical analysis of mutants with altered proficiency of recombination.

Genes whose products influence recombination have been identified by isolating, mapping, and phenotypically characterizing mutations which abolish, decrease, or increase recombination. Mutants altered in these genes are tested for recombination in different systems (such as conjugation, specialized and generalized transduction, transformation with plasmids, and DNA amplification; see Porter, this volume). Comparison of the effects of these mutations in different recombination systems helps in inferring the roles of the corresponding gene products in recombination.

Next, the gene products are identified, purified, and characterized with respect to their action on different DNA substrates. The in vitro properties of the proteins and the in vivo effects of the mutations can provide important information about recombinational steps catalyzed by the proteins in vivo.

Cloning of these genes and overproduction of their products have been useful in many respects. These include easier purification of large quantities of the corresponding proteins, study of the effects of excess intracellular amounts on recombination (Moreau, 1987), isolation of new mutations (Willis et al., 1985), study of the regulation of expression of these genes (Willis et al., 1985; Blanar et al., 1984b; Armengod and Lambiers, 1986), sequencing of wild-type and mutant alleles of the genes, and, at least in one case, identification of a new subunit (RecD) of the protein (Amundsen et al., 1986). In some cases, sequencing has been helpful in illuminating gene structure and expression (Blanar et al., 1984a; Wang and Tessman, 1986).

B. Isolation of Recombination-Deficient Mutants

Numerous screening schemes have identified *E. coli* mutants with altered recombinational proficiency. The reduced ability of colonies of (mutagenized) F⁻ cells to produce recombinants when mated with lawns of Hfr cells has been the most common screen used to identify Rec⁻ (hypo-Rec) mutants. This screen identified *recA* (Clark and Margulies, 1965), *recE* (Gillen et al., 1981), *recF* (Horii and Clark, 1973), *recJ* (Horii and Clark, 1973; Lovett and Clark, 1984), *recN* (Lloyd et al., 1983; Kolodner et al., 1985), and *recO* (Kolodner et al., 1985) mutations.

Other phenotypes associated with altered DNA metabolism identified numerous mutants subsequently found to be hypo-Rec in conjugation crosses. *recB* and *recC* mutants were initially identified by their extreme sensitivity to X rays and their reduced proficiency in Hfr × F⁻ crosses (Clark and Margulies, 1965; Howard-Flanders and Theriot, 1966). *recD* mutants were isolated for their ability to support good growth of phage T4 gene 2 mutants (Chaudhury and Smith, 1984a; Amundsen et al., 1986), to reduce plasmid stability (Biek and Cohen, 1986), or to raise the copy number of certain plasmids (Seelke et al., 1987). Other screens yielded mutations of the *recA*, *recB*, and *recC* genes that do not have a strong effect on recombination proficiency but do affect certain features of recombination (Tessman and Peterson, 1985; Schultz et al., 1983; Lundblad et al., 1984; see below).

ruv mutants were isolated for their sensitivity to mitomycin C (MitC) (Otsuji et al., 1974) and for their reduced recombination between duplicated *gal* genes in F′ merodiploids (Stacey and Lloyd, 1976). Later, Lloyd et al. (1984) observed that the *ruv* mutations were Rec⁻ in *recBC sbcB(C)* cells. *recQ* was first isolated as a thymineless death-resistant derivative and then shown to make *recBC sbcB(C)* cells Rec⁻ (Nakayama et al., 1984).

C. Hyper-Rec Mutants

Other mutants have been isolated for their increased recombination proficiency. Suppressor mutations, which increase the recombination proficiency of otherwise hypo-Rec mutants, include *sbcA*, *sbcB*, and *sbcC*, which increase viable recombinant production in *recBC* cells (Barbour et al., 1970; Kushner et al., 1971, 1972; Lloyd and Buckman, 1985), *srf*, which increases recombination proficiency of *recBC sbcB(C) recF* cells (Volkert and Hartke, 1984), and *srj*, which increases recombination proficiency of *recBC sbcB(C) recJ* cells (S. Lovett, personal communication). Increased recombinant formation in Hfr × F⁻ plate matings and increased resistance to MitC and UV light were the screens used to identify these mutants. The *sbcA* and *sbcB(C)* suppressors played an important role in the later identification of several *rec* genes (*recE*, *recF*, *recN*, *recO*, *recQ*, and *ruv*) which do not appreciably affect recombination in wild-type cells (see below).

Konrad (1977) identified hyper-Rec mutants in which increased recombination between two nontandem *lacZ* duplications on the *E. coli* chromosome was scored on indicator plates. With this screen several hyper-Rec mutants were isolated. Some of these had lesions in previously identified genes: *polA* (De Lucia and Cairns, 1969; Konrad and Lehman, 1974), *lig* (Gottesman et al., 1973), *mutU* (= *uvrD* = *uvrE* = *recL*) (Siegel, 1973, 1981), *dam* (Marinus 1973a, b; Marinus et al., 1983), and *dut* (= *dnaS* = *sof*) (Hochhauser and Weiss, 1978). Two previously isolated mutations, *lig-7* (Pauling and Hamm, 1968) and *polA1* (De Lucia and Cairns, 1969), were found by Konrad (1977) to be hyper-Rec in this assay. Others have used this assay to implicate additional DNA metabolism genes in recombination, including *uvrD*, *xthA*, *lexA*, and *rep* (Zieg et al., 1978), *xseA* (Chase and Richardson, 1977; Chase et al., 1986), and *rdgB* (Clyman and Cunningham, 1987).

Feinstein and Low (1986) used the criterion of simultaneous increased recombination in two intragenic regions to isolate six hyper-Rec mutants whose mutations were located in *mutS* (five mutations) and *mutL* (one mutation). They analyzed these and *mutH* (= *mutR*) and *mutU* for their effect on intragenic recombination in Hfr × F⁻ crosses and in an F′ *lac* chromosome homogenotization assay (Berg and Gallant, 1971). All were hyper-Rec, with some *mutU* alleles the most so. In Hfr × F⁻ crosses *mutU* reduced linkage between well-separated markers, while *mutH*, *mutL*, and *mutS* had smaller but measurable effects.

The identification of recombination genes by these indirect schemes presumably reflects the involvement of recombination in other DNA metabolic processes and/or its sharing of enzymes with some of these processes (Clark, 1971). These schemes have identified more than 30 *E. coli* genes, each of which influences recombination in one or more situations. Most of these genes, discussed below,

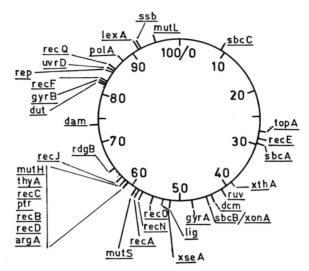

FIGURE 1. Map location of recombination genes on the standard *E. coli* map. For mapping references see the following: *recD* (Amundsen et al., 1986; Biek and Cohen, 1986); *recJ* (Lovett and Clark, 1984); *recN* (Lloyd et al., 1983); *recO* (Kolodner et al., 1985); *recQ* (Nakayama et al., 1984); *rdgB* (Clyman and Cunningham, 1987); other genes (Bachmann, 1983).

have been mapped (Fig. 1) and cloned, and some have been sequenced. The products of most of these genes have been identified. Many of these products have been purified and their activities have been characterized in vitro (see Table 1 for references).

Nearly all of the hypo-Rec mutations have been examined for effects on the proficiency of recombination in conjugation and other systems (see sections III. and IV.) and on the repair of UV damage and other phenotypes (see below). However, most of the hyper-Rec mutations, other than the suppressors of the *rec* mutations, have not been examined for their effects on recombination in different systems. Even in conjugation the effects of these mutations have been studied primarily only in the wild-type genetic background.

D. Characteristics of Recombinational Gene Mutations

Table 1 briefly describes the products of the recombinational genes identified by the schemes described above. Some additional comments on the mutations of these genes and the likely roles of their products in recombination are given below. Reviews by Clark and Low (1988) and Smith (1988) give additional information about specific genes and their products. Chapters by Radding, Taylor, and Radman in this book deal more extensively with *recA*, *recBCD*, and mismatch repair genes, respectively.

1. *recA*

recA is the most intensively studied recombination gene. Many *recA* mutations have been described (for details, see Templin et al., 1978; Walker, 1984;

TABLE 1
Genes affecting homologous recombination in *E. coli*

Gene	Protein product and approx size	Activities[a]	References[b]
recA	RecA protein, 37.5 kDa	Directly and indirectly regulates recombination; has DNA-dependent ATPase activity; rapidly and cooperatively binds to ssDNA, producing helical nucleoprotein filaments which bind dsDNA nonspecifically to generate large DNA networks; in these networks rapidly synapses homologous sequences and promotes ATP-dependent assimilation of ssDNA into homologous dsDNA; processively unwinds dsDNA; promotes pairing and strand exchange between dsDNAs with single-stranded ends to create Holliday junctions; anneals complementary ssDNA; in the presence of ssDNA and an adenine nucleotide cleaves (or stimulates autocleavage of) LexA repressor and λ repressor	A. Emmerson et al. (1981) B. Sancar et al. (1980) C. Emmerson and West (1977), Gudas and Mount (1977), McEntee (1977) D. Cox and Lehman (1987), Radding (this volume)
recB	Component of the β-subunit of RecBCD enzyme (ExoV), 140 kDa	Has dsDNA-dependent ATPase, ATP-dependent dsDNA and ssDNA exonuclease, and ATP-stimulated ssDNA endonuclease activities; has ATP-dependent unwinding activity on linear dsDNA with flush or nearly flush ends, producing long ssDNA and dsDNA with single-stranded tails; during unwinding from right to left makes single-stranded nicks near Chi sites (5′-GCTGGTGG-3′) 4 to 6 bp to the 3′ side; has limited exonucleolytic activity on RNA-DNA hybrids; has D-loop-cleaving activity	A. Hickson and Emmerson (1981), Sasaki et al. (1982), Dykstra et al. (1984), Amundsen et al. (1986) B. Finch et al. (1986a, b, c) C. Buttin and Wright (1968), Barbour and Clark (1970), Goldmark and Linn (1972), Lieberman and Oishi (1974), Taylor and Smith (1980), Hickson and Emmerson (1981), Ponticelli et al. (1985), Amundsen et al. (1986) D. Telander-Muskavitch and Linn (1981, 1982), Smith et al. (1984), Smith (1988), Taylor (this volume)
recC	Component of the β-subunit of RecBCD enzyme, 130 kDa		
recD	α-Subunit of RecBCD enzyme, 65 kDa		

Continued on following page

TABLE 1—*Continued*

Gene	Protein product and approx size	Activities[a]	References[b]
recE	Exonuclease VIII, 140 kDa	Has ATP-independent exonuclease activity; degrades dsDNA processively from the 5′ end, producing long 3′ tails; digests nicks or gaps in a duplex circle extremely slowly, but efficiently degrades DNA containing recessed 5′ termini (the otherwise analogous enzyme λ exonuclease lacks the last activity)	A. Clark et al. (1984), Willis et al. (1985), Mahajan and Clark (unpublished data) B. Chu, Mahajan, and Clark (unpublished data) C. Kushner et al. (1974), Gillen et al. (1977); Joseph and Kolodner (1983a, b)
recF	RecF protein, 40–42 kDa	Unknown	A. Ream and Clark (1983) B, C, and D. Blanar et al. (1984b), Armengod and Lambiers (1986), Krivonogov (1984), Ream et al. (1980)
recJ	RecJ protein, 53 kDa	Unknown	A and C. Lovett and Clark (1984, 1985)
recN	RecN protein, 62 kDa	Unknown	A. Picksley et al. (1984b) B. Rostas et al. (1987) C. Finch et al. (1985)
recO	Unknown	Unknown	D. Kolodner et al. (1985)
recQ	RecQ protein, 74 kDa	Unknown	A. Nakayama et al. (1985) B. Irino et al. (1986)
ruv	Ruv protein, 41 kDa	Unknown	A. Shurvinton et al. (1984) C. Attfield et al. (1985)
sbcA	SbcA protein, 86 kDa	Unknown	A, B, and C. Clark et al. (1984) D. Kaiser and Murray (1980)
sbcB	Exonuclease 1, 72 kDa	Attacks ssDNA at free 3′-OH ends, processively releasing 5′-mononucleotides; leaves the 5′-terminal dinucleotide intact	A. Vapnek et al. (1976) B. Phillips and Kushner (1987) C. Lehman and Nussbaum (1964), Kushner et al. (1972) D. Weiss (1981), Brody et al. (1986)
sbcC	Unknown	Unknown	D. Lloyd and Buckman (1985)
ssb	SSB protein, 19.5 kDa	Binds ssDNA selectively and cooperatively; facilitates RecA protein binding to ssDNA and hence RecA-promoted reactions; inhibits nuclease activities of RecBCD enzyme; stimulates DNA polymerases II and III; favors melting of DNA but can renature denatured DNA in the presence of polyamines	A and B. Sancar et al. (1981), C. MacKay and Linn (1976), Kowalczykowski et al. (1987), Kowalczykowski and Krupp (1987), Eigner et al. (1987) D. Chase and Williams (1986), Cox and Lehman (1987)

Continued on following page

TABLE 1—*Continued*

Gene	Protein product and approx size	Activities[a]	References[b]
lexA	LexA protein, 44 kDa	Represses *lexA*, *recA*, and other damage-inducible genes; undergoes autoproteolysis at the Ala[84]-Gly[85] dipeptide bond; this reaction is stimulated by RecA protein activated by ssDNA, ATP, $MgCl_2$, and NaCl	A. Emmerson et al. (1981) B. Horii et al. (1981) C. Little and Harper (1979), Little et al. (1980), Little (1984) D. Little and Mount (1982), Walker (1984)
polA	DNA polymerase I (PolI), 109 kDa	Has $5' \rightarrow 3'$ polymerase activity; has $3' \rightarrow 5'$ exonuclease activity on ssDNA with a 3'-OH deoxyribonucleotide terminus; has $5' \rightarrow 3'$ exonuclease activity on dsDNA with a base-paired or mismatched 5' end	A. Kelly et al. (1977) B. Joyce and Grindley (1982) C and D. Kornberg (1980, p. 101–166), Pyhtila and Syvaoja (1980), Lehman (1981)
lig	DNA ligase, 75 kDa	Seals nicks in dsDNA by making phosphodiester bonds between adjacent $5'$-PO_4 and 3'-OH ends.	A. Panasenko et al. (1977) C and D. Gottesman et al. (1973), Kornberg (1980, p. 261–276)
gyrA	Subunit A of DNA gyrase (GyrA), 105 kDa	Negatively supercoils closed circular duplex DNA in the presence of ATP; relaxes supercoiled circles in the absence of ATP; forms and resolves knotted and catenated duplexes; produces discrete double-strand breaks in DNA in the presence of oxolinic or nalidixic acid followed by addition of sodium dodecyl sulfate; has dsDNA-dependent ATPase activity	A, B, and C. Swenberg and Wang (1987), Mizuuchi et al. (1984)
gyrB	Subunit B of DNA gyrase (GyrB), 95 kDa		D. Gellert (1981), Kornberg (1980, p. 311–315), Drlica (1984)
topA	DNA topoisomerase I (TopI), 110 kDa	Nicks and closes duplex DNA; manifestations include relaxation of negatively supercoiled DNA, knotting and unknotting of single-strand rings, catenation and decatenation of a pair of dsDNA circles if one circle contains an ss region, and intertwining of complementary single-strand rings	A. Wang and Bechere (1983) C. Wang (1971) D. Wang (1981), Gellert (1981), Drlica (1984)
uvrD	Helicase II, 81 kDa	Causes ATP-dependent unwinding of dsDNA with an adjacent ssDNA region for the protein to initiate binding; stimulates UvrABC enzyme-catalyzed excision by releasing the UvrABC incised damaged fragment	A. Oeda et al. (1981) B. Finch and Emmerson (1984) C. Hickson et al. (1983), Kumara and Sekiguchi (1984), Kumara et al. (1985), Caron et al. (1985) D. Geider and Hoffman-Berling (1981)

Continued on following page

TABLE 1—*Continued*

Gene	Protein product and approx size	Activities[a]	References[b]
dam	DNA adenine methylase (Dam), 31 kDa	Methylates adenine residues in 5′-GATC-3′ in dsDNA to produce 6-methyladenine	A and B. Brooks et al. (1983) C. Geier and Modrich (1979), Herman and Modrich (1982) D. Marinus (1987), Claverys and Lacks (1986), Radman (this volume)
dcm	DNA cytosine methylase (Dcm), 23 kDa	Methylates the second cytosine in 5′-CC(A/T)GG-3′ in dsDNA to produce 5-methylcytosine	A. Bhagwat et al. (1986) D. Marinus (1987), Radman (this volume)
dut	Deoxyuridine triphosphatase, 16 kDa	Converts dUTP to dUMP, thereby reducing incorporation of the former into DNA	A. Taylor et al. (1980) C and D. Kornberg (1980, p. 62–64)
rep	Rep protein, 65 kDa	Has ssDNA-dependent ATPase activity; catalytically separates the strands of dsDNA in the presence of SSB protein and ATP, provided the dsDNA has an adjacent single-strand region in which Rep protein moves unidirectionally, 3′ → 5′, before invading the duplex	A and B. Bialkowska-Hobrzanska et al. (1985) C. Takahashi et al. (1979) D. Geider and Hoffman-Berling (1981)
xth	Exonuclease III	Has 3′ → 5′ exonuclease activity specific for dsDNA and for the RNA strand in an RNA-DNA hybrid (RNase H activity); has DNA-3′-phosphatase activity; has endonucleolytic activity on apurinic and apyrimidinic sites; converts nicked circles into gapped circles	A. Rogers and Weiss (1980) C. Low et al. (1984) D. Weiss (1981)
xseA	Subunit A of exonuclease VII, 52 kDa	The native enzyme has one large subunit and four small subunits; degrades linear ssDNA processively in both 5′ → 3′ and 3′ → 5′ directions to produce first large oligonucleotides and eventually small acid-soluble oligonucleotides	A and B. Chase et al. (1986)
xseB	Subunit B of exonuclease VII, 10.5 kDa		
mutH	MutH protein, 25 kDa	Hemi-methylated 5′-GATC-3′ binding protein	A. Lu et al. (1984) C. Lu et al. (1983), Lahue et al. (1987) D. Claverys and Lacks (1986), Meselson (1988), Radman (this volume)

Continued on following page

TABLE 1—*Continued*

Gene	Protein product and approx size	Activities[a]	References[b]
mutL	MutL protein, 70 kDa	Required in cell-free methyl-directed mismatch correction reaction	D. Radman (this volume)
mutS	DNA base pair mismatch binding protein, 97 kDa	Specifically binds DNA regions containing base pair mismatches	A and C. Su and Modrich (1986) D. Claverys and Lacks (1986), Lieb (1987), Meselson (1988), Radman (this volume)
rdgB	RdgB protein, 25 kDa	Unknown	D. Clyman and Cunningham (1987)

[a] For in vivo characteristics and mapping of these genes, see the text.
[b] References are divided into four parts: A, cloning; B, sequencing; C, Purification and characterization of the protein product; and D, reviews and other recent references.

Wang and Tessman, 1986). *recA* null mutations almost completely block recombination in most *E. coli* systems. A significant exception is the recombination of certain plasmids and bacteriophages promoted by the RecE pathway or analogous pathways encoded by phages λ and φ80 (see sections III. and IV. for references and details). *recA* null mutations also block phenotypes regulated by the RecA protein through its protease (Prt) activity, such as the induction of λ prophage by cleaving the λ repressor, induction of the SOS regulon genes by cleaving the LexA repressor (for a review, see Walker, 1984), *recBC*-dependent but *lexA*-independent induction of restriction alleviation by UV (Thoms and Wackernagel, 1984), *recBC*-dependent and *lexA*-regulated respiration inhibition by UV (Swenson and Norton, 1986), and induction of a *lexA*-independent gene required for UV mutagenesis of λ (Calsou et al., 1987). For a further description of *recA* phenotypes, see Radding (this volume).

With respect to the recombinational pathways, several significant observations on *recA* need special mention.

(i) The RecA protein-dependent steps take a much shorter time in the RecBCD pathway than in the RecF pathway. Elevated levels of RecA protein can accelerate the RecF pathway (section III.D.4.).

(ii) In addition to its direct role in homology matching and strand transfer (Radding, 1978, 1982, this volume), RecA protein may influence the RecF pathway of recombination indirectly by regulating the expression of the *recN*, *recQ*, and *ruv* genes. The expression of these three genes is blocked by the *lexA3*(Ind⁻) (Mount et al., 1972) mutation, indicating that their normal expression may be regulated by the activated RecA protein (RecA*). But the basal levels of these proteins, synthesized without RecA*-mediated induction, may suffice for recombination.

(iii) Some *recA* mutations, initially called *srfA* (Volkert and Hartke, 1984, 1987; Wang and Smith, 1986), suppress the recombination and repair deficiency of *recF* mutations (see above). Though some of these *recA* mutant alleles have been sequenced and the protein product of one has been partially characterized (M. V. V. S. Madiraju and A. J. Clark, personal communication), the mechanism of this suppression is not known.

(iv) Several *recA* mutations affect the recombinational and protease activities differentially. These mutations can be Rec$^+$ Prt^c (protease constitutive), Rec$^-$ Prt^c, or Rec$^+$ Prt$^-$ (for details and references, see Wang and Tessman, 1986; Walker, 1984). The Rec$^+$ Prt$^-$ and Rec$^+$ Prt^c mutants indicate that the recombinational and protease activities lie in different domains, whereas single-base-pair mutations affecting both phenotypes, e.g., Prt^c Rec$^-$ and Prt$^-$ Rec$^-$ (null mutations), indicate that these domains must partially overlap (Wang and Tessman, 1986).

(v) The activities of RecA protein on DNA can be separated into several stages such as homology search and formation of paranemic joints, formation of plectonemic joints, processive unwinding of double-stranded DNA (dsDNA) and branch migration (Table 1; Radding, this volume). Two of these activities, homology search and processive DNA unwinding, appear to reside in different parts of the protein: one monoclonal antibody blocks the former but not the latter (Makino et al., 1987). The rate-determining activity of RecA protein may differ in different Rec pathways.

2. *recB*, *recC*, and *recD*

Null mutations of *recB* and *recC* strongly reduce the proficiency of conjugational recombination (Clark and Margulies, 1965; Howard-Flanders and Theriot, 1966) and lack all of the known in vitro activities of the RecBCD enzyme (Table 1; Taylor, this volume). Since *recB* and *recC* null single mutants and *recB recC* double mutants have indistinguishable phenotypes, they are frequently denoted as *recBC* mutants, a practice followed in this review. Two commonly used null mutations are *recC22*, a nonsense mutation, and *recB21*, a 1.4-kilobase (kb) insertion that is polar on *recD* (Templin et al., 1978; Amundsen et al., 1986). Transposon insertion mutations and deletions of *recB* and *recC* have been isolated (Lloyd et al., 1987a, b; Chaudhury and Smith, 1984b).

In addition to recombination, the null mutations alter other aspects of DNA and cellular metabolism (see Taylor, this volume), such as the following: (i) enhanced sensitivity to DNA-damaging agents such as X rays, UV light, MitC, and nitrofurantoin (Howard-Flanders and Theriot, 1966; Schultz et al., 1983) and a block in the repair of dsDNA breaks (Sargentini and Smith, 1987); (ii) slow growth rate and segregation of a large fraction (up to 90%) of nonviable cells with associated DNA abnormalities (Capaldo-Kimball and Barbour, 1971; Capaldo et al., 1974), reduced oxygen consumption, and release of intracellular enzymes into the medium (Capaldo and Barbour, 1975; Miller and Barbour, 1977); (iii) nonviability in combination with mutations in other DNA metabolism genes such as *polA* and *dam* (Monk and Kinross, 1972; Strike and Emmerson, 1972; McGraw and Marinus, 1980); (iv) reduced DNA degradation following UV irradiation (Youngs and Bernstein, 1973); (v) altered growth of several phages including P1, P2, and certain mutants of λ and T4 (for references, see Chaudhury and Smith, 1984a); (vi) block in SOS induction by nalidixic acid, coumermycin, or UV-irradiated mini-F plasmids (Chaudhury and Smith, 1985; Bailone et al., 1985); (vii) block in UV or nalidixic acid induction of *lexA*-independent restriction alleviation (Thoms and Wackernagel, 1984); and (viii) absence of Chi activity, the stimulation of recombination in the neighborhood of the Chi sequence 5′-GCTGGTGG-3′ (Stahl, 1979; Smith et al., 1984; Smith, 1988; Taylor, this volume).

Some *recB* and *recC* mutants retain full or partial dsDNA nuclease activity and proficiency of conjugational recombination but have other special phenotypes. *recB* and *recC* mutations with the TexA phenotype (Lundblad et al., 1984) increase the frequency of excision of the inverted repeat transposons Tn5 and Tn10, reduce Chi activity, and increase UV sensitivity, but do not reduce viability. *recC** mutations (Schultz et al., 1983), which were selected as nitrofurantoin-resistant pseudorevertants of a null mutation, *recC73*, lack Chi activity but have fully or partially regained other wild-type phenotypes including viability, proficiency of conjugational recombination, and UV resistance (Schultz et al., 1983).

The *recBCD*‡ mutants are deficient in dsDNA nuclease, helicase, and Chi activity but have almost wild-type levels of conjugational recombination and UV resistance (Chaudhury and Smith, 1984a; Taylor, this volume). These mutations are in the *recC* and *recD* genes (Amundsen et al., 1986). *recD* mutants have increased plasmid instability (Biek and Cohen, 1986; Cohen and Clark, 1986), increased plasmid copy number (Seelke et al., 1987), and plate phage T4 gene 2⁻ (Chaudhury and Smith, 1984a).

Extragenic suppressors of *recBC* mutations, designated *sbcA*, *sbcB*, and *sbcC*, restore recombination and UV resistance but not RecBCD enzyme activity. These suppressors played a key role in elucidating pathways of recombination (see below and section III.).

In considering the role of RecBCD enzyme in recombination, one should keep in mind the following. RecBCD enzyme is multifunctional and may influence recombination at several steps. It seems to have a positive role in at least two pathways of recombination, only one of which is Chi dependent. It may have a negative role in other pathways by removing their substrates.

3. *recE* and *sbcA*

The *recE* gene is part of the defective lambdoid prophage rac present in the chromosome of most *E. coli* strains (Kaiser and Murray, 1979; Gillen et al., 1981) but not in AB1157, a strain whose derivatives have been extensively used in analysis of recombinational pathways other than the RecE pathway. *recE* encodes exonuclease VIII (ExoVIII), with activities similar to those of λ exonuclease (Little, 1967; Table 1). Normally, *recE* (together with other genes of the rac prophage) is repressed. However, its expression can be induced in several ways (for a review, see Clark et al., 1984).

(i) Mating an Hfr harboring a repressed rac prophage with a recipient devoid of rac transiently induces expression of *recE* by zygotic induction (Low, 1973).

(ii) *sbcA* mutations activate *recE* and were isolated as suppressors of the *recBC* recombination deficiency (Barbour et al., 1970). These can be large deletions, such as *sbcA8* (140 kb) and *sbcA81* (100 kb), which remove most of the rac prophage and some of the adjacent chromosomal regions, very small deletions or point mutations such as *sbcA1*, *sbcA2*, and *sbcA6*, or possibly duplications several kilobases long, such as *sbcA23* (Kaiser and Murray, 1980). Only for *sbcA8* is the nature of the induced expression known. *sbcA8* deletes 882 base pairs (bp) from the 5′ end of *recE* and fuses the remaining 3′ region of this gene to the 5′ region of a chromosomal gene; although the resulting fusion protein (150 kilodaltons [kDa]) is significantly larger than the ExoVIII (140 kDa) obtained from an *sbcA23* strain, it is fully active and constitutively expressed (S. K. Mahajan, C. C. Chu, and A. J. Clark, unpublished data).

(iii) Transposon insertions may provide a new promoter to *recE* (Fouts et al., 1983; Willis et al., 1983; Chu and Clark, unpublished data).

(iv) In some cases small deletions (≃1 kb) in the 5′ region of a cloned *recE* gene fuse the 3′ region of *recE* to the 5′ region of another rac gene (Willis et al., 1985; Chu and Clark, unpublished data).

The putative point mutation *recE159* and *recE101*::Tn*10* (Gillen et al., 1981; Fouts et al., 1983) inactivate ExoVIII and abolish recombination proficiency of *recBC sbcA* cells. Recombination in *recBC⁺ sbcA⁺* cells is not significantly influenced by *recE101*::Tn*10*.

In addition to *recE*, *sbcA* mutations seem to activate the expression of other genes with analogs in λ. These include a restriction alleviation gene, *ral* (Gillen et al., 1981; Kannan and Dharmalingam, 1987) and *int*- and *xis*-like genes (Clark and co-workers, unpublished data). An unidentified protein similar to the *bet* protein of λ (Kmiec and Holloman, 1981; Muniyappa and Radding, 1986), which can promote *recA*-independent recombination, may also exist (section IV.E.).

The RecE pathway seems to use linear duplex substrates efficiently (section IV.). ExoVIII may degrade one strand (5′→3′) to generate single-stranded regions with free 3′ ends (Joseph and Kolodner, 1983b), which can be efficient substrates for RecA protein-mediated synapsis. ExoVIII may also promote repair of single-strand nicks and gaps, though less efficiently than ExoV (Waldstein, 1979).

4. *sbcB*, *sbcC*, and *xonA*

sbcB mutations inactivate exonuclease I and were isolated as extragenic suppressors that restored recombination proficiency and MitC resistance to *recBC* mutants without restoring the RecBCD enzyme activity (Kushner et al., 1971, 1972; Templin et al., 1972). The original isolates apparently contained a mutation in another suppressor locus, *sbcC*, identified only recently (Lloyd and Buckman, 1985; also see Masters et al., 1984). Studies on the RecF pathway of recombination (sections III. and IV.) were believed to use *recBC sbcB* mutants but presumably used *recBC sbcB sbcC* mutants. These mutants are denoted *recBC sbcB(C)* to indicate the presumptive *sbcC* mutation.

The *sbcB* mutations make *recBC* cells UV resistant but increase their MitC resistance, growth rate, and viability only slightly; faster-growing *sbcB sbcC* mutants accumulate rapidly and have nearly wild-type levels of viability, MitC resistance and proficiency of conjugational recombination via the RecF pathway. It is noteworthy that the *sbcB* mutation is crucial for increasing recombination and repair in the *recBC* background; *sbcC* alone has little effect (Lloyd and Buckman, 1985). The activity of *sbcC* is unknown.

The mechanism of *sbcBC*-mediated recombination proficiency of *recBC* cells is not entirely clear. The degradation of single-stranded DNA (ssDNA) with free 3'-OH ends by ExoI (Lehman and Nussbaum, 1964) and the derepression of the *recN* gene in the *sbcB(C)* mutants (Picksley et al., 1984b) provide two clues (see section V.).

xonA mutations lie in the *sbcB* gene and also inactivate ExoI but fail to restore recombination proficiency to *recBC* cells though UV and MitC resistance are restored (Kushner et al., 1972; Yajko et al., 1974). The *xonA* mutations may not completely inactivate ExoI, or the mutants studied may not have acquired *sbcC* mutations.

5. *recF*, *recJ*, and *recO*

The *recF*, *recJ*, and *recO* genes, whose activities are unknown, seem to regulate an early step in conjugational recombination and may account for most of the recombination initiation in *recBC* mutants (Lloyd et al., 1987a, b). *recJ*⁺ is also required for recombination in *recD* mutants (see section III.B.). Mutations in these genes block plasmid recombination in wild-type genetic backgrounds (Kolodner et al., 1985) and increase the UV sensitivity of *recBC*, *recBC sbcA*, and *recBC sbcB(C)* cells (Horii and Clark, 1973; Lovett and Clark, 1984; Kolodner et al., 1985). In a *rec*⁺ genetic background *recF* and *recO*, but not *recJ*, mutations increase UV sensitivity (section IV.C.).

recF mutants have been more extensively characterized than *recO* and *recJ* mutants. *recF* mutations prevent chromosome mobilization by F' (Picksley et al., 1984a), reduce dsDNA break repair (Sargentini and Smith, 1987), reduce SOS induction by UV, nalidixic acid, and coumermycin (for references, see Thoms and Wackernagel, 1987), and increase intracellular levels of the RecA protein (Salles and Paoletti, 1983). The recombination deficiency and UV sensitivity of *recBC sbcB(C) recF* cells can be suppressed by special *recA* mutations (*tif-1*, *srfA*) (Volkert and Hartke, 1984, 1987; Wang and Smith, 1986).

A 74-kDa endonucleolytic protein, positively regulated by *recF*⁺, has been reported (Krivonogov, 1984). The *recF144* mutation significantly reduces the density of recombinational exchanges between selected and unselected markers, even in a wild-type genetic background (Krivonogov and Novitskaya, 1982). These observations indicate an indirect role for the RecF protein in (early stages of) recombination, though a direct role in modulating RecA-mediated reactions is not ruled out in view of the ability of the *srfA* allele of *recA* to suppress all *recF* alleles.

The *recF* gene forms part of an operon containing the *dnaN* and *gyrB* genes (Ream et al., 1980; Ream and Clark, 1983). The former lies upstream of *recF* with a 1-bp overlap, while the latter lies 28 bp downstream. Sequences which are strong negative regulators of *recF* exist within *dnaN* and within *recF* itself. As a consequence, only a few molecules of the RecF protein are found in each cell (Blanar et al., 1984b; Armengod and Lambiers, 1986).

Mutations (*srj*; see above) which suppress *recJ* phenotypes in *recBC sbcBC recJ* cells map to at least two different locations (S. Lovett, personal communication).

6. *recN*, *recQ*, and *ruv*

The *recN*, *recQ*, and *ruv* genes are essential for the RecF pathway of recombination and are repressed by *lexA*$^+$ (Lloyd et al., 1983; Irino et al., 1986; Shurvinton and Lloyd, 1982; Lloyd et al., 1987a). *ruv* may also contribute to certain kinds of recombination in wild-type cells (Stacey and Lloyd, 1976; Lloyd et al., 1987a; N. N. Pandit and S. K. Mahajan, unpublished data). The genes have been cloned and their products have been identified, but the activities are unknown.

recN$^+$ is required for repair of dsDNA breaks but seems to have no role in recombination repair by exchange of ssDNA (Picksley et al., 1984a; Sargentini and Smith, 1987). Consistent with this, *recN* mutants in a wild-type background are resistant to UV but sensitive to MitC and ionizing radiation, though they are sensitive to UV in a *recBC sbcB(C)* background. RecN protein seems to have no early role in recombination because *recN* mutation has no effect on β-galactosidase production in F′ *lacZ1* × F⁻ *lacZ2* crosses either in wild-type or *recB* backgrounds (Lloyd et al., 1987b) or on chromosome mobilization by an F′ (Picksley et al., 1984a; section III.D.3.). *recN* gene expression is derepressed by *sbcB(C)* mutations and is repressed by the LexA repressor (Picksley et al., 1984b).

recQ mutants are partially resistant to thymineless death and are UV sensitive in a *recBC sbcB(C)* background but UV resistant in a wild-type background (Nakayama et al., 1984, 1985).

In a *recBC sbcB(C)* background *ruv* strongly reduces F′ repliconation frequency though conjugal transfer, as seen by zygotic induction of λ prophage, is not affected (Lloyd et al., 1984). *ruv* mutants are sensitive to UV and γ rays in a wild-type background. Low doses of UV induce filamentation (Otsuji et al., 1974). Though all *ruv* mutations map to one locus, different alleles have different quantitative effects on recombination, nalidixic acid sensitivity, and UV sensitivity on minimal medium (Otsuji et al., 1974; Shurvinton et al., 1984; Attfield et al., 1985; Lloyd et al., 1987a, b). The reduction of F′ repliconation frequency is not seen in *ruv recA* recipients (R. G. Lloyd, personal communication). This observation has been interpreted to imply that recombination can be initiated in *ruv* mutants but that a late step, such as resolution of the interacting homologous molecules, is defective. *ruv*$^+$ is required for viability of *dam* mutants (Peterson et al., 1985); *ruv*$^+$ may be required for double-strand break repair (see *dam* below).

7. ssb

The *ssb* gene encodes a single-stranded DNA-binding (SSB) protein required for DNA replication and repair (for reviews, see Chase and Williams, 1986; Cox and Lehman, 1987). Two mutations, *ssb-1* and *ssb-113* (formerly *lexC113*), have been extensively characterized. *ssb-113* mutants are phenotypically similar to *lexA3*(Ind⁻) mutants in many respects, including increased sensitivity to radiation and methyl methanesulfonate, dominance over wild-type alleles, extensive DNA degradation following irradiation, failure to induce *recA* following UV irradiation, and recombination proficiency in conjugation or transduction crosses (Baluch et al., 1980; Chase and Williams, 1986; Mount et al., 1972). *ssb-113* enhances the precise excision of Tn5 and Tn10 (Lundblad and Kleckner, 1984) and reduces plasmidic recombination (Kolodner et al., 1985). On the other hand, *ssb-1* reduces recombination in conjugation and P1 transduction (Vinogradskaya et al., 1986; Glassberg et al., 1979), inversion of repeated elements in λ (Ennis et al., 1987), and λ *lacZ1* × λ *lacZ2* crosses (Golub and Low, 1983). Most *ssb-1* phenotypes are expressed only at high temperatures (41 to 42°C), while *ssb-113* effects are expressed at low temperature (32°C).

SSB protein facilitates RecA protein binding to ssDNA, thereby promoting the formation of presynaptic joint molecules, strand exchange, and branch migration, especially past long heterologies (Eigner et al., 1987; Radding, this volume). A role for SSB protein in modulating the protease and recombinational activities of the RecA protein has been suggested (Moreau, 1987). SSB protein may also inhibit the nuclease activity of the RecBCD enzyme and enhance its unwinding activity (MacKay and Linn, 1976; Cox and Lehman, 1987).

SSB may have a direct role in conjugational DNA transfer (see Porter, this volume). Plasmids derepressed for conjugation also show derepression of their *ssb* genes, which complement the *ssb-1* mutation (Golub and Low, 1986). *ssb* genes of these plasmids share extensive DNA homology (Golub and Low, 1986). These genes may synthesize SSB in the merozygotes soon after conjugal DNA is transferred, thereby protecting the donor ssDNA until its complement is synthesized or until it is coated with the RecA protein to initiate recombination.

8. topA

The *topA* gene encodes DNA topoisomerase I. *topA* mutants have high superhelicity of plasmid DNA, increased sensitivity to UV and methyl methanesulfonate, reduced growth rate, and reduced transposition and transcription (for reviews, see Sternglanz et al., 1981; Wang, 1981; Drlica, 1984). Plasmidic recombination is reduced 1,000-fold in *topA* deletion mutants (Fishel and Kolodner, 1984). Topological linkage of DNA molecules during strand transfer in regions lacking free ends is an obvious role for this enzyme during recombination (Cunningham et al., 1981) and is consistent with plasmidic recombination proceeding by a pathway that may involve pairing and strand transfer in the absence of free ssDNA ends (see section IV.E.).

9. *lexA*

The LexA protein indirectly affects recombination by regulating the expression of *recA* and of three genes, *recN*, *recQ*, and *ruv*, specific to the RecF pathway (for review, see Walker, 1984). *lexA*(Ind⁻) mutations make the LexA repressor noncleavable by autodigestion or by RecA* protease (Mount et al., 1972; Little, 1984), and reduce expression of some genes in the *lexA* regulon, whereas *lexA*(Def) mutations derepress these genes (Mount, 1977).

10. *polA* and *lig*

The *polA* and *lig* genes encode DNA polymerase I and DNA ligase, enzymes essential for DNA replication and repair. The essential function of *polA* for cell survival seems to be the $5' \rightarrow 3'$ exonuclease activity or its coordination with the polymerase activity; mutants in which the polymerase activity is reduced to about 1% of the wild-type level are viable (for review and references, see Kornberg, 1980). Curiously, mutations in *polA* or *lig* decrease the yield of recombinants in conjugation and the λ-inversion assay but increase it in the Konrad assay (Zieg et al., 1978; Ennis et al., 1987). These results have been interpreted to imply that these two enzymes can influence recombination at different steps in different systems (Smith, 1988).

11. *gyrA* and *gyrB*

DNA gyrase is encoded by the *gyrA* and *gyrB* genes. Inhibitors of this enzyme reduce recombination proficiency in two systems. Coumermycin, an inhibitor of the GyrB subunit, reduces fivefold the *recBC*-independent recombination between direct duplications in UV-irradiated lambda phage infecting homoimmune lysogens (Hays and Boehmer, 1978). Nalidixic acid, an inhibitor of the GyrA subunit, and coumermycin inhibit recombination between inverted duplications on the λ genome during its vegetative growth (Ennis et al., 1987). The possibility that DNA gyrase is indirectly involved as a result of induction of the *lexA* regulon (Little and Mount, 1982) seems unlikely because *lexA*(Def) and *lexA*(Ind⁻) mutations do not significantly alter either the inversion frequency or its inhibition by coumermycin (Ennis et al., 1987).

12. *rep*

Rep protein plays a nonessential but important role in replication of the *E. coli* chromosome. However, it is essential for replication of certain single-stranded phages (φX174, M13, etc.). *rep* mutants are viable but have a larger cell size, a faster-sedimenting nucleoid body containing more DNA, and a larger number of replication forks which move slowly (for a review, see Kornberg, 1980). The only reported recombination effects of *rep* mutations are a two- to threefold decrease in the yield of recombinants in conjugation and in the Konrad assay (Zieg et al., 1978).

13. *dam* and *uvrD*

dam and *uvrD* mutants lack DNA adenine methylase and helicase II, respectively, and are hyper-Rec for several systems, including conjugation (Zieg et al., 1978). The simplest explanation of this hyper-Rec character is that the wild-type products of these genes aid repair of potentially recombinogenic lesions, thereby removing substrates for recombination, and repair of mismatches in heteroduplex DNA, thereby eliminating potential recombinants (for reviews and references, see Claverys and Lacks, 1986; Radman, this volume).

dam mutants have phenotypes in addition to their hyper-Rec character. These include increased sensitivity to UV and 2-aminopurine, increased spontaneous mutability, elevated expression of several SOS genes (including *recA* and *lexA*), increased precise excision of Tn*5* and Tn*10*, increased single-strand breaks in DNA, and nonviability in combination with mutations of *recA*, *recB*, *recC*, *ruv*, *polA*, or *lexA*(Ind⁻) (for references, see Claverys and Lacks, 1986). But *dam recBC sbcA* and *dam recBC sbcB(C)* cells are viable (McGraw and Marinus, 1980). The nonviability of *dam recA* and *dam recB* cells is correlated with the inability of *recA* and *recB* mutants to repair double-strand breaks (Wang and Smith, 1987). These results are consistent with *dam* mutants accumulating single-strand breaks which, if unrepaired by the RecBCD, RecE, or RecF pathway, lead to double-strand breaks and cell death.

uvrD mutants have been isolated in other contexts and designated *uvrE*, *mutU*, or *recL* (Siegel, 1981).

The hyper-Rec character of *uvrD* mutants extends to intragenic recombination in F′ × F⁻ and Hfr × F⁻ crosses (Arthur and Lloyd, 1980; Lloyd, 1983; Feinstein and Low, 1986). In conjugation crosses, *uvrD* decreases linkage between well-separated markers (Feinstein and Low, 1986). Lloyd (1983) observed that the *lexA3*(Ind⁻) mutation abolishes the *uvrD*-induced hyper-Rec phenotype which seems to be controlled by an SOS gene other than *recA*. Howard-Flanders and Bardwell (1981) found a *uvrD* mutant to be hyper-Rec both for spontaneous and UV-induced recombination in λ phage × prophage crosses. These results are consistent with the role of helicase II (*uvrD* gene product; Table 1) in mismatch repair (Radman, this volume) and in double-strand break repair (Sargentini and Smith, 1987). Some *uvrD* mutations also enhance excision of Tn*5* and Tn*10* from the chromosome (Lundblad and Kleckner, 1984).

14. *dcm*

The *dcm* gene encodes DNA cytosine methylase and may be required for very-short-patch repair of mismatches in heteroduplex DNA (Lieb, 1987; Radman, this volume). No other recombination effects of this gene have been reported.

15. *mutH*, *mutL*, and *mutS*

mutH, *mutL*, and *mutS* are hyper-Rec (two- to sixfold increase) for intragenic recombination in conjugation (Feinstein and Low, 1986). Their hyper-Rec character presumably stems from their failure to repair mismatches in heteroduplex DNA; repair would eliminate potential recombinants, as for *dam* and *uvrD* mutants. All of these genes are involved in methyl-directed (long-patch) mismatch repair (see Claverys and Lacks, 1986; Radman, this volume) and enhance transposon excision (Lundblad and Kleckner, 1984). *mutL*[+] and *mutS*[+] are also required for full levels of very-short-patch mismatch repair (Lieb, 1987).

16. *dut*, *xth*, *xseA/xseB*, and *rdgB*

dut, *xth*, *xseA*, and *rdgB* mutants are hyper-Rec in the Konrad system but are not significantly affected in conjugational recombination. The hyper-Rec character (3- to 12-fold increase) of *dut* (also called *dnaS* or *sof*) may be due to increased incorporation of uracil into DNA and the subsequent transient gaps created during excision repair involving *N*-uracil glycosylase and ExoIII (Hochhauser and Weiss, 1978; Taylor and Weiss, 1982). Similarly, *xth* mutation may increase recombinogenic lesions due to a failure to efficiently repair uracil-containing DNA. *xth dut* double mutants are nonviable at high temperature. Mutations in *xseA*, but not in *xseB*, which encode subunits of ExoVII, are hyper-Rec in the Konrad assay (Chase and Richardson, 1977; Vales et al., 1983). *rdgB*[+] is required for growth and viability of *recA* mutants (Clyman and Cunningham, 1987). *rdgB* mutations block DNA synthesis and lead to extensive degradation of DNA in *recA200* cells but not in *recA*[+]. Their hyper-Rec character in the Konrad assay may be due to partial induction of the SOS regulon (Clyman and Cunningham, 1987).

III. MULTIPLE PATHWAYS OF RECOMBINATION IN *E. COLI* CONJUGATION

A. Original Evidence for Three Distinct Genetic Pathways of Recombination

The idea that general recombination in *E. coli* might proceed by more than one pathway was developed to explain some unusual characteristics of mutations that influenced the proficiency of conjugational recombination. *recA* mutations abolished the formation of viable recombinants in Hfr × F[−] crosses (Clark and Margulies, 1965; Low, 1968), whereas *recB* and *recC* mutations, including nonsense and double *recBC* mutations, permitted the formation of some recombinants (Willets and Mount, 1969; Willetts and Clark, 1969).

An ATP-dependent exonuclease, apparently the product of the *recBC* genes, was present in wild-type cells but missing from *recB* and *recC* mutants (Buttin and Wright, 1968; Oishi, 1969; Barbour and Clark, 1970; Tomizawa and Ogawa, 1972). With a view to establishing the correlation between this nuclease deficiency and the Rec[−] phenotype of the mutants, Clark and co-workers isolated three kinds of recombination-proficient (Rec[+]) revertants of *recBC* mutants (Barbour et al., 1970). One kind arose only in *recB* or *recC* single mutants, mapped to the *recBC* region of the chromosome, and restored the ATP-dependent nuclease activity.

These were, presumably, true revertants, or had acquired intragenic suppressors of the original mutations.

The other two kinds of Rec$^+$ revertants were still deficient in the ATP-dependent nuclease activity but had acquired unlinked suppressor mutations, named *sbcA* and *sbcB* [now *sbcB(C)*; see section II.D.4.]. The *recBC sbcA* cells had an ATP-independent exonuclease (ExoVIII) not present in the parental cells (Kushner et al., 1974; Joseph and Kolodner, 1983a, b; Table 1). The *sbcA* mutations arose frequently in most *E. coli recBC* strains but not in *recBC* derivatives of strain AB1157. *sbcB(C)* suppressors were recovered only from *recBC* derivatives of strain AB1157 and were associated with disappearance of exonuclease I from the cells (Kushner et al., 1971, 1972).

Clark (1971, 1973, 1974) hypothesized that three different pathways were responsible for recombination in *rec$^+$* (ExoV$^+$ ExoI$^+$ ExoVIII$^-$), *recBC sbcA* (ExoV$^-$ ExoI$^+$ ExoVIII$^+$), and *recBC sbcB(C)* (ExoV$^-$ ExoI$^-$ ExoVIII$^-$) genetic backgrounds. He named these the RecBC (now RecBCD; see below), RecE, and RecF pathways, respectively. *recE* denoted the gene encoding ExoVIII, and *recF* denoted a hypothetical gene required for recombination in *recBC sbcB(C)* mutants. Later, when several genes required for recombination in *recBC sbcB(C)* mutants were identified (Horii and Clark, 1973), the first one was named *recF*. It was also hypothesized that the residual recombination in *recBC (sbcB$^+$)* mutants utilized the RecF pathway, since the RecBC and RecE pathways were not expected to be available in the absence of ExoV and ExoVIII, and transfer of a *recF* mutation to the *recBC* cells further lowered the proficiency of recombination (Horii and Clark, 1973).

In support of the above hypothesis of three distinct pathways of recombination, several genes whose function is required for recombination by only the RecE and/or RecF pathways were subsequently identified (section II.). These are *recF* (Horii and Clark, 1973), *recJ* (Lovett and Clark, 1984), *recN* (Lloyd et al., 1983), *recO* (Kolodner et al., 1985), *recQ* (Nakayama et al., 1984), and *ruv* (Lloyd et al., 1984) for the RecF pathway; and *recE* (Gillen et al., 1981; Fouts et al., 1983), *recF* (Gillen et al., 1981), *recJ* (Lovett and Clark, 1984), *recO* (Kolodner et al., 1985), and probably *ruv* (Lloyd et al., 1987a; S. T. Lovett and A. J. Clark, unpublished data) for the RecE pathway. In addition, the *lexA3*(Ind$^-$) mutation (Mount et al., 1972) has little effect on recombination in *rec$^+$* cells but drastically reduces recombination in *recBC sbcB(C)* mutants, indicating that recombinant formation in the latter requires some SOS inducible genes (Lovett and Clark, 1983; Lloyd et al., 1987b; S. K. Mahajan and A. R. Datta, unpublished data). *recN*, *recQ*, and *ruv* are three of these inducible genes (Lloyd et al., 1983; Irino et al., 1986; Shurvinton and Lloyd, 1982; Picksley et al., 1984b). *lexA3*(Ind$^-$) also moderately reduces recombination by the RecE pathway (Lloyd et al., 1987a; N. N. Pandit and S. K. Mahajan, unpublished data). On the other hand, none of the above mutations, viz., *recE*, *recF*, *recJ*, *recN*, *recO*, *recQ*, *ruv*, and *lexA3*, modify the recombinational proficiency of wild-type (*rec$^+$*) cells by more than a factor of two or three (Horii and Clark, 1973; Lloyd et al., 1983, 1984; Lovett and Clark, 1984; Kolodner et al., 1985; Nakayama et al., 1984; Lloyd et al., 1987a). Figure 2 summarizes the effect of these mutations in different backgrounds. Other genes (in addition to

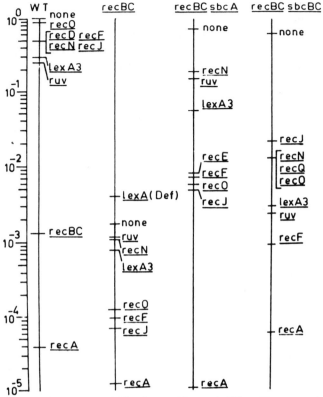

FIGURE 2. Effect of mutations on the yields of recombinants in Hfr × F⁻ crosses in wild-type cells (RecBCD pathway), *recBC* mutants (RecX pathway), *recBC sbcA* mutants (RecE pathway), and *recBC sbcB(C)* mutants (RecF pathway). None means no mutations other than those listed at the head of the column. The data are from several sources cited in the text (sections II. and III.).

recA) that may be required by one or more pathways include *ssb*, *polA*, *lig*, *topA*, *gyrA*, and *gyrB* (section II.). However, mutations in these genes have not been tested in all *recBC sbc* backgrounds.

B. Recent Developments: Evidence for Two More Pathways

Though the basic features of the hypothesis of multiple genetic pathways of recombination have stood the test of time, recent findings have necessitated some modifications.

(i) The name of the main pathway of recombination in wild-type cells has been changed from RecBC to RecBCD pathway (see Smith, 1988), as a result of the discovery of an additional subunit, RecD, of the RecBC enzyme (Amundsen et al., 1986). Mutations in the *recD* gene abolish the RecBCD pathway and activate a different pathway (see below).

(ii) The assumption that the residual recombination in *recBC* cells uses the RecF pathway needs revision in view of the following. First, at least three mutations, namely, *recN*, *recQ*, and *ruv*, which strongly reduce recombinational proficiency in *recBC sbcB(C)* cells have little effect in *recBC* cells (Lloyd et al., 1987a). Second, the viable recombinants produced in *recBC* and *recBC sbcB(C)* cells appear to arise by different mechanisms. For instance, in conjugation, *recBC sbcB(C)* recipients produce a higher frequency of nearby double exchanges than do *recBC* recipients (Mahajan and Datta, 1977; section III.D.2.). In specialized transduction the ratio of addition-type transductants (in which the transducing vector is integrated into the chromosome along with the selected bacterial gene) to substitution-type transductants (in which bacterial genes are exchanged between the chromosome and the transducing particle; see Porter, this volume) is significantly lower for *recBC sbcB(C)* than for *recBC* recipients (Wackernagel and Radding, 1974; Porter et al., 1982; Porter, 1983; see section IV.B.). In P1-mediated generalized transduction, *recBC* cells show greater marker discrimination than their *recBC sbcB(C)* counterparts (Masters et al., 1984; section IV.A.).

I suggest a new name, RecX, for the pathway operating in *recBC* cells. One would like to follow the practice of naming each recombinational pathway after a gene specific to it, but all the genes presently known to be required by this pathway (e.g., *recF*, *recJ*, *recO*, and *recA*) are also required by the RecE and RecF pathways. Indeed, according to this practice, even the RecF pathway needs renaming, as the *recF* gene product is required also by the RecE and RecX pathways. RecN might be a more appropriate name for the RecF pathway, but to avoid confusion in reading the past literature I shall continue to use RecF.

(iii) *recD* recipients produce viable recombinants at high frequency (about 50 to 300% of the wild type) (Chaudhury and Smith, 1984a; Biek and Cohen, 1986). This was interpreted to imply that the ATP-dependent dsDNA nuclease activity absent in these mutants is not required by the RecBCD pathway and that the essential recombinational activity of the RecBCD enzyme might be its unwinding activity (Chaudhury and Smith, 1984a). However, *recD* mutants seem to lack the unwinding activity as well (A. F. Taylor, quoted in Smith, 1988). A simpler explanation appears to be that the *recD* cells use a new recombinational pathway distinct from the RecBCD pathway of wild-type cells. In this interpretation, the wild-type RecBCD pathway uses the nuclease and unwinding activities, but the pathway operating in *recD* cells is independent of these activities.

Several lines of evidence seem to support this interpretation. First, the proficiency of conjugational recombination in *recD* recipients is depressed by *recJ* mutation (Lloyd et al., 1988; S. Lovett, C. Luisi-DeLuca, and R. Kolodner, *Genetics*, in press) which has no effect on the wild-type RecBCD pathway (Horii and Clark, 1973; Lovett and Clark, 1984). Second, *recD* mutants produce more exchanges between close markers than the wild type in biparental λ (Chi-free) vegetative crosses (Chaudhury and Smith, 1984a) and in conjugation crosses (S. H. Mangoli and S. K. Mahajan, unpublished data). Third, in λ vegetative crosses Chi sites stimulate recombination in wild-type hosts but not in *recD* mutants (Chaudhury and Smith, 1984a). Fourth, plasmids behave differently in *recD* mutants than in wild-type hosts; these behaviors include generation of linear multimers (Cohen and Clark, 1986), defective maintenance (Biek and Cohen, 1986), and increased copy number (Seelke et al., 1987) in *recD* mutants.

recD mutants are recombination proficient in Δ *recE recF143 sbcB⁺* backgrounds (Chaudhury and Smith, 1984a; Cohen and Clark, 1986; Ennis et al., 1987), suggesting that their recombination pathway is distinct from the RecE, RecF, and RecX pathways. I suggest the name RecY for this pathway.

The RecY recombination pathway may be a mixed pathway. It shares the *recJ* gene product with the RecE, RecF, and RecX pathways, perhaps for an early step. It shares the *recB* and *recC* gene products (perhaps the β subunit of RecBCD enzyme [Lieberman and Oishi, 1974; Amundsen et al., 1986; Lovett et al., in press]) with the RecBCD pathway, perhaps for a late step (see sections III.C. and III.D.1. below). The RecY pathway may also function in *recC*‡ mutants, which are phenotypically indistinguishable from *recD* mutants (Chaudhury and Smith, 1984a), and in *recC** and *recBC* (TexA) mutants (Schultz et al., 1983; Lundblad et al., 1984). The abolition of the hyper-Rec nature of the Chi-free λ recombination in *recD* cells in *recD⁺/recD* merodiploids (Chaudhury and Smith, 1984a) suggests that the RecY pathway is inhibited by the RecBCD enzyme (section IV.D.2.).

C. Objections to the Concept of Multiple Pathways

The genetically defined pathways described above could reflect pathways with significantly different DNA structures and steps in their interconversion. If so, what are these differences, and are they quantitative or qualitative? It is conceivable that the different pathways utilize the same DNA structures and interactions, i.e., proceed by a single pathway, and that the differential formation of viable recombinants in different mutants reflects differences in one or more of the following: (i) *prerecombinational steps*, such as the regulation of the concentrations of the substrates for recombination, whose nature is the same for all pathways; (ii) *levels of the catalysts or facilitators of recombination*, such as RecA or SSB proteins; and (iii) *postrecombinational steps*, such as the viability of merozygotes that have experienced recombination. For example, in certain mutants recombination might generate a replication inhibition signal or derepress a "killer" gene.

Physical analysis of the recombinational structures formed in different genetic backgrounds would appear to be the only way to decide whether the pathways differ at the level of DNA interactions. Very little work of this nature has been done in *E. coli* conjugation, though there is some evidence from other systems discussed in the next section. In conjugation the available evidence, based on phenotypic analysis of the recombinants, is primarily indirect, fragmentary, and complex, as a result of the cellular milieu and alternative DNA metabolism with which the recombining molecules have to contend. Below I review this evidence.

D. Additional Evidence for Multiple Pathways of Conjugational Recombinant Formation

1. Formation of Transcribable Recombinant DNA

Birge and Low (1974) studied the effects of *recA*, *recB*, and *recC* on recombination with an assay that monitored the production of a recombinant gene by its expression, even if the merozygotes failed to complete recombination and produce viable progeny. The assay involved mating Hfr and F⁻ cells harboring nonidentical and noncomplementing *lacZ* mutations and measuring the level of functional β-galactosidase (β-gal) produced. *recA* recipients produced no detectable β-gal. *recB21* recipients, which produced *lacZ⁺* recombinant clones at less than 1% of the wild-type level, produced β-gal at more than half of the wild-type level. They concluded that the RecBCD enzyme either acts at a late step in recombination (after the two *lacZ* alleles have interacted to produce a *lacZ⁺* gene) or is not at all involved in recombination but is required only for postrecombinational viability of the merozygotes (Capaldo-Kimball and Barbour, 1971; see section III.C. above). Lloyd and his co-workers have recently extended this assay to other genetic backgrounds and to F' *lacZ1* × F⁻ *lacZ2* crosses (Lloyd and Thomas, 1984; Lloyd et al., 1987b). *recF*, *recJ*, or *recO* recipients, or those with any combination of these mutations, produced β-gal at about the same level as *recB* recipients. However, introduction of *recB* into the *recF*, *recJ*, or *recO* recipients reduced β-gal production by a factor of 10 to 25. Hence, in these backgrounds the RecBCD enzyme may promote an early recombinational step, in addition to a late step (to explain the low recovery of the Lac⁺ clones in *recB* strains). (For an alternative interpretation, see Smith [1988].) Porter et al. (1978) also inferred an early recombinational role for the RecBCD enzyme in λ *lacZ1* × F' *lacZ2* specialized transduction crosses (section IV.B.).

Lloyd et al. (1987b) concluded that the RecBCD enzyme and a combination of the products of *recF*, *recJ*, and *recO* act in two alternative mechanisms for initiating conjugal recombination. I argue later for this interpretation.

2. Densities and Distribution of Recombinational Exchanges

Mahajan and Datta (1977) used another assay for recombination that is probably independent of differential viability of the recombinants in different genetic backgrounds. They determined the density and distribution of exchanges in different recipient strains among selected recombinant clones (i.e., among viable subpopulations) following conjugation. They measured a parameter, R1, defined as the frequency of recombination in a chromosomal region, 1, bounded by the selected donor marker a^+ and an origin-proximal nearby unselected marker *b* (see Fig. 3 and Table 2; S. K. Mahajan, Ph.D. thesis, University of Pennsylvania, Philadelphia, 1971). R1 was higher in *recBC* recipients than in *rec⁺* recipients for both intergenic and intragenic markers. They inferred that the *recBC* mutations increased the mean number of recombinational heteroduplex regions per unit length. The *recBC*-induced increase in R1 was not decreased by additional *sbcB(C)* mutations (Mahajan and Datta, 1977) and was decreased only slightly by

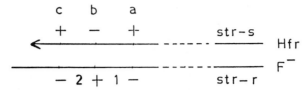

FIGURE 3. Standard three-factor conjugation cross. The arrowhead indicates the origin and direction of transfer of DNA from the Hfr donor to the F⁻ recipient; 1 is the genetic interval between the distal marker *a* and the linked marker *b*, while 2 is the interval between markers *b* and *c*. R = a⁺ Str^r recombinants per donor

$$R1 = \frac{a^+ \quad b^+ \quad Str^r}{a^+ \quad Str^r}$$

$$R2(1) = \frac{a^+ \quad b^+ \quad c^+ \quad Str^r}{a^+ \quad b^+ \quad Str^r}$$

sbcA mutations (S. K. Mahajan, A. R. Datta, and N. N. Pandit, unpublished data), both of which reverse *recBC*-induced nonviability (Capaldo-Kimball and Barbour, 1971; Masters et al., 1984) as well as recombination deficiency. This indicated that selective high viability of the recombinant subpopulation with high R1 was unlikely to be responsible for the observed increase in R1. *recD* mutants also give higher R1 values (Mangoli and Mahajan, unpublished data).

The higher R1 values in *recBC*, *recD*, *recBC sbcA*, and *recBC sbcB(C)* mutants than in *rec⁺* cells support the view that different pathways operate in the multiple mutants (RecX, RecY, RecE, and RecF, respectively). These alternate pathways appear to make little, if any, contribution in *recBCD⁺* cells. R1 is comparable in *rec⁺*, *recBC⁺ sbcB(C)*, and *recBC⁺ sbcA* strains (Mahajan and Datta, 1979; Mahajan et al., unpublished data). Similarly, the *lexA3*(Ind⁻) mutation has little effect in *recBCD⁺* cells but markedly reduces recombinant yields (R) in *recBC sbcB(C)* mutants (Table 2). The hyper-Rec character of *recD*

TABLE 2
Density and distribution of exchanges in *E. coli* Hfr × F⁻ crosses[a]

Recombination mutations in recipient cells	Pathway available	Relative recombination frequencies[b]		
		R	R1	R2(1)
recF143	RecBCD	1	1	1
recB21 recC22	RecX	0.01	4.5	1
recB21 sbcA8	RecE	0.3	3.2	1
recB21 recC22 sbcB15 sbcC201	RecF	0.4	4.5	2.2
lexA3(Ind⁻)	RecBCD	0.5	1	1
recB21 recC22 sbcB15 sbcC201 lexA3(Ind⁻)	RecX (?)	0.006	3.8	1

[a] All values are normalized relative to RecBCD pathway. The data are representative and are adapted from Mahajan and Datta (1977) and from unpublished results of Mahajan, Datta, and Pandit.
[b] See Fig. 3 and the text for definitions of R, R1, and R2(1).

mutants is absent in merodiploids producing wild-type RecBCD enzyme (Chaudhury and Smith, 1984a). *recE* (Gillen et al., 1981), *recF* (Horii and Clark, 1973), *recJ* (Lovett and Clark, 1984), and *recO* (Kolodner et al., 1985) have little effect on recombinant yields in $recBCD^+$ backgrounds. These results suggest that RecBCD enzyme reduces or abolishes the activities of the alternate pathways. In terms of the proposal of Lloyd et al. (1987b) that in the absence of RecBCD enzyme recombination is initiated by a mechanism dependent on RecF, RecJ, and RecO proteins, these results indicate that the latter mechanism(s) is inhibited by the wild-type RecBCD enzyme.

3. Chromosome Mobilization

F′ donor cells transfer the episomal DNA to F⁻ recipients with high efficiency and donor chromosome markers with much lower efficiency (Adelberg and Burns, 1960). Wilkins (1969) studied the mobilization of the chromosomal DNA by an F′ *lac* episome in two situations: when the F′ *lac* had been inside the donor cell for several generations, and when it had been transferred to the donor (from another donor) immediately before the (second) mating. He observed that the frequency of chromosome transfer from newly formed F′ *lac* cells harboring a *recB21* or *recC22* mutation was more than 50% of the wild-type level. The frequency was 10 to 20% of the wild-type level when well-established *recB* or *recC* F′ *lac* was used as a donor and 10^{-4}% if the host was *recA*. If it is assumed that covalent bond formation between single strands of homologous DNAs is a prerequisite for mobilization (Scaife and Gross, 1963), then *recB* or *recC* mutants can perform this act but the *recA* cannot. This would imply that either the RecBCD enzyme has no role in the (single-strand) recombination between the F′ and the chromosome or that a RecBCD-independent pathway exists in addition to the RecBCD-dependent pathway. The latter interpretation seems preferable because it can explain the difference between the RecBCD-independent mobilization frequencies of the freshly formed and old F′ *lac* cells if it is further postulated that the ssDNA is a better substrate of the RecBCD-independent pathway (section V.).

Picksley et al. (1984a) examined the F′ mobilization of chromosomal DNA from a *recBC sbcB(C)* background. They observed that these cells were about 60% as proficient as wild-type cells in the mobilization mediated by a well-established F′. Introduction of *recF143* mutation into this background reduced these mobilization frequencies by more than 30-fold, while *recN* mutations caused less than 2-fold reduction. These results would indicate that the steps leading to the covalent bonding between homologous single strands required for mobilization may use mostly RecX pathway functions and that up to this stage the RecF pathway specific gene, *recN*, has no role.

4. Kinetics of Recombination and Segregation

Several workers have compared the kinetics of recombination by the RecBCD and RecF pathways. Mahajan and Datta (1979) observed that no multiplication of the recombinants occurs for 4 to 5 h after mating in a *recBC sbcB(C)* recipient, while in a *rec*⁺ recipient there is no noticeable delay. They proposed that the delay reflected slow kinetics of heteroduplex formation in the RecF pathway. More definitive experiments using *recA*(Ts) mutants supported this interpretation. Completion of recombination by the RecBCD pathway requires functional RecA for a much shorter time than that by the RecF pathway; i.e., the RecA-mediated strand exchange takes longer in the RecF pathway than in the RecBCD pathway (Bresler et al., 1981). Recombination by the RecF pathway could be accelerated by increasing the intracellular RecA protein concentration by a *recAoc* mutation (Lloyd and Thomas, 1983).

5. Circular Intermediates in the *recBC sbcB(C)* Genetic Background

Bresler et al. (1981) have reported the presence of small circular DNA from exconjugants of an Hfr cross with a *recBC sbcB(C)* recipient. No such circles were seen with a *recBC*⁺ recipient.

6. Involvement of *lexA* Regulon Genes in the RecF Pathway

Three *lexA*-controlled genes (*recN*, *recQ*, and *ruv*) are required by the RecF but not the other pathways (sections II. and III.A. above). Of these, *recN* is essential for double-strand break repair (Picksley et al., 1984a) while *recQ*⁺ is required for thymineless death (Nakayama et al., 1984). Though there is no direct evidence that these genes are induced in *recBC sbcBC* merozygotes, even the basal level of *recN* gene expression is much higher in a *recBC sbcBC* background than in *recBC* or *rec*⁺. Apparently some proteins of the RecF pathway may act on specialized DNA structures created under different stress conditions; some of these structures may be intermediates of the RecF pathway of recombination.

7. Effect of Chi on Exchange Distribution

In conjugation between Hfr Gal⁻ Bio⁺ λ lysogens and F⁻ Gal⁺ Bio⁻ λ lysogens, a Chi sequence (Table 1) present on the donor λ prophage alters the distribution of exchanges in the prophage in *recB*⁺*C*⁺*D*⁺ recipients but not in isogenic *recBC sbcA* recipients (Dower and Stahl, 1981).

8. Differential Roles of SSB Protein in the RecBCD and RecF Pathways

Vinogradskaya et al. (1986) found that *ssb-1* affects recombination by the RecBCD and RecF pathways. The F factor, which encodes an ssDNA-binding protein (SSB-F), could complement this mutation to restore recombination by the RecBCD but not the RecF pathway, suggesting that SSB protein had a different, or an additional, role in recombination by the RecF pathway compared with the RecBCD pathway.

IV. MULTIPLE PATHWAYS IN OTHER *E. COLI* SYSTEMS

Mutations affecting conjugational recombination often affect recombination in other *E. coli* systems. The results reviewed below suggest that the efficiencies of different pathways differ in these systems.

A. Generalized Transformation and Transduction

In generalized transformation and transduction, small fragments of donor dsDNA are introduced into the recipients (see Porter, this volume). However, the yields of viable recombinants in the two cases differ in different mutants (Hoekstra et al., 1980; Margolin, 1987; Clark and Low, 1988). The main difference is that very few, if any, transformants are recovered in $recBCD^+$ cells, perhaps because the RecBCD enzyme readily degrades the transforming but not the transducing DNA molecules. The latter seem to be protected by proteins that can circularize the majority of the DNA molecules (Sandri and Berger, 1980). Introduction of *sbcA* or *sbcB* mutations into the *recBC* background increases recombinant yields in both transformation and transduction, indicating that recombination in these two systems may be mediated by the same genetic pathways that operate in conjugation. Though the effects of other mutations of the RecX, RecE, and RecF pathways on transformation have not been examined, in transduction they have qualitatively similar effects on the overall yields, as in conjugation (Zieg and Kushner, 1977; Young and Edlin, 1983; Lloyd et al., 1987a). *recBC* mutants, however, are much less hypo-Rec in P1 transduction (5 to 20 times) than in conjugation (50 to 1,000 times) (Emmerson and Howard-Flanders, 1967; Lloyd et al., 1987a).

Cosloy (1982) analyzed exchange distribution in P1 transduction in different recipients and concluded that *recBC sbcB(C)* strains integrated a larger number and longer pieces of donor DNA than either wild-type or *sbcB(C)* ($recBC^+$) strains. Hoekstra et al. (1980) obtained similar results for generalized transformation.

Both P1- and T4-mediated generalized transduction show marker-dependent variation in the relative yields of transductants (Masters, 1977; Young and Edlin, 1983). Masters et al. (1984) observed that the extent of the marker discrimination in P1 transduction is less in *recBC sbcB(C)* strains than in the wild type. The marker discrimination in wild-type cells is reduced by UV irradiation of the transducing phages or the recipient bacteria and to some extent by increasing the level of RecA protein in the recipients. All of these circumstances are likely to increase the contribution of the RecF pathway (Smith and Hays, 1985; Lloyd et al., 1983; Lloyd and Thomas, 1983). The *recB* mutation by itself failed to reduce the marker discrimination; this is further evidence that the RecX pathway (in *recBC* mutants) differs from the RecF pathway [in *recBC sbcB(C)* mutants].

Dower and Stahl (1981) showed that Chi is active in P1 transduction by the RecBCD pathway but not by the RecE pathway.

TABLE 3
Recombination in specialized transduction[a]

Cross	Recombination mutations in recipient cells	Relative β-gal activity[b]	Relative frequencies of Lac$^+$ recombinants[b]	% Addition type
λ *lacZ1* × F$^-$ *lacZ2*	*rec$^+$*	1	100	—
	recB	1	7	—
	recB sbcA	7	520	—
	recB sbcB(C)	20	100	—
λ *lacZ1* × F′ *lacZ2*	*rec$^+$*	1	100	56
	recB	0.04	<1	30
	recB sbcA	0.4	20	30
	recB sbcB(C)	2.2	22	13

[a] The data are adapted from Porter (1983).
[b] β-Gal activities and the frequencies of Lac$^+$ recombinants in each type of cross are separately normalized with respect to the *rec$^+$* recipients. The β-gal activity and the yield of Lac$^+$ recombinants in *rec$^+$* cells were nearly 50-fold greater in the second cross than in the first.

B. Specialized Transformation and Transduction

There are both qualitative and quantitative differences among λ- or φ80-mediated specialized transformants and transductants formed in different genetic backgrounds. Wackernagel and Radding (1974) and Basu and Oishi (1975) reported that more addition-type than substitution-type recombinants (for definitions, see section III.B. and Porter, this volume) were formed in wild-type strains than in *recBC sbcB(C)* strains.

Porter and co-workers (Porter et al., 1978, 1982; Porter, 1983; Seifert and Porter, 1984) have studied recombination between two *lacZ* alleles, one located on a specialized transducing λ *lacZ1* phage and the other on a second transducing λ *lacZ2* phage, the bacterial chromosome (F$^-$ *lacZ2*), an F′ *lacZ2* plasmid, or a P1 *dlacZ2* plasmid. In all cases, replication and expression of the infecting phage were blocked by a homoimmune λ prophage in the recipient. The production of β-gal and viable Lac$^+$ recombinants was monitored. The ability of different pathways to effect recombination between the two *lacZ* alleles was greatly influenced by the nature of the DNA molecules on which these alleles were located. For a given pair of molecules the yield and the nature of the recombinants differed in different pathways (Table 3). The following points are noteworthy.

(i) λ *lacZ1* recombined ≃50-fold more efficiently with F′ *lacZ2* than with F$^-$ *lacZ2*, whether β-gal or LacZ$^+$ recombinants were measured.

(ii) A *recB* mutation did not depress β-gal production in the λ *lacZ1* × F$^-$ *lacZ2* cross but strongly did so in the λ *lacZ1* × F′ *lacZ2* cross, suggesting an early recombinational role for the RecBCD enzyme in the latter cross but not in the former (section III.D.1.). However, *recB* reduced the number of Lac$^+$ recombinants in both cases, suggesting that RecBCD enzyme also had a late role in both crosses.

(iii) In crosses (called "conjuduction" by these workers) in which λ *lacZ1* recombined with F′ *lacZ2* soon after the latter entered F$^-$ Δ*lac* merozygotes, β-gal production was even higher, but the *recB* effect was very small. Perhaps freshly transferred (ss?)DNA can initiate recombination more frequently and in a *recB*$^+$-independent manner (sections III.D.2. and V.).

(iv) The high-level, *recB*$^+$-dependent recombination in λ *lacZ1* × F' *lacZ2* crosses required the elements involved in cutting the F DNA to initiate its conjugal transfer. The elements include the specific sequence *oriT*, present at the origin of transfer, and protein products of the F-factor *traY*$^+$ and *traZ*$^+$ genes, which act at *oriT* before the transfer begins (Willets and Wilkins, 1984; Ippen-Ihler and Minkley, 1986). These results have been interpreted to imply that the RecBCD enzyme enters F' *lac* at the cut *oriT* site to initiate recombination (for a discussion, see Smith, 1988).

(v) The presence of *sbcA* or *sbcB* mutations in *recB* cells enhanced the production of β-gal as well as viable Lac$^+$ recombinants in both λ *lacZ1* × F' *lacZ2* and λ *lacZ1* × F$^-$ *lacZ2* crosses.

(vi) In λ *lacZ1* × F' *lacZ2* crosses the majority (56%) of the RecBCD pathway *lac*$^+$ recombinants were of the addition type, but the RecX, RecE, and RecF pathways gave only 30, 30, and 13% addition-type recombinants, respectively.

(vii) λ *lacZ1* × λ *lacZ1* and λ *lacZ1* × P1 d*lacZ1* crosses gave results broadly similar to those of λ *lacZ1* × F$^-$ *lacZ2* crosses.

The above results permit two significant inferences. First, the RecBCD pathway efficiently functions only when the RecBCD enzyme can enter an end on one of the interacting duplexes. Second, the *recB*$^+$-independent pathways, particularly the RecX pathway, can use ssDNA or gapped DNA much more efficiently than dsDNA (sections III.D.2. and V.).

C. Repair of Damaged DNA

The relation of recombination and the repair of damaged DNA has been extensively studied by many groups (Rupp et al., 1971; Clark and Volkert, 1978; Hanawalt et al., 1979; Wang and Smith, 1983, 1986; Smith and Sharma, 1987). There is a strong correlation between the levels of recombination proficiency and UV resistance in various mutants. Thus, mutations in *recA*, *recB*, *recC*, *recE*, *recF*, *recJ*, *recN*, *recO*, *recQ*, *ruv*, *ssb*, and *lexA*(Ind$^-$) reduce UV resistance in the genetic backgrounds where they reduce recombination proficiency (Clark and Margulies, 1965; Howard-Flanders and Theriot, 1966; Horii and Clark, 1973; Glassberg et al., 1979; Hanawalt et al., 1979; Gillen et al., 1981; Lovett and Clark, 1984; Lloyd et al., 1984; Picksley et al., 1984a; Nakayama et al., 1985; Kolodner et al., 1985), while *sbcA*, *sbcB*, *sbcC*, and *srfA* increase UV resistance in the backgrounds where they increase recombination (Barbour et al., 1970; Kushner et al., 1971; Volkert and Hartke, 1984; Lloyd and Buckman, 1985). There are, however, some exceptions. In a wild-type background, mutations in *recF*, *recO*, and *ruv* have little effect on recombinant yields but reduce UV resistance significantly (Rothman et al., 1975; Kolodner et al., 1985; Shurvinton and Lloyd, 1982). In a *recBC* mutant background, mutations in *xonA* increase UV resistance without restoring recombination (Horii and Clark, 1973; Gopalakrishna and

Bhattacharjee, 1976); *xonA* mutations may be "leaky" *sbcB* alleles (see section II.).

Physical analysis of the molecular intermediates of repair has been carried out for some of the mutations, viz., *recA*, *recB*, *recC*, *recF*, *recJ*, *recN*, and *lexA3*(Ind⁻) (Wang and Smith, 1983, 1986; Picksley et al., 1984a; Smith and Sharma, 1987; Sargentini and Smith, 1987). These studies showed that *recA*⁺, *recB*⁺, and *recF*⁺ are involved in postreplication repair and some aspects of excision repair. Postreplication repair seems to proceed by two major *recA*⁺-dependent pathways. One pathway repairs most of the DNA daughter strand gaps by a *recF*⁺-dependent process, while the other repairs, by a *recB*⁺-dependent mechanism, double-strand breaks produced by the cleavage of unrepaired gaps (Wang and Smith, 1983, 1986). These results are consistent with the proposal that *recF*⁺-dependent recombination involves ssDNA exchanges, while *recBCD*⁺ involves dsDNA exchanges (Mahajan and Datta, 1979; section V.). The recombinational repair pathways may also be involved in *recA*⁺-dependent excision repair of partially replicated chromosomes (Smith and Sharma, 1987).

D. Amplification and Deamplification

Mahajan et al. (1984) studied the effect of different Rec backgrounds on the amplification of chloramphenicol resistance conferred by the transposon Tn9 located on a P1 prophage. They assumed that the level of chloramphenicol resistance reflected the number of copies of Tn9 and that this number changed by tandem amplification. They inferred that the RecBCD, RecE, and RecF pathways effectively promote amplification. Physical analysis of the prophage DNA showed that certain other rearrangements could also increase the P1 prophage copy number and significantly increase the subpopulation with enhanced chloramphenicol resistance. However, all mutations that reduced more than 10-fold the subpopulation resistant to 200 μg of chloramphenicol per ml significantly reduced tandem amplification. By this criterion the RecBCD and RecF pathways are highly proficient in amplification. In general, the mutations which blocked recombination by specific pathways also reduced amplification by those pathways.

A *recB sbcA8* strain seemed to have the two pathways, a *recA*-dependent pathway of high amplification and a *recA*-independent pathway of deamplification. Essentially similar results were obtained with a chromosomal insertion of Tn1721, whose drug (tetracycline) resistance level is a better indicator of tandem amplification (J. F. Sarkari and S. K. Mahajan, unpublished data; N. N. Pandit and S. K. Mahajan, manuscript in preparation). The RecX pathway weakly amplified either Tn9 or Tn1721.

E. Plasmidic Recombination

Plasmidic recombination has been monitored by the interconversion between circular monomers and larger circular oligomers (Bedbrook and Ausubel, 1976; Potter and Dressler, 1976; Kolodner, 1980; Fishel et al., 1981; Cohen and Clark,

1986; Biek and Cohen, 1986) and by the intra- or interplasmidic recombination between two mutant alleles of a gene to generate a wild-type gene (Laban and Cohen, 1981; Fishel et al., 1981; James et al., 1982; Cohen and Laban, 1983; Doherty et al., 1983; Laban et al., 1984; Fishel and Kolodner, 1984; Kolodner et al., 1985). Matfield et al. (1985) studied recombination between directly repeated duplications within a single gene to generate a wild-type gene. While the first type of recombination can be identified only by physical techniques, such as gel electrophoresis and electron microscopy, the latter types permit both genetic and physical analyses.

In wild-type cells most plasmidic recombination seems to proceed by a $recB^+$ $recC^+$-independent but $recA^+$-, ssb^+-, $recF^+$-, $recJ^+$-, and $recO^+$-dependent RecX-like pathway (Hall and Howard-Flanders, 1972; Laban and Cohen, 1981; James et al., 1982; Cohen and Laban, 1983; Kolodner et al., 1985), although the $recA^+$ requirement is not very stringent (especially for intraplasmidic recombination). In a wild-type background $lexA3$, ruv, and $recN$ have no detectable effect on plasmidic recombination (Kolodner et al., 1985), but $topA^+$ function is required (Fishel and Kolodner, 1984). $recAo281$, which produces high levels of RecA protein without inducing the $lexA$ regulon, has no effect, but the $lexA71$::Tn5(Def) mutation gives a several-fold increase (Kolodner et al., 1985). $recB$, $recC$, and $recD$ mutations stimulate plasmidic recombination and alter the distribution of exchanges (James et al., 1982; Matfield et al., 1985; Biek and Cohen, 1986). These mutations, as well as $sbcB(C)$, allow the formation of linear plasmid multimers and greatly reduce the stability of most plasmids (Cohen and Clark, 1986; Biek and Cohen, 1986). This has complicated the study of plasmidic recombination by the RecF pathway. Though oligomer formation may be to some extent $recBC^+$ dependent (James et al., 1982), the RecBCD pathway seems to contribute little to plasmidic recombination. Presumably, this is because the RecBCD pathway needs a linear duplex substrate, which is not present at high levels in $recB^+$ $recC^+$ $recD^+$ cells (Cohen and Clark, 1986).

$recBC$ $sbcA$ mutants promote plasmidic recombination considerably more efficiently than the wild type and in a $recE^+$-dependent but $recA^+$-, $recF^+$-independent manner (Fishel et al., 1981; Laban and Cohen, 1981; Doherty et al., 1983). Intraplasmidic, though not interplasmidic, recombination in $recBC$ mutants is $recF$ independent. $sbcA$ mutations increase intraplasmidic recombination in $recBC^+$ cells, though not as much as in $recBC$ mutants (Cohen and Laban, 1983). It appears that the linear multimers generated in $recBC$ mutants (Cohen and Clark, 1986) can recombine efficiently by a pathway independent of $recF^+$ but stimulated by ExoVIII.

Studies have shown that plasmids which contain intramolecular homologies and are linearized in vitro recombine after transformation (Symington et al., 1985; Abastado et al., 1987). Two important results emerged from these studies. (i) the RecE pathway can promote intramolecular recombination of linear DNA substrates in a $recA^+$-, $recF^+$-independent manner; in fact, mutations in these genes seem to slightly increase recombinant yields. (ii) A dsDNA end can, perhaps directly, recombine with an intramolecular homologous sequence in wild-type cells, though not as efficiently as in $recBC$ $sbcA$ strains. Even in the wild-type background this type of recombination is $recA^+$-, $recF^+$-, and $recJ^+$-independent but three- to fourfold reduced by $recB$ and $recC$ mutations.

The RecA independence of the RecE pathway and the less stringent dependence on RecA of the RecX pathway in plasmidic recombination raises the question of heteroduplex formation in plasmidic recombination. It seems likely that heteroduplex is formed in plasmidic recombination since "figure eight" molecules, presumed heteroduplex-containing intermediates of plasmidic recombination, have been observed in $recA^+$ and $recA$ mutant cells (Potter and Dressler, 1976; West et al., 1983; James et al., 1982; Mizuuchi et al., 1982). Furthermore, physical analysis of recombinant plasmids from $\Delta recE$ strains led Laban et al. (1984) to conclude that formation of these plasmids involved RecA-independent branch migration, though RecA protein could stimulate branch migration and polarize it.

There are two possibilities for heteroduplex formation in $sbcA$ $recA$ mutants. (i) The $sbcA$ mutations may activate, in addition to $recE$, a gene for an as yet unidentified annealing protein akin to the β protein of lambda (Kmiec and Holloman, 1981; Muniyappa and Radding, 1986). (ii) The requirement for RecA protein in other systems, e.g., conjugation, may be in the extension of heteroduplexes (branch migration). In plasmids, branch migration may be short (and nonenzymatic) (Meselson, 1972) or aided by superhelicity (Holloman et al., 1975).

The formation of linear plasmid multimers in $recBC$ mutants has suggested the rolling-circle mode of plasmid replication in these strains (Cohen and Clark, 1986). If true, then multiple recombinational pathways may reflect, at least partially, multiple pathways of replication in different genetic backgrounds, inasmuch as replication may generate substrates for recombination. Thus, the failure of RecBCD enzyme to contribute positively to plasmidic recombination in $sbcA^+$ cells and the suppression by RecBCD enzyme of the RecE pathway in $sbcA$ mutants may reflect the inhibition by this enzyme of rolling-circle replication (Unger and Clark, 1972; Cohen and Clark, 1986) which may generate linear duplex substrates for the RecBCD and RecE pathways.

Indeed, it appears that the circular duplexes present in the $recBCD^+$ cells are probably efficient substrates only for the RecX pathway. The absence of a major contribution by the RecF pathway is indicated by the neutrality of the $recN$, ruv, and $lexA$(Ind$^-$) mutations (Kolodner et al., 1985). This contribution can be considerably increased, however, by a $lexA$(Def) mutation. Constitutive expression of the $lexA$ regulon genes apparently either produces more lesions, which are specifically utilized by the RecF pathway, or makes the preexisting lesions more available to this pathway. An attractive possibility is that the RecF pathway is initiated only by free 3'-hydroxyl ends (which would explain its sensitivity to ExoI), whereas the RecX pathway uses locally melted single-stranded regions without nicks. The dependence of plasmidic recombination on the topoisomerase I function is consistent with this interpretation (Cunningham et al., 1981).

The linear multimers formed in $recBC$ mutants seem to be used more efficiently by the RecE than by the RecX pathway, perhaps because only the former can generate 3' tails (Symington et al., 1985; Cohen and Clark, 1986). All plasmidic recombination in $recBC$ mutants, including the formation of oligomeric circles, may use linear multimers and, therefore, be intramolecular. It must be remembered, however, that most plasmidic recombination takes place in the presence of multiple homologous molecules. Recombination is usually considered intramolecular whenever the selected recombinants can, in principle, arise by an interaction between two homologous sequences on a single molecule. But in all these cases intermolecular interactions can also produce the selected recombinants; in most cases the relative contributions of the two types of exchanges have not been assessed. Therefore, the terms intra- and intermolecular plasmidic recombination, as conventionally used, may not faithfully reflect the corresponding molecular events (for a discussion, see Ennis et al., 1987).

F. Lambda Recombination

Recombination in lambda, which is amenable to both genetic and physical analyses, provides perhaps the best evidence for the existence of multiple pathways (for reviews, see Stahl, 1979, 1984, 1986; Smith, 1983). General recombination in lambda can proceed by the phage-encoded Red pathway or by any of the host pathways, RecBCD, RecE, RecF, RecX, and RecY (Gillen and Clark, 1974; Stahl and Stahl, 1977; Smith, 1983; Chaudhury and Smith, 1984a). An additional host pathway, called Rpo because it requires RNA polymerase, has been suggested (Ikeda and Kobayashi, 1977; Matsumoto and Ikeda, 1983).

Recombination by the Red pathway is intimately connected with replication, while the RecBCD pathway can proceed without extensive replication (Stahl et al., 1974; Stahl and Stahl, 1986). Further, the Red pathway-mediated recombination is largely nonreciprocal, and the RecBCD pathway is largely reciprocal (Sarthy and Meselson, 1976).

Differential exchange distribution in recombinants produced by different pathways has been demonstrated by a combination of physical and genetic techniques (Stahl et al., 1974; Gillen and Clark, 1974). The lambda Red and the host RecE pathways seem to behave indistinguishably in most respects (Gillen and Clark, 1974), though recently Thaler et al. (1987) have suggested that these two pathways may also be significantly different. Lambda exonuclease and ExoVIII have similar activities, though they do not antigenically cross-react (Gillen et al., 1977; Table 1) and their sizes are different (λ Exo polypeptide is 24 kDa, while ExoVIII is 140 kDa [Little, 1967; Joseph and Kolodner, 1983a]). RecBCD pathway recombination is stimulated in the neighborhood of Chi, while the other pathways seem to be totally unresponsive to this sequence (Gillen and Clark, 1974; Stahl and Stahl, 1977; Mosig, this volume; Taylor, this volume).

G. Phage-Prophage Recombination

In phage-prophage crosses a lambda lysogen is infected with a lambda defective in replication, and the ability of the latter to rescue this replication function from the prophage by homologous recombination is monitored. If the infecting phage is homoimmune, its expression (including replication and recombination) is repressed even after the rescue, until the repressor is inactivated. If the infecting phage is heteroimmune, these functions are expressed, and replication can take place after the rescue, or even before the rescue, if a suppressor of the replication mutation exists in the host.

In homoimmune crosses the rescue frequencies are low but can be greatly enhanced by prior treatment of the superinfecting phage with UV and psoralen. The enhancement is traceable to UvrABC$^+$-dependent incision at the damage site (Lin et al., 1977). Both spontaneous and damage-induced phage-prophage recombination in these crosses is reduced only three- to fourfold by $recBC$ or $recB$ $recF$ double mutations and is increased about threefold by $recB$ $uvrD$ double mutations (Howard-Flanders and Bardwell, 1981). In heteroimmune crosses, the λ Red and the host RecF pathways promote phage-prophage recombination with high efficiency in a $recF^+$-dependent manner, while the host RecE pathway seems to make no contribution (Armengod, 1981).

H. Recombination between Tandem Duplications on Nonreplicating Phages

Hays and co-workers (Hays and Boehmer, 1978; Hays et al., 1984; Smith and Hays, 1985) have studied the loss of a tandem duplication (deamplification; see section IV.D. above) in λ (Bellet et al., 1971; Berg, 1971). The low spontaneous frequency of this recombination (~0.2%) under conditions of repressed replication and transcription can be increased to ~50% by UV irradiation of the phage before infection. This UV-stimulated recombination is strongly $recF^+$ dependent and partly $uvrABC^+$ dependent, but independent of $recBC$ or $recE$ mutations. These results, as well as those in section IV.G. above, are consistent with the proposal that the RecBCD and RecE pathways use linear duplex substrates. The nonreplicating λ in these crosses, being circular, is unable to recombine by these pathways and mostly uses the RecF pathway. Recombination in these crosses requires DNA gyrase (see section II.); coumermycin, an inhibitor of DNA gyrase, reduces the loss of duplications about fivefold in gyr^+ cells but not in a $gyrB$ (coumermycin-resistant) mutant (Hays and Boehmer, 1978).

I. Phage-Plasmid Recombination

In this system, cointegrates are formed between an infecting lambda phage and a plasmid harboring a part of lambda. The RecE, RecF, and RecX pathways are much less proficient than the RecBCD pathway in promoting this recombination, which is presumably reciprocal (Watt et al., 1985; King and Richardson, 1986; Shen and Huang, 1986). The proficiency of the Red pathway of λ was not examined. Interestingly, in $recBC^+$ cells, the $sbcA23$ mutation reduced the recovery of cointegrates by a factor of 20, while $recF143$ increased it 10-fold (King and Richardson, 1986). The $sbcA23$-promoted reduction in the cointegrate yield may reflect resolution of the cointegrates, similar to deamplification (see section IV.D. above). The $recF143$ effect may be due to the $recF^+$ gene product diverting most of the substrate molecules into substitution-type recombination, whose products were not monitored.

The minimum length of homology in the plasmid required for efficient recombination was lower for the RecBCD pathway (23 to 27 bp) than for the RecX or RecE pathways (40 to 90 bp) (Shen and Huang, 1986).

J. Inversions

Using a derivative of phage lambda in which recombination between two 1.4-kb inverted homologous repeats can be selected by simple phenotypes associated with inversion of the intervening region, Ennis et al. (1987) measured the frequencies of inversion in different Rec backgrounds. Only wild-type and *recD* cells promoted inversions efficiently. The lambda Red pathway and the host RecE, RecF, and RecX pathways were considerably less efficient in this process. Further, Chi sites located to the right of the invertible segments considerably enhanced inversion by the RecBCD pathway but not by the other pathways. Since inversion may be a reciprocal recombination event, these results appear to support the idea (see above) that the RecBCD pathway is more proficient in forming reciprocal exchanges than are the RecE and RecF pathways. Ennis et al. (1987) identified other genes whose products are required by the RecBCD pathway, including *ssb*, *gyrA*, *gyrB*, *polA*, and *lig*. In the absence of a Chi site, *recD* mutations enhanced inversions 8- to 12-fold above the *recD*$^+$ level in a *recF143* genetic background. Lambda × lambda recombination is also enhanced under these conditions (Chaudhury and Smith, 1984a).

V. COMPARISON OF DIFFERENT SYSTEMS: SOURCES OF COMPLEXITY AND COMMON FEATURES

The recombinational studies in the various *E. coli* systems described above are incomplete, as not all recombinational mutations have been examined in each system. Even so, the complexity of these results is obvious; a given mutation may have widely different, or even opposite, effects in different systems. The *recA* independence of lambda and plasmidic recombination in certain backgrounds, the *recBCD*$^+$ independence of several kinds of recombination (such as certain specialized transductions, conjuduction, plasmid crosses, and F'-mediated chromosome mobilization), and the reduction of phage-plasmid recombination by *sbcA* are some cases in point. Given these disparate results, is it reasonable to talk of genetic pathways of recombination as sets of genes whose products act in concert to promote recombination? The answer, it appears, is yes.

Below, I discuss some of the factors that may be responsible for the apparent inconsistencies of the results obtained from different systems and then describe some underlying common features. The latter permit the conclusion that genetically defined pathways of recombination have distinct identities with respect to the DNA structures involved and their interactions.

A. Sources of Complexity

Data obtained from phenotypic analysis of recombinants must be interpreted with caution, since recombinant frequencies can be influenced by many intracellular and extracellular factors. Recombining DNA molecules are concurrently subject to other aspects of DNA metabolism, including replication, transcription, and mismatch correction (see, e.g., Stahl and Stahl, 1986; Ikeda and Kobayashi, 1977; Claverys and Lacks, 1986; Mosig, this volume; Radman, this volume). Differential growth rates of different recombinant types (De Haan et al., 1972) and culture conditions (Bergmans and Hoekstra, 1977) can also influence the observed recombinant frequencies. These may affect one pathway or system more than others. Even when these effects are controlled, other complexities may remain.

1. Substrate Heterogeneity

Systems differ in the nature, size, and location of their recombinational substrates, which may be circular or linear, relaxed or supercoiled, single or double stranded, naked or bound to protective proteins. The gene in which recombination is monitored, such as *lacZ* or *tet*, may be present on a plasmid a few kilobases long, a P1 transducing fragment nearly 100 kb long, or the much longer *E. coli* chromosome. The two interacting genes may be on a single replicon or on separate replicons. Replicons may have special features, such as the *oriT* site (origin of transfer) of the F factor, or different modes of replication.

These factors can significantly and differentially influence the recombinational proficiency of different pathways. For instance, the RecBCD pathway seems to require an at least transiently double-stranded end. Thus, this pathway makes a major contribution to recombination in conjugation, P1-mediated transduction, λ Red⁻ Gam⁻ vegetative crosses, and F' crosses, but it makes little if any contribution to recombination of plasmids, DNA with single-stranded gaps generated by UvrABC-dependent excision of damaged bases, and repressed λ phage with a prophage.

2. Differences in the End Product of Recombination

The structure of the end product monitored may determine the kind of exchange and hence the pathway that can produce it. For instance, phage-plasmid fusions (Shen and Huang, 1986; King and Richardson, 1986) and inversions (Ennis et al., 1987) may arise primarily from reciprocal exchange of double-stranded segments; the RecBCD pathway appears to be more proficient in such exchanges than the RecE and RecF pathways. On the other hand, chromosome mobilization by F' factors (Scaife and Gross, 1963; Clowes and Moody, 1966; Wilkins, 1969) may involve nonreciprocal exchange of single-stranded segments and may be promoted primarily by the RecF and RecX pathways. Integration of nonhomologous DNA that is flanked by homologies might proceed readily by a pathway that effects double-stranded exchanges (the RecBCD pathway) but inefficiently by a pathway that transfers long single-stranded stretches (the RecF pathway), which may stall at the heterology (Laban et al., 1984).

3. Multiple Activities of the Recombinational Proteins

Several recombinational proteins have multiple activities. For example, RecA protein and RecBCD enzyme are complex proteins with multiple activities (section II.). Some of these activities seem to be directly involved in recombination or in regulating the supply of specific recombinational substrates, while others may influence the recombinant yields indirectly by regulating the activity of other recombinational genes.

RecA protein has direct roles in homology search and strand transfer. Both of these roles may be wholly or partially dispensable under special circumstances (sections IV.E. and IV.F.). The interaction of the RecBCD enzyme with the recombinational systems is even more complex as a result of the greater variety of its activities. RecBCD enzyme may have a direct role in producing unwound, linear single-stranded substrates for RecA protein. Under certain circumstances it may degrade the substrates of its own recombinational pathway, as seems to happen in the case of transforming DNA. In other cases, for instance, in λ and plasmids, it may inhibit the formation of linear duplex substrates for recombination (Unger and Clark, 1972; Cohen and Clark, 1986). It may also inhibit other (hyper-Rec) pathways (RecE, RecF, RecX, and RecY) by eliminating the DNA substrates or lesions used by the latter (sections II.D.3. and III.D.).

Other recombinational proteins may also have multiple effects on recombination. For instance, the *sbcB* mutation directly blocks the ExoI-mediated degradation of ssDNA (Kushner et al., 1971) and derepresses *recN* (Picksley et al., 1984b). The *recF* gene product regulates the expression of *recA* and other SOS genes (Karu and Belk, 1982; Salles and Paoletti, 1983) but may play a direct role, as well. PolA has multiple activities which may influence recombination at different stages (section II.).

4. Partial Overlap between Pathways

Although one pathway may make the major contribution to recombination, other pathways may also contribute. This may occur when two pathways can use the same starting substrate or when they use different available substrates to produce similar end products. For instance, single-strand exchanges can be produced by appropriate resolution of a Holliday junction formed between two duplexes or by strand transfer from a single-stranded donor to a double-stranded recipient.

The recombinant product of one pathway can also act as a substrate for another pathway which may even reverse the process. One example is the $recA^+$-independent reversal of tandem amplification by the RecE pathway (Mahajan et al., 1984). In addition, the RecE pathway may resolve phage-plasmid cointegrates, whose appearance is reduced by an *sbcA* mutation (Watt et al., 1985; Shen and Huang, 1986; King and Richardson, 1986).

B. Common Features

If the above sources of complexity are kept in mind, certain common threads in the different systems are discernible. In several systems, though not all, the proficiency of recombination shows similar qualitative dependence on the various *rec* mutations, suggesting a similar genetic structure of pathways in these systems as in conjugation. Mutations in *recA* reduce recombination in all systems tested except plasmid and lambda recombination by the RecE pathway (sections IV.E. and IV.F.). Mutations in *recB* and *recC* reduce recombination in all systems except plasmid recombination, chromosome mobilization by F' factors, conjuduction, and repressed phage × prophage crosses. Mutations in *sbcA* and *sbcB(C)* suppress *recBC*-associated recombination deficiency in all systems except phage-plasmid fusion and inversions. As discussed above, these exceptions may be due to a RecA-like protein being expressed in *sbcA* mutants (RecE pathway; section IV.E.), the absence of double-stranded ends in the systems in which RecBCD enzyme does not act (section II.D.2.), and the requirement for reciprocal exchange in plasmid-phage fusions and in inversions (sections IV.I. and IV.J.). Thus, it would appear that the concept of pathways, at least at the genetic level, is not merely a reflection of the peculiar properties of the conjugation system.

The RecBCD pathway is efficient in all systems in which at least one of the two interacting duplexes is linear, such as in P1 transduction or conjugation, or in which one of them appears to be transiently open to permit entry of the RecBCD enzyme, e.g., in λ *lacZ* × F' *lacZ2* specialized transduction, phage-plasmid recombination, λ vegetative crosses, or inversions. On the other hand, the RecBCD pathway does not efficiently recombine donor ssDNA in conjuduction or F'-mediated chromosome mobilization. Similarly, this pathway cannot recombine circular plasmid duplexes or repressed lambda, which may not have a double-stranded end even transiently.

In contrast, the RecE and RecF pathways, especially the latter, are inefficient in phage-plasmid fusions, in inversion, and in production of addition-type specialized transductants, which may require reciprocal double-strand exchanges. These pathways are more efficient in systems in which the selected recombinants can be produced by single-strand exchanges, e.g., F'-mediated chromosomal mobilization, production of substitution-type specialized transductants, conjuduction, and phage-prophage recombination.

The RecX, RecE, and RecF pathways also differ from one another with respect to their preferred substrates. The RecE pathway can efficiently use duplex substrates for intramolecular (*recA*$^+$-independent) recombination of linear plasmid multimers, but the RecF pathway cannot. Similarly, the RecX pathway can recombine circular plasmid DNA in *recBCD*$^+$ cells, whereas the RecF pathway seems to contribute very little (see section IV.E. above).

These observations imply that the genetic differences between the pathways are not recombinationally trivial. If the genetic backgrounds of the RecBCD, RecF, and RecE pathways were just different ways of increasing the levels of the same common recombinational enzymes above the RecX level, then it would be difficult to explain why only the RecBCD pathway appears to efficiently effect reciprocal double-strand exchanges to produce phage-plasmid cointegrates, inversions, or addition-type specialized transductants, or why only the RecE pathway gives high-level RecA-independent plasmidic recombination. Similarly, it is unlikely that their differences are confined to the generation of common precombinational substrates by different routes, because in that case their products should be similar.

VI. SUMMARY AND CONCLUDING REMARKS

More than 30 genes in which mutations influence homologous recombination have been identified in *E. coli*. Genetic and biochemical analysis of these genes and their products have led to two important inferences. (i) A central step common to all homologous recombination, namely, homology matching and heteroduplex formation (strand transfer), is mediated in most *E. coli* recombination systems by the RecA and SSB proteins. (ii) There are several distinct pathways for the preceding and succeeding steps. These steps include the generation of substrates for strand transfer, the processing of the intermolecular complexes resulting from strand transfer, the resolution of these intermolecular complexes, and the replication of the recombinant molecules thus formed.

Primarily defined in genetic terms, as groups of genes whose products act together to effect recombination, these pathways differ in the efficiencies with which they can use different kinds of substrates and produce different recombinant products. There seem to be at least five different $recA^+$-dependent pathways of recombination. These have been named the RecBCD, RecE, RecF, RecX, and RecY pathways. These are optimally functional in the wild type, *recBC sbcA*, *recBC sbcBC*, *recBC*, and *recD* genetic backgrounds, respectively. Specific genes, in addition to $recA^+$, whose products are essential for each of these pathways, are $recB^+$, $recC^+$, and $recD^+$ for the RecBCD pathway; $recE^+$, $recF^+$, $recJ^+$, and $recO^+$ for the RecE pathway; $recF^+$, $recJ^+$, $recO^+$, $recN^+$, $recQ^+$, and ruv^+ for the RecF pathway; $recF^+$, $recJ^+$, and $recO^+$ for the RecX pathway; and $recJ^+$, $recB^+$, and $recC^+$ for the RecY pathway. ssb^1, $polA^+$, lig^+, $gyrA^+$, and $gyrB^+$ may be required by all the pathways, but the effects of these genes have not been examined in all cases.

Mutations in several genes increase recombination between close markers. Some of these (e.g., *mutL* and *mutS*) seem to block mismatch correction within the heteroduplex regions formed by strand exchange. These mutations would not be expected to affect the amount of DNA exchanged between the recombining molecules, though they may influence linkage between marker pairs. Other mutations, such as *dam*, *uvrD*, *polA*, and *lig*, may increase the number of recombinogenic lesions. Double-strand lesions may provide substrates for the RecBCD pathway and single-strand lesions for the other pathways.

The RecBCD pathway, the dominant pathway in wild-type cells, differs from the rest in several important aspects. This pathway produces fewer recombinational events per unit chromosomal distance than the others and probably exchanges longer segments of DNA. The RecBCD pathway is stimulated by Chi sites and appears to require at least a transient double-stranded end for entry of the RecBCD enzyme. Detailed models for these recombinational pathways will require more information about their key proteins and the structures of the DNA intermediates.

The initial evidence for multiple pathways of recombination came from genetic analysis of recombination accompanying conjugation. Later studies indicated that the pathways extend to other *E. coli* systems. Distinguishing features of these pathways, apart from the specific genes that they use, include the nature of the initial DNA lesions, the density of exchanges per unit chromosomal length, the nature of the exchanges (whether single or double strand, reciprocal or nonreciprocal) produced by each, the time course of recombination, the structure of the product molecules, and their response to specific recombinogenic sequences such as Chi.

ACKNOWLEDGMENTS. I am indebted to A. J. Clark for several useful discussions while I was preparing the original outline of this review. My visit to his laboratory at the University of California, Berkeley, was supported by Public Health Service grant A 105371 from the National Institutes of Health.

The numerous contributions of K. A. V. David, Swapan Bhattacharjee, Niketan Pandit, Sandhya Suryavanshi, Jasmine Sarkari, Gayatri Sharma, and Suhas Mangoli in the preparation of this review are gratefully acknowledged. I am also thankful to Gerald Smith for many helpful suggestions.

LITERATURE CITED

Abastado, J.-P., S. Darche, F. Godeau, B. Cami, and P. Kourilsky. 1987. Intramolecular recombination between partially homologous sequences in *Escherichia coli* and *Xenopus laevis* oocytes. *Proc. Natl. Acad. Sci. USA* **84**:6496–6500.

Adelberg, E. A., and S. N. Burns. 1960. Genetic variation in the sex-factor of *Escherichia coli*. *J. Bacteriol.* **79**:321–330.

Amundsen, S. K., A. F. Taylor, A. M. Chaudhury, and G. R. Smith. 1986. *recD*: the gene for an essential third subunit of exonuclease V. *Proc. Natl. Acad. Sci. USA* **83**:5558–5562.

Armengod, M.-E. 1981. Role of the *recF* gene of *Escherichia coli* K-12 in λ recombination. *Mol. Gen. Genet.* **181**:497–504.

Armengod, M.-E., and E. Lambiers. 1986. Overlapping arrangements of *recF* and *dnaN* operon of *Escherichia coli*, positive and negative control sequences. *Gene* **43**:183–196.

Arthur, H. N., and R. G. Lloyd. 1980. Hyper-recombination in *uvrD* mutants of *Escherichia coli* K-12. *Mol. Gen. Genet.* **180**:185–191.

Attfield, P. V., F. E. Benson, and R. G. Lloyd. 1985. Analysis of the *ruv* locus of *Escherichia coli* K-12 and identification of the gene product. *J. Bacteriol.* **164**:276–281.

Bachmann, B. J. 1983. Linkage map of *Escherichia coli* K-12, edition 7. *Microbiol. Rev.* **47**:180–230.

Bailone, A., S. Sommer, and R. Devoret. 1985. Mini-F plasmid-induced SOS signal in *Escherichia coli* is RecBC dependent. *Proc. Natl. Acad. Sci. USA* **82**:5973–5977.

Baluch, J., J. W. Chase, and R. Sussman. 1980. Synthesis of *recA* protein and induction of bacteriophage lambda in single-strand deoxyribonucleic acid-binding protein mutants in *Escherichia coli*. *J. Bacteriol.* **144**:489–498.

Barbour, S. D., and A. J. Clark. 1970. Biochemical and genetic studies of recombination proficiency in *Escherichia coli*. I. Enzymatic activity associated with *recB*⁺ and *recC*⁺ genes. *Proc. Natl. Acad. Sci. USA* **65**:955–961.

Barbour, S. D., H. Nagaishi, A. Templin, and A. J. Clark. 1970. Biochemical and genetic studies of recombination proficiency in *Escherichia coli*. II. Rec⁺ revertants caused by indirect suppression of Rec⁻ mutations. *Proc. Natl. Acad. Sci. USA* **67**:128–135.

Basu, S. K., and M. Oishi. 1975. Factor which affects the mode of genetic recombination in *E. coli*. *Nature* (London) **253**:138–140.

Bedbrook, J. R., and F. M. Ausubel. 1976. Recombination between bacterial plasmids leading to the formation of plasmid multimers. *Cell* **9**:707–716.

Bellett, A. J. D., H. G. Busse, and R. L. Baldwin. 1971. Tandem genetic duplications in a derivative of phage lambda, p. 501–513. *In* A. D. Hershey (ed.), *The Bacteriophage Lambda*. Cold Spring Harbor Laboratory, Cold Spring Harbor, N.Y.

Berg, D. E. 1971. Regulation in phage with duplications of the immunity region, p. 677–678. *In* A. D. Hershey (ed.), *The Bacteriophage Lambda*. Cold Spring Harbor Laboratory, Cold Spring Harbor, N.Y.

Berg, D. E., and J. A. Gallant. 1971. Tests of reciprocality in crossingover in partially diploid F′ strains of *Escherichia coli*. *Genetics* **68**:457–472.

Bergmans, H. E. N., and W. P. M. Hoekstra. 1977. The fate of the donor DNA after conjugation in *Escherichia coli* K-12. Influence of the growth rate of the recipient. *Mol. Gen. Genet.* **158**:179–183.

Bhagwat, A. S., A. Sohail, and R. J. Roberts. 1986. Cloning and characterization of the *dcm* locus of *Escherichia coli* K-12. *J. Bacteriol.* **166**:751–755.

Bialkowska-Hobrzanska, H., C. A. Gilchrist, and D. T. Denhardt. 1985. *Escherichia coli rep* gene: identification of the promoter and N terminus of the Rep protein. *J. Bacteriol.* **164**:1004–1010.

Biek, D. P., and S. N. Cohen. 1986. Identification and characterization of *recD*, a gene affecting plasmid maintenance and recombination in *Escherichia coli*. *J. Bacteriol.* **167**:594–603.

Birge, E. A., and K. B. Low. 1974. Detection of transcribable recombination products following conjugation in Rec⁺, RecB⁻ and RecC⁻ strains of *Escherichia coli* K-12. *J. Mol. Biol.* **83**:447–457.

Blanar, M. A., D. Kneller, A. J. Clark, A. E. Karu, F. E. Cohen, R. Langridge, and I. D. Kuntz. 1984a. A model for the core structure of the *Escherichia coli* RecA protein. *Cold Spring Harbor Symp. Quant. Biol.* **49**:507–509.

Blanar, M. A., S. J. Sandler, M.-E. Armengod, L. W. Ream, and A. J. Clark. 1984b. Molecular analysis of the *recF* gene of *Escherichia coli*. *Proc. Natl. Acad. Sci. USA* **81**:4622–4626.

Bresler, S. E., I. Y. Goryshin, and V. A. Lanzov. 1981. The process of general recombination in *E. coli* K-12. Structure of intermediate products. *Mol. Gen. Genet.* **183**:139–143.

Brody, R. S., K. G. Doherty, and P. G. Zimmerman. 1986. Processivity and kinetics of the reaction of exonuclease I from *Escherichia coli* with polydeoxyribonucleotides. *J. Biol. Chem.* **261**:7136–7143.

Brooks, J. E., R. M. Blumenthal, and T. R. Gingeras. 1983. The isolation and characterization of *Escherichia coli* DNA adenine methylase gene. *Nucleic Acids Res.* **11**:837–851.

Buttin, G., and M. Wright. 1968. Enzymatic DNA degradation in *E. coli*: its relationship to synthetic processes at the chromosomal level. *Cold Spring Harbor Symp. Quant. Biol.* **33**:259–269.

Calsou, P., A. Villaverde, and M. Defais. 1987. Activated RecA protein may induce expression of a gene that is not controlled by the *lexA* repressor and whose function is required for mutagenesis and repair of UV-irradiated bacteriophage lambda. *J. Bacteriol.* **169**:4816–4821.

Capaldo, F. N., and S. D. Barbour. 1975. DNA content, synthesis and integrity in dividing and non-dividing cells of Rec⁻ strains of *Escherichia coli* K-12. *J. Mol. Biol.* **91**:53–66.

Capaldo, F. N., G. Ramsey, and S. D. Barbour. 1974. Analysis of the growth of recombination-deficient strains of *Escherichia coli* K-12. *J. Bacteriol.* **118**:242–249.

Capaldo-Kimball, F., and S. D. Barbour. 1971. Involvement of recombination genes in growth and viability of *Escherichia coli* K-12. *J. Bacteriol.* **106**:204–212.

Caron, P. R., S. R. Kushner, and L. Grossman. 1985. Involvement of helicase II (*uvrD* gene product) and DNA polymerase I in excision mediated by *uvrABC* protein complex. *Proc. Natl. Acad. Sci. USA* **82**:4925–4929.

Chase, J. W., B. A. Rabin, J. B. Murphy, K. L. Stone, and K. R. Williams. 1986. *Escherichia coli* exonuclease VII. Cloning and sequencing of the gene encoding the large subunit (*xseA*). *J. Biol. Chem.* **261**:14929–14935.

Chase, J. W., and C. C. Richardson. 1977. *Escherichia coli* mutants deficient in exonuclease VII. *J. Bacteriol.* **129**:934–947.

Chase, J. W., and K. R. Williams. 1986. Single-stranded DNA binding proteins required for DNA replication. *Annu. Rev. Biochem.* **55**:103–136.

Chaudhury, A. M., and G. R. Smith. 1984a. A new class of *Escherichia coli recBC* mutants: implications for the role of RecBC enzyme in homologous recombination. *Proc. Natl. Acad. Sci. USA* **81**:7850–7854.

Chaudhury, A. M., and G. R. Smith. 1984b. *Escherichia coli recBC* deletion mutants. *J. Bacteriol.* **160**:788–791.

Chaudhury, A. M., and G. R. Smith. 1985. Role of *Escherichia coli* RecBC enzyme in SOS induction. *Mol. Gen. Genet.* **201**:525–528.

Clark, A. J. 1971. Toward a metabolic interpretation of genetic recombination of *E. coli* and its phages. *Annu. Rev. Microbiol.* **25**:437–464.

Clark, A. J. 1973. Recombination deficient mutants of *E. coli* and other bacteria. *Annu. Rev. Genet.* **7**:67–86.

Clark, A. J. 1974. Progress toward a metabolic interpretation of genetic recombination of *Escherichia coli* and bacteriophage λ. *Genetics* **78**:259–271.

Clark, A. J. 1980. A view of the RecBC and RecF pathways of *E. coli* recombination. *ICN-UCLA Symp. Mol. Cell. Biol.* **19**:891–899.

Clark, A. J., and K. B. Low. 1988. Systems and pathways of homologous recombination in *Escherichia coli*, p. 155–215. *In* K. B. Low (ed.), *The Recombination of Genetic Material*. Academic Press, Inc., New York.

Clark, A. J., and A. D. Margulies. 1965. Isolation and characterization of recombination-deficient mutants of *Escherichia coli* K-12. *Proc. Natl. Acad. Sci. USA* **53**:451–459.

Clark, A. J., S. J. Sandler, D. K. Willis, C. C. Chu, M. A. Blanar, and S. T. Lovett. 1984. Genes of the RecE and RecF pathways of conjugational recombination in *Escherichia coli*. *Cold Spring Harbor Symp. Quant. Biol.* **49**:453–462.

Clark, A. J., and M. R. Volkert. 1978. A new classification of pathways repairing pyrimidine dimer damage in DNA, p. 57–72. *In* P. C. Hanawalt, E. C. Freidberg, and C. F. Fox (ed.), *DNA Repair Mechanisms*. Academic Press, Inc., New York.

Claverys, J. P., and S. A. Lacks. 1986. Heteroduplex deoxyribonucleic acid base mismatch repair in bacteria. *Microbiol. Rev.* **50**:133–165.

Clowes, R. C., and E. E. M. Moody. 1966. Chromosomal transfer from recombination-deficient strains of *Escherichia coli* K-12. *Genetics* **53**:717–726.

Clyman, J., and R. C. Cunningham. 1987. *Escherichia coli* K-12 mutants in which viability is dependent on *recA* function. *J. Bacteriol.* **169**:4203–4210.

Cohen, A., and A. J. Clark. 1986. Synthesis of linear plasmid multimers in *Escherichia coli* K-12. *J. Bacteriol.* **167**:327–335.

Cohen, A., and A. Laban. 1983. Plasmidic recombination in *Escherichia coli* K-12: the role of the *recF* gene function. *Mol. Gen. Genet.* **189**:471–474.

Cosloy, S. D. 1982. Analysis of genetic recombination by the RecBC and RecF pathways of *Escherichia coli* K-12, p. 261–273. *In* V. N. Streips, W. R. Guild, S. H. Goodgal, and G. A. Wilson (ed.), *Genetic Exchange*. Marcel Dekker, Inc., New York.

Cox, M. M., and I. R. Lehman. 1987. Enzymes of general recombination. *Annu. Rev. Biochem.* **56**: 229–262.

Cunningham, R. P., A. M. Wu, T. Shibata, C. DasGupta, and C. M. Radding. 1981. Homologous pairing and topological linkage of DNA molecules by combined action of *E. coli* RecA protein and topoisomerase I. *Cell* **24**:213–223.

De Haan, P. G., W. P. M. Hoekstra, and C. Verhoeff. 1972. Recombination in *Escherichia coli*. V. Genetic analysis of recombinants from crosses with recipients deficient in ATP-dependent exonuclease activity. *Mutat. Res.* **14**:375–380.

De Lucia, P., and J. Cairns. 1969. Isolation of an *E. coli* strain with a mutation affecting DNA polymerase. *Nature* (London) **224**:1164–1166.

Doherty, M. J., P. T. Morrison, and R. Kolodner. 1983. Genetic recombination of bacterial plasmid DNA. Physical and genetic analysis of the products of plasmid recombination in *Escherichia coli*. *J. Mol. Biol.* **167**:539–560.

Dower, N. A., and F. W. Stahl. 1981. χ activity during transduction-associated recombination. *Proc. Natl. Acad. Sci. USA* **78**:7033–7037.

Drlica, K. 1984. Biology of bacterial deoxyribonucleic acid toposiomerases. *Microbiol. Rev.* **48**:273– 289.

Dykstra, C. C., D. Prasher, and S. R. Kushner. 1984. Physical and biochemical analysis of cloned *recB* and *recC* genes of *Escherichia coli*. *J. Bacteriol.* **157**:21–27.

Eigner, C., E. Azhderiom, S. C. Tsong, C. M. Radding, and J. W. Chase. 1987. Effect of various single-stranded DNA-binding proteins on reactions promoted by RecA protein. *J. Bacteriol.* **169**: 3423–3428.

Emmerson, P. T, I. D. Hickson, R. L. Gordon, and A. E. Tomkinson. 1981. Cloning of *recA*[+] and *lexA*[+]

and some of their mutant alleles: an investigation of their mutual interaction, p. 281–285. *In* E. Seeburg and K. Kleppe (ed.), *Chromosome Damage and Repair*. Plenum Publishing Corp., New York.

Emmerson, P. T., and P. Howard-Flanders. 1967. Cotransduction with *thy* of a gene required for genetic recombination in *Escherichia coli. J. Bacteriol.* **93:**1729–1731.

Emmerson, P. T., and S. C. West. 1977. Identification of protein X of *Escherichia coli* as the *recA*⁺/ *tif*⁺ gene product. *Mol. Gen. Genet.* **155:**77–85.

Ennis, D. G., S. K. Amundsen, and G. R. Smith. 1987. Genetic functions promoting homologous recombination in *Escherichia coli*: a study of inversion in phage λ. *Genetics* **115:**11–24.

Feinstein, S. I., and K. B. Low. 1986. Hyper-recombining recipient strains in bacterial conjugation. *Genetics* **113:**13–33.

Finch, P. W., P. Chambers, and P. T. Emmerson. 1985. Identification of the *Escherichia coli recN* gene product as a major SOS protein. *J. Bacteriol.* **164:**653–658.

Finch, P. W., and P. T. Emmerson. 1984. The nucleotide sequence of the regulatory region of the *uvrD* gene of *Escherichia coli. Nucleic Acids Res.* **13:**5789–5799.

Finch, P.W., A. Storey, K. Brown, I. D. Hickson, and P. T. Emmerson. 1986a. Complete nucleotide sequence of *recD*, the structural gene for the α subunit of exonuclease V of *Escherichia coli. Nucleic Acids Res.* **14:**8583–8594.

Finch, P. W., A. Storey, K. E. Chapman, K. Brown, I. D. Hickson, and P. T. Emmerson. 1986b. Complete nucleotide sequence of the *Escherichia coli recB* gene. *Nucleic Acids Res.* **14:**8573–8582.

Finch, P. W., R. E. Wilson, K. Brown, I. D. Hickson, A. E. Tomkinson, and P. T. Emmerson. 1986c. Complete nucleotide sequence of the *Escherichia coli recC* gene and of the *thyA-recC* intergenic region. *Nucleic Acids Res.* **14:**4437–4451.

Fishel, R. A., A. A. James, and R. Kolodner. 1981. *recA*-independent genetic recombination of plasmids. *Nature* (London) **294:**184–186.

Fishel, R. A., and R. Kolodner. 1984. *Escherichia coli* strains containing mutations in the structural gene for topoisomerase I are recombination deficient. *J. Bacteriol.* **160:**1168–1170.

Fouts, K. E., T. Wasie-Gilbert, D. K. Willis, A. J. Clark, and S. D. Barbour. 1983. Genetic analysis of transposon-induced mutations of the Rac prophage in *Escherichia coli* K-12 which affect expression and function of *recE. J. Bacteriol.* **156:**718–726.

Fox, M. S. 1978. Some features of genetic recombination in procaryotes. *Annu. Rev. Genet.* **12:**47–68.

Geider, K., and H. Hoffman-Berling. 1981. Proteins controlling the helical structure of DNA. *Annu. Rev. Biochem.* **50:**233–260.

Geier, G. E., and P. Modrich. 1979. Recognition sequence of the *dam* methylase of *Escherichia coli* K-12 and mode of cleavage of *Dpn*I endonuclease. *J. Biol. Chem.* **254:**1408–1413.

Gellert, M. 1981. DNA topoisomerases. *Annu. Rev. Biochem.* **50:**879–910.

Gillen, J. R., and A. J. Clark. 1974. The RecE pathway of bacterial recombination, p. 123–136. *In* R. F. Grell (ed.), *Mechanisms in Recombination*. Plenum Publishing Corp., New York.

Gillen, J. R., A. E. Karu, H. Nagaishi, and A. J. Clark. 1977. Characterization of the deoxyribonuclease determined by lambda reverse as exonuclease VIII of *Escherichia coli. J. Mol. Biol.* **113:**27–41.

Gillen, J. R., D. K. Willis, and A. J. Clark. 1981. Genetic analysis of the RecE pathway of genetic recombination in *Escherichia coli* K-12. *J. Bacteriol.* **145:**521–532.

Glassberg, J., R. R. Meyer, and A. Kornberg. 1979. Mutant single-strand binding protein of *Escherichia coli*: genetic and physiological characterization. *J. Bacteriol.* **140:**14–19.

Goldmark, P. J., and S. Linn. 1972. Purification and properties of the *recBC* DNAase of *Escherichia coli* K-12. *J. Biol. Chem.* **247:**1849–1860.

Golub, E. I., and K. B. Low. 1983. Indirect stimulation of genetic recombination. *Proc. Natl. Acad. Sci. USA* **80:**1401–1405.

Golub, E. I., and K. B. Low. 1986. Unrelated conjugative plasmids have sequences which are homologous to the leading region of the F-factor. *J. Bacteriol.* **166:**670–672.

Gopalakrishna, K., and S. K. Bhattacharjee. 1976. Role of *rec* pathways on sensitivity of *Escherichia coli* to near ultraviolet and visible light. *J. Bacteriol.* **127:**1022–1023.

Gottesman, M. M., M. L. Hicks, and M. Gellert. 1973. Genetics and function of DNA ligase in *Escherichia coli. J. Mol. Biol.* **77:**531–547.

Gudas, L. J., and D. W. Mount. 1977. Identification of the *recA* (*tif*) gene product of *Escherichia coli*. *Proc. Natl. Acad. Sci. USA* **74:**5280–5284.

Hall, J. D., and P. Howard-Flanders. 1972. Recombinant F′ factors from *Escherichia coli* K-12 strains carrying *recB* or *recC*. *J. Bacteriol.* **110:**578–584.

Hanawalt, P. C., P. K. Cooper, A. K. Ganesan, and C. A. Smith. 1979. DNA repair in bacteria and mammalian cells. *Annu. Rev. Biochem.* **48:**783–836.

Hays, J. B., and S. Boehmer. 1978. Antagonists of DNA gyrase inhibit repair and recombination of UV-irradiated phage λ. *Proc. Natl. Acad. Sci. USA* **75:**4125–4129.

Hays, J. B., T. A. G. Smith, S. A. Friedman, E. Lee, and G. L. Coffman. 1984. RecF and RecBC function during recombination of non-replicating, UV-irradiated phage λ DNA and during other recombination processes. *Cold Spring Harbor Symp. Quant. Biol.* **49:**475–483.

Herman, G. E., and P. Modrich. 1982. *Escherichia coli dam* methylase. Physical and catalytic properties of the homologous enzyme. *J. Biol. Chem.* **257:**2605–2612.

Hickson, I. D., H. M. Arthur, D. Bramhill, and P. T. Emmerson. 1983. The *E. coli uvrD* gene product is DNA helicase II. *Mol. Gen. Genet.* **190:**265–270.

Hickson, I. D., and P. T. Emmerson. 1981. Identification of the *Escherichia coli recB* and *recC* gene products. *Nature* (London) **294:**578–580.

Hochhauser, S. J., and B. Weiss. 1978. *Escherichia coli* mutants deficient in deoxyuridine triphosphatase. *J. Bacteriol.* **134:**157–166.

Hoekstra, W. P. M., J. E. N. Bergmans, and E. N. Zuidweg. 1980. Role of recBC nuclease in *Escherichia coli* transformation. *J. Bacteriol.* **143:**1031–1032.

Holliday, R. 1964. A mechanism for gene conversion in fungi. *Genet. Res.* **5:**282–304.

Holloman, W. K., R. Wiegand, C. Hoessli, and C. M. Radding. 1975. Uptake of homologous single-stranded DNA by superhelical DNA: a possible mechanism for initiation of genetic recombination. *Proc. Natl. Acad. Sci. USA* **72:**2394–2398.

Horii, T., T. Ogawa, and H. Ogawa. 1981. Nucleotide sequence of the *lexA* gene of *E. coli*. *Cell* **22:**689–697.

Horii, Z. I., and A. J. Clark. 1973. Genetic analysis of the RecF pathway to genetic recombination in *Escherichia coli* K-12: isolation and characterization of mutants. *J. Mol. Biol.* **80:**327–344.

Howard-Flanders, P., and E. Bardwell. 1981. Effects of *recB21*, *recF143* and *uvrD152* on recombination in lambda bacteriophage-prophage and Hfr by F⁻ crosses. *J. Bacteriol.* **148:**739–743.

Howard-Flanders, P., and L. Theriot. 1966. Mutants of *Escherichia coli* K-12 defective in DNA repair and in genetic recombination. *Genetics* **53:**1137–1150.

Ikeda, H., and I. Kobayashi. 1977. Involvement of DNA-dependent RNA polymerase in a *recA*-independent pathway of genetic recombination in *Escherichia coli*. *Proc. Natl. Acad. Sci. USA* **74:**3932–3936.

Ippen-Ihler, K. A., and E. G. Minkley, Jr. 1986. The conjugation system of F, the fertility factor of *Escherichia coli*. *Annu. Rev. Genet.* **20:**593–624.

Irino, N., K. Nakayama, and H. Nakayama. 1986. The *recQ* gene of *Escherichia coli* K-12: primary structure and evidence for SOS regulation. *Mol. Gen. Genet.* **205:**298–304.

James, A. A., P. T. Morrison, and R. Kolodner. 1982. Genetic recombination of bacterial plasmid DNA. Analysis of the effect of recombination-deficient mutations on plasmid recombination. *J. Mol. Biol.* **160:**411–430.

Joseph, J. W., and R. Kolodner. 1983a. Exonuclease VIII of *Escherichia coli*. I. Purification and physical properties. *J. Biol. Chem.* **258:**10411–10417.

Joseph, J. W., and R. Kolodner. 1983b. Exonuclease VIII of *Escherichia coli*. II. Mechanism of action. *J. Biol. Chem.* **258:**10418–10424.

Joyce, C. M., and N. D. F. Grindley. 1982. Identification of two genes immediately downstream from the *polA* gene of *Escherichia coli*. *J. Bacteriol.* **152:**1211–1219.

Kaiser, K., and N. E. Murray. 1979. Physical characterisation of the "Rac prophage" in *E. coli* K-12. *Mol. Gen. Genet.* **175:**159–174.

Kaiser, K., and N. E. Murray. 1980. On the nature of *sbcA* mutations in *E. coli*. *Mol. Gen. Genet.* **179:**555–563.

Kannan, P. R., and K. Dharmalingam. 1987. Restriction alleviation and enhancement of mutagenesis of the bacteriophage T4 chromosome in *recBC sbcA* strains of *Escherichia coli*. *Mol. Gen. Genet*. **209**:413–418.

Karu, A. E., and E. D. Belk. 1982. Induction of *E. coli recA* protein via *recBC* and alternate pathways: quantitation by enzyme-linked immunosorbent assay (ELISA). *Mol. Gen. Genet*. **185**:275–282.

Kellenberger, G., M. L. Zichichi, and J. J. Weigle. 1961. Exchange of DNA in the recombination of bacteriophage λ. *Proc. Natl. Acad. Sci. USA* **47**:869–878.

Kellenberger-Gujer, G., and R. A. Weisberg. 1971. Recombination in bacteriophage lambda. I. Exchange of DNA promoted by phage and bacterial recombination mechanisms, p. 407–415. *In* A. D. Hershey (ed.), *The Bacteriophage Lambda*. Cold Spring Harbor Laboratory, Cold Spring Harbor, N.Y.

Kelly, W. S., K. Chalmers, and N. E. Murray. 1977. Isolation and characterization of a λ*polA* transducing phage. *Proc. Natl. Acad. Sci. USA* **74**:5632–5636.

King, S. R., and J. P. Richardson. 1986. Role of homology and pathway specificity for recombination between plasmids and bacteriophage λ. *Mol. Gen. Genet*. **204**:141–147.

Kmiec, E., and W. K. Holloman. 1981. β-Protein of bacteriophage λ promotes renaturation of DNA. *J. Biol. Chem*. **256**:12636–12639.

Kolodner, R. 1980. Genetic recombination of bacterial plasmid DNA: electron microscopic analysis of *in vitro* intramolecular recombination. *Proc. Natl. Acad. Sci. USA* **77**:4847–4851.

Kolodner, R., R. A. Fishel, and M. Howard. 1985. Genetic recombination of bacterial plasmid DNA: effect of RecF pathway mutations on plasmid recombination in *Escherichia coli*. *J. Bacteriol*. **163**:1060–1066.

Konrad, E. B. 1977. Method for the isolation of *Escherichia coli* mutations with enhanced recombination between chromosomal duplication. *J. Bacteriol*. **130**:167–172.

Konrad, E., and I. R. Lehman. 1974. A conditional lethal mutant of *Escherichia coli* K-12 defective in the 5'→3' exonuclease associated with DNA polymerase I. *Proc. Natl. Acad. Sci. USA* **71**:2048–2051.

Kornberg, A. 1980. *DNA Replication*. W. H. Freeman and Co., San Francisco.

Kowalczykowski, S. C., J. Clow, R. Somani, and A. Varghese. 1987. Effects of the *Escherichia coli* SSB protein on the binding of *Escherichia coli* RecA protein to single stranded DNA. Demonstration of competitive binding and the lack of specific protein-protein interaction. *J. Mol. Biol*. **193**:81–95.

Kowalczykowski, S. C., and R. A. Krupp. 1987. Effects of the *Escherichia coli* SSB protein on the single-stranded DNA-dependent ATPase activity of *Escherichia coli* RecA protein. Evidence that SSB protein facilitates the binding of RecA protein to regions of secondary structure within single-stranded DNA. *J. Mol. Biol*. **193**:97–113.

Krivonogov, S. V. 1984. The *recF*-dependent endonuclease from *Escherichia coli* K-12. Formation and resolution of pBR322 DNA multimer. *Mol. Gen. Genet*. **196**:105–109.

Krivonogov, S. V., and V. A. Novitskaya. 1982. A protein connected with the integrity of the *recF* gene in *Escherichia coli* K-12. *Mol. Gen. Genet*. **187**:302–304.

Kumara, K., and M. Sekiguchi. 1984. Identification of the *uvrD* gene product of *Escherichia coli* as DNA helicase II and its induction by DNA damaging agents. *J. Biol. Chem*. **259**:1560–1565.

Kumara, K., M. Sekiguchi, A. L. Steinum, and E. Seeberg. 1985. Stimulation of the UvrABC enzyme-catalyzed repair reactions by the UvrD protein (DNA helicaseII). *Nucleic Acids Res*. **13**:1485–1492.

Kushner, S. R., H. Nagaishi, and A. J. Clark. 1972. Indirect suppression of *recB recC* mutations by exonuclease I deficiency. *Proc. Natl. Acad. Sci. USA* **69**:1366–1370.

Kushner, S. R., H. Nagaishi, and A. J. Clark. 1974. Isolation of exonuclease VIII: the enzyme associated with the *sbcA* indirect suppressor. *Proc. Natl. Acad. Sci. USA* **71**:3593–3597.

Kushner, S. R., H. Nagaishi, A. Templin, and A. J. Clark. 1971. Genetic recombination in *Escherichia coli*: the role of exonuclease I. *Proc. Natl. Acad. Sci. USA* **68**:824–827.

Laban, A., and A. Cohen. 1981. Interplasmidic and intraplasmidic recombination in *Escherichia coli* K-12. *Mol. Gen. Genet*. **184**:200–207.

Laban, A., Z. Silberstein, and A. Cohen. 1984. The effect of non-homologous DNA sequences on interplasmidic recombination. *Genetics* **108**:39–52.

Lahue, R. S., S.-S. Su, K. Welsh, and P. Modrich. 1987. Analysis of methyl-directed mismatch repair *in vivo*, p. 125–134. *In* R. McMacken and T. J. Kelly (ed.), *DNA Replication and Recombination.* Alan R. Liss, Inc., New York.

Lehman, I. R. 1981. DNA polymerase I of *Escherichia coli*, p. 15–37. *In* P. D. Boyer (ed.), *The Enzymes*, vol. 14. Academic Press, Inc., New York.

Lehman, I. R., and A. L. Nussbaum. 1964. The deoxyribonucleases of *Escherichia coli*. V. On the specificity of exonuclease I (phosphodiesterase). *J. Biol. Chem.* **239:**2628–2636.

Lieb, M. 1987. Bacterial genes *mutL*, *mutS*, and *dcm* participate in repair of mismatches at 5-methylcytosine sites. *J. Bacteriol.* **169:**5241–5246.

Lieberman, R. P., and M. Oishi. 1974. The *recBC* deoxyribonuclease of *Escherichia coli*: isolation and characterization of the subunit proteins and reconstitution of the enzyme. *Proc. Natl. Acad. Sci. USA* **71:**4816–4820.

Lin, P.-F., E. Bardwell, and P. Howard-Flanders. 1977. Initiation of genetic exchanges in λ phage-prophage crosses. *Proc. Natl. Acad. Sci. USA* **74:**291–295.

Little, J. W. 1967. An exonuclease induced by bacteriophage λ. II. Nature of the enzymatic reaction. *J. Biol. Chem.* **242:**679–686.

Little, J. W. 1984. Autodigestion of *lexA* and phage λ repressors. *Proc. Natl. Acad. Sci. USA* **81:**1375–1379.

Little, J. W., S. H. Edmiston, L. Z. Pacelli, and D. W. Mount. 1980. Cleavage of the *Escherichia coli lexA* protein by the *recA* protease. *Proc. Natl. Acad. Sci. USA* **77:**3225–3229.

Little, J. W., and J. E. Harper. 1979. Identification of the *lexA* gene product of *Escherichia coli* K-12. *Proc. Natl. Acad. Sci. USA* **76:**6147–6151.

Little, J. W., and D. W. Mount. 1982. The SOS regulatory system of *Escherichia coli*. *Cell* **29:**11–22.

Lloyd, R. G. 1983. *lexA* dependent recombination in *uvrD* strains of *Escherichia coli*. *Mol. Gen. Genet.* **189:**157–161.

Lloyd, R. G., F. E. Benson, and C. E. Shurvinton. 1984. Effect of *ruv* mutations on recombination and DNA repair in *Escherichia coli* K-12. *Mol. Gen. Genet.* **194:**303–309.

Lloyd, R. G., and C. Buckman. 1985. Identification and genetic analysis of *sbcC* mutations in commonly used *recBC sbcB* strains of *Escherichia coli* K-12. *J. Bacteriol.* **164:**836–844.

Lloyd, R. G., C. Buckman, and F. E. Benson. 1987a. Genetic analysis of conjugational recombination in *Escherichia coli* K12 strains deficient in RecBCD enzyme. *J. Gen. Microbiol.* **133:**2531–2538.

Lloyd, R. G., N. P. Evans, and C. Buckman. 1987b. Formation of recombinant *lacZ*⁺ DNA in conjugational crosses with a *recB* mutant of *Escherichia coli* K12 depends on *recF*, *recJ* and *recO*. *Mol. Gen. Genet.* **209:**135–141.

Lloyd, R. G., S. M. Picksley, and C. Prescott. 1983. Inducible expression of a gene specific to the RecF pathway for recombination in *Escherichia coli* K-12. *Mol. Gen. Genet.* **190:**162–167.

Lloyd, R. G., M. C. Porton, and C. Buckman. 1988. Effect of *recF*, *recJ*, *recO* and *ruv* mutations on ultraviolet survival and genetic recombination in a *recD* strain of *Escherichia coli* K12. *Mol. Gen. Genet.* **212:**317–324.

Lloyd, R. G., and A. Thomas. 1983. On the nature of RecBC and RecF pathways of conjugal recombination in *Escherichia coli*. *Mol. Gen. Genet.* **190:**156–161.

Lloyd, R. G., and A. Thomas. 1984. A molecular model for conjugational recombination in *Escherichia coli* K-12. *Mol. Gen. Genet.* **197:**328–336.

Lovett, S. T., and A. J. Clark. 1983. Genetic analysis of the regulation of the RecF pathway of recombination in *E. coli* K-12. *J. Bacteriol.* **153:**1471–1478.

Lovett, S. T., and A. J. Clark. 1984. Genetic analysis of the *recJ* gene of *Escherichia coli*. *J. Bacteriol.* **157:**190–196.

Lovett, S. T., and A. J. Clark. 1985. Cloning of the *Escherichia coli recJ* chromosomal region and identification of its encoded proteins. *J. Bacteriol.* **162:**280–285.

Low, K. B. 1968. Formation of merodiploids in matings with a class of *rec*⁻ recipient strains of *Escherichia coli* K-12. *Proc. Natl. Acad. Sci. USA* **60:**160–167.

Low, K. B. 1973. Restoration by the *rac* locus of recombinant forming ability in *recB*⁻ and *recC*⁻ merozygotes of *Escherichia coli*. *Mol. Gen. Genet.* **122:**119–130.

Low, R. L., J. M. Kaguni, and A. Kornberg. 1984. Potent catenation of super-coiled and gapped DNA circles by topoisomerase I in the presence of a hydrophilic polymer. *J. Biol. Chem.* **259:**4576–4581.

Lu, A.-L., S. Clark, and P. Modrich. 1983. Methyl-directed repair of DNA base-pair mismatches *in vitro. Proc. Natl. Acad. Sci. USA* **80:**4639–4643.

Lu, A.-L., K. Welsh, S. Clark, S.-S. Su, and P. Modrich. 1984. Repair of DNA base-pair mismatches in extracts of *Escherichia coli. Cold Spring Harbor Symp. Quant. Biol.* **49:**589–596.

Lundblad, V., and N. Kleckner. 1984. Mismatch repair mutations of *Escherichia coli* enhance transposon excision. *Genetics* **109:**3–19.

Lundblad, V., A. F. Taylor, G. R. Smith, and N. Kleckner. 1984. Unusual alleles of *recB* and *recC* stimulate excision of inverted repeat transposons Tn*10* and Tn*5. Proc. Natl. Acad. Sci. USA* **81:** 824–828.

MacKay, V., and S. Linn. 1976. Selective inhibition of the DNAase activity of the *recBC* enzyme by the DNA binding protein from *Escherichia coli. J. Biol. Chem.* **251:**3716–3719.

Mahajan, S. K., and A. R. Datta. 1977. Nature of recombinants produced by the RecBC and the RecF pathways in *Escherichia coli. Nature* (London) **266:**652–653.

Mahajan, S. K., and A. R. Datta. 1979. Mechanism of recombination by the RecBC and the RecF pathways following conjugation in *Escherichia coli* K12. *Mol. Gen. Genet.* **169:**67–78.

Mahajan, S. K., N. N. Pandit, and J. F. Sarkari. 1984. Host functions in amplification and deamplification of Tn9 in *Escherichia coli* K-12: a new model for amplification. *Cold Spring Harbor Symp. Quant. Biol.* **49:**443–451.

Makino, O., S. Ikawa, Y.-I. Shibata, H. Maedo, T. Ando, and T. Shibata. 1987. recA protein-promoted recombination reaction consists of two independent processes, homologous matching and processive unwinding. A study involving an anti-recA protein monoclonal IgG. *J. Biol. Chem.* **262:**12237–12246.

Margolin, P. 1987. Generalized transduction, p. 1154–1168. *In* F. C. Neidhardt, J. L. Ingraham, K. B. Low, B. Magasanik, M. Schaechter, and H. E. Umbarger (ed.), *Escherichia coli and Salmonella typhimurium: Cellular and Molecular Biology.* American Society for Microbiology, Washington, D.C.

Marinus, M. G. 1973a. Isolation of deoxyribonucleic acid methylase mutants of *Escherichia coli* K-12. *J. Bacteriol.* **114:**1143–1150.

Marinus, M. G. 1973b. Location of DNA methylation genes on the *Escherichia coli* K-12 genetic map. *Mol. Gen. Genet.* **127:**47–55.

Marinus, M. G. 1987. DNA methylation in *Escherichia coli. Annu. Rev. Genet.* **21:**113–131.

Marinus, M. G., M. Carranway, A. Z. Frey, L. Brown, and J. A. Avraj. 1983. Insertion mutations in the *dam* gene of *Escherichia coli* K-12. *Mol. Gen. Genet.* **192:**288–289.

Masters, M. 1977. The frequency of P1 transduction of the genes of *Escherichia coli* as a function of chromosomal position: preferential transduction of the origin of replication. *Mol. Gen. Genet.* **155:** 197–202.

Masters, M., B. J. Newman, and C. M. Henry. 1984. Reduction of marker discrimination in transductional recombination. *Mol. Gen. Genet.* **196:**85–90.

Matfield, M., R. Badawi, and W. J. Brammar. 1985. *Rec*-dependent and *Rec*-independent recombination of plasmid-borne duplication in *Escherichia coli. Mol. Gen. Genet.* **199:**518–523.

Matsumoto, T., and H. Ikeda. 1983. Role of R loops in *recA*-independent homologous recombination of bacteriophage λ. *J. Virol.* **45:**971–976.

McEntee, K. 1977. Protein X is the product of the *recA* gene of *Escherichia coli. Proc. Natl. Acad. Sci. USA* **74:**5275–5279.

McGraw, B. R., and M. G. Marinus. 1980. Isolation and characterization of Dam$^+$ revertants and suppressor mutations that modify secondary phenotypes of *dam3* strains of *Escherichia coli* K-12. *Mol. Gen. Genet.* **178:**309–315.

Meselson, M. 1972. Formation of hybrid DNA by rotary diffusion during genetic recombination. *J. Mol. Biol.* **71:**795–798.

Meselson, M. 1988. Methyl-directed repair of DNA mismatches. *In* K. B. Low (ed.), *The Recombination of Genetic Material.* Academic Press, Inc., New York.

Meselson, M., and J. J. Weigle. 1961. Chromosome breakage accompanying genetic recombination in bacteriophage. *Proc. Natl. Acad. Sci. USA* **47:**857–868.

Miller, J. E., and S. D. Barbour. 1977. Metabolic characterization of the viable, residually dividing and non-dividing cell classes of recombination-deficient strains of *Escherichia coli*. *J. Bacteriol.* **130:** 160–166.

Mizuuchi, K., B. Kemper, J. Hays, and R. A. Weisberg. 1982. T4 endonuclease VII cleaves Holliday structures. *Cell* **29:**357–365.

Mizuuchi, K., M. Mizuuchi, M. H. O'Dea, and M. Gellert. 1984. Cloning and simplified purification of *Escherichia coli* DNA gyrase A and B proteins. *J. Biol. Chem.* **259:**9199–9201.

Monk, M., and J. Kinross. 1972. Conditional lethality of *recA* and *recB* derivatives of a strain of *Escherichia coli* K-12 with a temperature-sensitive deoxyribonucleic acid polymerase I. *J. Bacteriol.* **109:**971–978.

Moreau, P. L. 1987. Effects of overproduction of single-stranded DNA binding protein on RecA protein-dependent processes in *Escherichia coli*. *J. Mol. Biol.* **194:**621–634.

Mount, D. W. 1977. A mutant of *Escherichia coli* showing constitutive expression of the lysogenic induction and error-prone DNA repair pathways. *Proc. Natl. Acad. Sci. USA* **74:**300–304.

Mount, D. W., K. B. Low, and S. J. Edmiston. 1972. Dominant mutations (*lex*) in *Escherichia coli* K-12 which affect radiation sensitivity and frequency of ultraviolet light-induced mutations. *J. Bacteriol.* **112:**886–893.

Muniyappa, K., and C. M. Radding. 1986. The homologous recombination system of phage λ. Pairing activities of β-protein. *J. Biol. Chem.* **261:**7472–7478.

Nakayama, H., K. Nakayama, R. Nakayama, N. Irino, Y. Nakayama, and P. C. Hanawalt. 1984. Isolation and genetic characterization of a thymineless death-resistant mutant of *Escherichia coli* K12. Identification of a new mutation (*recQ1*) that blocks the RecF recombination pathway. *Mol. Gen. Genet.* **195:**474–480.

Nakayama, K., N. Irino, and H. Nakayama. 1985. The *recQ* gene of *Escherichia coli* K-12: molecular cloning and isolation of insertion mutants. *Mol. Gen. Genet.* **200:**266–271.

Oeda, K., T. Horiuchi, and M. Sekiguchi. 1981. Molecular cloning of the *uvrD* gene of *Escherichia coli* that controls ultraviolet sensitivity and spontaneous mutation frequency. *Mol. Gen. Genet.* **184:**191–199.

Oishi, M. 1969. An ATP-dependent deoxyribonuclease from *Escherichia coli* with a possible role in genetic recombination. *Proc. Natl. Acad. Sci. USA* **64:**1292–1299.

Oppenheim, A. B., and M. Riley. 1966. Molecular recombination following conjugation in *Escherichia coli*. *J. Mol. Biol.* **20:**331–357.

Otsuji, N., H. Iyehara, and Y. Hideshima. 1974. Isolation and characterization of an *Escherichia coli ruv* mutant which forms nonseptate filaments after low doses of ultraviolet light irradiation. *J. Bacteriol.* **117:**337–344.

Panasenko, S. M., J. R. Cameron, R. W. Davis, and I. R. Lehman. 1977. Five hundredfold overproduction of DNA ligase after induction of a lambda lysogen constructed *in vitro*. *Science* **196:** 188–189.

Pauling, C., and L. Hamm. 1968. Properties of a temperature-sensitive radiation-sensitive mutant of *Escherichia coli*. *Proc. Natl. Acad. Sci. USA* **60:**1495–1502.

Peterson, K. R., K. F. Wertman, D. W. Mount, and M. G. Marinus. 1985. Viability of *Escherichia coli* K-12 DNA adenine methylase (*dam*) mutants requires increased expression of specific genes in the SOS regulon. *Mol. Gen. Genet.* **201:**14–19.

Phillips, G. J., and S. R. Kushner. 1987. Determination of the nucleotide sequence for the exonuclease I structural gene (*sbcB*) of *Escherichia coli*. *J. Biol. Chem.* **262:**455–459.

Picksley, S. M., P. V. Attfield, and R. G. Lloyd. 1984a. Repair of DNA double-strand breaks in *Escherichia coli* K12 requires a functional *recN* product. *Mol. Gen. Genet.* **195:**267–274.

Picksley, S. M., R. G. Lloyd, and C. Buckman. 1984b. Genetic analysis and regulation of inducible recombination in *Escherichia coli* K-12. *Cold Spring Harbor Symp. Quant. Biol.* **49:**469–474.

Ponticelli, A. S., D. W. Schultz, A. F. Taylor, and G. R. Smith. 1985. Chi-dependent DNA strand cleavage by RecBC enzyme. *Cell* **41:**145–151.

Porter, R. D. 1983. Specialized transduction with λ*plac5*: involvement of the RecE and the RecF recombination pathways. *Genetics* **105:**247–257.

Porter, R. D., T. McLaughlin, and B. Low. 1978. Transduction versus "conjuduction": evidence for multiple roles for exonuclease V in genetic recombination in *Escherichia coli*. *Cold Spring Harbor Symp. Quant. Biol.* **43:**1043–1046.

Porter, R. D., R. A. Welliver, and T. A. Witkowski. 1982. Specialized transduction with λ p*lac5*: dependence on *recB*. *J. Bacteriol.* **150:**1485–1488.

Potter, H., and D. Dressler. 1976. On the mechanism of genetic recombination: electron microscopic observation of recombination intermediates. *Proc. Natl. Acad. Sci. USA* **73:**3000–3004.

Pyhtila, M. J., and J. E. Syvaoja. 1980. DNA polymerase I and *recBC* enzyme support the covalent closing of hydrogen-bonded λ DNA circles in extracts of *Escherichia coli* cells. *Eur. J. Biochem.* **112:**125–130.

Radding, C. M. 1978. Genetic recombination: strand transfer and mismatch repair. *Annu. Rev. Biochem.* **47:**847–880.

Radding, C. M. 1982. Homologous pairing and strand exchange in genetic recombination. *Annu. Rev. Genet.* **16:**405–437.

Ream, L. W., and A. J. Clark. 1983. Cloning and deletion mapping of the *recF dnaN* region of the *Escherichia coli* chromosome. *Plasmid* **10:**101–110.

Ream, L. W., L. Margossian, A. J. Clark, F. G. Hansen, and K. von Meyenburg. 1980. Genetic and physical mapping of *recF* in *Escherichia coli* K-12. *Mol. Gen. Genet.* **180:**115–121.

Rogers, S. G., and B. Weiss. 1980. Cloning of the exonuclease III gene of *Escherichia coli*. *Gene* **11:**187–196.

Rostas, K., S. J. Morton, S. M. Picksley, and R. G. Lloyd. 1987. Nucleotide sequence and LexA regulation of the *Escherichia coli recN* gene. *Nucleic Acids Res.* **15:**5041–5050.

Rothman, R. H., T. Kato, and A. S. Clark. 1975. The beginning of an investigation of the role of *recF* in the pathway of metabolism of ultraviolet irradiated DNA in *Escherichia coli*, p. 283–291. *In* P. C. Hanawalt and R. B. Setlow (ed.), *Molecular Mechanisms for Repair of DNA*, vol. 5, part A. Plenum Publishing Corp., New York.

Rupp, W. D., C. E. Wilde III, D. L. Reno, and P. Howard-Flanders. 1971. Exchanges between DNA strands in ultraviolet-irradiated *Escherichia coli*. *J. Mol. Biol.* **61:**25–44.

Salles, B., and C. Paoletti. 1983. Control of UV induction of *recA* protein. *Proc. Natl. Acad. Sci. USA* **80:**65–69.

Sancar, A., C. Stachelek, W. Konigsberg, and W. D. Rupp. 1980. Sequences of the *recA* gene and protein. *Proc. Natl. Acad. Sci. USA* **77:**2611–2615.

Sancar, A., K. R. Williams, J. W. Chase, and W. D. Rupp. 1981. Sequences of the *ssb* gene and protein. *Proc. Natl. Acad. Sci. USA* **78:**4274–4278.

Sandri, R. M., and H. Berger. 1980. Bacteriophage P1-mediated generalized transduction in *Escherichia coli*: structure of abortively transduced DNA. *Virology* **106:**30–40.

Sargentini, N. J., and K. C. Smith. 1987. Quantitation of the *recA*, *recB*, *recC*, *recF*, *recJ*, *recN*, *lexA*, *radA*, *radB*, *uvrD* and *umuC* genes in the repair of x-ray induced DNA double strand breaks in *Escherichia coli*. *Radiat. Res.* **107:**58–72.

Sarthy, P. V., and M. Meselson. 1976. Single burst study of *rec*- and *red*-mediated recombination in bacteriophage lambda. *Proc. Natl. Acad. Sci. USA* **73:**4613–4617.

Sasaki, M., T. Fujiyoshi, K. Shimada, and Y. Takagi. 1982. Fine structure of the *recB* and *recC* gene region of *Escherichia coli*. *Biochem. Biophys. Res. Commun.* **109:**414–422.

Scaife, J., and J. D. Gross. 1963. The mechanism of chromosome mobilization by an F-prime factor in *Escherichia coli* K-12. *Genet. Res.* **4:**328–331.

Schultz, D. W., A. F. Taylor, and G. R. Smith. 1983. *Escherichia coli* RecBC pseudorevertants lacking Chi recombinational hotspot activity. *J. Bacteriol.* **155:**664–680.

Seelke, R., B. Kline, R. Aleff, R. D. Porter, and M. S. Shields. 1987. Mutations in the *recD* gene of *Escherichia coli* that raise the copy number of certain plasmids. *J. Bacteriol.* **169:**4841–4844.

Seifert, J. S., and R. D. Porter. 1984. Enhanced recombination between λ *plac5* and F*42lac*: identification of *cis*- and *trans*-acting factors. *Proc. Natl. Acad. Sci. USA* **81:**7500–7504.

Shen, P., and H. V. Huang. 1986. Homologous recombination in *Escherichia coli*: dependence on substrate length and homology. *Genetics* **112:**441–457.

Shurvinton, C. E., and R. G. Lloyd. 1982. Damage to DNA induces expression of the *ruv* gene of *Escherichia coli*. *Mol. Gen. Genet.* **185:**352–355.

Shurvinton, C. E., R. G. Lloyd, F. E. Benson, and P. V. Attfield. 1984. Genetic analysis and molecular cloning of the *Escherichia coli ruv* gene. *Mol. Gen. Genet.* **194:**322–329.

Siddiqi, O. 1963. Incorporation of parental DNA into genetic recombinants of *E. coli. Proc. Natl. Acad. Sci. USA* **49:**589–592.

Siddiqi, O., and M. Fox. 1973. Integration of donor DNA in bacterial conjugation. *J. Mol. Biol.* **77:**101–123.

Siegel, E. C. 1973. Ultraviolet sensitive mutator strain of *Escherichia coli* K-12. *J. Bacteriol.* **113:**145–160.

Siegel, E. C. 1981. Complementation studies with the repair deficient *uvrD3*, *uvrE156* and *recL152* mutations in *Escherichia coli. J. Mol. Biol.* **121:**524–530.

Smith, G. R. 1983. General recombination, p. 175–205. *In* R. W. Hendrix, J. W. Roberts, F. W. Stahl, and R. A. Weisberg (ed.), *Lambda II.* Cold Spring Harbor Laboratory, Cold Spring Harbor, N.Y.

Smith, G. R. 1987. Mechanism and control of homologous recombination in *Escherichia coli. Annu. Rev. Genet.* **21:**179–201.

Smith, G. R. 1988. Homologous recombination in procaryotes. *Microbiol. Rev.* **52:**1–28.

Smith, G. R., S. K. Amundsen, A. M. Chaudhury, K. C. Cheng, A. S. Ponticelli, C. M. Roberts, D. W. Schultz, and A. F. Taylor. 1984. Role of RecBC enzyme and Chi sites in homologous recombination. *Cold Spring Harbor Symp. Quant. Biol.***49:**485–495.

Smith, K. C., and R. C. Sharma. 1987. A model for the *recA*-dependent repair of excision gaps in UV-irradiated *Escherichia coli. Mutat. Res.* **183:**1–9.

Smith, T. A. G., and J. B. Hays. 1985. Repair and recombination of nonreplicating UV-irradiated phage DNA in *E. coli.* II. Stimulation of RecF-dependent recombination by excision repair of cyclobutane pyrimidine dimers and other photoproducts. *Mol. Gen. Genet.* **201:**393–401.

Stacey, K. A., and R. G. Lloyd. 1976. Isolation of Rec⁻ mutants from an F-prime merodiploid strain of *Escherichia coli* K-12. *Mol. Gen. Genet.* **143:**223–232.

Stahl, F. W. 1979. Special sites in generalized recombination. *Annu. Rev. Genet.* **13:**7–24.

Stahl, F. W. 1984. A perspective on double strand cuts in genetic recombination, p. 429–440. *In* V. L. Chopra, B. C. Joshi, T. R. Sharma, and H. C. Bansal (ed.), *Genetics: New Frontiers.* Proceedings of the XV International Congress of Genetics, vol. 1. Oxford and IBH Publishing Co., New Delhi, India.

Stahl, F. W. 1986. Roles of double-strand breaks in generalized recombination. *Prog. Nucleic Acid Res. Mol. Biol.* **33:**169–194.

Stahl, F. W., K. D. McMilin, M. M. Stahl, J. M. Crasemann, and S. Lam. 1974. The distribution of crossovers along unreplicated lambda bacteriophage chromosomes. *Genetics* **77:**395–408.

Stahl, F. W., and M. M. Stahl. 1977. Recombination pathway specificity of Chi. *Genetics* **86:**715–725.

Stahl, F. W., and M. M. Stahl. 1986. DNA synthesis at the site of Red-mediated exchange in phage λ. *Genetics* **113:**1–12.

Sternglanz, R., S. diNardo, K. A. Volkel, Y. Nishimura, Y. Hirota, K. Becherer, L. Zumstein, and J. C. Wang. 1981. Mutations in genes coding for *Escherichia coli* DNA topoisomerase I affect transcription and transposition. *Proc. Natl. Acad. Sci. USA* **78:**2747–2751.

Strike, P., and P. T. Emmerson. 1972. Coexistence of *polA* and *recB* mutations of *Escherichia coli* in the presence of *sbc*, a mutation which indirectly suppresses *recB. Mol. Gen. Genet.* **116:**177–180.

Su, S.-S., and P. Modrich. 1986. *Escherichia coli mutS*-encoded protein binds to mismatched base pairs. *Proc. Natl. Acad. Sci. USA* **83:**5057–5067.

Swenberg, S. L., and J. C. Wang. 1987. Cloning and sequencing of the *Escherichia coli gyrA* gene coding for the A subunit of DNA gyrase. *J. Mol. Biol.* **197:**729–736.

Swenson, P. A., and I. L. Norton. 1986. RecBC enzyme activity is required for far-UV induced respirational shut-off in *Escherichia coli* K-12. *Mutat. Res.* **159:**13–21.

Symington, L. S., P. Morrison, and R. Kolodner. 1985. Intramolecular recombination of linear DNA catalyzed by the *Escherichia coli* RecE recombination system. *J. Mol. Biol.* **186:**515–525.

Takahashi, S., C. Homs, A. Chu, and D. T. Denhardt. 1979. The *rep* mutation. VI. Purification and properties of *Escherichia coli rep* protein, DNA helicase III. *Can. J. Biochem.* **57:**855–866.

Taylor, A., and G. R. Smith. 1980. Unwinding and rewinding of DNA by the RecBC enzyme. *Cell* **22:**447–457.

Taylor, A. F., P. G. Siliciano, and B. Weiss. 1980. Cloning of the *dut* (deoxyuridine triphosphatase) gene of *E. coli. Gene* **9:**321–336.

Taylor, A. F., and B. Weiss. 1982. Role of exonuclease III in the base excision repair of uracil-containing DNA. *J. Bacteriol.* **151:**351–357.

Telander-Muskavitch, K. M., and S. Linn. 1981. *recBC*-like enzymes: the exonuclease V deoxyribonucleases, p. 233–250. *In* P. D. Boyer (ed.), *The Enzymes*, vol. 14. Academic Press, Inc., New York.

Telander-Muskavitch, K. M., and S. Linn. 1982. A unified mechanism for the nuclease and unwinding activities of the *recBC* enzyme of *Escherichia coli. J. Biol. Chem.* **257:**2641–2648.

Templin, A., S. R. Kushner, and A. J. Clark. 1972. Genetic analysis of mutations indirectly suppressing *recB* and *recC* mutations. *Genetics* **72:**205–215.

Templin, A., L. Margossian, and A. J. Clark. 1978. Suppressibility of *recA*, *recB*, and *recC* mutations by nonsense suppressors. *J. Bacteriol.* **134:**590–596.

Tessman, E. S., and P. K. Peterson. 1985. Isolation of protease-proficient, recombination-deficient *recA* mutants of *Escherichia coli. J. Bacteriol.* **163:**688–695.

Thaler, D. S., M. M. Stahl, and F. W. Stahl. 1987. Tests of the double-strand-break repair model for Red-mediated recombination of phage λ and plasmid λdv. *Genetics* **116:**501–511.

Thoms, B., and W. Wackernagel. 1984. Genetic control of damage-inducible restriction alleviation in *Escherichia coli* K-12: an SOS function not repressed by *lexA. Mol. Gen. Genet.* **197:**297–303.

Thoms, B., and W. Wackernagel. 1987. Regulatory role of *recF* in the SOS response of *Escherichia coli*: impaired induction of SOS genes by UV irradiation and nalidixic acid in a *recF* mutant. *J. Bacteriol.* **169:**1731–1736.

Tomizawa, J. 1967. Molecular mechanisms of recombination in bacteriophage: joint molecules and their conversion to recombinant molecules. *J. Cell Physiol.* **70**(Suppl. 1):201–214.

Tomizawa, J., and H. Ogawa. 1972. Structural genes of an ATP-dependent deoxyribonuclease of *Escherichia coli. Nature* (London) *New Biol.* **239:**14–16.

Unger, R. C., and A. J. Clark. 1972. Interaction of the recombination pathways of bacteriophage λ and host *Escherichia coli*: effects on λ recombination. *J. Mol. Biol.* **70:**531–537.

Vales, L. D., B. A. Rabin, and J. W. Chase. 1983. Isolation and preliminary characterization of *Escherichia coli* mutants deficient in exonuclease VII. *J. Bacteriol.* **155:**1116–1122.

Vapnek, D., N. K. Alton, C. L. Bassett, and S. R. Kushner. 1976. Amplification in *Escherichia coli* of enzymes involved in genetic recombination: construction of hybrid ColE1 plasmids carrying the structural gene for exonuclease I. *Proc. Natl. Acad. Sci. USA* **13:**3492–3496.

Vinogradskaya, G. R., I. Yu. Goryshin, and V. A. Lantsov. 1986. SSB gene in the inducible and constitutive pathways of recombination in *Escherichia coli* K-12. *Dokl. Akad. Nauk USSR* **290**(1):228–231.

Volkert, M. R., and M. A. Hartke. 1984. Suppression of *Escherichia coli recF* mutations by *recA*-linked *srfA* mutations. *J. Bacteriol.* **157:**498–506.

Volkert, M. R., and M. A. Hartke. 1987. Effects of *Escherichia coli recF* suppression mutation, *rec801*, on *recF*-dependent DNA repair associated phenomena. *Mutat. Res.* **184:**181–186.

Wackernagel, W., and C. M. Radding. 1974. Transformation and transduction of *Escherichia coli*: the nature of recombinants formed by Rec, RecF and λRed, p. 111–121. *In* R. F. Grell (ed.), *Mechanisms in Recombination*. Plenum Publishing Corp., New York.

Waldstein, E. A. 1979. Role of exonucleases V and VIII in adenosine 5'-triphosphate and deoxynucleotide triphosphate-dependent strand break repair in toluenized *Escherichia coli* cells treated with X-rays. *J. Bacteriol.* **139:**1–7.

Walker, G. C. 1984. Mutagenesis and inducible responses to deoxyribonucleic acid damage in *Escherichia coli. Microbiol. Rev.* **48:**60–93.

Wang, J. C. 1971. Interaction between DNA and an *Escherichia coli* protein ω. *J. Mol. Biol.* **55:**523–533.

Wang, J. C. 1981. Type I DNA topoisomerases, p. 331–344. *In* P.D. Boyer (ed.), *The Enzymes*, vol. 14, part A. Academic Press, Inc., New York.

Wang, J. C., and K. Bechere. 1983. Cloning of the gene *topA* encoding DNA topoisomerase I and physical mapping of the *cysB-topA-trp* region of *Escherichia coli. Nucleic Acids Res.* **11:**1773–1790.

Wang, T. V., and K. C. Smith. 1983. Mechanisms for *recF*-dependent and *recB*-dependent pathways of postreplication repair in UV-irradiated *Escherichia coli uvrB. J. Bacteriol.* **156:**1093–1098.

Wang, T. V., and K. C. Smith. 1986. *recF* (Srf) suppression of *recF* deficiency in the postreplication repair of UV-irradiated *Escherichia coli* K-12. *J. Bacteriol.* **168:**940–946.

Wang, T. V., and K. C. Smith. 1987. Inviability of *dam recA* and *dam recB* cells of *Escherichia coli* is correlated with their inability to repair DNA double-strand breaks produced by mismatch repair. *J. Bacteriol.* **165:**1023–1025.

Wang, W., and E. S. Tessman. 1986. Location of functional regions of the *Escherichia coli* RecA protein by DNA sequence analysis of RecA protease-constitutive mutants. *J. Bacteriol.* **168:**901–910.

Watt, V. M., C. J. Ingles, M. S. Urdea, and W. J. Rutter. 1985. Homology requirements for recombination in *Escherichia coli*. *Proc. Natl. Acad. Sci. USA* **82:**4768–4772.

Weiss, B. 1981. Exodeoxyribonucleases of *Escherichia coli*, p. 203–231. *In* P. D. Boyer (ed.), *The Enzymes*, vol. 14. Academic Press, Inc., New York.

West, S. C., J. K. Countryman, and P. Howard-Flanders. 1983. Enzymatic formation of biparental figure-eight molecules from plasmid DNA and their resolution in *E. coli*. *Cell* **32:**817–829.

Whitehouse, H. L. K. 1963. A theory of crossing-over by means of hybrid deoxyribonucleic acid. *Nature* (London) **199:**1034–1040.

Wilkins, B. M. 1969. Chromosome transfer from F-*lac*⁺ strains of *Escherichia coli* K-12 mutant at *recA*, *recB*, or *recC*. *J. Bacteriol.* **98:**599–604.

Willets, N. S., and A. J. Clark. 1969. Characteristics of some multiply recombination-deficient strains of *Escherichia coli*. *J. Bacteriol.* **100:**231–239.

Willets, N. S., and D. W. Mount. 1969. Genetic analysis of recombination-deficient mutants of *Escherichia coli* K-12 carrying *rec* mutations cotransducible with *thyA*. *J. Bacteriol.* **100:**923–934.

Willets, N., and B. Wilkins. 1984. Processing of plasmid DNA during bacterial conjugation. *Microbiol. Rev.* **48:**24–41.

Willis, D. K., K. E. Fouts, S. D. Barbour, and A. J. Clark. 1983. Restriction nuclease and enzymatic analysis of transposon-induced mutations of the Rac prophage which affect expression and function of *recE* in *Escherichia coli* K-12. *J. Bacteriol.* **156:**727–736.

Willis, D. K., L. H. Satin, and A. J. Clark. 1985. Mutation-dependent suppression of *recB21 recC22* by a region cloned from the Rac prophage of *Escherichia coli* K-12. *J. Bacteriol.* **162:**1166–1172.

Yajko, E. M., M. C. Valentine, and B. Weiss. 1974. Mutants of *Escherichia coli* with altered deoxyribonucleases. II. Isolation and characterization of mutants for exonuclease I. *J. Mol. Biol.* **85:**223–243.

Young, K. K. Y., and G. Edlin. 1983. Physical and genetical analysis of bacteriophage T4 generalized transduction. *Mol. Gen. Genet.* **192:**241–246.

Youngs, D. A., and I. A. Bernstein. 1973. Involvement of the *recB-recC* nuclease (exonuclease V) in the process of X-ray-induced deoxyribonuclease acid degradation in radiosensitive strains of *Escherichia coli* K-12. *J. Bacteriol.* **113:**901–906.

Zieg, J., and S. R. Kushner. 1977. Analysis of genetic recombination between two partially deleted lactose operons of *Escherichia coli* K-12. *J. Bacteriol.* **131:**123–132.

Zieg, J., V. F. Maples, and S. R. Kushner. 1978. Recombination levels of *Escherichia coli* K-12 mutants deficient in various replication, recombination, or repair genes. *J. Bacteriol.* **134:**958–966.

Chapter 4

Mapping and Map Distortions in Bacteriophage Crosses

Gisela Mosig

I. INTRODUCTION

General or homologous recombination, exchange between segments of DNA that share extensive homology, is the basis for genetic mapping, a powerful tool in many biological analyses. Mapping experiments, if done well, allow correct ordering of mutated sites and, in many cases, estimates of the distances between them. Incidentally, they often reveal whether a change in sequence of a specific gene is the physiologically relevant mutation. Genetic linkage maps, like many other kinds of maps, can be constructed from data sets by applying logical rules, regardless of physical realities. The criteria by which such linkage maps are judged are additivity of distances and self-consistency of marker orders, but not necessarily congruence with distances in DNA molecules or chromosomes. On the other hand, many investigators want to use genetic mapping mainly to position certain mutations within their favorite chromosomes or DNA segments.

    Map distortions, i.e., the lack of correlation between genetic and physical distances in certain regions, became apparent when distances between mutated sites could be measured by methods other than recombination. They exposed the original assumption of equal probability of exchange per unit length of DNA and random distribution of exchanges as an oversimplification. Within short intervals multiple exchanges often appear highly correlated (high, or localized, negative interference). In other situations, there appears to be positive interference between exchanges. Both map distortions and interference of exchanges are contributing to our understanding of the mechanisms of recombination and of the enzymes that mediate its various steps. Incorporating the lessons learned into mapping strategies, in turn, results in better maps. This intimate relationship of theoretical and practical considerations in bacteriophage crosses is the topic of this chapter. It is probably inevitable that this view imposes some dichotomy on certain aspects of theoretical considerations that permeate map constructions. The reader will find an excellent treatment of the theories behind these considerations in the book by Stahl (1979a).

In discussing the rationale for mapping and possible reasons for map distortions, it is useful to relate them to models that describe the many pathways of recombination. As far as possible, I shall first deal with general aspects that apply to all phage crosses and then comment on certain aspects that appear unique to a specific phage.

Map distortions, i.e., lack of congruence between genetic and physical distances, can be caused by (i) natural hot spots of recombination in the wild-type DNA sequence, (ii) so-called marker effects, which are apparent only when certain sequence alterations are involved in the crosses, and (iii) phage packaging mechanisms, which can cut linear DNA from replicating concatemers, thereby separating genetic markers, in various ways.

In addition to such map distortions, so-called map expansion can be seen in certain situations; i.e., the summed recombination frequencies derived from crosses between several closely linked markers appear smaller than recombination frequencies derived from crosses between the outside markers of such a set (Holliday, 1964; Holliday, 1968; Stahl, 1979a).

I will use the words "hot spots," "marker effects," and "end effects" to distinguish among the first three kinds of map distortion and the term "map expansion" in the way originally defined by Holliday (1968), even though Weinstock and Botstein (1980) and Israel (1980) called the map distortion caused by the phage P22 and T1 *pac* sites "map expansion."

Obviously, different reasons for map distortions are more interdependent than any classification can imply since they are related to various pathways of recombination and the protein complexes that mediate them. In most phages several overlapping pathways operate, and their relative contributions can vary with the experimental protocols designed to investigate them; these situations can lead to confusion and apparent controversies.

II. OVERVIEW OF RECOMBINATION MODELS

As discussed elsewhere in this book, all current models involve the formation of "hybrid" or "heteroduplex" DNA as intermediates in recombination (Hershey, 1958). Complementary single-stranded segments of different parental origin

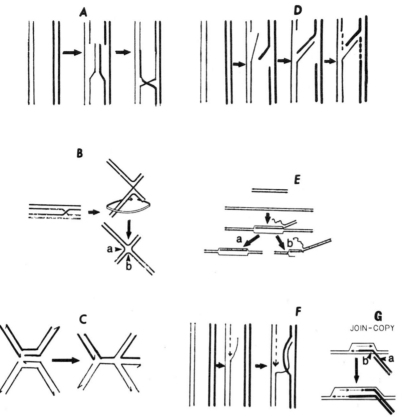

FIGURE 1. Models for initiation and propagation of heteroduplex regions in DNA. (A) Holliday (1964), (B) Sobell (1972) and Dressler and Potter (1982), (C) Broker and Lehman (1971), (D) Whitehouse (1963), (E) Fox (1966), (F) Meselson and Radding (1975) and Hotchkiss (1971), (G) Luder and Mosig (1982) and Skalka (1974). Cuts a and b, marked with arrowheads, would yield patch- and splice-type recombinants, respectively. (Modified from Mosig et al., 1979.)

are joined via base pairing in heteroduplex regions of variable lengths. When genetic markers are involved in the hybrid regions, so-called partial heterozygotes can be detected (Hershey and Chase, 1951). Extensive base pairing in the hybrid regions is responsible for the remarkably high fidelity of general recombination.

Whitehouse (1963) and Holliday (1964) proposed prototype molecular models for formation and resolution of hybrid DNA and implications for analyses of genetic fine structure. Current models differ mainly in the details of how hybrid DNA is formed and extended, and how the branched intermediates are resolved to yield final linear recombinant structures (Fig. 1). The models can be classified (somewhat arbitrarily) according to the way in which pairing of complementary strands is initiated. In one class (Fig. 1A and B), strands of the same polarity are exchanged at identical or closely spaced positions, e.g., by the concerted action of recombination enzymes. These models imply that some pairing occurs prior to cutting and joining, by binding of proteins, such as Z-DNA–binding proteins and topoisomerases (Kmiec and Holloman, 1986; Fishel et al., 1988), or by pairing of double strands in four-stranded structures, or both. Pairing of homologous double strands is stereochemically feasible (McGavin, 1971; Wilson, 1979; Hopkins, 1986). It is thought by some investigators to be insufficient for precise alignment during homologous recombination, but this assumption is not proven.

The other class of models (Fig. 1C through G) postulates that a single-stranded segment derived from one parental duplex pairs with the complementary strand from a second parental duplex, either within a single-stranded region (a gap) in the second parental duplex or by displacing one of the strands in a displacement loop ("D loop"). A possible role of triple-stranded DNA (Moser and Dervan, 1987) in recombination remains to be investigated.

Single-stranded termini on duplex DNA can be generated by (i) partial nucleolytic degradation (Broker and Lehman, 1971), (ii) partial unwinding (Fox, 1966), (iii) displacement synthesis from nicks in DNA (Meselson and Radding, 1975), or (iv) DNA replication (Watson, 1972; Broker, 1973; Luder and Mosig, 1982; Dannenberg and Mosig, 1983).

In all models, small heteroduplex regions formed at first are extended by branch migration. In addition, when the invading strand has a 3' terminus, DNA synthesis can be initiated from the recombinational intermediate (Skalka, 1974; Mosig et al., 1979; Stahl, 1986). In some phages, like T4, this mode of initiating replication forks is essential for viral growth (Luder and Mosig, 1982) since the transcriptional program inactivates initiation from primary origins (for review, see Mosig, 1987). The same is probably true for T1 (Pugh and Ritchie, 1984). In other phages, like λ, this mode, although it is not essential, augments viral DNA replication (Skalka, 1974; for reviews, see Furth and Wickner, 1983; Smith, 1983).

Eventually, the branched intermediates have to be resolved to nonbranched heteroduplexes (and eventually recombinants) by cutting across the H junction, by branch migration to the ends of the molecules and cutting of the resulting Y junction by endonuclease VII (gp49) (Minagawa and Ryo, 1979; Mizuuchi et al., 1982; Kemper et al., 1984; Jensch and Kemper, 1986), or by DNA replication, branch migration, or packaging (Mosig et al., 1984).

Most recombinational intermediates can be cut in two alternative ways which either leave outside markers (or DNA sequences) in the parental configuration (e.g., a in Fig. 1B, E, or G) or exchange outside markers (e.g., b in Fig. 1B, E, or G) (for review, see Dressler and Potter, 1982). These two types are called patch and splice heteroduplexes, respectively (Stahl, 1979a). (They have also been called insertion-type and crossover-type heteroduplexes, respectively.) Mismatches in the heteroduplex regions can be repaired by several different mechanisms. Unrepaired heteroduplexes will eventually segregate as homoduplexes after replication. If they are packaged before replication, they give rise to progeny that differ in one or more genetic markers.

On average, the various cuts implicit in these different steps of recombination are the basis for construction of linear recombination maps (Benzer, 1961). Marker effects and hot spots of recombination can affect any or several of these steps. Since recombination is thought to be mediated by complex protein machines (Alberts, 1984; Mosig et al., 1984), preferred entry sites for such machines can also affect recombination (Stahl, 1979a, b, 1986; Smith, 1988b; D. S. Thaler, E. Sampson, I. Siddiqi, S. M. Rosenberg, F. W. Stahl, and M. Stahl, *in* E. Friedberg and P. Hanawalt, ed., *Mechanisms and Consequences of DNA Damage Processing*, in press).

To circumvent problems associated with marker effects in fine-structure mapping, three-factor crosses are required. Because in phage crosses one usually scores only wild-type recombinants, two different crosses are done in which the markers are arranged in alternative ways

$$a^+ \quad b^+ \quad c \qquad \text{and} \qquad a^+ \quad b \quad c^+$$

$$\overline{\qquad\qquad\qquad} \qquad\qquad \overline{\qquad\qquad\qquad}$$

$$a \quad b \quad c^+ \qquad\qquad a \quad b^+ \quad c$$

$$\text{I} \qquad\qquad\qquad\qquad \text{II}$$

such that a single exchange (I) or double exchanges (II) are required to generate the selected $a^+b^+c^+$ recombinant. If exchanges occur truly at random, the double exchanges should be much less frequent than the single exchanges. In practice, phage recombinants requiring two exchanges between closely linked markers are approximately half as frequent as those requiring a single exchange (Steinberg and Edgar, 1962). Such a difference can, however, be sufficient to allow correct ordering of markers in many situations (cf. Fig. 5). Both patch-type recombination and heteroduplex repair contribute to the strong correlation of multiple exchanges (Beck, 1980), i.e., "high negative interference," to be discussed below.

In different phages, the idiosyncrasies of the different packaging mechanisms have important consequences for mapping (for reviews of individual phages, see R. Calendar, ed., *The Bacteriophages*, in press). Most phages package linear double-stranded DNA molecules from larger "concatemeric" intracellular DNA. All linear phage chromosomes examined contain some repeats at the ends, the so-called "terminal redundancies." Their lengths vary from the 12 bases of the cohesive ends of lambda to the approximately 10,000 base pairs in the terminal redundancy of P1. In some phages, like lambda, T5, and T7, both ends of mature chromosomes are genetically determined. Their heads package the amount of DNA between the left and right end cutting sequences (*cos*), regardless of small deletions or insertions. (Large deletions or insertions may be incompatible with successful packaging.) Other phages, like T4, T1, P1, and P22, package headfuls of DNA, larger than genome equivalents (Streisinger, 1966). At the extreme, in T4, packaging is initiated at nearly random positions on the circular map (Kalinski and Black, 1986; for minor exceptions, see Mosig et al., 1971; Grossi et al., 1983), most likely from recombinational intermediates (for review, see Mosig, 1987). In other phages with permuted ends (e.g., T1, P1, and P22), packaging can be initiated at specific *pac* sites, and a limited number of heads are subsequently filled from the same concatemer in a processive fashion. This mechanism results in nonrandom permutations such that ends are clustered in the vicinity of *pac* sites. Thus, depending on the phage, terminal redundancies can result in partial diploidy for unique, preferred, or random segments of the genome. Obviously, the diploid regions can carry different alleles of a given marker. If so, they are called "terminal redundancy heterozygotes" (Streisinger, 1966; Streisinger et al., 1967; Stahl, 1979a).

III. POPULATION GENETICS OF PHAGE CROSSES AND MAPPING FUNCTIONS

When bacteria are infected with several particles of two or more phage strains that differ in two or more genetic markers, the resulting progeny phage contain a certain fraction of recombinants. Such a mixed infection is considered to be the equivalent of a cross in cellular organisms, and the fraction of recombinants between a given marker pair is eventually converted to a measure of distance on the genetic map. The raw recombinant frequencies require corrections, however, mainly because a phage cross is, to some extent, an experiment in population genetics (Visconti and Delbrück, 1953). Phage genomes pair and exchange information repeatedly within efficient pairing regions of limited lengths (Doermann and Boehner, 1963; Doermann and Parma, 1967; Drake, 1967; Broker and Doermann, 1975) during a single infection cycle. The most dramatic consequence of such repeated "mating" is that parental markers from more than two parents can appear in a single recombinant progeny particle (Hertel, 1963, 1965). Another important consequence is the increase in recombinant frequencies during the latent period (Doermann, 1953) that occurs because the DNA packaged into the first progeny particles has experienced fewer matings than the DNA packaged last.

Several mapping functions have been devised to account for repeated matings and to convert the observed proportions of recombinants in mass lysates into map distances (for a thorough review, see Stahl, 1979a). These equations are based on the following considerations.

(i) The statistical principles that apply to probabilities of odd- versus even-numbered exchanges, first elaborated by Haldane (1919).

Consider a set of two- and three-factor crosses between the markers a, b, and c:

where the recombinant frequencies R_{ab}, R_{bc}, and R_{ac} are obtained in the crosses $ab^+ \times a^+b$, $b^+c \times bc^+$, and $ac^+ \times a^+c$, respectively. If genetic exchanges were statistically independent of one another (an assumption that is rarely correct), R_{ac} would equal $R_{ab} + R_{bc} - 2R_{ab} \cdot R_{bc}$ because double (or even numbers of) exchanges bring the outside markers back together and only odd numbers of exchanges contribute to R_{ac}. The Haldane (1919) formula corrects for the invisible even-numbered exchanges in a two-factor cross, without measuring smaller subintervals, by assuming that exchanges occur at random and follow essentially the Poisson distribution, $R = \frac{1}{2}(1 - e^{-2x})$, where x is the mean number of exchanges in that interval and R is the frequency of recombinants (x is a measure of genetic distance between the markers involved).

(ii) Depending on the average number of parental particles of different genotype per bacterium, individual bacteria are infected with different numbers and different ratios of parental genotypes. Recombinant frequencies are thus expected to differ among individual infected cells. A correction to the observed recombinant frequency in the mass lysate to account for this variation has been developed by Lennox et al. (1953). In practice, it is probably more reliable to do the crosses with equal input ratios and at controlled multiplicities of infection, sufficiently high (three to five of each parent) to ensure that most bacteria are infected with both parental types, since the increase in recombinant frequencies with increasing multiplicities in T4 is larger than the considerations of Lennox et al. (1953) predict (Mosig, 1962). This can now readily be explained by the increase in the number of different ends with increasing multiplicities (see below).

(iii) The average number of interactions (rounds of mating) in individual bacteria and the number of matings in the history of individual progeny particles are variable.

(iv) Each mating involves synapsis and multiple exchange only within limited regions of the genome, so-called "switch areas" (Chase and Doermann, 1958; Bresch, 1962; Drake, 1967; Amati and Meselson, 1965; Broker and Doermann, 1975).

(v) For genomes, like T4, where the gene sequences in different infecting chromosomes are circular permutations of one another (Streisinger et al., 1964; Streisinger, 1966), the linkage relationship of a given marker pair is different on different chromosomes. Thus, two distances must be considered. Obviously, this situation is more complex for those phages whose chromosomes are not random but are limited permutations of circular genomes, e.g., T1, P22, and P1 (see Israel, 1980).

(vi) Nonrandomness of multiple exchanges, interference, is usually measured by the coefficient of coincidence (c.o.c.) or interference index (i): $i = R_{abc}/R_{ab} \cdot R_{bc}$, where R_{abc} is the observed frequency of double exchange recombinants (those with an odd number of exchanges in the a–b interval and an odd number of exchanges in the b–c interval). The value of i is less than unity when one exchange interferes with another, i.e., when there is positive interference; it is larger than 1 when double exchanges are more frequent than expected, i.e., when there is negative interference. Interference is defined as $1 - i$.) In phage crosses within short intervals, there is so-called high or localized negative interference. With decreasing distances, i increases to values between 20 and 30 (Chase and Doermann, 1958; Amati and Meselson, 1965). These results are explained by multiple exchanges within limited switch areas. Both formation of patch-type heteroduplexes (Fig. 1) and mismatch repair within heteroduplex regions contribute to high (or true) negative interference in short intervals (Beck, 1980; for reviews, see Mosig, 1970; Broker and Doermann, 1975; Stahl, 1979a; Smith, 1988a, b).

Theoretical considerations imply, and experiments show, that i depends, among other things, on distances between markers, lengths of average pairing regions, marker effects, and activities of certain recombination and repair enzymes. Thus, no simple equation can correct for it. In practice, in various mapping functions, i, its distance dependence, and the lengths of pairing regions are estimated from best fits with available experimental results.

Part of the negative interference usually observed in phage crosses can be attributed to nonrandom mating and to exclusion of some infecting DNA molecules from the replicating-recombining pool. This aspect contributes to negative interference in all phage recombination regardless of distances beyond the lengths of pairing regions and accounts for i values on the order of 2 to 5 (Chase and Doermann, 1958; Amati and Meselson, 1965; Stahl, 1979a). It is often called low negative interference, but a better term is false negative interference (Whitehouse, 1982) since it is an artifact of phage growth.

Phage mapping functions incorporate terms (m) for the average number of matings and, when appropriate, for differences in mating experience between the first (m_1) and the last (m_2) encapsidated particles. Since it is now clear that vegetative DNA of most phages replicates and recombines as large concatemeric DNA, the term rounds of mating is, at best, an oversimplification.

Mapping functions are considered satisfactory if they convert fractions of recombinants into values that are additive, like true distances. Since many functions can be derived whose parameters have no obvious meaning, but which are successful in terms of the additivity criterion, exact values and parameters in these functions have little significance (Stahl and Steinberg, 1964; Stahl et al., 1964; Stahl, 1979a). Since different phages have different intrinsic recombination potential and other idiosyncrasies (see Calendar, in press), different phages would require different mapping functions. The guiding principles are discussed in an appendix by Stahl (1979a). Even for a given phage such mapping functions may need adjustments for different growth conditions. Differences in incubation temperature, concentrations of infected bacteria, the use of metabolic inhibitors, such as cyanide or fluorodeoxyuridine, the use of UV irradiation, or the inactivation of certain recombination enzymes or pathways by mutation influence recombination. Functions developed for mass lysates cannot be applied to individual single bursts.

As mentioned above, in addition to parental and recombinant types, there is a small proportion of progeny particles that are "partial heterozygotes" which carry alleles from both parents within limited regions of the genome (Hershey and Chase, 1951) as a result of packaging of heteroduplexes or of "terminal redundancy heterozygotes," or both (Hertel, 1963, 1965). It is a somewhat arbitrary decision whether to count all, half, or none of the heterozygotes as recombinants. In many experimental settings, they cannot be distinguished from true ("complete") recombinants. The potential error is usually negligible, except when certain markers are preferentially subject to heteroduplex repair or when the markers are very close in a gene that must be expressed before DNA replication can segregate the heteroduplex.

In phages for which mapping functions have been developed (Stahl, 1979a;

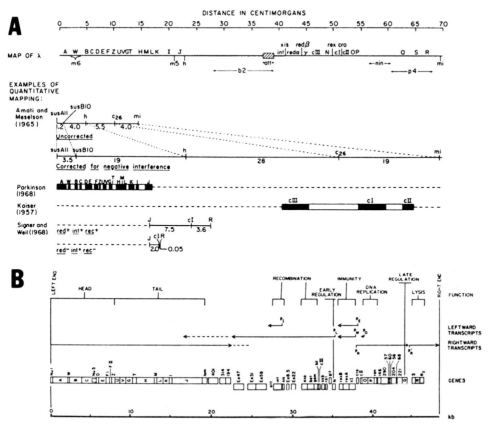

FIGURE 2. Comparison of the recombinational map (A) (Campbell, 1971) with the physical map (B) (Daniels et al., 1987) of phage lambda. (Reproduced by permission of the publisher.)

Stahl and Steinberg, 1964; Stahl et al., 1964; Israel, 1980; Gussin et al., 1980; Beck, 1980), the corrections give, for the most part, remarkable agreements with actual distances for both large and small intervals. The exceptions, due to true marker effects and hot spots, can then be considered with much more confidence.

A. A Mapping Function for the Linear Genome of Lambda

A mapping function for lambda (Stahl, 1979a) (equation 1) corrects for false (low) negative interference by incorporating a value of 0.4, and it assumes m (= rounds of mating) to be 1.

$$r = 0.4(1 - e^{-m/2(1-e^{-2x})}), \quad m = 1.0 \tag{1}$$

where r = recombinant frequency and x is the mean number of exchanges between the two markers in question. Without any additional corrections for correlations of exchanges, equation 1 gives the rectified genetic map of lambda shown in Fig. 2A, aligned for comparison with the physical map (Fig. 2B) based on the DNA sequence (Daniels et al., 1987).

Surprisingly, no striking marker effects are found in many λ crosses. Recombinant frequencies are proportional to distances and consistently additive, even for very closely linked markers, when recombinants for outside markers are selected (Gussin et al., 1980) (see Fig. 3).

At first glance, it may appear surprising that no correction for correlations of exchanges is required, since Amati and Meselson (1965) had found extremely high negative interference in some λ crosses involving short distances. In retrospect, the extremely high i values in λ crosses reported by Amati and Meselson (1965) are now known to be due to specific markers that are subject to very short patch repair, as discussed below (Lieb, 1981, 1983, 1985; Lieb et al., 1986; Raposa and Fox, 1987; Meselson, 1988; Lu and Chang, 1988; M. Fox, personal communication; P. Modrich, personal communication). Such markers do cause mapping anomalies (see Fig. 6), but these do not influence construction of a global map.

B. A Mapping Function for the Circular Genome of T4

A mapping function for phage T4 (equation 2) (Stahl and Steinberg, 1964; Stahl et al., 1964) takes into account the circularity of the T4 map and the different mating experience of particles released at different times of the latent period (Doermann, 1953).

$$R = 0.45\left\{(1 - D)\left[1 + \frac{e^{-m_2 p_D} - e^{-m_1 p_D}}{(m_2 - m_1)p_D}\right] + D\left[1 + \frac{e^{-m_2 p_{1-D}} - e^{-m_1 p_{1-D}}}{(m_2 - m_1)p_{1-D}}\right]\right\} \quad (2)$$

$$p_D = \frac{x}{2x + 1}\left[D + \frac{2Kx}{x + 1}(1 - e^{-(D/K)(2x+1)})\right] \quad (3)$$

where p_D is the probability of recombination per mating between two loci separated by a distance D; R is the recombination frequency; K is the mean length of synapsed regions of two chromosomes, 6.5×10^3 base pairs; x is the mean number of crossovers per switch region, 2; m_1 is the average number of matings among first matured chromosomes, 10; and m_2 is the average number of matings among last matured chromosomes, 55.

Again, the recombination map derived by using equation 2 shows remarkable congruence with the most recent physical map of T4 (Kutter et al., 1987) (Fig. 4), except for a major distortion of the gene 34-35 region discussed below.

This mapping function estimates that the total T4 map comprises approximately 2,000 map units. This would mean that each phage chromosome, on average, participates in approximately 20 recombinational exchanges, which may be largely clustered because a large proportion of recombinants are of the "patch" type (as defined by Stahl, 1979a), experience localized mismatch correction (for review, see Mosig, 1970), and participate in repeated exchanges within limited regions (Broker and Lehman, 1971; for review, see Broker and

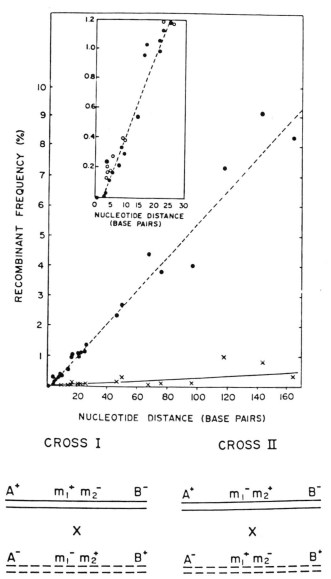

CROSS I

CROSS II

A^+ $m_1^+ m_2^-$ B^- A^+ $m_1^- m_2^+$ B^-

X   X

A^- $m_1^- m_2^+$ B^+ A^- $m_1^+ m_2^-$ B^+

FIGURE 3. Recombinant frequency as a function of nucleotide distance between markers in cy and cII. A^+B^+ recombinants were selected and scored for $m_1^+ m_2^+$ types. The dashed line is a linear regression fit to data from type I crosses; its slope is $5.6 \times 10^{-2}\%$ per nucleotide pair. The solid line ($\times$——$\times$) is a visual fit to data from type II crosses for which at least one turbid plaque was detected. The inset is an expansion of the plot of the type I data, illustrating the fact that the line intercepts the x axis at $+2$ nucleotides. Open circles in the inset are based on data from type I crosses involving $cy3002$ or $cy3004$, which are thought to coincide with $cy42$ and $cy3001$, respectively, but whose precise position is not known. (From Gussin et al., 1980. Reproduced by permission of the publisher.)

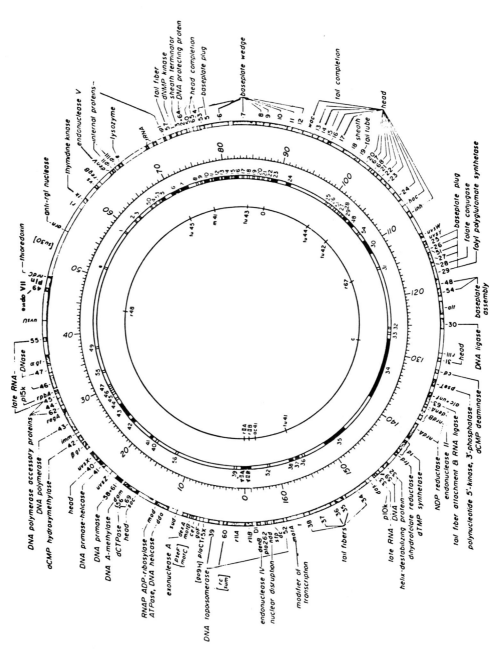

FIGURE 4. Comparison of the recombinational maps (inner circle and second double circle from Mosig, 1970) with the physical map (outer circle modified from Kutter et al., 1987).

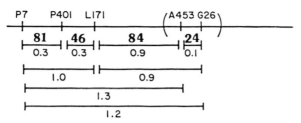

FIGURE 5. Map of gene 32 mutations of T4. The upper line shows the order deduced from three-factor crosses (Mosig et al., 1977). The lower lines show distances. Bold numbers above the lines indicate distances in nucleotides (Gauss et al., 1987). Numbers below the lines represent corrected percent recombinants (see text) obtained from two-factor crosses (Mosig et al., 1977).

Doermann, 1975). Doermann and Parma (1967) measured recombinant frequencies in a multifactor cross in all intervals defined by 15 markers distributed over the circular map. Twenty-one percent of the progeny were of parental type for all 15 markers, and an average of only 2.7 recombinational exchanges per progeny particle was observed. Since markers at such distances are exchanged predominantly by "splice"-type recombination, this result could imply that a large proportion of clustered exchanges were not detected. However, the special conditions of these crosses, especially the unequal input ratios and the multiple mutations, may also have influenced recombination.

The 2,000 map units also imply that there is, on average, approximately 0.01% recombination per nucleotide pair per infection cycle. This estimate is in remarkable agreement with observed intragenic recombinant frequencies. For example, the average frequency of recombination per nucleotide was found to be 0.014% in crosses within the T4 glutamine tRNA and the lysozyme (*e*) genes (Comer, 1977; Ravin and Artemiev, 1974). The recombinant frequencies within gene 32 (Mosig et al., 1977) can now be compared with the nucleotide distances between mutated sites (Gauss et al., 1987). Figure 5 shows that 0.01% recombination corresponds to 1 to 3 base pairs within this gene. The reasons for the slight variability (also apparent in the T4 data of Comer, 1977, and the λ data of Gussin et al., 1980) are not known. They could be related to differential initiation of recombination, heteroduplex repair (Kramer et al., 1982, 1984), or resolution of recombinational branches by endonuclease VII. This enzyme, mentioned above, has some limited preference for cutting of certain sequences in synthetic substrates (Kemper et al., 1984). Another possible source of error in the raw recombination data was corrected in Fig. 5. Since the gene 32 product, a single-stranded DNA-binding protein, is required for recombination (Tomizawa, 1967; see Radding, this volume) and different gene 32 mutants are defective in different steps (for review, see Mosig et al., 1984; Mosig, 1987), different mutant combinations affect recombination differently. These effects were determined in crosses between *r*II mutations in different gene 32 mutant backgrounds and were used to correct the raw recombinant frequencies (Mosig et al., 1977).

Less accurate comparisons, based on estimated molecular weights of amber peptides in gene 43 (Huang and Lehman, 1972) and gene 37 (Beckendorf et al.,

1973), are in general agreement with approximately 0.01% recombinants per nucleotide. Celis et al. (1973) estimated approximately half as many recombinants per nucleotide in gene 23.

IV. MAJOR MAP DISTORTIONS

Phages lambda and T4 provided the first examples of major map distortions in phage crosses beyond the statistical considerations that can be corrected by mapping functions (Jordan and Meselson, 1965; Mosig, 1966; Signer et al., 1968). Many more examples are now known.

A. Effects of Ends

The double-strand ends of infecting T4 chromosomes are highly recombinogenic. Because T4 ends are circularly permuted and because T4 packages a small proportion of incomplete genomes into small heads, it was possible to show in single-burst analyses that recombination between a given pair of markers is severalfold higher when they are located near an end than when the same pair is located in the interior of a molecule (Mosig, 1963; Doermann and Parma, 1967; Mosig et al., 1971). Both patch- and splice-type recombination is enhanced near ends. In contrast, the few "interstitial" recombinants are predominantly of the patch type (Mosig et al., 1971). The initiation of DNA replication from the invading single-stranded termini (Luder and Mosig, 1982) (Fig. 1G) eliminates the ends and amplifies negative interference associated with patch-type recombination and positive interference associated with splice-type recombination (Mosig et al., 1971). Results of experiments with terminal redundancy heterozygotes (Doermann and Boehner, 1963; Womack, 1963) can be interpreted in a similar way. Since the T4 ends are nearly randomly permuted, in T4 this end effect does not result in large overall map distortions. In phages with partially permuted chromosomes, such end effects distort the congruence of physical and genetic maps near the preferred packaging (*pac*) site, as found in phages T1 (Michalke, 1967) and P22 (Israel, 1980; Casjens et al., 1987). In phage P1 such an effect might be hidden by the strong recombination at the nearby *loxP* site, discussed below.

B. Hot Spots due to Protein-Mediated Site-Specific Recombination

The lambda attachment site is responsible for integration of lambda into the host chromosome. It also acts as a hot spot of recombination in lytic crosses between wild-type lambda chromosomes. This hot spot activity depends on the Int function of lambda (for review, see Weisberg and Landy, 1983) and might be considered simply as site-specific recombination. However, Int-dependent recombination, like general recombination, involves heteroduplex formation and limited branch migration (Enquist et al., 1979; Echols and Green, 1979; for review, see Weisberg and Landy, 1983). When lambda integrates into the host chromosome, branch migration beyond the attachment site is prohibited or at least greatly reduced by nonhomology of bacterial and phage sequences beyond that site. The nonhomology may, in fact, stimulate resolution of the intermediate, as it does in other situations (Lieb et al., 1984). In lambda × lambda crosses, however, a small but significant proportion of Int-dependent recombination branch migration does proceed beyond the attachment site. Thus, in this case the initiation of recombination, by a type I topoisomerase (the lambda Int protein), resembles the pattern postulated by Holliday (1964) (Fig. 1A).

Although the lambda chromosome circularizes soon after being injected into the host (Furth and Wickner, 1983), the genetic map of lambda is linear. The initiation sites for packaging (called *cos* in this phage) mark the ends of the genetic map. Since cutting at *cos* is obligatory for λ packaging, *R* and *Nul* are dissociated 100% of the time. There is considerable evidence that *cos* sites are also "recombinators"; i.e., they probably provide entry sites for λ Red and *Escherichia coli* RecBCD recombination complexes (Stahl, 1986; Smith, 1988a, b).

As in lambda, in P1 a site-specific crossover site, *loxP*, acts as a hot spot of recombination. This site is recognized by the P1 Cre protein, one of the best-understood site-specific recombination proteins (Hoess et al., 1987). This site-specific recombination system is responsible for the rare integration of P1 into the host chromosome. It is also thought to facilitate maintenance and segregation of monomeric P1 in the repressed plasmid state (Sternberg and Hoess, 1983; see also Calendar, in press). A small proportion of the exchanges initiated at *loxP* branch migrate into the adjacent region. This proportion is actually larger than that in the lambda *att* region (Sternberg, 1981). Frequent recombination at this site in lytic infections generates the apparent ends of the P1 map. Deletion of *loxP* confers linkage to markers bracketing this region (Sternberg and Hoess, 1983). Since the nearby *pac* site of P1 serves to cut only the first of a consecutive series of headfuls of DNA, it does not define the end of the P1 map.

Recombination across the remainder of the P1 genome appears not uniform, but divides the genetic map of P1 into several linkage groups. On the basis of special spot tests, considered as equivalents of two-factor crosses, Walker and Walker (1976) constructed linear maps of markers within each linkage group, but could not order markers between linkage groups. Since there are no physical discontinuities in the P1 chromosome, it is possible that the P1 genome contains several recombinational hot spots of unknown nature. It has been suggested that they may represent Chi sites (Sternberg and Hoess, 1983), but this possibility has not yet been tested. Another untested possibility is that these sites are recognized by the P1-encoded recombination-enhancing Ref protein (Windle and Hays, 1986).

Compared with other phages, recombination in single-stranded DNA phages is very rare. The binding site for the A protein in the φX174 genome acts like a hot spot of recombination. The hot spot is *recA* (but not *recBCD*) dependent. Most likely, the A-protein-induced nick, required for replication, can provide a target site for invasion mediated by RecA protein (for review, see Smith, 1988b).

C. Chi

Chi sites are the prototype of hot spots of recombination and serve as models for recognition of specific DNA sequences by general recombination enzymes (for reviews, see Stahl, 1979a, b; Smith et al., 1981b; Smith, 1983, 1987, 1988b). Chi sites do not respond to phage-encoded recombination systems, but they enhance recombination by the RecBCD enzyme of *E. coli* (and its relatives). However, Chi sites were first selected and characterized in phage lambda, under conditions in which concatemerization and packaging of lambda DNA required recombination by the RecBCD enzyme of the host (Lam et al., 1974; Henderson and Weil, 1975). All analyzed Chi sites have the sequence 5'-GCTGGTGG-3' (Smith et al., 1981a). Although they do not exist in wild-type lambda, they can be generated by mutations from similar sequences in lambda or by insertion of foreign DNA.

Chi stimulates recombination nearby and, with decreasing probability, at distances as far away as 10 to 20 kilobases, predominantly to the 5' side of that sequence. This stimulation of recombination at a distance is seen even when the phage chromosome with which it is recombining is grossly deleted or substituted over an interval that includes the Chi site (Stahl and Stahl, 1975; Stahl et al., 1980). In lambda crosses, Chi stimulation is orientation dependent. Elegant experiments by Kobayashi et al. (1982, 1984a, b) established that the orientation dependence reflects the requirement for a *cos* site, at which lambda terminase can generate a double-stranded end. Under special conditions the required double-stranded end can also be generated by a restriction enzyme (*Eco*RI) in vivo (Stahl et al., 1983). Regardless of the nature of the cut, recombination is stimulated predominantly to the distal side of Chi relative to the double-strand cut at *cos*.

The required double-stranded end is now understood to provide an entry site for the RecBCD enzyme which interacts in a not completely understood way with the Chi sequence. (For a review of the RecBCD enzyme, see Smith, 1988b; Taylor, this volume.) RecBCD enzyme can, under certain conditions, processively unwind DNA without cutting it. Under other conditions it has endo- and exonucleolytic activities. Purified RecBCD enzyme cuts preferentially near Chi (Ponticelli et al., 1985; Taylor et al., 1985; Taylor and Smith, 1985). Two models have been proposed for Chi action. In one model (Smith et al., 1981b), the in vitro cut by RecBCD enzyme near Chi occurs also in vivo and produces a single-stranded DNA end that initiates a strand exchange. In another model (Thaler et al., in press), Chi instructs the traveling RecBCD enzyme to assume a conformation competent to cut and resolve a recombinational intermediate, such as a Holliday junction, that it subsequently encounters. In one version of this model, Chi sites remove the D subunit from the traveling RecBCD enzyme, changing it from a nonrecombinogenic to a recombinogenic resolver of a recombinational intermediate (D. S. Thaler and F. W. Stahl, *Annu. Rev. Genet.*, in press).

Most experimental results are equally compatible with either model; some results (Rosenberg, 1987; *Genetics*, in press) appear to contradict the simplest version of the "nick 3' to Chi" model. Could it be that Chi sites can stimulate RecBCD enzyme to facilitate both initiation and resolution of intermediates, perhaps depending on whether the traveling recombination machine entered an unrecombined or a heteroduplex molecule?

D. Hot Spots Dependent on DNA Modification

Hot spots at which recombination exchanges are initiated are revealed not only by increases in recombination but also by increases in heterozygosity (Rottländer et al., 1967) and in marker rescue from irradiated DNA (Womack, 1965; Broker and Doermann, 1975). By these criteria, there are at least four recombinational hot spots in phage T4 (Womack, 1965; Rottländer et al., 1967; Broker and Doermann, 1975; Wu et al., 1984); a fifth one has been described in or near the *denV* gene (Radany et al., 1984). Only one of these (in or near gene 34) also grossly distorts the global map (Mosig, 1966, 1968; Beckendorf and Wilson, 1972). This could mean that most hot spots enhance predominantly patch-type recombination, but the one near gene 34 enhances both patch- and splice-type recombination.

The nature of these hot spots is unknown. Interestingly, the hot spot near gene 34 and another near *uvsY* correspond to tertiary origins of DNA replication (Kreuzer and Alberts, 1986; Kreuzer and Menkens, 1987). In contrast to primary T4 origins, which require functional unmodified host RNA polymerase, and to secondary initiation from recombinational intermediates mentioned above, tertiary origins are independent of RNA polymerase and of recombination functions, but they require the function and binding site of the T4 transcriptional activator protein, gp *mot* (for review of gp *mot*, see Brody et al., 1983). Kreuzer and Menkens (1987) proposed that at or near tertiary origins, replicative intermediates might also serve as preferred substrates for recombination.

The hot-spot activity near gene 34 requires the T4-specific DNA glucosylation; it does not occur in the closely related phage T2 (Russell, 1974), which has a similar gene sequence (Kim and Davidson, 1974) but a different glucosylation pattern. Hot-spot activity is also reduced in T4 mutants defective in the glucosylating enzymes (Levy and Goldberg, 1980). Possibly, glucosylation may affect the binding of T4 proteins required for tertiary initiation of replication or subsequent recombination, or both. T4 DNA topoisomerase, for example, has different activities on glucosylated versus nonglucosylated DNA (Kreuzer and Huang, 1983; Kreuzer and Alberts, 1984).

Conceivably, two different mechanisms contribute to the enhanced recombination in or near gene 34 which may depend on different, albeit closely linked, sites. The tertiary origin described by Kreuzer and Menkens (1987) lies within gene 34. In contrast, Beckendorf and Wilson (1972) concluded that a strong site for recombination initiation lies between genes 34 and 35, that strand exchange and branch migration are initiated there, and that termination of branch migration at variable distances from that site is responsible for the observed gradient of recombination activity within gene 34 and perhaps gene 35.

V. HETERODUPLEX REPAIR

A. Dam Methylation-Directed Mismatch Repair

Methyl-directed mismatch repair has been extensively reviewed (Radman et al., 1980; Radman and Wagner, 1986; Meselson, 1988; Radman, this volume). Artificially constructed heteroduplex chromosomes, containing different genetic markers in the two strands and differentially methylated, have been instrumental in demonstrating heteroduplex repair systems in *E. coli* (Wildenberg and Meselson, 1975) and the dependence of one repair system on DNA methylation by the Dam methylase (Marinus, 1987). This system converts markers in the unmethylated strand to the allele in the methylated strand of hemimethylated duplexes (Pukkila et al., 1983; Radman et al., 1980). In addition, the efficiency of repair is sequence and context specific (Kramer et al., 1982, 1984; Radman and Wagner, 1986; Jones et al., 1987a). The Dam-directed system generates long repair patches; several mismatches, separated by as many as 2,000 base pairs, can be co-corrected (Raposa and Fox, 1987).

B. Very Short Patch Mismatch Repair

Lieb (1981, 1983) first reported strong marker anomalies due to some mutations in the lambda *c*I repressor gene. These anomalies are caused by a very short patch repair that acts independently of Dam methylation. The map distortions due to these mutations are shown in Fig. 6. On the basis of the DNA sequences of the anomalous mutations, Lieb proposed that the very short patch repair system corrects only transition mutations that arose at the indicated cytosines in 5'-CC*(A/T)GG, C*AGG, and CC*AG. There is strong disparity in repair, favoring the wild-type sequence (Lieb, 1985). Jones et al. (1987b) have shown that T · G mispairs are corrected efficiently to C · G, but C · A mispairs are seldom corrected. The fact that the repair function is lacking in a *dcm* bacterial mutant suggested that the cytosine methylase participates in repair at its target site and at sites in related sequences (Lieb, 1987; Jones et al., 1987c; Zell and Fritz, 1987). However, the properties of certain *dcm* mutants indicate that methylation function is not required for very short patch repair. Since this repair system corepairs fewer than 10 bases on either side of the mutant sequence (Lieb et al., 1986), it results in frequent correlated double exchanges in multiply marked regions.

This model has received overwhelming support (Radman and Wagner, 1986; Jones et al., 1987c; Zell and Fritz, 1987; Raposa and Fox, 1987). It reconciles the absence of high negative interference in the crosses of Gussin et al. (1980), which contained no markers subject to very short patch repair, with the marker effects seen by Lieb et al. (1986) for certain other markers in the same region and with the very high negative interference seen in the early experiments of Amati and Meselson (1965). The *Pam80* mutation used by Amati and Meselson (1965), with the sequence 5'-CTAGG-3', can be restored in heteroduplexes to the wild-type sequence CCAGG. To explain some additional asymmetries of the data (Fig. 6),

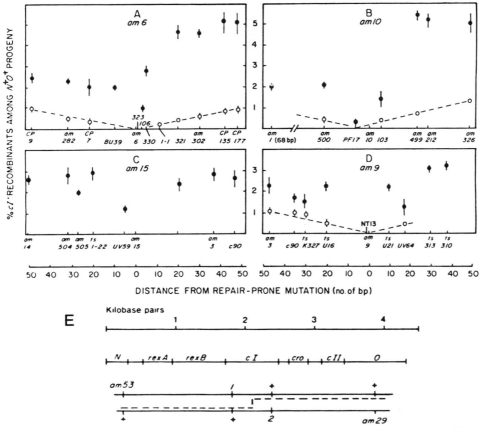

FIGURE 6. (A to D) Recombination frequencies from crosses of repair-prone cI amber mutations with nearby markers, among selected $N\,O$ recombinants as shown in panel E. Distances between mutations are indicated on the horizontal axes, with the repair-prone mutation in the center. (○) Average recombinant frequencies in control crosses; (●) average frequencies of recombinants in crosses with repair-prone amber mutations. Broken lines connecting the circles were fitted by eye. (From Lieb et al., 1986.)

Lieb et al. (1986) proposed that strand invasions, generating the heteroduplex substrates of repair, are formed only in the 3′ to 5′ direction (White and Fox, 1974).

Differences in recombinant frequencies due to specific markers similar in magnitude to those caused by Dcm target sequence mutants in lambda have been reported in the rII region of phage T4 (Tessman, 1965; Shcherbakov et al., 1982a, b). They might be caused by a similar mechanism. Since T4 DNA contains glucosylated hydroxymethylcytosine instead of cytosine, the T4 system probably recognizes a signal different from that of the very short patch repair system of $E.$ $coli.$

Recently, three different additional repair systems of $E.\ coli$ have been found (Lu and Chang, 1988; P. Radicella, E. Clark, and M. S. Fox, personal communication; Modrich, personal communication).

C. Repair of Large Heteroduplex Loops

Heteroduplex loops can be formed in vitro as well as in vivo during branch migration of DNA molecules containing homologous and heterologous regions (Lichten and Fox, 1984; Bianchi and Radding, 1983; Drake, 1966; G. Mosig and D. Powell, Abstr. Annu. Meet. Am. Soc. Microbiol. 1985, M4, p. 209). In packaged λ DNA few heteroduplex loops are found, partly because they are efficiently excised, generating homoduplex deletions. In addition, in λ crosses such heterologies generate excessive recombinational exchanges in their vicinity (Lieb et al., 1984), because the heterologies cause termination of branch migration or excision of the heteroduplex loops, or both. A minority of the heteroduplex loops, however, survive, and if they don't exceed 20 or so base pairs, they can even be packaged. Large deletion heteroduplexes do form, but they are neither repaired nor packaged in phage λ (Pearson and Fox, 1988a, b). The nature of the excising enzymes is unknown. Neither the Dam-dependent system nor the Dcm-dependent system is responsible (Dohet et al., 1987).

In phage T4, where heteroduplex loops up to 150 base pairs can be packaged, small heteroduplex loops (2 to 4 base pairs) can be excised by *denV* (endonuclease V) (Berger and Pardoll, 1976), whereas large heteroduplex loops (up to 150 base pairs) are efficiently excised in vivo (Mosig and Powell, Abstr. Annu. Meet. Am. Soc. Microbiol. 1985) and in vitro (Kleff and Kemper, 1988) by endonuclease VII (gene 49), which also cuts Holliday junctions, Y junctions, and cruciforms (Kemper et al., 1984). Excision of heteroduplex loops supports an earlier postulate to explain true negative interference in three-factor T4 crosses in which the central marker is a deletion. When wild-type recombinants are scored (Vigier, 1966; Berger and Warren, 1969), negative interference is reduced (compared with crosses in which the central marker is a point mutation). When triple mutant recombinants are scored (Doermann and Parma, 1967), negative interference is not reduced. In retrospect, this efficient excision of large heteroduplex loops is responsible for the earlier observations that in 49[+] phages deletion heterozygotes are mainly found in terminal redundancy heterozygotes and not in heteroduplex heterozygotes (Womack, 1963; Streisinger et al., 1967; Stahl, 1979a).

VI. EPILOGUE

Whereas all phages recombine their DNA and use similar mechanisms in the process, a comparison of maps and of recombination potential of different phages shows a bewildering diversity. Recombination mechanisms that generate only double exchanges will give circular maps, regardless of the physical structure of the chromosome (Stahl, 1979a). Almost all phages generate intracellular concatemeric DNA from the infecting linear genomes. The concatemeric DNA is frequently branched, largely as a result of recombination. Nevertheless, even among the few well-studied examples discussed here, there are both linear and circular maps, and recombination frequencies per unit length vary widely. In the T4 gene 32, there is approximately 0.1% recombination per 10 base pairs (Fig. 5).

In the gene 9 of P22, this frequency is approximately one-tenth of that of T4 (Fane and King, 1987), and in λ crosses the frequency as compared with T4 is lower by a factor of 100 (Gussin et al., 1980; Lieb et al., 1986). Mechanisms other than general recombination itself greatly influence these values: DNA replication, DNA packaging, and site-specific recombination are a few obvious contributors.

The overall frequency of recombination is largely dependent on the efficiency of recombination enzymes, the control of their activities and synthesis, and the generation of DNA ends. The detailed analysis of T4, discussed elsewhere (Mosig, 1987), provides a convincing example of how recombination is integrated into the life cycle and how the complexities of the transcriptional program and their influence on DNA replication might have selected for a high recombination potential in this phage.

ACKNOWLEDGMENTS. I thank Peggy Lieb, Frank Stahl, Mike Yarmolinsky, Nat Sternberg, and Gary Gussin for discussions, Gerry Smith for thoughtful editing, and Cindy Young for her expertise and patience in getting this review together.

This writing was supported by Public Health Service grant GM 13221 from the National Institutes of Health.

LITERATURE CITED

Alberts, B. M. 1984. The DNA enzymology of protein machines. *Cold Spring Harbor Symp. Quant. Biol.* **49**:1–12.

Amati, P., and M. Meselson. 1965. Localized negative interference in bacteriophage λ. *Genetics* **51**: 369–379.

Beck, B. N. 1980. High negative interference and recombination in bacteriophage T5. *Genetics* **96**:25–41.

Beckendorf, S. K., J. S. Kim, and I. Lielausis. 1973. Structure of bacteriophage T4 genes *37* and *38*. *J. Mol. Biol.* **73**:17–35.

Beckendorf, S. K., and J. H. Wilson. 1972. A recombination gradient in bacteriophage T4 gene *34*. *Virology* **50**:315–321.

Benzer, S. 1961. On the topography of the genetic fine structure. *Proc. Natl. Acad. Sci. USA* **47**:403–415.

Berger, H., and D. Pardoll. 1976. Evidence that mismatched bases in heteroduplex T4 bacteriophage are recognized in vivo. *J. Virol.* **20**:441–445.

Berger, H., and A. J. Warren. 1969. Effects of deletion mutants on high negative interference in T4D bacteriophage. *Genetics* **63**:1–5.

Bianchi, M. E., and C. M. Radding. 1983. Insertions, deletions and mismatches in heteroduplex DNA made by recA protein. *Cell* **35**:511–520.

Bresch, C. 1962. Replication and recombination in bacteriophage. A review of pertinent data and a molecular interpretation of the partial replica hypothesis. *Z. Vererbungsl.* **93**:476–490.

Brody, E., D. Rabussay, and D. H. Hall. 1983. Regulation of transcription of prereplicative genes, p. 174–183. *In* C. K. Mathews, E. M. Kutter, G. Mosig, and P. B. Berget (ed.), *Bacteriophage T4*. American Society for Microbiology, Washington, D.C.

Broker, T. R. 1973. An electron microscopic analysis of pathways for bacteriophage T4 DNA recombination. *J. Mol. Biol.* **81**:1–16.

Broker, T. R., and A. H. Doermann. 1975. Molecular and genetic recombination of bacteriophage T4. *Annu. Rev. Genet.* **9**:213–244.

Broker, T. R., and I. R. Lehman. 1971. Branched DNA molecules: intermediates in T4 recombination. *J. Mol. Biol.* **60:**131–149.

Campbell, A. 1971. Genetic structure, p. 13–44. *In* A. D. Hershey (ed.), *The Bacteriophage* λ. Cold Spring Harbor Laboratory, Cold Spring Harbor, N.Y.

Casjens, S., W. M. Huang, M. Hayden, and R. Parr. 1987. Initiation of bacteriophage P22 DNA packaging series. Analysis of a mutant that alters the DNA target specificity of the packaging apparatus. *J. Mol. Biol.* **194:**411–422.

Celis, J. E., J. D. Smith, and S. Brenner. 1973. Correlation between genetic and translational maps of gene *23* in bacteriophage T4. *Nature* (London) *New Biol.* **241:**130–132.

Chase, M., and A. H. Doermann. 1958. High negative interference over short segments of the genetic structure of bacteriophage T4. *Genetics* **43:**332–353.

Comer, M. M. 1977. Correlation between genetic and nucleotide distances in a bacteriophage T4 transfer RNA gene. *J. Mol. Biol.* **113:**267–271.

Daniels, D. L., J. L. Schroeder, W. Szybalski, F. Sanger, and F. R. Blattner. 1987. A molecular map of coliphage lambda, p. 1–21. *In* S. J. O'Brien (ed.), *Genetic Maps 1987.* Cold Spring Harbor Laboratory, Cold Spring Harbor, N.Y.

Dannenberg, R., and G. Mosig. 1983. Early intermediates in bacteriophage T4 DNA replication and recombination. *J. Virol.* **45:**813–831.

Doermann, A. H. 1953. The vegetative state in the life cycle of bacteriophage: evidence for its occurrence and its genetic characterization. *Cold Spring Harbor Symp. Quant. Biol.* **18:**3–11.

Doermann, A. H., and L. Boehner. 1963. An experimental analysis of bacteriophage T4 heterozygotes. I. Mottled plaques from crosses involving six *rII* loci. *Virology* **21:**551–567.

Doermann, A. H., and D. H. Parma. 1967. Recombination in bacteriophage T4. *J. Cell. Comp. Physiol.* **70**(Suppl. 1):147–164.

Dohet, C., S. Dzidic, R. Wagner, and M. Radman. 1987. Large nonhomology in heteroduplex DNA is processed differently than single base pair mismatches *Mol. Gen. Genet.* **206:**181–184.

Drake, J. W. 1966. Heteroduplex heterozygotes in bacteriophage T4 involving mutation of various dimensions. *Proc. Natl. Acad. Sci. USA* **55:**506–512.

Drake, J. W. 1967. The length of the homologous pairing region for genetic recombination in bacteriophage T4. *Proc. Natl. Acad. Sci. USA* **58:**962–966.

Dressler, D., and H. Potter. 1982. Molecular mechanisms in genetic recombination. *Annu. Rev. Biochem.* **51:**727–761.

Echols, H., and L. Green. 1979. Some properties of site-specific and general recombination inferred from Int-mediated exchanges by bacteriophage lambda. *Genetics* **93:**297–307.

Enquist, L. W., H. Nash, and R. A. Weisberg. 1979. Strand exchange in site-specific recombination. *Proc. Natl. Acad. Sci. USA* **76:**1363–1367.

Fane, B., and J. King. 1987. Identification of sites influencing the folding and subunit assembly of the P22 tailspike polypeptide chain using nonsense mutations. *Genetics* **117:**157–171.

Fishel, R. A., K. Detmer, and A. Rich. 1988. Identification of homologous pairing and strand-exchange activity from a human tumor cell line based on Z-DNA affinity chromatography. *Proc. Natl. Acad. Sci. USA* **85:**36–40.

Fox, M. S. 1966. On the mechanism of integration of transforming deoxyribonucleate. *J. Gen. Physiol.* **49**(no. 6, Suppl.):183–196.

Furth, M., and S. Wickner. 1983. Lambda DNA replication, p. 145–173. *In* R. W. Hendrix, J. W. Roberts, F. W. Stahl, and R. A. Weisberg (ed.), *Lambda II.* Cold Spring Harbor Laboratory, Cold Spring Harbor, N.Y.

Gauss, P., K. B. Krassa, D. S. McPheeters, M. A. Nelson, and L. Gold. 1987. Zinc (II) and the single-stranded DNA binding protein of bacteriophage T4. *Proc. Natl. Acad. Sci. USA* **84:**8515–8519.

Grossi, G. F., M. F. Macchiato, and G. Gialanella. 1983. Circular permutation analysis of phage T4 DNA by electron microscopy. *Z. Naturforsch. Teil C* **38:**294–296.

Gussin, G. N., E. D. Rosen, and D. L. Wulff. 1980. Mappability of very close markers of bacteriophage lambda. *Genetics* **96:**1–24.

Haldane, J. B. S. 1919. The combination of linkage values, and the calculation of distances between the loci of linked factors. *J. Genet.* **8:**299–309.

Henderson, D., and J. Weil. 1975. Recombination-deficient deletions in bacteriophage λ and their interactions with Chi mutations. *Genetics* **79:**143–174.

Hershey, A. D. 1958. The production of recombinants in phage crosses. *Cold Spring Harbor Symp. Quant. Biol.* **23:**19–46.

Hershey, A. D., and M. Chase. 1951. Genetic recombination and heterozygosis in bacteriophage. *Cold Spring Harbor Symp. Quant. Biol.* **16:**471–479.

Hertel, R. 1963. The occurrence of three allelic markers in one particle of phage T4. *Z. Vererbungsl.* **94:**436–441.

Hertel, R. 1965. Gene function of heterozygotes in phage T4. *Z. Vererbungsl.* **96:**105–115.

Hoess, R., A. Wierzbicki, and K. Abremski. 1987. Isolation and characterization of intermediates in site-specific recombination. *Proc. Natl. Acad. Sci. USA* **84:**6840–6844.

Holliday, R. 1964. A mechanism for gene conversion in fungi. *Genet. Res.* **5:**282–304.

Holliday, R. 1968. Genetic recombination in fungi, p. 157–174. *In* W. J. Peacock and R. D. Brock (ed.), *Replication and Recombination of Genetic Material.* Australian Academy of Science, Canberra, Australia.

Hopkins, R. C. 1986. A unique four-stranded model of homologous recombination intermediate. *J. Theor. Biol.* **120:**215–222.

Hotchkiss, R. D. 1971. Toward a general theory of genetic recombination in DNA. *Adv. Genet.* **16:**325–348.

Huang, W. M., and I.R. Lehman. 1972. On the direction of translation of the T4 deoxyribonucleic acid polymerase gene *in vivo. J. Biol. Chem.* **247:**7663–7667.

Israel, V. 1980. Genetic implications of limited circular permutation. *Virology* **106:**100–106.

Jensch, F., and B. Kemper. 1986. Endonuclease VII resolves Y-junctions in branched DNA *in vitro. EMBO J.* **5:**181–189.

Jones, M., R. Wagner, and M. Radman. 1987a. Repair of a mismatch is influenced by the base composition of the surrounding nucleotide sequence. *Genetics* **115:**605–610.

Jones, M., R. Wagner, and M. Radman. 1987b. Mismatch repair of deaminated 5-methyl-cytosine. *J. Mol. Biol.* **193:**155–159.

Jones, M., R. Wagner, and M. Radman. 1987c. Mismatch repair and recombination in *E. coli. Cell* **50:**621–626.

Jordan, E., and M. Meselson. 1965. A discrepancy between genetic and physical lengths on the chromosome of bacteriophage λ. *Genetics* **51:**77–86.

Kaiser, A. 1957. Mutations in a temperate bacteriophage affecting its ability to lysogenize *Escherichia coli. Virology* **3:**42–61.

Kalinski, A., and L. W. Black. 1986. End structure and mechanism of packaging of bacteriophage T4 DNA. *J. Virol.* **58:**951–954.

Kemper, B., F. Jensch, M. U. Depka-Prondzynski, H. J. Fritz, R. U. Borgmeyer, and K. Mizuuchi. 1984. Resolution of Holliday structures by endonuclease VII as observed in interactions with cruciform DNA. *Cold Spring Harbor Symp. Quant. Biol.* **49:**815–825.

Kim, J.-S., and N. Davidson. 1974. Electron microscope heteroduplex studies of sequence relation of T2, T4 and T6 bacteriophage DNAs. *Virology* **57:**93–111.

Kleff, S., and B. Kemper. 1988. Initiation of heteroduplex-loop repair by T4-encoded endonuclease VII *in vitro. EMBO J.* **7:**1527–1535.

Kmiec, E. B., and W. K. Holloman. 1986. Homologous pairing of DNA molecules by Ustilago Rec 1 protein is promoted by sequences of Z-DNA. *Cell* **44:**545–554.

Kobayashi, I., H. Murialdo, J. M. Crasemann, M. M. Stahl, and F. W. Stahl. 1982. Orientation of cohesive end site *cos* determines active orientation of Chi sequence in stimulating recA-recBC-mediated recombination in phage λ lytic infections. *Proc. Natl. Acad. Sci. USA* **79:**5981–5985.

Kobayashi, I., M. M. Stahl, E. R. Fairfield, and F. W. Stahl. 1984a. Coupling of packaging explains apparent nonreciprocality of Chi-stimulated recombination of bacteriophage lambda by RecA and RecB functions. *Genetics* **108:**773–794.

Kobayashi, I., M. M. Stahl, and F. W. Stahl. 1984b. The mechanism of the Chi-cos interaction in RecA-RecBC-mediated recombination in phage λ. *Cold Spring Harbor Symp. Quant. Biol.* **49:**497–506.

Kramer, B., W. Kramer, and H.-J. Fritz. 1984. Different base/base mismatches are corrected with different efficiencies by the methyl-directed DNA mismatch-repair system of *E. coli*. *Cell* **38**:879–881.

Kramer, W., K. Schughart, and H.-J. Fritz. 1982. Directed mutagenesis of DNA clones in filamentous phage: influence of hemimethylated GATC sites on marker recovery from restriction fragments. *Nucleic Acids Res.* **10**:6475–6485.

Kreuzer, K. N., and B. M. Alberts. 1984. Site-specific recognition of bacteriophage T4 DNA by T4 type II DNA topoisomerase and *Escherichia coli* DNA gyrase. *J. Biol. Chem.* **259**:5339–5346.

Kreuzer, K. N., and B. M. Alberts. 1986. Characterization of a defective phage system for the analysis of bacteriophage T4 DNA replication origins. *J. Mol. Biol.* **188**:185–198.

Kreuzer, K. N., and W. M. Huang. 1983. T4 DNA topoisomerase, p. 90–96. *In* C. K. Mathews, E. M. Kutter, G. Mosig, and P. B. Berget (ed.), *Bacteriophage T4*. American Society for Microbiology, Washington, D.C.

Kreuzer, K. N., and A. E. Menkens. 1987. Plasmid model systems for the initiation of bacteriophage T4 DNA replication, p. 451–471. *In* R. McMacken and T. J. Kelly (ed.), *DNA Replication and Recombination*. Alan R. Liss, New York.

Kutter, E., B. Guttman, W. Rüger, J. Tomaschewski, and G. Mosig. 1987. Bacteriophage T4, p. 22–37. *In* S. J. O'Brien (ed.), *Genetic Maps 1987*. Cold Spring Harbor Laboratory, Cold Spring Harbor, N.Y.

Lam, S. T., M. M. Stahl, K. D. McMilin, and F. W. Stahl. 1974. Rec-mediated recombinational hot spot activity in bacteriophage lambda. II. A mutation which causes hot spot activity. *Genetics* **77**:425–433.

Lennox, E. S., C. Levinthal, and F. Smith. 1953. The effect of finite input in reducing recombinant frequency. *Genetics* **38**:508–511.

Levy, J. N., and E. B. Goldberg. 1980. Region-specific recombination in phage T4. I. A special glucosyl-dependent recombination system. *Genetics* **95**:519–530.

Lichten, M., and M. S. Fox. 1984. Evidence for the inclusion of regions of non-homology in heteroduplex products of bacteriophage lambda. *Proc. Natl. Acad. Sci. USA* **81**:7180–7184.

Lieb, M. 1981. A fine structure map of spontaneous and induced mutations in the lambda repressor gene, including insertions of IS elements. *Mol. Gen. Genet.* **184**:364–371.

Lieb, M. 1983. Specific mismatch correction in bacteriophage lambda crosses by very short patch repair. *Mol. Gen. Genet.* **191**:118–125.

Lieb, M. 1985. Recombination in the λ repressor gene: evidence that very short patch (VSP) mismatch repair restores a specific sequence. *Mol. Gen. Genet.* **199**:465–470.

Lieb, M. 1987. Bacterial genes *mutL*, *mutS*, and *dcm* participate in repair of mismatches at 5-methylcytosine sites. *J. Bacteriol.* **169**:5241–5246.

Lieb, M., E. Allen, and D. Read. 1986. Very short patch mismatch repair in phage lambda: repair sites and length of repair tracts. *Genetics* **114**:1041–1060.

Lieb, M., M.-M. Tsai, and R. C. Deonier. 1984. Crosses between insertion and point mutations in λ gene cI: stimulation of neighboring recombination by heterology. *Genetics* **108**:277–289.

Lu, A.-L., and D.-Y. Chang. 1988. Repair of single base-pair transversion mismatches of Escherichia coli in vitro: correction of certain Ala mismatches is independent of *dam* methylation and host *mut* HLS gene functions. *Genetics* **118**:593–600.

Luder, A., and G. Mosig. 1982. Two alternative mechanisms for initiation of DNA replication forks in bacteriophage T4: priming by RNA polymerase and by recombination. *Proc. Natl. Acad. Sci. USA* **79**:1101–1105.

Marinus, M. G. 1987. DNA methylation in *Escherichia coli*. *Annu. Rev. Genet.* **21**:113–131.

McGavin, S. 1971. Models of specifically paired like (homologous) nucleic acid structures. *J. Mol. Biol.* **55**:293–298.

Meselson, M. 1988. Methyl-directed repair of DNA mismatches, p. 91–113. *In* K. B. Low (ed.), *The Recombination of Genetic Material*. Academic Press, Inc., New York.

Meselson, M., and C. M. Radding. 1975. A general model for genetic recombination. *Proc. Natl. Acad. Sci. USA* **72**:358–362.

Michalke, W. 1967. Erhöhte Rekombinations Hävfigkeit an den Enden des T1-chromosoms. *Mol. Gen. Genet.* **99**:12–33.

Minagawa, T., and Y. Ryo. 1979. Genetic control of formation of very fast sedimenting DNA of bacteriophage T4. *Mol. Gen. Genet.* **170**:113–115.

Mizuuchi, K., B. Kemper, J. Hays, and R. A. Weisberg. 1982. T4 endonuclease VII cleaves Holliday structures. *Cell* **29**:357–365.

Moser, H. E., and P. B. Dervan. 1987. Sequence-specific cleavage of double helical DNA by triple helix formation. *Science* **238**:645–650.

Mosig, G. 1962. The effect of multiplicity of infection on recombination values in bacteriophage T4D. *Z. Vererbungsl.* **93**:280–286.

Mosig, G. 1963. Genetic recombination in bacteriophage T4 during replication of DNA fragments. *Cold Spring Harbor Symp. Quant. Biol.* **28**:35–42.

Mosig, G. 1966. Distances separating genetic markers in T4 DNA. *Proc. Natl. Acad. Sci. USA* **56**: 1177–1183.

Mosig, G. 1968. A map of distances along the DNA molecule of phage T4. *Genetics* **59**:137–151.

Mosig, G. 1970. Recombination in bacteriophage T4, p. 1–53. *In* E. W. Caspari (ed.), *Advances in Genetics*, vol. 15. Academic Press, Inc., New York.

Mosig, G. 1987. The essential role of recombination in phage T4 growth. *Annu. Rev. Genet.* **21**:347–371.

Mosig, G., W. Berquist, and S. Bock. 1977. Multiple interactions of a DNA-binding protein *in vivo*. III. Phage T4 gene-*32* mutations differentially affect insertion-type recombination and membrane properties. *Genetics* **86**:5–23.

Mosig, G., R. Dannenberg, D. Ghosal, A. Luder, S. Benedict, and S. Bock. 1979. General genetic recombination in bacteriophage T4. *Stadler Genet. Symp.* **11**:31–55.

Mosig, G., R. Ehring, W. Schliewen, and S. Bock. 1971. The patterns of recombination and segregation in terminal regions of T4 DNA molecules. *Mol. Gen. Genet.* **113**:51–91.

Mosig, G., M. Shaw, and G. M. Garcia. 1984. On the role of DNA replication, endonuclease VII, and *r*II proteins in processing of recombinational intermediates in phage T4. *Cold Spring Harbor Symp. Quant. Biol.* **49**:371–382.

Parkinson, J. S. 1968. Genetics of the left arm of the chromosome of bacteriophage lambda. *Genetics* **59**:311–325.

Pearson, R. K., and M. S. Fox. 1988a. Effects of DNA heterologies on bacteriophage λ packaging. *Genetics* **118**:5–12.

Pearson, R. K., and M. S. Fox. 1988b. Effects of DNA heterologies on bacteriophage λ recombination. *Genetics* **118**:13–19.

Ponticelli, A. S., D. W. Schultz, A. F. Taylor, and G. R. Smith. 1985. Chi-dependent DNA strand cleavage by RecBC enzyme. *Cell* **41**:145–151.

Pugh, J. C., and D. A. Ritchie. 1984. The structure of replicating bacteriophage T1 DNA: comparison between wild type and DNA-arrest mutant infections. *Virology* **135**:189–199.

Pukkila, P. J., J. Peterson, G. Herman, P. Modrich, and M. Meselson. 1983. Effects of high levels of DNA adenine methylation on methyl-directed mismatch repair in *Escherichia coli*. *Genetics* **104**: 571–582.

Radany, E. H., L. Naumovski, J. D. Love, K. A. Gutekunst, D. H. Hall, and E. C. Friedberg. 1984. Physical mapping and complete nucleotide sequence of the *denV* gene of bacteriophage T4. *J. Virol.* **52**:846–856.

Radman, M., and R. Wagner. 1986. Mismatch repair in *Escherichia coli*. *Annu. Rev. Genet.* **20**:523–538.

Radman, M., R. E. Wagner, B. W. Glickman, and M. Meselson. 1980. DNA methylation, mismatch correction and genetic stability, p. 121–130. *In* M. Alacevic (ed.), *Progress in Environmental Mutagenesis*. Elsevier, Amsterdam.

Raposa, S., and M. S. Fox. 1987. Some features of base pair mismatch and heterology repair in *Escherichia coli*. *Genetics* **117**:381–390.

Ravin, V. K., and M. I. Artemiev. 1974. Fine structure of the lysozyme gene of bacteriophage T4B. *Mol. Gen. Genet.* **128**:359–365.

Rosenberg, S. M. 1987. Chi-stimulated patches are heteroduplex with recombinant information on the phage λ r chain. *Cell* **48**:855–865.

Rottländer, E., K. O. Hermann, and R. Hertel. 1967. Increased heterozygote frequency in certain regions of the T4 chromosome. *Mol. Gen. Genet.* **99**:34–39.

Russell, R. L. 1974. Comparative genetics of the T-even bacteriophages. *Genetics* **78**:967–988.

Shcherbakov, V. P., L. A. Plugina, E. A. Kudryashova, O. I. Efremova, S. T. Sizova, and O. G. Toompuu. 1982a. Marker-dependent recombination in T4 bacteriophage. I. Outline of the phenomenon and evidence suggesting a mismatch repair mechanism. *Genetics* **102**:615–625.

Shcherbakov, V. P., L. A. Plugina, E. A. Kudryashova, O. I. Efremova, S. T. Sizova, and O. G. Toompuu. 1982b. Marker-dependent recombination in T4 bacteriophage. II. The evaluation of mismatch repair abilities in crosses within indicator distances. *Genetics* **102**:627–637.

Signer, E., H. Echols, J. Weil, C. Radding, M. Shulman, L. Moore, and K. Manly. 1968. The general recombination system of bacteriophage lambda. *Cold Spring Harbor Symp. Quant. Biol.* **33**:711–714.

Signer, E. R., and J. Weil. 1968. Recombination in bacteriophage λ. I. Mutants deficient in general recombination. *J. Mol. Biol.* **34**:261–271.

Skalka, A. 1974. A replicator's view of recombination (and repair), p. 421–432. *In* R. F. Grell (ed.), *Mechanisms in Recombination.* Plenum Publishing Corp., New York.

Smith, G. R. 1983. General recombination, p. 175–209. *In* R. W. Hendrix, J. W. Roberts, F. W. Stahl, and R. A. Weisberg (ed.), *Lambda II.* Cold Spring Harbor Laboratory, Cold Spring Harbor, N.Y.

Smith, G. R. 1987. Mechanism and control of homologous recombination in *Escherichia coli. Annu. Rev. Genet.* **21**:179–201.

Smith, G. R. 1988a. Homologous recombination in procaryotes. *Microbiol. Rev.* **52**:1–28.

Smith, G. R. 1988b. Homologous recombination sites and their recognition, p. 115–154. *In* K. B. Low (ed.), *The Recombination of Genetic Material.* Academic Press, Inc., New York.

Smith, G. R., S. M. Kunes, D. W. Schultz, A. Taylor, and K. L. Triman. 1981a. Structure of Chi hotspots of generalized recombination. *Cell* **24**:429–436.

Smith, G. R., D. W. Schultz, A. Taylor, and K. Triman. 1981b. Chi sites, RecBC enzyme, and generalized recombination. *Stadler Genet. Symp.* **13**:25–37.

Sobell, H. M. 1972. Molecular mechanism for genetic recombination. *Proc. Natl. Acad. Sci. USA* **68**:2483–2487.

Stahl, F. W. 1979a. *Genetic Recombination. Thinking about It in Phage and Fungi.* W. H. Freeman, San Francisco.

Stahl, F. W. 1979b. Special sites in generalized recombination. *Annu. Rev. Genet.* **13**:7–24.

Stahl, F. W. 1986. Roles of double-strand breaks in generalized genetic recombination. *Prog. Nucleic Acid Res. Mol. Biol.* **33**:169–194.

Stahl, F. W., R. S. Edgar, and J. Steinberg. 1964. The linkage map of bacteriophage T4. *Genetics* **50**:539–552.

Stahl, F. W., and M. M. Stahl. 1975. Rec-mediated recombinational hot spot activity in bacteriophage λ. IV. Effect of heterology on Chi-stimulated crossing over. *Mol. Gen. Genet.* **140**:29–37.

Stahl, F. W., M. M. Stahl, R. E. Malone, and J. M. Crasemann. 1980. Directionality and nonreciprocality of Chi-stimulated recombination in phage λ. *Genetics* **94**:235–248.

Stahl, F. W., and C. M. Steinberg. 1964. The theory of formal phage genetics for circular maps. *Genetics* **50**:531–538.

Stahl, M. M., I. Kobayashi, F. W. Stahl, and S. K. Huntington. 1983. Activation of Chi, a recombinator, by the action of an endonuclease at a distant site. *Proc. Natl. Acad. Sci. USA* **80**:2310–2313.

Steinberg, C. M., and R. S. Edgar. 1962. A critical test of a current theory of genetic recombination. *Genetics* **47**:187–208.

Sternberg, N. 1981. Bacteriophage P1 site-specific recombination. III. Strand exchange during recombination at *lox* sites. *J. Mol. Biol.* **150**:603–608.

Sternberg, N., D. Hamilton, and R. Hoess. 1981. Bacteriophage P1 site-specific recombination. II. Recombination between loxP and the bacterial chromosome. *J. Mol. Biol.* **150**:487–507.

Sternberg, N., and R. Hoess. 1983. The molecular genetics of bacteriophage P1. *Annu. Rev. Genet.* **17**:123–154.

Streisinger, G. 1966. Terminal redundancy, or all's well that ends well, p. 335–340. *In* J. Cairns, G. S.

Stent, and J. D. Watson (ed.), *Phage and the Origins of Molecular Biology*. Cold Spring Harbor Laboratory, Cold Spring Harbor, N.Y.

Streisinger, G., R. S. Edgar, and G. H. Denhardt. 1964. Chromosome structure in phage T4. I. Circularity of the linkage map. *Proc. Natl. Acad. Sci. USA* **51**:775–779.

Streisinger, G., J. Emrich, and M. M. Stahl. 1967. Chromosome structure in phage T4. III. Terminal redundancy and length determination. *Proc. Natl. Acad. Sci. USA* **57**:292–295.

Taylor, A. F., D. W. Schultz, A. S. Ponticelli, and G. R. Smith. 1985. RecBCD enzyme nicking at Chi sites during DNA unwinding: location and orientation dependence of the cutting. *Cell* **41**:153–163.

Taylor, A. F., and G. R. Smith. 1985. Substrate specificity of the DNA unwinding activity of the RecBC enzyme of *Escherichia coli*. *J. Mol. Biol.* **185**:431–443.

Tessman, I. 1965. Genetic ultrafine structure in the T4 *r*II region. *Genetics* **51**:63–75.

Tomizawa, I. 1967. Molecular mechanisms of genetic recombination in bacteriophage: joint molecules and their conversion to recombinant molecules. *J. Cell. Physiol.* **70**(Suppl. 1):201–214.

Vigier, P. 1966. Role des heterozygotes internes dans la formation de genomes double-recombinants chez le bacteriophage T4. *C.R. Acad. Sci.* **263**:2010–2013.

Visconti, N., and M. Delbrück. 1953. Mechanism of genetic recombination in phage. *Genetics* **38**:5–33.

Walker, D. H., Jr., and J. T. Walker. 1976. Genetic studies of coliphage P1. III. Extended genetic map. *J. Virol.* **20**:177–187.

Watson, J. D. 1972. Origin of concatemeric T7 DNA. *Nature* (London) *New Biol.* **239**:197–201.

Weinstock, G. M., and D. Botstein. 1980. Genetics of bacteriophage P22. IV. Correlation of genetic and physical map using translocatable drug-resistance elements. *Virology* **106**:92–99.

Weisberg, R., and A. Landy. 1983. Site-specific recombination in phage lambda, p. 211–250. *In* R. Hendrix, J. Roberts, F. Stahl, and R. Weisberg (ed.), *Lambda II*. Cold Spring Harbor Laboratory, Cold Spring Harbor, N.Y.

White, R. L., and M. S. Fox. 1974. On the molecular basis of high negative interference. *Proc. Natl. Acad. Sci. USA* **71**:1544–1548.

Whitehouse, H. L. K. 1963. A theory of cross-over by means of hybrid deoxyribonucleic acid. *Nature* (London) **199**:1034–1039.

Whitehouse, H. L. K. 1982. *Genetic Recombination: Understanding the Mechanisms*. John Wiley & Sons, Inc., New York.

Wildenberg, J., and M. Meselson. 1975. Mismatch repair in heteroduplex DNA. *Proc. Natl. Acad. Sci. USA* **72**:2202–2206.

Wilson, J. H. 1979. Nick-free formation of reciprocal heteroduplexes: a simple solution to the topological problems. *Proc. Natl. Acad. Sci. USA* **76**:3641–3645.

Windle, B. E., and J. B. Hays. 1986. A phage P1 function that stimulates homologous recombination of the *Escherichia coli* chromosome. *Proc. Natl. Acad. Sci. USA* **83**:3885–3889.

Womack, F. C. 1963. An analysis of single-burst progeny of bacteria singly infected with a bacteriophage heterozygote. *Virology* **21**:232–241.

Womack, F. C. 1965. Cross-reactivation differences in bacteriophage T4D. *Virology* **26**:758–760.

Wu, J.-R., Y.-C. Yeh, and K. Ebisuzaki. 1984. Genetic analysis of *dar*, *uvs*W, and *uvs*Y in bacteriophage T4: *dar* and *uvs*W are alleles. *J. Virol.* **52**:1028–1031.

Zell, R., and H.-J. Fritz. 1987. DNA mismatch-repair in *Escherichia coli* counteracting the hydrolytic deamination of 5-methyl-cytosine residues. *EMBO J.* **6**:1809–1815.

Chapter 5

Mismatch Repair and Genetic Recombination

Miroslav Radman

I. INTRODUCTION

In this chapter I review the properties of two distinct DNA mismatch repair systems operating in *Escherichia coli* and *Streptococcus pneumoniae*. A hypothesis for the effects of these systems on genetic recombination is proposed and extended to eucaryotes. The essence of the hypothesis can be summarized in two points. (i) Long-patch mismatch repair (LPMR) conserves genetic information in the course of DNA replication (by repair directed to the newly synthesized strands) and in the course of genetic recombination (by reversing heteroduplex formation, thus aborting recombination intermediates). (ii) Very-short-patch mismatch repair (VSPMR) is highly specialized. It conserves certain sequences (e.g., cytosine methylation and similar sequences in *E. coli* and some different short sequences in *S. pneumoniae*) by very localized events. Thus, heteroduplexes containing these special mismatches recombine at high frequency. The antirecombination effect of LPMR and the hyperrecombination effect of VSPMR provide formal models for conservation, homogenization, and diversification of genes.

II. MISMATCHES, HETERODUPLEXES, AND THEIR REPAIR

A DNA base-pair mismatch can be defined as any noncomplementary base pair in the DNA duplex. Thus, unlike DNA damage, mismatches consist of chemically normal, but mispaired or unpaired bases in a DNA duplex which is then called heteroduplex DNA. A heteroduplex DNA molecule contains nonidentical genetic information on the two complementary strands, such that upon replication a single heteroduplex molecule segregates genetically; i.e., it yields mixed progeny consisting of DNA duplexes with sequences derived from both complementary strands. Mismatch repair transforms a heteroduplex molecule containing a single mismatch into a homoduplex which yields pure progeny of a single parental genotype. By this criterion, mismatch repair can result from events not triggered by the mismatch, e.g., "nick-translation" type of repair by excision and resynthesis of a strand, or simply asymmetric DNA replication in which one of the heteroduplex strands is lost. Keeping this in mind, we shall consider as biologically relevant mismatch repair only those processes that are provoked or triggered by the mismatch, i.e., processes that involve mismatch recognition as a necessary step in mismatch repair.

Mispaired or unpaired bases in a DNA duplex arise in vivo by at least three different mechanisms: (i) errors in DNA replication, (ii) genetic recombination by strand exchange between homologous but nonidentical DNA sequences, and (iii) deamination of 5-methylcytosine (5-meC) to thymine (T), generating a $G \cdot T$ mismatch. Deamination of cytosine (C) to uracil (U) will not be considered to form a DNA mismatch, since uracil in the $G \cdot U$ mismatch is a non-DNA base and is apparently recognized as a DNA lesion by a specific DNA uracil-N-glycosylase (for review, see Lindahl, 1982).

All three mismatch-generating events are rare and unpredictable, and thus it is difficult to isolate mismatched DNA intermediates in vivo in order to study their repair. Therefore, attempts have been made to mimic the in vivo situations by using suitable DNA heteroduplexes reconstituted in vitro. In studies of mismatch repair in E. coli, artificial heteroduplex DNAs isolated from bacteriophage lambda, M13, f1, or ϕX174 have provided important insights.

Two mismatch repair systems have been well characterized in the bacteria E. coli and S. pneumoniae (for reviews, see Radman and Wagner, 1984, 1986; Claverys and Lacks, 1986; Modrich, 1987; M. Meselson, in K. B. Low, ed., The Recombination of Genetic Material, in press), and although their respective molecular mechanisms and specificities are not identical, their basic mechanisms and effects on DNA replication and recombination appear quite similar. Therefore, to facilitate discussion of the involvement of the two mismatch repair systems in genetic recombination, I shall use the terms long-patch mismatch repair (LPMR) and very-short-patch mismatch repair (VSPMR) to distinguish the two basic mismatch repair systems in bacteria and, perhaps, eucaryotes.

A. LPMR

LPMR in *E. coli* requires either unmethylated 5'-GATC-3' sequences (Län-gle-Rouault et al., 1986, 1987; Lahue et al., 1987) or nicks (single-strand breaks) (Längle-Rouault et al., 1987) for its activity and is usually referred to as methyl-directed mismatch repair. The requirement for unmethylated 5'-GATC-3' sequences accounts for the activity of LPMR on the newly synthesized, tran-siently undermethylated DNA strands and therefore for correction of replication errors. *E. coli dam* mutants deficient in DNA adenine methylation (Marinus and Morris, 1975) (6-methyladenine in 5'-GATC-3' sequences) are proficient in mismatch repair but deficient in strand direction (Radman et al., 1980, 1981; Pukkila et al., 1983) and are therefore mutators. *E. coli mutH*, *mutL*, *mutS*, and *mutU* (formerly called *uvrE* or *uvrD*) (Cox, 1976) mutants are proficient in DNA methylation but deficient in mismatch repair (for review, see Radman and Wagner, 1986) and are therefore mutators. Over 99% of the spontaneous replica-tion errors are corrected by this mismatch repair system (Glickman and Radman, 1980). Long excision-resynthesis tracts (up to several kilobases) are hallmarks of the methyl-directed mismatch repair (Wildenberg and Meselson, 1975; Wagner and Meselson, 1976; Lu et al., 1984; Jones et al., 1987b). The functions of the *mut* gene products and their roles in this process are reviewed below.

The LPMR in *S. pneumoniae* has been studied by its effects on genetic recombination during transformation (see Lacks, this volume). This LPMR does not appear to be methyl directed (*S. pneumoniae* is devoid of Dam methylase activity [see Claverys and Lacks, 1986]). The unligated ends of the donor DNA single strand inserted into the recipient DNA seem to direct the removal of the strand bearing a recognizable mismatch. *S. pneumoniae* mismatch repair genes involved in LPMR are called *hex* (for review, see Claverys and Lacks, 1986).

B. VSPMR

VSPMR in *E. coli* repairs those G · T mismatches that originate by deami-nation of 5-meC (C*) to T in the sequence 5'-CC*(A or T)GG-3' or in the related, but presumably unmethylated, 5'-C(A or T)GG-3' sequences. By repairing such G · T mismatches exclusively to the G · C pair, using very short excision-resynthesis tracts, VSPMR apparently protects *E. coli* from the hydrolytic loss of 5-meC and "cools" considerably the 5-meC→T mutation "hot spots" (for review, see Radman and Wagner, 1986).

Sicard and colleagues (Lefevre et al., 1984) discovered a VSPMR in *S. pneumoniae* analogous in its mechanism to the *E. coli* VSPMR (see below), but the nature of the sequence acted on (5'-ATTAAT-3') and the mismatch repair pattern (G · A→G · C) do not inspire a specific hypothesis on the biological role of this VSPMR (Mostachfi and Sicard, 1987). It cannot be excluded that some VSPMR systems have evolved to correct some frequent sequence-specific (or mismatch-specific) replication errors which escape detection or repair by the LPMR system.

The effects of LPMR and VSPMR on genetic recombination and a situation in which recombination may be an essential part of a particular mode of mismatch repair are described below.

III. BASIC CONCEPTS DEFINED AROUND 1964

Mismatch repair was first postulated by Holliday (1964) as an integral part of his molecular model for genetic recombination. In the same year, Setlow and Carrier (1964) and Boyce and Howard-Flanders (1964) announced the discovery of the repair of DNA damage (UV light-induced pyrimidine dimers) by the excision of the damaged region of the DNA strand. Soon, Pettijohn and Hanawalt (1964) demonstrated that the gap left after excision is filled in by localized DNA repair synthesis.

Hybrid or heteroduplex DNA at the splice or patch site in genetic recombination was postulated in several organisms (Whitehouse, 1963; Meselson, 1965; Holliday, 1964). At that time, Lacks (1962) and Fox and Allen (1964) had shown that the structure of recombined DNA in pneumococcal transformation is a single-stranded insertion derived from the donor DNA which replaced, by strand exchange, the resident DNA strand, thus creating a "patch-type" heteroduplex (see Fig. 2). Ephrussi-Taylor and Gray (1966) then suggested that the wide difference in transformation efficiency between diverse genetic markers from the same genetic region reflects the repair efficiency of mismatches formed in the heteroduplex region. The nature of these mismatches is determined by the mutational nature of the genetic markers used. Mismatch repair in pneumococcal transformation was supposed to be directed to the donor strand, so as to exclude the incoming donor strand information (for review, see Claverys and Lacks, 1986). Thus, markers that did not provoke repair when mispaired with the wild type were high-efficiency transformation markers, whereas markers creating repaired mismatches were low-efficiency transformation markers (see Lacks, this volume).

Amati and Meselson (1965), working with bacteriophage lambda, demonstrated an apparent clustering of genetic exchanges between closely spaced markers (so-called negative interference [see Mosig, this volume]), but it was 10 years before White and Fox (1974) and Wildenberg and Meselson (1975) provided evidence that such clustering of genetic exchanges can be caused by a localized repair of mismatches in the heteroduplex region including two or more markers.

Thus, the formal concepts of the structure of recombined DNA and of the effects of genetic markers on the pattern and frequency of genetic exchanges were set two decades ago. Considerable genetic and molecular experimentation has resulted in a relatively coherent picture of the underlying molecular mechanisms in the bacteria *E. coli* and *S. pneumoniae*. That picture is summarized below.

IV. MISMATCH REPAIR OF REPLICATION ERRORS IN *E. COLI*: METHYL-DIRECTED LPMR

In order for an enzymatic process to be able to correct replication errors (without any interaction between homologous duplexes; see section X), it must detect the mismatch, distinguish newly synthesized strands from parental strands, and act selectively to preserve the parental sequence. In *E. coli*, a means of strand discrimination for the LPMR system is provided by adenine methylation in GATC sequences (for reviews, see Modrich, 1987; Radman and Wagner, 1986; Meselson, in press). Mismatch repair is greatly reduced in regions of DNA in which GATC sequences are fully adenine methylated and repair occurs preferentially, if not exclusively, on unmethylated strands of hemimethylated heteroduplexes (Radman et al., 1980, 1981; Pukkila et al., 1983; Lu et al., 1983, 1984). Because the newly synthesized strands are transiently undermethylated (Lyons and Schendel, 1984; Marinus, 1976), i.e., methylation lags somewhat behind replication, LPMR in *E. coli* is believed to occur primarily on newly synthesized strands immediately behind the replication fork. (Radioactive labeling experiments show that several minutes may elapse before the newly synthesized strands become fully methylated [Lyons and Schendel, 1984], but the length of the hemimethylated DNA at the replication fork has not been determined.)

Although much is known about this *E. coli* mismatch repair system, both from in vivo (Radman and Wagner, 1986; Meselson, in press) and in vitro (Modrich, 1987) studies, many specific aspects of the mechanism are not yet understood. Two different models are currently being considered and tested. Both models propose that the MutS protein (Su and Modrich, 1987) (and perhaps the MutL protein as well) is involved in mismatch recognition (Modrich, 1987; Radman and Wagner, 1986). In both models hemimethylated GATC sequences on either side of a recognized mismatch are nicked on the unmethylated strand by the MutH protein, and the strand segment containing the mispaired base is removed. Both models require some process of long-distance communication along the DNA to allow MutH protein to act only at adjacent unmethylated GATC sequences. The key difference between the models is that one proposes action at, and synthesis between, the mismatch and one adjacent GATC sequence (Lu et al., 1984; Modrich, 1987) whereas the other proposes action at, and synthesis between, two GATC sequences flanking the recognized mismatch (Längle-Rouault et al., 1987). The first model is supported by in vitro experiments (Lu et al., 1984), while the second model is supported by in vivo experiments (Längle-Rouault et al., 1987; Jones et al., 1987b).

MutH protein nicks the unmethylated GATC sequence (Lu et al., 1984; Modrich, 1987; Längle-Rouault, 1987; Welsh et al., 1987), whereas the MutU protein (helicase II) presumably melts the (nicked or to be nicked) strand, sometimes over several kilobases, to allow repair synthesis (presumably by the DNA polymerase III holoenzyme). DNA without GATC sequences is not mismatch repaired, and GATC sequences several kilobases away from a mismatch can stimulate its repair (Längle-Rouault et al., 1986, 1987; Lahue et al., 1987). The finding that a persistent nick in heteroduplex DNA can effectively substitute for both the MutH function and unmethylated GATC sequences (Längle-Rouault et al., 1987) brought together LPMR models for *E. coli* and *S. pneumoniae*. A weak but specific nicking of unmethylated GATC sequences was observed with the purified MutH protein (Welsh et al., 1987).

V. MOLECULAR SPECIFICITY OF THE LPMR SYSTEM

The two well-characterized LPMR systems, *E. coli* MutHLSU and *S. pneumoniae* HexAB, do not recognize and repair all mismatches with equal efficiency. Yet, with few exceptions, the two systems show striking similarities in their specificities of mismatch repair. For both systems, the extent of repair depends on the nature of the mismatch and its sequence environment (for reviews, see Claverys and Lacks, 1986; Radman and Wagner, 1986; Modrich, 1987; Meselson, in press). In general, it appears that transition mutation mismatches (G · T and A · C) are better repaired than transversion mutation mismatches (G · A, C · T, C · C, T · T, A · A, and G · G) (Claverys et al., 1981, 1983; Lacks et al., 1982; Lahue et al., 1987; Kramer et al., 1984; Dohet et al., 1985; Jones et al., 1987c). The *E. coli* LPMR efficiency for a given mismatch (in particular for G · A and C · T) increases with increasing G+C content in the neighborhood of the mismatch (Jones et al., 1987c).

The *E. coli* mismatch repair system can recognize and very efficiently repair heteroduplexes containing one-base deletion or addition frameshift mutations (Dohet et al., 1986). These heteroduplexes do not contain a mismatch, but rather an extra, and therefore unpaired, base. When such heteroduplexes are unmethylated, mismatch repair works equally well on either strand, and the nature of the unpaired base does not influence the extent of repair. The repair of such heteroduplexes can also be methyl directed; thus, the *E. coli* mismatch repair system can correct both addition and deletion frameshift mutations arising during replication (Dohet et al., 1986; Schaaper and Dunn, 1987). In *S. pneumoniae*, Hex-directed LPMR efficiently repairs heteroduplexes with one, two, and three unpaired bases, but four, five, or more unpaired bases are refractory to repair (Gasc et al., 1987).

Although heteroduplexes with a few unpaired bases are substrates for mismatch repair enzymes, heteroduplexes with a large single-stranded loop are not (Kramer et al., 1982; Dohet et al., 1987; Raposa and Fox, 1987). However, there appears to be some activity in mismatch repair-deficient cells (*mutH*, *mutL*, *mutS*, and *mutU* mutants) which can repair such looped structures in *E. coli* even in fully methylated DNA (Dohet et al., 1987). Heteroduplex regions with such large nonhomologies may not occur frequently during replication but do occur during bacteriophage lambda recombination (Lichten and Fox, 1984) between the wild type and deletion mutants.

The structural elements possibly responsible for the recognition of mismatches by the mismatch repair proteins and the effect of mismatch repair specificity on the nature of spontaneous mutations is discussed elsewhere (Fazakerley et al., 1986).

VI. MISMATCH REPAIR OF DEAMINATED 5-meC IN *E. COLI*: VSPMR

VSPMR acts only in specific sequences by the removal of very few specific bases and appears to preserve certain privileged sequences. 5-meC, the most common modified base in the DNA of many organisms, is the product of the reaction catalyzed by the cytosine methyltransferase enzymes (for review, see Marinus, 1987). 5-meC is also the most unstable base in DNA (Lindahl, 1982); it readily loses its amino group to become thymine. Therefore, there was little surprise when early sequencing of spontaneous mutations in *E. coli* showed hot spots of 5-meC→T (i.e., C · G→T · A transition) mutations (Coulondre et al., 1978). It was thought that there could be no way to discriminate the wild type from the mutant in the G · T mismatch that originated by the 5-meC→T deamination, since deamination presumably occurs irrespective of DNA replication. However, to account for certain anomalies in phage lambda crosses (see Mosig, this volume), Lieb (1983) was the first to propose that such G · T→G · C mismatch repair may occur by the excision-resynthesis of a few bases—hence, the name VSPMR. VSPMR was proposed to account for hyperrecombination observed with special markers (see below).

A direct demonstration of Lieb's hypothesis was provided by Jones et al. (1987a, b), who used in vitro-constructed phage lambda DNA heteroduplexes and found that indeed only the G · T mismatch at the position flanking the central A · T pair of the 5′-CC(A or T)GG-3′ sequence is subject to VSPMR; neither A · C mismatches at these or other positions nor G · T mismatches at other sites in, or outside, this Dcm sequence are subject to VSPMR. VSPMR operates on fully GATC (adenine) methylated DNA; it requires MutS, MutL, Dcm, and PolA proteins (S. Dzidic and M. Radman, submitted for publication), but not the MutH and MutU proteins which are required for LPMR (Jones et al., 1987a, b; Lieb, 1987). (Dcm protein is the cytosine methyltransferase of *E. coli* [for review, see Marinus, 1987].) The excision tract is indeed very short, and it may be that often just one base (T) is excised (Jones et al., 1987). Studies with M13 phage DNA heteroduplexes (Zell and Fritz, 1987) confirmed basic features of the VSPMR elaborated by Jones et al. (1987a, b). Thus, the VSPMR system appears to conserve the cytosine methylation sites by reversing the 5-meC→T change (Fig. 1). The system also recognizes G · T mismatches and repairs them to G · C pairs in tetranucleotide sequences that are part of the Dcm recognition sequence, i.e., C(A or T)GG (see Fig. 4; see Lieb et al., 1986).

VII. ANTIRECOMBINATION BY LPMR

LPMR excising one of the two parental strands and reconstituting (by repair synthesis) one parental sequence can destroy the potential recombinants created by strand-exchange heteroduplex formation. This will be true in particular when the "invading" strand is excised, as in the case of pneumococcal transformation (see Fig. 2). The observation, in this system, that the unrecognized and hence unrepaired mismatches produce high-efficiency transformation, i.e., recombination (irrespective of the presence of the pneumococcal Hex mismatch repair system), whereas the well-repaired mismatches (which are high-efficiency markers in *hex* mutants) produce low-efficiency transformation, is a clear demonstration that the Hex mismatch repair system is antirecombinogenic (for review, see

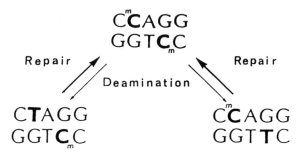

FIGURE 1. Generation and repair of mismatches produced by deamination of 5-meC. The *E. coli dcm* sequence has two 5-meCs which can deaminate to thymidine as shown. VSPMR acts on both deamination products: its sequence, site, mismatch (G · T), and base (T) specificity is such that it reconstitutes the original *dcm* sequence. (From Jones et al., 1987a.)

Claverys and Lacks, 1986). It appears that this mismatch repair system removes from the recombinant DNA the entire invading strand (about 5 to 10 kilobases long) (Mejean and Claverys, 1984), thus aborting the first step (hybrid DNA formation) in recombination (Fig. 2). Compared with very-high-efficiency transformation markers, low-efficiency markers escape mismatch repair and remain integrated only 1 to 10% of the time.

The antirecombinogenic effect of the *E. coli* LPMR (MutHLSU) system is implicated by the observation of Feinstein and Low (1986) that screening for hyperrecombination mutants in Hfr crosses yielded only mismatch repair-deficient mutants (*mutL* or *mutS*). Previously isolated mutations in any of the four *mut* genes (*mutH*, *mutL*, *mutS*, and *mutU*) involved in methyl-directed mismatch repair (see above) also showed the hyperrecombination effect (Feinstein and Low, 1986). In an earlier search for hyper-rec mutants using an intrachromosomal recombination system, Konrad (1977) found only *mutU* mutants as well as mutants that may initiate more recombination events by creating more nicks and gaps.

VIII. HYPOTHESIS FOR ANTIRECOMBINATION BY LPMR IN PNEUMOCOCCAL TRANSFORMATION AND IN GENETIC CROSSES IN *E. COLI*

I would like to propose that in bacterial recombination, during Hfr crosses or F' *lac* × chromosomal *lac* homogenotization crosses (Feinstein and Low, 1986), or intrachromosomal recombination (Konrad, 1977), the *E. coli* MutHLSU mismatch repair system may act analogously to the Hex system in pneumococci, i.e., to destroy the heteroduplex generated by strand exchange. How can the methyl-directed repair act on recombining, presumably fully GATC-methylated, DNA? Two pieces of information may help to generate a model to account for the observations of Feinstein and Low (1986).

(i) Längle-Rouault et al. (1987) showed that a persistent nick in heteroduplex DNA can effectively substitute for both unmethylated GATC sequences and the

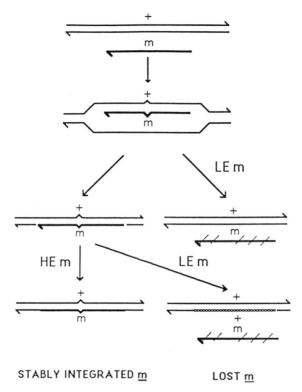

STABLY INTEGRATED m   LOST m

FIGURE 2. Mismatch repair and recombination during pneumococcal transformation. Donor DNA integrates as a single strand by displacing the recipient DNA strand (the three-stranded intermediate has not been demonstrated). A mismatched high-efficiency (HE) marker +/m is not recognized by the Hex LPMR system and is therefore stably integrated in the recipient DNA. A mismatched low-efficiency (LE) marker +/m is readily recognized, and the entire donor strand of 5 to 10 kilobases is aborted from the recipient DNA (Mejean and Claverys, 1984). (The fate of the aborted single-stranded DNA has not been determined.) For more information, see the text and Lacks (this volume).

MutH (GATC-nicking) function in repair of φX174 DNA heteroduplex. They suggested that the entire φX174 strand can be melted away by the helicase II (MutU function) in a mismatch-stimulated, *mutL*- and *mutS*-dependent (*mutH*-independent) process.

   (ii) Siddiqi and Fox (1973) analyzed the integration of donor DNA in bacterial Hfr conjugation by physical methods. They found that the donor DNA integrates as a single strand (as in pneumococcal transformation), preferentially into the newly synthesized recipient strands. Thus, the (presumably) methylated donor DNA strand may become covalently linked to the (presumably) undermethylated newly synthesized strands, and methyl-directed mismatch repair of the incoming donor strand marker may occur. (The parental recipient F⁻ strand is protected from mismatch repair by the methylation of GATC sequences.) In addition, any single-strand insertion heteroduplex involving a newly synthesized, and therefore presumably undermethylated, Hfr DNA (Ippen-Ihler and Minkley, 1986) could be mismatch repaired by methyl-directed LPMR, resulting in the loss of the Hfr marker. Such methyl-directed mismatch repair is less likely to operate in F′ *lac* homogenotization crosses. Indeed, the hyper-rec effect in Hfr crosses is the weakest for the *mutH* mutants, intermediate for *mutL* and *mutS* mutants, and strongest for the helicase II (*mutU*) mutants (Feinstein and Low, 1986). In F′ *lac* homogenotization crosses, *mutH* has no significant effect (1.2 times more recombinants in *mutH* than in *mut*$^+$ [Feinstein and Low, 1986]), whereas the helicase II mutants have a high hyper-rec effect in both types of crosses (factor of 26). Helicase II appears to be the most limiting in decreasing the frequency of recombinants. I propose that the decrease is due to the mismatch-stimulated and MutLS-requiring helicase II-catalyzed removal of the incoming donor strand from the parental recipient duplex (Fig. 3). Free ends of the incoming donor strand may be adequate for the mismatch-stimulated (*mutLS*-requiring) helicase II-catalyzed rejection of that strand. MutH function could further increase this rejection only if the invading Hfr DNA strand is undermethylated (i.e., newly synthesized in the F$^-$ cell) (for review, see Ippen-Ihler and Minkley, 1986) or if a methylated Hfr donor strand becomes covalently linked with the unmethylated GATC sequences of the newly synthesized recipient strand. Thus, the MutH function has a significant effect in Hfr crosses but little or no effect in F′ *lac* homogenotization crosses (Feinstein and Low, 1986).

According to this hypothesis, LPMR is expected to affect recombination involving heteroduplex DNA regions and to exhibit specific marker effects. That is, the antirecombinogenic effect of LPMR should be particularly effective for transition mutations, which produce G · T and A · C mismatches in heteroduplex DNA, and for small deletion or insertion frameshift mutations, but ineffective for some transversion and large heterology mutations (see section IV). Norkin (1970) found such pronounced marker effects in Hfr crosses between different point mutations in the *lacZ* gene: the recombination frequencies were inconsistent with the positions of the mutations deduced from deletion mapping. Using identified *lacI* mutations in F′ *lac* × chromosomal *lac* crosses, Coulondre and Miller (1977) observed 1,000-fold variations in recombination frequencies with equally separated close markers.

This hypothesis, which unifies conceptually the marker effects and mismatch repair in *S. pneumoniae* and *E. coli*, has one potential obstacle; the specificity of the *mutHLSU* mismatch repair system predicts that there should be no effect of these *mut* mutations on the recombination frequency between large deletion or insertion markers, since large loops of nonhomology are not recognized by the *mutHLSU* system (Dohet et al., 1987; Kramer et al., 1982; Raposa and Fox, 1987). Yet, Feinstein and Low (1986) did observe a significant, but small, hyperrecombination effect of *mut* mutants in crosses with deletion markers. (However, some genetically silent point mutation difference [''polymorphism''] between the two parents near the two deletions may have triggered occasional corepair of the deletion-wild type heteroduplex structure to regenerate parental sequences. Alternatively, special sequences at the loop heteroduplex junction may provide a signal for mismatch repair enzymes, as suggested previously [Radman and Wagner, 1986] to account for results of Fishel and Kolodner [1984].)

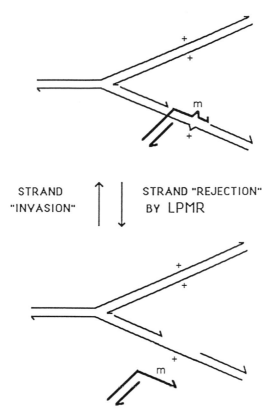

FIGURE 3. Model for antirecombinogenic effect of mismatch repair in *E. coli* Hfr crosses. Single-strand insertion of donor DNA into a replication fork of recipient DNA (Siddiqi and Fox, 1973) by recombination enzymes (for review, see Smith 1987; see also Mahajan, this volume; Radding, this volume; and Taylor, this volume) is shown arbitrarily with a double-stranded tail on the donor DNA. Abortion of the recombination intermediate (heteroduplex) is carried out by a mismatch-stimulated (*mutSL*-dependent) helicase II (MutU) activity. F⁻ recipient replicating DNA is represented by thin lines; Hfr donor DNA with a genetic marker (m) is represented by thick lines.

A. LPMR in Lambda Lytic Cycle Crosses

A most impressive antirecombinogenic effect of methyl-directed (MutHLSU) *E. coli* mismatch repair was observed in transfection experiments with pure heteroduplex DNA of phage lambda (Jones et al., 1987b): when heteroduplex DNA was unmethylated and only the MutHLSU system was active (e.g., in a *dcm* mutant deficient in VSPMR [Jones et al., 1987a]), not a single recombinant genotype (out of 200 molecules) was observed between two markers separated by 0.1 kilobase. Mismatch repair occurred on over 95% of the molecules, but all events corepaired the two mismatches on the same (but either) heteroduplex strand (Jones et al., 1987b). Thus, the LPMR system can conserve parental genetic information by the corepair of mismatched markers on one strand (see legend to Fig. 6). Whether the heteroduplex joints of crossover junctions can be

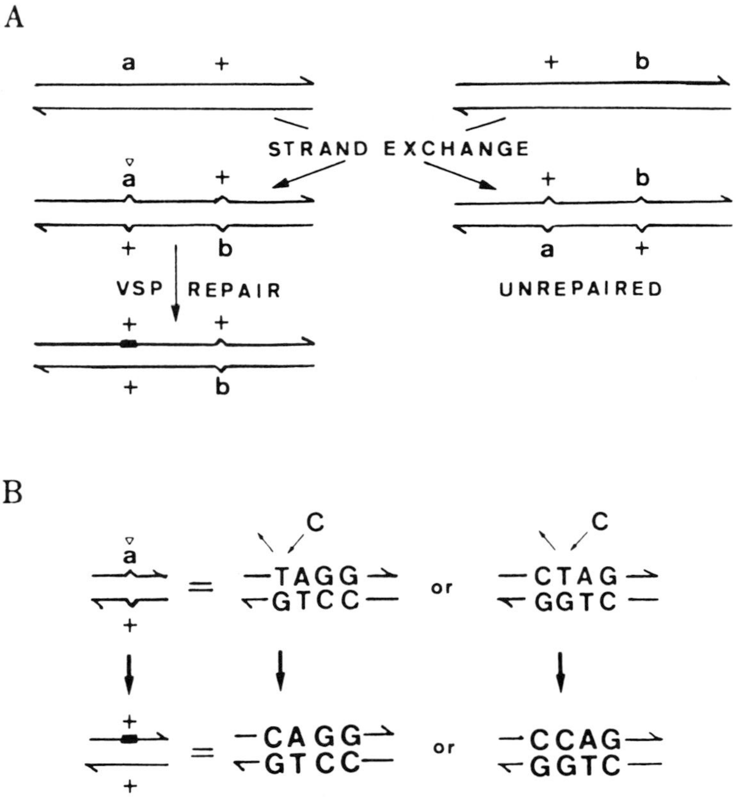

A

B

FIGURE 4. Hyperrecombinogenic effect of VSPMR in *E. coli*. (A) A genetic cross between two markers, a VSPMR marker (a) and a normal marker (b), is shown. Reciprocal strand exchange produces two heteroduplexes which differ only at the mismatch sites (e.g., if a/+ is a T/G mismatch, +/a is a C/A mismatch). VSPMR acts only on one, a/+ mismatch, whose chemistry is shown in B. This repair produces a ++ recombinant strand. For more information, see text and Jones et al. (1987b).

dissociated by LPMR is not known, but the work of Feinstein and Low (1986) suggests that they can. Such activity would also inhibit "splice" (crossover) recombination between distant markers only if the heteroduplex joint bears a recognized mismatch.

IX. HYPERRECOMBINATION BY VSPMR

The large marker effects in *E. coli* conjugational intragenic crosses and in pneumococcal transformation are not seen for most markers in phage lambda crosses (Gussin et al., 1981; Lieb, 1981). Genetic maps of phage lambda, both intra- and intergenic, are impressively congruent with the physical map (see Gussin et al., 1981; Lieb, 1981 [and references therein]; Mosig, this volume). It appears that LPMR does not interfere with genetic mapping either because it is inefficient on phage lambda or because it decreases recombination frequencies to similar extents in crosses with different markers (see Jones et al., 1987b). Notably, the hyper-rec effect of *mutL* mutations in conjugational recombination by the *E. coli* RecBCD pathway (Feinstein and Low, 1986) does not occur in RecBCD pathway recombination of phage lambda (J. Sawitzke and S. M. Rosenberg, personal communication; also see Rosenberg, 1987).

However, a class of genetic markers, initially identified by Coulondre and Miller (1977) in F' *lac* × *lac* crosses, misbehave also in phage lambda crosses in that they yield many more recombinants per physical distance than other nearby markers (Coulondre and Miller, 1977; Lieb, 1983, 1987; Lieb et al., 1986). This excess of recombination can be by a factor of 100 (Lieb, 1983) to 1,000 (Coulondre and Miller, 1977) in crosses with very close markers. Coulondre and Miller (1977) and Lieb (1983) noticed that the "hyper-rec" markers were due to C→T transitions in tetranucleotide sequences related to the cytosine methylation (Dcm) sequences 5'-CC(A or T)GG-3'. Lieb (1983) proposed that a VSPMR system acts as depicted in Fig. 4. Her hypothesis predicted that VSPMR generates recombinant genotypes on only one of the two possible heteroduplexes, i.e., on the G · T mismatch, which is repaired only to G · C, and not on the reciprocal A · C mismatch.

Lieb's hypothesis was tested by Jones et al. (1987b), who performed parallel genetic crosses and DNA transfection experiments with in vitro-constructed heteroduplex molecules of phage lambda using the same genetic markers: one normal marker (the temperature-sensitive *c*I857 mutation) and one hyper-rec marker (the *c*I am6). With this system, it was possible to detect all possible individual repair or corepair events. The results showed clearly that VSPMR indeed operates only on the G · T mismatch-bearing heteroduplex involving the am6 hyper-rec marker (Jones et al., 1987b). In addition, as predicted, the repair is unidirectional G · T→G · C. In the absence of LPMR (i.e., in a *mutH* mutant), VSPMR created recombinants in 60% of the heteroduplex molecules; in contrast, in the absence of VSPMR (i.e., in a *dcm* mutant), not a single recombinant was formed among 200 heteroduplex molecules (Jones et al., 1987b). Thus, it appears that those genetic markers forming with the wild type a G · T mismatched heteroduplex at sites flanking the A · T pair of the 5'-C(A or T)GG-3' sequence are detected as 5-meC→T deamination events and are repaired to restore the original G · C base pair (Fig. 1 and 4). (VSP may be as short as 1 or 2 bases but is always shorter than 10 bases [for references, see Jones et al., 1987b; Lieb et al., 1986].)

In the course of this highly localized repair event, a recombinant genotype is generated on the repaired strand. Apparent clustering of genetic exchanges among closely linked markers, called negative interference, can be largely accounted for by the action of VSPMR on appropriate genetic markers (Jones et al., 1987b; Raposa and Fox, 1987; Mosig, this volume; Meselson, in press). Classical studies on negative interference involving *P*am3 and *P*am80 markers of lambda (Amati and Meselson, 1965; White and Fox, 1974; Wildenberg and Meselson, 1975) can be explained by VSPMR since (i) *mutL*, *mutS* (Glickman and Radman, 1980), and *dcm* (M. F. Bourguignon-van Horen and M. Radman, unpublished data) mutations decrease, whereas *mutH* and *mutU* increase, their recombination (Glickman and Radman, 1980), (ii) both *P*am3 and *P*am80 mutations generate a VSPMR marker (by a 5'-CCAG-3'→5'-CTAG-3' change) (W. Reiser, doctoral dissertation,

University of Heidelberg, Heidelberg, Federal Republic of Germany, 1983), and (iii) transfections with *Pam3/Pam80* heteroduplex DNA strongly support the VSPMR mechanism (Fig. 4) as a major source of P^+ recombinants (Raposa and Fox, 1987; Bourguignon-van Horen and Radman, unpublished data; Meselson, in press).

Hyperrecombination of a special marker *amiA* in a special sequence (5'-ATTAAT-3') of *S. pneumoniae* occurs by a mechanism analogous to the *E. coli* VSPMR (Lefevre et al., 1984; Mostachfi and Sicard, 1987). Point mutations in this sequence decrease but do not abolish hyperrecombination by VSPMR (M. Sicard and A. M. Gasc, personal communication).

X. RECOMBINATION AS A MISMATCH REPAIR PROCESS: A HYPOTHESIS FOR SCE

Glickman and Radman (1980) suggested that the LPMR system can provoke double-strand breaks in unmethylated DNA (in *E. coli dam* mutants) as a result of simultaneous attack by mismatch repair enzymes on both DNA strands bearing a mismatch. This suggestion accounted for the lethality of *dam* mutations coupled with a mutation blocking recombinational repair of double-strand breaks (*recA*, *recB*, *polA*, or *lexA* [Ind⁻]) and the viability of a triple mutant, *dam recA mutH* (or *mutL* or *mutS*) (Glickman and Radman, 1980; McGraw and Marinus, 1980). Double-strand breakage of unmethylated DNA containing mismatches was demonstrated both genetically (as mismatch-stimulated killing of phage lambda) (Doutriaux et al., 1986) and physically (by analysis of bacterial DNA in neutral sucrose gradients) (Wang and Smith, 1986a). The 10-fold lower mutation rates in *dam* mutants than in *mut* mutants (*mutH*, *L*, *S*, or *U*) suggested that *dam rec⁺* bacteria repair, by recombination, about 90% of their DNA replication errors (Radman and Wagner, 1986; Wagner et al., 1984). Hastings (1984; this volume) adapted this model for meiotic recombination in *Saccharomyces cerevisiae*, in which mismatch repair appears to act unidirectionally, i.e., in favor of the recovery of the "invading" strand information. In Hastings' model, heteroduplex formation precedes the mismatch-stimulated recombination between homologs. The model presented below (Fig. 5) elaborates a mechanism for correction of replication errors by recombination between sisters, i.e., sister chromatid exchange (SCE). Mismatch correction of replication errors "in conjunction with SCE" was first suggested by Wagner and Meselson (1976), whereas mismatch repair by recombination via double-strand breaks was proposed by us (Glickman and Radman, 1980; Radman and Wagner, 1986; Wagner et al., 1984) and by Hastings (1984).

Although striking marker effects observed in fungal meiotic recombination (Leblon and Rossignol, 1973; White et al., 1985) argue in favor of some mismatch repair process, no direct evidence exists for the model presented in Fig. 5; however, the model can accommodate diverse observations, and it offers a mechanism and a biological role for the phenomenon of SCE in eucaryotes. The key features of the model are as follows (see Fig. 5). (i) Nonreplicating DNA is

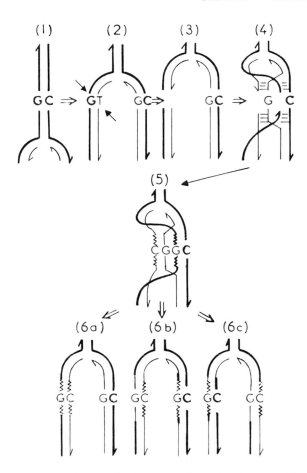

FIGURE 5. Mechanism for repair of replication error mismatches by recombination between sister molecules. Replication error is T mispaired with the template G (step 2). Double-stranded gap (step 3), resulting from the attack of mismatch repair enzymes on both strands, is recombinogenic; the two free ends invade the homologous sister molecule, forming short heteroduplexes (step 4) sufficient to prime DNA synthesis (step 5) from the two 3' ends. The resulting double Holliday junction can be resolved to liberate the recombined sister molecules by three mechanisms considered here: (6a) topoisomerase-mediated resolution leads to the segregation of two sisters parental for DNA flanking the repaired region (Thaler et al., 1987), (6b) noncrossover resolution also results from cleavage of pairs of either both crossing or both noncrossing strands, and (6c) crossover resolution results from cleavage of one pair of crossing strands and one pair of noncrossing strands (see Thaler et al., 1987 [and references therein]). Parental strands are thick lines, newly synthesized strands are thin lines, and the repair-synthesis tracts are zigzag lines.

kept, perhaps by some proteins in the chromatin, in a state inaccessible to recombination enzymes; only the nascent products of DNA replication, i.e., the two sister molecules, become transiently susceptible to recombination at the same locus, at the same time. Therefore, sister exchange (SCE) is much more frequent than nonsister (homolog) exchange. (In this model "chromatin" is an inhibitor of recombination.) (ii) DNA repair enzymes produce postreplicative double-strand breaks in one sister molecule at a site of a replication error (mismatch) or of damage. Single-strand-specific endonucleases, ubiquitous among eucaryotes (Shishido and Ando, 1982), may play a role in converting a damaged site to a double-strand break or gap, following its replication. (iii) Double-strand breaks in one of the two sister molecules are efficiently repaired by recombination using the intact sister molecule to restore the missing sequence by copying. This recombinational repair reestablishes the parental DNA sequence. (iv) Only crossing-over resolutions of the recombination repair intermediates will be detected as SCE by standard methods. The disparity between the number of expected initial replication errors in mammalian cells (my estimates are in the range of 10^2 to 10^4 per cell) and the number of SCEs (in the range of 3 to 10 per cell) (for review, see Latt, 1981) suggests that most resolutions in SCE recombination must be noncrossover, or that some other mismatch repair process removes the majority of replication errors. (v) Limiting recombination between sisters exclusively close to the replication fork (as suggested in i) may solve one major problem: when damage or mismatches occur in a repetitive DNA sequence, there is so much homology in the genome with which to recombine, often as many as 10^5 copies spread over many chromosomes, that deletions, deletions/amplifications (unequal SCE), translocations, and other mitotic homologous recombinants would be generated. These events are dangerous for mammalian cells, since they may lead to the loss of genetic information or to expression of recessive mutations (see Kinsella and Radman, 1978; Radman et al., 1982). In fact, agents that stimulate SCEs readily stimulate amplifications and rearrangements, although at a much lower frequency (Latt, 1981; Stark and Wahl, 1984). In several cases duplications or deletions in the human genome are associated with apparent crossovers involving different copies of a repeated sequence family (Mermer et al., 1987; Lehrman et al., 1987). This model, if applied to meiosis (Hastings, 1984), may account for the correlations between abundantly polymorphic regions and regions of high meiotic crossing-over frequency (Szabo et al., 1984).

The first three features of the model proposed above are supported by the following evidence. (i) Chromatin prevents certain repair enzymes from repairing damaged DNA (Mellon et al., 1986). The high mitotic recombination rates of small extrachromosomal DNAs (Waldman and Liskay, 1987 [and references therein]) and high recombination enzyme activities in vitro (Kucherlapati et al., 1985; Lopez et al., 1987), versus a near absence of mitotic recombination in mammalian chromosomes (Chasin, 1974), suggest that mitotic recombination of the chromosomes is inhibited. (ii) Restriction enzymes are potent inducers of SCEs in cultured mammalian cells but only if introduced during the S-phase (Natarajan et al., 1985). X ray-induced random double-strand breaks do not efficiently induce SCEs (for review, see Latt, 1981). These facts argue that only the newly replicated DNA is susceptible to restriction enzymes and that only double-strand breaks in the newly replicated DNA lead to SCEs. (iii) Double-strand breaks provoked by mismatch repair enzymes have been detected genetically (Doutriaux et al., 1986) and physically (Wang and Smith, 1986b, c). A single-strand-specific endonuclease isolated from the fungus *Ustilago maydis* produces mismatch-stimulated and mismatch-localized double-strand breaks in vitro (Pukkila, 1978). Post-UV irradiation, postreplication double-strand breaks and their repair have been detected by physical methods in bacteria and mammalian cells (Wang and Smith, 1986b, c). Molecular models proposed by Resnick and Martin (1976) and by Szostak et al. (1983) adequately account for the gap-directed gene conversion for the model in Fig. 5. Noncrossover resolutions of recombination intermediates by topoisomerases have been suggested by Thaler et al. (1987).

Several observations can be explained by this model. For example, there are three situations in which accelerated rates of replication errors correlate with increased SCEs. (i) In mammalian cells excess thymidine, which is mutagenic as a result of alteration of nucleotide pools (Bradley and Sharkey, 1978), also induces SCEs (Perry, 1983). (ii) Bromodeoxyuridine-induced pool bias (deficit of dCTP) is mutagenic and SCE inducing: deoxycytidine in the medium suppresses both the mutagenic and SCE-inducing effects (Davidson et al., 1980). (iii) Bloom's syndrome in humans is a mutator mutation (Warren et al., 1981) and can be diagnosed by the elevated levels of spontaneous SCEs (Chaganti et al., 1974) (it is not known whether the mutator phenotype results from replication errors without DNA damage).

XI. TWO MODES OF MISMATCH REPAIR CAN CONSERVE, HOMOGENIZE, AND DIVERSIFY DNA SEQUENCES IN THE COURSE OF GENETIC RECOMBINATION

LPMR of *E. coli* conserves genetic information in both replication and recombination: a DNA strand carrying a sequence different from the parental strand in replication (or different from the strand-assimilating molecule in recombination) is aborted from the DNA, and the single-strand gap is filled in by copying the parental strand (see section VII). In pneumococci, a similar process destroys a recombination intermediate (Fig. 2). Point mutation heterozygosities lower heteroduplex formation in meiosis of the fungus *Ascolobus immersus* (Nicolas and Rossignol, 1983), but (because of the lack of mismatch repair-deficient mutants in this system) we do not know whether this antirecombination effect is provoked by mismatch repair enzymes. A mismatch-stimulated antirecombination by LPMR, as described in Fig. 6, could act to prevent chromosomal aberrations caused by "illegitimate" recombination involving repetitive sequences and to prevent mitotic recombination (see also section X). Sequence polymorphisms among repetitive DNA sequences and between the two parental chromosomal lines in diploid mammalian cells may provide a sufficient number of mismatches in hybrid DNA regions to make LPMR a potent antirecombination process. Clearly, only the accurate SCE would not be prevented by LPMR as a result of sequence identity of sister molecules. Given the notorious insensitivity of the strand exchange process to base pair mismatches, including large heterologies (Lichten and Fox, 1984; review Smith, 1987), mismatch-stimulated antirecombination by LPMR appears to act as a general "proofreading" system, assuring the fidelity of homologous DNA recombination and thereby playing an important role in chromosomal stability. I suggest that the prevention of recombination by LPMR may be the mechanism of the genetic separation of newly arising species from

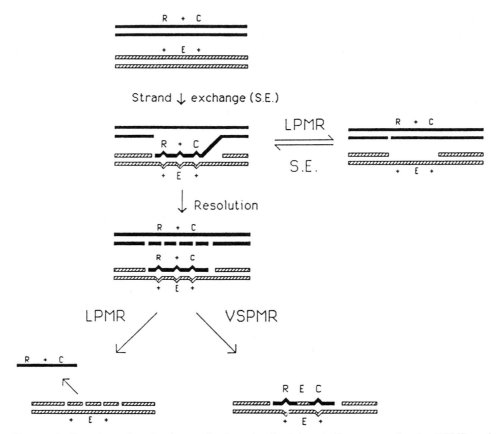

FIGURE 6. Overview of molecular mechanisms involved in genetic conservation by LPMR and genetic diversification by VSPMR in the course of genetic recombination. R/+, E/+, and C/+ symbolize point mutation heterozygosities between two homologous recombining genes. For simplicity, only one kind of strand exchange (asymmetric strand exchange) is shown. Heteroduplex DNA covers all three markers. LPMR may reverse the initial heteroduplex formation step before its resolution (shown at the right) or abort the inserted "patch" after resolution (lower left). If the heteroduplex is not aborted by LPMR, and if one or more mismatches are substrate for VSPMR (+/ E, in this figure), then VSPMR will generate a recombinant (REC) by localized and directed repair (+/ E to E/E). Interrupted lines symbolize repair-synthesized strand fragments.

their ancestors and therefore the primary cause of sterility in interspecific and intergeneric crosses.

In *E. coli*, VSPMR conserves cytosine methylation sites against hydrolytic 5-meC loss (see section VI), whereas in genetic recombination VSPMR generates recombinants and thus diversifies strand-exchanging parental DNA sequences (see section IX). These two opposing genetic effects of the LPMR and VSPMR systems in the course of genetic recombination (see Fig. 6) may have important impacts on the maintenance and evolution of genetic information, but in particular on the genetics of mammals, in which there are abundant, diverse repetitive sequences and diversified gene families. For example, sequence *conservation* is required to maintain the already diversified genes of the germ line repertoire of the immune system; sequence *homogenization* is required in some repetitive sequence families, such as ribosomal RNA genes or in genes coding for the constant immunoglobulin part; but rapid and extensive, highly localized sequence *diversification*, or somatic mitogenesis, is required to realize the somatic B-cell repertoire of immunoglobulins (see Engler and Storb, this volume). This mutation rate in the variable regions of the immunoglobulin G coding genes is about 10^{-5} mutations per base pair per replication (Wahl et al., 1987), whereas for the rest of the genome 10^{-11} to 10^{-9} is the postulated figure (Drake, 1969). VSPMR of heteroduplexes formed by recombinational strand exchange between diverse members of the V-gene family could easily make up to 10^{10} different sequence variants (Radman, 1983). Indeed, the repertoire of chicken B-cell immunoglobulins appears to consist of patchworks from diverse germ line genes (Reynaud et al., 1987). Furthermore, transfection of mouse cells with multiply mismatched heteroduplexes (involving mouse H2 histocompatibility gene sequences) yielded patchwork sequences, as expected for VSPMR (Abastado et al., 1984). The potential of VSPMR systems operating on recombinational heteroduplexes to generate localized preprogrammed somatic or germ line diversification in immunoglobulin, histocompatibility complex, and some other systems was discussed elsewhere (Bourguignon et al., 1982; Kourilsky, 1983, 1986; Radman, 1983).

Conservation and homogenization of DNA sequences could be realized during recombination by LPMR: conservation results if the repair is directed against the incoming strand (as in section VII); homogenization results if repair can act on either strand (conversion stops when no mismatches appear in different heteroduplexes). Thus, the properties of LPMR and VSPMR elucidated in bacteria (Fig. 6) seem sufficient to explain genetic conservation and diversification between repeated sequences in eucaryotes. It is not known whether these are the mechanisms employed or whether analogously conserving and diversifying mismatch repair systems exist in eucaryotes. However, the recent demonstrations of in vitro mismatch repair activity in *S. cerevisiae* (Muster-Nassal and Kolodner, 1986), *Xenopus* (P. Brooks, C. Dohet, M. Petranovic, and M. Radman, *UCLA Symp. Mol. Cell. Biol.* New Ser., in press), and HeLa (Brooks et al., in press; Glazer et al., 1987) cell extracts raise hope for experimental tests in the near future.

XII. MISMATCH REPAIR IN PALINDROMES: A GENETIC GAME?

Inverted repeated DNA sequences (palindromes) occur frequently in eucaryotic genomes. When imperfectly repeated, they provide a potential source of heteroduplex DNA within which mismatch repair systems might play curious genetic games. In an imperfect palindrome, the conformational switch rod⇌cruciform could produce heteroduplex DNA without breaking strands and, presumably, without involving the complete recombination machinery. Mismatch repair in the cruciform would produce new mismatches upon return to the rod configuration and vice versa, thus creating genetic instability in the form of jumping mutations. This genetic game of mismatch repair within palindromic sequences will cease at sites where mismatch repair has created a perfect palindromic sequence (e.g., when repair occurs on both cruciform arms and is directed to the same repeat sequence). On the basis of the preceding discussion, one would expect LPMR to prevent the mismatched cruciform formation and the VSPMR to perform mismatch repair on the cruciforms.

If LPMR and diverse sequence and mismatch-specific VSPMR systems act on multiple mismatches in palindromic sequences, fascinating possibilities can be imagined. For example, intrastrand gene conversions could act over very short distances and therefore also on single-copy haploid genes containing inverted repeats, thus enhancing both stable and unstable genetic diversity. Although mismatch repair in imperfect palindromes has not been tested experimentally, some complex mutations in yeast cells (Stewart and Sherman, 1974) and phage T4 (de Boer and Ripley, 1984) can be accounted for by such mismatch repair events.

ACKNOWLEDGMENTS. I enjoyed stimulation and criticism from Susan Rosenberg and David Thaler. Figure 4 is a by-product of discussions with Dadi Petranovic. I thank Gerry Smith for the critical reading of the manuscript. Bob Wagner, Christiane Dohet, Mady Jones, and Pascale Doutriaux are irreplaceable companions in the search for mismatch repair mechanisms.

The mismatch repair work in my laboratory was supported by grants from the Centre National de la Recherche Scientifique, the Association pour la Recherche sur le Cancer (Villejuif), the Ligue Nationale Française contre le Cancer, and the Commission of the European Communities.

LITERATURE CITED

Abastado, J. P., B. Cami, T. H. Dinh, J. Igolen, and P. Kourilsky. 1984. Processing of complex heteroduplexes in *Escherichia coli* and *Cos*-1 monkey cells. *Proc. Natl. Acad. Sci. USA* **81**:5792–5796.

Amati, P., and M. Meselson. 1965. Localized negative interference in bacteriophage lambda. *Genetics* **51**:369–379.

Bourguignon-Van Horen, F., A. Brotcorn, P. Caillet-Fauquet, W. P. Diver, C. Dohet, P. O. Doubleday, P. Lecomte, G. Maenhaut-Michel, and M. Radman. 1982. Conservation and diversification of genes by mismatch correction and SOS induction. *Biochimie* (Paris) **64**:559–564.

Boyce, R. P., and P. Howard-Flanders. 1964. Release of ultraviolet-light induced thymine dimers from DNA in E. coli. *Proc. Natl. Acad. Sci. USA* **51**:293–297.

Bradley, M. O., and N. A. Sharkey. 1978. Mutagenicity of thymidine to cultured chinese hamster cells. *Nature* (London) **274**:607–608.

Chaganti, R. S. K., S. Schonberg, and J. German. 1974. A manifold increase in sister chromatid exchanges in Bloom's syndrome lymphocytes. *Proc. Natl. Acad. Sci. USA* **71**:4508–4512.

Chasin, A. 1974. Mutations affecting adenine phosphoribosyl transferase activity in chinese hamster cells. *Cell* **2**:37–41.

Claverys, J. P., and S. A. Lacks. 1986. Heteroduplex deoxyribonucleic acid base mismatch repair in bacteria. *Microbiol. Rev.* **50**:133–165.

Claverys, J. P., V. Mejean, A. M. Gasc, F. Galibert, and A. M. Sicard. 1981. Base specificity of mismatch repair in *Streptococcus pneumoniae. Nucleic Acids Res.* **9**:2267–2280.

Claverys, J. P., V. Mejean, A. M. Gasc, and A. M. Sicard. 1983. Mismatch repair in *Streptococcus pneumoniae*: relationship between base mismatches and transformation efficiencies. *Proc. Natl. Acad. Sci. USA* **80**:5956–5960.

Coulondre, C., and J. H. Miller. 1977. Genetic studies of the *lac* repressor. III. Additional correlation of mutational sites with specific amino acid residues. *J. Mol. Biol.* **117**:525–575.

Coulondre, C., J. H. Miller, P. J. Farabaugh, and W. Gilbert. 1978. Molecular basis of base substitution hot spots in *Escherichia coli. Nature* (London) **274**:775–780.

Cox, E. C. 1976. Bacterial mutator genes and the control of spontaneous mutation. *Annu. Rev. Genet.* **10**:135–156.

Davidson, R. L., E. R. Kaufman, C. P. Dougherty, A. M. Ouellette, C. M. DiFolco, and S. A. Latt. 1980. Induction of sister chromatid exchanges by BudR is largely independent of the BUdR content of DNA. *Nature* (London) **284**:74–76.

de Boer, J. G., and L. S. Ripley. 1984. Demonstration of the production of frameshift base-substitution mutations by quasipalindromic DNA sequences. *Proc. Natl. Acad. Sci. USA* **81**:5528–5531.

Dohet, C., S. Dzidic, R. Wagner, and M. Radman. 1987. Large non-homology in heteroduplex DNA is processed differently than single base pair mismatches. *Mol. Gen. Genet.* **206**:181–184.

Dohet, C., R. Wagner, and M. Radman. 1985. Repair of defined single base pair mismatches in *E. coli. Proc. Natl. Acad. Sci. USA* **82**:503–505.

Dohet, C., R. Wagner, and M. Radman. 1986. Methyl-directed repair of frameshift mutations in heteroduplex DNA. *Proc. Natl. Acad. Sci. USA* **83**:3395–3397.

Doutriaux, M. P., R. Wagner, and M. Radman. 1986. Mismatch-stimulated killing. *Proc. Natl. Acad. Sci. USA* **83**:2576–2578.

Drake, J. W. 1969. Comparative rates of spontaneous mutation. *Nature* (London) **221**:1132–1133.

Ephrussi-Taylor, H., and T. C. Gray. 1966. Genetic studies of recombining DNA in pneumococcal transformation. *J. Gen. Physiol.* **49**:211–231.

Fazakerley, G. V., E. Quignard, A. Woisard, W. Guschlbauer, G. A. van der Marel, J. H. van Boom, M. Jones, and M. Radman. 1986. Structures of mismatched base pairs in DNA and their recognition by the *Escherichia coli* mismatch repair system. *EMBO J.* **5**:3697–3703.

Feinstein, S. I., and K. B. Low. 1986. Hyper-recombining recipient strains in bacterial conjugation. *Genetics* **113**:13–33.

Fishel, R. A., and R. Kolodner. 1984. An *Escherichia coli* cell-free system that catalyzes the repair of symmetrically methylated heteroduplex DNA. *Cold Spring Harbor Symp. Quant. Biol.* **49**:603–609.

Fox, M. S., and M. Allen. 1964. On the mechanisms of desoxyribonucleate integration in pneumococcal transformation. *Proc. Natl. Acad. Sci. USA* **52**:412–419.

Gasc, A. M., P. Garcia, D. Baty, and A. M. Sicard. 1987. Mismatch repair during pneumococcal transformation of small deletions produced by site-directed mutagenesis. *Mol. Gen. Genet.* **210**:369–372.

Glazer, P. M., S. N. Sarkar, G. E. Chisholm, and W. C. Summers. 1987. DNA mismatch repair detected in human cell extracts. *Mol. Cell. Biol.* **7**:218–224.

Glickman, B. W., and M. Radman. 1980. *Escherichia coli* mutator mutants deficient in methylation-instructed DNA mismatch correction. *Proc. Natl. Acad. Sci. USA* **77**:1063–1067.

Gussin, G. N., E. D. Rosen, and D. L. Wulff. 1981. Mapability of very close markers of bacteriophage lambda. *Genetics* **96**:1–24.

Hastings, P. J. 1984. Measurement of restoration and conversion; its meaning for the mismatch repair hypothesis of conversion. *Cold Spring Harbor Symp. Quant. Biol.* **49**:49–53.

Holliday, R. 1964. A mechanism for gene conversion in fungi. *Genet. Res.* **5**:282–304.

Ippen-Ihler, K. A., and E. G. Minkley, Jr. 1986. The conjugation system of F, the fertility factor of *Escherichia coli. Annu. Rev. Genet.* **20**:593–624.

Jones, M., R. Wagner, and M. Radman. 1987a. Mismatch repair of deaminated 5-methyl-cytosine. *J. Mol. Biol.* **194**:605–610.

Jones, M., R. Wagner, and M. Radman. 1987b. Mismatch repair and recombination in *E. coli. Cell* **50**:621–626.

Jones, M., R. Wagner, and M. Radman. 1987c. Repair of a mismatch is influenced by the base composition of the surrounding nucleotide sequence. *Genetics* **115**:605–610.

Kinsella, A. R., and M. Radman. 1978. Tumor promoter induces sister chromatid exchanges: relevance to mechanisms of carcinogenesis. *Proc. Natl. Acad. Sci. USA* **75**:6149–6153.

Konrad, E. B. 1977. Method for the isolation of *Escherichia coli* mutants with enhanced recombination between chromosomal duplication. *J. Bacteriol.* **130**:167–172.

single-stranded DNA, can be used to measure the uptake or displacement of labeled strands (Cox and Lehman, 1981b). S1 nuclease can also be used to measure the renaturation of complementary single strands (Cox and Lehman, 1981a). Gel electrophoresis serves to follow both the formation of joint molecules and the formation of new products by strand exchange (Cox and Lehman, 1981b; West et al., 1982b).

II. PRESYNAPTIC REACTION

A. Single-Stranded DNA

From the earliest observations on the pairing and strand exchange activities of purified RecA protein in vitro, it was apparent that single-stranded DNA was a key element in the reaction. Under physiological conditions, RecA protein is largely inert with duplex DNA, but is introduced to duplex DNA by its association with single-stranded DNA, present either in the form of separate single strands or as single-stranded regions of otherwise duplex molecules. The binding of RecA protein to single-stranded versus double-stranded DNA under a variety of conditions has recently been reviewed in detail (Cox and Lehman, 1987).

Single-stranded DNA in solution is, of course, in a highly folded configuration with secondary structure resulting from intrastrand base pairing of short runs of complementary sequences. One role of RecA protein is to organize the single-stranded DNA into an extended conformation, yet secondary structure in single strands impedes the binding of RecA protein, at least under the ionic conditions that are optimal for pairing and strand exchange (Muniyappa et al., 1984; Tsang et al., 1985b). But single-stranded DNA-binding proteins, including the single-stranded DNA-binding protein of *E. coli* (SSB) and the gene 32 protein of phage T4, remove secondary structure from single-stranded DNA and thereby favor the association of RecA protein with DNA (Muniyappa et al., 1984; Kowalczykowski and Krupp, 1987; Kowalczykowski et al., 1987b; Morrical et al., 1986). I will return to further consideration of the role of single-stranded DNA-binding proteins after I have discussed all three phases of the RecA reaction.

A second determinant of RecA reactions was recognized as a corollary to its cooperative binding to single-stranded DNA; the 38,000-dalton polypeptide acquires its pluripotency by polymerizing on single-stranded DNA to form helical nucleoprotein filaments that have multiple functions, as shown below. The filaments formed by RecA protein on single-stranded DNA have been called initiation complexes or presynaptic filaments to distinguish them from other filaments formed by RecA protein. In the following, I will use the term presynaptic filament.

RecA protein also forms filaments without DNA (Ogawa et al., 1978; Flory and Radding, 1982; McEntee et al., 1981; Cotterill and Fersht, 1983; Morrical and Cox, 1985; Register and Griffith, 1985a), and under certain experimental conditions, such as low pH or the presence of the nonhydrolyzable analog ATP-γ-S, RecA protein polymerizes on duplex DNA. As a natural concomitant of pairing and strand exchange, RecA protein invades duplex DNA from single-stranded regions (Ohtani et al., 1982a, b; Shibata et al., 1984; Cassuto and Howard-Flanders, 1986; Shaner and Radding, 1987; Shaner et al., 1987; J. E. Lindsley and M. M. Cox, *J. Biol. Chem.*, in press) and remains bound to heteroduplex DNA produced by strand exchange (Chow et al., 1986; Pugh and Cox, 1987b).

B. Structure of the Presynaptic Filament

A variety of observations revealed that the filament formed on single-stranded DNA is a key early intermediate (reviewed by Cox and Lehman, 1987). Presynaptic filaments that were isolated by gel filtration showed the behavior expected of an intermediate: such isolated filaments, when mixed with duplex DNA and ATP, formed joint molecules more rapidly and efficiently than equivalent concentrations of free RecA protein and single-stranded DNA (Flory et al., 1984; Tsang et al., 1985b).

Observations on the activity of isolated presynaptic filaments containing various amounts of bound RecA protein have indicated that saturation with RecA protein is not essential for activity: filaments that contained only 70% of the saturating amount of RecA protein were as active as saturated filaments, but filaments that retained only about one-third of their RecA protein were inactive (Tsang et al., 1985b). As explained further below, the formation of a joint at the end of duplex DNA is a stabilizing event; consequently, the assayable activity of a presynaptic filament depends more critically on the absence of secondary structure in the single-stranded DNA at the site corresponding to the end of duplex DNA than it does on coating of the entire single strand by RecA protein (Muniyappa et al., 1984; Tsang et al., 1985a).

Under conditions that minimize secondary structure in single-stranded DNA, presynaptic filaments formed in the presence of ATP-γ-S are morphologically indistinguishable from those formed in the presence of ATP (see below) and are nearly as active in homologous pairing (Honigberg et al., 1985).

Characterization of active presynaptic filaments that had been isolated by gel filtration revealed an apparently helical structure with a pitch of 90 Å (1 Å = 0.1 nm), containing about six molecules of RecA protein and 22 nucleotide residues per turn (Flory et al., 1984; Tsang et al., 1985b). The calculated axial separation of bases, about 4.1 Å, is 1.5 times that of duplex DNA observed on the same electron microscope grids (Flory et al., 1984; Williams and Spengler, 1986).

The apparent disparity between the spacing of bases in the RecA presynaptic filament and that in duplex DNA poses a quandary in understanding how RecA protein enables a single strand to recognize its complement in duplex DNA. The possibility that RecA protein increases the pitch of duplex DNA prior to homologous alignment is discussed below (see The partial unwinding of duplex DNA in section III.B.). However, there is at least one other, more compact form of the presynaptic filament seen in the absence of ATP. Flory et al. (1984) observed that removal of ATP from presynaptic filaments by gel filtration reduced the pitch from 90 to 69 Å. Williams and Spengler (1986) isolated by gel filtration, in the absence of ATP, a presynaptic filament whose pitch was 55 Å, which expanded to 93 Å upon addition of ATP-γ-S. Stasiak and Egelman (1986) observed a filament, formed on single-stranded DNA in the absence of nucleotide cofactor, whose pitch was 65 Å. If the presynaptic filament can alternate conformations rapidly, the axial separation of bases in one such conformation may match that in duplex DNA. The apparent disparity between the axial spacing in the presynaptic filament and that in duplex DNA could also be an artifact if nucleoprotein filaments and naked duplex DNA respond differently to stretching or shrinking forces during preparation for microscopy.

The location of single-stranded DNA in the presynaptic filament has not been determined. However, calculations based on the observed contour length of filaments and the maximal P-P spacing of nucleotide residues show that the DNA cannot be wrapped around the outside of protein beads, as in chromatin. If the DNA in the RecA presynaptic filament follows a helical path with the same pitch as the protein helix, the radius of the DNA path cannot exceed 20 Å, whereas the radius of the nucleoprotein filament is 45 to 50 Å (Flory et al., 1984; Williams and Spengler, 1986). Studies of oligodeoxynucleotides bound to RecA protein in the presence of ATP-γ-S have shown that the bases are available to methylation; by contrast, the phosphodiester backbone is protected from DNase, and ethylation of the oligonucleotide backbone interferes with binding (Leahy and Radding, 1986). On the basis of lessened methylation of adenine N3 in RecA complexes relative to naked single-stranded DNA, DiCapua and Müller (1987) suggested that RecA protein binds to a face of the polydeoxynucleotide chain corresponding to the minor groove of B-form DNA. Taken together, the observations above suggest that in the presynaptic filament, single-stranded DNA is located in a deep groove with its bases pointed outward.

The right-handed helicity of RecA nucleoprotein filaments was established rigorously by observations on another variant of such filaments, which forms with duplex DNA in the presence of ATP-γ-S (Stasiak et al., 1981; Stasiak and DiCapua, 1982; DiCapua et al., 1982). Since the parameters and morphologic appearance of these filaments are indistinguishable from those of presynaptic filaments, and since they are more regular in structure, they have served as models for structural analysis, including image reconstruction of electron micrographs. Such reconstructions reveal a filament with a deep but wide groove and six asymmetric units per turn. The latter indicates that the filament is a polymeric array of monomers of RecA protein. This filament is much less flexible than duplex DNA but much more flexible than other polymeric protein filaments such as F actin (Egelman and Stasiak, 1986). The detailed electron microscope analysis of the various RecA nucleoprotein filaments is described by Stasiak and Egelman (this volume).

III. SYNAPSIS

With model substrates, for example, circular single strands from a small DNA phage such as M13 and the corresponding duplex DNA, RecA protein at 37°C promotes homologous pairing at rates that have been estimated to be one or two orders of magnitude greater than the rate of thermal reannealing at $T_m - 20°C$. Under optimal conditions, RecA protein can pair 100% of DNA molecules in 1 or 2 min (Gonda et al., 1985; Kahn and Radding, 1984). As discussed in this section, after RecA protein binds to single-stranded DNA to form a presynaptic filament, the filament binds duplex DNA without regard to homology, and the resulting confinement of DNA to a limited volume facilitates homologous pairing. Still largely unknown, however, are the details of the interaction of the helical presynaptic filament with helical duplex DNA that proceeds from recognition of homology to stabilization of the joint and finally to an extensive exchange of strands.

A. Conjunction of DNA Molecules and the Search for Homology

The relative inability of RecA protein to bind directly to pure duplex DNA is remedied when single-stranded DNA is present, whether or not the latter is homologous (Shibata et al., 1979; Radding et al., 1980). Such observations and kinetics that resemble Michaelis-Menten enzyme kinetics led to the suggestion that homologous pairing proceeds via ternary complexes of RecA protein with single- and double-stranded DNA (Radding et al., 1980, 1981; Gonda and Radding, 1983; Gonda et al., 1985). The nature of these complexes and their role in homologous pairing in vitro were elucidated in studies of DNA aggregation and kinetics.

When single strands are incompletely coated by RecA protein, they form large aggregates that can be detected easily by their rapid sedimentation. Aggregation of single strands is suppressed, and even reversed, by conditions that favor the complete coating of single strands by RecA protein. Such conditions include a low concentration of Mg^{2+} (1 to 2 mM) and the presence of single-stranded DNA-binding proteins. However, completely coated single strands, i.e., presynaptic filaments that are saturated with RecA protein, participate in another significant interaction of multiple molecules. When naked duplex DNA is added to fully formed presynaptic filaments under conditions that are suitable for homologous pairing, all of the DNA, both single and double stranded, rapidly coaggregates (Tsang et al., 1985a; Rusche et al., 1985). This process has been called coaggregation, and its product has been called a coaggregate specifically to denote the inclusion of both single- and double-stranded DNA. In the following, the term aggregation is used in a limited sense, to signify the association of multiple single strands that are incompletely coated by RecA protein (Fig. 2).

As indicated above in section I.B., the pairing of complementary strands by RecA protein has requirements different from those of the strand exchange reaction. The differences between renaturation and the strand exchange reactions can be rationalized on the basis of the phenomena of aggregation and coaggregation (Fig. 2). The conditions that favor incomplete coating of single strands by RecA protein, and aggregation, correlate with those that favor renaturation (Fig. 1a), whereas the conditions that favor complete coating, and coaggregation, correlate with those that favor strand exchange reactions (Fig. 1b and c). Most notably, aggregation and renaturation are inhibited by SSB and by stoichiometric

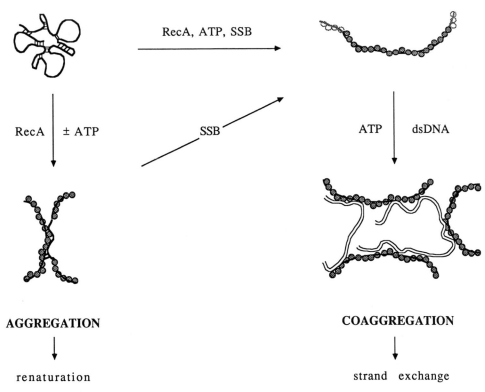

FIGURE 2. Colligative property of RecA protein and its relation to renaturation and strand exchange reactions. Under conditions that favor unfolding of single strands and stoichiometric binding of RecA protein, the resulting presynaptic filament forms large nucleoprotein networks by nonspecifically binding duplex DNA at many sites as it searches for homology. These networks are intermediates that speed homologous alignment by concentrating the DNA and lead to strand exchange reactions, as illustrated in Fig. 1b and c. When conditions do not favor stoichiometric association of RecA protein with single-stranded DNA, naked single-stranded regions can bind to the site normally occupied by duplex DNA. These conditions concentrate single-stranded DNA and thus favor the renaturation of single strands (Fig. 1a).

concentrations of RecA protein and do not absolutely require ATP, whereas the opposite is truc of coaggregation and strand exchange reactions. These correlations suggested that aggregation and coaggregation speed homologous pairing by concentrating DNA (Bryant and Lehman, 1985; McEntee, 1985; Tsang et al., 1985a).

Both aggregation and coaggregation appear to signal the existence of a second site on RecA protein that binds DNA. Since there is one molecule of RecA protein for every three or four nucleotide residues, the presynaptic filament is polyvalent with regard to this putative binding site. When duplex DNA is present, its weak binding to multiple sites cross-links presynaptic filaments and thus creates nucleoprotein networks. Aggregation of incompletely coated single strands probably arises in a similar fashion by the binding of naked regions of single-stranded DNA to the second site that would otherwise be occupied by duplex DNA. Further support for that interpretation comes from the observation that whereas two fully coated complementary strands do not renature, a single strand fully coated by RecA protein pairs perfectly well with a naked complementary strand (F. R. Bryant and I. R. Lehman, unpublished data, cited by Cox and Lehman, 1987).

Thus, as described further below, the mechanism by which RecA protein pairs complementary strands or homologous molecules appears to be based on two binding sites per molecule of RecA protein. The single strand that initially interacts with RecA protein can lie deep in a groove of the presynaptic filament, but the other DNA molecule, single or double stranded, must be naked and thus able to fit into a second binding site. The concept of two binding sites is also supported by studies on the binding of DNA (Bryant et al., 1985) and on the transfer of RecA protein from one molecule of DNA to another (Menetski and Kowalczykowski, 1987) and by modeling (Howard-Flanders et al., 1984a, b); however, direct identification of two binding sites per molecule of RecA protein has not yet been described.

Pairing within nucleoprotein networks that contain many complementary or homologous DNA molecules is reflected in the kinetics of both renaturation (Fig. 1a) and the initial formation of joint molecules (Fig. 1b and c). Unlike thermal renaturation or the renaturation that is promoted by single-stranded DNA-binding proteins, which are second-order reactions, the pairing of complementary strands by RecA protein is first order with respect to DNA (Bryant and Lehman, 1985; McEntee, 1985).

The pairing of naked duplex DNA with the presynaptic filament also follows a first-order rate law, which indicates that complexes containing both single-stranded DNA and duplex DNA are intermediates in homologous pairing (Julin et al., 1986; Gonda and Radding, 1983, 1986; Gonda et al., 1985). Much further evidence identified coaggregates as the instrumental intermediates that govern the first-order kinetics and accelerate homologous pairing. Under appropriate conditions, coaggregates include all of the DNA in a reaction mixture, they form faster than joint molecules, and they give rise to joint molecules faster than they can exchange with added DNA (Tsang et al., 1985a). A variety of conditions that enhance or inhibit coaggregation similarly affect homologous pairing (Chow and Radding, 1985). The stability of coaggregates was seen to depend on the length of duplex DNA. Further correlations were observed between the rate of formation of joint molecules, the yield of joint molecules, and the yield of coaggregates on the one hand and the length of duplex DNA on the other (Gonda and Radding, 1983, 1986). Such experiments involved duplex molecules with a homologous segment of constant length at the end of a variable heterologous region. Using duplex molecules that had no heterologous segment, Julin et al. (1986) observed a presumably related correlation between initial rate, the first-order rate constant, and the yield of joint molecules on the one hand and the length of duplex DNA on the other. They did not see an effect of length on the apparent rate constant when they varied the length of the heterologous portion of the duplex DNA. The reason for this difference has not been explained.

The bulk of the foregoing observations and correlations support the notion that double-stranded molecules bridge presynaptic filaments via multiple short-lived contacts and, accordingly, that longer duplex molecules make more stable coaggregates by making more contacts; however, the most convincing evidence that coaggregates are instrumental in homologous pairing was the observation that the suboptimal pairing of short duplex molecules with single strands is accelerated by the addition of long *heterologous* duplex molecules. This leads to the formation of more stable coaggregates which, moreover, absorb or retain shorter homologous molecules and thus accelerate their homologous pairing (Gonda and Radding, 1986).

Although the ability of RecA protein to bring DNA molecules into proximity accelerates the search for homology in vitro, this colligative property is not unique to RecA protein: polyamines, histones, histonelike proteins, and the β protein of phage lambda probably have similar effects (Cox and Lehman, 1981a; Hübscher et al., 1980; Kmiec and Holloman, 1981; Gonda and Radding, 1986; Muniyappa and Radding, 1986); moreover, in vivo, where macromolecules are much more concentrated, the role of such nucleoprotein networks is unclear. However, observations made on RecA protein in vitro have provided insights on how homologous pairing occurs in the seemingly implausible environment of concentrated long nucleoprotein filaments connected by multiple, albeit short-lived, intermolecular bridges (Honigberg et al., 1986).

Under optimal but commonly used experimental conditions, RecA protein appears to have uncommon abilities; it pairs all homologous molecules in 1 min and it enables a single-stranded probe to identify homologous duplex DNA even when the latter is only 1 molecule among nearly 1 million heterologous duplex molecules. Moreover, under the same optimal conditions, the rate of the search for homology does not limit the rate of formation of joint molecules, but rather some later step is slower than homologous alignment. These remarkable capacities, however, depend on a one-to-one search; to a first approximation, this means one presynaptic filament scanning one duplex molecule for homology. Experimentally, we can keep the respective concentrations of single-stranded and double-stranded DNA constant, and at the same time force each presynaptic filament to scan more duplex molecules, simply by making the homologous duplex DNA a small fraction of the total duplex DNA and by making the homologous probe a small fraction of the total single-stranded DNA. Such experiments showed that when each homologous presynaptic filament was forced to search more than about a dozen duplex molecules, the rate of the homologous search itself not only became limiting but also became diminishingly small. By inference, the mobility of presynaptic filaments and naked duplex DNA within the networks is limited. A mechanism based on manifold interconnected binding sites effectively concentrates the reactants, but incurs the disadvantage just described. We are left wondering how the mechanism works in vivo, where during either transformation or conjugation a single copy of DNA equivalent to a probe finds its target in a much larger chromosome at efficiencies ranging roughly from 10 to 50% (Lerman and Tolmach, 1957; Fox, 1966; Wollman et al., 1956).

B. Recognition

1. Homology

Homologous recombination is characterized by a requirement for some minimal length of homology, which was predicted by Thomas (1966). This was very clearly demonstrated in phage T4, where recombination frequency is proportional to length when the region of homology is greater than 50 base pairs. Recombination occurs, but at much lower frequencies, when the region of homology is less than 50 base pairs (Singer et al., 1982). The minimal recognition length (Thomas, 1966) or minimal efficient processing segment (Shen and Huang, 1986) has been variously estimated to be 20 to 100 base pairs in procaryotes (Singer et al., 1982; Shen and Huang, 1986; Watt et al., 1985; King and Richardson, 1986; also see Mahajan, this volume). Thomas (1966) postulated that a minimal length requirement would guard against adventitious recombination between short runs of homology that must occur on a stochastic basis. It may also guard against recombination between nonstochastic sequence repetitions. If the search for homology involves nonspecific binding of DNA in a second site, as described above, a minimal recognition length may be required simply to discriminate between heterologous and homologous contacts.

Closely related to the concept of minimal recognition length are questions about the effects of one or a few mismatches within a homologous region of some length. This relationship was clearly shown by Watt et al. (1985), who observed a 10-fold decrease in recombination resulting from a single mismatch within a homologous region of 53 base pairs. Similar sensitivities to mismatches were observed by Shen and Huang (1986).

The ability to observe homologous pairing in vitro provides an opportunity to look for the molecular bases for a minimal recognition length. Although a systematic study has not yet been made, there are some observations on the relationship between homologous pairing by RecA protein and the parameters related to homology. Gonda and Radding (1983) observed that the rate of formation of stable joint molecules by circular single strands and linear duplex DNA molecules fell off sharply between 100 and 30 base pairs of homology.

Circular single strands and linear duplex DNA of phages fd and M13, which differ at about 3% of their nucleotide residues, pair as well in mixed combinations as do the completely homologous pairs (DasGupta and Radding, 1982a; Bianchi and Radding, 1983). The DNAs of phages ϕX174 and G4 differ at about 30% of their nucleotide residues, yet RecA protein will pair even these two DNA molecules; in this case, however, the duplex DNA must be negatively superhelical (DasGupta and Radding, 1982b; Bianchi et al., 1983). This effect of negative superhelicity presumably results from stabilization of the joint molecules either in the reaction mixture or during the assay. Just as stability of mismatched joints or short joints can affect the apparent outcome in vitro, stability might have a similar determining role on the genetic outcome in vivo.

2. Topology

Paranemic joints and nascent joints. Recognition of homology takes place anywhere along the length of homologous molecules without regard to the location of ends in the involved DNA chains (Shibata et al., 1981; DasGupta et al., 1980; West et al., 1982b; Bianchi et al., 1983; Keener and McEntee, 1984; Riddles and Lehman, 1985a; Wu et al., 1983; Christiansen and Griffith, 1986; Register et al., 1987). For example, a circular single strand can pair with superhelical DNA to form a so-called paranemic joint. Strictly speaking, the term paranemic describes the overall topology of this union, since two circular structures cannot truly intertwine in the plectonemic fashion that is exemplified by the two strands of closed circular duplex DNA. When two circular structures are intertwined, every right-handed turn must be compensated by a left-handed turn (Fig. 3). If the pairing of closed circular duplex DNA with a circular right-handed helical RecA presynaptic filament involves the right-handed intertwining of duplex DNA and presynaptic filament, such right-handed turns must be compensated by an equal number of left-handed turns, which could be distant solenoidal turns in either partner that are not part of the actual base-paired joint. Other structures involving balanced right-handed and left-handed turns are also possible (Bianchi et al., 1983; Riddles and Lehman, 1985a), including Z-DNA-like structures (Kmiec and Holloman, 1984). Recently, Cox and his collaborators (Cox and Lehman, 1987; Cox et al., 1987; Schutte and Cox, 1987) proposed that a paranemic joint is a truly side-by-side structure in which duplex DNA and the helical presynaptic filament are not intertwined at all, but are extensively aligned via periodic homologous contacts as appropriate sequences come face to face on the two helices (see Fig. 3d and section IV.B.).

Since it lacks topoisomerase activity, RecA protein alone cannot convert a pair of circular substrates into a new plectonemic heteroduplex molecule plus a single strand derived from the original homoduplex DNA, but such an outcome can be accomplished when a topoisomerase is also present (Cassuto, 1984; Cunningham et al., 1981; Bianchi et al., 1983).

When RecA protein pairs a circular single strand with linear duplex DNA that has heterologous ends (Bianchi et al., 1983; Riddles and Lehman, 1985a, b), or pairs a circular duplex molecule with linear single strands that have heterologous ends (Christiansen and Griffith, 1986; Register et al., 1987) the reaction is similarly prevented from proceeding to the formation of heteroduplex DNA plus a displaced strand. The joint molecule remains some kind of three-stranded structure. Although there is no topological barrier to intertwining a linear molecule that has heterologous ends with a circular molecule, there is nonetheless a barrier to forming true heteroduplex DNA when the linear partner has heterologous ends (Fig. 3). Whatever their actual structure, joints formed from such substrates, which have also been termed paranemic joints, share the same instability as the true paranemic joints formed by the pairing of two circular molecules (Bianchi et al., 1983; Riddles and Lehman, 1985a; also see Role of ends, below).

To understand homologous recognition, one would like to know about the

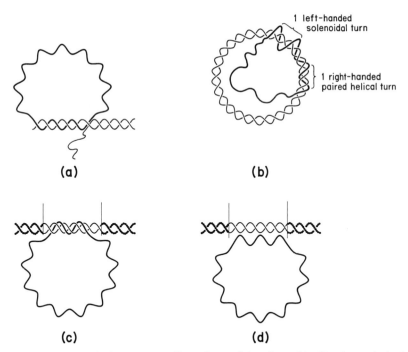

1 left-handed
solenoidal turn

1 right-handed
paired helical turn

(a)   **(b)**

(c)   **(d)**

FIGURE 3. Some hypothetical structures of homologous joints formed by RecA protein in the course of pairing a circular single strand with duplex DNA. (a) A plectonemic, heteroduplex joint formed from completely homologous molecules by displacement of one strand from the original duplex molecule and interwinding of the other two strands in a Watson-Crick duplex structure. (b) A paranemic joint formed from a circular single strand and closed circular duplex DNA. Putative right-hand turns must be compensated by an equal number of left-hand turns, which may be distant solenoidal turns that play no role in homologous alignment. For ease of illustration, the single strand exhibits the necessary gyrations, but the duplex DNA could follow a similar path around single-stranded DNA in the helical presynaptic filament. (c and d) Two structures that might be formed by circular single strands and duplex DNA with heterologous ends. Joints formed from such substrates are also called paranemic joints because they are barred from forming the plectonemic joint (illustrated in a) and because they share the instability of joints formed by circular single strands and circular duplex DNA (illustrated in b). (c) A three-stranded interwound structure formed in the region of homology. (d) A joint completely lacking any intertwining, but instead paired at periodic intervals as homologous sequences appear on apposing faces of helical duplex DNA and the helical presynaptic filament. In a, c, and d, the sinusoidal shape given to the single strand is meant to suggest its helical structure within the presynaptic filament.

formation and structure of nascent joints, the very first homologous complex (Wu et al., 1982; Bianchi et al., 1983). Since paranemic joints are blocked from becoming plectonemic joints (see Fig. 3), they have been used as a model for understanding early steps in pairing. Taking advantage of the sensitivity of paranemic joints to deproteinization, Riddles and Lehman (1985a) showed that even with completely homologous substrates, joints with the instability of a paranemic joint form more rapidly than plectonemic joints. This result is consistent with the idea that some form of three-stranded paranemic joint is an intermediate in the formation of a plectonemic joint. The conversion of a three-stranded joint to a plectonemic heteroduplex joint plus a displaced strand can also be blocked by the substitution of ATP-γ-S for ATP, or by the addition of ATP-γ-S to a reaction mixture containing ATP. No strand exchange can be detected in the presence of ATP-γ-S, and the joints formed have the sensitivity to deproteinization that is characteristic of paranemic joints (Cox and Lehman, 1981b; Riddles and Lehman, 1985a, b; Honigberg et al., 1985).

The length of paranemic joints has been estimated to be 400 to 700 base pairs (Bianchi et al., 1983; Christiansen and Griffith, 1986; Register et al., 1987). Electron micrographs show duplex DNA entering and leaving the RecA presynaptic filament, but have not yet revealed any details of structure within the joint (Christiansen and Griffith, 1986).

Role of ends. Although homologous recognition and the formation of a nascent joint can occur anywhere along homologous DNA molecules, ends play two distinct and important roles in stabilizing the joint. One role, already evident from the foregoing, is seen when DNA ends are located in a region of homology. When completely homologous linear duplex DNA and circular single-stranded DNA interact, pairing can take place at an end of the duplex DNA, where a nascent joint can be converted to a joint that is heteroduplex and plectonemic by displacement of the noncomplementary strand (see Fig. 3a). This stabilizing event converts a less stable three-stranded joint of unknown structure into Watson-Crick DNA plus a displaced strand, because the latter, having a free homologous end, can fully unwind and separate from the original duplex (Bianchi et al., 1983; Riddles and Lehman, 1985a, b).

There is a related but more subtle role of ends that is seen when nicks or double-strand breaks are located in regions of heterology (Honigberg et al., 1985). That role is to act as a swivel, not for complete strand separation but for a reduction in twist of the duplex DNA that is necessary for the formation of a nascent joint. RecA protein forms few detectable joints from circular single strands plus either positively superhelical DNA or relaxed closed circular DNA, but a nick or a double-strand break in a heterologous region of the closed circular duplex DNA changes it into a good substrate (K. Muniyappa and C. Radding, unpublished data). Not surprisingly, a topoisomerase can also provide the necessary swivel, and further analysis shows that pairing of a circular single strand with closed circular duplex DNA produces positive supertwists. It is the relaxation of these positive supertwists, either by a topoisomerase or by the cleavage of at least one strand in a heterologous region, that appears to be required for the formation of a paranemic joint molecule that can be detected by the filter assay (Muniyappa and Radding, unpublished data). The generation of positive supertwists and the requirement that they be relaxed indicate that pairing is accompanied by a partial unwinding or reduction in twist. Such partial unwinding might be important for homologous recognition itself, for stabilization of the paranemic joint, or for both, as discussed further in the following section.

Partial unwinding of duplex DNA. This section is an appropriate bridge between the sections on synapsis and strand exchange because questions about the unwinding of duplex DNA concern us from the moment of contact of the presynaptic filament with duplex DNA to the final separation of the displaced strand. The real subject is the conformation of DNA, the conformation that mediates recognition, and whatever conformations mediate strand exchange.

Experimentally, we detect unwinding in at least two different states, which I will call partial unwinding, or reduction in twist, and strand separation. The reduction in twist in duplex DNA has sometimes been detected by ligating a nick in circular duplex DNA while it is complexed with RecA protein and sometimes by treating such a complex with a topoisomerase. After either procedure, the topological state of the closed circular DNA, specifically the linking difference (usually assessed by gel electrophoresis), provides information about the effect of RecA protein on the conformation of the DNA. Topology here serves as a method to report on changes caused by RecA protein in the conformation of DNA. Strand separation, a more complete state of unwinding, is detected after the bases have exchanged pairing partners and some portion of the incoming strand has become resistant to S1 nuclease, which specifically degrades single-stranded DNA, while a corresponding portion of one strand of the parental duplex has become sensitive to S1 nuclease. In this section we are concerned principally with the reduction in twist.

Numerous reports have suggested that RecA protein binds stably to duplex DNA only after a nucleation event, which can occur in a variety of ways, including the interaction of the presynaptic filament with naked duplex DNA. However, nucleation can occur more slowly by the interaction of RecA protein alone with duplex DNA or more rapidly under certain other experimental conditions. Following nucleation, RecA protein extensively associates with duplex DNA, reduces its twist, and hydrolyzes ATP. Nucleation itself may be accompanied by a local reduction in twist (Muniyappa and Radding, unpublished data); the subsequent extensive association is accompanied by reduction in twist in a correspondingly greater portion of the duplex DNA. Arguably, the former is essential for homologous recognition; the latter, for strand exchange.

The ability of RecA protein to reduce the twist of duplex DNA was first seen in the presence of the nonhydrolyzable analog ATP-γ-S, which causes RecA protein to polymerize on duplex DNA in a reaction that is strongly stimulated by single-stranded DNA or even by oligonucleotides, whether or not they are homologous (Cunningham et al., 1979; Shibata et al., 1981). This stimulation by homologous or heterologous single-stranded DNA can be seen as the result of a nucleation event that leads to extensive polymerization of RecA protein on duplex DNA, with an attendant reduction in twist throughout the polymer. However, unwinding in the presence of ATP-γ-S can also be observed without the addition of a stimulating single-stranded cofactor, presumably by a slower reaction. In either event, unwinding is attributable to the formation of long, regular, stable, nucleoprotein filaments that are right-handed helices with 18.7 base pairs and 6.4 molecules of RecA protein per turn (Stasiak et al., 1981, 1983; Stasiak and DiCapua, 1982; DiCapua et al., 1982; see also Stasiak and Egelman, this volume). As already indicated, the parameters and morphology of this ATP-γ-S filament, which contains double-stranded DNA, are very similar to those of the presynaptic filament.

In the presence of ATP, RecA protein causes a similarly extensive untwisting of duplex DNA that is dependent upon the presence of homologous single-stranded DNA (Ohtani et al., 1982a, b; Wu et al., 1983). Shibata and his colleagues showed that this unwinding, which they called processive unwinding, is initiated by the formation of a joint molecule (Iwabuchi et al., 1983; Shibata et al., 1982a, b, 1984; Ohtani et al., 1982a, b). Processive unwinding and its attendant hydrolysis of ATP can also be initiated by the binding of RecA protein to superhelical DNA at low concentrations of Mg^{2+} in the absence of single strands, and processive unwinding plus hydrolysis of ATP continues even after homologous joints made by short single-stranded filaments have been caused to dissociate (Iwabuchi et al., 1983; Ohtani et al., 1982a, b). Processive unwinding reflects continued polymerization of RecA protein on duplex DNA, following the nucleation event that is stimulated by the formation of a joint, or by incubation under certain ionic conditions (Shibata et al., 1984). Nucleation leading to extensive association of RecA protein with duplex DNA and consequent ATP hydrolysis is also favored when RecA protein is initially bound to DNA at acid pH (Pugh and Cox, 1987a) or when the duplex DNA has a single-stranded region, such as a gap or a tail (see section V.C.) (West et al., 1980; Cassuto and Howard-Flanders, 1986; Shaner and Radding, 1987; Shaner et al., 1987). At neutral pH, there is a much slower interaction of RecA protein with duplex DNA that leads to the hydrolysis of ATP, an interaction that is favored by factors that lower the T_m of the DNA. These observations were interpreted as evidence of nucleation followed by extensive association (Roman and Kowalczykowski, 1986; Kowalczykowski et al., 1987a).

Experiments with monoclonal antibodies have provided an important insight into the unwinding reactions (Shibata et al., 1984; Makino et al., 1985, 1987). One monoclonal antibody permits homologous recognition but completely inhibits both processive unwinding and strand exchange. Further analysis of these experiments suggests that cooperative binding of RecA protein is essential for strand exchange but is not essential for homologous pairing and strengthens the distinction between a nucleating event and subsequent events.

The nature of the nucleation event is clearly critical. Naked B-form duplex DNA collides with the RecA presynaptic filament which contains the single strand in a complex binding site. The site is physically complex since it is embedded in a helical structure, and the site is chemically complex since it is composed partly of DNA with a highly specific sequence and partly of protein with a low degree of specificity that admits either a duplex molecule or another single strand. Recognition of homology follows collision. Does recognition require that the duplex unwind first?

Observations made independently in two laboratories indicate that there is an instantaneous homology-dependent unwinding that is more rapid than the subsequent strand exchange (Kahn and Radding, 1984; Shibata et al., 1984). This finding may be related to the observation that a reduction in twist is essential to form joint molecules that can be detected by a filter assay (see above). However, the requirement even for an early and rapid reduction in twist could merely reflect a degree of unwinding that is required to make a joint sufficiently stable to survive the assay.

IV. ASYMMETRIC STRAND EXCHANGE

RecA protein promotes two categories of strand exchange: *asymmetric*, in which one strand of a duplex molecule is transferred to the single strand originally embedded in the presynaptic filament to form a new heteroduplex molecule; and *symmetric*, in which two duplex molecules, one with an obligatory single-stranded region, exchange strands to form two new heteroduplex molecules. Symmetric exchange is discussed separately in section V.

Strand exchange promoted by RecA protein requires the hydrolysis of ATP (Cox and Lehman, 1981b) and proceeds directionally (see below), two characteristics that distinguish it from the random walk of branch migration, which takes place between naked DNA molecules that have been suitably joined (see Radding, 1978). The term strand exchange, as used here, refers to the kind of exchange that is actively promoted by RecA protein. Asymmetric strand exchange has been most thoroughly studied in reactions of circular single-stranded DNA and linear duplex DNA. When these substrates represent full genomes of one of the small DNA phages, the final products are linear single-stranded DNA and nicked circular heteroduplex DNA. The progress of strand exchange can be assayed by measuring the progressive change in sensitivity to S1 nuclease: the circular single strand becomes less sensitive as it is incorporated into heteroduplex DNA, whereas the strand displaced from the duplex DNA becomes more sensitive. This assay reflects the average extent of strand exchange in the population of molecules. Gel electrophoresis provides another useful assay that detects production of the final products.

A. Properties

Asymmetric strand exchange has three notable properties: it is slow, directional, and able to proceed past heterologous sequences to produce mismatched heteroduplex DNA.

1. Slowness

In the absence of other proteins, RecA protein promotes strand exchange at an average rate of only a few base pairs per second (Kahn and Radding, 1984). The addition of single-stranded DNA-binding proteins, such as SSB or T4 gene 32 protein, increases the rate about two to three times (see Radding, 1982). The reaction is asynchronous, however, since in the presence of SSB some complete products about 6 kilobases long can be detected as early as 5 min after the start of the reaction, which is equivalent to about 20 base pairs per second (Cox and Lehman, 1981b, 1987; Chow et al., 1988b). All of these rates are much below that of even the slowest helicases such as the RecBCD enzyme that unwinds DNA at about 300 base pairs per second (Taylor and Smith, 1980; see Taylor, this volume).

2. Directionality

Whereas pairing can take place at any homologous location between a single strand and duplex DNA, strand exchange is directional. The single strand in the presynaptic filament is an initiating strand whose polarity appears to determine the directionality of strand exchange, which is 5' to 3' with respect to the initiating strand (see Fig. 1) (Cox and Lehman, 1981c; Kahn et al., 1981; West et al., 1981b). RecA protein also polymerizes on single-stranded DNA in the same 5' to 3' direction (Register and Griffith, 1985b). The corresponding directionality of polymerization and strand exchange may or may not be causally related (see section IV.B.).

The directionality of strand exchange 5' to 3' with respect to the initiating strand suggests that the 5' end of a linear single strand should be the initiating end when such a strand pairs with circular duplex DNA. Partially duplex molecules with homology confined to a terminal portion of their single-stranded tails showed the expected properties: molecules with either 5' or 3' tails paired with circular duplex DNA, but only those with 5' ends formed stable joints, presumably because strand exchange could proceed 5' to 3' (DasGupta et al., 1981; Wu et al., 1982). Recent experiments carried out with fully linear single strands in the presence of SSB have challenged that explanation, however. Single strands with homology at their 3' ends formed stable joints, whereas those with homologous 5' ends did not (Konforti and Davis, 1987). At this writing, it is not clear whether the difference is attributable to preferential binding of SSB to 5' ends, as observed by Register and Griffith (1985b), or to some other cause (Konforti and Davis, 1987).

3. Production of mismatched heteroduplex DNA

One of the most interesting, and perhaps most surprising, properties of RecA protein is its ability to propagate or extend strand exchange even when sizable heterologous sequences stand in the way. Strand exchange can pass thymine dimers (Livneh and Lehman, 1982) and every kind of base pair mismatch. It can take place between M13 and fd DNA, which have about 3% mismatch base pairs. Even more surprisingly, strand exchange can pass insertions of extra sequences located either in the single strand or in the duplex DNA. When the insertion was in single-stranded DNA, its length, up to about 1,000 nucleotide residues, made little difference to the yield of heteroduplex DNA, which ranged between 30 and 40%. However, the yield of product was strongly dependent on length when the insertion was in duplex DNA (Bianchi and Radding, 1983). Nonetheless, the efficiency with which RecA protein drives strand exchange through small insertions in duplex DNA is noteworthy: the apparent yield of mismatched heteroduplex DNA was one-third of the homologous control when the duplex substrate contained a heterologous insertion of 140 base pairs. Such observations show not only that the mechanism of strand exchange is insensitive to sequence homology but that its ability to separate strands, i.e., to remove 360° of twist per 10.4 base pairs, is not strictly or temporally coupled to the formation of stable heteroduplex DNA. This conclusion is made surprising by observations which show that RecA protein has very limited helicase activity in the absence of strand exchange (Bianchi et al., 1985; West et al., 1981c). Bianchi et al. (1985) found that RecA protein was unable to separate an oligonucleotide longer than 30 nucleotide residues from a single strand to which it was annealed.

B. Mechanics of Strand Exchange

Asymmetric strand exchange is a disassembly of one helix and the concerted assembly of a new helix that occurs via an intermediate in which the original duplex DNA and some form of helical RecA nucleoprotein filament are joined (see Fig. 1b). Various studies have identified at least the following possible contributing mechanisms: (i) *treadmilling*, the disassembly of filament at a trailing end and reassembly at a leading end (Cox et al., 1984; Howard-Flanders et al., 1984b); (ii) *polarized polymerization*, a variation on treadmilling that involves only a growing end (Shibata et al., 1984; Kowalczykowski et al., 1987a); (iii) *rotation* of duplex DNA and nucleoprotein filament in relation to each other (Cox et al., 1984; Cox and Lehman, 1987); and (iv) *conformational changes* (Dunn et al., 1982; Shaner and Radding, 1987; Egelman and Stasiak, 1988).

Strand exchange is carried out by a nucleoprotein filament that seems deceptively simple and at the same time enigmatic. It should be clear already that our understanding of strand exchange falls short. Therefore, several models that have been proposed are especially valuable. In the following, I will briefly describe those models, followed in each case by discussion of various data that seem relevant, but which do not yet constitute a critical scrutiny of the interesting ideas embodied in the models.

According to a model proposed by Howard-Flanders et al. (1984a, b, 1987), duplex DNA and the helical nucleoprotein filament intertwine to form a single multistrand nucleoprotein filament of some undetermined length, but one that could be hundreds or even thousands of nucleotide residues in length (also see Stasiak and Egelman, this volume). During asymmetric strand exchange three strands of DNA are included in the filament, the single strand originally in the presynaptic filament plus the newly incorporated duplex molecule. Strand exchange results from the exchange of pairing partners within the three-stranded helix by a simple rotational motion of the bases. In an early version of this model (Howard-Flanders et al., 1984a, b) progression of this exchange was linked to treadmilling. Separation of the noncomplementary strand from the parental duplex molecule is envisaged as a passive process: as the protein scaffold dissolves, it leaves behind a new heteroduplex molecule and a displaced strand loosely wound around the heteroduplex a minimum of once about every 20 nucleotide residues. Eventually, the displaced strand would disentangle, presumably by rotary diffusion in vitro.

Cox et al. (1987) have categorized models of strand exchange that involve treadmilling as type I models, and those that do not, as type II models.

Treadmilling, in its classical form, involves a continuous protein filament from which subunits at the tail end dissociate as they hydrolyze ATP and reassociate at the head end (Cleveland, 1982). Observations made on the ATPase activity of RecA protein show that ATP is not hydrolyzed exclusively at the ends of presynaptic filaments, but rather must be hydrolyzed throughout the filament (Brenner et al., 1987; Shaner and Radding, 1987; Kowalczykowski and Krupp, 1987; Kowalczykowski et al., 1987a). Thus, the data do not support the simplest form of a type I or treadmilling model.

What is the evidence for a nucleoprotein filament that contains within it three strands in a helical conformation? Electron microscopic observations of fixed specimens have revealed molecules in which linear duplex DNA appears to have been incorporated into a nucleoprotein filament starting from one end of the duplex and continuing for thousands of nucleotides without evidence of a displaced single strand until late in the process (Stasiak et al., 1984; Register et al., 1987). These observations have been interpreted as evidence that strand separation is a delayed event which occurs only after a three-strand filament has formed that is thousands of nucleotide residues in length. Biochemical data bearing on the existence of a three-stranded helix within the nucleoprotein filament are mixed. On the one hand, observations made in the absence of SSB provide evidence that strand displacement is an early event: within a few minutes of the beginning of strand exchange, the 5' end of the displaced strand is able to participate in a second round of pairing. On the other hand, under optimal conditions in the presence of SSB, there is evidence that displaced strands remain attached in some way to duplex molecules for as long as 20 to 30 min after strand exchange has gone to completion, as judged by the S1 nuclease assay (see Fig. 8 of Chow et al., 1988b).

DNase protection studies have been done to assess how much of each strand of the DNA is in close contact with the RecA filament. Chow et al. (1986) studied DNase protection during strand exchange promoted by RecA protein in the absence of SSB (Fig. 4A). Using the method of footprinting, with each DNA strand individually labeled, they found that the single strand in the presynaptic filament remained largely protected while its homolog became increasingly protected concomitant with its displacement from the duplex DNA. The complementary strand in newly formed heteroduplex DNA was sensitive to DNase I during the entire reaction. At least in the absence of SSB, RecA protein appeared to detach first from the proximal end of heteroduplex DNA (pattern iii in Fig. 4). Other experiments which involved DNA with a long heterologous region showed that when strand exchange reached this block, protection extended some 50 residues into the heterologous region, but not more (pattern iii in Fig. 4A) (Chow et al., 1988a).

Pugh and Cox (1987b) studied DNase protection in the presence of SSB. They used uniformly labeled DNA and measured conversion of DNA to acid-soluble products, which requires about 10 times as much DNase as footprinting. They found that concomitant with strand exchange the incoming single strand became more sensitive to DNase with time, whereas about half of the label in duplex DNA became increasingly resistant. In this study, the increasing resistance of duplex

PATTERNS OF DNAse PROTECTION

Type of Strand Exchange:

A) Asymmetric   B) Symmetric

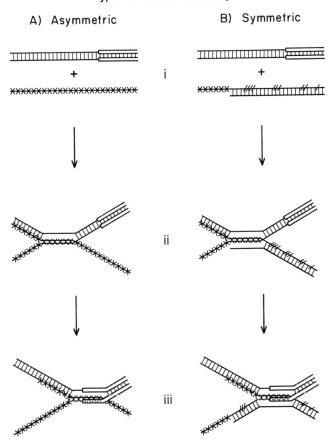

FIGURE 4. Patterns of protection from DNase I and restriction endonucleases observed during asymmetric and symmetric strand exchange. (×××) Regions of strong protection; (////) regions of weak protection. The unfilled bars represent long blocks of heterologous DNA. The length of the central region where strand switching occurs is uncertain. (From Chow et al., 1988a.)

DNA to DNase could not be assigned to the displaced strand or to the complementary strand incorporated into heteroduplex DNA.

These two studies of DNase protection are difficult to compare directly because SSB was used in one, but not the other, and because the methodology was quite different. However, both studies showed that production of some part of duplex DNA went hand in hand with the progress of strand exchange. These observations on shielding from DNase cleavage do not provide evidence that duplex DNA is extensively incorporated in a three-stranded structure, but they also do not exclude short three-stranded structures or structures in which one strand is more exposed near the surface of the nucleoprotein filament.

A quite different model has been proposed recently by Cox and his colleagues (Cox et al., 1987; Cox and Lehman, 1987; Schutte and Cox, 1987) based in part on the following observations. Schutte and Cox (1987) observed that a few minutes after the addition of duplex DNA to an otherwise complete reaction mixture, a new, lower steady-state rate of ATP hydrolysis was established. This new steady state reached a limiting value when the ratio of double-stranded DNA to single-stranded DNA was 1:1, and moreover the limiting value was inversely proportional to the length of homologous duplex DNA. They proposed that this steady state reflects a complete one-to-one homologous alignment of duplex DNA with the presynaptic filament via a paranemic structure in which there are only periodic contacts of homologous sequences in the two helical partners, as appropriate sequences come face to face on the two helices, namely, the presynaptic filament and duplex DNA (see Fig. 3d). This idea is further elaborated as a model for strand exchange in which the long axis of the duplex DNA is repeatedly passed around the long axis of the presynaptic filament, dissolving a turn of the paired duplex and establishing a turn of new heteroduplex for each rotation of the longitudinal axes (Cox et al., 1987; Cox and Lehman, 1987).

Given the helical nature of the reactants and the helical nature of both products (when sufficient RecA protein is present), some kind of rotation of duplex DNA and RecA filament must occur to effect strand exchange. Cox et al. (1987) have pointed out that the movement of a branch point in strand exchange might entail either the movement of duplex DNA around the presynaptic filament (or vice versa) or the axial rotation of duplex DNA and the presynaptic filament each about its own longitudinal axis. Honigberg and Radding (1988) devised substrates to test these various motions and found that most strand exchange is attributable to simultaneous rotation of duplex DNA and the RecA filament, each about its longitudinal axis. When strand exchange was initiated by a plus strand at one end of a duplex molecule and simultaneously by a minus strand at the other end, so that the exchanges proceeded head to head toward each other, the two exchanges were mutually inhibitory, but the inhibition was relieved by a nick in the middle of the duplex molecule (Fig. 5). These results suggest that the nick relieved torsional strain resulting from opposing rotations generated by head-to-head strand exchange in the same duplex molecule. Since strand exchange takes place in complex nucleoprotein networks, this kind of axial rotation may be favored over grosser motions of these very large macromolecules. These experimental observations seem to argue against models that absolutely require motions other than axial rotation.

No detailed model has yet incorporated explicit conformational changes, but various suggestions have been published (Dunn et al., 1982; Stasiak and Egelman, 1986; Shaner and Radding, 1987). Conformational changes might be expected in RecA protein, in the nucleoprotein filaments, or in the DNA within the filaments (see Partial unwinding of duplex DNA in section III.B. above). Recent observations of bundles of RecA protein nucleoprotein filaments have begun to reveal different conformations that monomers of RecA protein can adopt (Egelman and Stasiak, 1988; Stasiak and Egelman, this volume). Different conformations of the presynaptic filament have been noted in the presence and absence of ATP (see

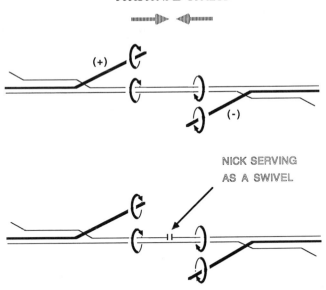

FIGURE 5. Diagram of the evidence that strand exchange involves the rotation of the presynaptic filament and of duplex DNA, each about its own longitudinal axis. When head-to-head strand exchange is arranged by pairing a plus strand at one end and a minus strand at the other, torsional stress inhibits strand exchange unless it is relieved by a nick acting as a swivel (Honigberg and Radding, 1988).

section II.B. above) (Flory et al., 1984; Stasiak and Egelman, 1986; Egelman and Stasiak, 1988; Williams and Spengler, 1986). The change of double-stranded DNA from B form to a partially unwound conformation in RecA filaments formed in the presence of ATP-γ-S was the first conformational change observed (Cunningham et al., 1979; Stasiak et al., 1981). It seems likely that the conformation of DNA in those filaments is a static form of the extensive filament formed by RecA protein on duplex DNA following nucleation by a variety of paths (see Partial unwinding of duplex DNA in section III.B. above).

C. Role of Single-Stranded DNA-Binding Proteins

Single-stranded DNA-binding proteins bind specifically to single-stranded DNA, and some proteins of this class are essential for replication and play a role in genetic recombination and DNA repair (for review, see Chase and Williams, 1986; Mahajan, this volume). The gene 32 protein of phage T4 and SSB, the single-stranded DNA-binding protein of E. coli, promote the renaturation of DNA in vitro (Alberts and Frey, 1970; Christiansen and Baldwin, 1977), an action that is attributable to their ability to disrupt secondary structure in single-stranded

DNA. By unfolding single strands, they expose bases for interstrand pairing by a reaction that is second order with respect to DNA. Efforts to detect the pairing of a single strand with duplex DNA by single-stranded DNA-binding proteins were unsuccessful, but soon after the discovery that RecA protein does so, an auxiliary or stimulating effect of both *E. coli* SSB and phage T4 gene 32 protein on that particular reaction was observed (McEntee et al., 1980; Shibata et al., 1980; West et al., 1982a).

Almost without exception, the effects of *E. coli* SSB on various aspects of the RecA reaction have been produced equally well by the genetically unrelated T4 gene 32 protein (Shibata et al., 1980; Kowalczykowski and Krupp, 1987; Egner et al., 1987). In limited examples, other DNA-binding proteins have shown similar actions (Muniyappa et al., 1984; Egner et al., 1987). This lack of specificity with respect to the primary structure of the DNA-binding proteins show that their auxiliary effects do not require specific protein-to-protein contacts with RecA protein. Therefore, in this chapter, descriptions of interactions of single-stranded DNA-binding proteins with RecA protein refer to functional interactions, not specific protein contacts. West et al. (1982a, b) found no effect of SSB on symmetric strand exchange, and most observations have been made on the effects of SSB on pairing and strand reactions involving circular single strands and linear duplex DNA.

SSB has a molecular weight of 19,000 and binds to single-stranded DNA as a tetramer (Chase and Williams, 1986). The recent discovery that it has several modes of binding to single-stranded DNA has greatly clarified its interactions with RecA protein (Lohman and Overman, 1985; Lohman et al., 1986; Bujalowski and Lohman, 1986; Griffith et al., 1984). For purposes of the present discussion, we can focus on two binding modes, one that is highly cooperative, with a site size of 33 nucleotide residues, which occurs at low concentrations of salt (below 10 mM) or in the absence of Mg^{2+} ions, and one that is not highly cooperative, with a site size of 65 nucleotide residues, which occurs at concentrations of salt above 0.2 M or at a Mg^{2+} concentration of 12 mM (Lohman and Overman, 1985; Griffith et al., 1984). In the low-salt mode, the filament formed by SSB on single-stranded DNA has a smooth appearance when visualized by electron microscopy, whereas in the high-salt mode the filament appears beaded (Griffith et al., 1984). In the low-salt mode, SSB excludes RecA protein from single-stranded DNA and thereby inhibits its ability to promote the cleavage of repressor and to promote homologous pairing (Griffith et al., 1984; Morrical et al., 1986; K. Muniyappa, K. R. Williams, J. W. Chase, and C. M. Radding, submitted for publication). Under the ionic conditions that are required for pairing reactions promoted by RecA protein, SSB favors the equilibrium binding of RecA protein to single-stranded DNA even when the DNA is initially coated with SSB. Inhibitory effects that have been seen when SSB was added to a reaction mixture before RecA protein have a kinetic basis; in such a case, the formation of an active presynaptic filament requires 10 min or more (Morrical et al., 1986; Muniyappa et al., submitted).

SSB affects both the presynaptic and postsynaptic phases of the RecA reaction, as described below, but has no direct effect on synapsis (Egner et al., 1987). Pairing and strand exchange reactions of RecA protein require a concentration of Mg^{2+} ions of 10 mM or greater. At this concentration of Mg^{2+}, the formation of a presynaptic filament limits the rate of the reaction and results in a short lag in the production of joint molecules (Kahn and Radding, 1984). As seen by effects of temperature, concentration of Mg^{2+} ions, and nucleotide sequence, secondary structure in single-stranded DNA impedes the binding of RecA protein and is responsible for the slowness of the presynaptic step. SSB, gene 32 protein, the SSB protein of the *E. coli* F factor, and the β protein of phage λ all overcome that barrier and favor the rapid formation of active presynaptic filaments (Muniyappa et al., 1984; Kowalczykowski et al., 1987b; Kowalczykowski and Krupp, 1987; Morrical et al., 1986; Egner et al., 1987). There is no evidence for any other important role of single-stranded DNA-binding proteins in the presynaptic phase. If presynaptic filaments are formed by preincubation of single-stranded DNA with RecA protein in 1 mM Mg^{2+}, the filament formed is nearly as active as that formed in the presence of SSB (Tsang et al., 1985b).

SSB, however, also stabilizes RecA presynaptic filaments, which otherwise decay slowly (over a period of 20 to 30 min), under conditions that are suitable for homologous pairing and strand exchange (Kahn and Radding, 1984; Kowalczykowski et al., 1987b; S. W. Morrical and M. M. Cox, unpublished data, cited by Cox and Lehman, 1987). Since the formation of joint molecules is so fast (a matter of a few minutes) under optimal conditions, this stabilizing effect of SSB presumably contributes little to the initial formation of joint molecules; on the other hand, since strand exchange, the next phase of the reaction, is so slow, stabilization of filaments by SSB could make a significant contribution to that process. Indeed, the formation of filaments by preincubation of single strands with RecA protein does not increase the steady-state rate of strand exchange or diminish the stimulatory effect of SSB (Kahn and Radding, 1984; Chow et al., 1986). Moreover, in the absence of SSB, the yield of nicked circular heteroduplex molecules is reduced by fivefold or more (Cox and Lehman, 1981b; Roman and Kowalczykowski, 1986; Chow et al., 1988b).

Recent experiments have shown that SSB and gene 32 protein counteract a back reaction that contributes to the slowness and incompleteness of strand exchange when RecA protein acts alone (Chow et al., 1988b). Within a few minutes of the beginning of strand exchange, the 5' end of the displaced strand invades the heteroduplex ends of other joint molecules and creates macromolecular complexes. Since the latter do not enter an agarose gel, their formation reduces the yield of free nicked circular duplex product. The formation of these complexes requires homology at the 5' end of the displaced strand, but the complexes persist even when RecA protein is removed by one of several different treatments, consistent with the formation of large heteroduplex joints. Either SSB or T4 gene 32 protein effectively suppresses this reinitiation. Further examination reveals that in the presence of SSB, the heteroduplex product is inert for a second round of pairing and exchange, whereas the displaced strand is active. Correspondingly, observations of Pugh and Cox (1987b) indicate that in the presence of SSB, RecA protein remains on the heteroduplex product for up to 1 h after the completion of strand exchange.

Thus, in the presence of a single-stranded DNA-binding protein, the immediate product of strand exchange is not nicked circular duplex DNA, but rather is a nucleoprotein filament which, moreover, is inert for a second round of exchange. A related example of this phenomenon was the report of Shibata et al. (1982a, b) that superhelical DNA is inactivated for a second round of homologous pairing by RecA protein remaining from a first round; in that case, for reasons presumably related to the closed circular state of the duplex DNA, inactivation occurred in the absence of SSB. The precise role of SSB or gene 32 protein in the production of inactive heteroduplex DNA is not yet clear. By some means, the presence of a single-stranded DNA-binding protein leads to a RecA nucleoprotein filament on heteroduplex DNA that is more saturated with RecA protein, more stable, or altered in some other way.

Finally, we come to a question that has elicited divergent views. What is the nature of the presynaptic filament that is made when both RecA protein and SSB are present? Does that filament contain both proteins or not? Under two conditions there is an unequivocal answer; in the absence of ATP or in the low-salt binding mode, SSB displaces RecA protein (Morrical et al., 1986; Kowalczykowski et al., 1987b). Of course, neither of these conditions is suitable for pairing and strand exchange. When we limit our consideration to conditions that promote homologous pairing and strand exchange, the data are more complex.

Electron microscopic observations by Flory et al. (1984) showed that active presynaptic filaments formed in 1 mM Mg^{2+} and shifted to 13 mM Mg^{2+} were identical in appearance and measured parameters to those formed in 13 mM Mg^{2+} in the presence of SSB. On the basis of fluorescence quenching measurements, Morrical et al. (1986) concluded that there is a continuous and stoichiometric association of SSB with presynaptic filaments. Using the same methodologic approach, Kowalczykowski et al. (1987b) concluded that SSB and RecA protein compete for the same binding sites. Using different experimental approaches, namely, immunoprecipitation and digestion with micrococcal nuclease, Muniyappa et al. (submitted) found evidence favoring the view that SSB associates with the presynaptic filament in stoichiometric amounts, 1 molecule of SSB per 20 to 22 nucleotide residues, a value that agrees with the finding of Morrical et al. (1986). At the same time, the presynaptic filament appears to retain its full complement of RecA protein (Morrical et al., 1986; Muniyappa et al., submitted).

Finally, Williams and Spengler (1986) observed that in the presence of ATP-γ-S, 4 mM Mg^{2+}, SSB, RecA protein, and single-stranded DNA, a filament is formed with a unique electron microscopic appearance, and they suggested that this filament might contain both proteins in some interspersed arrangement.

If, as suggested by some investigators, there exists a presynaptic filament containing 1 molecule of RecA protein per 4 nucleotide residues and 1 molecule of SSB per 20 nucleotide residues, its structure and functional significance remain to be shown. The single-stranded DNA-binding proteins clearly play important and fascinating roles in the recombination reactions of RecA protein, but our understanding of those interactions is still incomplete.

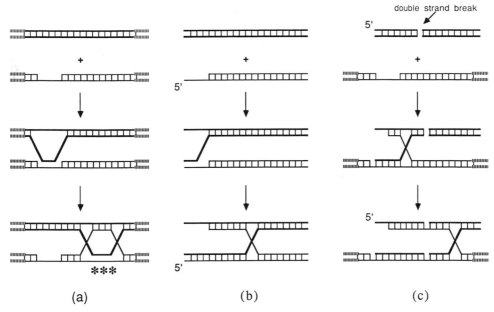

FIGURE 6. Reactions of partially single-stranded molecules with fully duplex DNA, sometimes called four-strand interactions. (a) Circular duplex molecules with a gap in one strand, pairing with closed circular duplex DNA. (b) Duplex DNA with a single-stranded tail pairing with linear duplex DNA. (c) Circular duplex molecules with a gap in one strand pairing with linear duplex DNA with a double-strand break. Interrupted lines at ends designate circular molecules. The asterisks in panel a indicate a region corresponding to the original gap, where for topological reasons the paired strands cannot intertwine.

V. REACTIONS OF DUPLEX MOLECULES THAT HAVE SINGLE-STRANDED REGIONS: SYMMETRIC STRAND EXCHANGE

Whereas reactions of single strands with duplex DNA provide amenable model systems, the more complex interactions of partially single-stranded molecules with fully duplex DNA probably simulate more closely the configuration of DNA molecules in vivo. These two kinds of reactions, sometimes referred to as three-strand and four-strand interactions, respectively, are closely related (see Fig. 1b and c), and study of the latter has provided additional insights into the action of RecA protein.

A. Substrates, Reactions, and Products

Substrates used to study four-strand interactions have included circular duplex molecules with a gap of various sizes in one strand and linear duplex molecules with a single-stranded end (Fig. 6). In each case, the other partner in the reaction is a completely duplex molecule, sometimes circular and sometimes linear. Such pairing reactions in general are slower and have a lower yield than the pairing of fully single-stranded DNA with duplex DNA. Given our understanding of the role of coaggregation in three-strand interactions (see Fig. 2), one is led to suspect that partially single-stranded molecules pair less rapidly simply because they have less single-stranded DNA and therefore form less stable coaggregates.

The simplest four-strand exchange reaction to consider is the pairing of a tailed duplex molecule with the linear duplex from which it was derived (Fig. 6b). The reaction begins in the same way as asymmetric strand exchange (also see Fig. 1b and c), but the exchange continues into the duplex regions and becomes symmetric, reproducing in vitro the classical recombination intermediate postulated by Holliday (1964; see DasGupta et al., 1981; West et al., 1981a, b).

The reaction of a gapped circular molecule with closed circular duplex DNA, illustrated in Fig. 6a, is more complicated (Cunningham et al., 1980; Cassuto et al., 1980; West et al., 1982b). Pairing takes place in the gap, followed by the complex strand exchange shown. Since only one of the four strands involved has an interruption, a complete strand exchange is impossible. Instead, a kind of D loop is formed which migrates away from the gap, driven by RecA protein. As indicated in Fig. 6a, true intertwining of strands in heteroduplex DNA can occur in only one of the two possible heteroduplex regions. (The region of nonintertwined strands is indicated by asterisks in Fig. 6a; see Cunningham et al., 1980.)

B. Properties of Symmetric Strand Exchange

Insofar as we know them, the properties of symmetric strand exchange are similar to those of asymmetric exchange. The rate of symmetric exchange appears to be of the same magnitude as that of asymmetric exchange, only a few base pairs per second (West et al., 1982b). As in the case of substrates that undergo asymmetric strand exchange, the single-stranded region of DNA can pair with homologous sequences in duplex DNA whether these be located in the middle of a molecule or at an end (see Fig. 6) (West et al., 1982b; Cunningham et al., 1980; DasGupta et al., 1981; Wu et al., 1982). Moreover, symmetric strand exchange and asymmetric exchange appear to share the same polarity. Although pairing can occur at any site in the fully duplex partner, it appears more efficient when the appropriate end of the duplex DNA falls within the single-stranded region (West et al., 1982b). Thus, joint molecules, which are Holliday structures in this case, can form whether the strand that pairs with the original single-stranded region has a 3' end or a 5' end (see Fig. 6b and c), but complete strand exchange, all the way to the distal end of the paired molecules, is strongly favored when the strand transferred initially to its single-stranded complement has a 3' end (West et al., 1982b; also see Wu et al., 1982). Polarity is determined by the 5' to 3' orientation of single-stranded DNA in the gap.

Since the *recA* gene is essential for the recombinational mechanism called postreplication repair, which remedies gaps caused by pyrimidine dimers in replicating DNA, the in vitro pairing of gapped DNA with intact duplex DNA is of special interest (Rupp et al., 1971; West et al., 1981d, 1982b; Cunningham et al., 1980). West et al. (1981d) showed that the polarity of the RecA reaction causes the 3' end at a nick preferentially to be transferred into the single-stranded region of a gapped molecule.

C. Mechanics of Four-Strand Interactions

Under the usual conditions of pairing reactions, single-stranded DNA rapidly activates the hydrolysis of ATP by RecA protein, but duplex DNA does not. However, West et al. (1980) observed that duplex DNA with single-stranded ends or gaps supported far more ATPase activity than could be attributed to the single-stranded regions themselves, and these authors inferred that RecA protein might bind cooperatively to the duplex portions of such molecules. Recent studies have shown that, starting from a single-stranded region, RecA protein rapidly associates with the much longer duplex region of the molecule and hydrolyzes ATP at a steady-state rate that is proportional to the length of the entire molecule (Shaner and Radding, 1987; Shaner et al., 1987). When the single-stranded DNA has a 5′ tail, the rate of ATP hydrolysis reaches its steady state after a lag of about 1 min. When the tail has a 3′ end, the lag is 4 to 5 min longer and the steady-state rate is smaller but still reaches more than half of that seen with 5′ tails. In a study in which an ATP regenerating system was not used, RecA protein dissociated slowly and reassociated via the single-stranded tail (Shaner and Radding, 1987). A similar directional association of RecA protein with tailed molecules is also seen in the presence of ATP-γ-S (Cassuto and Howard-Flanders, 1986).

The extensive association of RecA protein with the duplex portion of gapped or tailed DNA in the presence of ATP confers only weak protection from the action of DNase I: principally, the strand with a tail or gap is protected. However, the single-stranded end of a tailed molecule is more strongly protected (see pattern i in Fig. 4B) (Shaner et al., 1987; Chow et al., 1988a). Chow et al. (1988a) studied the pattern of DNase protection during symmetric strand exchange occurring between the substrates diagrammed in Fig. 6b. The stronger protection of the single-stranded tail extended along that strand into newly formed heteroduplex DNA (Fig. 4B). The newly created single-stranded tail was also strongly protected, but by contrast its adjacent heteroduplex region was not (pattern iii in Fig. 4B). As in the case of asymmetric strand exchange, RecA protein appeared to detach first from the proximal end of the initiating strand (pattern iii in Fig. 4B). No protection of the complementary strand in either duplex was detected.

Comparison of the patterns of DNase protection during asymmetric and symmetric strand exchange (Fig. 4) supports the generally held view that these occur via the same mechanism. The initiating single strand on which RecA protein polymerizes clearly dominates the reaction.

D. Duplex-Duplex Pairing

In all of the foregoing, we have discussed pairing reactions in which a single strand coated with RecA protein pairs with another naked single strand or with a naked duplex molecule. The discovery by West et al. (1980) of the ability of tailed or gapped DNA to activate the ATPase activity of RecA protein supported the idea that RecA protein may move into the duplex portion of such a molecule and promote pairing between a duplex region coated with RecA protein and the homologous region of another completely duplex molecule that is free from any

RecA protein. Two independent studies tested this idea by examining the pairing of a duplex molecule with a partially single-stranded molecule in which the single-stranded DNA was in a heterologous region. Cassuto et al. (1982) found evidence for direct pairing of duplex regions; DasGupta et al. (1981) did not.

West and Howard-Flanders (1984) discovered the remarkable reaction diagrammed in Fig. 6c: in substrates undergoing symmetric strand exchange, some intermediate could react with a distal duplex fragment that had been separated by restriction cleavage and exchange strands with that fragment. In effect, symmetric strand exchange passed a double-strand break. Consistent with our understanding of how RecA protein gets onto duplex DNA, this reaction did not occur if the double-strand break was made in the gapped molecule. Less clear is the significance of the observation that removal or addition of a few nucleotides at the double-strand break reduced the rescue of the distal fragment. Subsequently, Rusche et al. (1985) and Register and Griffith (1986) reported that RecA protein promotes the juxtaposition of DNA ends. The ability of RecA protein to juxtapose ends might be important in reestablishing strand exchange between homologous molecules at a double-strand break.

VI. CONCLUDING REMARKS

From the perspective provided by studies on *E. coli* RecA protein, the enzymology of homologous pairing and strand exchange is unique. It resembles little the enzymology of replication and transcription; rather, the interactions of protein with DNA seem more like those involved in chromosome structure. The ordered filaments formed by RecA protein on DNA are specialized minichromosomes assembled in a way that efficiently promotes homologous pairing and subsequent strand exchange. Genetic purposes may be served by the greater requirement for perfect homology in the initial pairing and the lower requirement for homology in strand exchange. The relationship of the specialized *recA* minichromosome to the normal chromosome remains to be explored, as does the relationship of RecA protein to the RecBCD nuclease and other enzymes of recombination (see Taylor, this volume; Mahajan, this volume).

Finally, proteins like RecA protein are widely distributed in bacteria (Pierre and Paoletti, 1983; Lovett and Roberts, 1985; West et al., 1983; Eitner et al., 1982; Better and Helinski, 1983; Kokjohn and Miller, 1985; Koomey and Falkow, 1987), and proteins with pairing and strand exchange activities have been discovered in fungi (Kmiec and Holloman, 1982; Kolodner et al., 1987), in the lily plant (Hotta et al., 1985), and in vertebrates (Kenne and Ljungquist, 1984; Hsieh et al., 1986; Kucherlapati et al., 1985; Cassuto et al., 1987). However, as described by Kucherlapati and Moore (this volume), other proteins that promote pairing and strand exchange show significant differences in their apparent mechanisms, including the lack of a requirement for ATP by some. Thus, more work is required before we understand the relatedness or diversity of these proteins.

LITERATURE CITED

Alberts, B. M., and L. Frey. 1970. T4 bacteriophage gene 32: a structural protein in the replication and recombination of DNA. *Nature* (London) **227**:1313–1318.

Better, M., and D. R. Helinski. 1983. Isolation and characterization of the *recA* gene of *Rhizobium meliloti*. *J. Bacteriol.* **155**:311–316.

Bianchi, M., C. DasGupta, and C. M. Radding. 1983. Synapsis and the formation of paranemic joints by *E. coli* recA protein. *Cell* **34**:931–939.

Bianchi, M. E., and C. M. Radding. 1983. Insertions, deletions and mismatches in heteroduplex DNA made by RecA protein. *Cell* **35**:511–520.

Bianchi, M., B. Riboli, and G. Magni. 1985. *E. coli* recA protein possesses a strand separating activity on short duplex DNA. *EMBO J.* **4**:3025–3030.

Brenner, S. L., R. S. Mitchell, S. W. Morrical, S. K. Neuendorf, B. C. Schutte, and M. M. Cox. 1987. RecA protein-promoted ATP hydrolysis occurs throughout recA nucleoprotein filaments. *J. Biol. Chem.* **262**:4011–4016.

Bryant, F. R., and I. R. Lehman. 1985. On the mechanism of renaturation of complementary DNA strands by the RecA protein of *Escherichia coli*. *Proc. Natl. Acad. Sci. USA* **82**:297–301.

Bryant, F. R., A. R. Taylor, and I. R. Lehman. 1985. Interaction of the RecA protein of *Escherichia coli* with single-stranded DNA. *J. Biol. Chem.* **260**:1196–1202.

Bujalowski, W., and T. M. Lohman. 1986. *Escherichia coli* single-strand binding protein forms multiple, distinct complexes with single-stranded DNA. *Biochemistry* **25**:7799–7802.

Cassuto, E. 1984. Formation of covalently closed heteroduplex DNA by the combined action of gyrase and RecA protein. *EMBO J.* **3**:2159–2164.

Cassuto, E., and P. Howard-Flanders. 1986. The binding of RecA protein to duplex DNA molecules is directional and is promoted by a single stranded region. *Nucleic Acids Res.* **14**:1149–1157.

Cassuto, E., L.-A. Lightfoot, and P. Howard-Flanders. 1987. Partial purification of an activity from human cells that promotes homologous pairing and the formation of heteroduplex DNA in the presence of ATP. *Mol. Gen. Genet.* **208**:10–14.

Cassuto, E., S. C. West, and P. Howard-Flanders. 1982. Can recA protein promote homologous pairing between duplex regions of DNA? *EMBO J.* **1**:821–825.

Cassuto, E., S. C. West, J. Mursalim, S. Conlon, and P. Howard-Flanders. 1980. Initiation of genetic recombination: homologous pairing between duplex DNA molecules promoted by RecA protein. *Proc. Natl. Acad. Sci. USA* **77**:3962–3966.

Chase, J. W., and K. R. Williams. 1986. Single-stranded DNA binding proteins required for DNA replication. *Annu. Rev. Biochem.* **55**:103–136.

Chow, S. A., S. M. Honigberg, R. J. Bainton, and C. M. Radding. 1986. Patterns of nuclease protection during strand exchange. *J. Biol. Chem.* **261**:6961–6971.

Chow, S. A., S. M. Honigberg, and C. M. Radding. 1988a. DNase protection by RecA protein during strand exchange: asymmetric protection of the Holliday structure. *J. Biol. Chem.* **263**:3335–3347.

Chow, S. A., and C. M. Radding. 1985. Ionic inhibition of formation of recA nucleoprotein networks blocks homologous pairing. *Proc. Natl. Acad. Sci. USA* **82**:5646–5650.

Chow, S. A., B. J. Rao, and C. M. Radding. 1988b. Reversibility of strand invasion promoted by RecA protein and its inhibition by *Escherichia coli* single-stranded DNA-binding protein or phage T4 gene 32 protein. *J. Biol. Chem.* **263**:200–209.

Christiansen, C., and R. L. Baldwin. 1977. Catalysis of DNA reassociation by the *Escherichia coli* DNA binding protein. *J. Mol. Biol.* **115**:441–454.

Christiansen, G., and J. Griffith. 1986. Visualization of the paranemic joining of homologous DNA molecules catalyzed by the RecA protein of *Escherichia coli*. *Proc. Natl. Acad. Sci. USA* **83**:2066–2070.

Clark, A. J. 1973. Recombination deficient mutants of *E. coli* and other bacteria. *Annu. Rev. Genet.* **7**:67–86.

Clark, A. J., and A. D. Margulies. 1965. Isolation and characterization of recombination deficient mutants of *E. coli* K12. *Proc. Natl. Acad. Sci. USA* **53**:451–459.

Cleveland, D. W. 1982. Treadmilling of tubulin and actin. *Cell* **28**:689–691.

Cotterill, S. M., and A. R. Fersht. 1983. RecA filaments in solution. *Biochemistry* **22**:3525–3531.

Cox, M. M., and I. R. Lehman. 1981a. Renaturation of DNA: a novel reaction of histones. *Nucleic Acids Res.* **9**:389–400.

Cox, M. M., and I. R. Lehman. 1981b. RecA protein of *Escherichia coli* promotes branch migration, a kinetically distinct phase of DNA strand exchange. *Proc. Natl. Acad. Sci. USA* **78**:3433–3437.

Cox, M. M., and I. R. Lehman. 1981c. Directionality and polarity in RecA protein promoted branch migration. *Proc. Natl. Acad. Sci. USA* **78**:6018–6022.

Cox, M. M., and I. R. Lehman. 1987. Enzymes of general recombination. *Annu. Rev. Biochem.* **56**:229–262.

Cox, M. M., S. W. Morrical, and S. K. Neuendorf. 1984. Unidirectional branch migration promoted by nucleoprotein filaments of RecA protein and DNA. *Cold Spring Harbor Symp. Quant. Biol.* **49**:525–533.

Cox, M. M., B. F. Pugh, B. C. Schutte, J. E. Lindlsey, J. Lee, and S. W. Morrical. 1987. On the mechanism of recA protein-promoted DNA branch migration, p. 597–607. *In* R. McMacken and T. J. Kelly (ed.), *DNA Replication and Recombination*. Alan R. Liss, Inc., New York.

Cunningham, R. P., C. DasGupta, T. Shibata, and C. M. Radding. 1980. Homologous pairing in genetic recombination: RecA protein makes joint molecules of gapped circular DNA and closed circular DNA. *Cell* **20**:223–235.

Cunningham, R. P., T. Shibata, C. DasGupta, and C. M. Radding. 1979. Homologous pairing in genetic recombination: single strands induce RecA protein to unwind duplex DNA. *Nature* (London) **281**:191–195. (Erratum, **282**:426.)

Cunningham, R. P., A. M. Wu, T. Shibata, C. DasGupta, and C. M. Radding. 1981. Homologous pairing and topological linkage of DNA molecules by combined action of E. coli RecA protein and topoisomerase I. *Cell* **24**:213–223.

DasGupta, C., and C. M. Radding. 1982a. Polar branch migration promoted by RecA protein: effect of mismatched base pairs. *Proc. Natl. Acad. Sci. USA* **79**:762–766.

DasGupta, C., and C. M. Radding. 1982b. Lower fidelity of RecA protein catalysed homologous pairing with a superhelical substrate. *Nature* (London) **295**:71–73.

DasGupta, C., T. Shibata, R. P. Cunningham, and C. M. Radding. 1980. The topology of homologous pairing promoted by RecA protein. *Cell* **22**:437–446.

DasGupta, C., A. M. Wu, R. Kahn, R. P. Cunningham, and C. M. Radding. 1981. Concerted strand exchange and formation of Holliday structures by E. coli RecA protein. *Cell* **25**:507–516.

DiCapua, E., A. Engel, A. Stasiak, and T. Koller. 1982. Characterization of complexes between RecA protein and duplex DNA by electron microscopy. *J. Mol. Biol.* **157**:87–103.

DiCapua, E., and B. Müller. 1987. The accessibility of DNA to dimethylsulfate in complexes with RecA protein. *EMBO J.* **6**:2493–2498.

Dunn, K., S. Chrysogelos, and J. Griffith. 1982. Electron microscopic visualization of recA-DNA filaments: evidence for a cyclic extension of duplex DNA. *Cell* **28**:757–765.

Egelman, E. H., and A. Stasiak. 1986. Structure of helical recA-DNA complexes. *J. Mol. Biol.* **191**:677–697.

Egelman, E., and A. Stasiak. 1988. Structure of helical recA-DNA complexes. II. Local conformational changes visualized in bundles of recA-ATP-Gamma-S filaments. *J. Mol. Biol.* **200**:329–349.

Egner, C., E. Azhderian, S. S. Tsang, C. M. Radding, and J. W. Chase. 1987. Effects of various single-stranded-DNA-binding proteins on reactions promoted by RecA protein. *J. Bacteriol.* **169**:3422–3428.

Eitner, G., B. Adler, V. A. Lanzov, and J. Hofemeister. 1982. Interspecies RecA protein substitution in *Escherichia coli* and *Proteus mirabilis*. *Mol. Gen. Genet.* **185**:481–486.

Flory, J., and C. M. Radding. 1982. Visualization of RecA protein and its association with DNA: a priming effect of single strand binding protein. *Cell* **28**:747–756.

Flory, J., S. S. Tsang, and K. Muniyappa. 1984. Isolation and visualization of active presynaptic filaments of RecA protein and single-stranded DNA. *Proc. Natl. Acad. Sci. USA* **81**:7026–7030.

Fox, M. 1966. On the mechanism of integration of transforming deoxyribonuclease. *J. Gen. Physiol.* **49**:183–196.

Gonda, D. K., and C. M. Radding. 1983. By searching processively RecA protein pairs DNA molecules that share a limited stretch of homology. *Cell* **34**:647–654.

Gonda, D. K., and C. M. Radding. 1986. The mechanism of the search for homology promoted by RecA protein. *J. Biol. Chem.* **261:**13087–13096.

Gonda, D. K., T. Shibata, and C. M. Radding. 1985. Kinetics of homologous pairing promoted by RecA protein: effects of ends and internal sites in DNA. *Biochemistry* **24:**413–420.

Griffith, J. D., L. D. Harris, and J. Register III. 1984. Visualization of SSB-ssDNA complexes active in the assembly of stable recA-DNA filaments. *Cold Spring Harbor Symp. Quant. Biol.* **49:**553–559.

Holliday, R. 1964. A mechanism for gene conversion in fungi. *Genet. Res.* **5:**282–304.

Honigberg, S. M., D. K. Gonda, J. Flory, and C. M. Radding. 1985. The pairing activity of stable nucleoprotein filaments made from RecA protein, single-stranded DNA, and adenosine 5'-(γ-thio)triphosphate. *J. Biol. Chem.* **260:**11845–11851.

Honigberg, S. M., and C. M. Radding. 1988. The mechanics of winding and unwinding helices in recombination: torsional stress associated with strand transfer promoted by RecA protein. *Cell* **54:** 525–532.

Honigberg, S. M., B. J. Rao, and C. M. Radding. 1986. Ability of RecA protein to promote a search for rare sequences in duplex DNA. *Proc. Natl. Acad. Sci. USA* **83:**9586–9590.

Horii, T., T. Ogawa, and H. Ogawa. 1980. Organization of the *recA* gene of *Escherichia coli. Proc. Natl. Acad. Sci. USA* **77:**313–317.

Hotta, Y., S. Tabata, R. A. Bouchard, R. Piñon, and H. Stern. 1985. General recombination mechanisms in extracts of meiotic cells. *Chromosoma* (Berlin) **93:**140–151.

Howard-Flanders, P., S. C. West, E. Cassuto, T.-R. Hahn, and E. Egelman. 1987. Structure of recA spiral filaments and their role in homologous pairing and strand exchange in genetic recombination, p. 609–617. *In* R. McMacken and T. J. Kelly (ed.), *DNA Replication and Recombination.* Alan R. Liss, Inc., New York.

Howard-Flanders, P., S. C. West, J. R. Rusche, and E. H. Egelman. 1984a. Molecular mechanisms of general genetic recombination: the DNA-binding sites of RecA protein. *Cold Spring Harbor Symp. Quant. Biol.* **49:**571–580.

Howard-Flanders, P., S. C. West, and A. Stasiak. 1984b. Role of RecA protein spiral filaments in genetic recombination. *Nature* (London) **309:**215–220.

Hsieh, P., M. S. Meyn, and R. D. Camerini-Otero. 1986. Partial purification and characterization of a recombinase from human cells. *Cell* **44:**885–894.

Hübscher, U., H. Lutz, and A. Kornberg. 1980. Novel histone H2A-like protein of *Escherichia coli. Proc. Natl. Acad. Sci. USA* **77:**5097–5101.

Iwabuchi, M., T. Shibata, T. Ohtani, M. Natori, and T. Ando. 1983. ATP-dependent unwinding of the double helix and extensive supercoiling by *Escherichia coli* recA protein in the presence of topoisomerase. *J. Biol. Chem.* **258:**12394–12404.

Julin, D. A., P. W. Riddles, and I. R. Lehman. 1986. On the mechanism of pairing of single- and double-stranded DNA molecules by the recA and single-stranded DNA-binding proteins of *Escherichia coli. J. Biol. Chem.* **261:**1025–1030.

Kahn, R., R. P. Cunningham, C. DasGupta, and C. M. Radding. 1981. Polarity of heteroduplex formation promoted by *Escherichia coli* recA protein. *Proc. Natl. Acad. Sci. USA* **78:**4786–4790.

Kahn, R., and C. M. Radding. 1984. Separation of the presynaptic and synaptic phases of homologous pairing promoted by RecA protein. *J. Biol. Chem.* **259:**7495–7503.

Keener, S. L., and K. McEntee. 1984. Homologous pairing of single-stranded circular DNAs catalyzed by RecA protein. *Nucleic Acids Res.* **12:**6127–6139.

Kenne, K., and S. Ljungquist. 1984. A DNA-recombinogenic activity in human cells. *Nucleic Acids Res.* **12:**3057–3068.

King, S. R., and R. P. Richardson. 1986. Role of homology and pathway specificity for recombination between plasmids and bacteriophage lambda. *Mol. Gen. Genet.* **204:**141–147.

Kmiec, E., and W. K. Holloman. 1981. β protein of bacteriophage λ promotes renaturation of DNA. *J. Biol. Chem.* **256:**12636–12639.

Kmiec, E., and W. K. Holloman. 1982. Homologous pairing of DNA molecules promoted by a protein from Ustilago. *Cell* **29:**367–374.

Kmiec, E. B., and W. K. Holloman. 1984. Synapsis promoted by Ustilago rec1 protein. *Cell* **36:**593–598.

Kobayashi, I., and H. Ikeda. 1978. On the role of *recA* gene product in genetic recombination: an analysis by *in vitro* packaging of recombinant DNA molecules formed in the absence of protein synthesis. *Mol. Gen. Genet.* **166**:25–29.

Kokjohn, T. A., and R. V. Miller. 1985. Molecular cloning and characterization of the *recA* gene of *Pseudomonas aeruginosa* PAO. *J. Bacteriol.* **163**:568–572.

Kolodner, R., D. H. Evans, and P. T. Morrison. 1987. Purification and characterization of an activity from *Saccharomyces cerevisiae* that catalyzes homologous pairing and strand exchange. *Proc. Natl. Acad. Sci. USA* **84**:5560–5564.

Konforti, B. B., and R. W. Davis. 1987. 3′ homologous free ends are required for stable joint molecule formation by the recA and single-stranded binding proteins of *Escherichia coli*. *Proc. Natl. Acad. Sci. USA* **84**:690–694.

Koomey, J. M., and S. Falkow. 1987. Cloning of the *recA* gene of *Neisseria gonorrhoeae* and construction of gonococcal *recA* mutants. *J. Bacteriol.* **169**:790–795.

Kowalczykowski, S. C. 1987. Mechanistic aspects of the DNA strand exchange activity of *E. coli* recA protein. *Trends Biochem. Sci.* **12**:141–145.

Kowalczykowski, S. C., J. Clow, and R. A. Krupp. 1987a. Properties of the duplex DNA-dependent ATPase activity of *Escherichia coli* recA protein and its role in branch migration. *Proc. Natl. Acad. Sci. USA* **84**:3127–3131.

Kowalczykowski, S. C., J. Clow, R. Somani, and A. Varghese. 1987b. Effects of the *Escherichia coli* SSB protein on the binding of *Escherichia coli* recA protein to single-stranded DNA. *J. Mol. Biol.* **193**:81–95.

Kowalczykowski, S. C., and R. A. Krupp. 1987. Effects of *Escherichia coli* SSB protein on the single-stranded DNA-dependent ATPase activity of *Escherichia coli* recA protein. *J. Mol. Biol.* **193**:97–113.

Kucherlapati, R. S., J. Spencer, and P. D. Moore. 1985. Homologous recombination catalyzed by human cell extracts. *Mol. Cell. Biol.* **5**:714–720.

Leahy, M. C., and C. M. Radding. 1986. Topography of the interaction of RecA protein with single-stranded deoxyoligonucleotides. *J. Biol. Chem.* **261**:6954–6960.

Lerman, L. S., and L. J. Tolmach. 1957. Genetic transformation. I. Cellular incorporation of DNA accompanying transformation in *Pneumococcus*. *Biochim. Biophys. Acta* **26**:68–82.

Little, J. W., and D. W. Mount. 1982. The SOS regulatory system of Escherichia coli. *Cell* **29**:11–22.

Livneh, Z., and I. R. Lehman. 1982. Recombinational bypass of pyrimidine dimers promoted by the RecA protein of *Escherichia coli*. *Proc. Natl. Acad. Sci. USA* **79**:3171–3175.

Lohman, T. M., and L. B. Overman. 1985. Two binding modes in *Escherichia coli* single strand binding protein-single stranded DNA complexes. *J. Biol. Chem.* **260**:3594–3603.

Lohman, T. M., L. B. Overman, and S. Datta. 1986. Salt-dependent changes in the DNA binding co-operativity of *Escherichia coli* single strand binding protein *J. Mol. Biol.* **187**:603–615.

Lovett, C. M., Jr., and J. W. Roberts. 1985. Purification of a RecA protein analogue from *Bacillus subtilis*. *J. Biol. Chem.* **260**:3305–3313.

Makino, O., S. Ikawa, Y. Shibata, H. Maeda, T. Ando, and T. Shibata. 1987. Rec-A protein-promoted recombination reaction consists of two independent processes, homologous matching and processive unwinding. *J. Biol. Chem.* **262**:12237–12246.

Makino, O., Y. Shibata, H. Maeda, T. Shibata, and T. Ando. 1985. Monoclonal antibodies with specific effects on partial activities of RecA protein of *Escherichia coli*. *J. Biol. Chem.* **260**:15402–15405.

McEntee, K. 1985. Kinetics of DNA renaturation catalyzed by the RecA protein of *Escherichia coli*. *Biochemistry* **24**:4345–4351.

McEntee, K., and W. Epstein. 1977. Isolation and characterization of specialized transducing bacteriophages for the *recA* gene of *Escherichia coli*. *Virology* **77**:306–318.

McEntee, K., J. E. Hesse, and W. Epstein. 1976. Identification and radiochemical purification of the RecA protein of *Escherichia coli* K-12. *Proc. Natl. Acad. Sci. USA* **73**:3979–3983.

McEntee, K., G. M. Weinstock, and I. R. Lehman. 1979. Initiation of general recombination catalyzed *in vitro* by the RecA protein of *E. coli*. *Proc. Natl. Acad. Sci. USA* **76**:2615–2619.

McEntee, K., G. M. Weinstock, and I. R. Lehman. 1980. RecA protein-catalyzed strand assimilation: stimulation by *Escherichia coli* single-stranded DNA-binding protein. *Proc. Natl. Acad. Sci. USA* **77**:857–861.

McEntee, K., G. M. Weinstock, and I. R. Lehman. 1981. Binding of the RecA protein of *E. coli* to single- and double-stranded DNA. *J. Biol. Chem.* **256**:8835–8844.

McKay, D. B., T. A. Steitz, I. T. Weber, S. C. West, and P. Howard-Flanders. 1980. Crystallization of monomeric RecA protein. *J. Biol. Chem.* **255**:6662.

Menetski, J. P., and S. C. Kowalczykowski. 1987. Transfer of RecA protein from one polynucleotide to another. *J. Biol. Chem.* **262**:2085–2092.

Morrical, S. W., and M. M. Cox. 1985. Light scattering studies of the RecA protein of *Escherichia coli*: relationship between free recA filaments and the recA-ssDNA complex. *Biochemistry* **24**:760–767.

Morrical, S. W., J. Lee, and M. M. Cox. 1986. Continuous association of *Escherichia coli* single-stranded DNA binding protein with stable complexes of RecA protein and single-stranded DNA. *Biochemistry* **25**:1482–1494.

Muniyappa, K., and C. M. Radding. 1986. The homologous recombination system of phage λ: pairing activities of β protein. *J. Biol. Chem.* **261**:7472–7478.

Muniyappa, K., S. L. Shaner, S. S. Tsang, and C. M. Radding. 1984. Mechanism of the concerted action of RecA protein and helix-destabilizing proteins in homologous recombination. *Proc. Natl. Acad. Sci. USA* **81**:2757–2761.

Ogawa, T., H. Wabiko, T. Tsurimoto, T. Horii, H. Masukata, and H. Ogawa. 1978. Characteristics of purified RecA protein and the regulation of its synthesis *in vivo*. *Cold Spring Harbor Symp. Quant. Biol.* **43**:909–915.

Ohtani, T., T. Shibata, M. Iwabuchi, K. Nakagawa, and T. Ando. 1982a. Hydrolysis of ATP dependent on homologous double-stranded DNA and single-stranded fragments by rec-A protein of *Escherichia coli*. *J. Biochem.* **91**:1767–1775.

Ohtani, T., T. Shibata, M. Iwabuchi, H. Watabe, T. Iino, and T. Ando. 1982b. ATP-dependent unwinding of double helix in closed circular DNA by RecA protein of *E. coli*. *Nature* (London) **299**: 86–89.

Pierre, A., and C. Paoletti. 1983. Purification and characterization of RecA protein from *Salmonella typhimurium*. *J. Biol. Chem.* **258**:2870–2874.

Pugh, B. F., and M. M. Cox. 1987a. Stable binding of RecA protein to duplex DNA. *J. Biol. Chem.* **262**: 1326–1336.

Pugh, B. F., and M. M. Cox. 1987b. RecA protein binding to the heteroduplex product of DNA strand exchange. *J. Biol. Chem.* **262**:1337–1343.

Radding, C. M. 1978. Genetic recombination: strand transfer and mismatch repair. *Annu. Rev. Biochem.* **47**:847–880.

Radding, C. M. 1982. Homologous pairing and strand exchange in genetic recombination. *Annu. Rev. Genet.* **16**:405–437.

Radding, C. M., T. Shibata, R. P. Cunningham, C. DasGupta, and L. Osber. 1980. RecA protein of *E. coli* promotes homologous pairing of DNA molecules by a novel mechanism, p. 863–870. *In* B. Alberts and C. F. Fox (ed.), *Mechanistic Studies of DNA Replication and Genetic Recombination*. Academic Press, Inc., New York.

Radding, C. M., T. Shibata, C. DasGupta, R. P. Cunningham, and L. Osber. 1981. Kinetics and topology of homologous pairing promoted by *Escherichia coli recA*-gene protein. *Cold Spring Harbor Symp. Quant. Biol.* **45**:385–390.

Register, J. C., III, G. Christiansen, and J. Griffith. 1987. Electron microscopic visualization of the recA protein-mediated pairing and branch migration phases of DNA strand exchange. *J. Biol. Chem.* **262**:12812–12820.

Register, J. C., III, and J. Griffith. 1985a. 10 nm RecA protein filaments formed in the presence of Mg^{2+} and ATPγS may contain RNA. *Mol. Gen. Genet.* **199**:415–420.

Register, J. C., III, and J. Griffith. 1985b. The direction of RecA protein assembly onto single strand DNA is the same as the direction of strand assimilation during strand exchange. *J. Biol. Chem.* **260**: 12308–12312.

Register, J. C., III, and J. Griffith. 1986. RecA protein filaments can juxtapose DNA ends: an activity that may reflect a function in DNA repair. *Proc. Natl. Acad. Sci. USA* **83**:624–628.

Riddles, P. W., and I. R. Lehman. 1985a. The formation of paranemic and plectonemic joints between DNA molecules by the recA and single-stranded DNA-binding proteins of *Escherichia coli*. *J. Biol. Chem.* **260**:165–169.

Riddles, P. W., and I. R. Lehman. 1985b. The formation of plectonemic joints by the RecA protein of *Escherichia coli*: requirements for ATP hydrolysis. *J. Biol. Chem.* **260**:170–173.

Roberts, J. W., C. W. Roberts, N. L. Craig, and E. M. Phizicky. 1978. Activity of the *Escherichia coli* recA-gene product. *Cold Spring Harbor Symp. Quant. Biol.* **43**:917–920.

Roman, L. J., and S. C. Kowalczykowski. 1986. Relationship of the physical and enzymatic properties of *Escherichia coli* recA protein to its strand exchange activity. *Biochemistry* **25**:7375–7385.

Rupp, W. D., C. E. Wilde III, D. L. Reno, and P. Howard-Flanders. 1971. Exchanges between DNA strands in ultraviolet-irradiated *Escherichia coli*. *J. Mol. Biol.* **61**:25–44.

Rusche, J. R., W. Konigsberg, and P. Howard-Flanders. 1985. Isolation of altered recA polypeptides and interaction with ATP and DNA. *J. Biol. Chem.* **260**:949–955.

Sancar, A., C. Stachelek, W. Konigsberg, and W. D. Rupp. 1980. Sequences of the *recA* gene and protein. *Proc. Natl. Acad. Sci. USA* **77**:2611–2615.

Schutte, B. C., and M. M. Cox. 1987. Homology-dependent changes in adenosine 5'-triphosphate hydrolysis during RecA protein promoted DNA strand exchange: evidence for long paranemic complexes. *Biochemistry* **26**:5616–5625.

Shaner, S. L., J. Flory, and C. M. Radding. 1987. The distribution of *E. coli* recA protein bound to duplex DNA with single-stranded DNA. *J. Biol. Chem.* **262**:9220–9230.

Shaner, S. L., and C. M. Radding. 1987. Translocation of *Escherichia coli* RecA protein from a single-stranded tail to contiguous duplex DNA. *J. Biol. Chem.* **262**:9211–9219.

Shen, P., and H. V. Huang. 1986. Homologous recombination in *Escherichia coli*: dependence on substrate length and homology. *Genetics* **112**:441–457.

Shibata, T., C. DasGupta, R. P. Cunningham, and C. M. Radding. 1979. Purified *Escherichia coli recA* protein catalyzes homologous pairing of superhelical DNA and single-stranded fragments. *Proc. Natl. Acad. Sci. USA* **76**:1638–1642.

Shibata, T., C. DasGupta, R. P. Cunningham, and C. M. Radding. 1980. Homologous pairing in genetic recombination: formation of D-loops by combined action of RecA protein and a helix-destabilizing protein. *Proc. Natl. Acad. Sci. USA* **77**:2606–2610.

Shibata, T., C. DasGupta, R. P. Cunningham, J. G. K. Williams, L. Osber, and C. M. Radding. 1981. Homologous pairing in genetic recombination: the pairing reaction catalyzed by *Escherichia coli* recA protein. *J. Biol. Chem.* **256**:7565–7572.

Shibata, T., O. Makino, S. Ikawa, T. Ohtani, M. Iwabuchi, Y. Shibata, H. Maeda, and T. Ando. 1984. Roles of processive unwinding in recombination reactions promoted by RecA protein of *Escherichia coli*: a study using a monoclonal antibody. *Cold Spring Harbor Symp. Quant. Biol.* **49**:541–551.

Shibata, T., T. Ohtani, P. K. Chang, and T. Ando. 1982a. Role of superhelicity in homologous pairing of DNA molecules promoted by *Escherichia coli* recA protein. *J. Biol. Chem.* **257**:370–376.

Shibata, T., T. Ohtani, M. Iwabuchi, and T. Ando. 1982b. D-loop cycle: a cycle of sequential reactions which comprise pairing of homologous DNA, dissociation of joint molecules and regeneration of substrates promoted by rec-A protein of *Escherichia coli*. *J. Biol. Chem.* **257**:13981–13986.

Singer, B. S., L. Gold, P. Gauss, and D. M. Doherty. 1982. Determination of the amount of homology required for recombination in bacteriophage T4. *Cell* **31**:25–33.

Slilaty, S. N., and J. W. Little. 1987. Lysine-156 and serine-119 are required for LexA repressor cleavage: a possible mechanism. *Proc. Natl. Acad. Sci. USA* **84**:3987–3991.

Stasiak, A., and E. DiCapua. 1982. The helicity of DNA in complexes with RecA protein. *Nature* (London) **299**:185–186.

Stasiak, A., E. DiCapua, and T. Koller. 1981. Elongation of duplex DNA by RecA protein. *J. Mol. Biol.* **151**:557–564.

Stasiak, A., E. DiCapua, and T. Koller. 1983. Unwinding of duplex DNA in complexes with RecA protein. *Cold Spring Harbor Symp. Quant. Biol.* **47**:811–820.

Stasiak, A., and E. H. Egelman. 1986. Structure and dynamics of recA protein-DNA complexes as determined by image analysis of electron micrographs. *Biophys. J.* **49**:5–7.

Stasiak, A., A. Z. Stasiak, and T. Koller. 1984. Visualization of recA-DNA complexes involved in consecutive stages of an *in vitro* strand exchange reaction. *Cold Spring Harbor Symp. Quant. Biol.* **49**:561–570.

Taylor, A., and G. R. Smith. 1980. Unwinding and rewinding of DNA by the recBC enzyme. *Cell* **22**:447–457.

Thomas, C. A., Jr. 1966. Recombination of DNA molecules. *Prog. Nucleic Acid Res.* **5**:315–337.

Tsang, S. S., S. A. Chow, and C. M. Radding. 1985a. Networks of DNA and RecA protein are intermediates in homologous pairing. *Biochemistry* **24**:3226–3232.

Tsang, S. S., K. Muniyappa, E. Azhderian, D. K. Gonda, C. M. Radding, J. Flory, and J. W. Chase. 1985b. Intermediates in homologous pairing promoted by RecA protein. *J. Mol. Biol.* **185**:295–309.

Walker, G. C. 1985. Inducible DNA repair systems. *Annu. Rev. Biochem.* **54**:425–457.

Watt, V. M., C. J. Ingles, M. S. Urdea, and W. J. Rutter. 1985. Homology requirements for recombination in *Escherichia coli. Proc. Natl. Acad. Sci. USA* **82**:4768–4772.

Weinstock, G. M., K. McEntee, and I. R. Lehman. 1979. ATP-dependent renaturation of DNA catalyzed by the RecA protein of *E. coli. Proc. Natl. Acad. Sci. USA* **76**:126–130.

Weinstock, G. M., K. McEntee, and I. R. Lehman. 1981. Hydrolysis of nucleoside triphosphates catalyzed by the RecA protein in *Escherichia coli*. Hydrolysis of UTP. *J. Biol. Chem.* **256**:8856–8858.

West, S. C., E. Cassuto, and P. Howard-Flanders. 1981a. RecA protein promotes homologous-pairing and strand-exchange reactions between duplex DNA molecules. *Proc. Natl. Acad. Sci. USA* **78**:2100–2194.

West, S. C., E. Cassuto, and P. Howard-Flanders. 1981b. Heteroduplex formation by RecA protein: polarity of strand exchanges. *Proc. Natl. Acad. Sci. USA* **78**:6149–6153.

West, S. C., E. Cassuto, and P. Howard-Flanders. 1981c. Homologous pairing can occur before DNA strand separation in general genetic recombination. *Nature* (London) **290**:29–33.

West, S. C., E. Cassuto, and P. Howard-Flanders. 1981d. Mechanism of *E. coli* RecA protein directed strand exchanges in post-replication repair of DNA. *Nature* (London) **294**:659–662.

West, S. C., E. Cassuto, and P. Howard-Flanders. 1982a. Role of SSB protein in recA promoted branch migration reactions. *Mol. Gen. Genet.* **186**:333–338.

West, S. C., E. Cassuto, and P. Howard-Flanders. 1982b. Postreplication repair in *E. coli*: strand exchange reactions of gapped DNA by RecA protein. *Mol. Gen. Genet.* **187**:209–217.

West, S. C., E. Cassuto, J. Mursalim, and P. Howard-Flanders. 1980. Recognition of duplex DNA containing single-stranded regions by RecA protein. *Proc. Natl. Acad. Sci. USA* **77**:2569–2573.

West, S. C., J. K. Countryman, and P. Howard-Flanders. 1983. Purification and properties of the RecA protein of *Proteus mirabilis. J. Biol. Chem.* **258**:4648–4654.

West, S. C., and P. Howard-Flanders. 1984. Duplex-duplex interactions catalyzed by RecA protein allow strand exchanges to pass double-strand breaks in DNA. *Cell* **37**:683–691.

Williams, R. C., and S. J. Spengler. 1986. Fibers of RecA protein and complexes of RecA protein and single-stranded φX174 DNA as visualized by negative-stain electron microscopy. *J. Mol. Biol.* **187**:109–118.

Wollman, E. L., F. Jacob, and W. Hayes. 1956. Conjugation and genetic recombination in *Escherichia coli* K-12. *Cold Spring Harbor Symp. Quant. Biol.* **21**:141–162.

Wu, A. M., M. Bianchi, C. DasGupta, and C. M. Radding. 1983. Unwinding associated with synapsis of DNA molecules by RecA protein. *Proc. Natl. Acad. Sci. USA* **80**:1256–1260.

Wu, A. M., R. Kahn, C. DasGupta, and C. M. Radding. 1982. Formation of nascent heteroduplex structures by RecA protein and DNA. *Cell* **30**:37–44.

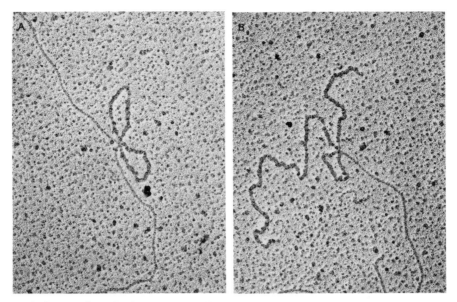

FIGURE 1. Intermediates in the unwinding of duplex DNA by ExoV. Electron micrographs of the reaction of duplex T7 DNA, in the presence of SSB protein, with cell-free extracts of an *E. coli* strain overproducing the RecBCD enzyme. (A) A twin-loop structure. (B) A loop-tail structure. Reaction conditions were as described previously (Taylor and Smith, 1980a), and the products were spread by the formamide/cytochrome *c* technique after glutaraldehyde fixation. Approximately 10 kb of DNA is shown in each micrograph. The thinner lines are duplex DNA, and the thicker lines are single strands of DNA coated with SSB protein. The contour lengths of single strands are reduced approximately twofold, relative to that of duplex DNA, by their binding SSB protein. (Reproduced with permission from Smith et al., 1984.)

Use of the polylysine electron microscope spreading technique produced more irregular-looking structures, but did reveal the presence of a small circular structure (presumably an ExoV molecule) at the junction of single- and double-stranded DNA in the structures (Muskavitch and Linn, 1982).

"Twin-loop" structures (Fig. 1A) consist of two single-stranded DNA loops of equal size, which appear thicker than duplex DNA as a result of their coating of SSB protein, flanked by duplex DNA. "Loop-tail" structures (Fig. 1B) consist of two single-stranded tails and a single-stranded loop, all emerging from a single point at the end of the duplex DNA. The sum of the contour lengths of the loop and the shorter tail frequently equals that of the longer tail, implying that the loop and the shorter tail are from the same strand of the DNA. Another prominent class of unwinding structure, termed a "fork," appears like Fig. 1B, but without the single-stranded loop. Heating the reaction products before fixation and spreading produces a decrease in the frequency of loop-tails and an increase in that of forks, showing loop-tails to be related to forks (Muskavitch and Linn, 1982).

While loop-tails and twin loops appear very different in structure, they are in fact closely related. Both types of structure travel along DNA at the same rate (~300 nucleotides per s), and their loops enlarge at the same rate (~100

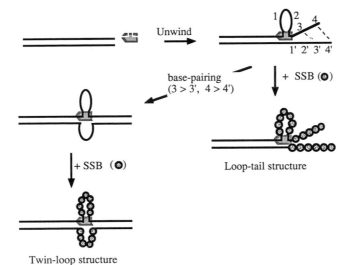

FIGURE 2. Simple model for the unwinding of duplex DNA by ExoV. This model is described in the text. The numbers on the loop and tails denote complementary sequences, to emphasize that the tails emerging from the enzyme are not immediately complementary. (Modified from Taylor and Smith, 1980a.)

nucleotides per s). Their topologies and similarities suggest the model shown in Fig. 2. Their constant rate of progress along the DNA also implies that the ATP-driven enzyme moves unidirectionally along the duplex DNA.

In the model, ExoV, pictured as the stippled box, interacts with only one strand of the DNA. It takes up that strand at 300 bases per s and releases it at 200 bases per s, hence causing a loop of single-stranded DNA to travel with it as it moves unidirectionally through the DNA and leaving two single-stranded tails behind itself. Where they emerge from the rear of the enzyme, the two tails are not complementary (because of the loop) and do not immediately base pair. The unwinding will thus always initiate as a loop-tail, but its subsequent fate will depend on whether the single-stranded tails encounter their partners before they encounter SSB protein. If the tails encounter SSB protein first, they will be completely covered with SSB protein, as the binding of SSB protein to single-stranded DNA is cooperative (Sigal et al., 1972). The loop-tail structure will be maintained and will travel as such down the DNA. If complementary regions of the two single-stranded tails find each other and base pair, all available base pairs will form, as base pairing is also a cooperative reaction. This leaves two single-stranded loops, one held by the enzyme and the other (its complement) maintained by the surrounding base pairs. Subsequent binding of SSB protein to the loops will produce a twin-loop structure, which will travel through the DNA as such.

The model thus predicts that the relative frequency of loop-tails and twin loops, which it claims results from a competition between base pairing of and SSB protein binding to the tails, should be a function of SSB protein concentration.

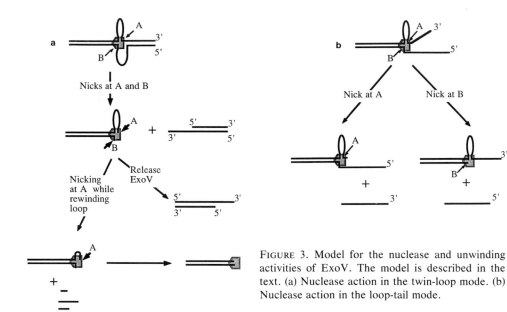

FIGURE 3. Model for the nuclease and unwinding activities of ExoV. The model is described in the text. (a) Nuclease action in the twin-loop mode. (b) Nuclease action in the loop-tail mode.

Low concentrations of gene 32 protein, the phage T4 homolog of SSB protein, did indeed favor twin loops, whereas high concentrations favored loop-tails (Taylor and Smith, 1980b). In the absence of SSB protein the structures were all internal (as the two tails inevitably base paired) and typically appeared as though the two complementary single-stranded loops had base paired to a certain extent: topological constraints prevented their complete base pairing. Twin-loop and loop-tail structures have also been observed with *Eco* ExoV under conditions similar to those used above but which allowed some nuclease activity, although the relative frequency of twin loops was greatly reduced (Muskavitch and Linn, 1982).

Twin-loop structures were also seen with *Hind* ExoV, but loop-tails were not, although occasional Y forms did occur (Taylor and Smith, 1980a; Taylor, unpublished data). The movement and growth of the loops have not been measured in an electron microscope for *Hind* ExoV. Biochemical experiments, described below, imply that the rates of DNA unwinding and rewinding for *Hind* ExoV are very similar to those of *Eco* ExoV.

C. Model for the Nuclease Activities of ExoV

To account for the nuclease activities of ExoV, at least one contact must occur between the enzyme and the second strand of the DNA (the "lower" strand in Fig. 3; the one not touched by the enzyme in Fig. 2), and the enzyme must have at least one single-stranded endonuclease activity for each strand of the DNA. In Fig. 3 the nuclease activity labeled A, that on the upper (loop-containing) strand, is pictured as cutting the strand emerging from the rear of the enzyme, while the activity labeled B cuts the lower strand during its only contact with the enzyme. In this version of the model, the loop held by the enzyme is on the strand of DNA whose 3' terminus entered the enzyme.

The predictions of the model differ depending on whether the enzyme is in its loop-tail or twin-loop mode of unwinding. In the twin-loop mode (see Fig. 3a, in which only the results of nicks at both A and B are shown), a nick at A causes the other loop to be temporarily disrupted and converted into a gap behind the enzyme. If the second strand is also nicked at B before the twin loop is reformed, a duplex DNA fragment with a 3'-terminated single-stranded tail will be released, as shown in Fig. 3a. If the twin loops reform (by base pairing, as in Fig. 2) without a second nick being made, the result will be a nick left behind the enzyme. If the first nick is at B, the final consequences are the same, except that the intermediate structure produced by the first nick resembles a loop-tail flanked by duplex DNA.

Small, wholly single-stranded fragments can be released if, after nicks at A and B and the release of the partially duplex intermediate, the enzyme releases the large loop as small fragments by rewinding it while repeatedly nicking at A until the loop is consumed. The enzyme, still attached to the end of the DNA, can then continue unwinding; it unwinds initially in loop-tail mode, switches to twin-loop mode, and repeats the cycle.

If the enzyme is constrained to the loop-tail mode, then the consequences of nicking are simple, as shown in Fig. 3b. A nick at A will produce a duplex DNA molecule with a 5'-terminated single-stranded tail, while a nick at B will produce a similar molecule with a 3'-terminated tail. In either case the material released will be wholly single stranded.

D. "Degradative" Activities

Investigations using different DNA substrates revealed several different degradative activities of the enzyme. I will consider the nucleolytic activities of the enzyme in ascending order of their rates and probable biological importance.

1. Single-stranded DNA endonuclease

Eco ExoV can cleave single-stranded DNA circles and render them sensitive to digestion by single-strand-specific exonucleases. This single-strand endonuclease activity is stimulated about sevenfold by, but is not completely dependent on, the presence of ATP (Goldmark and Linn, 1972). The ATP-independent activity is probably not a contaminant in the preparation, as its inhibition by the γ protein (a phage λ-encoded inhibitor of ExoV activity; see section II.F.) is similar to that of the ATP-dependent exonuclease and endonuclease activities of the enzyme (Karu et al., 1975).

The single-strand endonuclease appears to be only a minor activity of the enzyme. The turnover number for the ATP-stimulated single-strand endonuclease is 0.2 phosphodiester bonds cut per min, while the corresponding numbers for the single- and double-strand exonuclease activities are 500 and 2,000 bond scissions per min, respectively (calculated from data reported by Eichler and Lehman [1977] and Goldmark and Linn [1972]). Transfection experiments suggest that the single-strand endonuclease activity is negligible in vivo (Benzinger et al., 1975).

Unlike the exonuclease activities of the enzyme, the single-strand endonuclease activity is dependent on bovine serum albumin for its continued activity (Goldmark and Linn, 1972). The double-strand exonuclease activities and the single-strand endonuclease activity are also reported to cease after ~5 min of incubation unless the reaction contains either bovine serum albumin or KCl (Hermanns and Wackernagel, 1977).

*Hin*d ExoV also has an ATP-dependent single-strand endonuclease activity, which is greatly inhibited by the reaction conditions under which the enzyme was studied and has not been examined further (Friedman and Smith, 1972b).

2. Single-stranded DNA exonuclease

Single-stranded DNA (but not RNA) is rapidly degraded by *Eco* ExoV to oligonucleotides of the same length (ca. five nucleotides) as the limit product of the double-strand exonuclease reaction (Karu et al., 1973; Goldmark and Linn, 1972). The single-strand exonuclease activity is dependent on ATP but little influenced by its concentration. Molecules with 3'- or 5'-phosphate or hydroxyl termini are treated identically (Karu et al., 1973).

Under some reaction conditions 3' tails inhibit the degradation of duplex DNA 10-fold (Prell and Wackernagel, 1980), but under other conditions they cause only 2-fold inhibition (unpublished data). Presumably, the difference is due to the ability of the enzyme to digest the single-stranded tails under the different conditions. It is therefore difficult to evaluate the differences between the relative activities of single- and double-strand exonucleases seen with ExoV from different species. The turnover number for *Eco* ExoV single-strand exonuclease is about one-quarter that of the double-strand exonuclease (calculated from data reported by Eichler and Lehman [1977]), while that of *Hin*d ExoV is about 1/10 that of the double-strand exonuclease (Friedman and Smith, 1972a). An ATP-dependent double-strand exonuclease with a molecular weight of ~300,000 purified from *Pseudomonas aeruginosa* has no detectable single-strand DNase activity (Miller and Clark, 1976).

The single-strand exonuclease activity is clearly involved in the degradation of duplex DNA: all the data on reaction intermediates show that large single-stranded intermediates are released early in the reaction and subsequently converted to acid-soluble material by the single-strand exonuclease activity. However, little work has been done on the mechanism of this subsequent degradation. Attempts to determine the direction of attack on single-stranded DNA failed to detect release of either 5' or 3' termini from single-stranded DNA before the release of internal label. The enzyme may make initial endonucleolytic cuts, or it may attack DNA molecules processively with initiation as the rate-limiting step (Karu et al., 1973). ATP is essential for the exonucleolytic degradation of single-stranded DNA, but its role in the reaction is unclear; perhaps the enzyme uses the energy of ATP hydrolysis to move along single-stranded (as well as double-stranded) DNA, making endonucleolytic cuts as it passes, and reduces the DNA to short oligonucleotides by a number of passes through the intermediates.

3. DNA-dependent ATPase

ATP hydrolysis is essential for the double-strand exonuclease activity of both *Eco* and *Hin*d ExoV: neither α,β-methylene ATP nor β,γ-methylene ATP substitutes for ATP, nor do these analogs inhibit either enzyme (Smith and Friedman, 1972; unpublished results cited by Karu et al., 1973). ATP is hydrolyzed to ADP and P_i, but there is no detectable enzyme-catalyzed exchange between ATP and ADP or P_i (Goldmark and Linn, 1972; Smith and Friedman, 1972). Single-stranded linear or circular DNA (Goldmark and Linn, 1972) can act as a DNA cofactor of the *Eco* ExoV ATPase (there is no DNA-independent ATPase activity), but there are no reports as to whether ATP hydrolysis is essential for the nucleolytic activities of the enzyme on single-stranded DNA. Supercoiled or nicked circular duplex DNAs, which are not degraded by the enzyme, do not act as cofactors for the ATPase (Goldmark and Linn, 1972; Karu et al., 1973).

However, there are two nonhydrolyzable duplex substrates that can serve as cofactors for the ATPase: duplex DNA cross-linked with psoralen or a DNA-RNA hybrid (Karu and Linn, 1972; Karu et al., 1973; Orlosky and Smith, 1976). These latter substrates have termini similar to those of duplex DNA. With either of these two substrates, the DNase and ATPase activities are uncoupled; the enzyme can hydrolyze ATP seemingly indefinitely without any concomitant hydrolysis of DNA.

During the double-strand exonuclease reaction, either *Eco* or *Hin*d ExoV consumes about 20 ATP molecules per phosphodiester bond cleaved (Smith and Friedman, 1972; Goldmark and Linn, 1972). In a special mutant, *rorA* (Van Dorp et al., 1975), the ratio of ATPase to DNase activity is several times that of the wild type (section III.A.). Phosphodiester bond cleavage is an exothermic reaction (most nucleases do not require ATP), and the energy in one ATP molecule could melt about seven base pairs (Friedman and Smith, 1973); the need for the large amount of ATP hydrolysis remains a mystery.

4. Double-stranded DNA exonuclease

Linear duplex DNA is rapidly digested by *Eco* or *Hin*d ExoV to a limit digest of 3'-hydroxyl-, 5'-phosphate-terminated oligonucleotides with an average chain length of about five (Wright et al., 1971; Goldmark and Linn, 1972; Friedman and Smith, 1972a). The reaction is absolutely dependent on the presence of a nucleoside triphosphate; although any common ribo- or deoxyribonucleoside triphosphate can serve with varying efficiencies, ATP is most commonly used. ATP is required for the release of even a single nucleotide from the 3' end of duplex DNA (Eichler and Lehman, 1977).

Mg^{2+} is required for the reaction, with maximal rates of acid solubilization at 5 to 20 mM (Wright et al., 1971; Friedman and Smith, 1972a). Different activities of *Hin*d ExoV show differing sensitivities to lowered $MgCl_2$ concentrations (Wilcox and Smith, 1976b). The production of acid-soluble products is undetectable at ≤ 0.5 mM $MgCl_2$, but the ATPase activity and production of single-stranded DNA (as measured by S1 nuclease sensitivity) are virtually unaffected at these lower concentrations; the enzyme presumably can still unwind DNA but cannot hydrolyze it.

Effect of ATP concentration. The concentration of ATP has a strong effect on the amount of double-stranded DNA rendered acid soluble, after extended incubation, by *Eco* ExoV (Goldmark and Linn, 1972; Oishi, 1969; Wright et al., 1971; Eichler and Lehman, 1977). The double-strand exonuclease activity is maximal at ~30 μM ATP, and is inhibited several-fold by millimolar concentrations (Wright et al., 1971). The amount of single-stranded DNA solubilized, by contrast, increases with ATP concentration up to ~200 μM ATP and is very slightly inhibited at higher ATP concentrations. The initial rate of release (for the first 2 to 5 min) of acid-soluble nucleotides in the double-strand exonuclease reaction is unaffected by ATP concentration, but at concentrations above 200 μM the enzyme soon ceases to release acid-soluble material, while at lower concentrations the enzyme continues to do so (Eichler and Lehman, 1977). At high ATP concentrations the double-strand exonuclease activity of the enzyme seems to be irreversibly inactivated (though the ATPase activity is unaffected). Adding more enzyme to a reaction produces another round of reaction, but reducing the ATP concentration in a reaction which has ceased releasing acid-soluble products produces no more acid-soluble products.

It is tempting to equate the short time of rapid solubilization at high ATP concentrations with the time needed for the enzyme to pass once through the DNA substrate and to speculate that the enzyme is unable to attack more than one duplex DNA molecule under those conditions.

Reaction intermediates. *(i) Hind ExoV. Hin*d ExoV can bind to duplex DNA, without any DNA degradation, in the absence of ATP or $MgCl_2$ (Wilcox and Smith, 1976a). The binding is very tight, rapid, and specific for linear DNA, with two binding sites per duplex DNA molecule, presumably at the termini of the DNA. The binding reaction allows synchronization of nuclease reactions and, by allowing measurement of the number of active enzyme molecules in a preparation, allows such reactions to be carried out with one enzyme molecule per DNA terminus.

Such reactions were carried out at very low $MgCl_2$ concentrations, and the recycling of enzyme molecules was prevented by the addition of excess carrier DNA. Under such conditions the enzyme produced single-stranded DNA (but no acid-soluble material), but the release of single-stranded material ceased abruptly after a time proportional to the length of the substrate (Wilcox and Smith, 1976b). These observations are most compatible with the enzyme unwinding the DNA in the twin-loop mode (only 35% of the DNA becomes single stranded), but stopping after colliding with another enzyme molecule which entered the other end of the DNA. The duration of the reaction implies that *Hin*d ExoV travels at about 300 bases per s. The rate of twin-loop growth, as measured from the rate of production of single-stranded DNA, is about 100 bases per s. Both numbers are similar to those for *Eco* ExoV.

Short, partially duplex reaction products, whose duplex region was ~1,500 base pairs, were visible by electron microscopy after reaction for 15 s, and long, partially duplex intermediates were seen after 45 s of incubation (Wilcox and Smith, 1976b).

In previous experiments *Hin*d ExoV reaction intermediates were fractionated on sucrose gradients, and the sizes of the duplex and single-stranded regions in each fraction were estimated (Friedman and Smith, 1973). The proportion of the DNA in each fraction that was single stranded was estimated by CsCl buoyant density and was found to be equivalent to the fraction of the DNA sensitive to exonuclease I, an exonuclease specific for 3'-hydroxyl-terminated single strands. The tentative conclusion was that most of the molecules were duplex with 3'-terminated tails. The model in Fig. 3a predicts the release of short duplex fragments behind the traveling enzyme, with 3'-terminated tails of about half the length of the duplex region (from the relative rates of unwinding and rewinding; section II.B.). Such molecules with duplex regions of ~1,500 base pairs (the length seen by electron microscopy) and single-stranded regions of ~750 bases would have a sedimentation coefficient of 11.5S (calculated from a formula given by MacKay and Linn [1974]), very similar to the average value for the broad peak of reaction intermediates seen in *Hin*d ExoV intermediates (Friedman and Smith, 1973).

About half of the large single-stranded material produced by the enzyme in its initial interaction with duplex DNA was found to have no double-stranded DNA attached (Wilcox and Smith, 1976b). In the model in Fig. 3a, the enzyme is imagined to change its mode of action after releasing the partially duplex intermediate; it then rewinds the large single-stranded loop that remains, and (by occasional nicking of the tail it is extruding) releases it as large single-stranded fragments. The initial flush end is then regenerated, and the cycle of reaction can repeat. Long, completely single-stranded fragments would be released if some of the ExoV molecules were in the loop-tail mode (Fig. 3b). However, the tentative conclusion (Friedman and Smith, 1973) that the single-stranded tails on duplex intermediates were all at 3' termini would imply that, in terms of Fig. 3b, nicking could occur at B but not at A in the loop-tail mode.

(ii) Eco ExoV. Investigations of intermediates of the double-strand exonuclease activity of *Eco* ExoV have not been as straightforward as those with *Hin*d ExoV because of the lack of conditions that allow the *E. coli* enzyme to bind to the ends of duplex DNA without further reaction. The conditions (Taylor and Smith, 1985) that allow RecBCD enzyme-dependent retention of duplex linear DNA on a nitrocellulose filter have been found to allow very slow reaction of enzyme with DNA (Taylor, unpublished data). The lack of a filter binding assay also prevented the quantitation of active enzyme molecules.

Under normal reaction conditions *Eco* ExoV acts rapidly and processively, and no intermediates are detectable in the reaction (MacKay and Linn, 1974). Intermediates were detected in brief (45-s) reactions at reduced temperatures, with high salt and ATP concentrations (Karu et al., 1973) but with high $MgCl_2$, and with enzyme concentrations approximately equal to the number of duplex substrate ends (Karu et al., 1973; MacKay and Linn, 1974). Intermediates were purified by sucrose density gradient sedimentation, and their single-strand nature was investigated with single-strand-specific exo- and endonucleases. The association of single- and double-stranded material was estimated by equilibrium density gradient centrifugation. Partially duplex intermediates were found with both 3'- and 5'-terminated single-stranded tails (but very few with internal single-stranded regions); the single-stranded regions ranged from ~2,000 to ~4,000 nucleotides long, with the peak fraction having ~6,000 base pairs of duplex DNA and ~3,000 bases of single-stranded DNA. No fully single-stranded DNA fragments longer than a few hundred nucleotides were found.

The 3'-terminated intermediates and the single-stranded fragments are easily explained by Fig. 3a (as detailed above). Both 5'- and 3'-terminated strands are predicted by the scheme in Fig. 3b, in which unwinding in the loop-tail mode is followed by nicking and release of either strand, leaving a tail of the opposite polarity. The production of long (3,000-nucleotide) tails on duplex DNA but no observed single strands of that length may be due to the latter being attacked by free enzyme molecules after their release from the duplex DNA (but see MacKay and Linn [1974] for an alternative explanation).

The mode of unwinding by *Eco* ExoV in the absence of SSB, under reaction conditions similar to those above, but at a lower $MgCl_2$ concentration, is a function of salt concentration (unpublished data): at NaCl concentrations of 100 mM or higher rewinding predominates, while at lower concentrations the strands of the DNA are permanently separated. Interpretation of the reaction intermediates described above thus may be complicated by the occurrence of a mixture of two modes of enzyme reaction.

E. Substrate Specificity

1. Damaged DNA

recB and *recC* null mutants are sensitive to UV and X rays, implying that the enzyme is involved in repair of radiation damage. Heavily UV-irradiated linear duplex DNA is degraded by *Eco* ExoV but at a reduced rate (Karu et al., 1973; Tanaka and Sekiguchi, 1975). Pyrimidine dimers in duplex circular DNA are not attacked by the enzyme even after nicking on the 5' side of the dimers by T4 endonuclease V (Tanaka and Sekiguchi, 1975). X-irradiated linear DNA is digested at the same rate and to the same extent as unirradiated DNA (Karu et al., 1973). Thus, even though ExoV is implicated in the repair of damaged DNA, it has no enhanced reactivity with it.

2. DNA topology

The degradation of double-stranded DNA by the enzyme is, even though the limit digest products are oligonucleotides, strictly an exonucleolytic reaction. (Exonucleases are defined in terms of their substrate specificity rather than their reaction products. "An endonuclease can attack a covalently circular strand, but an exonuclease cannot because it requires a chain terminus" [Weiss, 1981].) Neither *Eco* nor *Hind* ExoV can digest covalently closed or nicked duplex circles (Friedman and Smith, 1972b; Goldmark and Linn, 1972; Karu et al., 1973). *Hin*d

ExoV is unable to digest duplex circular molecules with an average of one gap of 600 nucleotides per molecule (Friedman and Smith, 1972b).

Duplex circular molecules with gaps of from 10 to 770 nucleotides, made by annealing of restriction fragments, are not unwound by *Eco* ExoV (Taylor and Smith, 1985). Such gapped DNA is solubilized, under appropriate conditions, but the rate of cutting is very slow: ~0.005 cuts per min per enzyme for a gap of 770 nucleotides, with smaller gaps cut at correspondingly slower rates. The cutting of gapped circles is thus a very slow process and is abolished by the presence of SSB protein (MacKay and Linn, 1976). It is therefore probably of little biological relevance.

The degradation of duplex DNA by either *Hin*d or *Eco* ExoV is inhibited by single-stranded tails on the duplex DNA. The rate of digestion by *Hin*d ExoV is reduced 10-fold by either 5' or 3' tails of 3,000 nucleotides (Friedman and Smith, 1972b). Under appropriate conditions degradation of duplex DNA by *Eco* ExoV is abolished by 3' tails of 145 nucleotides (Prell and Wackernagel, 1980). Duplex DNA with short 3' or 5' tails can be unwound by *Eco* ExoV, but unwinding is prevented by tails of >50 nucleotides (Taylor and Smith, 1985).

To unwind or degrade duplex DNA, ExoV must apparently interact with the termini of both strands of one end of the duplex DNA.

3. Cross-linked DNA

Psoralens are photochemically activated agents that make random interstrand cross-links in duplex DNA. When incubated with psoralen cross-linked duplex DNA, either *Eco* or *Hin*d ExoV produces limited digestion (presumably from the end of the DNA to the first cross-link); the nuclease activity ceases, but the ATPase activity continues indefinitely (Karu and Linn, 1972; Orlosky and Smith, 1976). *Hin*d ExoV remains tightly bound to cross-linked molecules, sedimenting with them in sucrose gradients, and is unable to digest exogenous DNA (Orlosky and Smith, 1976). The *E. coli* enzyme can, however, leave the cross-linked molecule and degrade other duplex DNA molecules (Karu et al., 1973). The tighter binding of *Hin*d ExoV to cross-linked DNA is in accord with its tight binding to duplex DNA and may result from *Hin*d ExoV unwinding DNA in the twin-loop mode and *Eco* ExoV unwinding DNA in the loop-tail mode (Fig. 3).

F. Factors That Inhibit ExoV Activity

1. SSB protein

In the presence of SSB protein, single-stranded linear and circular DNAs are resistant to digestion by *Eco* ExoV. At low ATP concentrations, digestion of duplex DNA is stimulated by low levels of SSB protein but inhibited by saturating levels (MacKay and Linn, 1976). At high ATP concentrations, very little duplex DNA is made acid soluble, but most of the DNA is made single stranded and sediments as large fragments. The enzyme must be typically trapped in the loop-tail mode (Fig. 3b) and can release long single-stranded fragments by nicking (at A or B) but is unable to reduce those fragments to acid-soluble material because of their coating of SSB protein.

2. RecA protein

Much of the degradation of damaged DNA in vivo is by ExoV. This degradation is reduced if the cell is first induced for RecA protein synthesis, suggesting that large amounts of RecA protein (or a protein whose synthesis is controlled by RecA protein) may inhibit the activity of the RecBCD enzyme in vivo (Marsden et al., 1974). RecA protein, in vitro, inhibits the activity of *Eco* ExoV on linear or circular single-stranded DNA (Williams et al., 1981). The inhibition is stronger in the presence of adenosine 5'-O-(3-thiotriphosphate) (ATP-γ-S), an analog of ATP which stabilizes the binding of RecA protein to both single- and double-stranded DNA. The inhibition seems to result from RecA protein binding to the single-stranded DNA rather than to the enzyme. Inhibition of the double-strand exonuclease activity is weak and is seen only in the presence of ATP-γ-S. Strong inhibition of the digestion of duplex DNA with long single-stranded tails is seen even in the absence of ATP-γ-S. Long (~1,500-base) gaps in duplex DNA are also protected from ExoV degradation by RecA protein in the absence of ATP-γ-S. The strong inhibition of the single-strand exonuclease activity by RecA protein in the presence of ATP-γ-S seems at variance with the much weaker inhibition of the double-strand exonuclease activity, as the double-strand exonuclease produces large single-stranded fragments as reaction intermediates. It is possible that ExoV can attack the newly released single-stranded fragments before they can be bound by RecA protein.

3. Phage-encoded ExoV inhibitors

A number of phages encode proteins that serve to protect them from the action of ExoV (Sakaki, 1974). The phage T4 gene 2 product apparently binds to the ends of the DNA in the phage and protects it from ExoV attack on subsequent infection (Oliver and Goldberg, 1977). A seemingly similar protein has been isolated from phage Mu-infected cells and characterized in vitro: it binds to the ends of linear duplex DNA and protects it from ExoV (Williams and Radding, 1981; Akroyd and Symonds, 1986). Single-stranded DNA is not protected.

A second class of inhibitors function by binding to, and inactivating, ExoV. The best characterized is the γ protein of phage λ (Sakaki et al., 1973), which inhibits all the nuclease activities and the DNA-dependent ATPase of ExoV (Karu et al., 1975). The phage T7-encoded inhibitor (Pacumbaba and Center, 1975), by comparison, inhibits all the nuclease activities but not the ATPase activity of RecBCD enzyme; perhaps it also leaves the unwinding activity active. The γ protein functions best if it is incubated with RecBCD enzyme and ATP before the addition of substrate and has no effect on an ongoing *Eco* ExoV reaction.

The fact that various phages protect themselves from ExoV action implies that the degradative activities of ExoV are indeed functional within the cell (see also section IV.B.), though it provides no evidence as to which activity of ExoV is responsible for the demise of the phages.

4. Calcium ions

The nuclease, but not the ATPase, activities of RecBCD enzyme are greatly inhibited by the addition of Ca^{2+} ions to the reaction mixture (Rosamond et al., 1979). The inhibition is not total: while T7 DNA molecules which have been exposed to RecBCD enzyme under such conditions remain wholly duplex and full length, they do suffer approximately five nicks per DNA duplex.

G. "Recombinational" Activities

1. Chi cutting

Chi sites stimulate the rate of RecA-RecBCD-mediated homologous genetic recombination in their vicinity. (See Smith and Stahl [1985] and Smith [1987] for references and a more extensive review of the properties of Chi.) They were first identified in phage λ, and their action was studied in mutants of λ defective in the *red* and γ genes, in which λ × λ recombination occurs by the RecBCD pathway. The sites are also present in the *E. coli* chromosome and are active in P1-mediated transduction and in conjugation (Dower and Stahl, 1981). The sites are determined by the unique asymmetric sequence 5'-GCTGGTGG-3'. Elegant experiments by Stahl's group, using genetic crosses of derivatives of phage λ, have identified many features of the recombination-stimulating activity of Chi sites. Genetic studies with mutants of RecBCD enzyme, discussed in section III, identify RecBCD enzyme as the entity interacting directly with Chi sites.

Experiments with purified RecBCD enzyme and end-labeled linear DNA molecules show that, in vitro, Chi sites cause RecBCD enzyme to make single-strand interruptions in linear (but not circular) duplex DNA in the vicinity of the Chi sites (Ponticelli et al., 1985; Taylor et al., 1985). Under the reaction conditions used, the substrate was completely unwound during the reaction (unpublished data), and the labeled reaction product was released as a single-stranded fragment, regardless of whether the label was to the 5' or 3' side of the Chi site. The reaction, like other nuclease activities of ExoV, requires Mg^{2+} and ATP and is inhibited by Ca^{2+} ions. Chi cutting is seen with (almost) homogeneous RecBCD enzyme.

Extracts of recombination-proficient *recC** mutants (alleles *C1001*, *C1002*, *C1003*, and *C1004* overexpressed in plasmid pBR322) which ignore Chi in vivo (see section III) fail to cut at Chi in vitro, while extracts of wild-type cells do cut at Chi (Ponticelli et al., 1985; McKittrick, personal communication). An extract of a TexA mutant, *recC343*, which showed reduced Chi activity in vivo showed reduced Chi cutting in vitro (Ponticelli et al., 1985; McKittrick, personal communication). RecBCD enzyme must thus be directly involved in the cutting.

The cutting is seen only with linear duplex DNA and produces a nick with 5'-phosphate and 3'-hydroxyl termini, occurring on the strand of the DNA containing 5'-GCTGGTGG-3'. With the four Chi sites examined nicks occur as shown below, with different cuts being more prominent with different Chi sites (Taylor et al., 1985; K. Cheng, personal communication; Taylor, unpublished data).

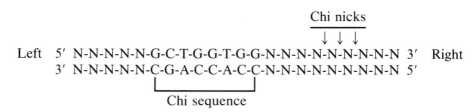

Left and right in this diagram refer to the orientation of phage λ DNA, as conventionally written. All active Chi sites in λ are in the orientation shown here.

Some sequences one nucleotide removed from that of Chi show reduced Chi activity in vivo (Schultz et al., 1981; Cheng and Smith, 1984) and show a degree of cutting in vitro that is correlated with their activity in vivo (Cheng and Smith, 1987).

Chi sites in phage λ are maximally active only when present in a particular orientation with respect to the *cos* site of λ (Kobayashi et al., 1984). *cos* is the site at which multimeric λ DNA suffers a double-strand cut to allow packaging of DNA into this phage head. This double-strand break allows RecBCD enzyme to enter the previously circular DNA. Only one end of the DNA (the right end of λ as conventionally written) is accessible to the enzyme, as the left end is presumably bound by the phage head assembly. Other genetic experiments support the notion that RecBCD enzyme enters the λ chromosome at the right end and travels leftwards (Stahl et al., 1986). Thus, Chi is active only if RecBCD enzyme encounters it while traveling from "right" to "left," as in the diagram above.

When RecBCD enzyme can approach Chi only from the left in vitro, no cutting at Chi is seen (Taylor et al., 1985). However, when the single-stranded tail blocking the entry of the enzyme at the right (see section II.E.) is removed, cutting at Chi is seen. Thus, RecBCD enzyme can cut at Chi in vitro only if it encounters the Chi, as drawn above, while it is traveling from right to left. In summary, RecBCD enzyme nicks DNA in vitro in the vicinity of Chi sites, but does so only if it encounters the Chi site in the orientation that leads to Chi activity in vivo.

2. D-loop cleavage

RecA protein can catalyze the synapsis of a single strand of DNA with a complementary strand in a duplex DNA molecule, making a three-stranded structure termed a displacement loop or D loop (see Radding, this volume). Cleavage of the displaced strand in such a structure can, as shown in the simple model in Fig. 4, lead to the formation of a "Holliday" recombination intermediate structure.

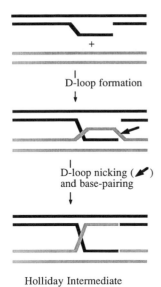

D-loop formation

D-loop nicking ( ) and base-pairing

Holliday Intermediate

FIGURE 4. Formation and cleavage of D loops to form a Holliday structure. The center panel shows a D loop arising from invasion of a single strand into a duplex DNA molecule. The upper panel depicts how such a structure could arise from a partially unwound, nicked, duplex DNA molecule. The lower panel shows the result of an appropriate nick in the D loop and annealing of complementary strands to form a Holliday recombination intermediate.

D loops were made between supercoiled circular DNA and a complementary linear single-stranded fragment, using thermal energy to form the complex. The displaced strand in the D loops was cut by RecBCD enzyme at a rate that was 20 times that of cleavage of single-strand circles, assuming the rate of cleavage to be linearly dependent on the length of the single-stranded DNA (Wiegand et al., 1977). RecA protein can dissociate D loops, but the reaction is blocked by the ATP analog ATP-γ-S, which does not inhibit RecBCD enzyme. In the presence of ATP (for the RecBCD enzyme) and ATP-γ-S, RecA protein inhibits the cleavage of D loops by RecBCD enzyme (Williams et al., 1981). SSB also prevents RecBCD enzyme from cutting D loops.

It is thus unclear whether RecBCD enzyme can cut D loops in vivo, though it should be borne in mind that these experiments detected cutting by exogenous ExoV, while in some models of recombination the RecBCD enzyme is already "riding on" one of the DNA duplexes involved in the D-loop reaction.

III. GENETIC AND PHYSICAL CHARACTERIZATION OF THE *recBCD* REGION

A. *recBCD* Mutants

1. Null mutants

Point mutants. Most *recB* and *recC* null mutants were isolated on the basis of X-ray or UV sensitivity rather than by their inability to recombine (Willetts and Mount, 1969). Several of the mutants were subsequently shown to be UGA suppressible (Templin et al., 1978). Amber (Templin et al., 1978) and temperature-sensitive (Tomizawa and Ogawa, 1972) mutants have also been isolated. It is perhaps worth noting that neither of the two mutants in common use is a missense mutant: *recC22* is UGA suppressible (Templin et al., 1978), while *recB21* contains an insertion of ~1.4 kilobases (kb) into its coding region and is polar on *recD* (Amundsen et al., 1986). *recC73*, however, seems to be a bona fide missense mutant: it is not suppressible by any nonsense suppressor tried (Templin et al., 1978), and Rec$^+$ revertants of it have been isolated (Schultz et al., 1983). By the same criteria, several *recB* mutants also seem to be missense. Despite these differences, all *recC* and *recB* null mutants show identical phenotypes: UV sensitivity and recombination deficiency. Only *recB21* and *recC22* were tested in vitro and they were both found to be deficient in all nuclease activities tested.

The null mutants, because they abolish all of the activities of ExoV, do not address the question of which of the several activities of RecBCD enzyme is most important for the various cellular activities in which the enzyme engages (see section IV).

The temperature sensitivity, in vitro, of ExoV from the *recB*(Ts) mutant showed *recB* to be a structural gene of the enzyme (Tomizawa and Ogawa, 1972). The single *recB*(Ts) or *recC*(Ts) mutants appear wild type at the permissive temperature (30°C), but become UV and mitomycin C sensitive at the nonpermissive temperature (43°C), although they are not very recombination deficient at 43°C. A double [*recB*(Ts) *recC*(Ts)] mutant was slightly recombination deficient (but very UV and mitomycin C sensitive) at 30°C, but resembled a null mutant at 43°C (Kushner, 1974b). The implication is that recombination requires less ExoV activity than does radiation resistance.

ExoV purified from the single and double temperature-sensitive mutants lacks the double-strand exonuclease but retains the single-strand exo- and endonuclease activities (Kushner, 1974a). The double-strand exonuclease activity of the *recC*(Ts) mutant is much less temperature sensitive than that of the *recB*(Ts) mutant, although their in vivo phenotypes are identical. A deficiency in unwinding activity would be a simple explanation for the greater effects of the mutations on the double-strand exonuclease activity than on the single-strand nuclease activities. One inference from the results is that the double-strand exonuclease (or unwinding) activity may be more important for recombination and repair than the nuclease activities on single-stranded DNA. This is in accord with transfection experiments showing the single-strand exo- and endonuclease activities to be very weak and almost undetectable, respectively, inside the cell (Benzinger et al., 1975).

A somewhat similar motif is shown by the *rorA* mutant. This mutant, whose lesion maps in *recB*, is X-ray sensitive but UV resistant and recombination proficient (Glickman, 1979; Glickman et al., 1971). The specific activity of RecBCD enzyme in the mutant is the same as that in the wild type, but the enzyme has an increased turnover of ATP during the degradation of double-stranded DNA (Van Dorp et al., 1975).

Deletion mutants. The low viability (section IV.A.) of *recB* and *recC* null mutants led to the suspicion that a total lack of ExoV would be lethal and that

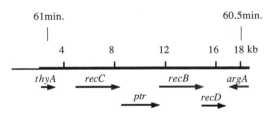

FIGURE 5. Map of the 18.5-kb *Bam*HI fragment of *E. coli* DNA containing the *recB*, *recC*, and *recD* genes and the flanking *thyA* and *argA* genes. The horizontal line represents the 18.5-kb *Bam*HI chromosomal fragment of *E. coli* DNA used in many of the studies described here, with locations in kilobases marked above the line. The thicker portion of the line shows the portion of the fragment whose DNA sequence has been established. The arrows represent the positions, sizes, and transcription directions of the known genes encoded by the fragment. The approximate location (in minutes) of the genes on the *E. coli* genetic map is noted. See text for references.

existing mutants survived because of their leakiness. However, deletions of the *recBCD* region were obtained as a result of aberrant excision of transposon Tn*10* inserted into *argA* or of a derivative of phage Mu inserted into the *thyA* gene (Chaudhury and Smith, 1984a). Some of these mutations deleted the whole region from *thyA* to *argA* (Fig. 5) and hence are RecB⁻, RecC⁻, and RecD⁻. The deletions have phenotypes identical to those of previously isolated null mutations. Therefore, ExoV activity is dispensable for *E. coli* growth (though the reason for the poor viability is unknown), and the previously isolated null mutants may well be devoid of all ExoV activity.

2. Rec⁺ Chi⁻ mutants

The involvement of RecBCD enzyme in the activation of Chi sites was first inferred from the pathway specificity of the action of Chi (Stahl and Stahl, 1977): Chi is active in the RecBCD pathway of recombination but not in other pathways that utilize RecA (see Mahajan, this volume). *recBCD* mutants that lack Chi activity but by all other criteria are RecBCD⁺ were isolated by the two protocols described below.

Certain pseudorevertants, designated *recC**, of the presumed missense mutation *recC73* have regained, fully or partially, all of the known activities of RecBCD enzyme except for Chi activity (Schultz et al., 1983). Three of the mutants (*recC1002*, *C1003*, and *C1004*) are very similar: they are partially Rec⁺, somewhat UV sensitive, recessive to wild type, and (except for *recC1002*, which is somewhat deficient) have about wild-type levels of ExoV activity in crude extracts. The exceptional mutant, *recC1001*, appears fully Rec⁺, is barely more UV sensitive than wild type, and is partially dominant over wild type for Chi activity. It has wild-type ExoV activity in extracts and is hyper-*rec* in crosses of phage λ lacking Chi. The mutants are thus qualitatively, rather than quantitatively (except for *recC1002*, which may be both), altered in their RecBCD enzyme activity. An appealing hypothesis to explain both classes of mutants is that *recC1002*, *C1003*, and *C1004* have lost the ability to recognize Chi and hence are somewhat depressed for recombination and recessive, while *recC1001* has lost Chi recognition but gained another specificity (for a sequence present in phage λ) and hence is fully recombination proficient, hyper-*rec*, and partially dominant.

Additional *recBCD* mutants of *E. coli*, designated "Tex" (for *t*ransposon *ex*cision), have an enhanced frequency of precise excision of transposons Tn*5* and Tn*10* from the chromosome (Lundblad et al., 1984). Complementation (Hickson and Emmerson, 1983) and mapping (Lundblad et al., 1984) identified one mutant as a *recB* mutant (*recB344*) and another as a *recC* mutant (*recC343*). A third mutant (*rec-345*) appeared to be in *recB*. The mutants were also somewhat UV sensitive (unpublished data) and showed wild-type levels of recombination and of ExoV activity in extracts, but were reduced for Chi activity in vivo (though not as reduced as the *recC** mutants discussed above). One of them (*recC343*) had reduced, but detectable, Chi cutting in vitro (Ponticelli et al., 1985). The *recC1002*, *C1003*, and *C1004* mutants also showed the Tex phenotype, though the hyper-*rec* mutant, *recC1001*, did not (Schultz et al., 1983).

Thus, two different approaches have produced mutants of RecBCD enzyme, five in *recC* and two in *recB*, with reduced or abolished Chi activity in vivo, but (with one exception) full ExoV nuclease activity in vitro, and with partial or full recombination ability by other measures. In all cases studied, the degree of Chi cutting in vitro mirrored the degree of Chi activity in vivo.

3. Rec+ nuclease-deficient mutants

Again, two separate lines of research led to the isolation *recBCD* mutants with a novel phenotype: recombination proficient but devoid of all activities of ExoV yet tested in vitro. The phenotype is dubbed ‡.

Chaudhury and Smith (1984b) isolated ‡ mutants by their ability to support the growth of phage T4 gene 2 mutants (section II.F.). On examination of their phenotypes it became clear why ‡ mutants had not been reported previously: they are both radiation resistant and recombination proficient, yet are devoid of ATP-dependent double-strand exonuclease activity. One mutation, *rec-1010*, is in *recC*, while the remainder, which were initially identified as being in *recB*, are in a previously unidentified gene, *recD* (Amundsen et al., 1986). The initial misclassification was due to the *recB* allele used for complementation tests, *recB21*, being an insertion that was polar on *recD*. The *recD* mutants (as detailed in section III.F.) lack both the activity and the polypeptide of a third subunit of ExoV. The *recD* mutations appear to be nonsense mutants: none showed a *recD* polypeptide of wild-type length in "maxicells" and some produced a truncated peptide. The solitary *recC*‡ mutant makes normal *recD* polypeptide, but an altered *recC* polypeptide that presumably cannot interact with *recD* polypeptide. *recD* mutants were also isolated by their ability to reduce the stability of certain plasmids (Biek and Cohen, 1986). Mutants isolated by either procedure appear identical in phenotype. These latter mutants are all Tn*10* insertion mutants.

By many criteria, ‡ mutants appear identical to wild-type cells: they are UV resistant, fully viable, and recombination proficient; yet they have little or no detectable ATP-dependent double-strand exonuclease activity either in vivo (measured as "reckless degradation"; see section IV.B.) or in extracts (Chaudhury and Smith, 1984b). They are also deficient in the single-strand nuclease activities of ExoV (S. K. Amundsen and A. M. Chaudhury, personal communication). However, their recombination ability is both qualitatively and quantitatively different from that of wild-type *E. coli.* ‡ mutants are hyper-recombinogenic for recombination of phage λ lacking Chi, but do not show Chi-mediated enhancement of recombination. They are also proficient for conjugal recombination and P1-mediated transduction. These two latter forms of recombination are dependent on an active *recJ* gene product (a gene in the RecF pathway of recombination; see Mahajan, this volume) in a *recD* mutant strain but not in a *recD*⁺ strain (S. T. Lovett, C. Luisa-DeLuca, and R. D. Kolodner, *Genetics*, in press). These results imply that recombination in *recD* mutants occurs by a pathway at least partly different from that in which wild-type RecBCD enzyme acts. However, *recD* mutant-mediated phage λ × λ (Red⁻ Gam⁻) recombination does not require a *recJ* function (D. Thaler, personal communication).

Despite the genetic evidence that *recD* mutants perform some active function, none of the known activities of RecBCD enzyme are detectable in extracts of *recD* mutants, although they are easily measured in wild-type extracts. These assays are particularly sensitive with the alleles, on a 18.5-kb *Bam*HI chromosomal fragment, inserted into plasmid pBR322. Wild-type enzyme is overproduced about 25-fold by such plasmids (Dykstra et al., 1984b). By the in vitro α-β complementation assay for RecBCD subunits (section III.F.), the ‡ clones overproduce their polypeptides 25-fold, yet there is no detectable nuclease or unwinding activity in the extracts (Amundsen et al., 1986; unpublished data).

A notable exception to these observations is *recD1009* (the nonsense mutant with the longest truncated peptide), which appears to be a somewhat leaky mutant. As a haploid it shows no Chi hot-spot activation, yet when overproduced it shows both limited Chi activity in vivo and Chi cutting and significant double-strand exonuclease activity in vitro (Amundsen et al., 1986; S. K. Amundsen, personal communication). It can also catalyze a very limited unwinding of duplex DNA, which seems more active on short DNA (unpublished data). The other *recD* nonsense mutants (*recD1011, 1013,* and *1014*) have no detectable nuclease, Chi cutting, or unwinding activity even in overproduced extracts. They are probably not merely unstable in extracts, as addition of purified *recD* polypeptide to the extracts produces full levels of active ExoV (Amundsen et al., 1986).

The *recD* mutants thus remain an enigma. They are fully recombinogenic, though qualitatively altered, yet are devoid of any of the known activities of ExoV in vitro.

Mutants similar to *recD* mutants may also have been isolated in *H. influenzae.* *Hin*d ExoV contains three polypeptides, with sizes similar to those of *Eco* ExoV (K. W. Wilcox, M. Orlosky, E. A. Friedman, and H. O. Smith, *Fed. Proc.* **34**:515, 1975). Eight *Hin*d ExoV (*add*) mutants were divided into three complementation groups (A, B, and C) by in vitro complementation (Kooistra et al., 1976). The three mutants in group A retain DNA-dependent ATPase activity and may correspond to *E. coli recD* mutants (although the latter have not been tested for ATPase activity). Four of the five group B and C mutants lack DNA-dependent ATPase, as do *E. coli recB* and *recC* null mutants. The exceptional mutant, in group C, may be similar to *E. coli recC1010*. The genetic mapping data for the *H. influenzae* mutants suggest the order C-B-A, which corresponds to the known order, *recC-recB-recD*, for the *E. coli* genes.

B. Map Location of Genes for ExoV

The *E. coli recB* and *recC* mutations were localized by P1 transduction (see Fig. 5) to a region of the *E. coli* map between *thyA* and *argA* (Willetts and Mount, 1969). The third gene, *recD*, was shown by deletion analysis (as discussed below and shown in Fig. 5) to map between *recB* and *argA*.

The apparent orderly clustering of the genes for ExoV is, however, interrupted by an interloper, *ptr*, the gene for protease III. Protease III is a periplasmic endopeptidase which is most active on small polypeptide fragments (Cheng and Zipser, 1979) but whose biological function is unclear, as null mutants in its structural gene (*ptr*) have no discernible phenotype (Cheng et al., 1979). The mutants were initially mapped to the *argA* region and were subsequently shown to be located between the *recC* and *recB* genes (Dykstra et al., 1984b). Neither Tn*1000* insertion mutations nor deletion of the *ptr* gene affected either the viability of the cell or the activity of ExoV: the connection between *ptr* and ExoV thus remains obscure.

C. Cloning of the Genes for ExoV

The strategies initially used to clone the genes were based on the presumption that overproduction of ExoV would be detrimental to the cell.

Emmerson and co-workers cloned chromosomal *Bam*HI fragments into a λ vector and selected for lysogens of a *recC22* mutant strain which had acquired mitomycin C and UV resistance (Hickson and Emmerson, 1981; Hickson et al., 1982). They obtained a defective phage carrying *recC*$^+$ and, by aberrant excision, obtained from it a defective λ phage carrying both *recC*$^+$ and *recB*$^+$. However, it was subsequently shown (Dykstra et al., 1984b) that the strain they used contained a duplication, at a remote location, of at least part of the *recBCD* region and that the restriction map (Hickson and Emmerson, 1981) of their *recC*$^+$ and *recB*$^+$ regions (when subcloned into plasmids) did not agree with that obtained by other workers (Sasaki et al., 1982; Dykstra et al., 1984b). The strains (*E. coli* AB1157 or its derivatives) used by other groups contained only a single *Bam*HI chromosomal fragment hybridizing to the cloned *recBCD* region, which suffered no detectable rearrangements on cloning (Fujiyoshi et al., 1983; Dykstra et al., 1984b).

Dykstra et al. (1984b) cloned the *argA*$^+$ gene on a plasmid and used it to identify, and subsequently clone into a λ vector, an 18.5-kb *Bam*HI fragment of *E. coli* DNA coding for *thyA*$^+$, *recC*$^+$, *recB*$^+$, and *argA*$^+$. Subsequent transfer of the *Bam*HI fragment into plasmids showed that the cell was unharmed by a ~25-fold overproduction of ExoV. Fujiyoshi et al. (1983) independently cloned the 18.5-kb

*Bam*HI fragment directly into a cosmid vector, by selecting for the linked *thyA*⁺ gene, and subcloned it into plasmid pBR322. Both wild-type and mutant ExoV genes have been cloned directly by ligating *Bam*HI fragments of *E. coli* DNA into pBR322 and selecting for Thy⁺ or Thy⁺ Arg⁺ transformants (Ponticelli et al., 1985).

The restriction maps obtained by the three groups (all using derivatives of *E. coli* AB1157) are essentially identical (Amundsen et al., 1986; Dykstra et al., 1984b; Sasaki et al., 1982).

D. DNA Sequence Analysis of the *recBCD* Region

Emmerson and his colleagues have determined the DNA sequence of the *recC* (Finch et al., 1986d), *recB* (Finch et al., 1986b), and *recD* genes (Finch et al., 1986a) and flanking regions. They have also determined the sequence of the *ptr* gene (Finch et al., 1986c), which is located between *recC* and *recB*, and that of the *argA* gene (Brown et al., 1987). Combining these sequences with that of the *thyA* gene (Belfort et al., 1983) produces a 16.4-kb stretch of DNA of known sequence in the *recBCD* region, extending ~2.1 kb from the *Bam*HI site near *thyA* to the *Bam*HI site near *argA* (see Fig. 5).

Most of the published sequence data were obtained with DNA isolated from *E. coli* AB1157. The portions (*recC* and part of *ptr*) sequenced from the partially diploid strain mentioned above have been confirmed with AB1157 DNA (P. T. Emmerson, personal communication). The sequences of the *recB* and *recC* genes of strain AB1157 have also been determined by H. Yoshioka and Y. Takagi (personal communication), who found sequences identical to those published.

E. Genes and Transcripts

The *recB* gene had been identified, before the region was cloned, as the structural gene for ExoV by the thermolability in vitro of the enzyme from a *recB*(Ts) mutant (Tomizawa and Ogawa, 1972). Deletion, subcloning, and Tn*1000* insertion analysis of the cloned *recBCD* region (Hickson and Emmerson, 1981; Sasaki et al., 1982; Dykstra et al., 1984b; Amundsen et al., 1986), together with mapping of the transcripts in the region (Sasaki et al., 1982) have identified locations and transcription direction of the genes (Fig. 5).

Analysis of the DNA sequence of the *recBCD* region allowed confirmation and extension of these results. A unique reading frame was identified for *recC* and confirmed by peptide analysis of the amino terminus of the RecC polypeptide (Finch et al., 1986d). The start of the *recC* transcript was mapped, but no strong consensus promoter sequence was identified, consistent with the low abundance (~10 copies per cell) of RecBCD enzyme (Finch et al., 1986d).

The *recB* sequence has two possible initiation codons (in the same reading frame) but no obvious ribosome binding site sequence (Finch et al., 1986b). The larger of the predicted polypeptides is consistent with the size of the RecB polypeptide. Mapping of transcription start sites also shows that the *recB* mRNA starts in the region of the larger predicted polypeptide (Sasaki et al., 1982).

A single long open reading frame is present in the *recD* region, but it contains three possible initiation codons (Finch et al., 1986a). The initiation codon favored by Finch et al. (1986a) is preceded by a sequence which is homologous to the consensus ribosome binding site, but which overlaps the termination codon of the *recB* gene by one nucleotide. It also predicts a 67-kilodalton (kDa) polypeptide, while the other two possible initiation codons (which are not preceded by obvious ribosome binding sites) predict polypeptides of 56 and 55 kDa. The reported molecular size of purified *recD* polypeptide is 58 kDa (Amundsen et al., 1986) or 63 kDa (unpublished results cited by Finch et al., 1986a).

The carboxyl terminus of the sequence for *ptr* overlaps the start of the *recB* coding region by eight nucleotides (Finch et al., 1986c). *ptr*, *recB*, and *recD* are transcribed in the same direction, and (as discussed above) *recB* and *recD* may also overlap. It is thus possible that the three genes are all part of one operon. Evidence from Tn*1000* insertions indicates that *recB* and *recD* are part of one operon but that *recD* may possess a minor promoter of its own (Amundsen et al., 1986).

F. Gene Products

1. RecB and RecC polypeptides

RecBCD enzyme purified to homogeneity was reported to contain two polypeptides, with sizes of approximately 130 and 140 kDa (Goldmark and Linn, 1972; Eichler and Lehman, 1977). The long-felt temptation to equate these gene products with the *recB* and *recC* genes was elegantly justified by Emmerson's group (Hickson and Emmerson, 1981). As mentioned above, they cloned the *recB* and *recC* genes onto separate plasmids. The proteins encoded by the two plasmids were radiolabeled by the "maxicell" technique and examined by sodium dodecyl sulfate-polyacrylamide gel electrophoresis. The *recB*$^+$ plasmid coded for a ∼135-kDa polypeptide which was abolished by Tn*1000* insertions that rendered the plasmid RecB$^-$. The *recC* gene product was similarly identified as a polypeptide of ∼125 kDa.

2. RecD polypeptide

The work of Lieberman and Oishi (1973, 1974), described below, strongly suggested that ExoV had three subunits, but recombination-deficient mutants defective in ExoV activity fell into only two complementation groups. The recent discovery of mutations in *recD*, the structural gene for the third subunit, explains this apparent conundrum: mutants in *recD* are deficient in ExoV activity and in one of the fractions isolated by Lieberman and Oishi (1974), but remain recombination proficient and radiation resistant and hence were not isolated in the schemes that produced *recB* and *recC* null mutants (Chaudhury and Smith, 1984b; Amundsen et al., 1986).

Lieberman and Oishi (1973) found that incubation of RecBCD enzyme with 4 M NaCl caused dissociation of the enzyme and loss of enzyme activity, and that removal of the salt restored ExoV activity. Chromatography of the dissociated enzyme produced two fractions, neither of which showed ExoV activity, but which would regenerate ExoV activity when mixed (Lieberman and Oishi, 1974). The activity in one of the fractions, termed β, had a molecular size (as estimated by sucrose gradient sedimentation) of about 170 kDa and was undetectable in either *recB* or *recC* mutant extracts. The activity in the other fraction, termed α, had an estimated molecular size of about 60 kDa and was present in both *recB* and *recC* mutant extracts (Lieberman and Oishi, 1974). It was unclear whether the α fraction contained a subunit of the enzyme or a protein (present in the ExoV preparation) necessary for the assembly of ExoV from its dissociated RecB and RecC subunits (the presumed active components of the β fraction).

Cloning and overproduction of ExoV greatly facilitated purification of the enzyme and enabled the detection, by sodium dodecyl sulfate-polyacrylamide gel electrophoresis, of a third polypeptide (of molecular size ~60 kDa) in highly purified preparations of the enzyme (Dykstra et al., 1984a; Amundsen et al., 1986). The β fraction, prepared from the pure enzyme (Amundsen et al., 1986), consisted of the RecB and RecC polypeptides, as expected from previous work (Lieberman and Oishi, 1974). The only detectable polypeptide in the α fraction was the ~60-kDa polypeptide found in the pure enzyme; α must therefore be a component of the active enzyme. Deletion analysis of the 18.5-kb *thyA-argA* *Bam*HI fragment located the gene for the 60-kDa protein (named the *recD* gene) to a 1.6-kb fragment located between the *recB* and *argA* genes (Amundsen et al., 1986). As described above, screens for loss of two apparently unrelated phenotypes led to mutations in the *recD* gene (Chaudhury and Smith, 1984b; Biek and Cohen, 1986). Maxicell analysis showed loss of the 60-kDa polypeptide in extracts of *recD* mutants (Amundsen et al., 1986). The extracts also lacked α in vitro complementation activity. The *recD* mutants have, in vitro, none of the biochemical activities of ExoV, though DNA-dependent ATPase has not yet been measured. The reported reconstitution of ExoV activity from purified RecB and RecC polypeptides (Hickson et al., 1985) presumably results from slight contamination of the RecB polypeptide with RecD polypeptide (the specific activity of the reconstituted enzyme was only 10% of that of pure ExoV [Eichler and Lehman, 1977]). Mixtures of extracts of overproducing clones of the *recB* and *recC* genes also yielded ExoV activity; the *recB* clone used in those experiments contains the DNA segment coding for *recD* (Umeno et al., 1983).

In summary, the RecBCD enzyme of *E. coli* consists of three polypeptides, with molecular sizes of about 140, 130, and 60 kDa, coded for by the *recB*, *recC*, and *recD* genes, respectively. The *H. influenzae* ExoV also contains three polypeptides, of sizes very similar to those of the *E. coli* enzyme (Wilcox et al., *Fed. Proc.* **34**:515, 1975).

3. Properties of individual polypeptides

The β fraction, which contains the RecB and RecC polypeptides, shows DNA-dependent ATPase activity but little or no nuclease activity (Lieberman and Oishi, 1974). The α fraction, now known to be RecD polypeptide, showed no

ATPase activity, though both the RecB and RecD polypeptides have recently been shown to have an affinity for ATP (Julin and Lehman, 1987).

It has recently been reported that purified RecBC enzyme (i.e., purified RecBCD enzyme with the RecD subunit removed) retains single-strand DNA endonuclease activity, DNA-dependent ATPase activity, and DNA unwinding activity (K. M. Palas and S. R. Kushner, *Fed. Proc.* **46:**2210, 1987). The report seems at variance with the failure to detect unwinding in extracts of *recD* mutants, which contained active RecB and RecC polypeptides as shown by α-β complementation (unpublished data).

IV. BIOLOGICAL FUNCTIONS

A. Cell Viability

Up to ~80% of the microscopically visible cells in cultures of *recB* or *recC* mutants cannot form colonies on agar (Capaldo et al., 1974), while only ~20% of the cells in cultures of *recA* strains are nonviable. The nonviable *rec* cells in *recA*, *recB recC*, or *recA recB recC* strains (identified as those cells that cannot form colonies on agar) are of two types: those that cannot divide and are metabolically inactive, and those that are still capable of several rounds of division but will eventually cease division ("residually dividing" cells) and are metabolically active (Capaldo et al., 1974; Miller and Barbour, 1977). Nondividing *rec* cells are also defective in DNA synthesis. *recA* nondividing cells contain no DNA, while the nondividing cells of *recB recC* or *recA recB recC* strains contain normal amounts of DNA (Capaldo and Barbour, 1975). More single-strand nicks were found in old than in newly synthesized DNA of a *recA recB recC* strain, but no extra nicks were found in old DNA of *rec*[+] or *recB recC* strains (Capaldo and Barbour, 1975). It thus appears that the RecA protein and RecBCD enzyme are involved in the repair of DNA chain breaks that arise during cell growth (Capaldo and Barbour, 1975). The lack of such repair in *recA* strains leads to DNA degradation (presumably by RecBCD enzyme) or to the accumulation of nicks in the DNA in *recA recB recC* strains. In the absence of RecBCD enzyme, some repair must proceed by another *recA*-dependent pathway (probably via the RecF pathway; see Mahajan, this volume), but it is not sufficient to allow full viability.

recD mutants are fully viable (Chaudhury and Smith, 1984b) but have no detectable in vitro ExoV activities. It seems likely that their high viability results from repair via the RecF pathway of recombination: *recD recJ* double mutants are much more sensitive to UV light than either single mutant (Lovett et al., in press).

B. Degradation of Foreign or Damaged DNA

Foreign DNA entering *E. coli* is cut into large fragments by the restriction-modification system of the cells if the incoming DNA is not appropriately modified. The subsequent solubilization of that restricted DNA is principally by the RecBCD enzyme (Simmon and Lederberg, 1972).

recA mutants degrade their chromosomal DNA after UV irradiation. This "reckless degradation" is effected by the RecBCD enzyme (Willetts and Clark, 1969). Presumably, DNA with double-strand breaks, produced during the repair of the UV lesions, cannot be repaired by recombination in the absence of RecA protein and is instead degraded by ExoV. Alternatively, UV irradiation may induce, in wild-type but not *recA* mutant strains, an inhibitor of RecBCD enzyme, such as an SOS-controlled analog of the γ protein of phage λ.

C. Repair of Damaged DNA

In *recA*$^+$ cells, the RecBCD enzyme helps to repair the damaged DNA rather than degrading it, as noted above. A *recA*$^+$ *recB* mutant is considerably more UV sensitive than its *recB*$^+$ parent, especially if the RecF pathway of recombination and repair is inactivated. A *recB recF* mutant is almost as UV sensitive as a *recA* mutant (Taylor, unpublished data). This result implies that *E. coli* has two pathways of repair, both requiring RecA protein; one requires RecBCD enzyme and the other RecF function. The RecBCD enzyme seems to mediate repair of double-strand breaks (Wang and Smith, 1983, 1986a). The entry of RecBCD enzyme into such a break could be the start of a recombinational repair of the damage.

D. SOS Induction

Damage to the DNA of *E. coli* causes the induction of a set of genes, called SOS genes, involved in the repair of that damage. The damaged DNA has to be processed to produce the as yet unidentified inducer of the SOS response, which activates the RecA protein to aid in the autocatalytic digestion of LexA protein, the repressor of the SOS genes (Little and Mount, 1982). RecBCD enzyme is needed to convert damage from certain agents, such as the DNA gyrase inhibitor nalidixic acid, into an active inducer of SOS (McPartland et al., 1980).

The observation that nalidixic acid induces SOS in ‡ mutants, which are devoid of nuclease activity, led to the suggestion that unwinding (and hence the production of single-stranded DNA) was the function of ExoV important for SOS induction (Chaudhury and Smith, 1985). However, both the failure to find unwinding in vitro in extracts of ‡ mutants and the observation that *recB2109* (a recently isolated mutant which is unwinding proficient, nuclease deficient, and recombination deficient; S. Thibodeaux, personal communication) cannot induce SOS after nalidixic acid treatment (G. Braedt, personal communication) cast doubt on that hypothesis.

The SOS induction in ‡ mutants thus gives further evidence that these mutants retain some activity of RecBCD enzyme in the cell but leaves unclear the nature of that function.

E. Genetic Recombination

1. Recombination pathways

The RecBCD pathway of recombination is the major pathway of conjugal and transductional recombination acting in wild-type *E. coli* (see Mahajan, this volume) and is the only pathway in which Chi recombination hot spots are active (Stahl and Stahl, 1977). There is no evidence for any RecBCD-mediated recombination in the absence of RecA function.

2. RecBCD pathway and Chi sites

The extensive information amassed on both the genetic and biochemical properties of Chi sites, RecA protein, and RecBCD enzyme has prompted a number of models of genetic recombination.

One such model (Smith et al., 1984; Smith, 1987) imagines the recombination reaction to be initiated by RecBCD enzyme binding to a double-strand end (the transiently cut *cos* site in the case of $\lambda \times \lambda$ recombination) and subsequently unwinding the DNA, as shown in Fig. 2. The enzyme is imagined to typically be in the twin-loop mode of unwinding within the cell, there being insufficient SSB in the cell to support extensive loop-tail unwinding (Taylor and Smith, 1980a). The enzyme is pictured in the model as nicking (in the twin-loop mode) at A, but not at B, in vivo after encountering a Chi site. The continued leftward travel of the enzyme would then result in the extrusion of a 3'-terminated single strand of DNA, resulting either from the release of the loop held by the enzyme or from its return to the loop-tail mode of unwinding. This tail can then be utilized by RecA protein to form a D loop, which can then be processed into Holliday junctions, as shown in Fig. 4, and thence into complete recombinants. RecBCD enzyme may be involved in these latter steps as well.

The model is supported by the observation in vitro of the unwinding and Chi cutting reactions and is bolstered by the good correlation between the in vivo and in vitro properties of mutant RecBCD enzymes and mutant Chi sites. Recent very elegant in vivo experiments question the simple model presented above (Rosenberg, 1987). The experiments examined "patch" recombinants, such as would be produced by separating the two DNA molecules at the bottom of Fig. 4 by a "horizontal" cut. It was inferred genetically that the short single strand of information transferred came from the strand complementary to that (the one cut at Chi in vitro) predicted by the model discussed above, though the origin of the strand bias observed was not established. Rosenberg (1987) presents a number of alternate models to explain the data, including a model in which the function of Chi sites in vivo is to stimulate the postulated Holliday junction resolving activity of RecBCD enzyme.

V. CONCLUSIONS

Despite nearly 20 years of effort, ExoV continues to amaze and baffle us with its myriad activities. The general nuclease activities of the enzyme are used to degrade damaged or foreign DNA. It was unclear how such DNA degradation could aid in the recombination of DNA molecules. The unwinding activity is a much more plausible activity for a recombination enzyme. However, break-join recombination requires that some phosphodiester bonds be broken. The ‡ mutants initially suggested that none of the nuclease activities of RecBCD enzyme are essential for recombination. That idea was recently undermined both by the discovery that ‡ mutants may not function by the wild-type RecBCD pathway (Lovett et al., in press) and also by the discovery of a *recB* mutant that is unwinding proficient and nuclease deficient, but is deficient in recombination (S. Thibodeaux, personal communication). RecBCD enzyme has been shown, by both in vivo and in vitro tests, to be responsible for the enhancement of recombination caused by Chi hot spots. The ability of the enzyme to nick DNA, in vitro, in the vicinity of Chi sites suggests a very plausible way to initiate genetic recombination, though such a reaction has not yet been demonstrated in vivo.

Nuclease activities are also needed to convert D loops into Holliday junctions (Fig. 4) and to resolve Holliday junctions into completed recombinants. There is some evidence that the cutting of D loops may be effected by ExoV, but there is as yet no published evidence as to whether ExoV may resolve Holliday junctions.

ACKNOWLEDGMENTS. I am most grateful to the people mentioned in this review who so kindly communicated their unpublished results to me. I also thank the members of G. R. Smith's laboratory for their many helpful comments on the manuscript.

My research is supported by Public Health Service research grants GM31693 and GM32194 from the National Institutes of Health to Gerald R. Smith.

LITERATURE CITED

Akroyd, J., and N. Symonds. 1986. Localization of the gam gene of bacteriophage mu and characterization of the gene product. *Gene* **49:**273–282.

Amundsen, S. K., A. F. Taylor, A. M. Chaudhury, and G. R. Smith. 1986. *recD*: the gene for an essential third subunit of exonuclease V. *Proc. Natl. Acad. Sci. USA* **83:**5558–5562.

Barbour, S. D., and A. J. Clark. 1970. Biochemical and genetic studies of recombination proficiency in *Escherichia coli*. I. Enzymatic activity associated with *recB*+ and *recC*+ genes. *Proc. Natl. Acad. Sci. USA* **65:**955–961.

Belfort, M., G. Maley, J. Pedersen-Lane, and F. Maley. 1983. Primary structure of the *Escherichia coli thyA* gene and its thymidylate synthase product. *Proc. Natl. Acad. Sci. USA* **80:**4914–4918.

Benzinger, R., L. W. Enquist, and A. Skalka. 1975. Transfection of *Escherichia coli* spheroplasts. V. Activity of RecBCD nuclease in rec+ and rec− spheroplasts measured with different forms of bacteriophage DNA. *J. Virol.* **15:**861–871.

Biek, D. P., and S. N. Cohen. 1986. Identification and characterization of *recD*, a gene affecting plasmid maintenance and recombination in *Escherichia coli*. *J. Bacteriol.* **167:**594–603.

Brown, K., P. W. Finch, I. D. Hickson, and P. T. Emmerson. 1987. Complete sequence of the *Escherichia coli argA* gene. *Nucleic Acids Res.* **15:**10586.

Buttin, G., and M. Wright. 1968. Enzymatic DNA degradation in *E. coli*: its relationship to synthetic processes at the chromosome level. *Cold Spring Harbor Symp. Quant. Biol.* **33:**259–269.

Capaldo, F. N., and S. D. Barbour. 1975. DNA content, synthesis and integrity in dividing and non-dividing cells of rec− strains of *Escherichia coli* K12. *J. Mol. Biol.* **91:**53–66.

Capaldo, F. N., G. Ramsey, and S. D. Barbour. 1974. Analysis of the growth of recombination-deficient strains of *Escherichia coli* K-12. *J. Bacteriol.* **118**:242–249.

Chaudhury, A. M., and G. R. Smith. 1984a. *Escherichia coli recBC* deletion mutants. *J. Bacteriol.* **160**: 788–791.

Chaudhury, A. M., and G. R. Smith. 1984b. A new class of *Escherichia coli recBC* mutants: implications for the role of RecBC enzyme in homologous recombination. *Proc. Natl. Acad. Sci. USA* **81**:7850–7854.

Chaudhury, A. M., and G. R. Smith. 1985. Role of *Escherichia coli* RecBC enzyme in SOS induction. *Mol. Gen. Genet.* **201**:525–528.

Cheng, K. C., and G. R. Smith. 1984. Recombinational hotspot activity of Chi-like sequences. *J. Mol. Biol.* **180**:371–377.

Cheng, K. C., and G. R. Smith. 1987. Cutting of Chi-like sequences by the RecBCD enzyme of *Escherichia coli*. *J. Mol. Biol.* **194**:747–750.

Cheng, Y. S., and D. Zipser. 1979. Purification and characterization of protease III from Escherichia coli. *J. Biol. Chem.* **254**:4698–4706.

Cheng, Y. S., D. Zipser, C. Y. Cheng, and S. J. Rolseth. 1979. Isolation and characterization of mutations in the structural gene for protease III (*ptr*). *J. Bacteriol.* **140**:125–130.

Dower, N. A., and F. W. Stahl. 1981. χ activity during transduction-associated recombination. *Proc. Natl. Acad. Sci. USA* **78**:7033–7037.

Dykstra, C. C., K. M. Palas, and S. R. Kushner. 1984a. Purification and characterization of exonuclease V from *Escherichia coli* K-12. *Cold Spring Harbor Symp. Quant. Biol.* **49**:463–467.

Dykstra, C. C., D. Prasher, and S. R. Kushner. 1984b. Physical and biochemical analysis of the cloned *recB* and *recC* genes of *Escherichia coli* K-12. *J. Bacteriol.* **157**:21–27.

Eichler, D. C., and I. R. Lehman. 1977. On the role of ATP in phosphodiester bond hydrolysis catalyzed by the RecBC deoxyribonuclease of *Escherichia coli*. *J. Biol. Chem.* **252**:499–503.

Finch, P. W., A. Storey, K. Brown, I. D. Hickson, and P. T. Emmerson. 1986a. Complete nucleotide sequence of *recD*, the structural gene for the alpha subunit of Exonuclease V of *Escherichia coli*. *Nucleic Acids Res.* **14**:8583–8594.

Finch, P. W., A. Storey, K. E. Chapman, K. Brown, I. D. Hickson, and P. T. Emmerson. 1986b. Complete nucleotide sequence of the *Escherichia coli recB* gene. *Nucleic Acids Res.* **14**:8573–8582.

Finch, P. W., R. E. Wilson, K. Brown, I. D. Hickson, and P. T. Emmerson. 1986c. Complete nucleotide sequence of the *Escherichia coli ptr* gene encoding protease III. *Nucleic Acids Res.* **14**:7695–7703.

Finch, P. W., R. E. Wilson, K. Brown, I. D. Hickson, A. E. Tomkinson, and P. T. Emmerson. 1986d. Complete nucleotide sequence of the *Escherichia coli recC* gene and of the *thyA-recC* intergenic region. *Nucleic Acids Res.* **14**:4437–4451.

Friedman, E. A., and H. O. Smith. 1972a. An adenosine triphosphate-dependent deoxyribonuclease from *Hemophilus influenzae* Rd. I. Purification and properties of the enzyme. *J. Biol. Chem.* **247**: 2846–2853.

Friedman, E. A., and H. O. Smith. 1972b. An adenosine triphosphate-dependent deoxyribonuclease from *Hemophilus influenzae* Rd. III. Substrate specificity. *J. Biol. Chem.* **247**:2859–2865.

Friedman, E. A., and H. O. Smith. 1973. Production of possible recombination intermediates by an ATP-dependent DNAase. *Nature* (London) *New Biol.* **241**:54–58.

Fujiyoshi, T., M. Sasaki, K. Ono, T. Nakamura, K. Shimada, and Y. Takagi. 1983. Construction of a lambda packageable ColE1 vector which permits cloning of large DNA fragments: cloning of *thyA* gene of *Escherichia coli*. *J. Biochem.* (Tokyo) **94**:443–450.

Glickman, B. W. 1979. *rorA* mutation of *Escherichia coli* K-12 affects the *recB* subunit of exonuclease V. *J. Bacteriol.* **137**:658–660.

Glickman, B. W., H. Zwenk, C. A. van Sluis, and A. Rörsch. 1971. The isolation and characterization of an X-ray-sensitive ultraviolet-resistant mutant of *Escherichia coli*. *Biochim. Biophys. Acta* **254**: 144–154.

Goldmark, P. J., and S. Linn. 1970. An endonuclease activity from *Escherichia coli* absent from certain rec⁻ strains. *Proc. Natl. Acad. Sci. USA* **67**:434–441.

Goldmark, P. J., and S. Linn. 1972. Purification and properties of the RecBC DNase of *Escherichia coli* K-12. *J. Biol. Chem.* **247**:1849–1860.

Hermanns, U., and W. Wackernagel. 1977. The RecBC enzyme of *Escherichia coli* K12: premature cessation of catalytic activities *in vitro* and reactivation by potassium ions. *Eur. J. Biochem.* **76**:425–432.

Hickson, I. D., K. E. Atkinson, and P. T. Emmerson. 1982. Construction of recombinant lambda phages that carry the E. coli *recB* and *recC* genes. *Mol. Gen. Genet.* **185**:148–151.

Hickson, I. D., and P. T. Emmerson. 1981. Identification of the *Escherichia coli recB* and *recC* gene product. *Nature* (London) **294**:578–580.

Hickson, I. D., and P. T. Emmerson. 1983. Involvement of *recB* and *recC* genes of *Escherichia coli* in precise transposon excision. *J. Bacteriol.* **156**:901–903.

Hickson, I. D., C. N. Robson, K. E. Atkinson, L. Hutton, and P. T. Emmerson. 1985. Reconstitution of RecBC DNase activity from purified *Escherichia coli* RecB and RecC proteins. *J. Biol. Chem.* **260**:1224–1229.

Julin, D. A., and I. R. Lehman. 1987. Photoaffinity labelling of the RecBCD enzyme of *Escherichia coli* with 8-azidoadenosine 5′-triphosphate. *J. Biol. Chem.* **262**:9044–9051.

Karu, A. E., and S. Linn. 1972. Uncoupling of the RecBC ATPase from DNase by DNA crosslinked with psoralen. *Proc. Natl. Acad. Sci. USA* **69**:2855–2859.

Karu, A. E., V. MacKay, P. J. Goldmark, and S. Linn. 1973. The RecBC deoxyribonuclease of *Escherichia coli* K-12. Substrate specificity and reaction intermediates. *J. Biol. Chem.* **248**:4874–4884.

Karu, A. E., Y. Sakaki, H. Echols, and S. Linn. 1975. The gamma protein specified by bacteriophage gamma. Structure and inhibitory activity for the RecBC enzyme of *Escherichia coli*. *J. Biol. Chem.* **250**:7377–7387.

Kobayashi, I., M. M. Stahl, and F. W. Stahl. 1984. The mechanism of the Chi-cos interaction in RecA-RecBC-mediated recombination in phage λ. *Cold Spring Harbor Symp. Quant. Biol.* **49**:497–505.

Kooistra, J., G. D. Small, J. K. Setlow, and R. Shapanka. 1976. Genetics and complementation of *Haemophilus influenzae* mutants deficient in adenosine 5′-triphosphate-dependent nuclease. *J. Bacteriol.* **126**:31–37.

Kushner, S. R. 1974a. Differential thermolability of exonuclease and endonuclease activities of the RecBC nuclease isolated from thermosensitive *recB* and *recC* mutants. *J. Bacteriol.* **120**:1219–1222.

Kushner, S. R. 1974b. In vivo studies of temperature-sensitive *recB* and *recC* mutants. *J. Bacteriol.* **120**:1213–1218.

Lieberman, R. P., and M. Oishi. 1973. Formation of the *recB-recC* DNase by in vitro complementation and evidence concerning its subunit nature. *Nature* (London) *New Biol.* **243**:75–77.

Lieberman, R. P., and M. Oishi. 1974. The RecBC deoxyribonuclease of *Escherichia coli*: isolation and characterization of the subunit proteins and reconstitution of the enzyme. *Proc. Natl. Acad. Sci. USA* **71**:4816–4820.

Little, J. W., and D. W. Mount. 1982. The SOS regulatory system of *Escherichia coli*. *Cell* **29**:11–22.

Lundblad, V., A. F. Taylor, G. R. Smith, and N. Kleckner. 1984. Unusual alleles of *recB* and *recC* stimulate excision of inverted repeat transposons Tn10 and Tn5. *Proc. Natl. Acad. Sci. USA* **81**:824–828.

MacKay, V., and S. Linn. 1974. The mechanism of degradation of duplex deoxyribonucleic acid by the RecBC enzyme of *Escherichia coli* K-12. *J. Biol. Chem.* **249**:4286–4294.

MacKay, V., and S. Linn. 1976. Selective inhibition of the dnase activity of the RecBC enzyme by the DNA binding protein from *Escherichia coli*. *J. Biol. Chem.* **251**:3716–3719.

Marsden, H. S., E. C. Pollard, W. Ginoza, and E. P. Randall. 1974. Involvement of *recA* and *exr* genes in the in vivo inhibition of the RecBC nuclease. *J. Bacteriol.* **118**:465–470.

McPartland, A., L. Green, and H. Echols. 1980. Control of *recA* RNA in E. coli: regulatory and signal genes. *Cell* **20**:731–737.

Miller, J. E., and S. D. Barbour. 1977. Metabolic characterization of the viable, residually dividing and nondividing cell classes of recombination-deficient strains of *Escherichia coli*. *J. Bacteriol.* **130**:160–166.

Miller, R. V., and A. J. Clark. 1976. Purification and properties of two deoxyribonucleases of *Pseudomonas aeruginosa*. *J. Bacteriol.* **127**:794–802.

Muskavitch, K. M. T., and S. Linn. 1981. RecBC-like enzymes: exonuclease V deoxyribonucleases, p. 234–250. *In* P. D. Bower (ed.), *The Enzymes*, vol. 9. Academic Press, Inc., New York.

Muskavitch, K. M., and S. Linn. 1982. A unified mechanism for the nuclease and unwinding activities of the RecBC enzyme of *Escherichia coli. J. Biol. Chem.* **257:**2641–2648.

Oishi, M. 1969. An ATP-dependent deoxyribonuclease from *Escherichia coli* with a possible role in genetic recombination. *Proc. Natl. Acad. Sci. USA* **64:**1292–1299.

Oliver, D. B., and E. B. Goldberg. 1977. Protection of parental T4 DNA from a restriction exonuclease by the product of gene 2. *J. Mol. Biol.* **116:**877–881.

Orlosky, M., and H. O. Smith. 1976. Action of ATP-dependent DNase from *Hemophilus influenzae* on cross-linked DNA molecules. *J. Biol. Chem.* **251:**6117–6121.

Pacumbaba, R., and M. S. Center. 1975. Partial purification and properties of a bacteriophage T7 inhibitor of the host exonuclease V activity. *J. Virol.* **16:**1200–1207.

Ponticelli, A. S., D. W. Schultz, A. F. Taylor, and G. R. Smith. 1985. Chi-dependent DNA strand cleavage by RecBC enzyme. *Cell* **41:**145–151.

Prell, A., and W. Wackernagel. 1980. Degradation of linear and circular DNA with gaps by the RecBC enzyme of *Escherichia coli.* Effects of gap length and the presence of cell-free extracts. *Eur. J. Biochem.* **105:**109–116.

Rosamond, J., K. M. Telander, and S. Linn. 1979. Modulation of the action of the *recBC* enzyme of *Escherichia coli* K12 by Ca^{2+}. *J. Biol. Chem.* **254:**8646–8652.

Rosenberg, S. M. 1987. Chi-stimulated patches are heteroduplex, with recombinant information on the phage lambda r chain. *Cell* **48:**855–865.

Sakaki, Y. 1974. Inactivation of the ATP-dependent DNase of *Escherichia coli* after infection with double stranded DNA phages. *J. Virol.* **14:**1611–1612.

Sakaki, Y., A. E. Karu, S. Linn, and H. Echols. 1973. Purification and properties of the gamma-protein specified by bacteriophage lambda: an inhibitor of the host RecBC recombination enzyme. *Proc. Natl. Acad. Sci. USA* **70:**2215–2219.

Sasaki, M., T. Fujiyoshi, K. Shimada, and Y. Takagi. 1982. Fine structure of the RecB and *recC* gene region of *Escherichia coli. Biochem. Biophys. Res. Commun.* **109:**414–422.

Schultz, D. W., and G. R. Smith. 1986. Conservation of Chi cutting activity in terrestrial and marine enteric bacteria. *J. Mol. Biol.* **189:**585–595.

Schultz, D. W., J. Swindle, and G. R. Smith. 1981. Clustering of mutations inactivating a Chi recombinational hotspot. *J. Mol. Biol.* **146:**275–286.

Schultz, D. W., A. F. Taylor, and G. R. Smith. 1983. *Escherichia coli* RecBC pseudorevertants lacking chi recombinational hotspot activity. *J. Bacteriol.* **155:**664–680.

Sigal, N., H. Delius, T. Kornberg, M. L. Gefter, and B. Alberts. 1972. A DNA-unwinding protein isolated from *Escherichia coli*: its interaction with DNA and DNA polymerases. *Proc. Natl. Acad. Sci. USA* **69:**3537–3541.

Simmon, V. F., and S. Lederberg. 1972. Degradation of bacteriophage lambda deoxyribonucleic acid after restriction by *Escherichia coli* K-12. *J. Bacteriol.* **112:**161–169.

Smith, G. R. 1987. Mechanism and control of homologous recombination in *Escherichia coli. Annu. Rev. Genet.* **21:**179–201.

Smith, G. R., S. K. Amundsen, A. M. Chaudhury, K. C. Cheng, A. S. Ponticelli, C. M. Roberts, D. W. Schultz, and A. F. Taylor. 1984. Roles of RecBC enzyme and chi sites in homologous recombination. *Cold Spring Harbor Symp. Quant. Biol.* **49:**485–495.

Smith, G. R., and F. W. Stahl. 1985. Homologous recombination promoted by Chi sites and RecBC enzyme of *Escherichia coli. BioEssays* **2:**244–249.

Smith, H. O., and E. A. Friedman. 1972. An adenosine triphosphate-dependent deoxyribonuclease from *Hemophilus influenzae* Rd. II. Adenosine triphosphatase properties. *J. Biol. Chem.* **247:**2854–2858.

Stahl, F. W., I. Kobayashi, D. Thaler, and M. M. Stahl. 1986. Direction of travel of RecBC recombinase through bacteriophage lambda DNA. *Genetics* **113:**215–227.

Stahl, F. W., and M. M. Stahl. 1977. Recombination pathway specificity of Chi. *Genetics* **86:**715–725.

Tanaka, J. I., and M. Sekiguchi. 1975. Action of exonuclease V (the RecBC enzyme) on ultraviolet-irradiated DNA. *Biochim. Biophys. Acta* **383:**178–187.

Taylor, A. F., D. W. Schultz, A. S. Ponticelli, and G. R. Smith. 1985. RecBC enzyme nicking at Chi sites during DNA unwinding: location and orientation-dependence of the cutting. *Cell* **41:**153–163.

Taylor, A., and G. R. Smith. 1980a. Unwinding and rewinding of DNA by the RecBC enzyme. *Cell* **22:** 447–457.

Taylor, A., and G. R. Smith. 1980b. Unwinding and rewinding of DNA by Exonuclease V, p. 909–917. *In* B. Alberts (ed.), *Mechanistic Studies of DNA Replication and Genetic Recombination.* Academic Press, Inc., New York.

Taylor, A. F., and G. R. Smith. 1985. Substrate specificity of the DNA unwinding activity of the RecBC enzyme of *Escherichia coli. J. Mol. Biol.* **185:**431–443.

Templin, A., L. Margossian, and A. J. Clark. 1978. Suppressibility of *recA*, *recB*, and *recC* mutations by nonsense suppressors. *J. Bacteriol.* **134:**590–596.

Tomizawa, J., and H. Ogawa. 1972. Structural genes of ATP-dependent deoxyribonuclease of *Escherichia coli. Nature* (London) *New Biol.* **239:**14–16.

Umeno, M., M. Sasaki, M. Anai, and Y. Takagi. 1983. Properties of the *recB* and *recC* gene products of *Escherichia coli. Biochem. Biophys. Res. Commun.* **116:**1144–1150.

Van Dorp, B., R. Benne, and F. Palitti. 1975. The ATP-dependent DNAase from *Escherichia coli rorA:* a nuclease with changed enzymatic properties. *Biochim. Biophys. Acta* **395:**446–454.

Wang, T. C., and K. C. Smith. 1983. Mechanisms for *recF*-dependent and *recB*-dependent pathways of postreplication repair in UV-irradiated *Escherichia coli uvrB. J. Bacteriol.* **156:**1093–1098.

Wang, T. C., and K. C. Smith. 1986a. Inviability of *dam recA* and *dam recB* cells of *Escherichia coli* is correlated with their inability to repair DNA double-strand breaks produced by mismatch repair. *J. Bacteriol.* **165:**1023–1025.

Wang, T. C., and K. C. Smith. 1986b. Postreplicational formation and repair of DNA double-strand breaks in UV-irradiated *Escherichia coli uvrB* cells. *Mutat. Res.* **165:**39–44.

Weiss, B. 1981. Exodeoxyribonucleases of *Escherichia coli*, p. 203–231. *In* P. D. Bower (ed.), *The Enzymes*, vol. 9. Academic Press, Inc., New York.

Wiegand, R. C., K. L. Beattie, W. K. Holloman, and C. M. Radding. 1977. Uptake of homologous single-stranded fragments by superhelical DNA. III. The product and its enzymic conversion to a recombinant molecule. *J. Mol. Biol.* **116:**805–824.

Wilcox, K. W., and H. O. Smith. 1976a. Binding of the ATP-dependent DNase from *Hemophilus influenzae* to duplex DNA molecules. *J. Biol. Chem.* **251:**6122–6126.

Wilcox, K. W., and H. O. Smith. 1976b. Mechanism of DNA degradation by the ATP-dependent DNase from *Hemophilus influenzae* Rd. *J. Biol. Chem.* **251:**6127–6134.

Willetts, N. S., and A. J. Clark. 1969. Characteristics of some multiply recombination-deficient strains of *Escherichia coli. J. Bacteriol.* **100:**231–239.

Willetts, N. S., and D. W. Mount. 1969. Genetic analysis of recombination-deficient mutants of *Escherichia coli* K-12 carrying *rec* mutations cotransducible with *thyA. J. Bacteriol.* **100:**923–934.

Williams, J. G., T. Shibata, and C. M. Radding. 1981. *Escherichia coli* recA protein protects single-stranded DNA or gapped duplex DNA from degradation by RecBC DNase. *J. Biol. Chem.* **256:**7573–7582.

Williams, J. G., and C. M. Radding. 1981. Partial purification and properties of an exonuclease inhibitor induced by bacteriophage Mu-1. *J. Virol.* **39:**548–558.

Wright, M., G. Buttin, and J. Hurwitz. 1971. The isolation and characterization from *Escherichia coli* of an adenosine triphosphate-dependent deoxyribonuclease directed by rec B, C genes. *J. Biol. Chem.* **246:**6543–6555.

Chapter 8

Visualization of Recombination Reactions

Andrzej Stasiak and Edward H. Egelman

I. INTRODUCTION

A. Introduction to the Method: Electron Microscopy—
Its Advantages and Drawbacks

Electron microscopy is one of several powerful methods for studying recombination reactions, as has been shown in the analysis of in vitro recombination reactions promoted by the *Escherichia coli* RecA protein or by the bacteriophage T4 UvsX protein. This method allows one to study the low-resolution molecular structure of DNA-protein complexes involved in recombination. One can observe different states of DNA and protein components during various reactions. Moreover, electron micrographs taken during sequential stages of a recombination reaction allow for an understanding of the multistage processes involved and will contribute to the eventual molecular models of DNA recombination which will emerge in the future. However, the interpretation of recorded images requires caution and expertise since important information can be lost or distorted during the required steps of specimen preparation.

1. Routinely used preparation techniques—how samples are prepared and what one observes

Two main types of specimen preparation are currently used for viewing DNA-protein complexes by electron microscopy: shadowing and negative staining (Fig. 1). For a review and detailed preparation protocols, see Sogo et al. (1987).

In the shadowed preparations the DNA-protein complexes, after adsorption to supporting film, are washed, dehydrated, and coated with evaporated heavy metal (e.g., platinum or tungsten) to provide sufficient contrast for viewing in an electron microscope. With this technique one observes surface features due to a relief made of a thin metal film deposited on the specimen (Fig. 1, right panel).

In negatively stained preparations DNA-protein complexes, after adsorption to a supporting film, are embedded in a thin layer of a heavy metal salt solution (e.g., 2% uranyl acetate). This solution dries to an amorphous electron-dense glass which surrounds biological matter (i.e., proteins and nucleic acids) that is almost transparent to electrons. With the negative stain technique one observes in projection an image of the cavity within the electron-dense stain due to the presence of the biological material (Fig. 1, left panel).

Shadowed and negative stain techniques have different advantages and limitations so the results obtained should be used in concert to complement each other. While the resolution in shadowed specimens is relatively low (approximately 40 to 50 Å [1 Å = 0.1 nm] with conventional techniques), the high contrast and stability of the metal film to an electron beam allow a thorough search for objects of interest directly in an electron microscope. Negatively stained speci-

TEM   TEM

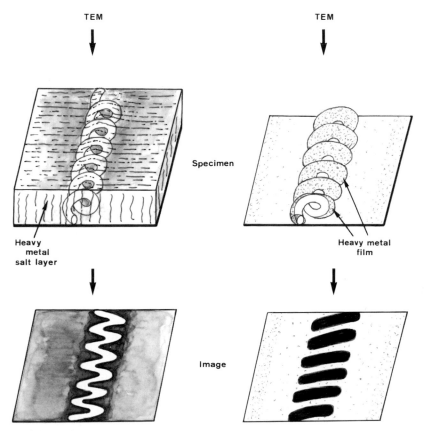

Specimen

Heavy
metal
salt layer

Heavy metal
film

Image

FIGURE 1. Schematic representation of helical protein-nucleic acid complexes after negative staining (left panel) and metal shadowing (right panel), and the images which these different preparations would give rise to in projection in a transmission electron microscope (TEM).

mens offer higher resolution (approximately 15 to 20 Å) since stain embedding is more accurate and precise than the deposition of metal film on the specimens which are being examined. However, these negatively stained specimens are rather sensitive to the electron beam. Even short viewing under an electron microscope causes not only the disintegration of the biological material (this happens in shadowed specimens as well) but also the rearrangement of the heavy metal salt layer, thus destroying the otherwise faithful imprints of biological objects which can be observed with this technique. Since the photographic recording of images requires much lower electron doses than extended direct observation in an electron microscope, achieving the full resolution possible in negative stain occurs at the price of taking pictures at once without an evaluation of the objects that are being recorded. Therefore, it is possible with negative staining to miss structures that occur rarely.

2. Distortions from native structure induced by routine specimen preparation

One aim of electron microscopy in biology is to observe such composite enzymatic reactions as DNA replication, transcription, and recombination during their in vivo occurrence, or at least under conditions established for in vitro assays. However, the conventional microscopy described so far requires that an original three-dimensional arrangement of molecules be brought into a two-dimensional form during adsorption to the supporting film. In addition to the interaction with the supporting film, distortions of observed molecules can be caused by specimen washing and drying. Fixation procedures used for stabilization of unstable protein-DNA complexes can also significantly change the structure of such complexes. Further, image contrast is generated in both shadowing and negative staining by heavy metals and not by the objects themselves.

We will describe here some typical examples of artifacts (distortions from the native structure) induced by routine steps of specimen preparation.

Different types of RecA protein-DNA complexes provide a textbook example of differential sensitivity to the routine fixation procedure of glutaraldehyde protein cross-linking (Sogo et al., 1987; Flory et al., 1984).

Figure 2 shows negatively stained preparations of three different types of RecA protein-DNA complexes prepared in an unfixed state (a, c, and e) and fixed with glutaraldehyde (b, d, and f). Complexes of RecA protein with double-stranded DNA (RecA-dsDNA complexes) (a and b) formed in the presence of ATP-γ-S (slow hydrolyzable analog of ATP) seem not to be changed by glutaraldehyde cross-linking, having the same contour length (for a given size DNA) and the same internal structure as unfixed complexes. Complexes of RecA protein with single-stranded DNA (RecA-ssDNA complexes) [phage ϕX174 (+) strand DNA] formed in the absence of nucleotide cofactor (c and d) show an intermediate sensitivity toward glutaraldehyde fixation; while the contour length of the complexes is not visibly affected, the fine regular internal structure of these complexes is blurred by fixation. A high sensitivity toward glutaraldehyde fixation is shown by RecA-ssDNA complexes formed in the presence of ATP (e and f); fixation decreases the contour length of these complexes by about 35% and also changes their internal structure.

This example shows that even in closely related biological objects the effects of glutaraldehyde fixation can be very different. Of course, one function of such fixation can be to maintain the integrity and not necessarily the structure of certain complexes. Unfixed RecA-DNA complexes formed in the presence of ATP would completely dissociate during the routine washing and drying procedures required for shadowed preparations. The negative stain technique does not require the washing and dehydration of specimen before embedding in stain; thus, if the stain components do not induce changes of unfixed biological structure, one has a good chance to observe and compare unfixed and fixed preparations of even such unstable objects as RecA-ssDNA complexes formed in the presence of ATP. The use of different preparative techniques (which do not involve fixation) can reveal possible artifacts introduced by fixation.

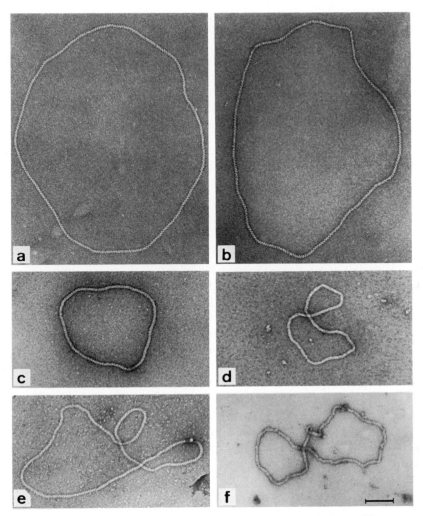

Figure 2. Electron micrographs of negatively stained preparations of three different types of RecA-DNA complexes. (a, c, and e) Unfixed preparations; (b, d, and f) fixed with glutaraldehyde. (a and b) RecA-dsDNA complexes formed in the presence of ATP-γ-S; (c and d) RecA-ssDNA complexes formed in the absence of nucleotide cofactor; (e and f) RecA-ssDNA complexes formed in the presence of ATP. Bar, 0.1 μM. (Based on Sogo et al., 1987, Fig. 2.)

Another example of an artifactual change from native structure is the shrinkage of dsDNA molecules in the routinely used ethanol dehydration or in air drying of specimens. This shrinkage (up to 25%) is probably caused by the adsorbed DNA molecules undergoing a transition from the hydrated B-DNA structure to the dehydrated A-DNA structure (Vollenweider et al., 1978). In B-DNA the axial rise per base pair (as measured in X-ray fiber diffraction) is 3.4 Å, while in A-DNA it is only 2.6 Å (Arnott and Chandrasekaran, 1981). Depending on how strongly DNA molecules are restricted in their movement by the adsorption to the supporting film and the extent of drying, the resulting shrinkage of DNA molecules can vary in different preparation procedures (Vollenweider et al., 1978).

3. Frozen-hydrated microscopy: a potentially nondistorting preparation technique

The preceding examples of artifacts lead to the question of how to prepare samples without distortions from the native structure. The frozen-hydrated technique (for review, see Stewart and Vigers, 1986; Dubochet et al., 1988), in which molecules are rapidly frozen in a thin film of reaction buffer and observed frozen in the vitreous ice layer formed, seems to be an ideal way to avoid artifacts: specimens are examined without any adsorption, fixation, staining, or drying. This increasingly popular technique obtains image contrast by taking advantage of the difference in electron density between an unstained biological object and the slightly less dense vitreous ice layer surrounding it. Since both the biological object (whether protein or nucleic acid) and the ice are quite transparent to electrons, the images must be formed by means of a phase-contrast mechanism generated by defocusing the microscope. Like negative stain, the images formed correspond to the projection of a three-dimensional mass distribution onto two dimensions. Unlike negative stain, the image contrast is generated directly by the structure and not by a heavy metal stain. Unfortunately, contrast in unstained frozen-hydrated preparations is rather poor, and the directly recorded (or observed) image can suffer from artifacts introduced by the phase-contrast optics; computed restoration of the images is frequently needed for the proper interpretation of fine detail. However, even without image analysis some general features of DNA-protein complexes seen directly in frozen-hydrated preparations can be compared with features visible in conventional negatively stained or shadowed preparations.

In the case of bundles of RecA-dsDNA complexes formed in the presence of ATP-γ-S (see below), we see the same helical parameters (a mean 95-Å pitch and a mean twist of about 6.15 units per turn) in frozen-hydrated preparations (E. H. Egelman and X. Yu, manuscript in preparation) as we do in conventional techniques. Further, the variable pitch and helical disorder present in conventional images of free RecA-dsDNA filaments in this state are also present in frozen-hydrated images, indicating that negatively stained or shadowed preparations of this complex are fairly faithful representations of the solution structure. It is always possible (although extremely unlikely) that the very different preparative procedures for negative stain, shadowing, and frozen-hydrated microscopy all yield the same artifactual structures, which differ in some significant way from the solution structure of these molecules. However, completely independent structural evidence, obtained from solution X-ray scattering (Egelman and Stasiak, 1986), provides an external check that this is not the case, at least for RecA-dsDNA complexes formed in the presence of ATP-γ-S.

4. Preparations with artifacts: recovery of usable information

Different methods of electron microscopy preparation should be checked to determine which step of preparation may cause a substantial change in the structure being studied. A knowledge of the extent of distortion can allow the use of distorting preparation steps, if required, for special purposes. For example, to follow the progress of a recombination reaction, it is necessary to remove the unbound phosphocreatine kinase, used for ATP regeneration, from the RecA-DNA complexes and naked DNA molecules. This purification is achieved by passing the reaction mixture over a Sepharose column, but the RecA-ssDNA-ATP complexes must be fixed with glutaraldehyde to stay together during this column purification. In observing pictures of such purified complexes (see section III), one must remember that after fixation these complexes are 67% shorter than they were before fixation, and they have lost the visible 95 Å pitch helical structure (Fig. 2e and f). Thus, images of these fixed structures may be misleading from the viewpoint of the substructure of the RecA-DNA complexes, but at the same time these images are invaluable in following the gross structure and topology of the recombination reaction.

B. Introduction to Reactions of RecA Protein with DNA Substrates In Vitro

General genetic recombination within bacterial cells is a highly complicated reaction. Many different enzymes, such as ssDNA-binding proteins, endonucleases, polymerases, topoisomerases, and ligases, help specialized recombination enzymes like RecA execute their function.

The aim of studying in vitro recombination reactions is ultimately to provide a detailed description and explanation of the complete in vivo reaction. Unfortunately, a recombination reaction occurring within a bacterial cell is very difficult to study by biochemical or microscopic techniques. Although thin sections from a bacterial cell can be analyzed under an electron microscope, many other processes involving bacterial chromosomes are also taking place concurrently with a recombination reaction, making the assignment of specific functions to different enzymes involved in the recombination reaction virtually impossible.

It is much easier to study and observe under an electron microscope recombination reactions which occur in vitro, and much effort has been invested in this regard. The first systems which were used involved the pairing of ssDNA molecules and D-loop formation, both promoted by RecA protein (McEntee et al., 1979; Weinstock et al., 1979; Shibata et al., 1979). These reactions do not involve a complete strand exchange. More appropriate model systems were then developed in which the complete strand exchange between DNA molecules occurs and does not depend upon the release of topological stress, cutting, or ligation (DasGupta et al., 1980; West et al., 1981; Cox and Lehman, 1981a). In such in vitro systems a recombination reaction can be mediated exclusively by specialized recombination enzymes like RecA protein without the help of auxiliary proteins. The use of simple model systems allows one to approach directly the crucial questions in recombination, namely, the mechanisms of homologous recognition and strand exchange. While the mechanism and the function in recombination of certain auxiliary proteins such as DNA polymerase I and DNA ligases are relatively well understood, the mechanism by which RecA brings together two DNA molecules at a region of homology has not yet been determined.

1. Possible mechanisms of homologous recognition

There are two diametrically different stereochemical possibilities for the recognition of homology between interacting dsDNA molecules or between a dsDNA and an ssDNA molecule. The first possibility, which can be called "strand separation before pairing," is based on the formation of intermolecular Watson-Crick hydrogen bonding between potentially complementary strands belonging to different DNA molecules (Radding, 1978). This possibility requires that strands in duplex DNA molecules be separated (at least locally) by recombination-promoting proteins to allow the formation of intermolecular base pairing.

The second possibility, called "pairing before strand separation," is based on the postulated additional hydrogen bonds that can be formed between the major grooves of two duplex DNA molecules (McGavin, 1971, 1977) or between the major grooves of one duplex DNA molecule and the equivalent side of a homologous single strand (Howard-Flanders et al., 1984, 1987). Recombination-promoting proteins acting in accordance with this second possibility would have to bring two interacting molecules into contact along their "major groove sides." Only when additional hydrogen bonds are formed between interacting molecules (homologous pairing) would the recombination-promoting proteins catalyze the second step of the recombination reaction, strand switching. This second step would result in the separation of the originally base-paired strands and the formation of a new duplex region containing the exchanged strands. In this model, the strand switch would involve a rotation of participating bases, allowing the conversion of the additional hydrogen bonds which are formed initially into a regular Watson-Crick hydrogen bonding (McGavin, 1977; Howard-Flanders et al., 1984, 1987).

Detailed studies of general genetic recombination reactions may soon answer the question of whether pairing occurs before or after strand separation. It is also possible that both mechanisms can operate within living cells. The mechanism of homologous recognition will be discussed again in the context of interpreting micrographs of recombining molecules (see section III).

II. STRUCTURE OF COMPLEXES OF DNA WITH RECOMBINATION PROTEINS STUDIED BY ELECTRON MICROSCOPY

A. Structure of RecA-DNA Complexes

The RecA-mediated recombination reaction, even in simple in vitro model systems, is a highly dynamic process driven by ATP hydrolysis. RecA monomers probably undergo movements and conformational changes (Howard-Flanders et al., 1984) during the sequential steps of binding to DNA, establishing homologous contacts between recombining DNA molecules, and promoting unidirectional branch migration. The analysis of micrographs from strand exchange reactions, in which RecA subunits may coexist in different states and interact with different types of DNA molecules, will be greatly facilitated by the knowledge gained from the observation of steady-state and equilibrium complexes formed with dsDNA and ssDNA.

1. RecA-DNA complexes formed in the presence of ATP-γ-S

Structure of dsDNA complexed with RecA protein. The first analyses of RecA-DNA interactions by electron microscopy were stimulated by biochemical evidence that in the presence of ATP-γ-S (an ATP analog hydrolyzed by RecA at a much lower rate than ATP [Weinstock et al., 1981]) RecA protein forms high-molecular-weight complexes which contain DNA (West et al., 1980). The electron micrographs clearly showed that RecA protein, in the presence of ATP-γ-S, cooperatively interacts with dsDNA. With limited amounts of RecA protein, some DNA molecules are covered by a seemingly continuous RecA polymer (covering up to several thousand base pairs), while other DNA molecules in the reaction mixture remain naked (West et al., 1980; see also Fig. 3). This "cooperativity" results from nucleation being rate limiting: the rate at which stable RecA polymer nuclei form is much less than the rate at which subunits add to the end of an existing polymer or nucleus. The complex of RecA protein with dsDNA formed in the presence of ATP-γ-S (RecA–dsDNA–ATP-γ-S) has been the central object of structural study in RecA-DNA interactions because this complex appears the most stable. The complex possesses several remarkable properties.

Figure 3 shows three nicked circular DNA molecules with 4,961 base pairs (bp) each. One molecule (c) is completely covered with RecA protein, the second (b) is partially covered, and the third (a) is naked. The length difference between the naked molecule and the RecA-covered one is striking. In fact, RecA protein in the presence of ATP-γ-S stretches DNA to 150% of its native length, increasing the average axial spacing between base pairs from 3.4 Å in the native B-DNA structure to 5.1 Å in RecA-complexed DNA (Stasiak et al., 1981; Dunn et al., 1982). This unprecedented protein-induced stretching of dsDNA indicates that DNA structure is greatly modified by RecA protein. A precise characterization of the RecA-induced DNA structure, determined crystallographically or spectroscopically, will be crucial for understanding the mechanism of recombination.

The second striking feature seen in shadowed preparations of the RecA–dsDNA–ATP-γ-S complex is the very regular cross striations (Fig. 3). In negatively stained preparations these striations are replaced by a visible zigzag pattern (Fig. 2a and b), which indicates that the RecA–dsDNA–ATP-γ-S complex has a structure of a single helix: in projection (negative stain image) this helix gives a zigzag pattern, while in a surface view (shadowed preparation) one sees only the top half of the striations (Fig. 1). From the shadowed images it is apparent

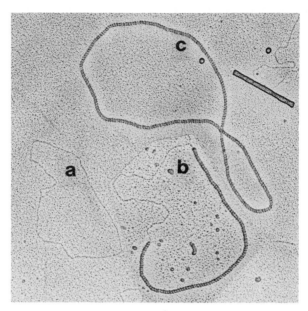

FIGURE 3. Electron micrographs of shadowed preparations of nicked circular dsDNA molecules (each 4,961 bp). The molecules are shown naked (a), partially covered by RecA (b), and completely covered by RecA (c). Since all molecules came from the same reaction mixture (in which there was a large excess of RecA), it is striking that some molecules are completely covered while complex formation has not yet begun on other molecules. This shows that under these conditions the rate of nucleation of polymers is much slower than the subsequent rate of growth of polymers. Notice the greatly increased length of the RecA-covered molecule (c) compared with the naked DNA molecule (a). Samples were prepared as described by Egelman and Stasiak (1986).

that the RecA–dsDNA–ATP-γ-S complex forms a right-handed helical filament. The striations or helical turns of the complex are so pronounced that it is possible to count them along the whole circular DNA molecule. The dsDNA used here had 4,961 bp, and the number of striations counted (each striation being one helical turn) was always either 267 or 268, indicating that 18.6 bp of DNA were present in each visible turn of the RecA-DNA complex (DiCapua et al., 1982). From these micrographs alone it is impossible to conclude anything directly about the helicity of DNA in the complex with RecA protein. Both the shadowed and negatively stained preparations show a clear protein coat on the DNA, while the DNA itself is hidden within the complex. Simple structural arguments suggest that if RecA protomers are bound continuously along the same "phase" of the DNA helix (e.g., along the minor groove), then the stretched DNA helix in the complex would have to adapt its pitch to the observable pitch of the filamentous RecA polymer. In this case 18.6 bp of the DNA contained within one turn of the RecA–dsDNA–ATP-γ-S complex would just form one complete helical repeat of this stretched DNA helix with 95-Å pitch and 18.6 bp per turn. Alternatively, RecA protomers could be bound to the DNA not in phase of DNA helix, with RecA polymer running transversally across the grooves of the DNA and approaching alternatively the major and minor groove of the DNA helix. In this case the DNA helix would not follow the visible helicity of the filamentous RecA polymer; the DNA could just have then one of many thinkable DNA structures which will result in having 18.6 bp per axial distance of 95 Å. (We will continue to use the conventional polymer nomenclature that an individual RecA molecule within a filament is a protomer, while the term monomer is reserved for a free molecule in solution.)

Figure 4 shows the method of electron microscopy used to determine the DNA helicity within a RecA-DNA complex (Stasiak and DiCapua, 1982; Stasiak et al., 1983). This method is based on the observation that circular DNA molecules can be completely covered with RecA protein only when the DNA is nicked.

When nicked and subsequently ligated (covalently closed relaxed) DNA molecules are used as substrates for RecA polymerization, only a part of each molecule is covered, and the molecules give the visual impression that high torsional stress accumulates in what was previously relaxed DNA (Fig. 4e). Circular DNA molecules are collapsed, with a folded and supercoiled appearance. The RecA-covered region is folded back on itself, showing two helical filaments coiling about each other, while the unreacted DNA region is visible as a highly supercoiled DNA tail. The fact that previously relaxed DNA molecules are converted to a highly supercoiled form by even partial covering with RecA protein proves that the RecA protein dramatically changes the DNA helicity.

Interestingly, all covalently closed relaxed DNA molecules incubated with an excess of RecA protein were covered to the same extent, indicating that the binding of RecA protein proceeded to the point where the energy of torsional deformation (imposed by the topological constraint) equals the energy to be gained by further RecA-DNA binding. Thus, in such partially covered molecules the DNA within the RecA-covered regions has one well-defined structure while the DNA in the uncovered supercoiled tails has a different well-defined structure.

The DNA linking number (α) in such DNA molecules (which consist of two different domains) can be described by the equation

$$\alpha = \frac{f_c n + f_t(N - n)}{360°} \tag{1}$$

where f_c is the composite rotation angle per base pair in the complexed DNA portion, n is the number of base pairs complexed with RecA, f_t is the composite rotation angle per base pair in the uncovered DNA portion (visible as the supercoiled tail), and N is the total number of base pairs in the molecule. Since both the RecA-covered region and the naked portion are apparently supercoiled, f_c and f_t are composite rotation angles because they arise as a result of two orders of helicity. In the case of f_t (the uncovered tail) the first order of helicity is most likely the native B-DNA helix, although strong topological stress can distort native B-DNA structure or induce transition to a different DNA structure, while the second order of helicity is a tight positive supercoiling.

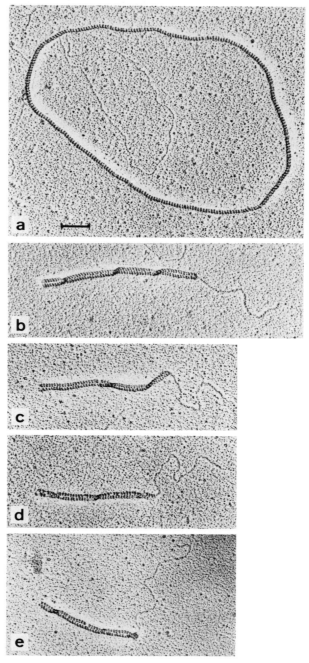

FIGURE 4. Shadowed micrographs of a nicked circular DNA molecule (4,961 bp long) which is completely covered with RecA protein (a) and of topoisomers of different linking numbers which have been covered by RecA protein to the point where further binding is precluded by the accumulated torsional stress. The linking number α is (b) 432, (c) 445, (d) 459, and (e) 472. Note that in b through e the rodlike striated structures arise from the side-by-side aggregation of two segments of the same circular RecA-DNA complex. These pairs of helical filaments are supercoiling with a pitch of about one supercoil for every 36 turns of the RecA helix (see text). The thin filaments which emerge from the paired RecA filaments are the highly positively supercoiled segments not covered by RecA. (From Stasiak and DiCapua, 1982.)

In the case of f_c (the covered region) the first order of helicity is given by the specific DNA structure in the complex with RecA protein, while the second order of helicity is visible as a regular left-handed coiling arising from the side-by-side interaction of RecA-covered complexes. This second-order helicity can be ignored since it occurs with only one turn per 36 striations or 670 bp.

When one complexes DNA molecules which have different degrees of supercoiling with RecA protein, one observes that the more negatively super-coiled the DNA, the longer are the portions of the molecules covered with RecA protein (Fig. 4e, d, c, and b). This observation demonstrates that the DNA in the complex with RecA protein undergoes a substantial unwinding and that the naked tail accumulates compensatory positive supercoils. It is apparent that if one could use more negatively supercoiled DNA (having an even lower linking number), one could obtain molecules which would be completely covered with RecA protein. A covalently closed supercoiled DNA molecule which would allow complete covering with RecA protein would have the same linking number (α) as the number of DNA turns in nicked molecules completely covered with RecA protein (Fig. 4a). For such molecules ($n = N$), equation 1 would be reduced to $\alpha = f_c N/360°$. Thus, by knowing the linking number (α) and the number of base pairs (N) in molecules which could be completely covered with RecA protein, one can determine the helicity of DNA in the RecA-DNA complex.

By forming RecA-DNA complexes with preparations of circular DNA with the same number of base pairs but with varying amounts of negative supercoils, it is possible to relate the extent of RecA covering to the linking number of the DNA molecules used (Fig. 4 and 5). Unfortunately, a precise measurement of the number of supercoils is possible only when the supercoil density (the ratio of the number of supercoils to the number of primary DNA turns in a molecule) is low, and this limits the experimentally accessible range. However, since there is a linear relation between the linking number of a DNA molecule and the number of base pairs covered (from equation 1, $\Delta\alpha = \Delta nc$, where c is a constant), a linear extrapolation can be used to find the value of α when $n = N$ (the linking number needed for completely covered molecules). A linear extrapolation from the experimental results to $n = N$ for the DNA molecules used (4,961 bp) indicates that the DNA linking number which would allow complete covering by RecA protein is 267 ± 19 (Fig. 5). As has already been shown, the number of visible striations on completely covered nicked circular DNA of the same size is also about 267. This clearly indicates that DNA within the RecA-DNA complex follows the helical arrangement of RecA protomers in the complex, with 18.6 bp per 95-Å-pitch helical turn. The average rotation per base pair in the RecA complex is thus 19.4° (360°/18.6), significantly untwisted from the approximately 34.3° per base pair in B-DNA.

Another electron microscopy approach led to a similar conclusion. Chrysogelos et al. (1983) allowed well-characterized supercoiled DNA molecules to react with RecA protein to such an extent that the uncomplexed part of the DNA was relaxed. The extent of the RecA-DNA complex necessary for relaxation of the uncomplexed DNA suggested that one supercoil is lost for every 31.5 bp bound by a RecA protein. This indicates that RecA protein changes three 10.5-bp turns of

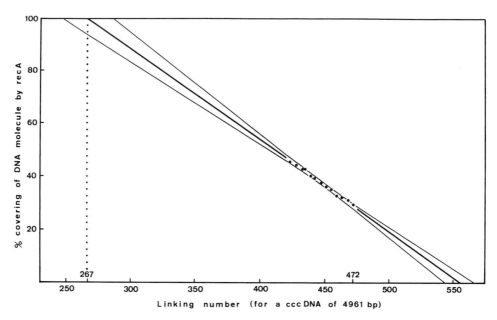

FIGURE 5. Relationship between the linking number of circular DNA molecules (4,961 bp long) and the extent of their covering by RecA protein (Stasiak and DiCapua, 1982). The extrapolation of experimentally determined points (obtained from micrographs such as those shown in Fig. 4b through e) to complete covering of the DNA shows that the linking number for completely covered covalently closed DNA molecules (267 ± 19) is consistent with the number of visible helical striations in completely covered nicked DNA molecules (267). The thin lines represent the 95% confidence interval. (From Stasiak et al., 1983.)

B-DNA into two 16-bp turns of an untwisted DNA. Since under the conditions of specimen preparation used by these authors one helical repeat of the RecA-DNA complex occurred for every 17 ± 1 bp, their results also show that the helicity of a complexed DNA is the same as the visible helicity of the RecA polymer formed on DNA.

The electron microscopy results have been confirmed by a physicochemical analysis of ethidium bromide fluorescence of partially RecA-covered DNA molecules. This method reveals that the helicity of DNA in the RecA-DNA complex is about 18.8 bp per turn (Dombroski et al., 1983).

In conclusion, different laboratories have determined the DNA helical parameters within the RecA–DNA–ATP-γ-S complex, showing that the DNA structure is changed by its interaction with RecA protein to about 18.6 bp per turn and an average axial spacing between base pairs of 5.1 Å, resulting in a DNA helix with a pitch of about 95 Å (see Table 1). These results strongly indicate that RecA protomers are continuously bound along the same "phase" of the DNA helix. Indeed, DiCapua and Müller (1987) demonstrated that the DNA within the RecA–dsDNA–ATP-γ-S complex is characteristically protected from dimethyl sulfate methylation as if RecA protomers would be bound continuously along the minor groove of the DNA helix. Interestingly, the structure of DNA which has

TABLE 1
Comparison of the native B-DNA structure with the DNA
structure in the RecA–dsDNA–ATP-γ-S complex

Type of structure	No. of base pairs per turn	Avg rise/base pair (Å)	Pitch (Å)	Rotation angle/base pair (°)
B-DNA	10.4	3.4	35.4	+34.6
RecA–dsDNA–ATP-γ-S complex	18.6	5.1	95	+19.4

been cooperatively intercalated with ethidium bromide seems to have similar parameters to DNA complexed with RecA protein. The average axial rise per base pair in such ethidium bromide-intercalated DNA is 5.1 Å, and the number of base pairs per turn is about 16 (Sobell, 1980). This similarity suggests that RecA protein might stretch DNA by intercalation of its aromatic residues between DNA bases. On the other hand, ethidium bromide intercalation leads to a nearest-neighbor exclusion, with ethidium bromide present only between every other base pair. The axial rise in this structure is thus alternating 3.4-Å rises with 6.8-Å rises, yielding an average rise of 5.1 Å (Sobell, 1980). Such a mode presents a problem for RecA intercalation, since the stoichiometry in this complex is 3 bp per RecA protomer (see below). This would mean that one RecA protomer would be involved in two intercalations, the next RecA protomer in one, the next in two, etc., breaking the symmetry between every RecA subunit. An alternative which cannot be excluded is that RecA protein intercalates between every base pair, resulting in a local axial rise per base pair of 6.8 Å. Since what is measured in an electron microscope, 5.1 Å per base pair, is the axially projected rise per base pair (with respect to the RecA filament axis), and not the local separation along the phosphate backbone, there is not necessarily any conflict. The separation between base pairs along the phosphate backbone would be equal to 5.1 Å only if the DNA were located exactly on the RecA filament axis and not coiling about it. If there actually were intercalation between every base pair, the path of hydrogen bonding between base pairs in the DNA would need to be at a radius of about 15 Å from the RecA filament axis for the observed projected rise per base pair to be 5.1 Å. Given that the position of DNA within the complex has not yet been determined (see below), this possibility cannot be excluded.

Arrangement of RecA protein molecules in the RecA–DNA–ATP-γ-S complex. A basic question about the structure of RecA-DNA complexes is the number of RecA protomers per helical turn of the complex. Since the molecular weight of the RecA monomer is known to be 37,824 (Sancar et al., 1980), one would need to know only the molecular mass of one turn of the RecA-DNA complex to tell how many RecA monomers it contains. Scanning transmission electron microscopy is a very sensitive method of molecular mass determination for unstained biological specimens. The number of electrons elastically scattered by objects of known molecular mass such as tobacco mosaic virus (coprepared on the same grid) can be directly compared with the scattering of RecA-DNA filaments to calculate the mass per turn of the RecA-DNA complex.

From such an experiment DiCapua et al. (1982) calculated the molecular weight of a 100-Å-long stretch of a RecA–DNA–ATP-γ-S complex to be 264,000. Taking the value of 94.7 Å as the average distance between two successive striations of RecA complex, a molecular weight of 250,000 can be obtained for one helical turn of a RecA-DNA complex. From this value one must subtract the mass of 18.6 bp of DNA to find that the RecA protein component of one turn has a mass of 238,000. Thus, the experiment using scanning transmission electron microscopy shows slightly less than 6.3 RecA monomers per turn of the helix.

Computed image analysis of electron micrographs of negatively stained side-by-side assemblies of RecA-DNA complexes has yielded the highest precision in determining the number of RecA protomers per turn of the RecA helix. The mean value found for the RecA–dsDNA–ATP-γ-S complex was 6.151 ± 0.003 units per turn, and that for the RecA–ssDNA–ATP-γ-S complex was 6.00 ± 0.03 units per turn (Egelman and Stasiak, 1988).

Given that there are 18.6 bp per turn, the results from both scanning transmission electron microscopy and image analysis are consistent with a RecA-DNA stoichiometry of one RecA per 3 bp, while other integral stoichiometries such as one RecA per 2 or 4 bp are excluded.

A stoichiometry of one RecA per 3 bp in the RecA–DNA–ATP-γ-S complex has also been demonstrated by using ethidium fluorescence assays and sedimentation analysis of complexes (Dombroski et al., 1983).

Three-dimensional image reconstruction of RecA–dsDNA–ATP-γ-S complexes. New perspectives in the investigation of the structure of RecA–DNA–ATP-γ-S complexes were opened by image analysis of negatively stained preparations (Egelman and Stasiak, 1986). Since a helix consists of identical subunits, related by translations along and rotations around a common helical axis, a single image of a helix in projection contains many different views of the component subunit, the different views being projections of the subunit from different angles. These different projections can be mathematically combined to synthesize the three-dimensional density distribution of the component subunit and the polymeric filament (DeRosier and Klug, 1968). Figure 6 shows such a three-dimensional image reconstruction of the RecA–dsDNA–ATP-γ-S complex which reveals that the complex forms a deeply grooved right-handed helical filament with close to six asymmetric units (presumably RecA protomers) per turn of the helix. Although boundaries between RecA protomers are not visible in a low-resolution reconstruction (one would have to trace the polypeptide chain to unambiguously separate protomers within the filament), it is clear that part of each monomer contributes to a continuous helical backbone at a radius of about 20 to 25 Å. The rest of each subunit is visible as an axially elongated domain which is about 50 Å long and appears to protrude axially and radially from the backbone. As all protruding domains point in the same axial direction, the RecA-DNA filament is clearly polar. Since the dsDNA in the complex has no net polarity, the polarity of a RecA filament on dsDNA must be a random consequence of the initial nucleation. The springlike appearance of the RecA-DNA filament corresponds

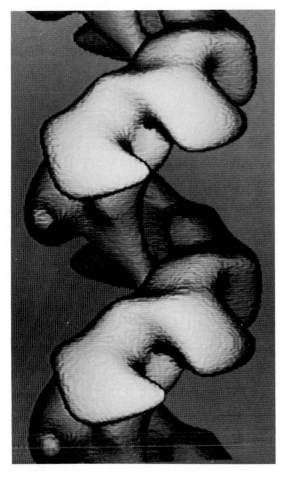

FIGURE 6. Three-dimensional reconstruction of a RecA filament which was formed on dsDNA with ATP-γ-S (Egelman and Stasiak, 1986). This particular filament had parameters which were each near one extreme: about 5.1 RecA protomers per turn of the helix (the average being about 6.15 under these conditions), a pitch of 100 Å (the average being about 95 Å), and a diameter of about 100 Å (the average being about 115 Å).

quite well with the observed elasticity of the complex. Although the RecA polymer has about 50% more mass per unit length than an F-actin polymer, RecA is about 10 times more flexible than F-actin, as determined by quantifying the fluctuations in curvature of molecules (Egelman and Stasiak, 1986). In the isolated single filaments which were analyzed, by the number of subunits per turn ranged from 5.1 to 6.8 while the pitch ranged from 90 to 100 Å (Egelman and Stasiak, 1986).

Three-dimensional image reconstruction of RecA–dsDNA–ATP-γ-S complexes in negative stain yields little information about the position of DNA within the complex. Since DNA accounts for less than 5% of the mass of the complex and since it is not known whether the DNA in the complex sequesters or excludes stain, the position of DNA in the three-dimensional reconstructions is ambiguous. However, a simple consideration of path length dictates that DNA must be located near the filament axis. The maximum observed axial rise per base pair in free DNA is 7.6 Å (Arnott and Chandrasekaran, 1981), and since this value is quite close to what one would predict purely from stereochemical grounds, we will use this value as the upper limit for local DNA base separation within the complex. Taking the measured axial rise of 5.1 Å and 18.6 bp per turn, one finds the maximal radius for a DNA helical path to be 17 Å. Since the RecA-DNA complex is at least 100 Å in diameter, the DNA must be located within a small volume around the filament axis.

Simple structural arguments of contact equivalency between each RecA monomer and 3 bp of DNA demand that the DNA within the RecA filament cannot be a coiled coil with two orders of helicity. Since a RecA–DNA–ATP-γ-S complex has a helical symmetry in which consecutive RecA monomers are related to each other by a rotation around the complex axis by 58° (360°/6.2) coupled with a translation along the filament axis by 15.3 Å (95 Å/6.2), consecutive triplets of base pairs must also be related to each other by the same simple rotation and translation.

A great deal of information has also emerged from the analysis of aggregates of RecA-DNA filaments (shown in Fig. 7), whose formation is induced by raising the cation concentration. Initial structures were found by raising the Mg^{2+} levels to 6 mM or greater (Egelman and Stasiak, 1986, 1988), at which point a highly cooperative transition occurs between filaments which are basically free in solution and bundles of filaments in which specific bonds are being made between neighboring RecA filaments. The transition is so sharp in this aggregation process that virtually no bundles are seen in an electron microscope at 5 mM Mg^{2+}, whereas large numbers are found at 6 mM. Similar structures can be induced by other cations, such as Ca^{2+} and Ba^{2+} (Egelman and Yu, in preparation). Because the interactions between filaments in the most ordered bundles which have been analyzed involve specific bonds, with the precise alignment of filaments, rather than a nonspecific aggregation, the mean twist of the RecA filaments can be discerned with great precision. If filaments had a mean twist of exactly six RecA protomers per turn, filaments could form bonds in a parallel manner and maintain equivalent specific contacts along their length. Instead, a supercoiling of the helical RecA filaments around a common axis is observed in the bundles, arising from the deviation from exactly six protomers per turn. An analysis of three different forms of aggregates (two forms contained six filaments, while the third form contained three filaments) showed that the mean twist of RecA protein in these bundles was 6.151 ± 0.003 protomers per turn (Egelman and Stasiak, 1988). All of these bundles contained RecA filaments which had initially been formed on linear dsDNA. A stoichiometry of 3 bp of DNA per RecA protomer would lead to 18.5 bp per turn of the RecA helix, in excellent agreement with the mean 18.6 bp per turn found in circular nicked dsDNA molecules covered with RecA protein.

Three-dimensional reconstructions of these bundles have revealed that not all RecA protomers are in the same conformation: filament-filament interactions within these bundles have broken the structural equivalence of each RecA

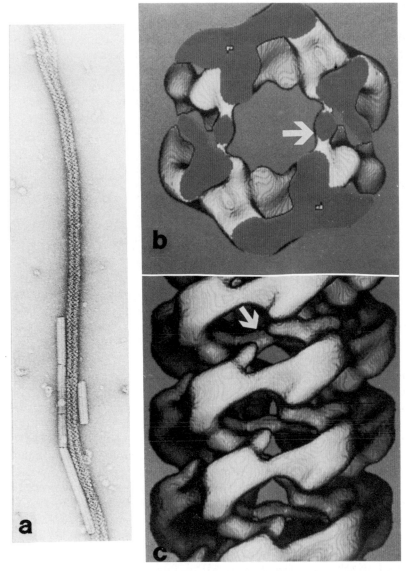

FIGURE 7. (a) Negatively stained bundle of six RecA-dsDNA filaments (prepared with ATP-γ-S). Tobacco mosaic virus particles, shown next to the bundle, are used as magnification standards. The six filaments are supercoiled about a hollow core, with each filament axis about 100 Å in radius from the bundle axis. (b and c) Three-dimensional reconstruction of such a bundle (Egelman and Stasiak, 1988). The side view (c) shows a 300-Å length of the bundle, containing slightly more than three turns of the RecA helix within each of the six component filaments of the bundle. Each filament makes a long-pitch supercoil about the axis of the bundle, and the pitch of the supercoil is about 3,600 Å. In the axial view (b) one can see the hollow core. The main bond between the filaments (arrow in b and c) is a bridge of density formed by a large rotation of at least one of every six RecA protomers which are present per turn of each filament.

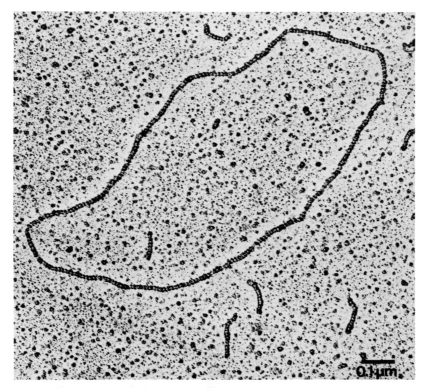

FIGURE 8. Electron micrograph of a shadowed preparation of RecA on an ssDNA circle, prepared with ATP-γ-S (Koller et al., 1983). One can see the clear right-handed 95-Å-pitch striations on the top surface of the filament. (From Koller et al., 1983.)

protomer and induced large conformational changes to occur within certain RecA protomers. While the relationship between these conformational changes and those that occur within an entire filament as a result of ATP hydrolysis is not yet known, it is quite possible that these local conformational changes are related to those which occur during the strand exchange reaction mediated by RecA protein. In this instance, however, these conformational changes are induced by filament-filament interactions, and there is no evidence that the RecA-mediated strand exchange involves more than one RecA polymer.

Complexes of RecA protein with ssDNA formed in the presence of ATP-γ-S. The structure of RecA–ssDNA–ATP-γ-S complexes seems to be very similar to that of the corresponding complexes formed with dsDNA. Figure 8 shows a 5,386-nucleotide-long ssDNA molecule completely covered with RecA protein. This complex, like the RecA–dsDNA–ATP-γ-S complex, has prominent striations arising from a helical structure with a pitch of about 95 Å and a diameter of about 100 Å (Koller et al., 1983). This suggests that the filament structure is probably not very different from the complex on dsDNA, also having about six RecA protomers per helical turn of the complex. The naked ssDNA under native conditions adopts a rather compact form with many internal base-paired regions and would appear on such micrographs as a small globular structure.

Although the structure of the RecA-ssDNA complex is slightly less regular than that of the RecA-dsDNA complex, making it difficult to count every striation in the circular molecule, the overall circumference of a complexed ssDNA circle can be measured and it corresponds to an average axial rise (with respect to the RecA filament axis) per nucleotide of about 5 Å. Since the pitch in regular stretches is about 95 Å, one helical turn of the complex must contain between 18 and 19 nucleotides (Koller et al., 1983). These results suggest that the stoichiometry of RecA protein to DNA in the RecA–ssDNA-ATP-γ-S complex is one RecA protomer per three nucleotides (in complexes with dsDNA the stoichiometry is one RecA monomer per 3 bp).

The increased disorder found in the RecA–ssDNA–ATP-γ-S complex (in comparison with the same complex on dsDNA) can be explained kinetically. We have stated that electron microscopy of RecA–dsDNA–ATP-γ-S complexes shows that nucleation must be rate limiting. With ssDNA this no longer appears to be the case: the relative rate of nucleation increases substantially, while the rate of polymer addition probably remains the same. Pugh and Cox (1987), using solution kinetic techniques, arrived at the same conclusion. Thus, multiple independent initiation events probably occur on a single ssDNA molecule, at random separations which in general will not be a multiple of three nucleotides apart. The subsequent polymer addition reactions will fail to completely fill the gaps between initiation sites (in two of three cases), leaving one- and two-nucleotide-long gaps in the protein coat. An even larger disorder would result from the possibility that even if two nucleation events occur an integral multiple of three nucleotides apart, one or more RecA monomers would be unable to be inserted as one growing polymer reaches the second nucleation site, producing gaps of RecA protomers in the helix. The less regular structure of the RecA–ssDNA–ATP-γ-S complex makes three-dimensional image reconstruction much more difficult than it is for complexes on dsDNA. However, unlike dsDNA, which has no net polarity, complexes on ssDNA should display a polarity with respect to the underlying polarity of the single strand. By allowing exonuclease III-treated linear dsDNA molecules (producing long, protruding, single-stranded tails with 5′ ends) to react with RecA protein it has been possible to relate the polarity of the three-dimensional reconstruction of the complex to the polarity of RecA-covered ssDNA (A. Stasiak, E. H. Egelman, and P. Howard-Flanders, *J. Mol. Biol.*, in press).

One main point should be stressed when drawing conclusions from the structure we have described for the RecA–DNA–ATP-γ-S complexes, namely, that both ssDNA and dsDNA can be induced by RecA protein to adopt highly stretched related structures with the same twist and axial extension. Without interaction with the RecA protein, dsDNA (B-DNA) and ssDNA (compact coils) differ in their structures so strongly that recognition of homology which requires extended contacts between homologous or complementary bases is difficult to achieve. This must be seen in light of the fact that recombination reactions mediated by RecA protein are usually initiated by the interaction of ssDNA with a homologous duplex region. As discussed later (section III), the observation of recombination intermediates suggests that both interacting DNA molecules (ssDNA molecule and dsDNA molecule) can be enveloped coaxially into the same RecA filament, thus allowing the formation of extended contacts between strands being kept in the same type of structure.

2. RecA-DNA complexes formed in the presence of ATP

Complexes of DNA with RecA protein formed in the presence of ATP-γ-S are convenient to study because of their stability and regularity. However, the complete recombination reaction can occur only in the presence of ATP, and that is why complexes formed with ATP are potentially more interesting, since they are more physiological. Unfortunately, these complexes rapidly hydrolyze ATP and are unstable. ATP hydrolysis and the accumulation of ADP lead to a conformational change of RecA monomers and eventually to a dissociation of RecA protein from DNA (Menetski and Kowalczykowski, 1985). To overcome this, an ATP regeneration system and an excess of ATP can be used to obtain reproducible preparations of such complexes which show a regular internal structure (Fig. 2e). Contour length measurements of RecA-ssDNA-ATP complexes formed on defined circular ssDNA molecules do show small variations in length, probably reflecting local dissociations of RecA protein monomers from the complex. However, the longest molecules (presumably those with the least net dissociation) closely resemble complexes formed in the presence of ATP-γ-S (Flory et al., 1984; Sogo et al., 1987). Image analysis from the most ordered stretches of RecA-ssDNA-ATP complexes gives a low-resolution mass distribution which is nearly identical with that obtained from complexes formed in the presence of ATP-γ-S (Egelman and Stasiak, 1986).

The above findings allow us to conclude that if RecA protomers have either a bound ATP or a bound ADP-P_i the structure of the complex is identical to that of complexes formed in the presence of ATP-γ-S. Therefore, the data in Table 1 which describe parameters of the RecA–DNA–ATP-γ-S complex most likely apply to RecA-DNA-ATP complexes as well.

As discussed later, RecA protein in the presence of ATP-γ-S is able to mediate only the first step of a recombination process, namely, the pairing reaction, while RecA protein in the presence of ATP mediates pairing, the strand switch, and unidirectional branch migration. Since the structure of the RecA-DNA-ATP complex is so similar to the structure of complexes formed in the presence of ATP-γ-S, and since only the ATP complex can mediate the entire strand exchange reaction, it is probable that a conformational change of RecA protomers driven by the ATP hydrolysis and their eventual dissociation from DNA are required to traverse further steps of the recombination reaction.

An interesting property of RecA protein in the presence of ATP is the difference in its binding to ssDNA and dsDNA. At low magnesium concentrations (1 to 2 mM) both ssDNA and dsDNA molecules are complexed with RecA protein into filaments, while at higher magnesium concentrations (5 to 15 mM) only ssDNA is incorporated into complexes (Iwabuchi et al., 1983; Stasiak et al., 1984). Since the in vitro RecA-mediated recombination reaction occurs only in the higher magnesium concentrations, during the initiation phase of the reaction contacts occur between ssDNA regions covered with RecA protein and regions of naked dsDNA (Cox and Lehman, 1982; Stasiak et al., 1984).

As already shown in section I, the structure of the RecA-ssDNA complex formed in the presence of ATP is greatly changed by glutaraldehyde fixation. Fixation causes a shrinkage of these complexes to an average axial rise per nucleotide of 3.4 Å from about 5 Å in the absence of fixation, and their helical structure becomes less visible (Flory et al., 1984; Koller et al., 1983). This effect of fixation should be kept in mind while analyzing samples which must be fixed with glutaraldehyde for preparative reasons. One of the most characteristic properties of RecA-ssDNA-ATP complexes is the necessity of saturation as a condition for maximal recombination activity. ssDNA must be completely covered with RecA protein without leaving uncomplexed cruciforms or other forms of internal pairing, which would otherwise impede the homologous aligning reaction with dsDNA or block branch migration (Muniyappa et al., 1984; Tsang et al., 1985).

3. RecA-DNA complexes formed in the absence of nucleotide cofactor or in the presence of ADP

RecA protein will form filamentous complexes with ssDNA in the absence of nucleotide cofactors. In the presence of 1 mM Mg^{2+}, complexes form readily and appear as right-handed helical filaments (Fig. 2c) with a pitch of about 64 Å (Stasiak and Egelman, 1986). The axial rise per nucleotide (with respect to the filament axis) in these complexes is about 2.1 Å (Koller et al., 1983), indicating about 30 nucleotides per helical turn. Williams and Spengler (1986) reported very similar parameters for this type of complex.

The stoichiometry of RecA-ssDNA complexes formed in the absence of a nucleotide cofactor is less clear than it is for the complexes formed in the presence of ATP-γ-S and ATP. Structural evidence, however, is consistent with a conserved stoichiometry of one RecA protomer for every three nucleotides. The mean cross-sectional area of such filaments is about 250% greater than that found for the ATP and ATP-γ-S filaments, as one would expect if the stoichiometry were conserved. If the stoichiometry is conserved, it is still a question whether these complexes can remain intact while undergoing the large conformational change induced by ATP or ATP-γ-S which would result in the stretching and winding of preformed filaments. If the stoichiometry of RecA protein to DNA is changed during the interconversion of complex types, it is more likely that RecA protomers dissociate from DNA and rebind to a new conformation.

When RecA-ssDNA complexes are formed in the presence of 1 to 2 mM ADP, the complexes are very similar to the filaments formed in the absence of nucleotide cofactor, suggesting either that the ADP is not bound or that the binding of ADP does not lead to a significantly different conformation from that of the filament formed in the absence of nucleotide cofactors (Stasiak and Egelman, unpublished data).

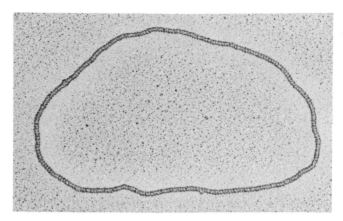

FIGURE 9. Electron micrograph of a shadowed preparation of the complex formed by *P. mirabilis* RecA protein on a nicked dsDNA circle in the presence of ATP-γ-S.

4. Different states of RecA-DNA complexes: coexistence and interchange leading to full activity

One may generalize that there are two main structural classes of RecA-DNA complexes: a class of stretched complexes formed with ATP or ATP-γ-S and either ssDNA or dsDNA, in which the average axial rise per nucleotide is 5.1 Å; and a compact class formed on ssDNA in the absence of nucleotide cofactor or in the presence of ADP, in which the average axial rise per nucleotide is 2.1 Å. Complexes formed in the presence of ATP hydrolyze ATP, with a concomitant conformational change in the complex. Since ATP is an essential cofactor in the recombination process, and since the change from ATP to ADP cofactors makes a dramatic difference in the filament conformation, it is quite probable that large local conformational changes are required for strand transfer and directional branch migration during the recombination process (Dunn et al., 1982; Stasiak and Egelman, 1986). It is possible that these conformational changes involve switching between a stretched and compressed form of the filament.

B. Structure of Nucleoprotein Complexes Formed by Proteins Analogous to RecA

Despite the fact that genes analogous to the *E. coli recA* gene have been found in many other procaryotes, only in a few cases has the appropriate gene product been isolated (West et al., 1983; Lovett and Roberts, 1985). Figure 9 shows a micrograph of a *Proteus mirabilis* RecA protein complexed with DNA in the presence of ATP-γ-S. The complex appears indistinguishable from complexes formed with the *E. coli* RecA protein.

Complexes formed with phage T4 UvsX gene product, which is essential for DNA repair and general recombination, are clearly distinguishable from the RecA protein complexes with DNA (Griffith and Formosa, 1985). Although very regular helical turns are also visible, the dimensions seem to be different from those of RecA-DNA complexes. One helical turn of a UvsX-dsDNA-ATP filament has a pitch of 140 Å and contains about 42 bp and 9 to 12 UvsX protein protomers (Griffith and Formosa, 1985). Interestingly, the complexes of UvsX protein formed with ssDNA and dsDNA appear to have the same structural parameters (Griffith and Formosa, 1985). This was also a characteristic of complexes formed by RecA protein with ssDNA and dsDNA in the presence of ATP-γ-S. As already discussed, the ability to bring ssDNA and dsDNA into the same configuration seems to be a prerequisite for homologous alignment between DNA molecules which would otherwise have a different structure.

Although only three examples have been discussed, it is striking that all three proteins act stoichiometrically, and not catalytically, by building a helical polymer along DNA. This polymer serves as a recombinational scaffold (Griffith and Formosa, 1985) within which the actual DNA pairing and strand exchange process between two homologous DNA molecules can occur (see section III).

C. SSB Protein and Its Interaction with DNA and RecA-DNA Complexes

Although the RecA protein alone is able to mediate the entire strand exchange reaction, some auxiliary proteins, like single-strand binding protein (SSB protein), can significantly improve the yield of the ssDNA-dsDNA strand exchange reaction (Cox and Lehman, 1982; Muniyappa et al., 1984; Stasiak et al., 1984). The same auxiliary protein, SSB, has little effect on the ds-gapped–dsDNA strand exchange reaction (West et al., 1982).

1. Structure of SSB-ssDNA complexes

Detailed electron microscopic studies of complexes between SSB protein and ssDNA revealed two different morphological forms of these complexes in low salt. One type of complex, which forms at a lower SSB-to-DNA ratio, has a beaded appearance and measures about 20% of the length of the corresponding protein-free dsDNA; the second type, which forms at a high SSB-to-DNA ratio, has a smooth appearance and measures about 40% of the length of the corresponding dsDNA (Griffith et al., 1984).

Under conditions suitable for a strand exchange reaction (high ionic strength), only beaded complexes form (Fig. 10), regardless of the SSB-to-DNA ratio used (Griffith et al., 1984).

Biochemical studies using fluorescence and gel retardation assays have shown two main binding modes of SSB to ssDNA: in one mode about 65 nucleotides interact with each functional tetramer of SSB, while in the second mode only about 33 nucleotides interact with the SSB tetramer (Lohman et al., 1986). The biochemical assays also showed, in agreement with electron microscopy, that at high salt conditions (levels comparable to strand exchange reaction conditions) only one binding mode is observed. This mode has 65 nucleotides per

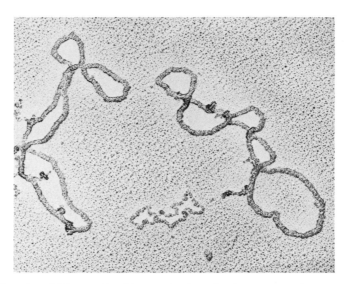

FIGURE 10. Complex of SSB protein with ssDNA (short circular "beaded" filament in the middle of the micrograph) coprepared with RecA-ssDNA-ATP complexes (the two long twisted circular filaments). ssDNA circles of the same length were used in both complexes.

tetramer, presumably giving rise to short beaded filaments (Lohman et al., 1986; Bujalowski and Lohman, 1987).

2. SSB protein: probable action at two different steps in strand exchange reactions

Figure 2e shows an extended RecA-ssDNA-ATP complex. To obtain such extended molecules, it is necessary to keep the magnesium concentration low (1 to 2 mM). At high magnesium concentrations, such as 10 mM, i.e., conditions required for a strand exchange reaction, a complex formed with the same ssDNA molecule is several times shorter and rather inactive in the strand exchange reaction. The addition of SSB protein restores the extended configuration of RecA-ssDNA-ATP complexes even at higher magnesium concentrations and makes such complexes highly reactive in strand exchange (Flory et al., 1984). Apparently, RecA protein alone is not able to open the secondary structures of ssDNA stabilized by higher magnesium concentrations, while the SSB protein has the ability to cooperate with RecA protein in opening such duplex regions (Tsang et al., 1985). As already discussed, ssDNA must be saturated with RecA protein to be active in strand exchange; SSB clearly helps to achieve such saturation since it helps to unfold ssDNA.

The second step of a reaction in which SSB protein plays a role is in binding to the displaced strand and sequestering it from repeated rounds of the reaction (Flory et al., 1984; Register et al., 1987; see also description of recombination intermediates in section III).

D. RecBCD Enzyme and Its Interaction with DNA

A RecA protein can initiate strand exchange between two DNA molecules only when a single-stranded region of one molecule is homologous with a duplex region of the second molecule. Such situations cannot occur between two fully duplex DNA molecules. However, proteins which possess a helicase activity can facilitate the pairing reaction by separating strands of at least one molecule. The RecBCD enzyme is a multisubunit complex with helicase, exonuclease, and endonuclease activities (see Taylor, this volume). Mutants lacking this enzyme are recombination deficient (Clark, 1973; Mahajan, this volume).

The helicase activity of the RecBCD enzyme is unique. The enzyme binds to one strand of linear dsDNA and moves in a 3' to 5' direction, coupled with hydrolysis of ATP. The "front" of the enzyme takes in ssDNA more rapidly than the "back" of the enzyme passes it through, producing a moving single-stranded loop that grows. ssDNA leaving the "back" of the enzyme can freely anneal with the complementary strand of the same duplex, but only to the point where the enzyme is in contact with one end of the ssDNA loop. As a result of this, a sister loop appears on the strand not interacting with RecBCD helicase, and this grows and moves with the same speed as the loop actively produced by the helicase. ssDNAs in sister loops cannot anneal freely, since the loops behave like topologically closed and unlinked circles, which restricts their annealing ability.

When a moving loop on a strand bound by a RecBCD enzyme approaches the Chi sequence (5'-GCTGGTGG-3'), the enzyme executes its endonucleolytic activity by cutting the loop at sites located four to six nucleotides to the 3' side of the Chi sequence (Taylor et al., 1985). After the Chi-induced cutting, the enzyme continues its action of DNA unwinding.

An isolated helicase present in catalytic amounts does not have long-lasting effects on DNA, since a helicase moves like a wedge along DNA and strands can anneal immediately behind the moving enzyme. However, in the presence of proteins having a high affinity for ssDNA, like SSB protein and RecA protein, the effect of a helicase action on DNA can be long lasting (see Fig. 2 in the review by Taylor, this volume). Within cells, strands separated by a helicase would probably be bound by these proteins and could therefore be active in a pairing reaction with a homologous duplex region of another molecule.

The fact that Chi sequences are recombination hot spots, dependent upon *recA* and *recBCD*, in *E. coli* suggests that RecA protein does indeed use RecBCD enzyme-separated strands to initiate a strand transfer reaction.

III. DIRECT VISUALIZATION OF IN VITRO RECOMBINATION REACTIONS

A. Main Model System: Reaction between Completely Homologous Linear dsDNA and Circular ssDNA

The complete in vitro strand exchange reaction between a linear dsDNA molecule and a homologous circular ssDNA molecule, which leads to the

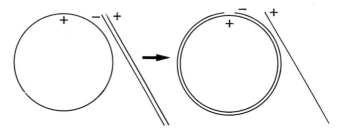

FIGURE 11. Schematic representation of the strand exchange reaction between a linear dsDNA molecule and a completely homologous circular ssDNA molecule, which gives rise to a nicked circular dsDNA molecule and a displaced ssDNA molecule.

formation of a nicked circular dsDNA and a displaced ssDNA linear molecule (Fig. 11), was the first example of homologous pairing followed by a full-length strand exchange accomplished with a purified RecA protein (Cox and Lehman, 1981a). This reaction is easy to assay by DNA gel electrophoresis, since intermediates and circular dsDNA products migrate slower than the linear double-stranded substrate. Electron microscopic analysis of such in vitro reactions requires conditions under which interacting molecules do not form large aggregates. A high concentration of ATP and an ATP regeneration system reduce aggregation and allow RecA protein to stretch ssDNA to achieve complete saturation (Stasiak et al., 1984). The ssDNA fully saturated with RecA protein is the active presynaptic filament. The addition of SSB protein can serve similar purposes in facilitating the complete saturation of the ssDNA by RecA protein (Stasiak et al., 1984; Register et al., 1987). The use of conditions which avoid aggregation does not affect the molecular mechanism of the reaction, as shown by the appearance in DNA gel electrophoresis of identical patterns of intermediates from reactions made under aggregating and nonaggregating conditions (Stasiak, unpublished data). The yield of a reaction under conditions which do not promote aggregation is lower, which is most likely a consequence of the reduced probability of molecules interacting when they are in the total volume of the reaction mixture in contrast to the effectively reduced volume of aggregates.

1. Microscopy at the nucleoprotein level

Reaction mediated by RecA only. The analysis of samples from different time points in the strand exchange reaction reveals sequential stages in this reaction (Stasiak et al., 1984; Stasiak and Egelman, 1987; Register et al., 1987). Figure 12 shows molecules fixed, by cross-linking proteins to DNA with glutaraldehyde, during the early phase of the reaction (up to 5 min). In the micrographs the ssDNA circles are always completely covered with RecA protein, while the linear dsDNA molecules have no bound RecA protein. These images are consistent with the fact, discussed earlier, that RecA protein does not bind to dsDNA under conditions suitable for strand transfer.

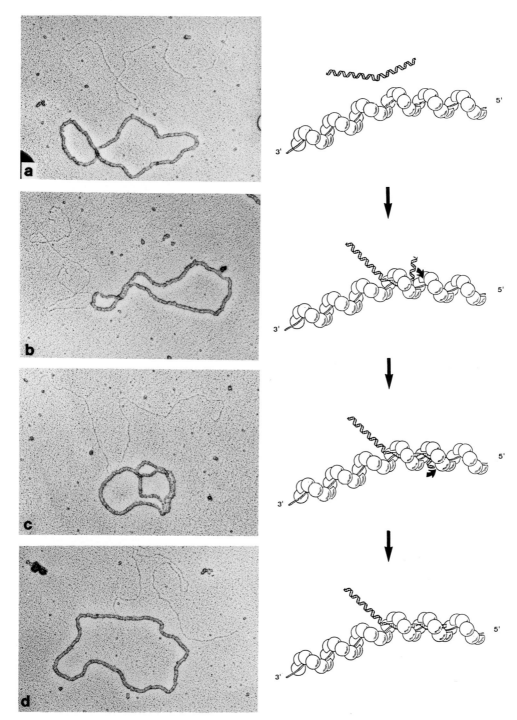

FIGURE 12. Electron micrographs and accompanying models of the early phase (up to 10 min) of the strand exchange reaction mediated by RecA protein. A circular ssDNA molecule, completely covered with RecA protein, interacts with a homologous naked dsDNA. The contact can occur anywhere within the deep groove of the RecA filament, and when an homologous juxtaposition occurs, the initial contacts can be extended, leading to the envelopment of the duplex DNA by the complex of RecA protein on ssDNA. Because the reaction was stopped at various time points by fixing samples with glutaraldehyde, a compaction of the filaments has occurred (see Fig. 2). The models (right panels) show the filaments as extended helical structures, as they would exist in solution before fixation.

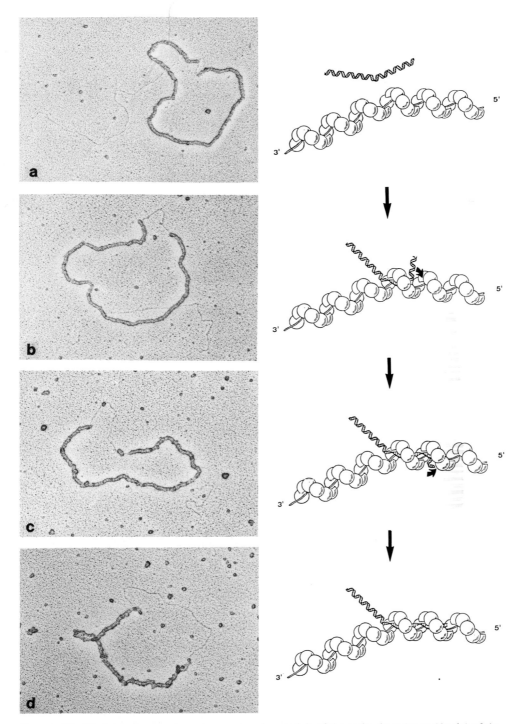

Figure 12. Electron micrographs and accompanying models of the early phase (up to 10 min) of the strand exchange reaction mediated by RecA protein. A circular ssDNA molecule, completely covered with RecA protein, interacts with a homologous naked dsDNA. The contact can occur anywhere within the deep groove of the RecA filament, and when an homologous juxtaposition occurs, the initial contacts can be extended, leading to the envelopment of the duplex DNA by the complex of RecA protein on ssDNA. Because the reaction was stopped at various time points by fixing samples with glutaraldehyde, a compaction of the filaments has occurred (see Fig. 2). The models (right panels) show the filaments as extended helical structures, as they would exist in solution before fixation.

In the earliest phase of the reaction most of the RecA-covered single-stranded circles do not interact with naked dsDNA molecules (Fig. 12a). The first joint molecules appear within a few minutes of the initiation of the reaction, and the dsDNA in such molecules appears to have a short fused junction with the RecA-covered single-stranded circles (Fig. 12b and c). Both ends of the dsDNA are visible, and the "internal" junctions seem to be randomly placed with respect to the ends of the duplex DNA molecules (Howard-Flanders et al., 1987; Stasiak and Egelman, 1987; Register et al., 1987). In the reaction shown here, a linear duplex DNA has been obtained by a unique cleavage of circular DNA of phage ϕX174 with restriction endonuclease *Pst*1. Thus, all linear double-stranded molecules in this reaction are identical with respect to the sequence and the position of the ends. Despite the random localization of the initial junctions, there is a compelling reason to believe that the observed junction occurs at a point of homology: no junctions occur when nonhomologous ssDNA and dsDNA molecules are used (Stasiak and Egelman, 1987).

With increasing reaction time (5 to 10 min), a new type of association between RecA-covered ssDNA circles and naked linear dsDNA molecules appears (Fig. 12d). Only one end of the dsDNA visibly protrudes from the circular complex, and the other end of the linear duplex DNA has presumably been drawn into the complex (Stasiak et al., 1984; Howard-Flanders et al., 1987; Stasiak and Egelman, 1987; Register et al., 1987). Length measurements of the visible part of dsDNA molecules indicate that fragments longer than 2,000 bp can be placed within the RecA complex formed on a single-stranded circle (Stasiak et al., 1984; Register et al., 1987). Thus, three DNA strands are in homologous alignment within the confines of the RecA-DNA complex. An understanding of the precise structure of this three-stranded DNA complex will be crucial in explaining the mechanism of homologous recognition and pairing.

From the order of appearance of joint molecules, one can conclude that molecules with a short internal fused junction, which are seen at early times in the reaction, are converted into molecules with a long "terminal" junction, in which one end of the linear dsDNA is taken into the complex. Presumably, this occurs by wrapping one branch of a linear duplex into the groove of a helical RecA-DNA complex (see model, Fig. 12). The wrapping probably occurs rapidly in a 3′ to 5′ direction relative to the covered circular ssDNA, which would then allow the subsequent strand exchange to proceed from the ends of the duplex DNA in the 5′ to 3′ direction relative to the circular ssDNA, as has been shown for this type of reaction (Cox and Lehman, 1981b; Kahn et al., 1981; West et al., 1981). The fact that at late times in the reaction a second branch of linear duplex is also taken into the complex (Fig. 13) suggests that "wrapping" also occurs in the 5′ to 3′ direction with respect to the circular ssDNA, but at a much slower rate than the 3′ to 5′ wrapping (see model, Fig. 13).

Shortly after the appearance of molecules with long terminal junctions, another type of joint molecule appears (Fig. 13) (Stasiak et al., 1984; Stasiak and Egelman, 1987; Register et al., 1987). These molecules have a characteristic protein-free gap on the circumference of the circular RecA-DNA filament. These gaps, seen more frequently when ATP regeneration is limiting, appear to start at

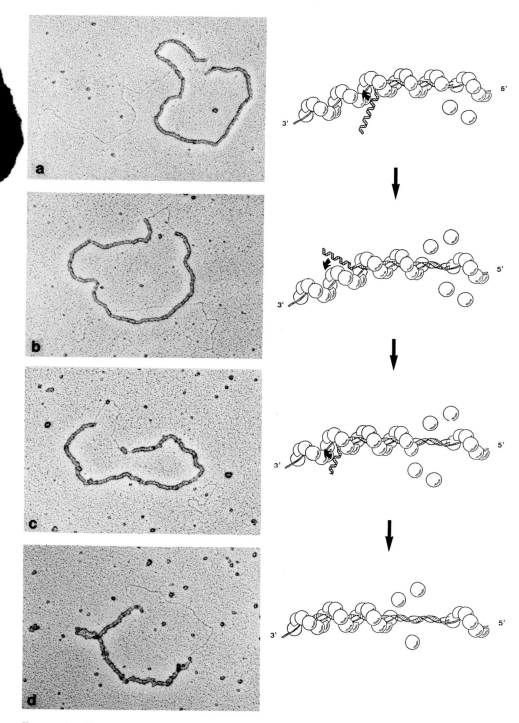

FIGURE 13. Electron micrographs and models of the middle phase (10 to 30 min) of the strand exchange reaction. The branch of the duplex DNA which contains the 5' end of the strand to be displaced becomes enveloped by the complex, and RecA protomers start to dissociate from this end, leaving a gap in the RecA covering of the circular DNA. The other branch of the duplex DNA, containing the 3' end of the strand to be displaced, is also taken into the complex as the reaction proceeds in time.

Genetic Recombination

Editors: Raju Kucherlapati and Gerald R. Smith
American Society for Microbiology, Washington, D.C., November 1988

Erratum

Pages 293, 295, and 297: The left panels of Figures 12, 13, and 14 were inadvertently disarranged.

The left panels appearing in Figure 12 should actually be the left panels of Figure 13.

The left panels appearing in Figure 13 should actually be the left panels of Figure 14.

The left panels appearing in Figure 14 should actually be the left panels of Figure 12.

Corrected pages are attached.

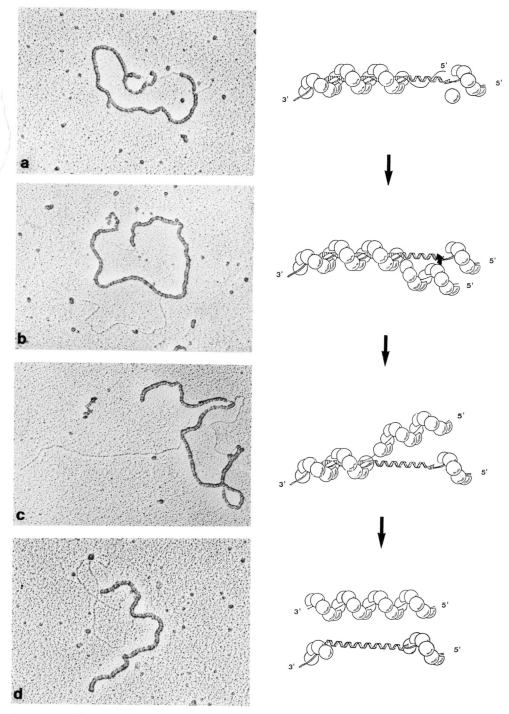

FIGURE 14. Electron micrographs and models of the late phase (30 to 60 min) of the strand exchange reaction. The displaced strand, after separating from the newly formed duplex region, starts becoming covered with RecA protein. The unidirectional branch migration proceeds with further RecA binding to the displaced strand and with the associated release of RecA protomers from the newly formed duplex region.

297

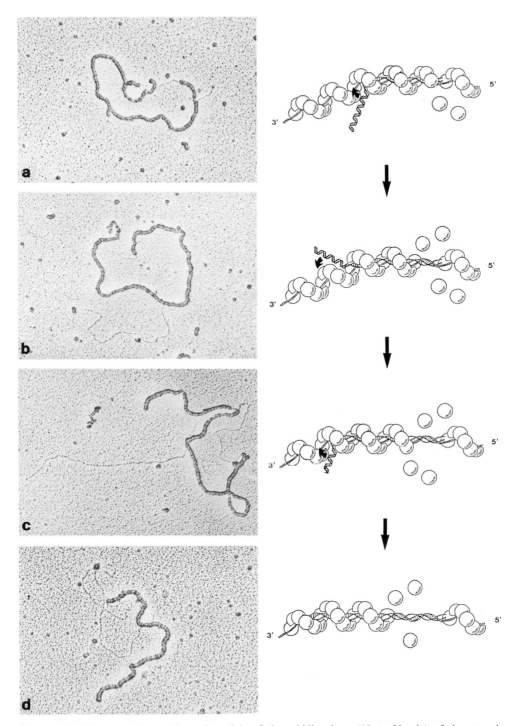

FIGURE 13. Electron micrographs and models of the middle phase (10 to 30 min) of the strand exchange reaction. The branch of the duplex DNA which contains the 5′ end of the strand to be displaced becomes enveloped by the complex, and RecA protomers start to dissociate from this end, leaving a gap in the RecA covering of the circular DNA. The other branch of the duplex DNA, containing the 3′ end of the strand to be displaced, is also taken into the complex as the reaction proceeds in time.

the 5′ end of the strand within the original linear dsDNA which will be displaced (Stasiak et al., 1984; Register et al., 1987). The protein gaps uncover naked DNA in these regions. The DNA filament visible there probably contains all three DNA strands taking part in the reaction. A displaced ssDNA would be visible as a RecA-covered filament, as is clearly the case later in the reaction (see Fig. 14). If RecA protein mediates the strand switching process while it is dissociating (Howard-Flanders et al., 1984, 1987), then the plectonemic arrangement of the three strands restricts the unraveling of the displaced strand as long as its end is bound by RecA protein. The model presented (Fig. 14) shows such a situation, and the micrographs are consistent with the possibility that the end of the ssDNA which is to be displaced is still bound by RecA protein adjacent to the gap.

Pictures taken at a late stage in the reaction (30 to 60 min) show joint molecules with a partially displaced strand covered with RecA protein (Fig. 14). Most likely, it is the end of the displaced strand which becomes accessible for RecA nucleation, and the subsequent polymerization of RecA protein rapidly proceeds from the displaced 5′ end toward the 3′ end, removing the strand to be displaced from the three-stranded region (see Fig. 14). This is consistent with the known directionality of RecA polymerization on ssDNA, from 5′ to 3′ (Register and Griffith, 1985; Cassuto and Howard-Flanders, 1986).

The fact that the RecA-covered protein of a displaced strand never overlaps with the RecA-covered region of homologous alignment but is contiguous with it proves that the strand to be displaced is kept within the RecA complex and is not freed from the RecA-covered region of homologous alignment until the RecA protein dissociates from this region.

In joint molecules with a RecA-covered, partially displaced single strand, one frequently sees naked dsDNA not completely taken into the region of the alignment (Fig. 14a, b, and c), but as the reaction nears its completion the whole duplex DNA is drawn into the complex (Fig. 14d). When the reaction nears completion, the displaced strand and the newly formed naked duplex region encompass almost the complete dsDNA circle. Such molecules still have a short RecA-covered region where ssDNA and dsDNA are aligned (Fig. 14d). At the end of the reaction both linear ssDNA molecules (displaced strands), which are completely covered with RecA protein, and naked dsDNA circles are observed (not shown).

A characteristic of the in vitro strand exchange reaction is the high asynchrony of the process in different joint molecules, with some molecules having finished the strand exchange reaction while others have not yet started. Since one cannot follow a particular joint molecule during the entire reaction, conclusions about the progress of the reaction must be based on the order of appearance of different types of joint molecules and not on the frequency with which different types of joint molecules appear at subsequent time points in the reaction.

In addition to this asynchrony, the interpretation of micrographs is complicated by the fact that the initial homologous contacts occur at random points on the substrates. The molecules which formed initial homologous junctions close to the 5′ end of the strand to be displaced are likely to progress through the reaction

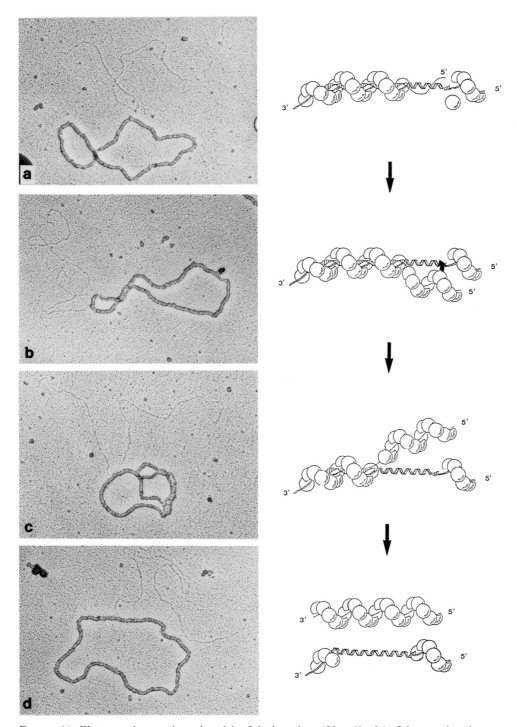

FIGURE 14. Electron micrographs and models of the late phase (30 to 60 min) of the strand exchange reaction. The displaced strand, after separating from the newly formed duplex region, starts becoming covered with RecA protein. The unidirectional branch migration proceeds with further RecA binding to the displaced strand and with the associated release of RecA protomers from the newly formed duplex region.

297

differently from the molecules which paired initially at a region close to the 3′ end of the strand to be displaced. Intermediate circular molecules without a displaced strand protruding from the gap in the complex and with a linear duplex completely drawn into complex (Fig. 13d; see also Stasiak et al., 1984; Register et al., 1987) probably arise from molecules which paired initially at a region close to the 3′ end of the strand to be displaced. In such molecules even a slow wrapping of the duplex in the 5′ to 3′ direction (with respect to ssDNA in the complex) can bring the 3′ end of the strand to be displaced into the complex before the 5′ end will become visible as a displaced RecA-covered strand.

It is interesting to speculate on the arrangement of three strands when they are enclosed within a RecA helical filament, and what causes RecA protein to dissociate from these three strands, leaving the gap in the complex. As already discussed in section I, a homologous recognition between ssDNA and dsDNA can occur either before or after strand separation of duplex DNA. To correspond with the micrographs presented here, pairing after strand separation would require that every time a segment of the duplex DNA enters into a groove of the complex, it is actively opened by RecA protein and one strand of the duplex is tested against the resident single strand. If there is no match (no homology), the duplex leaves the complex. If the contact occurs in a homologous position, then intermolecular Watson-Crick hydrogen bonds stabilize this initial interaction and tend to elongate the paired region over the entire homologous stretch.

The possibility of pairing before strand separation requires the formation of additional hydrogen bonds, thus implying that duplex DNA approaching the groove of the complex is placed by RecA protein in such a relation to the resident single strand that specific additional hydrogen bonds could be formed between two molecules while maintaining the original Watson-Crick base pairing of the incoming duplex DNA (McGavin, 1977; Howard-Flanders et al., 1984). If the contact between molecules is not in a region of homology, additional hydrogen bonds cannot form and the duplex is not retained within the complex. In the event that there is homology at the point of contact, additional hydrogen bonds can form, tending to elongate the paired region as far as there is homology, and as long as there are no topological obstacles. The micrographs shown here do not allow one to directly draw conclusions about the arrangement of three DNA strands enclosed within the same RecA helical filament.

Microscopy has provided evidence that DNA molecules within the complex are homologous (Stasiak and Egelman, 1987; Register et al., 1987). An interesting question is whether dsDNA can be enveloped into the complex beyond the region of homology, and this seems to be answered with the help of electron microscopy. Radding et al. (1983) studied an in vitro strand exchange reaction in which only a part of a linear dsDNA molecule, the part adjacent to the 5′ end of the strand to be displaced, was homologous to an interacting circular ssDNA. A micrograph of this reaction shows that the dsDNA does not enter the RecA-ssDNA complex beyond the region of homology of the two interacting DNA molecules (Radding et al., 1983). Similarly, Register et al. (1987) observed that if homology between linear dsDNA and circular ssDNA was limited to an internal portion of the linear dsDNA, then the pairing process never extended from this internal region of the linear dsDNA, ending up with characteristic molecules having two protein-free dsDNA branches protruding from the circular RecA-DNA filament (like "internal" junctions shown in Fig. 12b and c).

These results suggest that homology is not only required for the initial junction formation but is also necessary for driving the incorporation of a duplex DNA into the complex formed on an ssDNA. Thus, over the whole region where ssDNA and dsDNA are enclosed within the same RecA filament, a specific mechanism for the recognition of homology must operate. Interestingly, in the pairing reaction presented here the region within a RecA complex which contains all three DNA strands extends initially in both directions until the ends of the dsDNA are reached. Only after the end of a dsDNA which contains the 5' end of the strand to be displaced becomes enveloped by a RecA filament does the dissociation of RecA protein from the region containing all three DNA strands begin. The dissociation of RecA protomers probably reflects the strand switch reaction being completed in this region.

The two different pairing mechanisms discussed here could give rise to quite different strand arrangements within the paired region. If a recognition of homology is achieved by Watson-Crick hydrogen bonding between one of the separated strands of duplex DNA and the ssDNA resident in the complex, then such an arrangement of strands should still be present within the complex, which contains all three strands. One should expect within the complex the newly paired hydrogen-bonded strands and a displaced strand which is not hydrogen bonded.

If the recognition of homology occurred by the formation of additional hydrogen bonds between the ssDNA resident in the complex and an incoming duplex DNA, there would be no need to maintain such hydrogen bonding within the complex which encloses all three participating strands. Since such a pairing stage must be followed by a strand switch stage, it is likely that the triple-stranded region within the complex triggers the RecA-mediated strand switch process. In this case, pairing before strand separation can result in an arrangement of the three strands within the RecA-DNA complex identical to the arrangement produced by pairing after strand separation. However, it is also possible that triple-stranded DNA produced by pairing before strand separation could persist within the complex (especially in the presence of ATP-γ-S; see section III.C.).

Reaction mediated by RecA and SSB proteins. As already discussed, RecA protein alone can mediate the complete strand exchange reaction between a linear dsDNA molecule and a circular ssDNA molecule. However, the addition of SSB protein to RecA protein increases the yield and rate of the strand exchange reaction (Cox and Lehman, 1982; Radding et al., 1983). Since the distinction between RecA- and SSB-covered DNA regions is strikingly clear (Fig. 10), one can analyze joint molecules in micrographs and pinpoint those regions interacting with RecA protein and those interacting with SSB protein.

Micrographs from reactions supplemented with SSB protein show clearly that the displaced strand is always covered with SSB protein (Flory et al., 1984; Register et al., 1987). It appears that SSB protein sequesters the strand to be displaced from the RecA-free gap in the complex much more quickly than does free RecA protein. Further, SSB protein seems to pull the strand to be displaced from the RecA-covered region of DNA alignment so quickly that frequently such regions of alignment are not seen after several minutes of reaction (Fig. 15). Apparently, the speed of SSB-stimulated strand displacement is faster than the speed of wrapping of the linear duplex into the RecA-DNA filament (in the 5' to 3' direction with respect to the resident circular ssDNA). This increase in the speed of strand displacement and strand envelopment leads to a quick elimination of RecA protein in the region of homologous alignment. In the reaction mediated by RecA protein alone, a RecA covering of the region of homologous alignment always preceded the point of visible strand displacement (Fig. 14). Since such joint molecules, with partially displaced and partially exchanged strands, are kept in homologous register by the strand being exchanged, there is no need for the maintenance of homologous alignment of the three strands. When molecules have reached such a state, strand switching, which now can be unambiguously called strand transfer, can occur directly from the paired original linear duplex to the newly paired circular duplex (Fig. 15c and d) without proceeding through an ambiguous intermediate state of the reaction within the RecA-covered region of homologous alignment. Branch migration therefore seems to be driven by SSB protein pulling out the strand being displaced from the RecA complex and by a RecA-mediated annealing of the strand being exchanged with the resident strand in the RecA-ssDNA complex (Fig. 15, model). A RecA protein annealing activity was proposed to have an important function during strand exchange reactions (Cox and Lehman, 1987). After annealing, the RecA protein dissociates from the new duplex (Fig. 15). In the reaction made in the presence of RecA and SSB proteins, the final products of the strand exchange reaction are a naked dsDNA circle and a linear displaced strand covered with SSB protein (not shown).

The micrographs and the model presented for the strand exchange reaction mediated by RecA and SSB proteins require an explanation of why at the beginning of the reaction RecA protein wins the competition with SSB protein for binding to ssDNA while at the end of the reaction the opposite situation is observed. Probably the most important factor controlling the relative affinity of RecA and SSB proteins for ssDNA is the ATP/ADP ratio in the reaction mixture. While SSB binding to an ssDNA is rather unaffected by the accumulation of ADP in the reaction mixture, the RecA affinity for ssDNA decreases in the presence of ADP (Menetski and Kowalczykowski, 1985). Although in vitro strand exchange reactions usually contain an ATP regeneration system, at long incubation times there is a significant accumulation of ADP. It is possible that in vivo a reaction consuming large amounts of ATP with time will locally decrease the ATP/ADP ratio, favoring the binding of SSB protein to ssDNA.

An alternate explanation for the entire phenomenon is the nonequivalence of the ssDNA bound early in the reaction by RecA protein (a circle with no free ends) and that bound late in the reaction by SSB protein (a single-stranded free end). It is possible that the relative affinity of SSB protein for the free end is much greater than that of RecA protein.

Reaction mediated by RecA protein in the presence of ATP-γ-S. In the

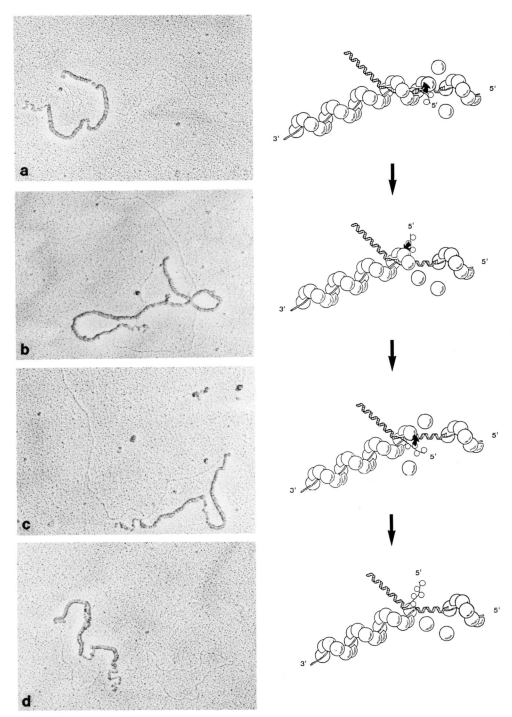

FIGURE 15. Electron micrographs and models of the strand exchange reaction mediated by both RecA protein and SSB protein. Notice that in this reaction the displaced strand becomes covered with SSB protein. In these models RecA protomers are represented by large balls and SSB protomers by small balls.

presence of ATP-γ-S, an analog of ATP, RecA protein is able to mediate the pairing reaction but not subsequent steps of recombination (Honigberg et al., 1985; Riddles and Lehman, 1985). Honigberg et al. (1985) demonstrated that the displaced strand of duplex DNA is resistant to S1 nuclease in molecules homologously paired by RecA protein acting in the presence of ATP-γ-S. This is in contrast to joint molecules formed in the presence of ATP, which clearly show the appearance of a displaced strand from the original duplex. The existence of joint molecules between circular ssDNA and linear dsDNA where homologous recognition occurs without a detectable strand separation of the duplex DNA suggests that pairing can occur before strand separation of the duplex DNA. Probably only after pairing occurs does RecA protein hydrolyze ATP, which is then coupled to a strand switch process which changes the triple-stranded DNA into a new duplex and a displaced strand.

The microscopy of a reaction made in the presence of ATP-γ-S shows RecA-covered ssDNA molecules in complexes with linear dsDNA molecules. Gaps in the RecA-covered region of alignment or displaced strands covered with RecA protein, both frequently seen in the ATP reaction, are not observed in the reaction made in the presence of ATP-γ-S (Stasiak, unpublished data).

2. Microscopy of deproteinized recombination intermediates

The microscopy preparations which have been described show recombination intermediates as they exist in the form of complexes between DNA and RecA or SSB protein. To elucidate the DNA-DNA interactions leading to pairing, deproteinized samples of recombination intermediates may be more useful, since proteins frequently obscure regions of DNA-DNA contacts, such as the regions of homologous alignment containing all three DNA strands within the RecA filament.

Electron microscopy preparations aimed at visualizing nicely spread ssDNA must be made in the presence of DNA denaturing agents, such as formamide, which may destroy the DNA-DNA interactions present in initial junctions. Register et al. (1987) noticed that paired molecules which were not able to convert the paired stage to the strand exchange stage were usually separated by the preparation technique which was used. Thus, even if a fragile triple-stranded DNA structure within the RecA-covered region of homologous alignment is stabilized with additional hydrogen bonds (Howard-Flanders et al., 1984), classical electron microscopy preparations are likely to melt such structures.

The microscopy of DNA intermediates from the deproteinized strand exchange reaction (DasGupta et al., 1980; Cox and Lehman, 1981b) may thus show a simplified and perhaps somewhat misleading picture of a strand exchange reaction. Figure 16 shows such a micrograph, after removal of RecA, where double- and single-stranded regions of a joint molecule are clearly distinguishable by their relative thicknesses. One can easily observe the branch point where the strand being exchanged passes from the original linear duplex to the new circular duplex. It is also easy to see which part of the circular ssDNA is paired with the exchanged strand and to see the single-stranded character of the strand being

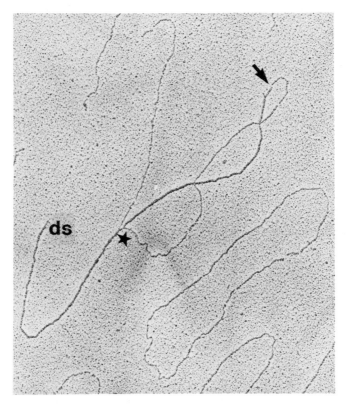

FIGURE 16. A deproteinized DNA sample from a RecA-mediated strand exchange reaction. A double-stranded segment of the joint DNA molecule is visible as a heavy line (marked at one end by ds), while all the other narrower lines are due to ssDNA. The star indicates the branch point in the joint molecule, where one strand of the duplex DNA is being displaced and where the second strand is being paired with a circular ssDNA molecule. The arrow shows where the second strand of the dsDNA ends. A circular ssDNA molecule which has not yet begun the strand exchange reaction is shown to the right and below the joint molecule. The sample was prepared by the hypophase method (Sogo et al., 1987).

displaced. Such pictures are therefore in perfect agreement with pictures of the reaction made in the presence of both RecA and SSB proteins, where such typical branch points are also visible (Fig. 15c and d).

Given that the reaction which was subsequently deproteinized was initially conducted with RecA protein alone, there is a discrepancy between the deproteinized pictures and the pictures of RecA-covered joined molecules, which show long regions of alignment within the RecA filament. This region of alignment is destroyed by the preparation procedure, and the ambiguity exists as to where strands were exchanged in the RecA-covered region of alignment. In this regard it is interesting that Register et al. (1987) showed micrographs of deproteinized paired molecules which are consistent with a triple-stranded structure in the paired regions.

B. In Vitro Recombination Reactions Mediated by Phage T4 UvsX Protein

UvsX protein is a gene product of the bacteriophage T4 and is essential for the general genetic recombination of bacteriophage DNA molecules in infected bacterial cells. Purified UvsX protein, aided with SSB protein, is able in an ATP-dependent reaction to promote a complete strand exchange between a linear dsDNA molecule and a homologous circular ssDNA (Yonesaki and Minagawa, 1985; Formosa and Alberts, 1986), in a manner very similar to RecA protein. Harris and Griffith (1987) visualized homologous pairing of DNA molecules catalyzed by UvsX protein. Their micrographs showed that UvsX protein interacts with homologous naked dsDNA, leading to the envelopment of the dsDNA by the helical UvsX-DNA complex. In their pictures the UvsX-DNA helical filament can contain within it dsDNA and ssDNA homologously aligned over a stretch ranging up to 600 bp. However, in the case of a UvsX-promoted reaction the formation of such extended junctions is dependent on the presence of a free DNA end in the region of pairing (Harris and Griffith, 1987). This requirement is in clear contrast to the situation observed in reactions mediated by RecA protein, in which extended junctions are formed independent of the presence of free DNA ends in the region of pairing (Register et al., 1987). The presence of free DNA ends in the region of pairing enables relatively easy strand separation of dsDNA. The UvsX requirement for a free DNA end in the region of pairing might indicate that UvsX protein causes strand separation of dsDNA before pairing; this is very interesting with regard to the mechanism of pairing. The function of the UvsX protein is to pair and recombine phage T4 DNA molecules in which all the cytosines are glucosylated. Glucosylation makes the formation of additional specific hydrogen bonds between homologous dsDNA and ssDNA structurally impossible. Thus, while there are two structural possibilities for the recognition of homology for a nonglucosylated DNA (normal cellular DNA), the first involving the formation of a transient triple-stranded DNA structure (pairing before strand separation) and the second involving Watson-Crick hydrogen bonding between a separated strand of duplex DNA and an ssDNA (pairing after strand separation), for glucosylated DNA only the second possibility can exist. The fact that RecA protein does not need any free DNA ends to form extended homologous junctions supports the view that RecA protein can mediate pairing before strand separation.

IV. CONCLUDING REMARKS

Electron microscopy has provided us with information about the structure of RecA-DNA complexes. Micrographs of recombining molecules are essential to developing molecular models of DNA recombination. However, many interesting questions about the molecular mechanisms of homologous recognition and strand exchange, such as whether pairing can occur before or only after strand separation, will never be definitively answered solely with the microscopic approach. Electron microscopy and image analysis will provide the framework into which high-resolution structural data from crystallography, spectroscopy, and biochemical techniques must be integrated.

ACKNOWLEDGMENTS. We thank Paul Howard-Flanders and Theo Koller for their support and encouragement.

This work was supported in part by Public Health Service grant GM35269 from the National Institutes of Health.

LITERATURE CITED

Arnott, S., and R. Chandrasekaran. 1981. Fibrous polynucleotide duplexes have very polymorphic secondary structures, p. 99–122. *In* R. Sarma (ed.), *Biomolecular Stereodynamics I.* Academic Press, Inc., New York.

Bujalowski, W., and T. M. Lohman. 1987. Limited co-operativity in protein-nucleic acid interactions, a thermodynamic model for the interactions of *Escherichia coli* single strand binding protein with single-stranded nucleic acids in the ''beaded'', (SSB)65 mode. *J. Mol. Biol.* **195:**897–907.

Cassuto, E., and P. Howard-Flanders. 1986. The binding of RecA protein to duplex DNA molecules is directional and is promoted by a single stranded region. *Nucleic Acids Res.* **14:**1149–1157.

Chrysogelos, S., J. C. Register III, and J. Griffith. 1983. The structure of RecA protein-DNA filaments. *J. Biol. Chem.* **258:**12624–12631.

Clark, A. J. 1973. Recombination deficient mutants of *E. coli* and other bacteria. *Annu. Rev. Genet.* **7:**67–86.

Cox, M. M., and I. R. Lehman. 1981a. RecA protein of *E. coli* promotes branch migration, a kinetically distinct phase of DNA strand exchange. *Proc. Natl. Acad. Sci. USA* **78:**3433–3437.

Cox, M. M., and I. R. Lehman. 1981b. Directionality and polarity in RecA protein-promoted branch migration. *Proc. Natl. Acad. Sci. USA* **78:**6018–6022.

Cox, M. M., and I. R. Lehman. 1982. RecA protein-promoted DNA strand exchange. *J. Biol. Chem.* **257:**8523–8532.

Cox, M. M., and I. R. Lehman. 1987. Enzymes of general genetic recombination. *Annu. Rev. Biochem.* **56:**229–262.

DasGupta, C., T. Shibata, R. P. Cunningham, and C. M. Radding. 1980. The topology of homologous pairing promoted by RecA protein. *Cell* **22:**437–446.

DeRosier, D. J., and A. Klug. 1968. Reconstructions of three-dimensional structures from electron micrographs. *Nature* (London)**217:**130–134.

DiCapua, E., A. Engel, A. Stasiak, and T. Koller. 1982. Characterization of complexes between RecA protein and duplex DNA by electron microscopy. *J. Mol. Biol.* **157:**87–103.

DiCapua, E., and B. Müller. 1987. The accessibility of DNA to dimethylsulfate in complexes with recA protein. *EMBO J.* **6:**2493–2498.

Dombroski, D. F., D. G. Scraba, R.D. Bradley, and A. R. Morgan. 1983. Studies of the interaction of RecA protein with DNA. *Nucleic Acids Res.* **11:**7487–7504.

Dubochet, J., M. Adrian, J. Chang, J. Homo, J. Lepault, A. W. McDowall, and P. Schultz. 1988. Cryo-electron microscopy of vitrified specimens. *Q. Rev. Biophys.* **21:**129–228.

Dunn, K., S. Chrysogelos, and J. Griffith. 1982. Electron microscopic visualization of RecA-DNA filaments: evidence for a cyclic extension of duplex DNA. *Cell* **28:**757–765.

Egelman, E. H., and A. Stasiak. 1986. The structure of helical RecA-DNA complexes. I. Complexes formed in the presence of ATP-γ-S or ATP. *J. Mol. Biol.* **191:**677–697.

Egelman, E. H., and A. Stasiak. 1988. Structure of helical RecA-DNA complexes. II. Local conformational changes visualized in bundles of RecA-ATP-γ-S filaments. *J. Mol. Biol.* **200:**329–349.

Flory, J., S. S. Tsang, and K. Muniyappa. 1984. Isolation and visualization of active presynaptic filaments of RecA protein and single-stranded DNA. *Proc. Natl. Acad. Sci. USA* **81:**7026–7030.

Formosa, T., and B. Alberts. 1986. Purification and characterization of the T4 bacteriophage uvsX protein. *J. Biol. Chem.* **261:**6107–6118.

Griffith, J., and T. Formosa. 1985. The uvsX protein of bacteriophage T4 arranges single-stranded and double-stranded DNA into similar helical nucleoprotein filaments. *J. Biol. Chem.* **260:**4484–4491.

Griffith, J., L. D. Harris, and J. C. Register III. 1984. Visualization of SSB-ssDNA complexes active in the assembly of stable RecA-DNA filaments. *Cold Spring Harbor Symp. Quant. Biol.* **49:**553–559.

Harris, L. D., and J. Griffith. 1987. Visualization of the homologous pairing of DNA catalyzed by the bacteriophage T4 UvsX protein. *J. Biol. Chem.* **262:**9285–9292.

Honigberg, S. M., D. Gonda, J. Flory, and C. M. Radding. 1985. The pairing activity of stable nucleoprotein filaments made from RecA protein, single-stranded DNA, and adenosine 5'-(γ-thio)triphosphate. *J. Biol. Chem.* **260:**11845–11851.

Howard-Flanders, P., S. C. West, E. Cassuto, T.-R. Hahn, E. H. Egelman, and A. Stasiak. 1987. Structure of RecA spiral filaments and their role in homologous pairing and strand exchange in genetic recombination, p. 609–617. *In* T. Kelly and R. McMacken (ed.), *DNA Replication and Recombination.* Alan R. Liss, New York.

Howard-Flanders, P., S. C. West, and A. Stasiak. 1984. Role of RecA protein spiral filaments in genetic recombination. *Nature* (London) **309:**215–220.

Iwabuchi, M., T. Shibata, T. Ohtani, M. Natori, and T. Ando. 1983. ATP-dependent unwinding of the double helix and extensive supercoiling by *Escherichia coli* RecA protein in the presence of topoisomerase. *J. Biol. Chem.* **258:**12394–12404.

Kahn, R., R. P. Cunningham, C. DasGupta, and C. M. Radding. 1981. Polarity of heteroduplex formation promoted by *E. coli* RecA protein. *Proc. Natl. Acad. Sci. USA* **78:**4786–4790.

Koller, T., E. DiCapua, and A. Stasiak. 1983. Complexes of RecA with single stranded DNA, p. 723–729. *In* N. Cozzarelli (ed.), *Mechanisms of DNA Replication and Recombination.* Alan R. Liss, New York.

Lohman, T. M., L. B. Overman, and S. Datta. 1986. Salt-dependent changes in the DNA binding co-operativity of *Escherichia coli* single strand binding protein. *J. Mol. Biol.* **187:**603–615.

Lovett, C. M., and J. W. Roberts. 1985. Purification of a recA protein analogue from *Bacillus subtilis.* *J. Biol. Chem.* **260:**3305–3313.

McEntee, K., G. M. Weinstock, and I. R. Lehman. 1979. Initiation of general recombination catalyzed *in vitro* by the *recA* protein of *Escherichia coli. Proc. Natl. Acad. Sci. USA* **76:**2615–2619.

McGavin, S. 1971. Models of specifically paired like (homologous) nucleic acid structures. *J. Mol. Biol.* **55:**293–298.

McGavin, S. 1977. A model for the specific pairing of homologous double-stranded nucleic acid molecules during genetic recombination. *Heredity* **39:**15–25.

Menetski, J. P., and S. C. Kowalczykowski. 1985. Interaction of RecA protein with single-stranded DNA, quantitative aspects of binding affinity modulation by nucleotide cofactors. *J. Mol. Biol.* **181:**281–295.

Muniyappa, K., S. L. Shaner, S. S. Tsang, and C. M. Radding. 1984. Mechanism of the concerted action of RecA protein and helix destabilizing proteins in homologous recombination. *Proc. Natl. Acad. Sci. USA* **81:**2757–2761.

Pugh, B. F., and M. Cox. 1987. Stable binding of RecA protein to duplex DNA. *J. Biol. Chem.* **262:**1326–1336.

Radding, C. M. 1978. Genetic recombination: strand transfer and mismatch repair. *Annu. Rev. Biochem.* **47:**847–880.

Radding, C. M., J. Flory, A. Wu, R. Kahn, C. DasGupta, D. Gonda, M. Bianchi, and S. S. Tsang. 1983. Three phases in homologous pairing: polymerization of RecA protein on single-stranded DNA, synapsis, and polar strand exchange. *Cold Spring Harbor Symp. Quant. Biol.* **47:**821–828.

Register, J. C., III, G. Christiansen, and J. Griffith. 1987. Electron microscopic visualization of the RecA protein-mediated pairing and branch migration phases of DNA strand exchange. *J. Biol. Chem.* **262:**12812–12820.

Register, J. C., III, and J. Griffith. 1985. The direction of RecA protein assembly onto single strand DNA is the same as the direction of strand assimilation during strand exchange. *J. Biol. Chem.* **260:**12308–12312.

Riddles, P. W., and I. R. Lehman. 1985. The formation of plectonemic joints by the RecA protein of *Escherichia coli*: requirements for ATP hydrolysis. *J. Biol. Chem.* **260**:170–173.

Sancar, A., C. Stachelek, W. Konigsberg, and W. D. Rupp. 1980. Sequences of RecA gene and protein. *Proc. Natl. Acad. Sci. USA* **77**:2611–2615.

Shibata, T., C. DasGupta, R. P. Cunningham, and C. M. Radding. 1979. Purified *Escherichia coli recA* protein catalyzes homologous pairing of superhelical DNA and single-stranded fragments. *Proc. Natl. Acad. Sci. USA* **76**:1638–1642.

Sobell, H. M. 1980. Structural and dynamic aspects of drug intercalation into DNA and RNA, p. 289–323. *In* H. Sarma (ed.), *Nucleic Acid Geometry and Dynamics*. Pergamon Press, Inc., Elmsford, N.Y.

Sogo, J., A. Stasiak, W. De Bernardin, R. Losa, and T. Koller. 1987. Binding of protein to nucleic acids, p. 61–79. *In* J. Sommerville and U. Scheer (ed.), *Electron Microscopy in Molecular Biology, a Practical Approach*. IRL Press, Oxford.

Stasiak, A., and E. DiCapua. 1982. The helicity of DNA in complexes with RecA protein. *Nature* (London) **299**:185–186.

Stasiak, A., E. DiCapua, and T. Koller. 1981. The elongation of duplex DNA by RecA protein. *J. Mol. Biol.* **151**:557–564.

Stasiak, A., E. DiCapua, and T. Koller. 1983. Unwinding of duplex DNA in complexes with RecA protein. *Cold Spring Harbor Symp. Quant. Biol.* **47**:811–820.

Stasiak, A., and E. H. Egelman. 1986. The structure and dynamics of RecA protein-DNA complexes as determined by image analysis of electron micrographs. *Biophys. J.* **49**:5–7.

Stasiak, A., and E. H. Egelman. 1987. RecA protein-DNA interactions in recombination, p. 619–628. *In* T. Kelly and R. McMacken (ed.), *DNA Replication and Recombination*. Alan R. Liss, New York.

Stasiak, A., A. Z. Stasiak, and T. Koller. 1984. Visualization of RecA-DNA complexes involved in consecutive stages of an *in vitro* strand exchange reaction. *Cold Spring Harbor Symp. Quant. Biol.* **49**:561–570.

Stewart, M., and G. Vigers. 1986. Electron microscopy of frozen-hydrated biological material. *Nature* (London) **319**:631–636.

Taylor, A. F., D. W. Schultz, A. S. Ponticelli, and G. R. Smith. 1985. RecBC enzyme nicking at Chi sites during DNA unwinding: location and orientation-dependence of the cutting. *Cell* **41**:153–163.

Tsang, S. S., K. Muniyappa, E. Azhderian, D. K. Gonda, C. M. Radding, J. Flory, and J. W. Chase. 1985. Intermediates in homologous pairing promoted by RecA protein: isolation and characterization of active presynaptic complexes. *J. Mol. Biol.* **185**:295–309.

Vollenweider, H. J., A. James, and W. Szybalski. 1978. Discrete length classes of DNA depend on mode of dehydration. *Proc. Natl. Acad. Sci. USA* **75**:710–714.

Weinstock, G. M., K. McEntee, and I. R. Lehman. 1979. ATP-dependent renaturation of DNA catalyzed by the recA protein of *E. coli*. *Proc. Natl. Acad. Sci. USA* **76**:126–130.

Weinstock, G. M., K. McEntee, and I. R. Lehman. 1981. Hydrolysis of nucleoside triphosphates catalyzed by the RecA protein of *Escherichia coli*. *J. Biol. Chem.* **256**:8845–8849.

West, S. C., E. Cassuto, and P. Howard-Flanders. 1981. Heteroduplex formation by RecA protein—polarity of strand exchanges. *Proc. Natl. Acad. Sci. USA* **78**:6149–6153.

West, S. C., E. Cassuto, and P. Howard-Flanders. 1982. Role of SSB protein in recA promoted branch migration reactions. *Mol. Gen. Genet.* **186**:333–338.

West, S. C., E. Cassuto, J. Mursalim, and P. Howard-Flanders. 1980. Recognition of duplex DNA containing single-stranded regions by RecA protein. *Proc. Natl. Acad. Sci. USA* **77**:2569–2573.

West, S. C., J. K. Countryman, and P. Howard-Flanders. 1983. Purification and properties of the recA protein of *Proteus mirabilis*. *J. Biol. Chem.* **258**:4648–4654.

Williams, R. C., and S. J. Spengler. 1986. Fibers of RecA protein and complexes of RecA protein and single-stranded φX174 DNA as visualized by negative-stain electron microscopy. *J. Mol. Biol.* **187**:109–118.

Yonesaki, T., and T. Minagawa. 1985. T4 phage gene uvsX product catalyzes homologous DNA pairing. *EMBO J.* **4**:3321–3327.

Chapter 9

Illegitimate Recombination in Bacteria

Neca D. Allgood and Thomas J. Silhavy

I. INTRODUCTION

Illegitimate recombination is a broad, imprecise term frequently used to refer to any DNA rearrangement that leads to the covalent joining of nonhomologous linear DNA segments which previously were nonadjacent (Franklin, 1971; Weisberg and Adhya, 1977). Accordingly, an illegitimate event leads to the formation of a novel DNA joint(s), a term commonly used for the site of the fusion between the DNA segments in question. Viewed in this manner, a deletion or a tandem duplication leads to the formation of a single novel joint, and an inversion leads to the formation of two novel joints. In pioneering work, Franklin (1967) demonstrated that these events occur by a mechanism(s) that is independent of homologous recombination functions such as *recA*, and this remains a hallmark of illegitimate recombination. For the purpose of this review, we add the further distinction that the event does not involve functions specified by transposable genetic elements, nor does it involve any identifiable site-specific recombination system, such as the *int* function of bacteriophage λ or the *hin* function of *Salmonella* spp.; these topics are considered by Syvanen (this volume) and Hatfull and Grindley (this volume).

Although little is known about the mechanism(s) of illegitimate recombination, it has been used extensively as an experimental tool in bacterial genetics. These studies have provided important data for consideration and have also raised a more practical motive for studying illegitimate recombination. Consider the case of recombinant DNA. During the past decade, recombinant DNA has had a major impact on biology. In part, this impact stems from the ability it provides the experimentalist to create novel DNA joints. With this technology, structural genes can be moved from their genomic location to another replicon, genetic fusions can be created which place a given structural gene under the control of a different regulatory system or which result in genes that specify hybrid proteins, or specific DNA sequences can be targeted for deletion or for insertion at another chosen location. Construction of recombinant DNA in vitro is analogous to illegitimate recombination in vivo. For each of the in vitro examples cited, there is a counterpart for illegitimate recombination in bacterial genetics. The low frequency and the inherently random nature of illegitimate recombination have, however, generally prevented its widespread use. If the frequency can be increased and if the events can be more precisely defined and controlled, illegitimate recombination could become a legitimate rival to recombinant DNA. Indeed, the successful harnessing of transposons, another method of producing novel DNA joints, has revolutionized the practice of bacterial genetics.

II. GENETIC REARRANGEMENTS CAUSED BY ILLEGITIMATE RECOMBINATION

This section describes genetic rearrangements thought to be the products of illegitimate recombination. Not surprisingly, more data are available than can be discussed adequately. However, many of these data are limited in that novel joints were defined by genetic mapping only. At this level of analysis, many of the relevant features of novel joints and the sites of illegitimate recombination are difficult to detect. Accordingly, we will focus primarily on studies including nucleotide sequence analysis.

In an attempt to categorize genetic rearrangements, we have divided the events into two major classes. This division is useful in considering the possible reaction mechanisms. We stress, however, that this is only a conceptual classification. It is not meant to imply the existence of separate mechanisms, although that may be the case. This classification is best described by comparison with the familiar process of homologous recombination.

Homologous recombination involves the breakage and rejoining of genetically marked, but otherwise identical, parent molecules. It is an accurate and faithful process that depends on extensive nucleotide sequence homology and generally occurs more frequently than illegitimate recombination. In addition, homologous recombination can be either reciprocal or nonreciprocal, depending on whether both of the possible products are recovered. Similarly, in certain illegitimate events both products are recovered, and therefore the event is reciprocal. Examples include intramolecular inversions and the joining of two circular DNA molecules to produce a larger circle, as postulated by Campbell (1962) for the integration of phage λ into the *Escherichia coli* chromosome. In contrast, other illegitimate events such as deletions and duplications need not be reciprocal. In these cases, one of the products of the reaction can be sacrificed without affecting recovery of the other. Considered in this light, the mechanism(s) involved in the formation of deletions or duplications may differ from those involved in the formation of inversions or intermolecular recombinants.

III. GENETIC REARRANGEMENTS THAT MAY OCCUR BY A NONRECIPROCAL MECHANISM

A. Deletions

Deletions create a single novel joint by fusing two previously distant genetic elements. They are prized genetic events both because they result in the loss of information and because they can create interesting chimeric loci. For these reasons, deletion formation has been well studied.

Although deletion formation initially seemed haphazard, careful genetic analysis showed that this process is not totally random, but rather is influenced by local chromosome structure. For example, Coukell and Yanofsky (1971) observed that the frequency of deletions in the *tonB-trp* region varies considerably in different strains of *E. coli*. By performing transductional crosses, they demonstrated that these differences in frequency are due to differences in local chromosome structure among the various strains. Changing the *tonB-trp* region of a particular strain altered both the frequency and pattern of deletions in that region. Relevant features of chromosome structure, however, remained elusive until the development of DNA sequencing technology.

Franklin (1971) defined illegitimate recombination as lacking any "requirement for homology," but subsequent nucleotide sequence analysis revealed that small direct or inverted repeats are common features of sites at which deletions are formed. Farabaugh et al. (1978) examined nine novel joints formed by spontaneous deletions of 13 to 123 base pairs within the *lacI* gene of *E. coli*. By comparing the nucleotide sequence of the deletion joint with that of the wild type, they found that four of the nine deletions occurred at sites of "microhomology" consisting of 5 to 8 base pairs in direct repeat; each deletion removed one of these repeats and all DNA between the two. Such microhomologies are common, but not essential, features of sites of deletion formation. Albertini et al. (1982) examined deletions of 700 to 1,000 base pairs that produced *lacI-lacZ* fusions. Direct repeats of 5 to 17 base pairs were found at 8 of the 12 sites, although some of these microhomologies were imperfect. Berman and Jackson (1984) analyzed 10 deletions of 300 to 1,000 base pairs which fused *ompR* to *lacZ*; 3 had associated microhomologies. Similar results have been obtained with additional *lacI-lacZ* fusions (Brake et al., 1978), other small deletions in several regions of the *E. coli* chromosome (Emr et al., 1980; Ghosal and Saedler, 1979; Post and Nomura, 1980; Wu et al., 1980), and deletions in phages T4 and T7 (Owen et al., 1983; Pribnow et al., 1981; Studier et al., 1979).

There are, however, deletions lacking microhomology at the site of novel joint formation. For example, Benson and Bremer (1987) analyzed seven deletions of 600 to 1,000 base pairs which fused *lamB* to *lacZ*. Only one of the seven novel joints was formed in a region of limited (50%) microhomology. The other six deletions were between regions that exhibited only 20 to 35% homology. These authors compared the identified novel joints with other possible homology alignments that could have generated a fusion by an in-frame deletion. For each of the six utilized sites in *lacZ* there are between 7 and 50 in-frame matches in *lamB* with greater homology. Thus, these deletions did not occur between the most extensive microhomologies. Even in the collections of Farabaugh et al. (1978) and Albertini et al. (1982), where microhomologies are clearly important, only about half of the deletion sites contain small repeat sequences. Thus, microhomologies are clearly not obligatory for the formation of deletion endpoints.

Glickman and Ripley (1984) proposed that deletion termini may also occur at sites that flank small repeat sequences in inverted orientation. These repeats, termed quasipalindromes, could form transient hairpin or cruciform structures, bringing the endpoints into close proximity at the moment of deletion formation. Support for this model was obtained from the data collected by Farabaugh et al. (1978); quasipalindromes could be identified at five of the nine deletion sites. Three of these five sites also have direct repeat microhomology between the two fused sequences, and consequently, both types (or either type) of repeat sequence may participate in deletion formation in these cases. Deletions at the two remaining sites (which could form palindromes using 5 of 6 base pairs or 8 of 12 base pairs) could involve quasipalindrome structures but not direct repeats. Still, three of the deletion sites in the Farabaugh collection remain unexplainable by either model. Moreover, only two of the seven deletions of Benson and Bremer (1987) have quasipalindromes near the site of deletion formation, and neither of these quasipalindromes would correctly align the ends of the novel joint. Thus, for these deletions as well, neither direct repeats nor quasipalindromes appear to be necessary.

Other more subtle features of chromosome structure or specific DNA sequences may also be involved in deletion formation. Glickman et al. (1986) reported an over-representation of *lacI* mutations, including deletions, at sites containing the sequence 5'-GATC-3'. This sequence is the site of strand scission for the methyl-instructed mismatch repair system (Lu et al., 1984), and reactions such as this may well be important for initiating deletion formation (see below).

1. Deletion hot spots correspond to microhomologies

While one cannot, on the basis of nucleotide sequence, accurately predict sites for illegitimate events, one can infer from deletion "hot spots" that microhomologies can be preferred locations for this illegitimate event. If deletions were formed at random, the probability of isolating the same deletion more than once would be exceedingly small. However, there are deletion hot spots that repeatedly lead to the formation of identical deletions. To date, such hot spots invariably exhibit significant microhomology in direct repeat. For example, in the collection of Farabaugh et al. (1978), three of the four deletions associated with microhomology were isolated twice, whereas the five without microhomology were each isolated only once. In the study by Albertini et al. (1982), 14 of the 37 deletions localized by restriction analysis involve the largest available microhomology, and nucleotide sequence analysis confirmed that 10 of these 14 are identical. The other repeatedly isolated deletions also involve microhomologies, and the number of occurrences corresponds well with the extent of the microhomology. About 20% of the deletions isolated by Berman and Jackson (1984) occurred at a single hot spot containing a 9-of-10 base pair microhomology between *ompR* and *lacZ* (N. Allgood, unpublished data). Thus, there is substantial circumstantial evidence supporting the involvement of microhomology in direct repeat in deletion formation.

Albertini et al. (1982) have provided convincing evidence that short direct repeats can be hot spots for deletion formation by clever exploitation of the powerful *lacI* genetic system. Known amber mutations were recombined into the *lacI* gene to reduce the extent of the microhomology at the hot spot mentioned above by a single base. Each of two different single-base changes reduced the frequency of deletions at the hot spot by an order of magnitude. A third single-base change, which was located within the repeat sequence, but outside the microhomology, had no effect on the frequency. Therefore, the degree of homology between two short direct repeats is an important determinant for the frequency with which those two sequences participate in an illegitimate event such as deletion formation.

2. Cellular components that may participate in deletion formation

Mutants with alterations in various aspects of DNA metabolism have been tested for their effect on deletion formation. With few exceptions, these tests have been negative; this result is not surprising since these mutants were identified by selections or screens unrelated to illegitimate recombination. The enzymes involved in deletion formation therefore remain largely unknown.

Although Franklin's conclusion that illegitimate events such as deletion formation occur independently of RecA has been verified repeatedly, Albertini et al. (1982) found that deletion formation in their system was decreased 25-fold by *recA* mutations. A plausible explanation for this difference has been proposed by Syvanen et al. (1986). Albertini et al. (1982) selected *lacI* deletions in F′128, and Syvanen et al. (1986) showed that conjugal transfer during matings within F′ populations leads to elevated RecA activity. They propose that the RecA-dependent stimulation observed is a consequence of elevated RecA levels, which serve to stabilize an important intermediate. Thus, the observed effect of *recA* on the frequency of deletion formation may be related to F′128 and not to the direct involvement of RecA in the illegitimate event. Further support for this view comes from the fact that the spectrum of deletions Albertini et al. (1982) obtained in *recA* mutants is identical to that obtained in the wild type.

Coukell and Yanofsky (1970) reported that *polA* mutants exhibit an increased frequency of deletion formation in the *tonB-trp* region. This result suggests that this DNA polymerase may play an important role in deletion formation. Recently, Fix et al. (1987) have found that the endpoints of deletions obtained in a *polA1* mutant preferentially occur at or near the sequence 5'-GTGG-3'. A similar site preference is observed in wild-type strains, although to a much lower degree. The simplest explanation for these results is that deletion formation can occur by multiple pathways and *polA* mutants affect only a subset of these.

B. Precise Excision of Transposable Genetic Elements

Transposable genetic elements cause insertion mutations, and therefore their reversion requires a deletion event. Because transposable elements can be easily manipulated genetically and because revertants of some mutations can be selected, precise excision has received considerable attention as a model for illegitimate recombination. Although precise excision is a decidedly nonrandom event, these studies have revealed several features of deletion formation that are likely to be of general importance.

Prompted by early work by McClintock (1950) on mutable loci in maize which suggested that reversion may require some functions also required for transposition, Bukhari (1975) analyzed the excision of the transposable bacteriophage Mu. Starting with a strain containing a polar Mu insertion in the *lacZ* gene, he sought mutants regaining expression of a gene (*lacY*) distal to the insertion. Most of the resulting mutants were not true revertants since they remained *lacZ*; only a minority were *lacZ*+, resulting from precise excision. Khatoon and Bukhari (1981) showed that precise excision of Mu required *recA* gene function and one component of the Mu transposition machinery, the Mu *A* gene product. Another function required for transposition, the Mu *B* gene product, however, must be absent. Because of these requirements, and because excision is not correlated with transposition to a different chromosomal location, the excision event was termed an "abortive transposition."

Recently, Nag and Berg (1987) carried out a similar, but more extensive, analysis of Mu excision using a hybrid λ Mu phage, λ *plac*Mu3, which was inserted in the nontranslated leader region of a plasmid-borne *tet* gene. Again, revertants were sought by selecting for relief of polarity, i.e., the restoration of tetracycline resistance. DNA sequence analysis of these revertants showed that the deletion endpoints occur preferentially at or near directly repeated microhomologies, including the 5-base-pair direct repeats that are generated upon Mu insertion. The involvement of microhomology may reflect a common mechanism for deletion formation observed in other test systems. Nevertheless, the requirement of the Mu A and RecA proteins would appear to make Mu excision different in some aspects.

The transposons Tn*10* and Tn*5* have similar overall structures. Both are large

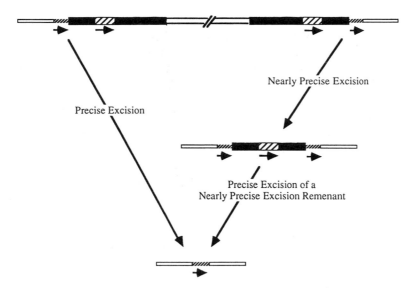

FIGURE 1. Three Tn*10* excision events. Thick lines represent the Tn*10*. Thin lines represent chromosomal DNA flanking the insertion. Thick black lines indicate the large inverted repeats at each end of the Tn*10*, and hatched regions with arrows underneath indicate short direct repeats. See text for discussion. (Adapted from Foster et al., 1981.)

(9,300 and 5,700 base pairs, respectively) and contain a central region with a drug resistance determinant (tetracycline and kanamycin, respectively) flanked by insertion elements (IS*10*, 1,329 base pairs, and IS*50*, 1,534 base pairs, respectively) in inverted repeat fashion. Upon insertion, both cause a small direct repeat of 9 base pairs. Precise excision of these elements therefore involves deletion of one of the direct repeats (microhomologies) plus all of the DNA between them. In addition, the insertion elements adjacent to these direct repeats are palindromes which potentially could form cruciform or hairpin structures.

Studies of Tn*5* have focused on precise excision. This event involves the direct repeats at the site of insertion and results in reversion to wild type, i.e., true reversion. With Tn*10*, however, three types of excision events have been characterized (Foster et al., 1981) (Fig. 1). The first is precise excision. The second, nearly precise excision, involves a 24-base-pair direct repeat which occurs within and very near the outside ends of the Tn*10* element. This reaction results in the excision of all but 50 base pairs of the Tn*10*, including one of the 24-base-pair repeats. The final reaction is the excision of the 50-base-pair remnant, restoring the original wild-type sequence. Precise excision and excision of the remnant result in true reversion, whereas nearly precise excision relieves the polar effects of the original insertion on the expression of distal genes.

Early work on the excision of both Tn*5* and Tn*10* has established the following facts. First, excision is not correlated with transposition to a new site, nor does it involve any of the transposition functions specified by the elements. In addition, it occurs independently of identified host recombination functions such as RecA (Egner and Berg, 1981; Foster et al., 1981; Lundblad and Kleckner, 1982). Accordingly, transposon excision is an example of illegitimate recombination. It is worth noting in this regard that precise excision is significantly more frequent than other examples of illegitimate recombination considered in this review. Whether this reflects the participation of a distinct, additional mechanism(s) is not yet clear.

The characteristic DNA structures associated with transposon insertions clearly contribute to the deletion potential. Reducing the length of the flanking inverted repeats or reversing their orientation markedly decreases the excision frequency (DasGupta et al., 1987; Egner and Berg, 1981; Foster et al., 1981). This observation argues convincingly that palindromes can promote deletion formation. Although no direct evidence has been presented, it seems likely that the short direct repeats that are produced upon insertion are important features as well. Nearly precise excision of Tn*10*, which involves a 24-base-pair direct repeat, is much more frequent than precise excision, an event which involves a 9-base-pair direct repeat (Foster et al., 1981). Thus, deletion frequency appears to be proportional to the size of the directly repeated sequence.

Analysis of *lacI* deletions that appear to involve microhomologies (see above) suggests that small deletions (<200 base pairs) are formed more frequently than larger deletions. The palindromes that flank transposons may bring the direct repeats at the insertion site into close proximity via the formation of cruciform structures and thereby effectively shorten the distance between the direct repeats. As for *lacI* deletions noted above, precise excision is stimulated when the transposable element is present on an F' plasmid. Syvanen et al. (1986) have shown that this stimulation is the result of homosexual transfer of the plasmid between cells in the population and that it occurs in the recipient. Since conjugation involves the transfer of single-stranded DNA to the recipient and since the formation of cruciform structures should be enhanced under single-stranded conditions, this hypothesis seems attractive.

Features of the nucleotide sequence other than the large inverted repeats and the flanking short direct repeats appear important as well, since reversion frequencies for different insertions within the same gene vary widely. For example, DasGupta et al. (1987) have examined the reversion of a very large (45,000-base-pair) Tn*5*-related transposon which was inserted at different sites within the *amp* gene of plasmid pBR322. In addition to reaffirming the importance of the size of the inverted repeat, these authors found variations in the reversion frequencies as large as 500-fold among the different insertion sites. This demonstrates the importance of local DNA sequence for the formation of deletions; however, the mechanistic implications of these observations are unclear.

To identify the enzymes participating in excision, mutants with altered frequencies of precise excision have been sought. Hopkins et al. (1983) modified the papillation screening technique of Konrad (1978) to search for mutants exhibiting an increased frequency of Tn*5* excision: mutagenized colonies of a *lacZ*::Tn*5* strain were screened by scoring for increased frequencies of lactose-positive papillae growing out of the lactose-negative colony after several days of incubation. Mutations at the *uup* locus of *E. coli* stimulate precise excision of Tn*5* by as much as 600-fold and of other elements, such as Tn*10*, to a lesser degree. In addition, they reduce growth of phage Mu. These mutations map to 21.3 min on the *E. coli* chromosome, a location that does not correspond to a previously identified gene.

Using a similar technique, Lundblad and Kleckner (1982) have isolated a number of mutations (called *tex* for Tn*10 ex*cision) that increase the frequency of Tn*10* excision. About half of these mutants have altered DNA metabolism; the lesions in these *tex* mutants are in genes specifying components of the methylation-instructed pathway for DNA repair (*dam*, *mutH*, *mutL*, *mutS*, or *uvrD*) or in the *recB* or *recC* gene (see Taylor, this volume). The phenotype of *tex* alleles that alter components of the mismatch repair pathway is the same as that of previously isolated, and presumably null, alleles (Lundblad and Kleckner, 1985). The effect of these mutations in Tn*10* excision would seem to implicate a mechanism involving the mismatch repair pathway (however, see below). Presumably, *tex* mutants altered in the *recB* or *recC* gene produce an altered RecBCD protein (Lundblad et al., 1984). Since the excision of Tn*10* does not normally depend on RecA or RecBCD (*recA* and *recBCD* null mutants excise at wild-type frequencies), the latter mutations appear to provide an additional excision pathway.

The *tex* mutations also stimulate excision of Tn*5* insertions, but the magnitude of the effect depends upon the particular Tn*5* (or Tn*10*) mutation studied (Lundblad and Kleckner, 1985; Lundblad et al., 1984). Presumably, this variation reflects some contribution of the local DNA sequence. Interestingly, the *tex* mutations increase precise excision and nearly precise excision of Tn*10*, but not the precise excision of the nearly precise excision remnant (Lundblad and Kleckner, 1982). Recall that the remnant is only 50 base pairs long (Fig. 1). This distinction may reflect the participation of mismatch repair and the altered form of RecBCD enzyme in deletion events involving substantial palindromic regions but not small microhomologies.

Lundblad and Kleckner (1982, 1985) examined the effects of mutations in other genes that specify functions involved in DNA metabolism on the frequency of Tn*10* excision. Most of these mutations, including several alleles of *polA* (see above), had no effect, but *mutD* (*dnaQ*) and *ssb* caused modest stimulation (3- to 15-fold). The finding that *polA* mutations do cause an increase in the frequency of excision of the precise excision remnant adds further evidence that this reaction may occur by a different mechanism. Certain mutations, such as mutations in the genes (*him* and *hip*) specifying host integration factor for bacteriophage λ, decrease the frequency of precise excision by 1,000-fold (Miller and Friedman, 1980).

C. Prophage Excision and the Formation of Specialized Transducing Phage

As noted above, chromosomal deletions need not occur by a reciprocal mechanism, but if they do, the second product would be a covalently closed circular DNA molecule composed of the deleted sequences. Normally, this molecule would be lost because it could not be propagated successfully. However, if the deleted region contains an origin of replication, it could be recovered.

Bacteriophages, such as λ, which form lysogens by chromosomal integration contain a functional origin and thus provide a means to maintain the sequences that were deleted. Prophage excision can also be used to study nonreciprocal deletion formation. In this case, of course, one would characterize the sequences deleted; the chromosome that suffered the deletion event would be lost.

Specialized transducing phages are produced by aberrant excision events thought to be illegitimate because they occur independently of RecA and phage-encoded functions (Gingery and Echols, 1968; Gottesman and Yarmolinsky, 1968). Such transducing phages have been studied extensively, and they provide a wealth of data relevant to illegitimate recombination. These data are especially valuable because the deletion which generates the transducing phage must be approximately 50 kilobase pairs, a size considerably larger than that examined in other experimental systems. Therefore, the analysis of transducing phage formation could address directly the important question of size constraint.

In an earlier review of illegitimate recombination, Weisberg and Adhya (1977) discussed prophage excision and transducing phage formation extensively. On the basis of genetic data available at that time, they concluded that there was weak, but suggestive, evidence in favor of preferred endpoints for aberrant excision. The nature of these preferred sites, however, was not known. More information has become available, but it remains primarily genetic, and hot spots are not always visible at this level of analysis. Also, microhomologies and quasipalindromes require DNA sequence analysis for detection. No doubt, with the rapidly increasing number of known DNA sequence, this will soon be remedied.

D. Duplications and Amplifications

Like deletions, tandem duplications create a single novel joint. However, certain duplications can arise at frequencies as high as 3% of the population (Anderson and Roth, 1981); this frequency is much higher than that observed for deletion formation at even the most active of hot spots. Some of this difference in frequency may occur because, unlike duplications, deletions are limited sharply in size since they may not extend into any essential sequences. Duplications have no such size constraint, and some have been estimated to carry more than 25% of the chromosome. Indeed, the only apparent constraint on duplicated sequences is that they may not include the termination site for chromosome replication (Anderson and Roth, 1981).

Because duplications create a novel joint, they can create operon or gene fusions. Anderson and Roth (1978a) exploited this fact to obtain duplications in the *his* region of the *Salmonella* chromosome. In particular, they selected for duplications that fuse the *hisD* gene to a distant, unrelated promoter. Because of obvious constraints imposed by the selection, duplications found by this method are rare (approximately one duplication per 10^9 cells per generation). Further analysis revealed at least two mechanisms for formation of these duplications. The first is RecA dependent and forms duplications that are identical. These duplications appear to result from unequal crossovers between homologous DNA sequences termed REP (for *re*petitive *e*xtragenic *p*alindromic; REP sequences are highly conserved inverted repeats of approximately 38 base pairs present at 500 to 1,000 copies per chromosome [Gilson et al., 1984; Higgins and Ames, 1981; Stern et al., 1984]). One REP sequence is located immediately upstream of *hisD*, and the other is found within the transcribed region near *argA* (J. Roth, personal communication). The second mechanism for duplication formation, operating at a sixfold lower frequency, is RecA independent and forms duplications with varying novel joints (Anderson and Roth, 1978a). These duplications, which occur by an illegitimate event, can be as large as 12% of the genome, but additional information that might shed light on the mechanism of this event is still lacking.

Two other, much less stringent selections were employed by Anderson and Roth (1978b, 1981) to detect duplications. Without the demand for operon fusion, the frequency increased to 0.001 to 3% of the population, depending on the region of the chromosome examined. However, this increase resulted primarily from the RecA-dependent mechanism. These duplications are formed by unequal cross-overs between homologous DNA sequences of substantial size, such as *rrn* operons, that are arranged as direct repeats on the chromosome. Such a mechanism could not generate viable deletions because deletions of this size would certainly be lethal.

An insight into the mechanisms of duplication came from the studies of Edlund and Normark (1981). In their analysis of chromosomal *ampC* duplications, they discovered that the junction point of one duplication occurred at a 12-base-pair sequence that was repeated on each side of the *ampC* gene, 10 kilobase pairs apart. This suggested the involvement of microhomology in duplication formation.

The importance of microhomology for the formation of duplications has recently been confirmed and extended by the work of Whoriskey et al. (1987), who studied *lacZ* gene amplifications. Starting with a strain containing an F′ plasmid carrying a *lacI-lacZ* fusion with a *lacI* promoter deletion, they selected for mutants able to grow on lactose. As with the *his* duplications of Anderson and Roth (1978a), the *lac* duplications can fuse the *lacI-lacZ* hybrid to a distant, unrelated promoter. In this case, however, the duplication, by itself, does not suffice. In all cases examined, the level of gene expression from a single copy of the duplication was not sufficient to allow growth on lactose. Rather, amplifications which increased the copy number of the duplication 40- to 200-fold were observed. These amplifications are the product of multiple unequal crossovers (a RecA-dependent event) between the initial duplication that generated the novel joint. Whoriskey et al. (1987) have determined the nucleotide sequence of 30 novel joints. By comparing these data with the known nucleotide sequence of the *lac* operon and surrounding regions, they found that seven of the nine different joints had occurred between sequences with 5 to 18 base pairs of homology. As is the case with deletions, the microhomologies correspond to hot spots. The two novel joints lacking microhomology were each isolated only once, while the most extensive microhomology was at a novel joint common to 12 of the 30 duplications. The striking similarity of DNA sequences that are prone to participate in both deletion and duplication formation suggests that these two events may occur by similar mechanisms. However, as noted above, small deletions (200 base pairs) are formed more frequently than larger ones, yet the most commonly isolated duplication was 9 kilobase pairs.

Because the *lacI* system of Whoriskey et al. (1987) depends on amplification (a process which requires RecA), it was not possible to determine whether RecA is required for formation of the original duplication. They argued that since none of these joints are at homologies larger than 20 base pairs, these events appear to be truly illegitimate rather than unequal crossovers.

IV. GENETIC REARRANGEMENTS THAT APPARENTLY OCCUR BY A RECIPROCAL MECHANISM

A. Inversions

An inversion mutation changes the orientation of the sequences affected relative to outside markers, with the concurrent generation of two novel joints. For example, an inversion in the sequence A B C D could produce the sequence A C B D, in which case the novel joints would be located between A and C and between B and D. If we assume that the substrate for inversion is a single DNA molecule, then the mechanism responsible must be reciprocal, since both products of the recombination remain in the same molecule. Numerous inversions in bacteria have been characterized. Indeed, a large inversion, covering 15% of the chromosome, provides a classic and prominent difference between the chromosomes of *E. coli* and *S. typhimurium* (Casse et al., 1973; Sanderson and Hall, 1970). Perhaps the most striking feature of inversions is their exceeding rarity.

Schmid and Roth (1983a, b) have studied the formation of inversions in *Salmonella* spp. The experimental system employed involves selection for a gene or operon fusion that restores expression of a promoterless *hisD* gene. In theory, inversions should create an appropriate novel joint at roughly the same frequency as duplications since the expected target sizes are comparable. However, this is not the case: deletions occur about 20 to 100 times more frequently than inversions (Schmid and Roth, 1983a). Nonetheless, endpoints of the inversions characterized to date are remarkably similar to those of duplications. In particular, the novel joint which is formed adjacent to *hisD* is located in the same region just upstream of the gene, most likely in the REP sequence. On the basis of this fact, these authors proposed that most inversions are formed in a RecA-dependent fashion by a crossover between homologous DNA sequences, such as REP, that are present as inverted repeats on the chromosome.

Inversions are commonly found in derivatives of phage λ, termed λdv. These defective phage are small circular DNA molecules that have lost all λ genes except *O*, *P*, and *cro*; they are therefore capable of autonomous replication. Chow et al. (1974) proposed that these molecules are formed by template switching during theta replication of λ. It is unlikely that such a mechanism generates the observed inversions in bacteria because this mechanism deletes most of the chromosome. However, Nag and Berg (1987) invoked a similar explanation to account for the appearance of short inverted sequences at the site of Mu excision. This reaction requires transposase. Indeed, most transposable elements are capable of catalyzing inversions (Kleckner and Ross, 1979).

B. Intermolecular Reactions

Intermolecular reactions fuse two circular DNA molecules into a single circular molecule generating two novel joints, each a junction of sequences previously in separate molecules. This is the mechanism proposed by Campbell (1962) for the integration of λ into the chromosome. Both reactants become part of the same molecule, and therefore, as with inversions, intermolecular reactions appear to be reciprocal. The most widely studied illegitimate intermolecular reactions are those between λ and pBR322-derived plasmids. These reactions produce phages capable of transducing plasmid-borne antibiotic resistance markers; such recombinant phages occur at a frequency of about 1 per 10^9 phages produced.

The structure of λ-pBR322 recombinants was studied by Pogue-Geile et al. (1980), who obtained two classes of recombinants at approximately equal frequency. The first class appears to be formed by λ integration into a pBR322 sequence similar to the *att* site on the *E. coli* chromosome. These molecules are probably due to Int-mediated site-specific recombination, but their elimination in *int* mutant infections was not tested. The second class does not appear to occur at a specific site on either molecule: since λ and pBR322 do not share large regions of homology, the recombinants apparently arise from illegitimate events. Sequence analysis of three members of the second class revealed that each fusion occurs between regions of microhomology 10 to 13 base pairs long. The two fusion joints in the recombinants are perfect reciprocal exchanges with no deleted or additional bases surrounding the joints (King et al., 1982).

In a *recA* mutant strain, phage-plasmid recombinants were isolated at a 10-fold lower frequency. Surprisingly, the structure of the resulting recombinant phages was also altered (Marvo et al., 1983). Each of the four recombinants sequenced appears to have deleted material surrounding the fusion joints concurrent with their fusion. One of the recombinants also contains a duplication of plasmid-derived sequences. These deletions and duplications differentiate the two fusion joints carried by each recombinant; each joint must be examined independently for important structural features. Of the eight fusion joints, only three are within microhomologies of 5 base pairs or more. Thus, microhomologies may be less important in the creation of recombinants in *recA* mutants. This difference may reflect simply the small numbers of recombinants analyzed, or it may reflect a genuine difference in mechanism.

Recombinants formed in $recA^+$ and *recA* mutant cells could arise by different classes of illegitimate events. Different mechanisms would account for the structural differences between the two types of products. In this case, however, one of the mechanisms would be greatly enhanced by RecA protein, which in other situations is not required for illegitimate recombination (reviewed by

Franklin, 1971; Weisberg and Adhya, 1977). As yet, no mutants are available which might clarify this difference.

Only intermolecular illegitimate recombination has been achieved in vitro. Ikeda et al. (1981) added plasmid DNA to an in vitro packaging system for λ DNA and isolated phage capable of transducing plasmid-borne resistance markers at 10^{-7} to 10^{-8} per total packaged phage. This frequency was not altered by *int* or *red* mutations in the lambda prophages or by *recA*, *recB*, *recC*, or *recF* mutations in the bacteria from which the extracts were prepared. This result indicates that the recombinants arise by illegitimate recombination.

E. coli DNA gyrase can both create and reseal double-strand breaks. Ikeda et al. (1981) found a role for DNA gyrase in their in vitro recombination system by studying the effect of oxolinic acid, which binds to the A subunit of gyrase and causes the enzyme to introduce single-stranded nicks in the DNA. Oxolinic acid increased the frequency of recombinant transducing phages as much as 15-fold in proportion to its concentration. With an extract from a *gyrA* mutant resistant to oxolinic acid, the enhancing effect of oxolinic acid was no longer observed. Further, coumermycin (a drug that binds to the B subunit of gyrase and inhibits its DNA supercoiling activity) partially reversed the effect of oxolinic acid with extracts of wild-type cells but not of a *gyrB* mutant, which is resistant to coumermycin. On the basis of these drug and mutant effects, Ikeda et al. (1981) proposed that DNA gyrase promotes these illegitimate recombination events. As further evidence for this proposal, they showed that increasing the concentration of *gyrA* protein or *gyrB* protein in their extracts increased the frequency with which recombinants were recovered.

By nucleotide sequence analysis, Ikeda and co-workers (Naito et al., 1984) found that four of the six recombinants analyzed had two novel joints, apparently produced by a simple reciprocal exchange. The other two recombinants contained deletions of parental DNA at the junction sites. None of the novel joints were at microhomologies of 5 base pairs or more. Comparison of these joints with preferred sites of gyrase cleavage in pBR322 showed that three of the five sites were at weak gyrase cleavage sites, but none were at the strongest gyrase cleavage sites (Ikeda et al., 1984).

Although the correspondence between gyrase cleavage sites and novel joints is not perfect, the effects of gyrase inhibitors and the addition of purified gyrase subunits suggest that DNA gyrase is involved in these illegitimate recombination events in vitro. It is not clear whether DNA gyrase is involved in illegitimate recombination in vivo. One suggestive piece of evidence that gyrase is involved in vivo is that the in vitro recombinants, particularly the two containing deletions (Naito et al., 1984), are structurally similar to those isolated in *recA* mutants by Marvo et al. (1983). From this similarity Marvo et al. (1983) proposed that the complex structure of their in vivo recombinants resulted from multiple cuts and ligations by DNA gyrase. Topoisomerase I has been proposed to mediate illegitimate recombination in mammalian cells by a similar mechanism (see Champoux and Bullock, this volume).

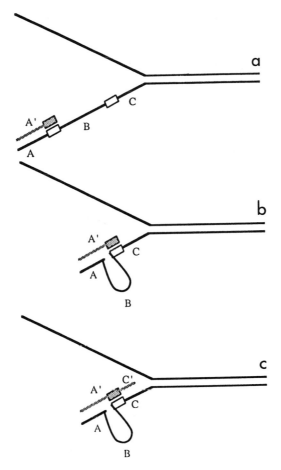

FIGURE 2. Slipped mispairing to create a deletion. (a) The two blocks indicate short homologous sequences in direct repeat (microhomologies). (b) Polymerase slippage could misalign parent and daughter strands. The structure in c could be resolved by further replication and segregation or by excision repair of the single-stranded loop. (Adapted from Albertini et al., 1982.)

V. POSSIBLE MECHANISMS OF ILLEGITIMATE RECOMBINATION

Franklin (1971) proposed two fundamentally different mechanisms for illegitimate recombination. The first, commonly termed slipped mispairing, involves errors in DNA replication in which the polymerase either fails to copy or repeatedly copies a given sequence to yield a deletion or a tandem duplication, respectively (Fig. 2). This model stems from the insights of Streisinger et al. (1967) on the mechanism of frameshift mutations: according to this view, a polymerase may occasionally misalign the daughter and parent strands at sequences that are partially homologous and produce a mutant daughter molecule. This model equates deletions and duplications with very large − and + frameshift mutations. The slipped mispairing model does not involve the cutting and rejoining of DNA molecules, and therefore it cannot yield reciprocal products.

The second mechanism for illegitimate recombination involves the cutting and rejoining of DNA molecules. These two reactions could happen either separately or in a concerted fashion. Such an event is reminiscent of homologous recombination, and accordingly, it offers the potential for reciprocal events. Enzymes that catalyze DNA scission and ligation are required by this model. The list of such enzymes is large but would include DNA repair enzymes, DNA gyrase, and DNA ligase.

Neither the slipped mispairing nor the cut and rejoin model can adequately account for all of the data available that characterize illegitimate recombination. Inversions and intermolecular events that result in the formation of a single DNA molecule appear to be the product of a reciprocal event, and therefore the slipped mispairing model cannot be invoked without major modification. The cut and rejoin model can be proposed for any of the observed events; however, this model fails to make specific predictions about the DNA sequences at hot spots and about proteins involved. In addition, it is quite possible that many illegitimate events involve important aspects of both models. For example, single-stranded DNA loops produced by slipped mispairing could be repaired by excision and religation to produce a deletion mutation. Nevertheless, these two models provide a useful framework in which the results that have been obtained can be considered.

A. Local DNA Sequence

The compelling data presented above indicate that local DNA sequence can influence illegitimate recombination, and although we do not understand all of the contributions of the local sequence, certain features, such as microhomology, figure prominently. Microhomology is characteristic of hot spots for the formation of deletions and duplications and for precise excision of transposable elements (see above). Recall, however, that microhomology is not absolutely required.

The slipped mispairing model can account for the involvement of direct repeats as sites at which the polymerase could realign after slippage (Fig. 1). In contrast, the cut and rejoin model makes no specific predictions about microhomology, although it could certainly be helpful. It has been suggested that low-level homologous recombination accounts for illegitimate events at regions with microhomology (Whoriskey et al., 1987). We think this unlikely. All of the known *E. coli* recombination systems operating in wild-type bacteria require RecA, and illegitimate events that are observed at sequences with microhomology do not require this protein. Because the cut and rejoin model makes no specific prediction about the enzymes that may participate, it is difficult to comment on the potential effect of microhomology.

Studies on precise excision of transposable elements demonstrate that large palindromes enhance the formation of deletions; quasipalindromes may also be important, although the available data are less convincing (see above). Palindromes could stabilize the single-stranded intermediates in the slipped mispairing model. Again, the cut and join model makes no clear predictions about the effects of inverted repeat sequences.

Many of the enzymes that catalyze DNA scission have a sequence preference. If these enzymes participate in illegitimate events, one would predict that their recognition sequences would frequently occur at or near novel joints. To date, no obvious sequence has emerged, and attempts to correlate the DNA sequence of novel joints with sites of gyrase cutting, for example, have been inconclusive. However, certain observations warrant further scrutiny. Glickman et al. (1986) suggested that the sequence 5'-GATC-3' is located at preferred sites for novel joint formation. The 5'-GATC-3' sequence is thought to be the site for strand scission during methyl-instructed DNA mismatch repair (Lu et al., 1984), and this scission may account for the genetic results discussed in the following section.

B. Proteins

The slipped mispairing model for illegitimate recombination predicts the direct involvement of a DNA polymerase in novel joint formation. In several different experimental systems, *polA* mutants do exhibit an increased frequency of illegitimate events (Coukell and Yanofsky, 1970; Fix et al., 1987). However, the mutations employed for these studies are invariably *polA1* or a similar allele. The *polA1* allele causes a dramatic decrease in polymerase function with little effect on the exonuclease activity of the enzyme (Joyce et al., 1985). Accordingly, it seems unlikely that the effect of this mutation is the direct result of slipped mispairing by the enzyme during DNA synthesis. Rather, the mutation may lead to an increase in gapped DNA present in the mutant cell, and it is this damaged DNA that leads ultimately to the generation of deletions. A similar explanation could be offered for mutations in *dnaQ*, the gene specifying the proofreading subunit of polymerase III (Scheurmann et al., 1983). Mutants with altered *dnaQ* function exhibit increased levels of base pair mismatches (Lundblad and Kleckner, 1985), which may also stimulate the activity of functions that lead to an illegitimate event. Thus, mutant analysis has produced no data that strongly favor the slipped mispairing model over alternative possibilities.

Conditions that increase or stabilize the amount of single-stranded DNA frequently enhance the formation of deletions. For example, precise excision of transposable elements is markedly more frequent when they are present on an F' plasmid rather than on the chromosome. This stimulation appears to be the result of conjugal transfer of the F', which involves single-stranded intermediates (Syvanen et al., 1985). Transposons contain large palindromes that could stabilize the single-stranded DNA molecule. Also, the *ssb-113* mutation, which decreases the activity of single-stranded DNA-binding protein (Meyer et al., 1982), stimulates precise excision of elements with substantial inverted repeats (Lundblad and Kleckner, 1985).

Mutants with increased frequencies of precise excision of transposable elements have been characterized. Genes thus identified fall into three groups: (i) genes that specify components of DNA mismatch repair, i.e., *dam*, *mutS*, *mutL*, *mutH*, and *uvrD* (Lundblad and Kleckner, 1982, 1985); (ii) genes for the RecBCD enzyme, i.e., *recB* and *recC* (Lundblad et al., 1984); and (iii) a gene for an unknown function, i.e., *uup* (Hopkins et al., 1983). Each of these is considered in turn.

The evidence implicating the enzymes of methyl-instructed DNA mismatch repair in the formation of deletions is indeed striking. Mutants defective in these enzymes show marked increases in the frequency of precise excision, and their site of action, the sequence 5'-GATC-3', is often found at or near novel joints (Glickman et al., 1986). Despite these findings, the mechanistic implications remain mysterious. One might expect *dam* mutations to increase precise excision, perhaps because they broaden mismatch repair activity (Radman et al., 1980; Glickman and Radman, 1980). However, mutations in *mutH*, *mutL*, *mutS*, or *uvrD* stimulate precise excision as well, and these mutations should reduce mismatch repair activity (Glickman and Radman, 1980). To further confound the issue, it has been shown that the effect of *dam* and each of the other mutations is additive (Lundblad and Kleckner, 1985). We cannot formulate a model to explain these findings, and we suspect, as did Lundblad and Kleckner, that the Dam methylase must have an additional function not associated with mismatch repair. In any event, the stimulatory effect of the mismatch repair mutations is abolished by reducing the size of the inverted repeats (Lundblad and Kleckner, 1985). Accordingly, the pathway for precise excision of transposons may not contribute significantly to the other illegitimate events discussed here.

The *recB* and *recC* alleles identified as mutations that stimulate precise excision (*tex* alleles) alter the RecBCD enzyme (see Taylor, this volume). These mutant enzymes have normal nuclease activity but reduced activity of Chi sites (the sites at which the wild-type enzyme nicks the DNA during DNA unwinding) (Ponticelli et al., 1985). The significance of this altered activity is not yet clear. In any event, since RecBCD enzyme is normally not required for precise excision, these mutations apparently activate an alternative pathway. Further study will provide insights into RecBCD enzyme function but may not shed light on the mechanism of precise excision.

As with the *tex* alleles, studies with *uup* are promising, but more work is clearly required (Hopkins et al., 1983). In particular, the product of the gene must be identified and some insights into its biological activity must be obtained. Data currently available do not allow mechanistic predictions.

DNA gyrase, which transiently introduces double-strand breaks in DNA, is a prime candidate for participation in the cut and join mechanism of illegitimate recombination. Ikeda and colleagues demonstrated that this enzyme can catalyze illegitimate events in vitro (Ikeda et al., 1981; Ikeda and Shiozaki, 1984; Naito et al., 1984) and proposed that gyrase mediates illegitimate recombination via a mechanism involving exchange of gyrase subunits (Ikeda et al., 1981). In this model gyrase binds and cuts both substrate DNAs, but before the cuts are resealed, subunits of the gyrase molecules dissociate from each other and exchange with those of another gyrase molecule. Resealing of the DNA ends bound to the exchanged subunits creates two reciprocal novel joints. Multiple exchanges such as those observed by Marvo et al. (1983) would generate more complex recombinants. There is only circumstantial evidence that gyrase functions in illegitimate recombination in vivo; the novel joints between λ and pBR322 produced by illegitimate recombinations in vivo resemble those produced in vitro. As discussed above, these in vivo reaction products are complex, and there does not appear to be strong dependence on microhomology. Accordingly, a gyrase-dependent mechanism may produce only a subset of the observed illegitimate events (for further discussion, see Champoux and Bullock, this volume).

VI. CONCLUSION

The application of DNA sequencing techniques to the study of illegitimate recombination has revealed important features of the local DNA structure at which these events occur. The preference for microhomology and large palindromes seems established. Given these results, the slipped mispairing model appears most attractive and can account for the most commonly occurring illegitimate events. Details of the reaction mechanism and the identity of the enzymes responsible, however, remain obscure. There is no direct evidence for a particular DNA polymerase, and additional enzymes may participate as well. For example, the single-stranded loop produced by slipped mispairing may be repaired by excision rather than simply segregating upon further replication. Although this model may currently be the most attractive, it cannot explain all of the observed reactions. Certain events occur without microhomology, and some events occur by a reaction that appears to be reciprocal. Accordingly, other mechanisms for illegitimate recombination such as the cut and join mechanism must operate as well. Many different enzymes could participate in such a cut and join reaction, and at least half a dozen have been implicated. Of these enzymes, gyrase is perhaps the most attractive; indeed, in vitro at least, this enzyme can promote illegitimate recombination. The reactions are complicated, however, and often produce more than a single novel joint.

The enzymes promoting illegitimate recombination may be identified by further mutant analysis. Additional selections or screens to detect mutants with altered frequencies of illegitimate recombination must be developed. Identification of the functions altered in these mutants could support one of the two mechanisms discussed here or suggest a new one. Screens such as the colony papillation technique have identified mutants with increased rates of precise of transposons; these techniques can be modified to search for mutants with altered rates of other illegitimate events. In this regard, we have devised a screen to identify mutants with increased rates of formation of small deletions fusing *lacZ* to *ompR* to produce a functional hybrid gene. Using this screen, we have identified strains which exhibit increased or decreased rates of deletion formation. In our analysis at least two classes have emerged: those which require microhomology and those which do not. Further analysis may shed light on the enzymes that participate in these two types of reactions.

ACKNOWLEDGMENTS. We thank S. Adhya, A. Campbell, N. Kleckner, R. Kolodner, M. Schmid, and M. Syvanen for helpful discussions, J. Roth for helpful discussion and for communicating unpublished results, and members of the laboratory staff for reviewing the manuscript.

This work was supported by Public Health Service grants to T.J.S. from the National Institutes of Health. N.D.A. is supported by a National Science Foundation Graduate Fellowship.

LITERATURE CITED

Albertini, A. M., M. Hofer, M. P. Calos, and J. H. Miller. 1982. On the formation of spontaneous deletions: the importance of short sequence homologies in the generation of large deletions. *Cell* **29:** 319–328.

Anderson, R. P., and J. Roth. 1981. Spontaneous tandem genetic duplications in *Salmonella typhimurium* arise by unequal recombination between rRNA (RRN) cistrons. *Proc. Natl. Acad. Sci. USA* **78:**3113–3117.

Anderson, R. P., and J. R. Roth. 1978a. Tandem chromosomal duplications in *Salmonella typhimurium*: fusion of histidine genes to novel promoters. *J. Mol. Biol.* **119:**147–166.

Anderson, R. P., and J. R. Roth. 1978b. Tandem genetic duplications in *Salmonella typhimurium*: amplification of the histidine operon. *J. Mol. Biol.* **126:**53–71.

Benson, S. A., and E. Bremer. 1987. In vivo selection and characterization of internal deletions in the *lamB::lacZ* gene fusion. *Gene* **52:**165–173.

Berman, M. L., and D. E. Jackson. 1984. Selection of *lac* gene fusions in vivo: *ompR-lacZ* fusions that define a functional domain of the *ompR* gene product. *J. Bacteriol.* **159:**750–756.

Brake, A. J., A. V. Fowler, I. Zabin, J. Kania, and B. Muller-Hill. 1978. β-Galactosidase chimeras: primary structure of a lac repressor-β-galactosidase protein. *Proc. Natl. Acad. Sci. USA* **75:**4824–4827.

Bukhari, A. I. 1975. Reversal of mutator phage Mu integration. *J. Mol. Biol.* **96:**87–99.

Campbell, A. M. 1962. Episomes. *Adv. Genet.* **11:**101–145.

Casse, F., M. C. Pascal, and M. Chippaux. 1973. Comparison between the chromosomal maps of *Escherichia coli* and *Salmonella typhimurium*. *Mol. Gen. Genet.* **124:**253–257.

Chow, L. T., N. Davidson, and D. Berg. 1974. Electron microscope study of the structures of lambda-dv DNAs. *J. Mol. Biol.* **86:**69–89.

Coukell, M. B., and C. Yanofsky. 1970. Increased frequency of deletions in DNA polymerase mutants of *Escherichia coli*. *Nature* (London) **228:**633–635.

Coukell, M. B., and C. Yanofsky. 1971. Influence of chromosome structure on the frequency of *tonB-trp* deletions in *Escherichia coli*. *J. Bacteriol.* **105:**864–872.

DasGupta, U., K. Weston-Hafer, and D. E. Berg. 1987. Local DNA sequence control of deletion formation in *Escherichia coli* plasmid pBR322. *Genetics* **115:**41–49.

Edlund, T., and S. Normark. 1981. Recombination between short DNA homologies causes tandem duplication. *Nature* (London) **292:**269–271.

Egner, C., and D. E. Berg. 1981. Excision of transposon Tn5 is dependent on the inverted repeats but not on the transposase function of Tn5. *Proc. Natl. Acad. Sci. USA* **78:**459–463.

Emr, S. D., J. Hedgpeth, J.-M. Clement, T. J. Silhavy, and M. Hofnung. 1980. Sequence analysis of mutations that prevent export of lambda receptor, an *Escherichia coli* outer membrane protein. *Nature* (London) **285:**82–85.

Farabaugh, P. J., U. Schmeissner, M. Hofer, and J. H. Miller. 1978. Genetic studies of the *lac* repressor. VII. On the molecular nature of spontaneous hotspots in the *lacI* gene of *Escherichia coli*. *J. Mol. Biol.* **126:**847–863.

Fix, D. F., P. A. Burns, and B. W. Glickman. 1987. DNA sequence analysis of spontaneous mutation in a *polA1* strain of *Escherichia coli* indicates sequence-specific effects. *Mol. Gen. Genet.* **207:**267–272.

Foster, T. J., V. Lundblad, S. Hanley-Way, S. M. Halling, and N. Kleckner. 1981. Three Tn10-associated excision events: relationship to transposition and role of direct and inverted repeats. *Cell* **23:**215–227.

Franklin, N. C. 1967. Extraordinary recombinational events in *Escherichia coli*. Their independence of the rec⁺ function. *Genetics* **55**:699–707.

Franklin, N. C. 1971. Illegitimate recombination, p. 175–194. *In* A. D. Hershey (ed.), *The Bacteriophage Lambda*. Cold Spring Harbor Laboratory, Cold Spring Harbor, N.Y.

Ghosal, D., and H. Saedler. 1979. IS2-*6l* and IS2-*6ll* arise by illegitimate recombination from IS2-*6*. *Mol. Gen. Genet.* **176**:233–238.

Gilson, E., J. M. Clement, D. Brutlag, and M. Hofnung. 1984. A family of dispersed repetitive extragenic palindromic DNA sequences in *E. coli*. *EMBO J.* **3**:1417–1422.

Gingery, R., and H. Echols. 1968. Integration, excision, and transducing particle genesis by bacteriophage lambda. *Cold Spring Harbor Symp. Quant. Biol.* **33**:721–727.

Glickman, B. W., P. A. Burns, and D. F. Fix. 1986. Mechanisms of spontaneous mutagenesis: clues from altered mutational specificity in DNA repair defective strains, p. 259–281. *In* D. M. Shankel, P. E. Hartman, and K. T. Hollaender (ed.), *Antimutageneses and Anticarcinogenesis, Mechanisms*. Plenum Publishing Corp., New York.

Glickman, B. W., and M. Radman. 1980. *Escherichia coli* mutator mutants deficient in methylation-instructed DNA mismatch correction. *Proc. Natl. Acad. Sci. USA* **77**:1063–1067.

Glickman, B. W., and L. S. Ripley. 1984. Structural intermediates of deletion mutagenesis: a role for palindromic DNA. *Proc. Natl. Acad. Sci. USA* **81**:512–516.

Gottesman, M. E., and M. B. Yarmolinsky. 1968. The integration and excision of the bacteriophage lambda genome. *Cold Spring Harbor Symp. Quant. Biol.* **33**:735–747.

Higgins, C. F., and G. F. Ames. 1981. Two periplasmic transport proteins which interact with a common membrane receptor show extensive homology: complete nucleotide sequences. *Proc. Natl. Acad. Sci. USA* **78**:6038–6042.

Hopkins, J. D., M. Clements, and M. Syvanen. 1983. New class of mutations in *Escherichia coli* (*uup*) that affect precise excision of insertion elements and bacteriophage Mu growth. *J. Bacteriol.* **153**: 384–389.

Ikeda, H., I. Kawasaki, and M. Gellert. 1984. Mechanism of illegitimate recombination: common sites for recombination and cleavage mediated by *E. coli* DNA gyrase. *Mol. Gen. Genet.* **196**:546–549.

Ikeda, H., K. Moriya, and T. Matsumoto. 1981. In vitro study of illegitimate recombination: involvement of DNA gyrase. *Cold Spring Harbor Symp. Quant. Biol.* **45**:399–408.

Ikeda, H., and M. Shiozaki. 1984. Nonhomologous recombination mediated by *Escherichia coli* DNA gyrase: possible involvement of DNA replication. *Cold Spring Harbor Symp. Quant. Biol.* **49**:401–409.

Joyce, C. M., D. M. Fujii, H. S. Laks, C. M. Hughes, and N. D. F. Grindley. 1985. Genetic mapping and DNA sequence analysis of mutations in the *polA* gene of *Escherichia coli*. *J. Mol. Biol.* **186**:283–293.

Khatoon, H., and A. I. Bukhari. 1981. DNA rearrangements associated with reversion of bacteriophage Mu-induced mutations. *Genetics* **98**:1–24.

King, S. R., M. A. Krolewski, S. L. Marvo, P. J. Lipson, K. L. Pogue-Geile, J. H. Chung, and S. R. Jaskunas. 1982. Nucleotide sequence analysis of in vivo recombinants between bacteriophage lambda DNA and pBR322. *Mol. Gen. Genet.* **186**:548–557.

Kleckner, N., and D. Ross. 1979. Translocation and other recombination events involving the tetracycline-resistance element Tn*10*. *Cold Spring Harbor Symp. Quant. Biol.* **43**:417–428.

Konrad, E. B. 1978. Isolation of an *Escherichia coli* K-12 *dnaE* mutation as a mutator. *J. Bacteriol.* **133**:1197–1202.

Lu, A. L., K. Welsh, S. Clark, S. S. Su, and P. Modrich. 1984. Repair of DNA basepair mismatches in extracts of *Escherichia coli*. *Cold Spring Harbor Symp. Quant. Biol.* **49**:589–596.

Lundblad, V., and N. Kleckner. 1982. Mutants of *Escherichia coli* K12 which affect excision, p. 245–258. *In* J. F. Lemontt and W. M. Generoso (ed.), *Molecular and Cellular Mechanisms of Mutagenesis*. Academic Press, Inc., New York.

Lundblad, V., and N. Kleckner. 1985. Mismatch repair mutations of *Escherichia coli* K12 enhance transposon excision. *Genetics* **109**:3–19.

Lundblad, V., A. F. Taylor, G. R. Smith, and N. Kleckner. 1984. Unusual alleles of *recB* and *recC* stimulate excision of inverted repeat transposons Tn*10* and Tn*5*. *Proc. Natl. Acad. Sci. USA* **81**: 824–828.

Marvo, S. L., S. R. King, and S. R. Jaskunas. 1983. Role of short regions of homology in intermolecular illegitimate recombination events. *Proc. Natl. Acad. Sci. USA* **80:**2452–2456.

McClintock, B. 1950. The origin and behavior of mutant loci in maize. *Genetics* **36:**344–355.

Meyer, R. R., D. C. Rein, and J. Glassberg. 1982. The product of the *lexC* gene of *Escherichia coli* is single-stranded DNA-binding protein. *J. Bacteriol.* **150:**433–435.

Miller, H. I., and D. I. Friedman. 1980. An *E. coli* gene product required for λ site-specific recombination. *Cell* **20:**711–719.

Nag, D. K., and D. E. Berg. 1987. Specificity of bacteriophage Mu excision. *Mol. Gen. Genet.* **207:**395–401.

Naito, A., S. Naito, and H. Ikeda. 1984. Homology is not required for recombination mediated by DNA gyrase of *Escherichia coli*. *Mol. Gen. Genet.* **193:**238–243.

Owen, J. E., D. W. Schultz, A. Taylor, and G. R. Smith. 1983. Nucleotide sequence of the lysozyme gene of bacteriophage T4: analysis of mutations involving repeated sequences. *J. Mol. Biol.* **165:**229–248.

Pogue-Geile, K. L., S. Dassarma, S. R. King, and S. R. Jaskunas. 1980. Recombination between bacteriophage lambda and plasmid pBR322 in *Escherichia coli*. *J. Bacteriol.* **142:**992–1003.

Ponticelli, A. S., D. W. Schultz, A. F. Taylor, and G. R. Smith. 1985. Chi-dependent DNA strand cleavage by RecBC enzyme. *Cell* **41:**145–151.

Post, L. E., and M. Nomura. 1980. DNA sequences from the *str* operon *Escherichia coli*. *J. Biol. Chem.* **255:**4660–4666.

Pribnow, D., D. C. Sigurdson, L. Gold, B. S. Singer, N. Napoli, J. Brosius, T. J. Dull, and H. F. Noller. 1981. rII cistrons of bacteriophage T4 DNA sequence around the intercistronic divide and positions of genetic landmarks. *J. Mol. Biol.* **149:**337–376.

Radman, M., R. E. Wagner, B. W. Glickman, and M. Meselson. 1980. DNA methylation, mismatch correction and genetic stability, p. 121–130. *In* M. Alacevic (ed.), *Progress in Environmental Mutagenesis*. Elsevier/North-Holland Biomedical Press, Amsterdam.

Sanderson, K. E., and C. A. Hall. 1970. F-prime factors of *Salmonella typhimurium* and an inversion between *S. typhimurium* and *Escherichia coli*. *Genetics* **64:**215–228.

Scheurmann, R., S. Tam, P. M. J. Burgers, C. Lu, and H. Echols. 1983. Identification of the e-subunit of *Escherichia coli* DNA polymerase III holoenzyme as the *dnaQ* gene product: a fidelity subunit for DNA replication. *Proc. Natl. Acad. Sci. USA* **80:**7085–7089.

Schmid, M. B., and J. R. Roth. 1983a. Selection and endpoint distribution of bacterial inversion mutations. *Genetics* **105:**539–557.

Schmid, M. B., and J. R. Roth. 1983b. Genetic methods for analysis and manipulation of inversion mutations in bacteria. *Genetics* **105:**517–537.

Stern, M. J., G. F. Ames, N. H. Smith, E. C. Robinson, and C. F. Higgins. 1984. Repetitive extragenic palindromic sequences: a major component of the bacterial genome. *Cell* **37:**1015–1026.

Streisinger, G. S., Y. Okada, J. Emrich, J. Newton, A. Tsugita, E. Terzaghi, and M. Inouye. 1967. Frameshift mutations and the genetic code. *Cold Spring Harbor Symp. Quant. Biol.* **31:**77–84.

Studier, F. W., A. H. Rosenberg, M. N. Simon, and J. J. Dunn. 1979. Genetic and physical mapping in the early region of bacteriophage T7 DNA. *J. Mol. Biol.* **135:**917–937.

Syvanen, M., J. D. Hopkins, T. J. Griffin, T. Y. Liang, K. Ippen-Ihler, and R. Kolodner. 1986. Stimulation of precise excision and recombination by conjugal proficient F′ plasmids. *Mol. Gen. Genet.* **203:**1–7.

Weisberg, R. A., and S. Adhya. 1977. Illegitimate recombination in bacteria and bacteriophage. *Annu. Rev. Genet.* **11:**451–473.

Whoriskey, S. K., V.-H. Nghiem, P.-M. Leong, J.-M. Masson, and J. H. Miller. 1987. Genetic rearrangements and gene amplification in *Escherichia coli*: DNA sequences at the junctures of amplified gene fusions. *Genes Dev.* **1:**227–237.

Wu, A. M., A. B. Chapman, and T. Platt. 1980. Deletions of distal sequence affect termination of transcription at the end of the tryptophan operon in *E. coli*. *Cell* **19:**829–836.

Chapter 10

Bacterial Insertion Sequences

Michael Syvanen

I. INTRODUCTION

Insertion sequences are highly specialized genetic elements that are often present in multiple copies in bacterial genomes. Unlike other bacterial genes, these elements have a highly variable genetic map location, even within populations of the same bacterial species (Brahma et al., 1982; Hartl et al., 1986). Though present in most, if not all, bacteria, insertion sequences do not confer obvious phenotypes to strains that carry them, other than the induction of mutations and the rearrangement of chromosomes, though they are often closely linked to highly selected accessory genes. There are indications that novel molecular mechanisms may be employed when these elements transpose to new sites or rearrange chromosomes. The search for molecular mechanisms associated with insertion sequences has stimulated so much research interest that over 20 different elements have now been scrutinized.

This review provides an overview of bacterial insertion sequences. A comprehensive overview of these elements was last given by Iida et al. (1983), though more specialized reviews have appeared since (Syvanen, 1984; Grindley

FIGURE 1. Prototype insertion sequence. The insertion sequence is indicated by the sequences between points a and b. The open arrows represent the recipient (target) sequences duplicated upon insertion. The solid arrows show the outside limits of the insertion sequence, whose ends are similar in sequence but in inverted orientation. Most insertion sequences have at least one promoter, p, that transcribes a gene, *tnpA*, that encodes a transposase which promotes insertion sequence movement.

and Reed, 1985; Deonier, 1987; Craig and Kleckner, 1987). After presenting a classification system and summary of the different kinds of insertion sequences and transposons that have been studied, I discuss a current view of the transposition mechanism with an emphasis on insights gained through studying the cointegration reaction. I also include an update on bacterial genes and external stimuli that influence both transposition mechanisms and the precise excision reaction, and a summary of the arguments favoring an evolutionary function for these elements.

Twenty-eight different, relatively well-studied insertion sequences are listed in Table 1. This partial listing includes those elements for which some sequence information is available. New elements are being reported monthly for a variety of different bacteria; e.g., pseudomonads appear to be well endowed with insertion sequences (Gertman et al., 1986; Scordilis et al., 1987). The property that unifies the insertion sequences is their ability to transpose. In addition, they have three salient common features (Fig. 1). First, the specific DNA sequences at their ends are important for movement; in most cases the sequences at the right and left ends are nearly identical but are in an inverted orientation. Second, as far as is known each element contains an open reading frame coding for proteins required for movement. These proteins are called transposases and presumably act on the specific end sequences. The size of the longest open reading frame or of the known transposase is listed in Table 1. And third, when an element moves to a new site, a small duplication of the recipient sequences is usually generated; the size of the duplication, which flanks the element at its site of insertion, is characteristic for a given element. These small duplications are thought to arise through the repair of staggered cuts that are made in the recipient sequences upon insertion.

The term transposon is commonly used to describe a transposable element that contains additional accessory genes, such as genes for antibiotic resistance and toxin production. Because these accessory genes do not directly affect transposition properties and because transposons do not differ from insertion sequences in their transposition mechanism, the terms transposons and insertion sequences are often used synonymously. In this review I maintain the original distinction; insertion sequences do not carry accessory genes, whereas transposons do. This convention is reflected in the nomenclature in Table 1.

II. CLASSIFICATION OF THE INSERTION SEQUENCES

At the level of nucleotide sequence, there are no conspicuous sequence homologies between any of the pairs of elements listed in Table 1. Mollet et al.

(1981) found a similarity in the end sequences of IS*1*, IS*26*, IS*903*, and, to a lesser degree, IS*50*. It is unclear whether these four sequences share other properties that distinguish them from rest of the elements. On the basis of a number of general criteria, I have divided the elements into four assemblages; the first and second are the "class 1" transposable elements, the third is the "class 2" elements (with "classes" as originally defined by Kleckner [1981]), and the fourth assemblage includes two site-specific transposons.

A. Assemblage 1

The 10 elements in assemblage 1 may share a common ancestor, though it is less clear whether they share common mechanisms of movement. Kathary et al. (1985) and Mahillon et al. (1985) found that a region of the transposase gene (or longest open reading frame) is related in the insertion sequences listed as assemblage 1a. The six elements display a similar 35-amino-acid sequence in their transposases (Fig. 2). In addition, members of this group code for a single transposase of approximately the same size, and most of these elements produce an 8- to 12-base-pair duplication in the recipient site upon insertion. The nature of these similarities in amino acid sequence is interesting to elaborate upon because the similarities suggest that this region constitutes at least part of the active site of the transposase.

The region of the transposase with the strongest amino acid sequence alignment (Fig. 2) displays not only a consensus primary structure but also a consensus secondary structure. Using Chou and Fasman (1978) pseudo-probability calculations on each of the assemblage 1a elements, I found the following pattern. In all of the sequences listed, residues −10 to 1 have a high value for the beta sheet, though there is little sequence alignment through this region. Position 6 defines the amino-terminal end of an alpha helix, designated Helix 1, that extends to about position 15 to 18. The region between 16 and 20 has no identifiable secondary structure, while in five of the six cases position 20 to 22 defines the amino-terminal end of another alpha helix, designated Helix 2. The Chou-Fasman algorithm can only weakly predict secondary structure for any given protein sequence; however, in Fig. 2 the assignment is based upon an average from six homologous but highly diverged sequences. This result should promote confidence in the validity of the pattern derived here.

Helix 2 has features that invite further speculation. If this alpha helix is real, then the highly conserved amino acids define one of its faces, i.e., Tyr-20, Arg-24, Glu-27, Arg- or Lys-31, and Lys-34. The common tyrosine at position 20 and glutamate at position 27 bring to mind some features of enzymes known to break and rejoin phosphodiester bonds (Wang, 1985). A nonspecific DNA-binding site could be suggested from the abundance of arginines and lysines along the face of Helix 2. A common binding site is supported by another line of reasoning. The complete lack of homology outside the Helix 1-Helix 2 region strongly suggests that the conserved residues reflect high functional constraint and not just common ancestry; it also implies that these regions perform a similar function. Because each of these transposases interacts with a highly specific DNA sequence (their

TABLE 1

Insertion sequences and transposons from bacteria[a]

Assemblage	Insertion sequence	Close relatives[b]	Direct repeat[c]	Amino acids[d]	Host[e]	References
1a	IS4		12	443	*E. coli* K-12	Iida et al. (1983)
	IS10	Tn10, Tn2921	9	404	Enteric bacteria	Iida et al. (1983), Navas et al. (1985)
	IS50	Tn5	9	478	Enteric bacteria	Iida et al. (1983)
	IS186		8–11	426	*E. coli* K-12	Kathary et al. (1985), Sengstad et al. (1986)
	IS231	Tn4430	11	478	*Bacillus thuringiensis*	Mahillon et al. (1985), Lereclus et al. (1986)
	ISH1		8	270	*Halobacter* sp.	Simsek et al. (1982)
1b	IS26	IS15, IS46, IS160, Tn2680, Tn1525	8	234	Enteric bacteria	Iida et al. (1983), Brown et al. (1984), Mollet et al. (1981), Nies et al. (1985), Trieu-Cuot and Courvalin (1984)
	IS52		4	236	*Pseudomonas* sp.	Yamada et al. (1986)
	IS903	IS102, Tn903, Tn602	9	308	Enteric bacteria	Iida et al. (1983), Stibbitz and Davies (1987)
2a	IS1	Tn9	8, 9, or 10	91, 124	*E. coli* K-12	Iida et al. (1983), Machida et al. (1984a), Iida and Hiestand-Naver (1986)
	IS5		4	326	*E. coli* K-12	Iida et al. (1983)
	IS66		8	258	*Agrobacterium tumefaciens*	Machida et al. (1984b)
	IS91[f]		0	?	*E. coli*	Diaz-Aroca et al. (1987)
	IS200[f]		0	?	*S. typhimurium*	Lam and Roth (1983)
	ISH2		10 or 20		*Halobacter* sp.	Iida et al. (1983), Simsek et al. (1982)
	ISH50		8	275	*Halobacter* sp.	Xu and Doolittle (1983)
	ISH51		3	?	*Halobacter* sp.	Hofman et al. (1986)
	ISM1		8	403	*Methanobrevibacter* sp.	Hamilton and Reeve (1985)

2b	IS2	E. coli K-12	301	5		H. J. Ronecker and B. Rak (Gene, in press)
	IS3	E. coli K-12	290	4		Iida et al. (1983), Timmerman and Tu (1985)
	IS136	A. tumefaciens	310	9		Vanderleyden et al. (1986)
2c	IS30	E. coli K-12	383	2		Dalrymple (1987), Dalrymple et al. (1984)
	IS4351	Bacteroides fragilis	326	3	Tn4351	Rasmussen et al. (1987)
3	Tn21[f]	Enteric bacteria	1000[g]	5	Tn4, Tn501, Tn1401, Tn1721, Tn2603, Tn2607, Tn2608, Tn2613, Tn4000	Hyde and Tu (1985), Tanaka et al. (1983)
	Tn3	Enteric bacteria	1017	5	Tn1, Tn2, IS101, Tn1720	Ishizaki and Ohtsubo (1985), Heffron (1983)
	Tn917	Streptococcus faecalis	560, 211	5	Tn551	Perkins and Youngman (1984), Shaw and Clewell (1985), Youngman et al. (1983)
4	Tn7[f]	Enteric bacteria	670, 420, 330[g]	5		Brevet et al. (1985), N. Craig (personal communication), Rogowsky et al. (1985)
	Tn554	Staphylococcus aureus	630, 361, 125	None	(IS21?)	Murphy and Lofdahl (1984), Murphy et al. (1985), Danilevitch and Kostyuchenko (1985)

[a] This list contains the elements for which partial or complete DNA sequences could be found in the literature. The criteria for assigning elements to the various assemblages are described in the text. Each insertion sequence or transposon is considered a distinct element if, at the nucleotide sequence level, no overall homology with any other element can be observed.

[b] Close relatives may fall into one of three categories: (i) an independent discovery of the given element, (ii) the composite transposon from which the given element was discovered or subsequently found, or (iii) a distinct but easily identifiable relative.

[c] Size of the direct repeat generated during insertion.

[d] Size, based on nucleic acid sequence, of the known transposase or the largest open reading frame.

[e] Host in which one given element naturally resides or is frequently found; for those listed as enteric bacteria, the natural host is not known and the element was discovered on broad-host-range plasmids found in numerous gram-negative bacteria.

[f] Complete sequence not available.

[g] Size based on protein molecular weight.

Assemblage 1a

```
        1       6       11      16      21      26      31      36
IS4    L L T S M T D A M R F P G G E M G D L Y S H R W E I E L G Y R E I K Q T

IS10   L A T N L P V E I R T P K Q L V - N I Y S K R M Q I E E T F R D L K S P

IS50   L L T G E P V E S L A Q A L R V I D I Y T H R W R I E E F H K A W K T G

IS186  L L T S L P - E D E Y S A E Q V A D C Y R L R W Q I E L A F K R L K S L

IS231  Y V S N T P - E G I V P M E Q I H D F Y S L R W Q I E I I F K T W K S L

ISH1   H R E Q T P L Q - K A H H A R M N E D Y N Q R W M S E T G F S Q L K E D
```

consensus:
1° structure   l - t - - p - e - - - - - - - - - - d - Y - - R w - i E - - f - - - K - -

2° structure ←——beta——→|←———alpha Helix 1———→| No Helix |←———alpha Helix 2———————→|

Assemblage 1b

```
IS26                    K T A Y A T I K G I E V M R A L R K G Q

IS52                    S V L Y R V K R K I E Y A K A Q L R A K

IS903                   T T D Y N R R S I A E T A M Y R V K Q L

orfC                    F N E Y V N E K G Y E L E R G T S K E V


                        t - Y - - - - - i E - - - - - - k
```

FIGURE 2. Amino acid sequence homology defining assemblage 1. The region shown is in the C-terminal one-third of the transposase or coding sequence for each of the elements. Boldface amino acids (single-letter code) below the lines show the consensus sequence; capital letters give invariant residues and lowercase letters give those that are highly conserved. For the assemblage 1a sequences, the consensus secondary structure was deduced by submitting each of the six respective amino acid sequences to the Chou-Fasman pseudo-probability calculations (Chou and Fasman, 1978). The assignments were made on the following basis. A high beta score was found for all but ISH1 between positions −10 and about +4. From the P at +6 to position 16 all six sequences show a high score (average maxima, 1.21) for an alpha helix (Helix 1). No alpha helix is predicted for any of the six at positions 18 to 20; there is no secondary structure in common for this region. Beginning with or soon after the highly conserved Y at position 20, an alpha helix (Helix 2) is predicted for all sequences except that of IS10. The criteria used for the alpha helix were a definition value of 1.12, a threshold value for four residues of 1.08, and a cutoff value of 1.00. Scores were also calculated for beta turns, omega loops, and hydrophobicity; no consensus on any of these was found. For the beta strand threshold value, a three-residue average of 1.25 and a cutoff value of 0.97 were used. The alignments and secondary structures were found using the programs LOCAL and PRSTRC at the Dana Farber Computer Facility. Assemblage 1b sequences are those that have the YE_7K_{14} signature, but only weakly align at the other positions compared with those of assemblage 1a. Secondary structure calculations were not performed on these sequences. OrfC is from a gene of unknown function on the plasmid pE194 (Horinouchi and Weisblum, 1982).

own end sequences) and a relatively nonspecific sequence (the recipient site), it seems reasonable to suggest that the nonspecific DNA-binding site would be highly conserved.

Besides the six elements that appear to be homologous, the transposases from IS903, IS52, and IS26 carry part of this conserved region. This piece contains a Tyr followed by a Glu 7 amino acids later and by a Lys 14 amino acids after the Tyr. This is the Y E_7 K_{14} signature. In a general homology search, in which I used a string containing, in part, the Y E_7 K_{14} signature, I found another protein in the data base on amino acid sequences maintained by NBRF. This is the 404-residue open reading frame (OrfC) found on the *Staphylococcus aureus* plasmid pE194 (Horinuchi and Weisblum, 1982). Though this putative protein was not contained within an identifiable insertion sequence, I did find a number of other similarities between it and the transposase from IS26 (Fig. 2), suggesting that the OrfC in pE194 is involved in some kind of transposition reaction.

B. Assemblage 2

Assemblage 2 contains the remaining insertion sequences; they lack the Y E_7 K_{14} signature in their transposase and are not characterized by similarities as strong as those in assemblage 1. Some of these elements are striking in their unusual gene organization. For example, IS1 contains a complex array of overlapping open reading frames, of which two seem to code for transposase (Gamas et al., 1987), while IS5 and IS4351 appear to code for proteins from overlapping genes on opposing strands. IS1 and ISH2 produce variably sized direct duplications of the recipient site, and in fact, the size of the duplication produced by IS1 can be altered by point mutations in IS1 (Machida and Machida, 1987).

I found that two groups from this assemblage share similarities. The first, designated 2b, contains IS2, IS3, and IS136. Homology between the transposases of IS2 and IS136 is high; there is an exact match of half of the residues over two-thirds of their length. Interestingly, both the sequences at their ends as well as the size of the direct duplication they generate upon insertion differ significantly. On the basis of this example and that of IS1 (mentioned above), it appears that the specific end sequences and the size of the direct duplication are probably not useful characters in assigning common ancestry. The transposase from IS3 aligns weakly to both IS2 and IS136. It is interesting that in the region of greatest similarity the three transposases from assemblage 2b share the amino acid sequence YN----HS-----SP near their C termini. The second group, designated 2c, contains IS30 and IS4351; the similarity between their transposases extends over 300 residues, and about one-third of their amino acids match exactly. Again, the end sequences and the sizes of their direct repeats differ.

The elements in assemblages 1 and 2 do not contain genes that confer obvious traits such as antibiotic resistance. However, when two of these elements flank another gene, the entire unit can become a transposable element, in which case it would be called a composite transposon (Kleckner, 1981). The next two assemblages differ in that the smallest transposable unit itself may also carry accessory genes.

C. Assemblage 3

The elements in assemblage 3 code for both a transposase and a resolvase. These elements transpose by means of a two-step pathway (Heffron, 1983). First, the transposase forms a cointegrate structure between donor and recipient molecules. The donor is then excised through the action of the site-specific resolvase. This latter reaction has been extensively studied (for reviews, see Heffron, 1983; Grindley and Reed, 1985; Craig and Kleckner, 1987). Tn3, Tn21, and Tn917 are homologous with respect to their end sequences, resolution sites, and resolvase genes (Hyde and Tu, 1985; Shaw and Clewell, 1985; Perkins and Youngman, 1984), and also at the level of the amino acid sequence of their transposases (this study). Interestingly, the large transposase of Tn3 appears to be related to two smaller transposases from Tn917 by a gene fusion. The amino-terminal third of the *tnpA* gene product of Tn3 is 28% similar to the shorter open reading frame of Tn917, and the carboxy-terminal half of the *tnpA* gene product of Tn3 contains 150 amino acids that align with the longer open reading frame of Tn917. This is the only group of bacterial insertion sequences for which a replicative mechanism of transposition is accepted (see below).

D. Assemblage 4

Tn7 and Tn544 are characterized by extreme site preference. Tn554 inserts into only one site in the *S. aureus* chromosome, and Tn7 prefers a single site in the *Escherichia coli* chromosome (Rogers et al., 1986; Craig and Kleckner, 1987). The unique feature of the assemblage 4 elements, however, is the apparent involvement of at least three genes coding for transposase functions. Like assemblage 3 elements, they carry accessory genes within their transposable units, but they neither form cointegrates nor possess resolvase activities. There are no indications that these two are homologous, though the complete sequence of Tn7 is not yet reported.

III. MECHANISM OF INSERTION SEQUENCE MOVEMENT

Cutting and rejoining phosphodiester bonds within both the donor and recipient sequences is obviously part of any transposition mechanism. The outstanding question is whether the transposition mechanism involves only these cut-and-insert reactions or involves extensive DNA synthesis as well. For example, when selection is made for transposition from a given donor site into a second site, the resulting clones invariably retain a copy of the element in the original donor site; this suggested that transposition results in duplication of the element. However, this kind of evidence only indirectly reflects upon the molecular mechanism and is no longer accepted as support for a replicative transposition mechanism. A second line of evidence is found in the ability of most of these elements to give rise to cointegrates. Assemblage 3 elements, for which cointegration is an integral part of the transposition reaction, are generally thought to form cointegrates in a direct replicative step (Heffron, 1983). However, because

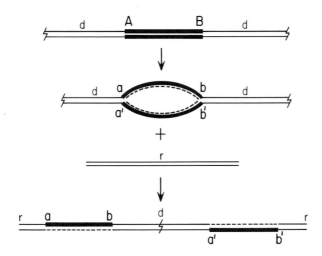

FIGURE 3. Possible replicative mechanism for the cointegration reaction. The insertion sequence (thick line) in a donor molecule (d) is specifically replicated in the first step to give an "eyelet" intermediate. The ends of the parental strands are designated a and a' and b and b'. The broken line shows newly synthesized DNA. The parental strands at a and b' are cut and inserted into the recipient (r) to yield a cointegrate (bottom line). Cutting at a and b or a' and b' (or double-strand cuts at each end) and insertion of the ends at r would produce a simple transposition event. This model for the relationship between transposition and cointegration is similar to that of Galas and Chandler (1981) and stresses a possible close relationship between replicative transposition (single-strand cutting) and cut-and-insert transposition (double-strand cutting). This formulation makes the idea of alternative pathways (Galas and Chandler, 1982; Hirschel et al., 1982) easier to visualize.

cointegrates are much rarer for the other elements, the same conclusion cannot be drawn. For these elements the question is whether the duplication event is a direct result of the transposition mechanism or a result of a fortuitous coupling of a simple cut-and-insert mechanism with normal chromosome replication.

A. Cointegration Reaction

As a means of orienting our thinking, a possible replicative pathway for cointegrate formation is shown in Fig. 3. This replication pathway was motivated by a desire to account for the possibility that transposition may operate to form either simple transposition products or cointegrates (Galas and Chandler, 1981, 1982; Biel and Berg, 1984; Grindley and Reed, 1985). Transposition may be achieved by a simple cut-and-insert mechanism where sites A and B are cut in Fig. 3, but it could also be achieved by ab or a'b' cutting in the eyelet. Cointegrates could, therefore, arise by ab' or a'b cutting of the eyelet in Fig. 3. I will discuss the validity of this model using IS50, but note that this discussion is relevant to all studies of the cointegration reaction promoted by assemblage 1 and 2 elements, especially IS1, IS21, IS26, and IS903 (for review see Grindley and Reed, 1985).

IS50 promotes cointegrate formation (Isberg and Syvanen, 1981), but the resulting cointegrates are not transposition intermediates (Isberg and Syvanen, 1985; but see Ahmed, 1986). When Hirschel et al. (1982) showed that cointegrate

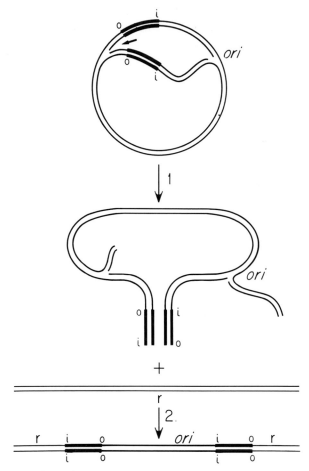

FIGURE 4. Cointegrate formation resulting from a cut-and-insert event acting on a monomeric replicating donor. Should transposition become fortuitously coupled with normal donor replication, a cointegrate could result. A theta replication intermediate is shown at the top of the figure (unidirectional replication is shown by the arrow; bidirectional replication would also be possible) where the insertion sequence (thick lines between o and i) has been replicated. In step 1 double-strand cuts (open arrows) occur at o on one daughter copy and at i on the other copy to produce two free double-strand ends. In step 2 these ends are inserted into the recipient (r). The other free double-strand ends (which contain donor sequences) need to be removed, and the replication fork needs to be blocked; exonuclease V would probably perform this task. This would result in the final cointegrate (bottom). (From Lichens-Park and Syvanen, 1988.)

formation by Tn5 is greatly stimulated by RecA function, whereas transposition by Tn5 is not, they suggested that cointegration and transposition may result from alternative pathways employing subtly different mechanisms, one of which requires RecA. However, Berg (1983) questioned this interpretation when he showed that DNA adjacent to the donor element transposes along with Tn5 from previously formed dimeric donor molecules. He suggested that RecA is required only to make these dimers and that cointegrates result from two insertion sequences in such a dimer acting in concert (as they do in composite transposons) to transpose one of the monomers. In other words, transposition and cointegration employ exactly the same mechanisms. However, Berg's scheme failed to explain the results of Isberg and Syvanen (1981), who found high-frequency cointegrate formation in *recA* mutant strains of *E. coli*. Thus, it appeared quite possible that IS*50* can form cointegrates by both the dimer donor and the alternative end product mechanisms (Fig. 3). The caveat to this conclusion is that the donor in the Isberg and Syvanen (1981) study was replication proficient; the possibility therefore remained that a cut-and-insert event coupled with normal donor replication created the cointegrate. This possibility was given strong support by Lichens-Park and Syvanen (1988), who showed, using a nonreplicating lambda::IS*50* derivative, that the ability of IS*50* to promote cointegrates is completely (>99%) dependent on (i) multiple copies of the lambda::IS*50* donor chromosome and (ii) functional RecA protein. It has been firmly established that normal IS*50* transposition requires none of these factors. Thus, the ability of IS*50* to promote cointegrates is dependent on normal donor replication, and we can no longer use the cointegration reaction as evidence for a replicative transposition mechanism.

This raises the question of how cut-and-insert transposition could be coupled to normal replication to produce cointegrates. Figure 4 outlines a possible mechanism whereby two insertion sequences found on the daughter duplexes in a theta replication intermediate undergo a concerted transposition event before the replication fork completes its replication of the donor replicon. One might think this model unlikely because the theta structure, as shown in Fig. 4, would have such a transient existence that the probability of a transposition event occurring during this brief period would be very low. However, the transposition frequency for several insertion sequences appears to be highest immediately after passage of a replication fork through the insertion sequence. The evidence for this possibility is that *dam* methylation sites at the ends of, for example, IS*10*, IS*50*, and IS*903*, would be hemimethylated in freshly replicated DNA, and as has been clearly shown for IS*10* and IS*50*, when these sites are hemimethylated transposition is enhanced (Roberts et al., 1985; Yin and Reznikoff, 1987; see below). Thus, the theta form as shown in Fig. 4 may be in a highly activated state for transposition. In addition to having the potential to produce a cointegrate, if the free o and i ends in the intermediate structure shown in Fig. 4 were to insert not into the recipient but rather into adjacent donor sequences, adjacent deletions or other rearranged donor molecules would be formed (putative pathways not shown). These arrangements would have the structures observed previously (e.g., with IS*1* and IS*903*) and interpreted as resulting from replicative transposition (for a description of these structures, see Grindley and Reed, 1985; Craig and Kleckner, 1987).

In summary, before a cointegrate can be used as evidence for a replicative transposition pathway, its formation should be observed when the donor is present in a single copy and its replication is blocked. This remains an open question. For example, Grindley and Reed (1985) interpret the rearrangements induced by IS*903* as evidence for at least some replicative transposition events induced by this insertion sequence, and IS*26* and IS*21* appear to give rise preferentially or exclusively to cointegrates (Willets et al., 1981; Reiss et al., 1983;

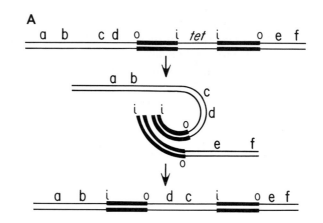

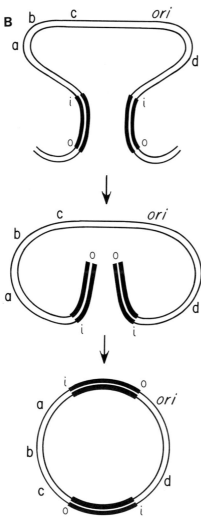

FIGURE 5. Intramolecular rearrangements induced by transposons Tn*10* and Tn*5* that support a simple cut-and-insert transposition mechanism. (A) Pathway suggested for the formation of the inversion-deletions induced by Tn*10* in the *his* operon from the *Salmonella typhimurium* chromosome (Kleckner et al., 1979). Tn*10* (shown at top as the sequences from o through o) induces the complex simultaneous loss of the tetracy-cline resistance element (*tet*) and inversion of one of the IS*10*s (thick lines between o and i) and the adjacent sequences c and d. Though the rearrangement appears complicated (compare the top and bottom lines), op-eration of a relatively simple cut-and-insert event is adequate to account for the rearrangement. In the first step the two IS*10* inside ends (i) are cut and the *tet* element is lost. In the second step the two free i ends insert into the adjacent sequences between loci b and c. Two possible products could be formed at this step, the inversion-deletion shown or the deletion of the *tet* element and the region i-o-d-c; deletions with this structure were also recovered (Ross et al., 1979). (B) Pathway suggested for the formation of the excision-inversion plasmids produced from an ancestral plas-mid that contains phage lambda fused to a Tn*5* deriv-ative where two IS*50*s (thick lines between o and i) flank an origin of plasmid replication (*ori*). As with the Tn*10* inversion-deletion event in A, two steps seem to be required, but a single cut-and-insert event will suffice. In the first step the two outside ends (o) are cut and the flanking sequences are lost. In the second step, the two o ends insert into the interior of the transposon between locus c and *ori*. We call this an "inposition" event. Two products could be formed at this step: the excision-inversion plasmid shown or an excision-dele-tion plasmid in which the region is deleted. Both products were recovered (Isberg and Syvanen, 1985).

Trieu-Cuot and Courvalin, 1985). Interestingly, for IS*21* Reimmann and Haas (1987) have convincingly shown that most of the cointegrates induced by IS*21* arise by means of a conservative pathway.

B. Direct Evidence for Cut-and-Insert Transposition Mechanisms

As summarized above, there is no direct evidence supporting replicative transposition for any of the assemblage 1 and 2 elements, but three lines of evidence support a cut-and-insert mechanism for IS*10* and IS*50* transposition.

(i) In the IS*10*::Lac heteroduplex transposition experiment, Bender and Kleckner (1986) showed that the genetic content from both strands of the donor insertion sequence DNA is found at the new recipient site. In this experiment, a heteroduplex transposon was constructed so that the fate of the genetic content for each DNA strand could be determined after a transposition event had occurred. By reference to Fig. 3 one can see that if a replicative transposition mechanism was used (say by ab or a'b' cutting) the movable unit would have the information from only one of the two strands. However, it was found that genetic markers from both strands moved to the new recipient, which means that extensive semiconservative replication does not occur during IS*10* transposition.

(ii) Certain complex intramolecular rearrangements induced by composite IS*10* or IS*50* transposons are easily explained by cut-and-insert mechanisms.

IS10. Kleckner et al. (1979) and Ross et al. (1979) characterized a Tn*10*-promoted deletion/inversion event that could be produced by cutting the inside ends of the IS*10*s in Tn*10* and inserting the excised DNA into the DNA immediately adjacent to the transposon. Figure 5A shows the resulting structures and a likely pathway.

IS50. Isberg and Syvanen (1985) described a Tn*5*-like transposon that produced an excision/inversion event; the structure of the product is easily explained by cutting the outside ends of the IS*50*s and inserting the excised DNA into the DNA between those ends, the so-called "inposition" reaction (Fig. 5B). In *recA recF* mutant hosts, only these excision/inversion plasmids were formed. No other rearranged plasmids (0 of 75) with structures predicted by replicative transposition were found.

(iii) The finding of circular forms of Tn*10* (Morisato and Kleckner, 1984) shows that double-strand breaks at the ends of Tn*10* do occur; this event requires the transposase of IS*10*. Circle formation in vitro, which is also transposase dependent, occurs without DNA synthesis (Morisato and Kleckner, 1987).

IV. FACTORS INFLUENCING TRANSPOSITION

A. Insertion Sequence Ends

One of the most conspicuous features of an insertion sequence element is the sequence defining its end; the two ends of an element are nearly identical but occur in inverted orientation. These inverted regions are usually between 8 and 20 base pairs in length but may be 50 or more base pairs long, and they almost certainly define sequence-specific cutting sites. If we generalize from IS*10*, the transposase is the cutting enzyme (Morisato and Kleckner, 1984). Detailed analysis of structure-function relationships between the ends and transposition has been begun only with IS*1*, IS*10*, and IS*50*.

A functional end requires more than the inverted repeat. This was shown by a deletion analysis of the ends of IS*1* (Gamas et al., 1985; Prentki et al., 1987), IS*10* (Way and Kleckner, 1984), and IS*50* (Johnson and Reznikoff, 1983). End sequence requirements have been most rigorously tested with IS*50*, in which the two ends are designated o and i. The terminal 9 base pairs of o and i define the inverted repeat, while the functional ends are 19 base pairs long. Two groups have recently shown by saturating the ends with point mutations that there are strong functional constraints on these sequences. At 34 of the 38 positions, substitutions result in a transposition deficiency; at the remaining 4 positions substitutions were neutral (Phadnis and Berg, 1987; J. Makris, P. Nordmann, and W. Reznikoff, submitted for publication). These results indicate that the sequences at the ends of IS*50* are highly constrained; in general, DNA sequences of this size have many more neutral positions. It is as if the information content through these sequences were unusually high. Within the ends of many insertion sequences there are functionally important GATC sites, i.e., the *dam* methylation sequence (see section IV.D.).

B. Insertion Sites

Early studies suggested that insertion sequences insert into target sites at random. More detailed analysis showed that many elements insert preferentially at certain target sequences. Near total site specificity is shown by IS*4* and ISH*1*, as well as by the assemblage 4 elements. A definite site bias is shown by IS*10* and IS*1*; each inserts preferentially into regions with weak homology to sequences found within the elements. Sengstad and Arber (1987) showed that phage P1 has at least two "attracting" sequences for IS*2*, and Pickens et al. (1984) mapped a plasmid site into which Tn*3* inserted preferentially. IS*50* and IS*1* have their own insertion "hot spots" within pBR322 (Prentki et al., 1987); for IS*1* this hot spot appears to be an IHF protein binding site and is most likely related to the apparent involvement of IHF protein in IS*1* transposition (as below, section IV.D.). IS*50* shows a preference for AT-rich DNA, while IS*186* has been found exclusively in the poly G/poly C tails used to clone foreign DNA into plasmids.

There is as yet no indication that differences in site specificity reflect significant differences in transposition mechanism. Assemblage 1 and 2 elements, for example, seem to run the gamut in site preference, from the highly promiscuous IS*1* and IS*50* to the sequence-specific IS*4* and ISH*1*; site preference may reflect differences in detail rather than in basic mechanism.

C. Regulation of Insertion Sequence Movement

Much of the pertinent information on regulation has been reviewed elsewhere (Syvanen, 1984; Craig and Kleckner, 1987). Here I will outline the overall patterns of transposition regulation without describing particular molecular mechanisms.

Insertion sequences encode not only transposases but also factors that impede movement. This makes sense, since insertion sequences are mutators and high mutational frequencies would lower host cell viability. Regulation of transposition is achieved through a variety of strategies. Assemblage 1 elements (at least for IS*10*, IS*50*, and IS*903*) encode diffusible inhibitors acting on all copies of the element in the cell. This *trans* inhibitor is probably important in modulating copy number (Johnson and Reznikoff, 1984; Kleckner et al., 1985), which may be necessary since transposition events result in duplication of the element and cause an increase in insertion sequence copy number. (Note that on the basis of the above discussion, it seems unlikely that the transposition mechanism is duplicative per se; however, this does not negate the result that insertion sequence movement leads to an increase in insertion sequence copy number in descendant cells.) Production of diffusible inhibitors by each element in a multicopy family would, in principle, serve to prevent an indefinite increase in the number of copies. In addition, the transposases are predominantly *cis* acting so that, for instance, one IS*50* can barely promote movement of a second IS*50* in the same cell. The combination of *cis*-acting transposase and *trans*-acting inhibitor would be expected to reduce transposition frequency with increasing copy number.

The constraint of the transposase to act only in *cis* may serve an additional purpose. In a cell containing multiple copies of an element, defective copies will not transpose. Thus, defective elements will be unlikely to spread to new sites, leaving only self-transposing elements to spread.

The assemblage 3 and 4 elements employ a different means of regulating copy number. These elements contain specific nucleotide sequences that block insertion of another copy in their vicinity on the same chromosome. This *cis*-acting inhibition is called transposition immunity (Arthur et al., 1984; Huang et al., 1986; Lee et al., 1983; Murphy, 1983). Copy number for these elements is therefore limited to the number of replicons in the cell.

D. Host Factors That Influence Insertion Sequence Movement

Insertion sequence transposition requires the transposase encoded by an insertion sequence. There are host factors, on the other hand, that stimulate movement of some elements, but none that are absolutely required. The best-characterized host factors involved in transposition are listed in Table 2. Since different elements utilize different host factors, no general principles have yet emerged.

Diverse evidence revealed the host factors involved in transposition. *E. coli* strains with mutations in *polA* (Syvanen et al., 1982; Sasakawa et al., 1981), *gyrA* and *gyrB* (Isberg and Syvanen, 1982), *dam* (Roberts et al., 1985; Yin and Reznikoff, 1987), and *rho* (Rosner, 1985) have significantly altered transposition frequencies. Involvement of the DnaA protein in IS*50* transposition and the IHF protein in IS*1* transposition was initially suggested by the presence of putative DnaA and IHF binding sites at the ends of IS*50* and IS*1*, respectively; these were in sequences known to be essential for transposition. In both cases the purified proteins bind to those end sequences (Johnson and Reznikoff, 1983; Gamas et al., 1987). Careful examination of *dnaA* mutants of *E. coli* strongly indicates a

TABLE 2
Host factors implicated in insertion sequence movement in *E. coli*

Factor	Function	Genetic locus	Known elements	References
DnaA	DNA replication	*dnaA*	IS*50*	Yin and Reznikoff (1987)
DNA gyrase	DNA supercoiling	*gyrA*, *gyrB*	IS*50*	Isberg and Syvanen (1982), Wang (1985)
DNA polymerase I	DNA repair	*polA*	IS*1*, IS*5*, IS*10*, IS*50*	Syvanen et al. (1982), Sasakawa et al. (1981), Joyce et al. (1985)
Dam	DNA repair	*dam*	IS*10*, IS*50*	Roberts et al. (1985), Yin et al. (1987)
HU	Nucleoid structure	?	IS*10*	Morisato and Kleckner (1987)
IHF	Phage integration, regulation of gene expression	*himA*, *himD*	IS*1*, IS*10*, IS*50*?	Morisato and Kleckner (1987), Gamas et al. (1987)
Rho	Transcription termination	*rho*	IS*1*	Datta and Rosner (1987)

stimulation by, but probably not a requirement for, DnaA protein in IS*50* movement (Yin and Reznikoff, 1987). Similarly, IHF and HU proteins may stimulate IS*10* movement (Morisato and Kleckner, 1987), since these proteins greatly stimulate, but are not required for, an in vitro IS*10* DNA cutting reaction dependent on the transposase of IS*10*. Host factors may affect transposition at various steps. IS*50* requires DNA gyrase because the recipient DNA molecule needs to be supercoiled (Isberg and Syvanen, 1982); similarly, IS*10* may require DNA gyrase since the in vitro transposase-dependent cutting reaction requires supercoiled DNA (Morisato and Kleckner, 1987). The *dam* methylase reduces transposition of IS*10* and IS*50* by methylating adenines in the multiple GATC sequences in these elements. Two mechanisms seem to be involved: (i) fully methylated GATC sites at the end presumably are not recognized by transposase (Roberts et al., 1985; Yin and Reznikoff, 1987), and (ii) fully methylated GATC sites in the promoters for the transposase and regulatory factor reduce the amount of active transposase synthesized (Roberts et al., 1985; Yin and Reznikoff, 1987; McCommas and Syvanen, 1988). Consequently, in *E. coli dam* mutants, IS*10* and IS*50* show higher transposition frequencies than they do in wild-type strains. DnaA protein (in the case of IS*50*) and IHF and HU proteins (in the case of IS*10*) apparently stimulate transposition by binding to the end sequences, not by altering transposase synthesis. DNA polymerase I seems to be directly involved in transposition, since the *polA* mutants defective in both the polymerase and the 5'-3' exonuclease activities are deficient in transposition, but the mechanism remains obscure. Perhaps transposition leaves the DNA nicked or gapped, and DNA polymerase I is required to repair these regions. This notion is supported by the observation that SOS functions (Little and Mount, 1982) are induced in those cells in which transposition has occurred (D. Roberts and N. Kleckner, *Proc.*

Natl. Acad. Sci. USA, in press). Transposition may be indirectly affected by *rho* mutations (Datta and Rosner, 1987) since Rho protein regulates the expression of DNA gyrase (Fassler et al., 1986), which in turn affects transposition indirectly, as noted previously.

Host mutations other than those listed in Table 2 affect insertion sequence movement, but the affected products are not yet characterized (Illyina et al., 1981; Ismailov et al., 1986; D. Roberts, Ph.D. thesis, Harvard University, Cambridge, Mass., 1986).

V. PRECISE EXCISION

Mutations induced by insertion sequences are generally quite stable; most revert to wild type with frequencies of $\leq 10^{-8}$ per cell per generation. On the other hand, mutations induced by insertion of composite transposons with inverted insertion sequences can be exceedingly unstable. These usually revert to wild type at frequencies of about 10^{-4} to 10^{-7}, but some revert at frequencies of greater than 10^{-3}. Reversion of these mutations requires loss of the transposon plus the small direct duplication created by the insertion. Such loss is called precise excision.

Precise excision of transposons is mechanistically unrelated to their transposition (Egner and Berg, 1981; Foster et al., 1981). The evidence is as follows: (i) transposition requires transposase, while precise excision does not; (ii) transposition occurs at similar frequencies whether the insertion sequences are in direct or inverted orientation, while precise excision occurs at 10^{3}- to 10^{7}-fold higher frequency when the insertion sequences are in the inverted orientation; and (iii) with a single exception, host mutations that affect transposition do not affect precise excision and vice versa.

The study of *E. coli* mutants that effect precise excision suggests that this process is influenced by factors used in other DNA processes (Table 3). Because new phenotypes for previously known loci and even new genes with pleiotropic phenotypes have been revealed during the isolation of these mutants, it appears that precise excision may be a useful tool for studying these other processes.

General recombination is one of these processes, and the current model for precise excision suggests how these two may be related. The model for precise excision of transposons with large inverted repeats invokes pairing of the repeated DNA sequences as an early step in the reaction pathway (Foster et al., 1981; Egner and Berg, 1981). This is not necessarily a cruciform in which single-stranded DNA snaps back on itself, but rather a pairing of the repeated duplex DNA stems which brings into proximity the two points that must eventually be united in a novel joint. If host factors aided in this pairing of the inverted stems in precise excision, it would not be unexpected for the same factors to promote general recombination as well.

This specific model has only indirect support, but analysis of the host mutants has shown that, though precise excision is independent of null mutations in *recA* and *recB recC*, there is some kind of linkage between precise excision and recombination (Hopkins et al., 1981). Lundblad et al. (1984) established the first

TABLE 3
Host factors affecting precise excision[a]

Genetic locus	Precise excision in mutants	Other phenotypes	References
recB recC (TexA)	High	General recombination	Lundblad et al. (1984)
F factor (wild type)	High	General recombination	Hopkins et al. (1980), Syvanen et al. (1986)
F factor traS	Very high	General recombination	
F factor tra	Normal	General recombination	
uvrD	High	Mismatch repair	Lundblad and Kleckner (1984)
mutH, S, L	High	Mismatch repair	Lundblad and Kleckner (1984)
dam	High	Mismatch repair	Lundblad and Kleckner (1984)
uup	High	Phage Mu growth	Hopkins et al. (1983)
drpA	Low	RNA synthesis	Lech et al. (1985)

[a] Precise excision was monitored by using Tn5 or Tn10, or both, in the strains of E. coli having the indicated genotypes. Tn5 and Tn10 contain flanking insertion sequences in inverted orientation.

specific linkage when they found unusual alleles of recB and recC, called texA, that cause an increase in precise excision. These mutations represent a new class of alleles affecting recB and recC, whose products form part of exonuclease V (Taylor, this volume); they are not recombination deficient but do change the interaction of exonuclease V with Chi sites and alter bacteriophage lambda growth. One of the texA mutants renders precise excision largely dependent upon functional RecA. In a different study (Syvanen et al., 1986), analysis of precise excision led to the discovery that F' plasmid transfer not only stimulates this process but also stimulates RecA-dependent recombination in E. coli. This latter study raised the possibility that there may be factors involved in some kinds of RecA-dependent recombination that have not yet been revealed by existing mutants.

There is an overlap between the sets of host factors involved in precise excision and methyl-directed mismatch repair. Lundblad and Kleckner (1984) isolated a large class of mutants that had enhanced precise excision of Tn10 (and to a lesser degree Tn5) and were defective in methyl-directed mismatch repair (Table 3) (see Radman, this volume). This class includes uvrD, mut, and dam mutants. In the uvrD but not the other mutants, enhanced precise excision is dependent upon functional RecA.

The drpA gene was found as a mutation that caused both a defect in precise excision at 30°C and an inability to synthesize RNA at 37°C (Lech et al., 1985). This gene does not code for any of the known transcription factors from E. coli.

VI. EVOLUTIONARY CONSIDERATIONS

A. Do External Factors Influence Insertion Sequence Movement?

The possibility that environmental stress may trigger insertion sequence movement is an attractive idea that has received serious attention because it bears directly on any hypothesis purporting evolutionary function for insertion sequences. One such formulation is the genomic stress hypothesis (McClintock, 1984), which postulates that DNA damage directly stimulates transposition. This notion seems quite plausible, since transposons use elaborate regulatory mechanisms to control their movement, and analogous regulatory mechanisms in bacteriophages and bacteria are known to respond to environmental stimulations. Furthermore, the existence of the UV-inducible system and the hypothesis of ''stress-triggered evolution'' (Echols, 1981) have provided a reasonable precedent for such speculations. However, despite a number of surveys, bacterial insertion sequence movement appears to be quite insensitive not only to DNA damage but also to any environmental stimulation. Because most experiments on this have given null results, they are, as is the custom, unpublished. Datta et al. (1983) and Rosner (1985) systematically tested over 100 chemical compounds for their effect on Tn9 transposition. They tested mutagens, UV light, antibiotics, nutrients, and various noxious agents and found none that influenced Tn9 transposition. My laboratory (unpublished data) has tested a variety of mutagens, UV light, and mitomycin C and found no effect on IS1, IS5, or IS50 transposition. Since these lists include many of the agents known to damage DNA (Taylor, this volume), it seems unlikely that genomic stress per se stimulates insertion sequence movement.

One environmental factor, low temperature, increases the frequency of transposition of a number of elements. This has been reported for IS1, Tn3 (Iida et al., 1983), and IS50 (Isberg and Syvanen, 1982). For these elements the temperature response may be adaptive, either to the host that carries them or to the elements themselves. Perhaps low temperature could be considered a stress, but this result must be considered weak evidence for any environmental stress hypothesis.

The report that transposition frequency is higher in stationary-phase cells than in exponentially growing cells (Read and Jaskunas, 1980; Arber and Iida, 1982) has been supported by casual observations of many workers who have found that strains harboring insertion sequences frequently pick up unselected secondary insertions during storage in room temperature stabs. This does indeed raise the possibility that stationary-phase growth activates movement. However, before this conclusion can be accepted, one must determine whether the transposition frequency is normally dependent upon the number of generations producing the population of cells or is dependent upon the age of that population. In the reports cited above, the number of generations leading to either the ''exponential'' or the ''stationary'' populations was controlled to be the same (where it is assumed that during stationary growth there were zero generations). This design does not permit time to be controlled. There seems to be little doubt that transposition events do, in fact, occur in stationary cells, and this is probably significant, given that so many other cellular processes are turned off under these conditions; but whether stationary cells show enhanced insertion sequence movement remains to be determined. One study failed to support this hypothesis; Edlin et al. (1986) showed that under a variety of growth-limiting conditions in chemostats, IS10 transposition was not stimulated.

Transposition of the assemblage 3 elements Tn21 and Tn917 can be externally stimulated by mercury ions (Lund and Brown, 1987) and erythromycin (Tomich et al., 1980), respectively. These agents cause derepression of transposase synthesis and, consequently, enhance transposition frequencies. Since Tn21 carries a mercury resistance gene and Tn917 carries an erythromycin resistance gene, and since these two elements engage in replicative transposition, these cases are the best examples of a possibly adaptive mutational event that is directly stimulated by the selective environment. They remain isolated examples.

B. Functional Role of Insertion Sequences in Adaptive Evolution

Because of the absence of clear phenotypic manifestations (other than their mutator activity), it was initially proposed that insertion sequences are sustained by direct selection acting on genetic variability (Cohen, 1976; Nevers and Saedler, 1977; Reanney, 1976). That is, the trait of genetic variability associated with the insertion sequences is their function, and they are selected each time a mutation induced by these elements is selected. I have defended this position as the "selection hypothesis" (Syvanen, 1984). The selection hypothesis was developed by bacteriologists who noted the very close relationship between insertion sequences and the highly selected antibiotic resistance genes that have appeared in bacterial populations in the past 35 years. In addition, direct evidence in its favor has been obtained for IS10 in chemostat competition experiments (Chao and McBroom, 1985). However, this hypothesis has been strongly resisted by evolutionists schooled in metazoan biology. For example, Mayr (1982) criticizes the selection hypothesis as a "teleological answer [where] seemingly functionless DNA is stored up in order to have it available in future times of need," and Hartl et al. (1986) wonder "how some future benefit accruing to such elements could be the driving force behind their evolution." While Mayr (1982) objects to the notion that a complex element could be maintained because it helps its host evolve, Hartl et al. (1986) seem to question both how insertion sequences could now contribute to the adaptive evolution of the hosts in which they reside and how they could have evolved to do this in the first place. The selection hypothesis does not attempt to answer the latter question. It makes no statement about how they arose and does not claim that they have served the same, if any, function throughout evolutionary history. The hypothesis does claim that these elements have contributed to adaptive evolution in the past and may still possess mutator activities with the potential to induce variants that can be fixed in the future. This hypothesis is not teleological. Stanley (1979) stressed a view of evolution as a race among evolving species; those species that give rise to successful variants at the highest rate will be favored. It seems reasonable to assume that a species that has given rise to successful variants will in some cases carry along with it the mutational mechanism that gave rise to that variant. Such a mechanism may give rise to future variants at a high rate. If the benefits of carrying these variants outweigh the disadvantage of carrying these mutators, we have a simple, nonteleological explanation for the maintenance of insertion sequences based on their function as mutators.

I would like to acknowledge that there is currently a climate of suspicion surrounding selectionist thinking in evolution. This healthy reaction against overly deterministic and often ad hoc evolutionary explanations has given rise to an important and undoubtedly correct emphasis on the role of random fluctuations and changes. With all these caveats in mind, a bacteriologist looking at drug resistance-determining plasmids and their associated insertion sequences is seeing natural selection working in one of its purest manifestations and must consider the possibility that this phenomenon is a model for the role of insertion sequences in evolution.

One of the most persistent objections to the selection hypothesis is that it relies on the assumption of group selection, and as has been mathematically argued by population geneticists, intergroup competition is weak compared with competition between individuals. From this it follows that the evolutionary function of insertion sequences cannot be genetic variability. This criticism rests primarily on the strength of mathematical models which must always rely on simplified assumptions; in this case I think questionable assumptions are used. It seems to me that the models under discussion are based upon populations at equilibrium in which competition with closely related individuals determines the outcome. These models do not treat situations in which there has been a catastrophic population collapse and survivors or new invaders are struggling to occupy an empty niche. The rise of antibiotic-resistant bacteria in hospitals represents such a scenario. Furthermore, these models do not take into account horizontal gene flow, for which the mobile elements are well adapted and through which a single successful adaptation may appear among multiple groups. It seems unwise to reject the selection hypothesis solely on the basis of these mathematical arguments.

For these various reasons the selection hypothesis has been rejected by many and has been succeeded by the alternative view that insertion sequences are genomic parasites. This view is epitomized by the "selfish DNA" hypothesis (Sapienza and Doolittle, 1981), which has gained support mainly among students of metazoans. According to this hypothesis the ability of self-replication and genome spread is sufficient to account for the existence of insertion sequences. One of the major appeals of the selfish gene hypothesis is that it is amenable to mathematical modeling, thus providing it with an abstract, if not biological, reality. The association of insertion sequences with genes that are highly selected is not viewed as support for the selection hypothesis; according to this perspective, if these genomic parasites contribute to adaptive changes, these are their effect and not their function (Maynard-Smith, 1982).

There is currently insufficient information to rigorously discriminate between the selfish DNA hypothesis and the selection hypothesis. The distribution of IS4 and IS5 in populations of *E. coli* was shown to be consistent with a variety of competing models (Sawyer and Hartl, 1986). The selfish DNA hypothesis has the advantage that it is easily incorporated into existing theory (especially as formulated mathematically by the population geneticists). The selection hypothesis, though perhaps more consistent with the facts it has been designed to explain, has implications that challenge widely accepted theory and may necessitate major changes in that theory.

ACKNOWLEDGMENTS. I thank Temple Smith at Dana Farber for his help and for the computer facility used in analyzing transposase sequences. I thank Alison Delong, Nancy Kleckner, and Sue Greenwald for their help in writing this paper.

LITERATURE CITED

Ahmed, A. 1986. Evidence for replicative transposition of Tn5 and Tn9. *J. Mol. Biol.* **191**:75–84.

Arber, W., and S. Iida. 1982. The involvement of IS elements of E. coli in the genesis of transposons and in spontaneous mutagenesis, p. 3–13. *In* S. Mitsohashi (ed.), *Drug Resistance in Bacteria*. Japan Scientific Society Press, Tokyo.

Arthur, A., E. Nimmo, S. Hettle, and D. Sherratt. 1984. Transposition and transposition immunity of transposon Tn-3 derivatives having different ends. *EMBO J.* **3**:1723–1730.

Bender, J., and N. Kleckner. 1986. Genetic evidence that Tn10 transposes by a nonreplicative mechanism. *Cell* **45**:801–815.

Berg, D. E. 1983. Structural requirement for IS50-mediated gene transposition. *Proc. Natl. Acad. Sci. USA* **80**:792–796.

Biel, S., and D. Berg. 1984. Mechanism of IS1 transposition in *Escherichia coli*: choice between simple insertion and cointegration. *Genetics* **108**:319–330.

Brahma, N., A. Schumacher, J. Cullum, and H. Saedler. 1982. Distribution of the *Escherichia coli* K12 insertion sequences IS1, IS2 and IS3 among other bacterial species. *J. Gen. Microbiol.* **128**:2229–2234.

Brevet, J., F. Faure, and D. Brorowski. 1985. Transposon Tn7-encoded proteins. *Mol. Gen. Genet.* **201**:258–264.

Brown, A. M., G. M. Coupland, and N. S. Willetts. 1984. Characterization of IS46, an insertion sequence found on two IncN plasmids. *J. Bacteriol.* **159**:472–481.

Chao, L., and S. M. McBroom. 1985. Evolution of transposable elements: an IS10 insertion increases fitness in *E. coli*. *Mol. Biol. Evol.* **2**:359–369.

Chou, P., and G. Fasman. 1978. Empirical predictions of protein conformation. *Annu. Rev. Biochem.* **17**:251–276.

Cohen, S. N. 1976. Transposable genetic elements and plasmid evolution. *Nature* (London) **263**:731.

Craig, N., and N. Kleckner. 1987. Transposition and site-specific recombination, p. 1054–1070. *In* F. C. Neidhardt, J. L. Ingraham, K. B. Low, B. Magasanik, M. Schaechter, and H. E. Umbarger (ed.), *Escherichia coli and Salmonella typhimurium: Cellular and Molecular Biology*. American Society for Microbiology, Washington, D.C.

Dalrymple, B. 1987. Novel rearrangements of IS30 carrying plasmids leading to reactivation of gene expression. *Mol. Gen. Genet.* **207**:413–420.

Dalrymple, B., P. Caspers, and W. Arber. 1984. Nucleotide sequence of the mobile element IS30. *EMBO J.* **3**:2145–2149.

Danilevich, V. N., and D. A. Kostyuchenko. 1985. Immunity to repeated transposition of insertion sequence IS21. *Mol. Biol.* (Moscow) **19(5)**:1242–1250.

Datta, A., B. Randolph, and J. Rosner. 1983. Detection of chemicals that stimulate Tn9 transposition in *E. coli* K12. *Mol. Gen. Genet.* **189**:245–250.

Datta, A., and J. Rosner. 1987. Reduced transposition in *rho* mutants of *Escherichia coli* K-12. *J. Bacteriol.* **169**:888–890.

Deonier, R. 1987. Locations of native insertion sequence elements, p. 982–989. *In* F. C. Neidhardt, J. L. Ingraham, K. B. Low, B. Magasanik, M. Schaechter, and H. E. Umbarger (ed.), *Escherichia coli and Salmonella typhimurium: Cellular and Molecular Biology*. American Society for Microbiology, Washington, D.C.

Diaz-Aroca, E., M. Mendiola, J. Zabala, and F. de la Cruz. 1987. Transposition of IS91 does not generate a target duplication. *J. Bacteriol.* **169**:442–443.

Echols, H. 1981. SOS functions, cancer and inducible evolution. *Cell* **25**:1–2.

Edlin, G., S. W. Lee, and M. M. Green. 1986. Tn10 transposition does not respond to environmental stress. *Mutat. Res.* **175**:159–164.

Egner, C., and D. Berg. 1981. Excision of transposon Tn5 is dependent on the inverted repeats but not on the transposase function. *Proc. Natl. Acad. Sci. USA* **78**:459–463.

Fassler, J., G. Arnold, and I. Tessman. 1986. Reduced superhelicity of plasmid DNA produced by the *rho-15* mutation in *E. coli*. *Mol. Gen. Genet.* **204**:424–429.

Foster, T., V. Lundblad, S. Hanley, M. Halling, and N. Kleckner. 1981. Three Tn10 associated excision events: relationship to transposition and role of direct and inverted repeats. *Cell* **23**:215–227.

Galas, D., and M. Chandler. 1981. On the molecular mechanisms of transposition. *Proc. Natl. Acad. Sci. USA* **78**:5848–5862.

Galas, D., and M. Chandler. 1982. Structure and stability of Tn9-mediated cointegrates: evidence for two pathways of transposition. *J. Mol. Biol.* **154**:245–272.

Gamas, P., M. Chandler, P. Prentki, and D. Galas. 1987. *E. coli* integration host factor binds specifically to the ends of the insertion sequence IS1 and to its major insertion hotspot in pBR322. *J. Mol. Biol.* **195**:261–272.

Gamas, P., D. Galas, and M. Chandler. 1985. DNA sequences at the end of IS1 required for transposition. *Nature* (London) **317**:458–460.

Gertman, E., B. N. White, D. Berry, and A. M. Kropinski. 1986. IS*222*, a new insertion element associated with the genome of *Pseudomonas aeruginosa*. *J. Bacteriol.* **166**:1134–1136.

Grindley, N., and R. Reed. 1985. Transpositional recombination in procaryotes. *Annu. Rev. Biochem.* **54**:863–896.

Hamilton, P. T., and J. N. Reeve. 1985. Structure of genes and an insertion element in the methane producing archaebacterium methanobrevibacter Smith II. *Mol. Gen. Genet.* **200**:47–59.

Hartl, D. L., M. Medhorn, L. Green, and D. E. Dyhuizen. 1986. The evolution of DNA sequences in *Escherichia coli*. *Philos. Trans. R. Soc. London B Biol. Sci.* **312**:191–204.

Heffron, F. 1983. Tn3 and its relatives, p. 223–260. *In* J. Shapiro (ed.), *Mobile Genetic Elements*. Academic Press, Inc., New York.

Hirschel, B. J., D. J. Galas, and M. Chandler. 1982. Cointegrate formation by Tn5, but not transposition, is dependent on *recA*. *Proc. Natl. Acad. Sci. USA* **79**:4530–4534.

Hofman, J. D., L. C. Schalkwyk, and W. F. Doolittle. 1986. ISH51: a large, degenerate family of insertion sequence-like elements in the genome of the archaebacterium, *Halobacterium volcanii*. *Nucleic Acids Res.* **14**:6983–7000.

Hopkins, J., M. Clements, T. Liang, R. Isberg, and M. Syvanen. 1981. Recombination genes on the *E. coli* sex factors specific for transposable elements. *Proc. Natl. Acad. Sci. USA* **77**:2814–2818.

Hopkins, J. D., M. Clements, and M. Syvanen. 1983. New class of mutations in *Escherichia coli* (*uup*) that affect precise excision of insertion elements and bacteriophage Mu growth. *J. Bacteriol.* **153**:384–389.

Horinouchi, S., and B. Weisblum. 1982. Nucleotide sequence and functional map of pE194, a plasmid that specifies inducible resistance to macrolide, lincosamide, and streptogramin type B antibiotics. *J. Bacteriol.* **150**:804–814.

Huang, C. J., F. Heffron, J. S. Tu, R. H. Scholoemer, and C. H. Lee. 1986. Analysis of Tn3 sequences required for transposition immunity. *Gene* **41**:23–32.

Hyde, D., and C. P. Tu. 1985. TnpM: a novel regulatory gene that enhances Tn21 transposition and suppresses cointegrate resolution. *Cell* **42**:629–638.

Iida, S., and R. Hiestand-Naver. 1986. Insertion element IS1 can generate a 10-base pair target duplication. *Gene* **45**:233–235.

Iida, S., J. Meyer, and W. Arber. 1983. Prokaryotic IS elements, p. 159–222. *In* J. Shapiro (ed.), *Mobile Genetic Elements*. Academic Press, Inc., New York.

Illyina, T., Y. Romanova, E. Necherva, and V. Smirnov. 1981. Isolation and mapping of *E. coli* K12 mutants defective for Tn9 transposition. *Mol. Gen. Genet.* **81**:384–389.

Isberg, R., and M. Syvanen. 1981. Replicon fusions promoted by the inverted repeats of Tn5: the right repeat is an insertion sequence. *J. Mol. Biol.* **150**:15–32.

Isberg, R., and M. Syvanen. 1982. DNA gyrase is a host factor required for transposition of Tn5. *Cell* **30**:9–18.

Isberg, R., and M. Syvanen. 1985. Tn5 transposes independently of cointegrate resolution: evidence for an alternative model for transposition. *J. Mol. Biol.* **182**:69–78.

Ishizaki, K., and E. Ohtsubo. 1985. Cointegration and resolution mediated by IS101 present in plasmid pSC101. *Mol. Gen. Genet.* **199:**389–395.

Ismailov, Z. F., S. Smirnov, and V. Tarasov. 1986. Characterization of *het* mutations in the chromosome of *Escherichia coli* causing increase in *Tn1* transposition frequency. *Genetika* **22:**777–786.

Johnson, R., and W. Reznikoff. 1983. DNA sequences at the ends of transposon Tn5 required for transposition. *Nature* (London) **304:**280–282.

Johnson, R. C., and W. S. Reznikoff. 1984. Role of the IS*50*R proteins in the promotion and control of Tn*5* transposition. *J. Mol. Biol.* **177:**645–661.

Joyce, C. M., D. M. Fujii, H. S. Laks, C. M. Hughes, and N. Grindley. 1985. Genetic mapping and DNA sequence analysis of mutations in the *polA* gene of *E. coli. J. Mol. Biol.* **186:**282–293.

Kathary, R., D. Jones, and E. Candido. 1985. IS*186*: an *Escherichia coli* insertion element isolated from a cDNA library. *J. Bacteriol.* **164:**957–959.

Kleckner, N. 1981. Transposable elements in prokaryotes. *Annu. Rev. Genet.* **15:**341–404.

Kleckner, N., K. Reichardt, and D. Botstein. 1979. Inversions and deletions of the Salmonella chromosome generated by transposon Tn10. *J. Mol. Biol.* **127:**89–115.

Kleckner, N., R. Simons, and J. Kittle. 1985. Mechanism of IS10-mediated multicopy inhibition. *J. Cell. Biochem.* **9**(Suppl. Part B):227.

Lam, S., and J. Roth. 1983. Mapping of IS200 copies in Salmonella typhimurium strain LT2. *Genetics* **105:**801–811.

Lech, K. F., C. H. Lee, R. R. Isberg, and M. Syvanen. 1985. New gene in *Escherichia coli* K-12 (*drpA*): does its product play a role in RNA synthesis? *J. Bacteriol.* **162:**117–123.

Lee, C. H., A. Bhagwat, and F. Heffron. 1983. Identification of a transposon Tn-3 sequence required for transposition immunity. *Proc. Natl. Acad. Sci. USA* **80:**6765–6769.

Lereclus, D., J. Mahillon, G. Menou, and M. M. Lecadet. 1986. Identification of transposon 4430: a transposon of *Baccillus thuringiensis* functional in E. coli. *Mol. Gen. Genet.* **204:**52–57.

Levesque, R. C., and G. A. Jacoby. 1988. Molecular structure and interrelationships of multiresistance β-lactamase transposons. *Plasmid* **19:**21–29.

Lichens-Park, A., and M. Syvanen. 1988. Cointegrate formation by IS50 requires multiple donor molecules. *Mol. Gen. Genet.* **211:**244–251.

Little, J. W., and D. W. Mount. 1982. The SOS regulatory system of *Escherichia coli. Cell* **29:**11–22.

Lund, P. A., and N. L. Brown. 1987. Role of mer-T and mer-P gene products of transposon Tn501 in the induction and expression of resistance to mercuric ions. *Gene* **52:**207–214.

Lundblad, V., and N. Kleckner. 1984. Mismatch repair mutations of *E. coli* enhance transposon excision. *Genetics* **109:**3–19.

Lundblad, V., A. Taylor, G. Smith, and N. Kleckner. 1984. Unusual alleles of recB and recC stimulate excision of inverted repeat transposons Tn10 and Tn5. *Proc. Natl. Acad. Sci. USA* **81:**824–828.

Machida, C., and Y. Machida. 1987. Base substitutions in transposable element IS1 cause DNA duplication of variable length at the target site for plasmid co-integration. *EMBO J.* **6:**1799–1804.

Machida, Y., C. Machida, and E. Ohtsubo. 1984a. Insertion element IS1 encodes two structural genes required for its transposition. *J. Mol. Biol.* **177:**229–240.

Machida, Y., M. Sakuraim, S. Kiyokawa, A. Ubasawa, Y. Suzuki, and J. E. Ikeda. 1984b. Nucleotide sequence of the insertion sequence found in the T-DNA region of mutant Ti plasmid pTiA66 and distribution of its homologues in octopine Ti plasmid. *Proc. Natl. Acad. Sci. USA* **81:**7495–7499.

Mahillon, J., J. Seurinck, L. Van Rempoy, J. Delcour, and M. Zabeau. 1985. Nucleotide sequence and structural organization of an insertion sequence (IS231) from *Bacillus thuringiensis* strain berliner 1715. *EMBO J.* **4:**3895–3899.

Maynard-Smith, J. 1982. Overview—unsolved evolutionary problems. *In* G. Dover and R. Flavel (ed.), *Genome Evolution.* Academic Press, Inc., New York.

Mayr, E. 1982. *The Growth of Biological Thought*, p. 579. Harvard University Press, Cambridge, Mass.

McClintock, B. 1984. The significance of responses of the genome to challenge. *Science* **226:**792–801.

McCommas, S. A., and M. Syvanen. 1988. Temporal control of transposition in Tn*5. J. Bacteriol.* **170:**889–894.

Mollet, B., S. Iida, J. Shepherd, and W. Arber. 1981. Nucleotide sequence of IS26, a new prokaryotic mobile genetic element. *Nucleic Acids Res.* **11:**6319–6330.

Morisato, D., and N. Kleckner. 1984. Transposase promotes double strand breaks and single strand joints at Tn10 termini *in vivo. Cell* **39:**181–190.

Morisato, D., and N. Kleckner. 1987. Tn10 transposition and circle formation in vitro. *Cell* **51:**101–111.

Murphy, E. 1983. Inhibition of T*n*554 transposition: deletion analysis. *Plasmid* **10:**260–269.

Murphy, E., L. Howler, and M. Freie Bastos. 1985. Transposon Tn554: complete nucleotide sequence and isolation of transposition-defective and antibiotic-sensitive mutants. *EMBO J.* **4:**3357–3365.

Murphy, E., and S. Lofdahl. 1984. Transposition of Tn554 does not generate a target duplication. *Nature* (London) **307:**292–294.

Navas, J., J. M. Garcia-Lobo, J. Leon, and J. M. Ortiz. 1985. Structural and functional analyses of the fosfomycin resistance transposon Tn*2921. J. Bacteriol.* **162:**1061–1067.

Nevers, P., and H. Saedler. 1977. Transposable genetic elements as agents of gene instability and chromosome rearrangements. *Nature* (London) **268:**109–115.

Nies, B. A., J. F. Meyer, and B. Wiedemann. 1985. Tn2440, a composite tetracycline resistance transposon with direct repeated copies of IS160 at its flanks. *J. Gen. Microbiol.* **131:**2443–2447.

Perkins, J., and P. Youngman. 1984. A physical and functional analysis of Tn917, a *Streptococcus* transposon in the Tn3 family that functions in *Bacillus. Plasmid* **12:**119–138.

Phadnis, S., and D. Berg. 1987. Identification of base pairs in IS50 o-end needed for IS50 and Tn5 transposition. *Proc. Natl. Acad. Sci. USA* **84:**9118–9122.

Pickens, R. N., A. J. Mazaitis, and W. K. Maas. 1984. High incidence of transposon Tn*3* insertions into a replication control gene of the chimeric R-Ent plasmid pCG86 of *Escherichia coli. J. Bacteriol.* **160:**430–433.

Prentki, P., P. Gamas, D. Galas, and M. Chandler. 1987. Functions of the ends of IS1, p. 719–734. *In* R. McMacken and T. J. Kelly (ed.), *DNA Replication and Recombination.* Alan R. Liss, Inc., New York.

Rasmussen, J., D. Odelson, and F. Macrina. 1987. Complete nucleotide sequence of insertion element IS*4351* from *Bacteroides fragilis. J. Bacteriol.* **169:**3573–3580.

Read, H. A., and S. R. Jaskunas. 1980. Isolation of E. coli mutants containing multiple transpositions of IS sequences. *Mol. Gen. Genet.* **180:**157–164.

Reanney, D. 1976. Extrachromosomal elements as possible agents of adaptation and development. *Bacteriol. Rev.* **40:**552–590.

Reimmann, C., and D. Haas. 1987. Mode of replicon fusion mediated by the duplicated insertion sequence IS21 in E. coli. *Genetics* **115:**619–626.

Reiss, G., B. Masephohl, and A. Puehler. 1983. Analysis of IS21-mediated mobilization of plasmid PACYC184 by R6845 in E. coli. *Plasmid* **10:**111–118.

Roberts, D., B. Hoopes, L. McClure, and N. Kleckner. 1985. IS10 transposition is regulated by DNA adenine methylation. *Cell* **43:**117–126.

Rogers, M., N. Ekgterinaki, E. Nimmo, and D. Sherratt. 1986. Analysis of Tn7 transposition. *Mol. Gen. Genet.* **205:**550–556.

Rogowsky, P., S. Halford, and R. Schmitt. 1985. Definition of 3 resolvase binding sites at the res loci of transposon Tn21 and Tn1721. *EMBO J.* **4:**2135–2142.

Rosner, J. L. 1985. Nonheritable resistance to chloramphenicol and other antibiotics induced by salicylates and other chemotactic repellents in *Escherichia coli* K 12. *Proc. Natl. Acad. Sci. USA* **82:**8771–8774.

Ross, B. G., J. Swan, and N. Kleckner. 1979. Physical structures of Tn10-promoted deletions and inversions: role of the 1400 bp inverted repetitions. *Cell* **16:**721–731.

Sapienza, C., and F. W. Doolittle. 1981. Genes are things you have whether you want them or not. *Cold Spring Harbor Symp. Quant. Biol.* **45:**177–182.

Sasakawa, C., R. Uno, and M. Yoshikawaz. 1981. The requirement for both DNA polymerase and 5′ to 3′ exonuclease activities of DNA polymerase I during Tn5 transposition. *Mol. Gen. Genet.* **182:**19–24.

Sawyer, S., and D. Hartl. 1986. Distribution of transposable elements in prokaryotes. *Theor. Popul. Biol.* **30:**1–16.

Scordilis, G., H. Reed, and T. Lessie. 1987. Identification of transposable elements which activate gene expression in *Pseudomonas cepacia. J. Bacteriol.* **169:**8–13.

Sengstad, C., and W. Arber. 1987. A cloned fragment from bacteriophage P1 enhances IS2 insertion. *Mol. Gen. Genet.* **206:**344–351.

Sengstad, C., S. Iida, R. Hiestand-Naver, and W. Arber. 1986. Terminal inverted repeats of transposable element IS186 which can generate duplications of variable length at an identical target sequence. *Gene* **49:**153–156.

Shaw, J., and D. Clewell. 1985. Complete nucleotide sequence of multiresistance transposon Tn*917* in *Streptococcus faecalis. J. Bacteriol.* **164:**782–796.

Simsek, M., S. Dassarma, V. Rajbuandary, and H. Khorana. 1982. A transposable element from *Halobacterium holobium* which inactivates the bacteriorhodospin gene. *Proc. Natl. Acad. Sci. USA* **79:**7268–7272.

Stanley, S. M. 1979. A theory of evolution above the species level. *Proc. Natl. Acad. Sci. USA* **72:**646–650.

Stibbitz, S., and J. E. Davies. 1987. Tn602: a naturally occurring relative of Tn903 with direct repeats. *Plasmid* **17:**202–209.

Syvanen, M. 1984. The evolutionary implications of mobile genetic elements. *Annu. Rev. Genet.* **18:**271–293.

Syvanen, M., J. Hopkins, and M. Clements. 1982. A new class of mutants in DNA polymerase I that affects gene transposition. *J. Mol. Biol.* **158:**203–212.

Syvanen, M., J. Hopkins, T. Griffin, T. Liang, K. Ippen-Ihler, and R. Kolodner. 1986. Stimulation of precise excision and recombination by conjugal proficient F plasmids. *Mol. Gen. Genet.* **203:**1–7.

Tanaka, M., T. Yamamoto, and T. Sawai. 1983. Evolution of complex resistance transposons from an ancestral mercury transposon. *J. Bacteriol.* **153:**1432–1438.

Timmerman, K., and C. P. Tu. 1985. Complete sequence of IS3. *Nucleic Acids Res.* **13:**2127–2139.

Tomich, P. K., Y. An, and D. B. Clewell. 1980. Properties of erythromycin-inducible transposon Tn*917* in *Streptococcus faecalis. J. Bacteriol.* **141:**1366–1374.

Trieu-Cuot, P., and P. Courvalin. 1984. Nucleotide sequence of the transposable element IS15. *Gene* **30:**113–120.

Trieu-Cuot, P., and P. Courvalin. 1985. Transposition behavior of IS15 and its progenitor IS15-V: are cointegrates exclusive end products? *Plasmid* **14:**80–89.

Vanderleyden, J., J. Desair, C. De-Meirsman, K. Michiels, A. P. Van-Gool, M. D. Chilton, and G. C. Jen. 1986. Nucleotide sequence of an insertion sequence (IS) element identified in the T-DNA region of a spontaneous variant of the Ti-plasmid pTiT37. *Nucleic Acids Res.* **14:**6699–709.

Wang, J. C. 1985. DNA topoisomerases. *Annu. Rev. Biochem.* **54:**665–698.

Way, J., and M. Kleckner. 1984. Essential sites at transposon Tn10 termini. *Proc. Natl. Acad. Sci. USA* **81:**3452–3456.

Willetts, N. S., C. Crowther, and B. W. Holloway. 1981. The insertion sequence IS21 of R68.45 and the molecular basis for mobilization of the bacterial chromosome. *Plasmid* **6:**30–52.

Xu, W. L., and W. F. Doolittle. 1983. Structure of the archaebacterial transposable element ISH50. *Nucleic Acids Res.* **11:**4195–9.

Yamada, T., P. Lee, and T. Kosoge. 1986. Insertion elements of *Pseudomonas savastanoi:* nucleotide sequence and homology with *Agrobacterium tumefaciens* transfer DNA. *Proc. Natl. Acad. Sci. USA* **83:**8263–8267.

Yin, J., M. Krebs, and W. Reznikoff. 1987. The effect of *dam* methylation on Tn*5* transposition. *J. Mol. Biol.* **199:**33–45.

Yin, J. C. P., and W. S. Reznikoff. 1987. *dnaA*, an essential host gene, and Tn*5* transposition. *J. Bacteriol.* **169:**4637–4645.

Youngman, P., J. Perkins, and R. Losick. 1983. Genetic transposition and insertional mutagenesis in *Bacillus subtilis* and *Streptococcus faecalis* transposon Tn917. *Proc. Natl. Acad. Sci. USA* **80:**2305–2309.

Chapter 11

Resolvases and DNA-Invertases: a Family of Enzymes Active in Site-Specific Recombination

Graham F. Hatfull and Nigel D. F. Grindley

I. INTRODUCTION

Site-specific recombination is a process of reciprocal exchange catalyzed by specialized recombinational proteins acting at specific DNA sites. No net loss or synthesis of DNA occurs in the systems examined thus far; the exchange occurs by a simple breakage-reunion mechanism. There are three possible outcomes of site-specific recombination which depend upon the arrangement of the recombination sites. Intramolecular events give either excision (resolution) or inversion, while intermolecular events give integration.

Numerous biological processes in procaryotes involve site-specific recombination. These include the integration and excision of certain bacteriophages, alternation of the expression of two mutually exclusive sets of genes (in bacteria and bacteriophages), plasmid stabilization, and formation of the final products of transposition by certain transposable elements. The proteins that mediate these events can be subdivided into two main groups on the basis of amino acid similarities: a group that includes the integrase (Int)-related proteins from lambdoid bacteriophages together with the Cre and FLP recombinases (Argos et al., 1986; Cox, this volume) and a group that includes the resolvases encoded by transposons of the Tn3 family and the DNA-invertases. In this chapter we discuss the latter group of recombination proteins, the two types of DNA rearrangement that they mediate, and our current understanding of these processes at the molecular level.

Class II transposable elements such as Tn3 and γδ (Tn1000) transpose via a two-stage process (see Fig. 1; for reviews, see Grindley, 1983; Heffron, 1983; Grindley and Reed, 1985). In the first stage, the donor molecule carrying the transposon is fused to the target molecule in a replicative reaction to form a cointegrate, which contains two copies of the transposable element in direct orientation; this step requires the action of transposase, the product of the tnpA gene of the transposon. In the second stage, resolution of this cointegrate molecule by a conservative site-specific recombination event generates the products of transposition, donor and target molecules each carrying one copy of the transposon; this event is catalyzed by a second transposon-encoded protein, resolvase, the product of the tnpR gene (see Fig. 1 and 2).

The DNA sequences of the resolvase genes from transposons Tn3, γδ, Tn501, Tn1721, Tn21, Tn2501, and Tn917 have been determined, and the amino acid sequences have been deduced; the protein sequences are clearly homologous (see Fig. 3). The resolvases can be classified into several groups. The Tn3 and γδ proteins have about 80% amino acid identity (Heffron et al., 1979; Reed et al., 1982) and are functionally interchangeable (Kostriken et al., 1981; Reed, 1981a; Kitts et al., 1982). The Tn501 and Tn21 resolvases have about 80% amino acid identity with each other (the Tn1721 and Tn501 resolvases differ by only a single amino acid [Rogowsky and Schmitt, 1985]) but only about 30% identity with the Tn3 group. Resolvases of the Tn501 subgroup are interchangeable with one another, but not with resolvases from the Tn3 group (Altenbuchner et al., 1981; Grinsted et al., 1982; Diver et al., 1983; Rogowsky and Schmitt, 1984, 1985; Halford et al., 1985). The Tn2501 resolvase (Michiels et al., 1987) has about 30%

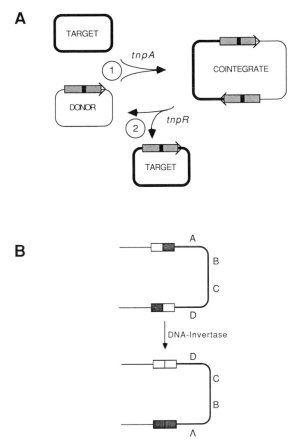

FIGURE 1. Transposition, cointegrate resolution, and DNA inversion. (A) Transposition of class II transposons (Kleckner, 1981) is a two-step process. The first step is fusion of the donor and target molecules to form a cointegrate in which the transposable element (shaded box) is duplicated. This step is mediated by transposase, the transposon *tnpA* gene product. The second step is a conservative site-specific recombination event which resolves the cointegrate into the final products of transposition. This step is catalyzed by resolvase, the product of the transposon *tnpR* gene, acting at a site within the transposon called *res* (black bar). (B) DNA inversion occurs as a result of the action of a protein, the DNA-invertase, at a site (box). Like cointegrate resolution, this is a conservative site-specific recombination event and results in inversion of a specific segment of DNA.

amino acid identity with each of the other two groups but cannot substitute for any other resolvase. The Tn*917* resolvase (the only member isolated from a gram-positive organism [Shaw and Clewell, 1985]) has no more than about 25% amino acid identity with any of the other resolvase proteins, and it is unlikely to be interchangeable with the other resolvases, although this remains to be determined.

Four different systems, the Hin, Gin, Pin, and Cin systems, have been described which affect biological processes by DNA inversion (see Fig. 1 and 4) (for reviews, see Silverman and Simon, 1983; Plasterk and van de Putte, 1984;

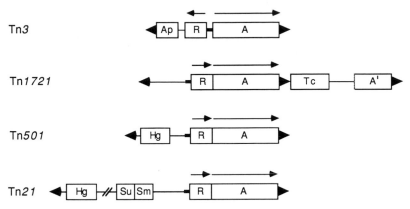

FIGURE 2. Organization of class II transposons. Four members of the class II transposons, of various complexity, are shown. The terminal repeats which are recognized by transposase, the *tnpA* gene product, are represented as a large arrowhead. The *tnpA* (A) and *tnpR* (R) genes are transcribed in the direction shown by the thin arrows over the genes. The recombination site, *res*, is represented as a short, thick line located next to the start of the *tnpR* gene. Genes coding for resistance to ampicillin (Ap), mercury (Hg), streptomycin (Sm), and sulfonamides (Su) are shown. (Adapted from Schmitt et al., 1985.)

Simon and Herskowitz, 1985). The Hin protein, responsible for flagellar phase variation in *Salmonella* spp., mediates inversion of a DNA segment containing the promoter for the H2 flagellin gene; in one orientation the gene is activated, whereas in the other it remains inactive (Zeig et al., 1977). Gin inverts a region of the bacteriophage Mu genome, resulting in alternation of expression between two sets of tail fiber genes (Kamp et al., 1978; Howe et al., 1979; van de Putte et al., 1980). Likewise, Cin controls the expression of the tail fiber genes of bacteriophage P1 (Iida, 1984; Kamp et al., 1984). The Pin inversion system is found associated with a defective e14 viral genome integrated in the *Escherichia coli* chromosome; its function is unknown (Plasterk et al., 1983a; Plasterk and van de Putte, 1985). These four DNA-invertases have approximately 60% amino acid identity (Zeig and Simon, 1980; Hiestand-Nauer and Iida, 1983; Plasterk et al., 1983a) and are functionally interchangeable with one another but not with any of the resolvases (Kutsukake and Iino, 1980; Kamp and Kahmann, 1981; Silverman et al., 1981; Iida et al., 1982).

In spite of the different biological consequences of resolution and inversion, the similarity of the amino acid sequences (about 13% of the amino acids are common to all members [see Fig. 3] [Simon et al., 1980; Newman and Grindley, 1984]) indicates that all 11 proteins belong to a single family, presumably descended from a common ancestor. The proteins are similar in size, ranging between 179 residues (Tn*917*) and 193 residues (Gin); the residues common to all members lie within the first 142 amino acids.

The ability to reproduce both resolution and DNA inversion in vitro has made it possible to analyze these recombinational events biochemically and topologically and to determine the requirements for the reactions. The requirements for

```
              10        20        30        40        50        60        70
               .         .         .         .         .         .         .
Tn 501   MQGHRI-GYVRVSSFDQNPERQ-----LEQTQ--VSKVFTDKASGK--DTQRPQLEALLSFVREGDTVVVHSMDRLA
Tn 21    MT.Q..-..I...T.......-----..GVK--.DRA.S.....--.VK.......I..A.T............

Tn 3     MRIFGYARVSTSQQSLDIQIRALKDAGVK--ANRIFTDKASGS--STDREGLDLLRMKVEEGDVILVKKLDRLG
γδ       M.L........V.........--...........--.S..K....................

Tn 2501  MSRVFAYCRVSTLEQTTENQRREIEAAGFAIRPQRLIEEHISGSVAASERPGFIRLLDRMENGDVLIVTKLDRLG

Tn 917   MIFGYARVSTDDQNLSLQIDALTHYG----IDKLFQEKVTGA--KKDRPQLEEMINLLREGDSVVIYKLDRIS

Hin      MATI-GYIRVSTIDQNIDLQRNALTSAN----CDRIFEDRISGK--IANRPGLKRALKYVNKGDTLVVWKLDRLG
Gin      ML.-..V....ND..TD.....VC.G----.EQ....KL..T--RTD.....RA.KRLQK...L.........
Pin      ML.-..V....ND..TD.....NC.G----.EL....KI..T--KSE.....KL.RTLSA..V.........
Cin      ML.-..V....NE..TA.....ES.G----.EL....KA..K--KAE.....KV.RMLSR...L.........


              80        90       100       110       120       130       140
               .         .         .         .         .         .         .
Tn 501   RNLDDLRRLVQKLTQRGVRIEFLKEGLVFTGEDSPMANLMLSVMGAFAEFERALIRERQREGITLAKQRGAYR
Tn 21    .......I..T......H...V..H.S.................A.......

Tn 3     RDTADMIQLIKEFDAQGVAVRFIDDGISTDGD---MGQMVVTILSAVAQAERRRILERTNEGRQEAKLKGIKF
γδ       ..............SI.........E---..K...........Q...........MA...VV.

Tn 2501  RNAMDIRKTVEQLASSDIRVHCLALGGVDLTS--AAGRMTMQVISAVAEFERDLLLERTHSGIARAKATGKRF

Tn 917   RSTKHLIELSELFEELSVNFISIQDNVDTSTS---MGRFFFRVMASLAELERDIIIERTNSGLKAARVRGKKG

Hin      RSVKNLVALISELHERGAHFHSLTDSIDTSS---AMGRFFFHVMSALAEMERELIVERTLAGLAAARAQGRLG
Gin      ..MKH.IS.VG...RE..IN.R.........S---P.........G.........I...M...AA..NK...I.
Pin      ..MRH.VV.VE...RE..IN.R.........T---P.........G.........V...K...ET..AQ...I.
Cin      ..MRH.VV.VE...RD..IN.R.........T---P.........G.........V...R...DA..AE...I.


             150       160       170       180
              .         .         .         .
Tn 501   GRKKALSDEQAATLRQRATAGEPKAQLAREFNISRETLYQYLRTDD
Tn 21    ...S..S.R1.E....VE...Q.TK.....G............Q

Tn 3     GRRRTVDRNVVLTLHQK---GTGATEIAHQLSIARSTVYKILEDERAS
γδ       ..K.KI..DA..NMW.Q---.L..SH.SKTMN.......VINESN

Tn 2501  GRPSALNEEQQLTVIARINAGISISAIAREFNTTRQTILRVKAGQQSS

Tn 917   GRPSKGKLSIDLALKMYDSKEYSIRQILDASKLKNNLLPLPQ

Hin      GRPRAINKHEQEQISRLLEKGHPRQQLAIIFGIGVSTLYRYFPASSIKKRMN
Gin      ..PPKLTKAEWE.AG..LAQ.I..KQV.L.YDVAL....KKH..KRAHIENDDRIN
Pin      ..RPKLTPEQWA.AG..IAA.T..QKV.I.YDVGV....KRF..GDK
Cin      ..RPKYQEETWQ.MR..LEK.I..KQV.I.YDVAV....KKF..SSFQS
```

FIGURE 3. Amino acid sequences of members of the family of resolvases and DNA-invertases. Members of subgroups that complement each other are grouped, and dots are shown where the amino acid is identical in all members of a subgroup. The amino acid sequences are aligned with padding (−) for maximum alignment. Amino acids in the shaded areas are present in all of the family members. The numbering used corresponds to the γδ sequence. The Tn501 and Tn21 sequences are from Diver et al. (1983), Tn3 and γδ are from Reed et al. (1982), Tn2501 is from Michiels et al. (1987), Tn917 is from Shaw and Clewell (1985), Pin and Gin are from Plasterk et al. (1983a), Cin is from Hiestand-Nauer and Iida (1983), and Hin is from Zeig and Simon (1980). The resolvase sequence from Tn1721 is identical to that from Tn501 except that position 26 (corresponding to position 29 in the γδ sequence) is a histidine (Rogowsky and Schmitt, 1985).

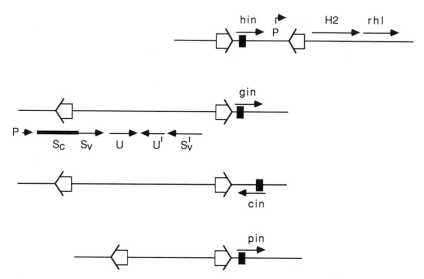

FIGURE 4. Four systems of site-specific DNA inversion. The genes for the DNA-invertases Hin, Gin, Cin, and Pin are shown as well as the direction in which they are transcribed. The invertible DNA segments are bracketed by inverted arrows. The systems shown are, from top to bottom: control of flagellar phase variation in *S. typhimurium* by Hin (H2 is a flagellin gene; *rhl* is a gene for a repressor of the unlinked flagellin H1 gene); variation of host specificity by inversion of the G loop and the C loop by Gin and Cin, respectively (S, S', U, and U' are tail fiber genes; S_c and S_v are constant and variable parts of the S gene, which is affected by inversion); and the Pin system associated with the e14 element in *E. coli*, whose function is unknown. The filled boxes represent the recombinational enhancers, *sis*, located within the genes for the DNA-invertases. P denotes transcriptional promoters of genes affected by DNA inversion.

the Tn*3*, γδ, Tn*1721*, and Tn*21* resolution systems are simple; only purified resolvase protein, a negatively supercoiled DNA cointegrate substrate, and a simple buffer are needed (Reed, 1981b; Kitts et al., 1983; Krasnow and Cozzarelli, 1983; Halford et al., 1985; Rogowsky and Schmitt, 1985). The Tn*3* and γδ systems also require Mg^{2+} for the complete reaction. In the absence of Mg^{2+} an intermediate accumulates; subsequent treatment with protease or sodium dodecyl sulfate releases the two "halves" of the DNA substrate which has been cleaved at both of the crossover points (Reed and Grindley, 1981). Mg^{2+} is apparently not essential for resolution by the Tn*21* protein (Castell et al., 1986). The DNA inversion systems are somewhat more complex and require additional protein factors in addition to the purified recombinase and the supercoiled DNA substrate (Mertens et al., 1984; Huber et al., 1985; Kahmann et al., 1985; Johnson and Simon, 1985; Johnson et al., 1986). There is no requirement for a high-energy cofactor in any of these reactions in vitro.

Interestingly, the DNA sites must be on the same molecule and in the appropriate orientation for efficient recombination. For example, resolution in vitro is very efficient when two *res* sites (the site of resolvase action) are in the same orientation in the cointegrate molecule. However, if the *res* sites are in inverted orientation, the products of inversion are barely detectable (Reed,

1981b). Likewise, the DNA-invertases mediate DNA inversion efficiently if the recombinase binding sites are in inverted orientation, but they resolve cointegrates with directly repeated sites rather poorly (Plasterk et al., 1983b; Johnson and Simon, 1985). The directional specificity of this family of proteins contrasts with the more promiscuous behavior of the lambda integrase family of recombination proteins, which can perform integration, excision, and inversion with equal facility (see Weisberg and Landy, 1983).

In addition to formation of the normal recombinant products in the in vitro reactions, the resolvases and the DNA-invertases convert their supercoiled substrate into topological isomers of reduced superhelicity. The substrate requirements for this topoisomerase activity are the same as those for recombination; thus, a molecule with a single recombination site is not a substrate for superhelical relaxation by resolvase (Krasnow and Cozzarelli, 1983). Likewise, a substrate with a single *res* site or with two inverted sites is not cleaved in the absence of Mg^{2+} by the $\gamma\delta$ resolvase (Reed and Grindley, 1981). Thus the recombination (and topoisomerase) potential of the protein-DNA complexes is apparently not activated unless the substrate contains two recombination sites in the appropriate orientation.

These site-specific recombination reactions can be thought of as a series of biochemical events. (i) The recombination proteins (and accessory proteins for DNA inversion) find and bind to specific DNA sites. (ii) The two protein-DNA complexes (possibly after assembly of each into a particular higher-order structure) come together to form a synaptic complex in which the recombinational potential of the recombination proteins is activated. (iii) The specific sugar-phosphate bonds are cleaved, and the strands are exchanged and ligated to the appropriate recombinational partner. (iv) The complex disassociates to give the products of recombination.

We first discuss the substrate requirements and the structure and function of the recombinase proteins. Next, we consider how the proteins recognize their DNA sites and mediate the changes in DNA structure that accompany the formation of nucleoprotein complexes. Finally, we discuss how two such nucleoprotein structures could interact to form a synaptic complex, in which the recombinational potential is activated, exchange of DNA strands occurs, and the products of recombination are released.

II. DNA SUBSTRATES FOR RECOMBINATION

A. *res* Sites

Analysis of Tn*3* mutants identified a region of the transposon that was necessary in *cis* for cointegrate resolution (Arthur and Sherratt, 1979; Heffron et al., 1979; Sherratt et al., 1981). The site of recombination was mapped more precisely by using the observation that resolvases within a subgroup are interchangeable; Kostriken et al. (1981) and Reed (1981a) made hybrid cointegrate substrates with one site from Tn*3* and one from $\gamma\delta$, resolved them in vivo with

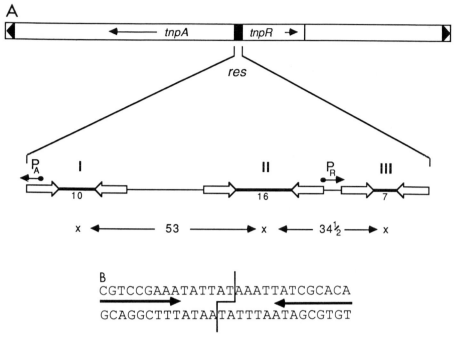

FIGURE 5. Organization of γδ showing details of the *res* site. (A) The open bar represents the γδ transposon, showing the terminal inverted repeats (filled arrowheads), the direction of transcription and location of the *tnpA* and *tnpR* genes, and the location of *res* (filled box). Below is a schematic representation of the *res* site, showing sites I, II, and III, each of which contains two related sequences in inverted orientation (open arrows) separated by a variable spacer (heavy line with length indicated in base pairs). The distances (in base pairs) between the centers of the sites are also shown. The transcription start sites from the P_A and P_R promoters are indicated. (B) The crossover point at the center of site I. Resolvase-mediated cleavage produces a two-base 3′ extension (⌐).

either the Tn*3* or γδ resolvase, and analyzed the DNA sequence of the recombinant joints. Recombination was found to occur within a 19-base-pair (bp) segment in the intercistronic region between the *tnpA* and *tnpR* genes of the transposons. This intergenic region of Tn*3* and γδ contains the promoters for both of these genes (Fig. 5), and resolvase acts not only as a recombinase but also as a transcriptional repressor (Chou et al., 1979a, b; Gill et al., 1979; Heffron et al., 1979; Reed et al., 1982). Similarly, a region upstream of the *tnpR* genes of Tn*501* and Tn*1721* is essential for cointegrate resolution (Altenbuchner and Schmitt, 1983; Diver et al., 1983), and the crossover region was further defined by resolution of hybrid cointegrates (Rogowsky and Schmitt, 1984). However, in this group of transposons the *tnpA* and *tnpR* genes are organized differently (see Fig. 2), although the recombination site is always close to the start of the *tnpR* gene.

Deletion studies (Wells and Grindley, 1984) demonstrated that a 120-bp segment of the *tnpA-tnpR* intergenic region is the minimum sequence required for recombination (Fig. 5 and 6); a cointegrate substrate containing two copies of the minimal 120-bp segment, termed *res*, is resolved efficiently in vivo and in vitro.

The Tn3 and γδ resolvases bind to three sites within this region, sites I, II, and III (Grindley et al., 1982; Kitts et al., 1983). As mentioned earlier, omission of Mg^{2+} from the γδ in vitro resolution reaction followed by digestion with proteinase K releases the two "halves" of the cointegrate, cleaved by double-strand breaks at each *res* site. These breaks have been mapped precisely and occur at the center of site I within the palindromic sequence 5'-TTATAA; the cuts are symmetrical and staggered to leave two-base 3' single-strand extensions (Fig. 5B; Reed and Grindley, 1981). This defines the actual crossover point for recombination. The Tn1721 and the Tn21 resolvases also bind to three regions within their respective *res* sites (Rogowsky et al., 1985), so that the overall organization of *res* is similar for each of these transposons, even though the primary DNA sequences are different (Fig. 6). By DNA sequence comparison a region corresponding to *res* can be identified in Tn501 and Tn2501. The precise point of recombination has not been determined for resolvases other than γδ, although it seems likely that similar cuts are made at the center of site I; the sequence of the resolved hybrid Tn501-Tn1721 cointegrate is consistent with this assumption. Moreover, the central dinucleotide ApT around which the cuts are made by the γδ resolvase is well conserved in each of the *res* sites (although it is apparently GpT in Tn2501; see Fig. 6). As we shall discuss later, the sequence of the central dinucleotide appears to play a critical role in recombination.

Each of the binding sites within the Tn3 and γδ *res* sites is composed of two 9-bp imperfectly repeated sequences related to the consensus 5'-TGTCYR^A_TTA (Y = pyrimidine; R = purine) in inverted orientation (Grindley et al., 1982; E. Falvey, Ph.D. thesis, Yale University, New Haven, Conn., 1988). These repeats are separated by a variable spacer which is 10 bp in site I, 16 bp in site II, and 7 bp in site III (Fig. 5 and 6). The corresponding binding sites within the Tn501, Tn1721, and Tn21 *res* sites also contain short repeats which are related to a consensus sequence, 5'-YGTCARRNTA (Rogowsky et al., 1985). These two consensus sequences are related and are composed of two motifs, YGTC and RNTA (N = any nucleotide), separated by a pyrimidine in Tn3/γδ and by AR in Tn501, Tn1721, and Tn21. In the Tn501 group the spacing within sites I and II (between the 10-bp repeats) is similar to those in Tn3 and γδ, 10 bp in site I and 14 to 15 bp in site II. The site III spacer is harder to define since a sequence that corresponds to the consensus cannot easily be identified for the left half of site III (Fig. 6). The short repeats in the Tn2501 *res* site correspond to a 9-bp consensus 5'-TGTNCGAAA, such that the spacers within sites I and II are 8 and 14 bp, respectively.

Another conserved feature of *res* is the distance between the centers of sites I and II (Fig. 5 and 6); in each of the *res* sites it is close to an integral number of helical turns of B-DNA, four (43 bp) in Tn2501 and five in the other transposons (53 bp in Tn3, γδ, and Tn501; 53.5 bp in Tn1721 and Tn21). As discussed below, the length of this intersite region is extremely important and must be close to an integral number of helical turns for *res* to be a substrate for efficient recombination.

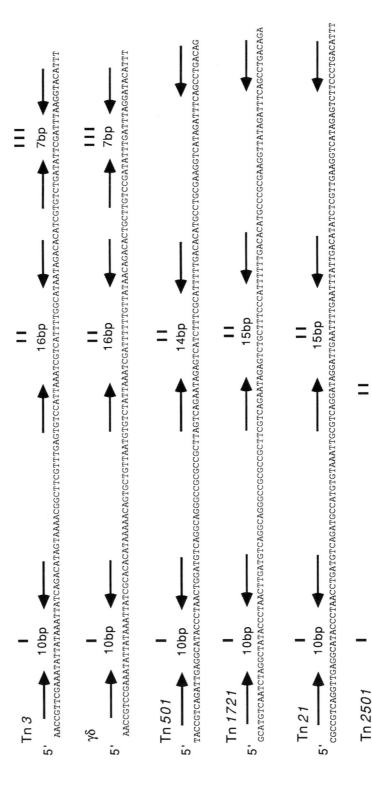

B. Sites for DNA Inversion

Whereas resolvase-mediated recombination requires just two DNA elements (the *res* sites), the substrates for DNA inversion contain three. Two of these are short inverted repeats which flank the invertible segment, are recognized by the DNA-invertases, and contain the crossover points for recombination (Fig. 4). By analogy to the *res* sites for resolution, we will refer to the recombination sites for inversion as *inv*; each one, however, has its own individual name. The third site (*sis*) has been called a recombinational enhancer sequence and is bound by an additional protein factor, FIS.

The *inv* sites at which Hin-mediated inversion occurs are called *hixL* (left) and *hixR* (right); those for Gin-mediated inversion are *gixL* and *gixR*; those for Pin-mediated inversion are *pixL* and *pixR*; and those for Cin-mediated inversion are *cixL* and *cixR*. One of each pair of *inv* sites is usually closely linked to the start of the recombinase gene, reminiscent of the *res-tnpR* linkage (see Fig. 2 and 4). These sites are small (approximately 25 to 30 bp) compared with the *res* sites and are very similar in sequence to one another (Fig. 7); this structural similarity is not surprising since the sites are functionally interchangeable (Plasterk et al., 1983a; Silverman and Simon, 1983). Each *inv* site consists of two 12-bp sequences in inverted orientation which are related to the consensus sequence $5'$-TT^A_CTCN^T_A AAACC and are separated from each other by two nucleotides: $5'$-ApA in the *hix*, *pix*, and *cix* sites, and $5'$-GpA in the *gix* sites (Fig. 7). Footprinting studies show that the *inv* sites are binding sites for the DNA-invertase (Johnson et al., 1984; Bruist et al., 1987b; Kahmann et al., 1987). A synthetic 26-bp consensus sequence is fully active as a recombination site and therefore represents the minimal sequence that is required at each end of the invertible segment of DNA (Johnson and Simon, 1985).

The two nucleotides between the 12-bp motifs play an important role in recombination. Not surprisingly, inversion is inhibited by insertion of an additional base pair between the repeats. More significantly, a mutant site that contains a $5'$-ApT dinucleotide instead of $5'$-ApA recombines efficiently in a substrate molecule with another mutant site, but recombination is reduced to less than 5% in a substrate with one mutant and one wild-type site (Johnson and

---

FIGURE 6. DNA sequence and organization of various *res* sites. The *res* sites from transposons Tn*3*, γδ, Tn*501*, Tn*1721*, Tn*21*, and Tn*2501* are shown, each containing three binding sites for their cognate resolvase. The crossover point for recombination is at the center of site I. Each binding site is composed of two imperfectly conserved inverted repeats (horizontal arrows) which are 9 bp long in the Tn*3*, γδ, and Tn*2501* sites and 10 bp long in the Tn*501*, Tn*1721*, and Tn*21* sites. The inverted repeats are separated by a short spacer which varies among the different sites. An inverted repeat that corresponds to the left half of site III in the Tn*501*, Tn*1721*, Tn*21*, and Tn*2501* sequences has not been identified. The distance between the centers of sites I and II is an important parameter for *res* function (see text) and is five helical turns except in Tn*2501*, where it is four helical turns. The Tn*917* sequence has not been included since, although there are many short inverted repeats immediately adjacent to the start of the *tnpR* gene, the location of *res* has not been determined (Shaw and Clewell, 1985). The Tn*2501* sequence is from Michiels et al. (1987); Tn*501*, Tn*1721*, and Tn*21* are from Diver et al. (1983); γδ is from Reed et al. (1982); and Tn*3* is from Heffron et al. (1979).

A

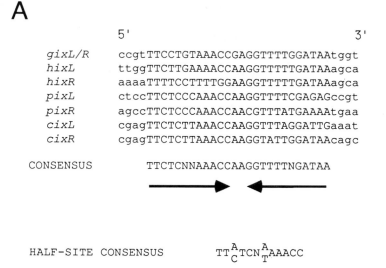

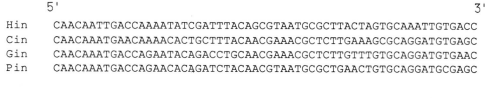

FIGURE 7. Sites for DNA inversion. (A) DNA sequence of *inv* sites. Each *inv* site shown in capital letters is composed of two imperfectly conserved 12-bp sequences in inverted orientation, separated by a 2-bp spacer. The *hix* sites are those found in the ON orientation (as shown in Fig. 4). The two *gix* sites are identical. The consensus derived from all 14 half-sites is shown. The crossover point for recombination is presumed to be at the center of each *inv* site. DNA sequences are from Plasterk and van de Putte (1985). (B) DNA sequence of *sis*. The sequences shown are from the early part of the *hin*, *pin*, *cin*, and *gin* genes; this region in *hin* has been shown by deletion analysis to be necessary for full enhancer function (Johnson et al., 1986). The location of a sequence that may be recognized by FIS (Johnson et al., 1987) is shown below; this sequence is present twice, indicating that there are two binding domains for FIS. DNA sequences are from Huber et al. (1985) and Johnson et al. (1987).

Simon, 1985). Although identity of the two central dinucleotides is clearly required, it is not known whether their sequence is irrelevant or whether only a subset of the 16 possible sequences will work efficiently. Inversion (rather than resolution) of the ApT substrate indicates that inversion specificity is not driven by the central dinucleotide sequences (the ApT sequence is palindromic). The requirement for identity suggests that these 2 bp lie between the staggered breaks made by the recombinase. This hypothesis is supported further by analysis of the recombination products from a substrate in which the two recombinational sites have a different central dinucleotide pair (Iida and Hiestand-Nauer, 1986). Although recombination was inefficient, products could be selected genetically and the sequences at the crossover sites could be determined; localized conversion was observed at the central nucleotides, suggesting that heteroduplexes were formed and subsequently repaired by the host. Klippel et al. (1988) have now demonstrated that Gin cleaves *gix* sites to give the expected two-base 3' single-strand extensions, and preliminary evidence suggests that Hin makes similar cleavages (R. Johnson, personal communication). The structure of each *inv* site is thus analogous to site I of *res*, and the strand cleavages made during recombination are similar to those made by resolvase.

The third DNA sequence element required for efficient DNA inversion is the recombinational enhancer (*sis*); in the absence of this element recombination mediated by a DNA-invertase occurs at the normal crossover sequences but 20- to 200-fold less efficiently (Huber et al., 1985; Johnson and Simon, 1985; Kahmann et al., 1985). The enhancer sequences (which like the DNA-invertases are interchangeable [Huber et al., 1985]) are located in the early portion of the recombinase genes but retain activity in almost any location and in either orientation, provided they are on the same DNA molecule as, and not too close to, the *inv* sites (the Hin enhancer is inactive when located only 30 bp from an *inv* site). These properties are similar to the transcriptional enhancers of eucaryotic gene expression (McKnight and Tijan, 1986). Deletion analyses indicate that a region of approximately 60 bp composed of two separate domains is required for full enhancer activity (Huber et al., 1985; Johnson and Simon, 1985; Kahmann et al., 1985). The domains are each about 20 bp long, lie in the same orientation, and are separated by about four and one-half turns of the DNA helix (see Fig. 7, Bruist et al., 1987a; Kahmann et al., 1987). A protein factor, FIS, is required for enhancer activity in vitro and binds independently to the two domains. A sequence related to the consensus 5'-CA$_G^A$NANNTGANC may be important for FIS recognition (Bruist et al., 1987a; Johnson et al., 1987). Interestingly, the length of the spacer between the domains is important; insertion of nonintegral numbers of helical turns into the spacer inactivates the enhancer, whereas an integral number of turns does not (Johnson et al., 1987). The possible role of this enhancer in recombination is discussed below.

III. RECOMBINASE PROTEINS

A. Two-Domain Recombinase

The resolvases and DNA-invertases are composed of two domains that are structurally and functionally distinct (see Fig. 8). The γδ resolvase is cleaved by chymotrypsin between residues phenylalanine 140 and glycine 141 to produce two proteolytic fragments, a 140-amino-acid N-terminal fragment and a 43-amino-acid C-terminal fragment (Abdel-Meguid et al., 1984).

The smaller C-terminal fragment has DNA-binding activity and binds to six

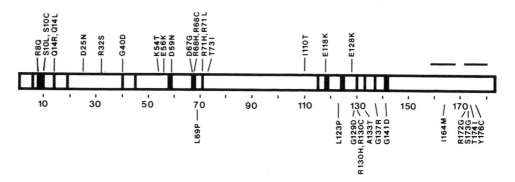

FIGURE 8. Mutants of the γδ resolvase. The horizontal open bar represents the primary sequence of resolvase; the C-terminal domain (amino acids 141 to 183) is shaded. Vertical bars represent amino acid residues that are present in all members of the family of related proteins (see Fig. 3). All the amino acid substitutions shown result in a recombination-defective phenotype; those above the line retain repressor activity, while those below the line do not. The nomenclature for the mutants is the wild-type amino acid followed by its position and the substituting amino acid; thus, R8Q is a change of arginine to glutamine at residue 8. The horizontal lines above the C-terminal domain represent a sequence that is thought to form a helix-turn-helix structure important for DNA binding.

sites within *res*, two sites within each of the three binding sites for the intact protein (see below). Within the C-terminal domain there is a region of approximately 20 amino acids (amino acids 161 to 180 in Tn*3* and γδ) with amino acid sequence similarity to the helix-turn-helix structural motif found in several other sequence-specific DNA-binding proteins, including the cyclic AMP binding protein and the Cro and CI repressors of bacteriophages lambda, 434, and P22 (Pabo and Sauer, 1984). The structure of the repressor-operator complex of bacteriophage 434 at high resolution (Anderson et al., 1987) confirms earlier proposals (see Pabo and Sauer, 1984) that this structural element provides the basis of DNA recognition by allowing interactions between amino acid side chains on the surface of the second helix and individual base pairs.

The binding affinity of the small proteolytic fragment of resolvase is different for each "half-site" and is approximately two orders of magnitude weaker than that of the intact protein for a whole site (K_d, ~2 × 10^{-8} M) (Abdel-Meguid et al., 1984). The large N-terminal proteolytic fragment has no DNA-binding activity. Since the intact protein binds cooperatively to both halves of an individual binding site, it is presumably active as a multimer (probably a dimer) whose formation is mediated by the N-terminal domain. Two lines of evidence suggest that the dimer is the active form. (i) Stoichiometric analyses showed that 12 monomers of the Tn*21* resolvase are required per cointegrate molecule for efficient resolution, presumably six dimers, three bound to each *res* site (Castell et al., 1986). (ii) Separation of the Tn*21* resolvase by gel filtration under conditions similar to those used to assay recombination produced several fractions, but the only fraction with recombination activity was that eluting at the position expected for a dimer (Halford et al., 1985).

In addition to its role in dimerization, the N-terminal domain is probably involved in interactions between resolvase dimers, both within a resolvase-*res* complex and between synapsed *res* sites. It also appears to contain the recombinational catalytic site (see below). In crystals of resolvase the C-terminal domain is disordered, suggesting that the two domains are not spatially fixed relative to each other but, rather, may be linked by a flexible hinge (Abdel-Meguid et al., 1984).

Bruist et al. (1987b) have synthesized the 52-amino-acid peptide of Hin that corresponds to the C-terminal proteolytic fragment of the γδ resolvase. This peptide has specific DNA-binding activity and interacts independently with two half-sites within *hixL* and *hixR*; its binding affinity for the inside half-site of *hixL* is about 50-fold less than the affinity of complete Hin for *hixL*.

The amino acid sequence similarities of the N-terminal domains and the demonstrated DNA binding activities of the C-terminal domains suggest that the resolvases and the DNA-invertases are structurally and functionally very similar.

B. Is Serine 10 an Active-Site Residue?

If Mg^{2+} is omitted from the in vitro resolution reaction, an intermediate accumulates in which resolvase is covalently attached to the 5' end of the DNA at the crossover point (Reed and Grindley, 1981). Resolvase conserves the energy of the phophodiester bond by forming a high-energy protein-DNA linkage in much the same way that topoisomerases do (for reviews, see Gellert, 1981; Wang, 1985), negating the requirement for a high-energy nucleotide cofactor; interestingly, resolvase and the DNA-invertases also have a site-specific topoisomerase activity (see below).

Reed and Moser (1984) have shown that the resolvase-DNA covalent linkage is via a phosphoserine bond, in contrast to the phosphotyrosine linkage used by other topoisomerases (Tse et al., 1980; Champoux, 1981; Rowe et al., 1984; Brougham et al., 1986; Horowitz and Wang, 1987; see also Cox, this volume; Champoux and Bullock, this volume) and Int-related recombinases (Gronostajski and Sadowski, 1985). It has proved difficult to show directly which serine residue is involved (Reed and Moser, 1984), although indirect evidence suggests that serine 10 is a good candidate (Hatfull and Grindley, 1986). First, an active-site residue should be highly conserved within a family of related proteins (the tyrosine that apparently participates in covalent linkage of the Int-related proteins is totally conserved [Argos et al., 1986]). There are 15 serine residues in the protein but only 2 of these, serine 10 and serine 39, are well conserved (see Fig. 3; residue 39 is a threonine in the resolvase of Tn*917*). Second, the behavior of mutant proteins with substitutions at position 10 is consistent with the idea that serine 10 is close to the DNA at the crossover point. A conservative substitution by a cysteine has relatively little effect on DNA binding, whereas substitution by a leucine (with a bulky hydrophobic side group) results in a reduced affinity specifically for site I DNA (which contains the crossover point). In both cases the mutant proteins are recombinationally inactive. By contrast, a cysteine substitution at position 39 of resolvase has no effect on resolution in vivo (V. Rimphanitchayakit, G. F. Hatfull, and N. D. F. Grindley, unpublished data). Substitution of a threonine for serine 10 of Hin results in loss of recombinogenic activity (Johnson, personal communication). The behavior of these mutants is that expected if serine 10 is an essential catalytic residue. Very recently, Klippel et al. (1988) have demonstrated that the covalent attachment of Gin to cleaved *gix* DNA also occurs by a phosphoserine linkage. Mutational studies similar to those described above for resolvase indicate that serine 9 of Gin (equivalent to serine 10 of resolvase) is the active residue.

The formation of a phosphoserine linkage was unexpected since the topoisomerases and the integrase family of site-specific recombinases form a phosphotyrosine bond during the reaction. This raises the possibility that the serine linkage observed by Reed and Moser (1984) was incidental and that the true intermediate involves a phosphotyrosine linkage; one possible scenario is that the absolutely conserved tyrosine at residue 6 (see Fig. 3) makes the biologically relevant covalent bond to the DNA but that linkage is transferred to the nearby serine 10 (which could be close to the DNA) in the experimental protocol used by Reed and Moser (1984). Direct biochemical evidence is needed to clarify this situation.

C. Mutant Resolvase Proteins

Newman and Grindley (1984) isolated a set of mutants of the γδ resolvase lacking recombination activity but retaining repressor activity. The mutations mapped in four conserved residues within the N-terminal domain. We have now collected a larger set of recombination-deficient mutants which includes both repressor-defective and repressor-active phenotypes (Fig. 8). In addition, we have constructed several mutants by oligonucleotide-directed mutagenesis (unpublished data). As shown in Fig. 8, approximately 60% of the mutations that map in the N-terminal domain alter totally conserved residues, even though these constitute only about 15% of the amino acid residues.

The distribution of mutations that affect repressor function is particularly interesting. Not surprisingly, repressor activity was severely impaired by several substitutions in the C-terminal domain, which has DNA-binding activity, as discussed above. In addition, however, the same phenotype was associated with substitutions in a highly conserved region at the C-terminal end of the N-terminal domain (residues 115 to 142). Interestingly, two additional substitutions in this same region (E118K and E128K) both resulted in mutant resolvases with altered DNA-binding properties (see below). There are three likely explanations for the role of this region of the protein in DNA binding. First, it could provide specific DNA-protein contacts; a comparison of the DNA contacts made by the C-terminal domain and the intact protein indicates that some specific contacts are dependent on the N-terminal domain (see below). It seems unlikely, however, that all of the residues in which substitutions have been isolated would be involved in specific DNA contacts. Second, these mutants could be defective in dimer formation; this also seems unlikely since preliminary experiments indicate that at least some of the mutations are partially dominant over the wild type in vivo, presumably as a result of forming inactive heterodimers (unpublished data). We prefer a third explanation, that this region is involved in positioning the C-terminal DNA-binding domains in the appropriate spatial locations relative to the dimerization domains, so that each of the binding sites can be recognized. This explanation is supported by the properties of the E128K mutant protein. E128K specifically fails to bind site III, but binding can be restored if site III is converted into a site I analog by increasing the intrasite spacer from 7 to 10 bp by duplication of the central 3 bp (unpublished data). Thus, E128K appears to have lost the ability to bind to the site III geometry, rather than having lost a specific DNA contact.

IV. DNA-PROTEIN INTERACTIONS

A. Recognition

1. DNA contacts made by intact resolvase

Contacts between intact resolvase and *res* DNA have been mapped by identifying the alkylations of phosphates and purines that inhibit binding. Native polyacrylamide gel electrophoresis provided a convenient way to separate resolvase-*res* complexes from free DNA (Falvey and Grindley, 1987; V. Rimphanitchayakit and N. D. F. Grindley, unpublished data). These studies showed that the protein recognizes a relatively large region of the DNA within each "half-site," with similar (but not identical) contacts in each of the six half-sites (see Fig. 9). Interestingly, there do not appear to be contacts with the crossover region at the center of site I (which can be probed by these modifications) that contribute to the binding energy; depurination of any of the central 4 bp does not adversely affect the binding of resolvase (Rimphanitchayakit and Grindley, unpublished data). Perhaps surprisingly, ethylation of the phosphates that form a covalent linkage to the protein does not affect binding, although it does inhibit resolution (Falvey and Grindley, 1987).

2. DNA contacts made by the C-terminal domain

In each half of a binding site, resolvase interacts with a region that includes both a major and a minor groove spanning approximately 15 bp (see Fig. 9). Interestingly, most (but not all) of the protein-DNA contacts are made by the C-terminal domain (Rimphanitchayakit and Grindley, unpublished data). This small peptide (43 amino acids) requires about 15 bp for efficient binding, which includes the 9-bp repeat, 2 bp outside it, and 4 bp on the inside (toward the center of the site) (see Fig. 9). Presumably, the putative helix-turn-helix motif in the C-terminal domain recognizes the major groove, as has been demonstrated for other repressors (Anderson et al., 1987). It is not clear which parts of this domain are interacting with the minor groove. The main difference between the contacts made by the C-terminal fragment and those of the intact resolvase involves the phosphate that lies at the inner edge of the contacted region; ethylation of this phosphate (at all sites except site II-R) interferes with binding of intact resolvase

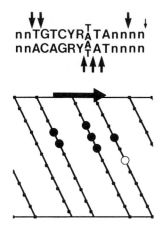

FIGURE 9. Contacts between the C-terminal domain of the resolvase and half of a binding site. The upper portion shows the consensus sequence for half a binding site (the center of the site is to the right); the 15-bp segment shown is the minimum length required for maximal binding of the C-terminal fragment. The lower portion shows a planar projection of the DNA helix. The phosphates which when ethylated interfere with binding of both intact resolvase and the C-terminal proteolytic fragment are shown by heavy arrows in the upper portion and filled circles in the lower. The phosphate which when ethylated inhibits the binding of only intact resolvase is indicated by the small arrow in the upper portion and the open circle in the lower. The large horizontal arrow represents the 9-bp segment that is related to the consensus sequence. Note that the segment of DNA in contact with the C-terminal fragment spans both a major and a minor groove.

but not that of the C-terminal fragment (Fig. 9). This result suggests either that a direct contact is made between this phosphate and the N-terminal domain or that the N-terminal domain acts indirectly on the positioning or folding of the C-terminal domain.

As mentioned above, a 52-amino-acid synthetic peptide (52-mer) of Hin (corresponding to the C-terminal domain of the resolvase) also has specific DNA-binding activity and recognizes the half-sites within *hixL*. An elegant experiment demonstrates that the N-terminal residue of this peptide is in the minor groove at the inside of the binding site. An EDTA moiety was covalently attached to the N terminus of the 52-mer, such that in the presence of Fe(II) and reducing agents, 52-mer–dependent cleavage at *hixL* was observed; cleavage occurred on both DNA strands but displaced to the 3' side on one strand relative to the other, indicating that the EDTA · Fe(II) moiety, and thus the N-terminal amino acid, is in the minor groove (Sluka et al., 1987).

The results obtained with the EDTA–52-mer Hin peptide, together with the chemical modification experiments with the C-terminal proteolytic fragment of γδ resolvase, demonstrate that this part of the recombinase proteins interacts with a major groove and an adjacent minor groove. The peptide chain must therefore at some point cross the phosphate backbone, perhaps making specific amino acid-phosphate contacts to it. A candidate for interaction with these phosphates (or with the phosphate on the other side of the minor groove; see Fig. 9) is the arginine residue at position 142 in the γδ sequence which is absolutely conserved within all members of the family of DNA recombinases (the only other absolutely conserved residue in the C-terminal domain is the glycine at position 141).

B. Changes in DNA Structure

The γδ resolvase-*res* complex appears by electron microscopy to be very compact (Benjamin and Cozzarelli, 1988; J. S. Salvo, Ph.D. thesis, Yale University, New Haven, Conn., 1987), suggesting that a highly ordered, complex nucleoprotein structure may be formed. What changes in DNA structure accompany the formation of such complexes, and are they important for recombination? As discussed below, formation of the resolvase-*res* structure involves considerable distortion of the DNA within each binding site, as a result of protein-DNA interactions, and between sites, as a result of protein-protein interactions. (We use three terms in the discussion of alterations in DNA structure, "bend," "kink," and "loop." We use the term bend to describe any change in DNA structure that results in altered mobility of DNA [or protein-DNA complexes] during polyacrylamide gel electrophoresis, as described by Wu and Crothers [1984]. This net bend results from one or more structural distortions of the DNA which may be gradual [involving several or many base pairs] or may be abrupt local distortions, in which case they are referred to as kinks. A DNA loop occurs between binding sites as a result of protein-protein interactions involving protein-DNA complexes at two or more binding sites.)

1. Resolvase induces bends at each binding site

The mobility of resolvase-*res* complexes during electrophoresis in native polyacrylamide gels is dependent on the location of the *res* site relative to the end of the fragment; for a set of fragments of the same size, the mobility is slowest when the *res* site is at the center of the fragment (J. J. Salvo and N. D. F. Grindley, *EMBO J.*, in press). This phenomenon is characteristic of a protein-induced (or fixed) DNA bend, as has been shown for the regulatory protein cyclic AMP binding protein (Wu and Crothers, 1984). Complexes of resolvase with DNA fragments that separately contain the three individual binding sites exhibit a similar position-dependent electrophoretic mobility, indicating that DNA bends are induced at each of sites I, II, and III (see Fig. 10; Hatfull et al., 1987; Salvo and Grindley, in press). DNA bending is probably a property of the intact protein since it seems unlikely that the C-terminal domain alone induces a substantial DNA bend. However, this is difficult to determine directly by gel electrophoresis because the mobility of the complexes is only slightly slower than that of the free DNA and, to be easily observed, very small DNA fragments and high-percentage acrylamide gels must be used (Rimphanitchayakit and Grindley, unpublished data).

2. DNA loop between sites I and II

Not only are the individual binding sites within *res* bent by resolvase, but also the DNA between sites I and II is bent (or looped) when all three sites are occupied (Salvo and Grindley, in press). The analysis was facilitated by a previous study of the site I-II spacer which demonstrated that additional DNA sequences could be inserted into this spacer without loss of recombination efficiency, provided the insertion was close to an integral number of helical turns (insertion or deletion of a nonintegral number of helical turns results in a dramatic reduction in recombination) (Salvo and Grindley, in press). The centers of sites I and II in the *res* site are naturally separated by five helical turns (assuming a helical repeat of 10.5 bp), so that the centers of these sites are on the same face of the DNA

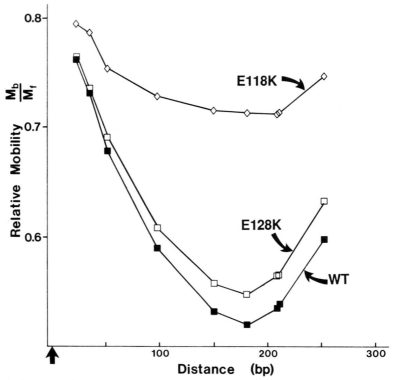

FIGURE 10. Resolvase-dependent bending of site I DNA. Complexes between the wild type (WT) or two mutants (E118K and E128K) of the resolvase and a family of 320-bp fragments containing site I were analyzed by native polyacrylamide gel electrophoresis. The relative mobility is the mobility of the bound DNA fragment (M_b) divided by the mobility of the free DNA fragments (M_f). The fragments are a circularly permuted set which differ from one another only in the location of site I relative to the ends of the DNA fragment. The distance shown on the abscissa is that from the center of site I (↑) through site I-R (the half-site normally proximal to site II) to the end of the fragment. The effect of site location on the mobility of the wild-type resolvase-site I complexes is characteristic of a DNA bend (Hatfull et al., 1987). The mobility of the E118K-site I complexes indicates a defect in the DNA bending. The mobilities of site I complexes with the E128K mutant protein (which has the same charge change as the E118K mutant) are also shown (from Hatfull et al., 1988).

helix. This relationship is also found in the other *res* sites, although the site I-II spacer in Tn*2501* is approximately one helical turn shorter (Fig. 6).

In several other systems, bending or looping of DNA alters its sensitivity to DNase I (Drew and Travers, 1985; Hochschild and Ptashne, 1986; Kramer et al., 1987). DNase I cutting is enhanced on the outside of the bend and suppressed on the inside; these effects are phased at 10- to 11-bp intervals, and the direction of the DNA bend can easily be mapped. The +10 and +21 *res* analogs (which have 10 and 21 bp, respectively, inserted between sites I and II) show this alternating pattern of suppressions and enhancements indicative of a DNA loop (see Fig. 11) (Salvo and Grindley, in press). The pattern is less clear in the wild type because there are relatively few DNase I cuts in this sequence. Interestingly, the +17

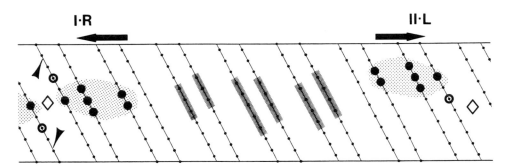

FIGURE 11. Direction of the DNA loop between sites I and II of *res*. A planar projection of the DNA is shown with the phosphates represented as small dots. The diamonds show the centers of sites I and II in a mutant *res* site that has an additional 21 bp between these sites; I-R and II-L indicate the right half of site I and the left half of site II, respectively. The resolvase-phosphate contacts are as shown in Fig. 9; the crossover point at the center of site I is shown by arrowheads. The shaded rectangles along the backbone represent regions of increased sensitivity to DNase I in the resolvase-*res* complex and thus the outside of the DNA loop. (Adapted from Salvo and Grindley, in press.)

analog (a nonintegral number of helical turns between sites I and II, and a poor substrate for recombination) shows none of these effects, nor does a substrate from which site III has been removed (Salvo and Grindley, in press). Thus, the looping of the DNA between sites I and II does not result simply from the binding of resolvase at these two adjacent sites but rather is the consequence of interactions between the protein subunits bound at *res*. However, the specific interactions that are involved remain to be determined.

3. Resolvase induces an unusual structure at the crossover site

The nature of the DNA distortion that results in a DNA bend at each site is unclear, although at site I of γδ *res* it appears to result in part from a kink at the center of the site, at the crossover point for recombination. The evidence for such a structure comes from footprinting experiments with the reagent MPE · Fe(II) and from the properties of a mutant protein, E118K.

Methidiumpropyl-EDTA · Fe(II) [MPE · Fe(II)] is a small molecule that intercalates into the DNA helix (via the minor groove) and cleaves the DNA backbone by hydroxyl radical attack (Dervan, 1986). When resolvase is bound at site I, the site is protected from MPE · Fe(II) cleavage except at its center, where cleavage is enhanced by the binding of resolvase. These enhanced cleavages can be accounted for by a single position of enhanced intercalation between the central 2 bp of site I (Hatfull et al., 1987).

Properties of the E118K mutant protein indicate that the protein-induced structural distortion that enhances MPE · Fe(II) cleavage correlates with the DNA bend; the mutant protein fails to induce the normal DNA bend at site I (Fig. 10) and also fails to promote enhanced cleavage by MPE · Fe(II). Our interpretation of these observations is that the structural distortion induced by resolvase at the center of site I is a kink (a localized distortion) toward the major groove which both opens up the minor groove for intercalation and imparts a substantial bend to the DNA (as detected by polyacrylamide gel electrophoresis). The enhanced cleavage (and thus the formation of the site I kink) is not dependent on the presence of sites II and III, and no enhanced cleavage is seen in MPE · Fe(II) footprinting of resolvase complexed to site II or site III (Hatfull et al., 1987).

Complementation studies in vitro with the E118K mutant protein are consistent with the hypothesis that its interaction with site I is recombinationally unproductive and, thus, that the structural deformation of the DNA is necessary for recombination (Hatfull et al., 1987). There could be several reasons for this; a kink could be important for appropriate docking of the resolvase active site with the crossover point, for inducing strain into the DNA to make strand cleavage energetically more favorable, or perhaps to allow direct DNA-DNA interactions in the synaptic complex.

4. Direction of the bend at site I

The MPE · Fe(II) footprinting experiment described above indicated that the site I kink is toward the major groove. This has been confirmed by determining the interbend distances at which the resolvase site I bend reinforces a known intrinsic (protein-independent) bend on the same DNA fragment (Fig. 12) (Salvo and Grindley, 1987). A segment of the kinetoplast (K-DNA) of *Leishmania tarentolae* was used to provide the intrinsic bend (Trifonov and Sussman, 1980; Wu and Crothers,1984; Koo et al., 1986; Ulanovsky and Trifonov, 1987; Zinkel and Crothers, 1987). The site I/K-DNA fragments, when complexed with resolvase, exhibited mobility minima when the center of site I was separated by an integral number of turns from the center of the K-DNA segment. Since the K-DNA bend is toward the major groove at its center (see Fig. 12; Zinkel and Crothers, 1987), the site I bend is also toward the major groove.

5. Is the resolvase-*res* structure a closed complex?

It is clear from the discussion above that the DNA within the resolvase-*res* complex is not a straight rod of B-DNA, but is bent and kinked in various ways. We know both the direction of the bend at site I and the direction of the loop between sites I and II. Interestingly, these curvatures are not in the same direction, as might be expected in a simple nucleosomelike structure with the DNA wrapped around a protein core, but rather oppose each other, so that the outside of the kink at the center of site I faces the inside of the site I-II loop. The directions of the other DNA bends are not yet known, but it seems likely from the DNase I sensitivities within site II (see Grindley et al., 1982) that the resolvase-induced bend at that site would continue in the same course as the site I-II loop. The T tracts both within site II and between sites I and II would assist bends in this direction (Burkhoff and Tullius, 1987; Zinkel and Crothers, 1987). (All the *res* sites shown in Fig. 6 contain a run of between three and six T residues adjacent to the center of site II.)

The looping of the DNA between sites I and II presumably results from

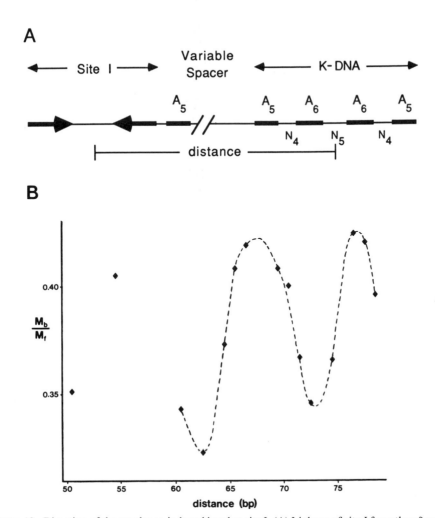

FIGURE 12. Direction of the resolvase-induced bend at site I. (A) Linkage of site I from the γδ *res* site and a K-DNA segment through variable spacer. Each of the four A tracts in the K-DNA segment and the A tract to the right of site I contribute to the intrinsic DNA bend (Salvo and Grindley, 1987). (B) Relative mobility (as defined in Fig. 10) of the resolvase-DNA complexes plotted as a function of the distance between the center of site I and the center of the K segment. (The K-DNA segment [from the kinetoplast of *Leishmania tarentolae*] has an effective bend at its center into the major groove. This has been shown in a similar experiment in which the DNA bend induced by cyclic AMP binding protein was used as a reference bend [Zinkel and Crothers, 1987] and is consistent with several models [Trifonov and Sussman, 1980; Wu and Crothers, 1984; Koo et al., 1986; Ulanovsky and Trifonov, 1987] and with hydroxyl radical cleavage data [Burkhoff and Tullius, 1987] which suggest that the individual A5-A6 tracts [four of which constitute the K-DNA segment] bend into the minor groove). The DNA fragments used contained an additional 170 bp to the left of site I and 90 bp to the right of the K-DNA segment. The lowest relative mobilities occur when an integral number of helical turns separates the centers of site I and the K-DNA segment (assuming a helical repeat of 10.5 bp), indicating that in these complexes the bend at site I is in the same direction as the overall K-DNA bend, that is, toward the major groove at the center of the site (from Salvo and Grindley, 1987).

protein-protein interactions within the resolvase-*res* complex and suggests that the DNA and protein together form a topologically closed (ringlike) structure with resolvase at site I interacting directly with the resolvase at site III (or at site II but in a site III-dependent manner) to complete the protein-DNA "circle." It seems reasonable that this aspect of the resolvase-*res* complex is relevant to the resolution reaction since substrates that fail to make the loop are also defective in resolution. However, it is probable that these same substrate defects (loss of site III, and inappropriate spacing between sites I and II) also interfere with formation of the required synaptic complex of two resolvase-bound *res* sites. One question raised by the apparent topological closure of a single complexed *res* site is what happens to it upon synapsis. Do two closed complexes persist, or do the intra-*res* protein-protein interactions dissociate to be replaced by equivalent inter-*res* interactions? As discussed below, this may be of significance when considering models for synapsis.

6. Role of the site I central dinucleotide

As described above, intercalation of MPE · Fe(II) occurs between the central ApT dinucleotide at the center of site I. During recombination these 2 bp contribute one strand to each of the product molecules and might be expected to play an important role in recombination. The ApT dinucleotide at the center of site I (the crossover point) is well conserved among the class II transposons (it is GpT in Tn*2501*; see Fig. 6). Analogs of the γδ site I have been constructed that have all possible dinucleotides substituted at the center, and we find that this sequence plays a role in all of the resolvase-site I interactions from binding to recombination (Hatfull et al., 1988; Hatfull and Grindley, unpublished data).

Although there appear to be no specific contacts between resolvase and any bases of the central 2 bp, most base substitutions dramatically reduce the binding of resolvase to site I (by at least 30-fold). For efficient binding it is necessary that (i) one of the base pairs is wild type (i.e., ApN or NpT) and (ii) neither of the 5' positions is T (thus, TpT and ApA, which fulfill the first requirement, are both inhibitory). It is the dinucleotide context that is important for binding rather than the specific base pair at each position. For example, resolvase binds efficiently to the GpT substrate and to its complement ApC substrate but binds poorly to the GpC substrate (with both 5'-G and 3'-C). It is difficult to satisfy these binding data with a model for specific resolvase-base pair contacts. Independent evidence against such contacts comes from the observation that depurination of any of the central 4 bp of site I does not inhibit complex formation. Indeed, resolvase binding is enhanced if any base of the central dinucleotide is removed (by depurination or depyrimidination) (Hatfull et al., 1988; Rimphanitchayakit and Grindley, unpublished data). We conclude that the sequence of the central dinucleotide has a pronounced effect on the ability of resolvase to deform (kink) the DNA at the crossover site and, moreover, that it requires less energy to make the kink if a base is missing than if the wild-type sequence is intact. All those sequence combinations that are bound efficiently by resolvase show enhanced cleavage by MPE · Fe(II) similar to the wild type, suggesting that they are similarly kinked.

Further study of these site I crossover point mutants has established the following. (i) Mutants that are tightly bound by resolvase do not necessarily recombine efficiently (indeed, none of the mutants recombines as well as the wild-type *res*). We have shown, in the case of one recombination-defective mutant (with a CpT central dinucleotide), that biochemical transactions are essentially limited to just one strand (the top strand), allowing superhelical relaxation but not recombination (Falvey et al., 1988). (ii) Some site I mutants that bind poorly are reasonably efficient recombination substrates (e.g., ApA). (iii) Both recombination sites must have the same central dinucleotide sequence to give recombination in vitro. However, the Mg^{2+}-independent cleavage of *res*, although requiring two *res* sites (correctly oriented on the same superhelical molecule) does not require that they have the same central dinucleotide. Thus, the requirement for sequence identity is after cleavage but before rejoining and presumably reflects the inability of resolvase to seal a mismatched recombinant joint.

7. DNA inversion systems

The *inv* site is relatively simple compared with *res*, containing a single recombinase binding site rather than three. There are both similarities and differences between the interaction of the DNA-invertases with *inv* and the resolvase-site I interaction. Studies of the Gin-*gix* interaction by MPE · Fe(II) footprinting show enhanced cleavage at the center of the binding site (Mertens et al., 1988), similar to that seen with resolvase. Presumably, a similar distortion of the DNA structure is induced in the vicinity of the crossover point. However, the central dinucleotides in the *inv* sites (ApA in *hix*, GpA in *gix*) are sequences that, at the center of site I of *res*, inhibit resolvase binding and (particularly in the case of GpA) reduce recombination efficiency. Thus, although the DNA-invertases require identity of the two sites, they exhibit much more tolerance than the resolvases toward variations in the sequence of the central dinucleotide. In support of this, the Cin recombinase uses secondary *inv* sites whose sequences are related to the normal sites; although there is a preference for the use of secondary sites in which the dinucleotides at the center are the same as the primary site, secondary sites with a variety of other dinucleotides are also used (Iida and Hiestand-Nauer, 1987).

It is probable that the DNA inversion systems also form complex protein-DNA structures during recombination analogous to the resolvase-*res* interactions (Fig. 13). The influence of the host factor HU on the rate of Hin-mediated inversion supports the idea that there is interaction between *sis* and *inv* or between the proteins bound at these sites. HU is a small protein found in most bacteria which binds nonspecifically to double-stranded DNA. In the absence of HU, the rate of DNA inversion drops approximately 10-fold; addition of enough HU to coat 20% of the entire molecule fully restores recombination (Johnson et al., 1986). However, this dependence on HU is greatest when the enhancer is within 350 bp of the closest *inv* site. Substrates in which *sis* is far from *inv* (up to 4 kilobases from the nearest site) recombine at similar rates in the presence and absence of HU. HU is thought to wrap DNA into nucleosomelike structures (for

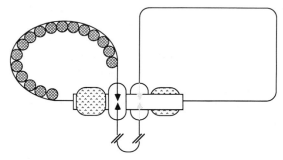

FIGURE 13. Schematic representation of the factors involved in DNA inversion. A substrate molecule (thin line) containing two *inv* sites (each represented by two arrows in inverted orientation) and the recombination enhancer, *sis* (open horizontal box), is shown. HU (stippled circles) may coat the DNA between one *inv* site and *sis*, allowing an interaction between FIS (stippled box) and the recombinase (open vertical box), when the intersite distance is small. The binding of FIS on opposite sides of the DNA at *sis* may result in the duplexes crossing to form two (−) nodes, as shown. The sign of these nodes would be changed to (+) by recombination. (Adapted from Johnson et al., 1987.)

review, see Drlica and Rouviere-Yaniv, 1987) and may facilitate interactions between FIS and the DNA-invertase, by wrapping up or bending the DNA between *sis* and *inv* (Fig. 13); this would be important when *inv* and *sis* are close (<350 bp) but less so when the sites are farther apart (a situation that would impart greater flexibility to the intervening DNA segment). There is as yet no direct evidence for interaction between FIS and the DNA-invertase, although such an interaction has been postulated (Johnson et al., 1987).

V. RECOMBINATIONAL PROCESS

We have discussed above the recognition of the DNA sites by the recombinational proteins and the formation of higher-order protein-DNA complexes prior to synapsis. How do these complexes align at synapsis, what determines the specificity of these reactions, and what is the nature of the strand exchange process? Topological analysis of the DNA inversion and the resolution reactions has provided an insight into many of these questions. Moreover, DNA inversion and resolution are topologically quite different from each other since cointegrate resolution yields two DNA molecules whereas DNA inversion does not alter the number of DNA molecules present. We assume that the apparent similarity between the proteins that mediate resolution and DNA inversion is likely to result in a similar mechanism of strand exchange and that any model for strand exchange that accounts for the observed topological changes during DNA inversion must also account for the changes during cointegrate resolution. First, we must explain the topological terms we will be using.

A. Topological Terms

The unit of the topological method is the node (Wasserman and Cozzarelli, 1986). If the DNA is represented on a flat surface, then a node is created when one

DNA duplex crosses another; the orientation of the DNA can be determined by following the path of the DNA molecule, and the direction of the DNA segments can be determined at the node. There are two ways in which the DNA duplexes can cross, producing either a $(-)$ node or a $(+)$ node. The sign is determined by the following rule: the topmost duplex is rotated not more than 180° so that it is aligned in the same direction with the underlying duplex; if a clockwise rotation is required, the node is $(-)$, and if a counterclockwise rotation is required, the node is $(+)$ (Cozzarelli et al., 1984). As suggested by others (see Cozzarelli et al., 1984), a ribbon model is a particularly helpful aide in understanding nodes and topological processes in general.

The recombinational crossover points separate the substrate molecule into two *domains*. A node formed by the crossing of two DNA segments from the same domain is an *intradomainal node*, and one in which the duplexes are from different domains is an *interdomainal node*. The number of the duplex nodes is the writhe, Wr; the number of Wr nodes in a substrate molecule resulting from the crossing of DNA duplexes from different domains is $^{ter}Wr^s$ ("ter" meaning interdomainal and "s" for substrate). This term is important, as specific numbers of interdomainal nodes are trapped in the substrate at synapsis.

The number of nodes introduced by the mechanism (Me) of the reaction is composed of two functions, the interdomainal nodes (^{ter}Me) and the intradomainal nodes (^{tra}Me). Assuming a similar biochemical mechanism for strand rotation, ^{tra}Me will be the same for both resolution and DNA inversion, and the sign of this node reflects the direction of rotation of the DNA strands within each duplex. If ^{tra}Me is $(+)$, the strands undergo a right-handed rotation. The numerical value of ^{ter}Me is also expected to be the same for resolution and inversion; however, the sign will be different, not because of an intrinsic difference in the mechanism of recombination but merely because of the direction of the path of the DNA duplexes that connect the recombination sites in the synaptic complex ($^{ter}Me^{res} = -^{ter}Me^{inv}$).

The total linkage change, ΔLg, is the change in the number of nodes after recombination has occurred. This term includes product nodes resulting from catenation and knotting as well as changes in the linking number (ΔLk) of the product DNA (which are detected as changes in superhelicity). (In discussions of DNA topology, it is assumed that the twist of the DNA duplex remains constant in proportion to the length of the DNA [i.e., about one turn per 10.5 bp]. Any change in twist introduced during recombination will therefore be translated into a change in writhe and will be detected experimentally as changes in superhelicity, catenation, or knotting.) Thus, $\Delta Lg = \Delta Lk + Ca^p + Kn^p$ (Ca^p, nodes contributing to product catenanes; Kn^p, nodes contributing to product knots) (Cozzarelli et al., 1984). For cointegrate resolution ΔLg is expected to be the same as the total mechanistic change, Me, regardless of the initial interdomainal nodes ($^{ter}Wr^s$). However, for DNA inversion this is not the case, and the total change in linkage is dependent on both Me and $^{ter}Wr^s$. The reason for this is simple. During DNA inversion all the interdomainal nodes trapped between the sites at synapsis change their sign as a result of the recombination, since the path of the DNA has reversed its original direction [thus, in Fig. 13 the initial two $(-)$ nodes will become two $(+)$

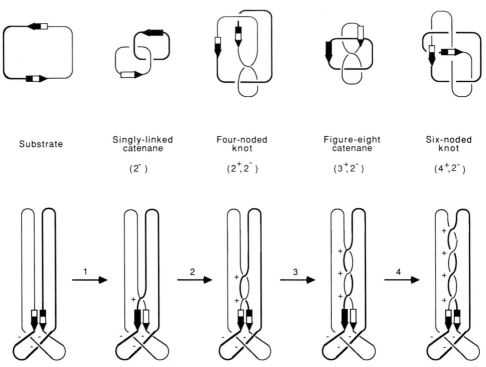

Substrate	Singly-linked catenane	Four-noded knot	Figure-eight catenane	Six-noded knot
	(2^-)	$(2^+,2^-)$	$(3^+,2^-)$	$(4^+,2^-)$

FIGURE 14. Topology of products formed by successive rounds of resolvase-mediated recombination. The major products of recombination are singly linked catenanes with two (−) nodes; further rounds of recombination without release of the substrate produce four-noded knots, figure-eight catenanes and six-noded knots, each with a unique topology, containing the number of (+) and (−) nodes indicated in parentheses. Each round of recombination introduces one additional (+) interdomainal node. Three (−) interdomainal nodes are trapped in the synaptic complex and are held there throughout each round of recombination. Intradomainal nodes are not shown. The two domains are represented as thick and thin lines, respectively. (Adapted from Wasserman et al., 1985.)

nodes following inversion, a change of +4]. In resolution, however, the sign of the initial nodes is retained following recombination (see Fig. 14). For inversion, therefore, the change in linkage (ΔLg) will include a term which is independent of Me and equals twice the number of interdomainal nodes.

B. Recombination Products Have a Defined Topology

Approximately 95% of the products of recombination are of a single topological form; resolvase-mediated resolution produces singly linked catenanes (Krasnow and Cozzarelli, 1983), while the DNA-invertases form unknotted circles (Kahmann et al., 1987; Kanaar et al., 1988). Wasserman and Cozzarelli (1985) determined that the catenanes produced by resolution are always linked by two (−) nodes. The invariant topology of the majority of the recombinant products is independent of the degree of supercoiling of the substrate, indicating that recombination does not result from simple random collision of the sites, which would trap supercoils and produce molecules of various topological forms as is seen with integrative recombination of bacteriophage lambda (Mizuuchi et al., 1980).

C. Topology of Cointegrate Resolution

When cointegrate resolution is performed in vitro, there is a gradual appearance of several minor products in addition to singly linked catenanes, the major products of resolution. The minor products are, first, a four-noded knot, then a five-noded figure-eight catenane, and a six-noded knot (see Fig. 14; Krasnow et al., 1983; Wasserman et al., 1985). These, too, are topologically unique, so that, for example, only one of many possible forms of six-noded knots is observed (Wasserman et al., 1985). These complex structures are assumed to occur by rare repetition of recombination in which additional rounds of strand exchange occur without release of the product molecules. All the product forms can be accounted for only if the enzyme introduces a single (+) interdomainal node as part of the recombination process (Fig. 14). This defines the term ^{ter}Me as a value of +1; that is, one positive interdomainal node is introduced (each time the recombinase performs an exchange of duplex strands).

As the topology of the resolution products has been determined and the interdomainal component of the mechanism is known, the number of interdomainal nodes at synapsis can be deduced. Since two (−) interdomainal nodes are found in the catenated products and one (+) interdomainal node is introduced during recombination, three (−) interdomainal nodes must be trapped at synapsis and $^{ter}Wr^s = -3$ (Fig. 14). Benjamin and Cozzarelli (1988) have demonstrated this directly.

The total change in linking number (ΔLk) during resolution is +4 (Boocock et al., 1987). Since two (−) nodes are trapped in the catenated products, both ΔLg and Me are +2 (for resolution, $\Delta Lg = Me$). Me is the sum of ^{ter}Me and ^{tra}Me, and since $^{ter}Me = +1$, ^{tra}Me must = +1 (see Fig. 15).

D. Topology of DNA Inversion

The products of DNA inversion are unknotted circles (Kanaar et al., 1988), indicating that, like resolvase, invertases also trap a specific number of interdomainal nodes at synapsis. Since the products are unknotted, the total change in linking number can be measured; this value for ΔLk is +4 (Kahmann et al., 1987; Kanaar et al., 1988). It has been argued that one particular set of topological parameters would produce both unknotted circles and a total linking number change of +4; these are that $^{ter}Me = -1$, $^{tra}Me = +1$, and $^{ter}Wr^s = -2$ (Kanaar et al., 1988). These values seem likely for the following reasons: (i) ^{ter}Me is expected to be the same for both resolution and DNA inversion but of a different sign, and for resolution ^{ter}Me is +1; (ii) we expect ^{tra}Me to be the same for both systems, and for resolution it is +1; and (iii) for parallel alignment of the sites, $^{ter}Wr^s$ should be an even number (or zero).

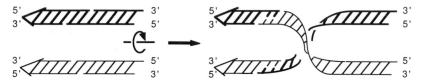

FIGURE 15. Double-strand cut-rotation model for strand exchange. The two duplexes are aligned with either the major grooves or the minor grooves of the two helices facing each other (but not the major groove of one facing the minor groove of the other). The recombinase makes double-strand breaks, to produce two-base, 3' extensions. One pair of duplexes (those to the right of the breaks in the figure) rotates 180° about the axis between them (the line with the circular arrow) in a right-handed direction (as shown on the right), introducing one interdomainal node. If the sites are in direct orientation (as in a substrate for resolution) and aligned in parallel, the sign of the node is (+); if the sites are connected in inverted orientation (as in a substrate for DNA inversion), the sign of the node is (−). Thus ^{ter}Me is +1 for resolution and −1 for DNA inversion. As shown in the right-hand panel, the strands within each duplex must rotate 180° around their helix axis in order for the 3' and 5' ends to be ligated. Thus, the sum of the intradomainal rotations is 360°, which is equivalent to one node, the sign is (+), and thus $^{tra}Me = +1$. This is not influenced by the way in which the strands are connected and is therefore true for both cointegrate resolution and DNA inversion. The total mechanistic change, Me, which is the sum of ^{ter}Me and ^{tra}Me, is therefore 0 for DNA inversion and +2 for cointegrate resolution.

E. Double-Strand Cut-Rotation Model for Recombination

What model for strand exchange would produce the observed topological changes for resolution and DNA inversion? Let us consider two types of model. In one model, a single-strand cut is made in each duplex, and the strands are exchanged to form a Holliday structure; this structure is then resolved by cleavage and exchange of the other pair of strands (Kikuchi and Nash, 1979). These single-strand exchanges can occur in several ways, depending on the alignment of the sites at synapsis and on the direction and extent of strand rotations needed to bring recombining ends together. Int-mediated recombination of phage lambda occurs by single-strand exchanges (see Kitts and Nash, 1987; Nash et al., 1987; Nunes-Duby et al., 1987). In the second model, double-strand cuts are made at both sites, partners are exchanged by a simple rotation, and the strands are ligated. The topology of both resolution and DNA inversion can be accounted for by this second model.

Figure 15 illustrates the double-strand cut-rotation model. The DNA duplexes align (or cross) such that either the major grooves or the minor grooves face each other (the alignment of one major groove of one duplex facing the minor groove of the other duplex seems unlikely from its inherent asymmetry and would not result in the observed topological changes). The recombinase makes concerted double-strand breaks with concomitant covalent linkage to the protein, and the duplexes rotate around each other; this introduces one interdomainal node. The value for ^{ter}Me is therefore 1. The sign is dependent on the way in which the strands are connected, which is different for resolution and DNA inversion. As shown in Fig. 15, $^{ter}Me = +1$ for cointegrate resolution and −1 for DNA inversion. In order for the two Watson strands to religate and likewise for the

Crick strands, the strands within each duplex must rotate 180° about the helix axis. (Note that if the duplexes are held within a protein-DNA complex, the 180° rotations [around the common axis] that generate the strand exchanges will automatically give the rotations about each helix axis necessary for aligning Watson and Crick strands.) Two such 180° rotations are made (one for each helix), forming one intradomainal node (each 180° rotation is equivalent to half a node). This node is (+) for the right-handed helical rotation shown in Fig. 15, so $^{tra}Me = +1$. The total Me for resolution will thus be +2, and for DNA inversion it will be zero.

The attractive feature of this model for strand exchange is that it can account for the observed topological changes in both cointegrate resolution and DNA inversion. Although the topological parameters for each system could perhaps be satisfied by a particular set of single-strand exchanges, there does not appear to be any one way of making single-strand exchanges that would produce the requisite topological changes in both systems. However, additional experimental evidence is needed to confirm the hypothesis that the mechanism for strand exchange is the same for both the resolution and inversion systems. Further support for the involvement of concerted duplex breaks comes from the finding that protease treatment of $\gamma\delta$ resolvase "minus Mg^{2+}" complexes yields molecules with double-strand cuts at both crossover sites. This suggests that in the synaptic complex both strands become covalently linked to resolvase before the occurrence of any strand exchange.

It is important to note that the topological changes observed during superhelical relaxation by resolvase are consistent with either model of exchange, since reduction of superhelicity in steps of one can result from either single- or double-strand cuts and rotation. The observation that resolvase relaxes supercoiled cointegrate substrates by steps of one (Krasnow and Cozzarelli, 1983), as do the type I topoisomerases, is not inconsistent with a mechanism involving concerted double-strand breaks. Double-strand cuts result in relaxation by steps of two only if another DNA duplex is passed through before the break is resealed (as has been postulated for the type II topoisomerases [Liu et al., 1980]).

An obvious way in which a site-specific recombinase might act to reduce substrate superhelicity would be for the protein to make the normal cuts in the substrate and start on the rotations, but then, instead of stopping (and ligating) when the recombinant strands are encountered, continue through a second round of rotations until the parental joints are reformed (this would mean a 360° rotation in Fig. 15, rather than the 180° rotation depicted). Although this would indeed change the linking number (reducing the negative superhelicity if the rotations are in the appropriate direction), this cannot be the way by which resolvase effects simple relaxation of its substrate. This is true because the process just described is equivalent to a double round of recombination which, as discussed above, yields a four-noded knot (Wasserman et al., 1985). To achieve relaxation without knotting, the synaptic complex must dissociate at least partially after the strand cleavage step, to allow strand rotations within a single resolvase-*res* complex. (Note that the topoisomerase activity must be initiated within a normal two-*res* synaptic complex since a single *res* or inverted *res* sites are not substrates for resolvase-mediated relaxation [Krasnow and Cozzarelli, 1983].) Simple relaxation will result from rotations of cleaved strands in one of two ways: (i) if both strands are cut, one duplex end rotates 360° relative to the other about the helix axis and the two are rejoined, or (ii) if one strand is cut, one free end rotates 360° around the unbroken strand and the nick is sealed. Both mechanisms change the substrate linking number by one (i.e., a type I topoisomerase activity) even though in the first both strands are broken. There are reasons to believe that resolvase might be able to use both pathways (see Falvey et al., 1988).

F. Models for Synapsis

The resolvases and the DNA-invertases both trap a specific number of interdomainal nodes at synapsis, as defined by $^{ter}Wr^s$. This means that productive synapsis does not occur by simple random collision of the two sites, which would trap various numbers of interdomainal nodes. As discussed above, resolvase traps three (−) interdomainal nodes at synapsis, and there are two general ways of doing this (Fig. 16). Each *res* could make a toroidal wrap around the protein (as in a nucleosome) to form one (−) interdomainal node within each of the resolvase-*res* complexes. The third node would then result from aligning the two sites in parallel; theoretically, this node could be (−) or (+), but it would be (−) in a negatively supercoiled substrate (Fig. 16, pathway 1). Alternatively, the two *res* sites could be plectonemically interwound to trap three (−) interdomainal nodes (as shown in Fig. 16, pathway 2).

An early model, the "tracking model," for synapsis attempted to explain the exclusion of random interdomainal interlinks and the orientation specificity (Krasnow and Cozzarelli, 1983). This model proposed that resolvase binds to one *res* site and then searches along the DNA (or loops it through the complex) until another *res* site is found in the appropriate orientation, excluding super-coils as it searches. In addition, the tracking complex would possibly recognize another site (or complex) only in the appropriate orientation, thus accounting for the specificity of site orientation. In support of this model, Benjamin et al. (1985) demonstrated that there was a preference for recombination between neighboring sites in a substrate that had a total of four *res* sites all in the same orientation. To test this model, Benjamin et al. (1985) devised the elegant "reporter ring" experiment, in which two small circles were catenated with a cointegrate substrate, which was then resolved in vitro. The tracking model predicts that these two rings would be segregated to just one of the two product molecules; however, such asymmetric segregation of the reporter rings was not observed. Thus, in its simplest form the tracking model does not seem to be appropriate.

A modification of the tracking model is the so-called "slithering" model proposed by Benjamin and Cozzarelli (1986). This model suggests that the structure of supercoiled DNA itself may pair sites in the appropriate alignment by plectonemic slithering of the DNA duplexes past each other until the sites come together. Synapsis is DNA directed, rather than protein directed as in the tracking

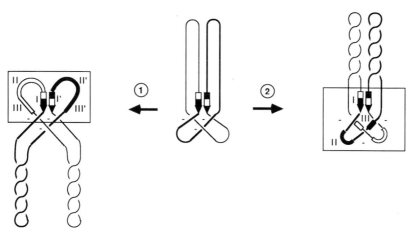

FIGURE 16. Trapping of three interdomainal nodes by toroidal or plectonemic wrapping of DNA. The central diagram shows the alignment of two recombination sites with three trapped (−) interdomainal nodes (as drawn in Fig. 15) (Cozzarelli et al., 1984). The three nodes can be trapped in two ways by the resolvase-*res* complex. Note that both 1 and 2 are simply redrawn from the central figure, and in both cases the boxed regions indicate the resolvase-determined topology of the synaptic complex. In 1 *res* occupies the upper loops of DNA, each resolvase-*res* complex entrapping one (−) toroidal node; a third node results from parallel alignment of the two recombination sites. In 2 *res* occupies the lower loops of DNA; thus, the two *res* sites are wrapped plectonemically about each other. Sites II and III of one *res* site (solid boxes as labeled) are paired antiparallel with sites II and III of the other (open boxes, unlabeled), but the two site I's are aligned in parallel. This is the structure proposed as the productive synaptic complex in the ''two-step'' model of synapsis (Boocock et al., 1986, 1987).

model. If the sites are in the ''incorrect'' orientation relative to each other, they would be brought together, but in an unproductive manner since the sites would not be in parallel alignment. In the slithering model, the trapped interdomainal nodes could arise either from toroidal wrapping at each *res* site or by plectonemic interwrapping of the sites. As discussed below, there is now evidence that the sites are plectonemically interwrapped at synapsis (H. Benjamin and N. Cozzarelli, personal communication).

The two-step model proposes a quite different pathway to synapsis (Boocock et al., 1986, 1987). Rather than being DNA directed or protein directed, synapsis occurs initially via random collision of the two sites, to produce a set of molecules with different initial synaptic geometries (i.e., the alignment of sites and trapping of interdomainal nodes). However, only a subset of these initial geometries would be topologically competent to proceed further in the reaction. Unproductive initial geometries would presumably disassemble, and another set of initial geometries would be formed. The concept of plectonemic interwrapping of the *res* sites with antiparallel alignment of sites II and III was initially proposed as part of the two-step model. Synapsis would initially involve collision of the *res* sites and interwrapping of sites II and III (step I), followed by pairing of the two site I's (step 2) to give the final structure shown in Fig. 16 (pathway 2). If this is accomplished, recombination can proceed. The collision of the sites occurs randomly, but the topology of this initial complex dictates whether the interwrapping of sites II and III and the subsequent pairing of the two site I's can occur. This process has been termed a "topological filter." For example, if additional (−) interdomainal nodes are trapped in the synaptic complex, then compensatory (+) nodes would have to be introduced in order to appropriately interwrap the *res* sites. The introduction of (+) nodes would not be favored in a negatively superhelical substrate. (For further discussion see Gellert and Nash, 1987.)

A model similar to the two-step model for synapsis has been proposed by Craigie and Mizuuchi (1986) for Mu transposition, in which the Mu ends are normally in inverted orientation. Experimentally, they showed that there is neither an absolute requirement for the sites to be on the same DNA molecule nor a requirement for them to be in inverted orientation, provided that the global conformation of the substrate allows appropriate interwrapping of the sites. Thus, a DNA substrate, in which the two recombination sites are on different DNA molecules which are catenated with multiple links, is a good substrate for recombination. Preliminary experiments indicate that multiply linked catenanes (but not singly linked catenanes) are also substrates for resolvase-mediated recombination. Moreover, the distribution of the catenane nodes after recombination indicates that all the nodes are plectonemic at synapsis (Benjamin and Cozzarelli, personal communication). This important result rules out the possibility that the interdomainal nodes trapped at synapsis arise from toroidal wrapping of the DNA at each *res* site. While the sites do seem to be plectonemically interwrapped at synapsis, it is not clear whether this is achieved by slithering of the DNA or by random collision with a "topological filter." Moreover, the biochemical experiments described above suggest that each resolvase-*res* complex is a closed structure, with a DNA loop held together by protein-protein interactions. If this is the case, interwrapping of the two *res* sites of necessity would involve partial disassembly of each individual complex, with intrasite protein-protein interactions presumably replaced by intersite interactions.

The DNA inversion systems trap only two interdomainal (−) nodes at synapsis, and the recombinational enhancer may play a role in this. An attractive possibility is that the nodes are introduced by interactions between FIS bound at the enhancer and the recombinase bound at *inv*, such that the DNA strands cross. Analysis of the Hin enhancer shows how this scheme could work (see Fig. 13; Bruist et al., 1987a; Johnson et al., 1987). The enhancer contains two binding sites for FIS that are separated by approximately four and one-half turns of the DNA helix, so that FIS would bind on opposite sides of the DNA helix. The relative positioning of these sites is critical since enhancer function is not impaired if the length of the spacer between the two FIS binding sites in *sis* is increased by integral numbers of helical turns but is inactivated by addition of nonintegral numbers of helical turns (Johnson et al., 1987). With FIS bound on opposite faces of the DNA helix, two interdomainal nodes of the same sign would naturally result from the interaction of one DNA-invertase-*inv* complex with each FIS.

VI. GOALS AND PERSPECTIVES

Our understanding of the recombination processes discussed above has been greatly enhanced both by the ability to reproduce them in vitro and by the application of a variety of experimental approaches including genetic, biochemical, topological, and structural methodologies. We now have a reasonable understanding of the primary interactions between the proteins and their DNA substrates and are beginning to learn more about the changes in DNA structure that occur in the process of building recombinogenic protein-DNA complexes. Our understanding of the proteins and their DNA interactions would be greatly enhanced if high-resolution structural data were available. Although crystals of the γδ resolvase have been available for some time (Abdel-Meguid et al., 1984), it has not proved easy to perform the crystallographic tricks necessary to extract the appropriate information from the X-ray diffraction pattern. In addition, crystallographic disorder of the C-terminal domain means that only the N-terminal domain is currently amenable to solution. One problem in solving the structure of this domain has been the difficulty in obtaining isomorphous heavy-metal derivatives; however, the introduction of cysteine residues by oligonucleotide-directed mutagenesis (G. F. Hatfull, M. Sanderson, P. Freemont, T. Steitz and N. D. F. Grindley, unpublished data) has now provided a solution to this problem.

It is clearly important to understand the nature of the protein-DNA complex formed at synapsis. This, too, may be approached by a crystallographic analysis, and attempts to grow crystals of the resolvase bound to the entire *res* site are in progress (Grindley, unpublished data). Unfortunately, the complexes formed in the DNA inversion systems at synapsis may be more difficult to analyze since two different DNA sites (*inv* and *sis*) as well as two different proteins (DNA-invertase and FIS) are probably required. However, an understanding of the interactions in this complex and how they compare to interactions within the resolvase-*res* complex is an important goal.

One of the intriguing properties of the two classes of recombination proteins discussed in this chapter is their mutually exclusive directionality. What is the origin of this directionality? Why do the resolvases catalyze only resolution and the DNA-invertases only DNA inversion? We suggest that the answer to this question is integral to understanding these reactions and that our comprehension will be incomplete until we can predict how to turn one system into the other. The problem is a complex one, since the specificity may be determined by the substrates, by the proteins, and by the interactions between them. For example, would a resolvase with two copies of site I and a suitably positioned *sis* become a DNA-invertase, or is there something about the resolvase-site I complex that is unique to the resolvases? Are there specific interactions between resolvase dimers in a synaptic complex (distinct from the interactions between dimers at the crossover sites) that are not found with the invertases? If the DNA-invertases interact directly with FIS at *sis*, are the resolvases capable of a similar interaction or is this interaction specific for the DNA-invertases?

Finally, we would like to point out that the rapid advance in our understanding of these site-specific recombination events has been in part due to parallel studies undertaken by many laboratories on the various members of this family of recombination systems. The similarities and differences between these systems have been, and will continue to be, a rich and fruitful source of information.

ACKNOWLEDGMENTS. We are particularly grateful to all those who sent us unpublished information and manuscripts in preparation or in press. We also express thanks to the members of our laboratory for their continued help and support.

The work in this laboratory is supported by Public Health Service grant GM28470 to N.D.F.G. from the National Institutes of Health.

LITERATURE CITED

Abdel-Meguid, S. S., N. D. F. Grindley, N. S. Templeton, and T. A. Steitz. 1984. Cleavage of the site-specific recombination protein, $\gamma\delta$ resolvase: the smaller of the two fragments binds DNA specifically. *Proc. Natl. Acad. Sci. USA* **81**:2001–2005.

Altenbuchner, J., C.-L. Choi, J. Grinsted, R. Schmitt, and M. H. Richmond. 1981. The transposons Tn*501* (Hg) and Tn*1721* (Tc) are related. *Genet. Res.* **37**:285–289.

Altenbuchner, J., and R. Schmitt. 1983. Transposon Tn*1721*: site-specific recombination generates deletions and inversions. *Mol. Gen. Genet.* **190**:300–308.

Anderson, J. E., M. Ptashne, and S. C. Harrison. 1987. Structure of the repressor-operator complex of bacteriophage 434. *Nature* (London) **326**:846–852.

Argos, P., A. Landy, K. Abremski, J. B. Egan, E. Haggard-Ljungquist, R. H. Hoess, M. J. Kahn, W. Kalionis, S. V. L. Narayana, L. S. Pierson, N. Sternberg, and J. M. Leong. 1986. The integrase family of site-specific recombinases: regional similarities and global diversity. *EMBO J.* **5**:433–440.

Arthur, A., and D. J. Sherratt. 1979. Dissection of the transposition process: a transposon-encoded site-specific recombination system. *Mol. Gen. Genet.* **175**:267–274.

Benjamin, H. W., and N. R. Cozzarelli. 1986. DNA-directed synapsis in recombination: slithering and random collision of sites. *Proc. Robert A. Welch Found. Conf. Chem. Res.* **29**:107–129.

Benjamin, H. W., and N. R. Cozzarelli. 1988. Isolation and characterization of the Tn*3* resolvase synaptic intermediate. *EMBO J.* **7**:1897–1905.

Benjamin, H. W., M. M. Matzuk, M. A. Krasnow, and N. R. Cozzarelli. 1985. Recombination site selection by the Tn*3* resolvase: topological tests of a tracking mechanism. *Cell* **40**:147–158.

Boocock, M. R., J. L. Brown, and D. J. Sherratt. 1986. Structural and catalytic properties of specific complexes between the Tn*3* resolvase and the recombination site, *res. Biochem. Soc. Trans.* **14**:214–216.

Boocock, M. R., J. L. Brown, and D. J. Sherratt. 1987. Topological specificity in Tn*3* resolvase catalysis, p. 703–718. *In* T. J. Kelley and R. McMacken (ed.), *DNA Replication and Recombination*. Alan R. Liss, Inc., New York.

Brougham, M. J., T. C. Rowe, and W. K. Holloman. 1986. Topoisomerase from *Ustilago mavdis* forms covalent complex with single-stranded DNA through a phosphodiester bond to tyrosine. *Biochemistry* **25**:7362–7368.

Bruist, M. F., A. C. Glasgow, R. C. Johnson, and M. I. Simon. 1987a. Fis binding to the recombinational enhancer of the Hin DNA inversion system. *Genes Dev.* **1**:762–772.

Bruist, M. F., S. J. Horvath, L. E. Hood, T. A. Steitz, and M. I. Simon. 1987b. Synthesis of a site-specific DNA-binding peptide. *Science* **235**:777–780.

Burkhoff, A. M., and T. D. Tullius. 1987. The unusual conformation adopted by adenine tracts in kinetoplast DNA. *Cell* **48**:935–943.

Castell, S. E., S. L. Jordan, and S. E. Halford. 1986. Site-specific recombination and topoisomerization by Tn*21* resolvase: role of metal ions. *Nucleic Acids Res.* **14**:7213–7226.

Champoux, J. J. 1981. DNA is linked to the rat liver DNA nicking-closing enzyme by a phosphodiester bond to tyrosine. *J. Biol. Chem.* **256**:4805–4809.

Chou, J., M. J. Casadaban, P. Lemaux, and S. N. Cohen. 1979a. Identification and characterization of a self-regulated repressor of translocation of the Tn3 element. *Proc. Natl. Acad. Sci. USA* **76**:4020–4024.

Chou, J., P. Lemaux, M. J. Casadaban, and S. N. Cohen. 1979b. Transposition protein of Tn3: identification and characterisation of an essential repressor-controlled gene product. *Nature* (London) **282**:801–806.

Cozzarelli, N. R., M. A. Krasnow, S. P. Gerrard, and J. H. White. 1984. A topological treatment of recombination and topoisomerases. *Cold Spring Harbor Symp. Quant. Biol.* **49**:383–400.

Craigie, R., and K. Mizuuchi. 1986. Role of DNA topology in Mu transposition: mechanism of sensing the relative orientation of two DNA segments. *Cell* **45**:793–800.

Dervan, P. B. 1986. Design of sequence-specific DNA-binding molecules. *Science* **232**:464–471.

Diver, W. P., J. Grinsted, D. C. Fritzinger, N. L. Brown, J. Altenbuchner, P. Rogowsky, and R. Schmitt. 1983. DNA sequences of and complementation by the *tnpR* genes of Tn501 and Tn1721. *Mol. Gen. Genet.* **191**:189–193.

Drew, H. R., and A. A. Travers. 1985. DNA bending and its relation to nucleosome positioning. *J. Mol. Biol.* **186**:773–790.

Drlica, K., and J. Rouviere-Yaniv. 1987. Histonelike proteins of bacteria. *Microbiol. Rev.* **51**:301–319.

Falvey, E., and N. D. F. Grindley. 1987. Contacts between γδ resolvase and the γδ *res* site. *EMBO J.* **6**:815–821.

Falvey, E., G. F. Hatfull, and N. D. F. Grindley. 1988. Uncoupling of the recombination and topoisomerase activities of the γδ resolvase by a mutation at the crossover point. *Nature* (London) **332**:861–863.

Gellert, M. 1981. DNA topoisomerases. *Annu. Rev. Biochem.* **40**:879–910.

Gellert, M., and H. Nash. 1987. Communication between segments of DNA during site-specific recombination. *Nature* (London) **325**:401–404.

Gill, R., F. Heffron, and S. Falkow. 1979. Identification of the protein encoded by the transposable element Tn3 which is required for transposition. *Nature* (London) **282**:797–801.

Grindley, N. D. F. 1983. Transposition of Tn3 and related transposons. *Cell* **32**:3–5.

Grindley, N. D. F., M. R. Lauth, R. G. Wells, R. J. Wityk, J. J. Salvo, and R. R. Reed. 1982. Transposon-mediated site-specific recombination: identification of three binding sites for resolvase at the *res* sites of γδ and Tn3. *Cell* **30**:19–27.

Grindley, N. D. F., and R. R. Reed. 1985. Transpositional recombination in prokaryotes. *Annu. Rev. Biochem.* **54**:863–896.

Grinsted, J., F. de la Cruz, J. Altenbuchner, and R. Schmitt. 1982. Complementation of transposition of *tnpA* mutants of Tn3, Tn21, Tn501, and Tn1721. *Plasmid* **8**:276–286.

Gronostajski, R. M., and P. D. Sadowski. 1985. The FLP recombinase of the *Saccharomyces cerevisiae* 2 plasmid attaches covalently to DNA via a phosphotyrosyl linkage. *Mol. Cell. Biol.* **5**:3274–3279.

Halford, S. E., S. L. Jordan, and E. A. Kirkbride. 1985. The resolvase protein from the transposon Tn21. *Mol. Gen. Genet.* **200**:169–175.

Hatfull, G. F., and N. D. F. Grindley. 1986. Analysis of γδ resolvase mutants *in vitro*: evidence for an interaction between serine-10 of resolvase and site I of *res*. *Proc. Natl. Acad. Sci. USA* **83**:5429–5433.

Hatfull, G. F., S. M. Noble, and N. D. F. Grindley. 1987. The γδ resolvase induces an unusual DNA structure at the recombinational crossover point. *Cell* **49**:103–110.

Hatfull, G. F., J. J. Salvo, E. E. Falvey, V. Rimphanitchayakit, and N. D. F. Grindley. 1988. Site-specific recombination by the γδ resolvase. *Symp. Soc. Gen. Microbiol.* **43**:149–181.

Heffron, F. 1983. Tn3 and its relatives, p. 223–260. *In* J. A. Shapiro (ed.), *Mobile Genetic Elements*. Academic Press, Inc., New York.

Heffron, F., B. J. McCarthy, H. Ohtsubo, and E. Ohtsubo. 1979. DNA sequence analysis of the transposon Tn3: three genes and three sites involved in transposition of Tn3. *Cell* **18**:1153–1163.

Hiestand-Nauer, R., and S. Iida. 1983. Sequence of the site-specific recombinase gene *cin* and its substrates serving in the inversion of the C segment of bacteriophage P1. *EMBO J.* **2**:1733–1740.

Hochschild, A., and M. Ptashne. 1986. Cooperative binding of λ repressors to sites separated by integral turns of the DNA helix. *Cell* **44**:681–687.

Horowitz, D. S., and J. C. Wang. 1987. Mapping the active site tyrosine of *Escherichia coli* DNA gyrase. *J. Biol. Chem.* **262:**5339–5344.

Howe, M. M., J. M. Schumm, and A. L. Taylor. 1979. The S and U genes of bacteriophage Mu are located in the invertible G segment of Mu DNA. *Virology* **92:**108–124.

Huber, H. E., S. Iida, W. Arber, and T. A. Bickle. 1985. Site-specific DNA inversion is enhanced by a DNA sequence element in *cis*. *Proc. Natl. Acad. Sci. USA* **82:**3776–3780.

Iida, S. 1984. Bacteriophage P1 carries two related sets of genes determining the host range in the invertible C segment of its genome. *Virology* **134:**421–434.

Iida, S., and R. Hiestand-Nauer. 1986. Localized conversion at the crossover sequences in the site-specific DNA inversion system of bacteriophage P1. *Cell* **45:**71–79.

Iida, S., and R. Hiestand-Nauer. 1987. Role of the central dinucleotide at the crossover site for the selection of quasi sites in DNA inversion mediated by the site-specific recombinase of phage P1. *Mol. Gen. Genet.* **208:**464–468.

Iida, S., J. Meyer, K. E. Kennedy, and W. Arber. 1982. A site-specific, conservative recombination system carried by bacteriophage P1. Mapping the recombinase gene *cin* and the crossover sites *cix* for the inversion of the C segment. *EMBO J.* **1:**1445–1453.

Johnson, R. C., M. F. Bruist, M. B. Glaccum, and M. I. Simon. 1984. *In vitro* analysis of Hin-mediated site-specific recombination. *Cold Spring Harbor Symp. Quant. Biol.* **49:**751–760.

Johnson, R. C., M. F. Bruist, and M. Simon. 1986. Host protein requirements for *in vitro* site-specific DNA inversion. *Cell* **46:**531–539.

Johnson, R. C., A. C. Glasgow, and M. I. Simon. 1987. Spatial relationship of the Fis binding sites for Hin recombinational enhancer activity. *Nature* (London) **329:**462–465.

Johnson, R. C., and M. I. Simon. 1985. Hin-mediated site-specific recombination requires two 26bp recombination sites and a 60bp recombinational enhancer. *Cell* **41:**781–791.

Kahmann, R., G. Mertens, A. Klippel, B. Brauer, F. Rudt, and C. Koch. 1987. The mechanism of G inversion, p. 681–690. *In* T. J. Kelley and R. McMacken (ed.), *DNA Replication and Recombination*. Alan R. Liss, Inc., New York.

Kahmann, R., F. Rudt, C. Koch, and G. Mertens. 1985. G inversion in bacteriophage Mu DNA is stimulated by a site within the invertase gene and a host factor. *Cell* **41:**771–780.

Kamp, D., and R. Kahmann. 1981. The relationship of two invertible segments in bacteriophage Mu and *Salmonella typhimurium* DNA. *Mol. Gen. Genet.* **174:**564–566.

Kamp, D., R. Kahmann, D. Zipser, T. R. Broker, and L. T. Chow. 1978. Inversion of the G DNA segment of phage Mu controls phage infectivity. *Nature* (London) **271:**577–580.

Kamp, D., E. Kardas, W. Ritthaler, R. Sandulache, R. Schmucker, and B. Stern. 1984. Comparative analysis of invertible DNA in phage genomes. *Cold Spring Harbor Symp. Quant. Biol.* **49:**301–311.

Kanaar, R., P. van de Putte, and N. R. Cozzarelli. 1988. Gin-mediated DNA inversion: product structure and the mechanism of strand exchange. *Proc. Natl. Acad. Sci. USA* **85:**752–756.

Kikuchi, Y., and H. A. Nash. 1979. Nicking-closing activity associated with bacteriophage *int* gene product. *Proc. Natl. Acad. Sci. USA* **76:**3760–3764.

Kitts, P., and H. Nash. 1987. Homology-dependent interactions in phage λ site-specific recombination. *Nature* (London) **329:**346–348.

Kitts, P. A., L. S. Symington, M. Burke, R. R. Reed, and D. Sherratt. 1982. Transposon-specified site-specific recombination. *Proc. Natl. Acad. Sci. USA* **79:**46–50.

Kitts, P. A., L. S. Symington, P. Dyson, and D. J. Sherratt. 1983. Transposon-encoded site-specific recombination: nature of the Tn*3* DNA sequences which constitute the recombination site *res*. *EMBO J.* **2:**1055–1060.

Kleckner, N. 1981. Transposable elements in prokaryotes. *Annu. Rev. Genet.* **15:**341–404.

Klippel, A., G. Mertens, T. Patschinsky, and R. Kahmann. 1988. The DNA invertase Gin of phage Mu: formation of a covalent complex with DNA via a phosphoserine at amino acid position 9. *EMBO J.* **7:**1229–1237.

Koo, H. S., H.-M. Wu, and D. M. Crothers. 1986. DNA bending at adenine-thymine tracts. *Nature* (London) **320:**501–506.

Kostriken, R., C. Morita, and F. Heffron. 1981. Transposon Tn*3* encodes a site-specific recombination system: identification of essential sequences, genes, and actual site of recombination. *Proc. Natl. Acad. Sci. USA* **78:**4041–4045.

Kramer, H., M. Niemoller, M. Amouyal, B. Revet, B. von Wilcken-Bergmann, and B. Muller-Hill. 1987. *lac* repressor forms loops with linear DNA carrying two suitably spaced *lac* operators. *EMBO J.* **6:** 1481–1491.

Krasnow, M. A., and N. R. Cozzarelli. 1983. Site-specific relaxation and recombination by the Tn*3* resolvase: recognition of the DNA path between oriented *res* sites. *Cell* **32:**1313–1324.

Krasnow, M. A., A. Stasiak, S. J. Splengler, F. Dean, T. Koller, and N. R. Cozzarelli. 1983. Determination of the absolute handedness of knots and catenanes of DNA. *Nature* (London) **304:** 559–560.

Kutsukake, K., and T. Iino. 1980. A transacting factor mediates inversion of a specific DNA segment in flagellar phase of variation *Salmonella. Proc. Natl. Acad. Sci. USA* **77:**7338–7341.

Liu, L. F., C.-C. Liu, and B. M. Alberts. 1980. Type II DNA topoisomerases: enzymes that can unknot a topologically knotted DNA molecule via a reversible double-strand break. *Cell* **19:**697–707.

McKnight, S., and R. Tijan. 1986. Transcriptional selectivity of viral genes in mammalian cells. *Cell* **46:**795–805.

Mertens, G., A. Hoffman, H. Blocker, R. Frank, and R. Kahmmann. 1984. Gin-mediated site-specific recombination in bacteriophage Mu DNA: overproduction of the protein and inversion *in vitro. EMBO J.* **3:**2415–2421.

Mertens, G., A. Klippel, H. Fuss, H. Blocker, R. Frank, and R. Kahmann. 1988. Site-specific recombination in bacteriophage Mu: characterization of binding sites for the DNA invertase Gin. *EMBO J.* **7:**1219–1227.

Michiels, T., G. Cornelis, K. Ellis, and J. Grinsted. 1987. Tn*2501*, a new component of the lactose transposon Tn*591*, is an example of a new category of class II transposable elements. *J. Bacteriol.* **169:**624–631.

Mizuuchi, K., M. Gellert, R. A. Weisberg, and H. A. Nash. 1980. Catenation and supercoiling in the products of bacteriophage λ integrative recombination *in vitro. J. Mol. Biol.* **141:**485–494.

Nash, H. A., C. E. Bauer, and J. F. Gardner. 1987. Role of homology in site-specific recombination of bacteriophage λ: evidence against joining of cohesive ends. *Proc. Natl. Acad. Sci. USA* **84:**4049–4053.

Newman, B. J., and N. D. F. Grindley. 1984. Mutants of the γδ resolvase: a genetic analysis of the recombination function. *Cell* **38:**463–469.

Nunes-Duby, S. E., L. Matsumoto, and A. Landy. 1987. Site-specific recombination intermediates trapped with suicide substrates. *Cell* **50:**779–788.

Pabo, C. O., and R. T. Sauer. 1984. Protein-DNA recognition. *Annu. Rev. Biochem.* **53:**293–321.

Plasterk, R. H., A. Brinkman, and P. van de Putte. 1983a. DNA inversions in the chromosome of *Escherichia coli* and in bacteriophage Mu: relationship to other site-specific recombination systems. *Proc. Natl. Acad. Sci. USA* **80:**5355–5358.

Plasterk, R. H., T. A. M. Ilmer, and P. van de Putte. 1983b. Site-specific recombination by Gin of bacteriophage Mu: inversions and deletions. *Virology* **127:**24–36.

Plasterk, R. H. A., and P. van de Putte. 1984. Genetic switches by DNA inversions in prokaryotes. *Biochim. Biophys. Acta* **782:**111–119.

Plasterk, R. H., and P. van de Putte. 1985. The invertible P-DNA segment in the chromosome of *Escherichia coli. EMBO J.* **4:**237–242.

Reed, R. R. 1981a. Resolution of cointegrates between transposons γδ and Tn*3* defines the recombination site. *Proc. Natl. Acad. Sci. USA* **78:**3428–3432.

Reed, R. R. 1981b. Transposon-mediated site-specific recombination: a defined *in vitro* system. *Cell* **25:**713–719.

Reed, R. R., and N. D. F. Grindley. 1981. Transposon-mediated site-specific recombination *in vitro*: DNA cleavage and protein-DNA linkage at the recombination site. *Cell* **25:**721–728.

Reed, R. R., and C. D. Moser. 1984. Resolvase-mediated recombination intermediates involve a serine-DNA linkage. *Cold Spring Harbor Symp. Quant. Biol.* **49:**245–249.

Reed, R. R., G. I. Shibuya, and J. A. Steitz. 1982. Nucleotide sequence of the resolvase gene and demonstration that its gene product acts as a repressor of transcription. *Nature* (London) **300:**381–383.

Rogowsky, P., S. E. Halford, and R. Schmitt. 1985. Definition of three resolvase binding sites at the *res* loci of Tn*21* and Tn*1721. EMBO J.* **4:**2135–2141.

Rogowsky, P., and R. Schmitt. 1984. Resolution of a hybrid cointegrate between transposons Tn*501* and Tn*1721* defines the recombination site. *Mol. Gen. Genet.* **193**:162–166.

Rogowsky, P., and R. Schmitt. 1985. Tn*1721*-encoded resolvase: structure of the *tnpR* gene and its *in vitro* functions. *Mol. Gen. Genet.* **200**:176–181.

Rowe, T., K. Tewey, and L. Liu. 1984. Identification of the breakage-reunion subunit of T4 DNA topoisomerase. *J. Biol. Chem.* **259**:9177–9181.

Salvo, J. S., and N. D. F. Grindley. 1987. Helical phasing between DNA bends and the determination of bend direction. *Nucleic Acids Res.* **23**:9771–9779.

Satchwell, S. C., H. R. Drew, and A. A. Travers. 1986. Sequence periodicities in chicken nucleosome core DNA. *J. Mol. Biol.* **191**:659–675.

Schmitt, R., P. Rogowsky, S. E. Halford, and J. Grinsted. 1985. Transposable elements and evolution, p. 91–104. *In Evolution of Prokaryotes.* Academic Press, Inc. (London) Ltd., London.

Shaw, J. H., and D. B. Clewell. 1985. Complete nucleotide sequence of macrolide-lincosamide-streptogramin B-resistance transposon Tn*917* in *Streptococcus faecalis. J. Bacteriol.* **164**:782–796.

Sherratt, D., A. Arthur, and M. Burke. 1981. Transposon-specified, site-specific recombination systems. *Cold Spring Harbor Symp. Quant. Biol.* **45**:275–281.

Silverman, M., and M. Simon. 1983. Phase variation and related systems, p. 537–557. *In* J. A. Shapiro (ed.), *Mobile Genetic Elements.* Academic Press, Inc., New York.

Silverman, M., J. Zeig, G. Mandel, and M. Simon. 1981. Analysis of the functional components of the phase variation system. *Cold Spring Harbor Symp. Quant. Biol.* **45**:17–26.

Simon, M., and I. Herskowitz (ed.). 1985. *Genome Rearrangement.* Alan R. Liss, Inc., New York.

Simon, M., J. Zeig, M. Silverman, G. Mandel, and R. Doolittle. 1980. Phase variation: evolution of a controlling element. *Science* **209**:1370–1374.

Sluka, J. P., S. J. Horvath, M. F. Bruist, M. I. Simon, and P. B. Dervan. 1987. Synthesis of a sequence specific DNA-cleaving peptide. *Science* **238**:1129–1132.

Trifonov, E. N., and J. L. Sussman. 1980. The pitch of chromatin DNA is reflected in its nucleotide sequence. *Proc. Natl. Acad. Sci. USA* **77**:3816–3820.

Tse, Y.-C., K. Kirkegaard, and J. C. Wang. 1980. Covalent bonds between protein and DNA. *J. Biol. Chem.* **255**:5560–5565.

Ulanovsky, L. E., and E. N. Trifonov. 1987. Estimation of wedge components in curved DNA. *Nature* (London) **326**:720–722.

van de Putte, P., S. Cramer, and M. Giphart-Gassler. 1980. Invertible DNA determines host specificity of bacteriophage Mu. *Nature* (London) **286**:218–222.

Wang, J. C. 1985. DNA topoisomerases. *Annu. Rev. Biochem.* **54**:665–697.

Wasserman, S. A., and N. R. Cozzarelli. 1985. Determination of the stereo structure of the product of Tn*3* resolvase by a general method. *Proc. Natl. Acad. Sci. USA* **82**:1079–1083.

Wasserman, S. A., and N. R. Cozzarelli. 1986. Biochemical topology: applications to DNA recombination and replication. *Science* **232**:951–960.

Wasserman, S. A., J. M. Dungan, and N. R. Cozzarelli. 1985. Discovery of a predicted DNA knot substantiates a model for site-specific recombination. *Science* **229**:171–174.

Weisberg, R. A., and A. Landy. 1983. Site-specific recombination in phage lambda, p. 211–250. *In* R. W. Hendrix, J. W. Roberts, F. W. Stahl, and R. A. Weisberg (ed.), *Lambda II.* Cold Spring Harbor Laboratory, Cold Spring Harbor, N.Y.

Wells, R. G., and N. D. F. Grindley. 1984. Analysis of the γδ *res* sites required for site-specific recombination. *J. Mol. Biol.* **179**:667–687.

Wu, H.-M., and D. M. Crothers. 1984. The locus of sequence-directed and protein-induced DNA bending. *Nature* (London) **309**:509–513.

Zeig, J., M. Silverman, M. Hilmen, and M. Simon. 1977. Recombinational switch for gene expression. *Science* **196**:170–172.

Zeig, J., and M. Simon. 1980. Analysis of the nucleotide sequence of an invertible controlling element. *Proc. Natl. Acad. Sci. USA* **77**:4196–4200.

Zinkel, S. S., and D. M. Crothers. 1987. DNA bend direction by phase sensitive detection. *Nature* (London) **328**:178–181.

Chapter 12

Conversion Events in Fungi

P. J. Hastings

I. INTRODUCTION

A. Conversion and Crossing-Over

Recombination, the production of new combinations of alleles, occurs in two distinct ways in both mitosis and meiosis: conversion and crossing-over. A reciprocal breakage and joining process is called crossing-over. A crossover causes all alleles on one side to be recombined with all alleles on the other, unless further crossovers intervene, and the two participating chromatids are reciprocally recombined. The term conversion is used for any nonreciprocal transfer of information.

The occurrence of the two processes is correlated in both meiosis and mitosis. For this reason, models of recombination mechanisms seek to explain reciprocal and nonreciprocal recombination as two consequences of the same process, which I shall call a recombination event. Since determination of reciprocity requires the recovery of all products of a recombination event, much of the information about recombination events comes from the study of meiotic tetrads in fungi.

B. Two Mechanisms of Conversion

There is substantial evidence that conversion in fungi occurs by two distinct mechanisms: (i) the formation and repair of heteroduplex DNA and (ii) double-strand gap filling. The relative roles of the two processes are not delimited, and at least in some cases, both may operate within the same recombination event.

It was postulated long ago that there is a heteroduplex intermediate in recombination in fungi, which is subject to mismatch repair (Holliday, 1962, 1964; Whitehouse, 1963; Meselson, 1965). Survival through meiosis of heterozygosity within a chromatid, seen as postmeiotic segregation (PMS), is taken as evidence of the occurrence of heteroduplex DNA. That this is a precursor to conversion, in which both DNA chains locally acquire the genotype of the homolog, was shown for *Ascobolus immersus* by Leblon and Rossignol (1973). They found that conversion of one allele is epistatic to PMS for another nearby allele. This experiment has also been performed in the yeast *Saccharomyces cerevisiae* (Fogel et al., 1979; Hastings, 1984). The logic of this experiment is discussed in more detail in the next section. The conclusion that much meiotic conversion results from heteroduplex repair is further supported by the isolation of yeast mutants defective in this repair (White et al., 1985) and by the detection of mismatch repair activity in vitro using yeast cell extracts (Muster-Nassal and Kolodner, 1986).

In contrast, it has been shown very clearly that conversion can occur in *S. cerevisiae* without the formation of a heteroduplex intermediate at the site where the conversion will occur. This was achieved by Orr-Weaver et al. (1981) in experiments involving the integration of plasmids by homologous recombination. It was found that a plasmid containing a double-strand gap was integrated as efficiently as a plasmid which had a single cut in the same region and that the gap was filled with a copy of the homologous information carried on the chromosome. Thus, the chromosomal information has been transferred nonreciprocally into the plasmid, and the process must be regarded as conversion. This conversion process is called double-strand gap filling.

These observations form the basis of our understanding of the mechanisms of recombinational repair of DNA, at least of double-strand damage, and of mating-type switching in *S. cerevisiae*, where progression from double-strand cut to double-strand gap followed by gap filling has been established (for review and references, see Strathern, this volume). Conversion by double-strand gap filling has been used as the basis of a model of meiotic recombination by Szostak et al. (1983), who suggested that the initiating event was the formation of a double-strand gap. It has also been suggested that double-strand gap filling may occur during repair of mismatches in DNA occurring in heteroduplex (Hastings, 1984) or resulting from polymerase errors (Wagner et al., 1984; Hastings, 1984).

II. MEIOTIC RECOMBINATION EVENTS

A. General Description of Meiotic Recombination

The subject of meiotic recombination has been reviewed in recent years by several authors (for example, Whitehouse, 1982; Szostak et al., 1983; Roman, 1985; Hastings, 1986a). A meiotic recombination event consists of lengths of one or two chromatids which show conversion or PMS. Both conversion and PMS may occur in the same chromatid in the same event, and lengths of conversion and PMS may be interrupted by lengths showing the parental genotype of the chromatid. It is the endings of these lengths which give rise to most intragenic recombination. In many situations, it seems that one end of the length in which conversion and PMS occur is outside the gene, while the other end is within the gene at a variable position. In association with these intragenic nonreciprocal events, there may be a crossover, detected by reciprocal recombination of markers flanking the region in which nonreciprocal recombination occurs. Approximately half of the nonreciprocal events are associated with crossovers. The associated crossover occurs between the chromatids which were involved in nonreciprocal events whenever this can be determined. The position of the crossover is ambiguous, but it does appear to be within the gene where the recombination event occurs.

B. Parameters Relatable to Heteroduplex and Heteroduplex Correction

1. Conversion spectra

Recombination events are usually detected by deviation from Mendelian (two wild-type:two mutant) segregation of a marker. "Even ratios" show conversion of a chromatid. These are 3 wild type:1 mutant (3:1) and 1:3, which are also expressed as 6:2 and 2:6 when seen in eight-spored tetrads in which each meiotic product has divided by mitosis. "Odd ratios" are 5:3 and 3:5, in which one spore pair, related by mitosis, is mixed and, therefore, showing PMS. Two mixed spore pairs indicate an aberrant 4:4 ratio in the tetrad. "Wider ratios" also occur—7:1, 1:7, 8:0, and 0:8. The wider ratios result from conversion or PMS, or both, for a marker on two sister chromatids. These events occur at the frequency expected for multiple events occurring at the same site. The pattern of ratios shown by a marker is known as the conversion spectrum of that marker.

The relative proportions of the different ratios for a marker can be accounted

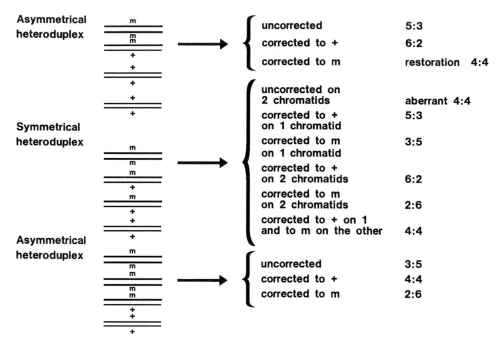

FIGURE 1. Origin of various non-Mendelian segregation ratios on the mismatch correction hypothesis of conversion. The number of wild-type spores is listed first in the ratios.

for by assuming that heteroduplex is formed on one or two chromatids at a given frequency and that the resulting mismatch is corrected to mutant or to wild type, or not corrected, with given probabilities. This exercise was first performed by Whitehouse and Hastings (1965). The assumptions in the algebra have been the subject of some debate (Emerson, 1966; Gutz, 1971), but the simplest form of the algebra was developed and applied to data from *A. immersus* with great success by Paquette and Rossignol (1978). Figure 1 shows how the simple ratios may be related.

The conversion spectrum of a marker is a mixture of allele-specific, position-specific, and locus-specific parameters. Leblon (1972a, b) classified mutations of two spore color loci in *A. immersus* according to the conversion spectra of the alleles. Some markers show little or no PMS. These markers show a strong disparity (6:2 ≠ 2:6). A-type alleles are converted to wild type more often than to mutant genotype (6:2 > 2:6), while B-type alleles have the opposite disparity (2: 6 > 6:2). Other alleles show a larger proportion of PMS ratio tetrads among non-Mendelian segregations. These C-type alleles may show parity, or they may show a disparity which is often less pronounced than that shown by A-type or B-type mutations. Leblon (1972a, b) found that A-, B-, and C-type alleles occur in both the *b1* and *b2* loci. The only parameter which was found to be locus specific was the frequency with which non-Mendelian segregation occurs. This parameter is referred to as the basic frequency of conversion. It was found that the class of mutation (A, B, or C type) depended not on the position but on the mutagenic origin of the marker. Leblon made a case that the A- and B-type mutations are frameshift mutations of opposite sign (insertion or deletion), while the C-type alleles are substitution mutations. These observations led to the very important conclusion that some conversion parameters reflect the nature of the difference in nucleotide sequence between the mutant and the wild-type DNA and that, therefore, they reflect the nature of the mismatch occurring in heteroduplex DNA. Approximately the same relationship between mutagenic origin and conversion spectrum was found in *Sordaria brevicollis* by Yu-Sun et al. (1977).

In yeasts, we do not find that the conversion spectrum depends upon the nature of the mismatch in the same way. Almost all known alleles of *S. cerevisiae* (Fogel et al., 1981) and *Schizosaccharomyces pombe* (Thuriaux et al., 1980) show the same spectrum—parity of 3:1 and 1:3 ratios, with a low proportion (a few percent) of aberrant tetrads showing PMS. There is a rare class of markers in *S. cerevisiae* in which a very high proportion of the non-Mendelian segregation tetrads are seen as PMS. It has been suggested that these high-PMS alleles have a common mismatch in heteroduplex (White et al., 1985). Other exceptional conversion spectra in yeasts have been attributed to the creation by mutation of sites at which recombination is initiated. The best known example is the *M26* allele in the *ade6* locus of *Schizosaccharomyces pombe* (Gutz, 1971). A similar explanation was favored by Pukkila et al. (1986) for disparity in conversion of a large deletion in *S. cerevisiae*. DiCaprio and Hastings (1976) reported that an apparent disparity of conversion of an auxotrophic mutation (*met10-4* in *S. cerevisiae*) was caused by differential survival of mutant and wild-type ascospores. Fink and Styles (1974) showed that deletion heterozygotes in *S. cerevisiae* show parity of 3:1 and 1:3 tetrads. Large heterologies in the *b2* locus of *A. immersus* behave in a way which is very similar to almost all yeast alleles in that they show parity of 6:2 and 2:6, and very low PMS (J. L. Rossignol, A. Nicholas, H. Hamza, and A. Kalogeropoulos, *in* B. Low, ed., *The Recombination of Genetic Material*, in press). A further extraordinary parallel between heterologies in *A. immersus* and normal alleles in yeasts is described in section B.4.

Some aspects of the conversion spectrum of an allele depend upon the position of the allele within a locus. It is these parameters which reveal the shape of the event. They are discussed in section C.

2. Coconversion

When a marker is converted, it is very common to find that adjacent alleles have been converted with it on the same chromatid. This phenomenon is called coconversion and is taken to imply that the conversion process involves a length of a chromatid rather than a point. This phenomenon is important for understanding several aspects of intragenic recombination.

It is also found that PMS concerns a length of a chromatid. When two adjacent alleles show PMS on the same chromatid, they retain their parental linkage relationship in the subchromatids, as would be expected if PMS represented heterozygosity in a length of heteroduplex DNA surviving through meiosis.

3. Interaction of conversion spectra

When there are two sites of heterozygosity within a recombinant event, the conversion spectra of the two alleles interact. The nature of interactions is extremely informative and provides the best genetic evidence that conversion occurs by repair of mismatches in heteroduplex DNA.

The original experiments were performed by Leblon and Rossignol (1973), using alleles of the *b2* locus of *A. immersus*. They found, first, that when an A-type allele (low PMS, 6:2 > 2:6) is placed beside a B-type allele (low PMS, 6:2 < 2:6), both markers retain their characteristic conversion spectrum in those tetrads in which conversion affected only one allele, but that when the two alleles are coconverted, the conversion spectrum is an average of that of the two individual alleles, as though some coconversion is under the direction of one allele and some is under the direction of the other. They then showed that this averaging of conversion spectra does not occur when a C-type allele (high PMS) is placed beside a B-type allele. In this case, the coconversion imposes the B-type spectrum on the C alleles, giving conversion instead of PMS, with the conversion showing the disparity characteristic of the B allele. Conversely, the C allele does not impose its high-PMS characteristic on the B allele. Thus, conversion is epistatic to PMS, as it would be if PMS and conversion represented products from the same series of events, analogous to a biochemical pathway, with PMS representing an earlier step in the pathway than conversion. If PMS represents heteroduplex DNA surviving through meiosis to be resolved by a postmeiotic mitotic DNA replication, then conversion appears to be the meiotic resolution of the same substrate; that is, the mismatch in heteroduplex DNA was removed before the meiotic divisions occurred. The same conclusion was suggested by the finding that the conversion spectrum depends upon the mutagenic origin of the allele. This requires that the nucleotide sequences of the parents are compared but that the resulting heteroduplex does not survive. It is, rather, removed by a process which causes the loss of a copy of one allele and the gain of another copy of the homologous allele, i.e., conversion.

In yeasts, where, for most alleles, there is no variation in conversion spectrum which could be attributed to the interaction of a specific mismatch with the mismatch repair mechanism, one might imagine that conversion had a different cause (Stahl, 1969, 1979; Szostak et al., 1983). However, experiments after the style of the Leblon and Rossignol (1973) experiment give the same result in *S. cerevisiae* as in *A. immersus*: a normal allele coconverts a high-PMS allele nearby so that much less PMS is seen for that allele (Fogel et al. 1979; Hastings, 1984). This seems to force one to conclude that in *S. cerevisiae*, as in *A. immersus*, there is a heteroduplex precursor to conversion.

Other evidence is accumulating to encourage belief in a heteroduplex precursor to conversion in *S. cerevisiae*. The isolation of mutants (called *pms* mutants) which cause normal alleles to behave like high-PMS alleles (Fogel et al., 1981) is most easily understood in this way, although Szostak et al. (1983) have shown that other interpretations are possible. The achievement of mismatch repair in vitro by Muster-Nassal and Kolodner (1986), using extracts from yeast cells, shows that *S. cerevisiae* is capable of this reaction. These authors also reported that *pms1* strains lack the in vitro activity. In addition, it has been found that high-PMS alleles in *S. cerevisiae* have base sequence characteristics in common (White et al., 1985).

A new dimension in the interaction of base sequence with recombination is revealed by the study of Moore et al. (1988). They reevaluated the accumulated data on meiotic and mitotic recombination frequencies in the *cyc1* gene of *S. cerevisiae* in the context of a precise knowledge of the position and nature of the mutations which are used as markers. It was found that mutations that lead to G/G or C/C mismatches have up to 1,000-fold-higher recombination frequency per base pair than other mutations at the same site. This marker effect was not seen when the mutations were closer than 4 base pairs, but outside this limit, it was strongest over short distances. The authors suggested that G/G or C/C mismatches block the extension of cocorrection tracts, thereby increasing recombination by enhancing independent correction. This would occur if an enzyme bound to G/G or C/C did not then proceed to process it.

That the nature of the heterozygosity should have such effects is very strong evidence for the formation of mismatched sequences and, therefore, of heteroduplex formation as an intermediate in the recombination process.

4. Restoration

The hypothesis that conversion results from mismatch repair at a point of heterozygosity in heteroduplex DNA supposes that the mismatched base pair is recognized, and one chain is cut and excised, so that polymerization using the other chain as a template will restore homozygosity within the DNA molecule. If the mismatched base is excised from the strand remaining from the invaded DNA molecule, conversion will result. If it is the invading information which is excised, the mismatch repair process will restore the parental genotype and give a Mendelian ratio in a tetrad, as shown in Fig. 1.

When an allele shows parity in conversion to mutant and to wild type, this is taken to mean that the mutant and wild-type sequences are equally likely to be excised from heteroduplex. Thus, on any one chromatid, conversion and restoration are equally likely.

If an allele shows disparity (6:2 ≠ 2:6), this is because one genotype is more likely to be excised than the other. Thus, it will be seen that conversion predominates when the heteroduplex forms on one chromatid and restoration predominates when it forms on the other. When there is a disparity in the direction of conversion, there should be an equal and opposite disparity in restoration.

This prediction was first tested for frameshift mutations in the *b2* locus of *A. immersus* by Hastings et al. (1980). The method used was to select for heteroduplex spanning the *b2* locus by the use of two well-spaced C-type markers (showing high PMS) and then to study the segregation of an intermediate phenotype marker showing low PMS situated within the length of heteroduplex. The result of the experiment was very close to the expectation based on the model explained above, namely, that conversion and restoration are equally frequent and show equal and opposite disparities on the two chromatids. However, this simple outcome was not found when, using the same system, large heterologies were studied instead of small frameshift mutations (Rossignol et al., 1984; Hamza et al., 1986). For a large deletion, conversion is much more frequent than restoration, and conversion in the two directions shows parity. In *S. cerevisiae* the latter result has been inferred for point mutations, using arguments based on the relative proportions of different classes of recombinant tetrad (Savage and Hastings, 1981; Hastings and Savage, 1984). A direct demonstration of restoration deficiency in *S. cerevisiae* has not yet been achieved.

From the *Ascobolus* data, Hamza et al. (1986) concluded that there are two mechanisms of conversion operating in the *b2* locus. The mechanism of conversion of at least some point mutations has the properties expected of excision repair in which excision of one genotype of a mismatch may be preferred over excision of the other. The second mechanism applies to conversion of large heterologies and has the properties of double-strand break repair. The experimental design selected for heteroduplex spanning the site of the heterology, so that, if double-strand break repair is involved, it is subsequent to heteroduplex formation. This sort of conversion, therefore, looks as if it results from double-strand cutting at a mismatch, as suggested by Hastings (1984). The double-strand gap does not appear to have been a primary gap originating from the initiation of the event as predicted by Szostak et al. (1983).

C. Parameters Relatable to Heteroduplex Distribution

Several aspects of a recombination event are locus specific and may also vary according to position within a locus. The discovery that recombination parameters varied dramatically from locus to locus within an organism and between organisms seemed, at one time, to paint a picture of extreme confusion and multiplicity of mechanism. The variation can, however, be reduced to very few parameters of the heteroduplex distribution along the length of the gene and between chromatids and of the distribution of correction tracts within the lengths of heteroduplex (see Hastings, 1986a).

1. Polarity

Recombination events in meiosis are not randomly distributed throughout the genome. It was mentioned above that one can see from the work of Leblon (1972a) that different loci have different and characteristic ranges of conversion frequency. This can be seen in the data from any well-studied organism. Conversion frequencies for 30 alleles in *S. cerevisiae* vary from 0.6 to 18%, but within the *arg4* locus, for example, the range is 2.5 to 8.3% (Fogel et al., 1979, 1981), while in *his1* the range is from 1.0 to 2.9% (E. A. Savage, Ph.D. thesis, University of Alberta, Edmonton, Alberta, Canada, 1979). In *A. immersus*, Leblon (1972a) reported that conversion frequencies ranged up to 6.4% in the *b1* locus, but up to 28% at the *b2* locus.

Even on the scale of a single locus, there is a nonrandom distribution of the probability of conversion. This is seen as polarity in conversion. Polarity can be seen in three ways. It was first discovered in *A. immersus* that, in some loci, intragenic recombination, detected by the occurrence of wild-type spores from a heteroallelic cross, occurred by conversion of the marker on one particular side (the minority allele) coupled with Mendelian segregation of the other marker (the majority allele). This was true for any two-point cross throughout a large part of the gene (Lissouba et al., 1962). It was found by nonselective tetrad analysis that both markers were being converted, but that when the majority allele was converted, it was always coconverted with the minority allele so that no recombination occurred. Thus, the recombination is occurring because the length of conversion ends between the markers. When this happens, the same marker is always the one converted, as though all conversion lengths come from the same side of the gene.

The same phenomenon can be seen in another way in some *Ascobolus* data: there is a gradient in conversion frequency (basic frequency of conversion, or BFC, which is the frequency of all non-Mendelian segregations) of individual markers seen in one-point crosses (i.e., mutant × wild type), showing that the markers near one end of the gene are converted more often than those in the middle or at the other end. There are very few loci in which this can be seen, probably because allele-specific influences on BFC are of the same magnitude as the gradient. The gradient is, however, very clear in the best-known locus, *b2*, in the data of Paquette and Rossignol (1978). They corrected the BFC of the markers for those components which are allele specific and related to heteroduplex correction. The resulting estimate of the distribution of heteroduplex within the locus gave a clearer picture of polarity than did BFC.

The third way to see polarity applies to the analysis of recombinant meiotic products with markers flanking the locus in which the recombination was selected. Murray (1963) selected for prototrophs from heteroallelic crosses in the *me-2* locus of *Neurospora crassa*. Those prototrophic meiotic products which were not recombinant for flanking markers showed the combination of flanking markers which entered the cross with the right-hand allele more often than the combination which entered the cross with the left-hand allele. This observation applied to every two-point cross over the length of the gene which was studied. Since tetrad analysis in many organisms shows that almost all intragenic recombination occurs by conversion, the simplest interpretation of Murray's observation is that most recombinants arise by conversion of the allele on the right-hand side of each cross. Thus, polarity can be seen in selected meiotic products with flanking markers in a way which is equivalent to the minority or majority parent criterion in *Ascobolus* tetrads.

The most common interpretation of polarity is that the length in which conversion occurs, presumably a length of heteroduplex, has one end outside the gene (the "high-conversion end") while the other end is at a variable position within the gene (Hastings and Whitehouse, 1964). The map length in which recombination is polarized has been described as a polaron (Lissouba et al., 1962).

The polaron seems to be related to the unit of function, and this has caused speculation and discussion concerning the possibility of common components in the control of function and of recombination (e.g., Hastings and Whitehouse, 1964; Thuriaux, 1977; Holliday, 1984; Catcheside, 1967, 1986). There are, however, many two-ended polarons reported in the literature—genes in which polarity is high at both ends and low in the middle (see Whitehouse and Hastings, 1965). This may imply that initiation of recombination occurs at both ends of the gene or that events may enter a gene from distant initiation points so that they run over several genes. Coconversions covering several genes do occur (DiCaprio and Hastings, 1976) and can now be seen more clearly by the use of restriction site heterozygosity to provide markers between genes (Borts and Haber, 1987; Symington and Petes, 1988).

2. Symmetry

In meiosis, crossing-over involves two of the four chromatids. When conversion or intragenic recombination is found in association with a crossover, it is the crossover chromatids which show conversion or recombination (Savage and Hastings, 1981). Conversion and PMS may be found on one or two chromatids. This parameter is known as the symmetry of the event, an asymmetrical event being confined to one chromatid, while a symmetrical event involves two.

Symmetry is measured by the occurrence of certain classes of tetrad. One such class is the aberrant 4:4 ratio which is presumed to represent uncorrected heteroduplex at the site of the allele on two chromatids. However, an allele must show a very high frequency of PMS to remain uncorrected on both chromatids. If flanking markers are included in the cross, it is possible to detect conversion occurring opposite to PMS in a class of tetrad called a tetratype 5:3 (see Fig. 2). From the cross $\dfrac{M\ 1\ N}{m\ +\ n}$, if heteroduplex forms on two chromatids to give $\dfrac{M\ 1/+\ N}{m\ +/1\ m}$, correction of one of these would give equality of the tetratype 5:3 $\dfrac{M\ +\ N}{m\ +/1\ n}$ with the tritype 5:3 $\dfrac{M\ 1/+\ N}{m\ +\ n}$. Asymmetrical heteroduplex will give only tritype 5:3. Hence, an excess of tritype over tetratype 5:3 is a measure of the relative frequency of asymmetrical heteroduplex.

Within *A. immersus*, asymmetry is more common than symmetry, but there is variation from locus to locus. Stadler and Towe (1971) found predominantly asymmetrical events in the *w3* locus, while the *b2* locus shows extensive symmetry as well as asymmetry (Paquette and Rossignol, 1978). In *S. cerevisiae*, there is no evidence of symmetrical heteroduplex (Fogel et al., 1979, 1981). Although aberrant 4:4 tetrads are seen (DiCaprio and Hastings, 1976), they occur at the frequency expected if the two occurrences of PMS were independent events.

Only in the *b2* locus of *A. immersus* do we know the distribution of heteroduplex within a single event. Paquette and Rossignol (1978) found that there was a gradient in the relative frequency of aberrant 4:4 tetrads that ran counter to the gradient in BFC, which showed polarity. Calculation of the relative proportion

Asymmetrical
heteroduplex

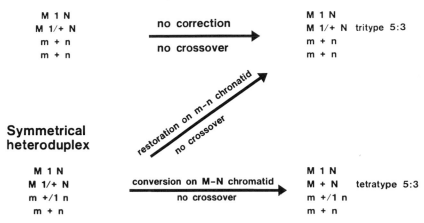

FIGURE 2. Determination of symmetry by the relative frequency of tritype and tetratype 5:3 segregations. The diagram shows why tritype 5:3 segregations in excess of the observed tetratype 5:3 segregations can be attributed to asymmetrical heteroduplex.

of symmetrical and asymmetrical heteroduplex confirms not only that there is a higher proportion of asymmetrical events at the high-BFC end than at the low-BFC end of the locus but also that there is an increase in the absolute frequency of symmetrical heteroduplex for alleles further toward the low-BFC end of the locus. By showing that aberrant 4:4 segregation at the low-BFC end of the locus is correlated with 5:3 segregation of a marker at the high-BFC end of the locus in the same tetrad, Hamza et al. (1981) showed that, at least in some cases, symmetrical and asymmetrical events are two phases of the same event.

This picture of some recombination events entering the gene at the high-BFC end, and ending at a variable position within the gene, may then be modified to include the idea that the event is asymmetrical when it enters the gene, becomes symmetrical at a variable position, and ends at a variable position. This is the pattern of events predicted by the model of Meselson and Radding (1975). The direction of travel will be discussed below, as will the position of the crossover which may occur in association with this event.

It is also necessary to discuss the possibility that not all events have this pattern. In the experiments described above which measure restoration in *A. immersus* (Hastings et al., 1980; Rossignol et al., 1984), it is apparent that the equality of restoration and conversion would not be likely to be seen if the length of heteroduplex were commonly discontinuous. A failure of heteroduplex to cover the middle marker when it covered those to either side would be indistinguishable from restoration. At least in this system it seems that almost all, if not all, heteroduplex is continuous. In contrast, there is evidence which suggests very strongly that the heteroduplex length is not always continuous into the intragenic region at the high-conversion end of the gene. Catcheside and Angel (1974) studied recombination within a gene of *N. crassa* (*his3*) which was heterozygous for an interchange breakpoint. They were also able to control the *cis*-dominant *cog* site which controls recombination within *his3*. It is difficult, from their data, to avoid the conclusion that a significant proportion (perhaps 10%) of events are not established as lengths of heteroduplex until they have passed the interchange breakpoint in the middle of the gene. Some theoretical implications of this were discussed by Catcheside (1986).

Upstream (high-conversion end) discontinuity in heteroduplex lengths was also reported by Kalogeropoulos and Rossignol (1988) in the *b2* locus of *A. immersus*. In this experiment, they used three C-type (high-PMS) alleles which spanned the length of the locus and found that there was much more Mendelian segregation of the allele nearest to the high-conversion end of the gene than could be explained by restoration from heteroduplex. They consider two hypotheses: that there are initiation sites within *b2* and that initiation at sites outside *b2* at the high-conversion end may be delayed in its expression as heteroduplex, as discussed above for *his3* in *N. crassa*. A third possibility is that there are some events at the *b2* locus which enter the locus at the low-conversion end, so that *b2* is bipolar. It should also be noted that discontinuity between the conversion length and the position of the crossover has been detected in the *buff* locus of *Sordaria brevicollis* (Sang and Whitehouse 1979, 1983; MacDonald and White-house, 1979; Theivendirarajah and Whitehouse, 1983). The method here was the same: markers in the middle of the event show more Mendelian segregation than is expected from restoration from heteroduplex. This expectation is derived from the amount of non-Mendelian segregation shown by the same markers and from the assumption that conversion and restoration should be equal. A discontinuity in the recombination event between conversion lengths and the crossover is surprising if one expects the crossover to arise by resolution of the event which gives rise to heteroduplex, as in all extant models of recombination (e.g., Holliday, 1964; Meselson and Radding, 1975; Szostak et al., 1983). Evidence on the position and nature of the crossover is presented in a later section.

The symmetry of an event is subject to marker effects (Nicolas and Rossignol, 1983; Rossignol et al., 1984; Hamza et al., 1987). In the *b2* locus of *A. immersus*, the presence of large heterologies causes an almost complete block to the formation of symmetrical heteroduplex toward the low-conversion end. This is inferred from a large decrease in the relative frequency of aberrant 4:4 segregations when a large heterology is present. Asymmetrical heteroduplex is unaffected, so that the frequency of non-Mendelian segregations is almost unchanged. Point mutations have the same effect, but it is much less pronounced (Nicolas and Rossignol, 1983).

The lack of symmetrical events for crosses containing heterologies extends the parallel between the conversion of heterologies in *A. immersus* and conversion of almost all markers in *S. cerevisiae*: a close parity in frequency of conversion to mutant and to wild type, a probable excess of conversion over restoration, and an absence of symmetry.

3. Direction of propagation of heteroduplex

We do not know a priori whether recombination events enter the gene at the high-conversion end as asymmetrical heteroduplex, becoming symmetrical and terminating at a variable position, or whether they come from the low-conversion end, being established as symmetrical at a variable position, becoming asymmetrical at a variable position, and then running off the high-conversion end. The only system in which the initiation site has been described is the *cog* control of recombination in *his3* of *N. crassa*. Catcheside and Angel (1974) used *cog* in conjunction with an interchange breakpoint in *his3*. It was apparent that some influence could cross the breakpoint to establish heteroduplex and crossing-over beyond the breakpoint, but this system is not sufficiently defined to equate the observations with the different phases of the event as seen in *A. immersus*.

In the *b2* locus of *A. immersus*, Rossignol and his group described polar influences on the frequency of both asymmetrical heteroduplex (Nicolas and Rossignol, 1983) and symmetrical heteroduplex (Hamza et al., 1981). Both appear, from the direction of the polar effects, to be propagated from the high-conversion end toward the low-conversion end. Rossignol et al. (1984) also found that crossovers are trapped on the high-conversion-end side of a heterology, as though they too travel from the high-conversion end.

In contrast to this, influences believed to be related to the correction process are effective in both directions. This was reported for the epistasis of conversion over the PMS shown by C-type alleles (Leblon and Rossignol, 1973) and for the epistasis of the pattern of conversion of heterologies over the conversion spectra of nearby point mutations (Hamza et al., 1987).

4. Intragenic recombination

Reciprocal intragenic recombination does occur in fungi but rarely amounts to 10% of the recombinants, and for some crosses it is not seen. Presumably, it could arise by conversion in opposite directions on two chromatids (reciprocal conversion), which would give an apparent two-strand double crossover when it was not associated with crossing-over. It seems reasonable to assume that most intragenic reciprocal recombination is caused by the occurrence of a crossover between the alleles. The absence of conversion in such tetrads does not mean that no alleles were included in heteroduplex, since restoration or reciprocal conversion may have occurred.

In the majority of cases of intragenic recombination, tetrad analysis reveals that the formation of the recombinant product was caused by conversion or PMS of one allele and not of another. Thus, it is necessary that the conversion of PMS length ended between the alleles. On the hypothesis of mismatch repair in heteroduplex, these endings may be either the endings of heteroduplex lengths between the alleles, so that one allele was not involved in the event, or the endings of correction tracts within continuous heteroduplex, so that the alleles differed in the conversion, restoration, or PMS outcome of resolution of the mismatch.

This secondary relationship of conversion to intragenic recombination means that there is no uniform quantitative relationship between the two processes. For example, in the *b2* locus of *A. immersus*, the frequencies of conversion and recombination differ by only one order of magnitude (Rossignol et al., in press), but in the *SUP6* locus of *S. cerevisiae* they differ by four orders of magnitude (DiCaprio and Hastings, 1976). Fincham and Holliday (1970) described the characteristics of a genetic map that would be obtained from recombination frequencies derived from the endings of two sorts of length: the heteroduplex lengths and the length of correction tracts within the heteroduplex. They demonstrated that the map will show expansion if correction tracts are shorter than heteroduplex lengths. That is, the long distances will exceed the sum of the short distances of which they are composed. This phenomenon was described by Holliday (1968) from data for many loci and several organisms. In the algebra of Fincham and Holliday (1970), the map will be additive over short physical distances because correction tracts will show the minimal length (x). In this circumstance all recombination will result from heteroduplex lengths (h) ending between the markers. When the physical distance between two markers exceeds x, a new source of recombination, the ends of correction lengths, is effective, so that extra recombinants appear for longer distances. The map distance is expanded when one compares a length which includes x endings with its constituent lengths which are too short to include x endings.

This model of the generation of intragenic recombinants has been extended to explain other phenomena of intragenic recombination (Hastings, 1975). If one allows that the relationship of h to x is variable, but characteristic of a locus, one can explain why some loci show map expansion, and some (e.g., series *46* in *A. immersus* in the data of Lissouba et al. [1962]) do not. If x is very long compared with h, all recombination will be related to h endings, and the map will be additive. At the same time, this explains the very strong polarity shown by series *46* since two sites in the same length of heteroduplex will always show coconversion. If all heteroduplex tracts enter from the same side and $x \gg h$, only the upstream marker can show independent conversion.

At the other end of the spectrum, the recombination parameters of some loci can be interpreted to mean that all recombination results from x endings. In other words, when heteroduplex occurs, it covers the whole locus, showing insufficient h endings to have an impact on the recombination parameters. Such loci will show no polarity (since polarity presumably results from h endings), no map expansion (since map expansion results from a mixture of x and h endings), and no mappability (since the ability to produce a map also results from h endings, as discussed below). Examples of immappable loci are *mtr* in *N. crassa* (Stadler and Kariya, 1969) and *SUP6* in *S. cerevisiae* (DiCaprio and Hastings, 1976). One wonders whether others have not been discovered and forgotten, since such amorphous data are not conducive to publication. The ability to produce a map from recombination frequencies depends upon recombination in a long length being greater than that in the constituent short lengths. Ideally, a map would be additive, where the long length is the sum of the constituent lengths. If the longest distance derived from one triad of crosses is not consistent with decisions as to which members are farthest apart from other related triads, no map can be

From the cross $\frac{M\ 1\ N}{m\ 2\ n}$, if $\underline{1}$ is to the left of $\underline{2}$:

$$m\ \text{++}\ N\ >\ M\ \text{++}\ n$$

$$R_1\ >\ R_2$$

because:

$$\begin{array}{l} M\ 1\ +\ N \\ \hspace{1.2cm} X \hspace{1cm} \text{gives}\ m\ +\ +\ N \\ m\ +\ 2\ n \end{array}$$

if $\underline{1}$ is to the right of $\underline{2}$:

$$M\ \text{++}\ n\ >\ m\ \text{++}\ N$$

$$R_1\ >\ R_2$$

because:

$$\begin{array}{l} M\ +\ 1\ N \\ \hspace{1.2cm} X \hspace{1cm} \text{gives}\ M\ \text{++}n \\ m\ 2\ +\ n \end{array}$$

FIGURE 3. R_1:R_2 method for determining the order of allelic mutations relative to the flanking markers. Where 1 and 2 are heteroalleles and M/m and N/n are flanking markers, the flanking marker configuration of the most common class of crossover prototrophic product reveals the order of the alleles if it is assumed that the simplest event is a crossover between the alleles. If a single exchange leads to the most common configuration (R_1), it would take a triple exchange to give the R_2 configuration.

produced. If the probability of the ending of a correction tract shows little distance dependence, recombination resulting from such endings will not lead to a consistent map. In continuous heteroduplex lengths, there must be some distance dependence of the probability of a correction length coming to an end, but there is evidence that allele-specific influences on correction are both diverse and strong, as will be discussed below, so that marker effects in specific crosses will mask any distance-dependent component in recombination.

There is an alternative method for placing alleles of a gene in their map order. When flanking markers are present, meiotic products showing intragenic recombination may be parental or recombinant for the flanking markers, depending on whether or not the intragenic recombination event had an associated crossover. All four combinations of flanking markers are found in recombinant products. They are approximately equally common. Inequality of the two parental classes was presented above as a way to detect polarity. Inequality of the two crossover classes provides a mapping method, as shown in Fig. 3. The more common crossover configuration of flanking markers in a recombinant wild-type meiotic product is designated R_1. If it is assumed that R_1 arose by a single exchange between the alleles, then the less common crossover class, R_2, required a triple exchange. Thus, the order of any two alleles relative to the flanking markers is established by placing them in the order which would give the R_1 combination by a single exchange. The map produced by this method is generally in close agreement with the map obtained by use of recombination frequencies.

Tetrad analysis shows that R_1 recombinants actually arise most frequently by conversion to wild type of the allele closest to the apparent position of the crossover, while the R_2 class shows the crossover to be separated from the converted allele by an allele showing Mendelian segregation (Fig. 3) (see Whitehouse and Hastings, 1965; Savage and Hastings, 1981). In some cases, the proportion of R_2 recombinants is consistent with the idea that the unconverted allele has been restored to parental genotype by correction of heteroduplex (Savage and Hastings, 1981), but as mentioned earlier, in some cases the crossover appears to have been physically separated from the converted segment (Sang and Whitehouse, 1979). If R_2 tetrads usually arise from a length of restoration between the conversion and the crossover, we see that the intragenic recombination was caused by the ending of a correction tract in continuous heteroduplex. The R_1 class, too, may arise from a correction tract ending when the marker nearest the crossover is converted while the other is restored but can also be caused by short heteroduplex so that the heteroduplex ending is between the markers, as shown in Fig. 3. If there are no heteroduplex endings, R_1 will be equal to R_2, so that a locus which is immappable by the use of recombination frequency data is also immappable by the $R_1:R_2$ method.

Deletion mapping, which has been so successfully used for bacteria and bacteriophages (see Mosig, this volume), has not been widely used in fungi, though it was used with the *b2* locus of *A. immersus* to support a map based on recombination frequencies (Rossignol et al., 1979).

D. Meiotic Crossing-Over

1. Amount of crossing-over

Crossovers between flanking markers occur in association with approximately half the events detected by conversion or by intragenic recombination. The crossovers that are associated with intragenic events affect the chromatids on which conversion and recombination occur (Savage and Hastings, 1981). The crossover is close to the recombination event, since, when allowance has been made for unrelated crossovers, the distance of the flanking markers does not affect the association. The position of the crossover will be discussed below.

However, it must be stressed that the amount of associated crossing-over is not the 50% which would be expected from free isomerization of a cross-strand exchange (Sigal and Alberts, 1982) except in one very special circumstance described later. The amount of associated crossing-over shows some locus specificity (compare Murray [1970] with Murray [1963], for example), varies for events in different parts of a gene (see, for example, Murray, 1963; Savage and Hastings, 1981), and is different for different aberrant ratios in the outcome of the event (Kitani and Whitehouse, 1974; Sang and Whitehouse, 1979; Nicolas, 1982). Nicolas (1982) studied tetrads of *A. immersus* showing non-Mendelian segregation at the *b8* locus and segregating for other markers flanking the *b8* locus. He reported that only 19% of 5:3 asci had an associated crossover. In 6:2 asci, the frequency of crossing-over is higher, but still below 50%. It is only in the aberrant 4:4 asci (PMS on two chromatids) that it actually reaches a value which is not statistically different from 50%. The same observation had been made earlier from data for *Sordaria fimicola* (Whitehouse, 1974).

These very interesting observations appear to tell us that when heteroduplex is symmetrical, and remains uncorrected on both chromatids, the cross-strand exchange is indeed equally likely to be cut in either isomeric form, to give equality of crossover to noncrossover events, as expected from the models of Holliday (1964) and Meselson and Radding (1975). The reasons we do not normally see the 50% relationship can now be explained. First, the mismatch correction process interferes with the resolution of the event, and, second, many events which we detect as asymmetrical do not have a symmetrical phase elsewhere and do not form crossovers. This possibility was first raised by Stadler and Towe (1971) and was discussed in detail by Hastings (1986b), where it was developed into a model to explain crossover position interference. However, not all evidence conforms with this view of the nature of the crossover, as will be discussed below.

2. Position of the crossover

When a crossover is associated with a length of conversion, it is not possible to determine where, with respect to the conversion, the crossover occurs. Two methods have been used to infer the position of the crossover. One (Savage and Hastings, 1981) uses the relative frequency of tetrad classes in which the crossover is separated from the conversion by a marker showing Mendelian segregation. This method is explained in Fig. 4. The other method makes use of PMS in tetrads where the heteroduplex was believed to be asymmetrical and is illustrated in Fig. 5 (Fogel et al., 1981; Nicolas, 1982). Savage and Hastings (1981) concluded that, in the *his1* locus of *S. cerevisiae*, the crossovers usually occur at the variable, low-conversion end of the event, within the gene. Sang and Whitehouse (1979, 1983) saw that a significant proportion of events in the *buff* locus of *Sordaria brevicollis* have the crossover separated from the event beyond the low-conversion end by a length which was not converted. In contrast, the PMS method shows that the crossover may be to either side of the site showing PMS. In the data of Fogel et al. (1981), using the *arg4* locus of *S. cerevisiae*, two-thirds of the crossovers appear to be upstream (i.e., toward the high-conversion end) of the high-PMS allele *arg4-16*. Nicolas (1982) saw that crossovers were equally often to either side of a C-type allele of the *b8* locus in *A. immersus*. Thus, it seems that, in the region of PMS, the crossover does not have a definable position. Savage and Hastings (1981) suggested that a cross-strand exchange structure can migrate back into the event when no mismatch is being corrected. Where heteroduplex is asymmetrical, this would not change the distribution of heteroduplex, but it would change the final position of a crossover which was formed from a cross-strand exchange which initially occurred at the low-conversion end of the event, as expected from the models of Holliday (1964) and Meselson and Radding (1975).

Crossovers have been seen to occur within the *b2* locus of *A. immersus*. Rossignol et al. (1984) found that, when there is a large heterology in the middle of the locus, crossovers accumulate to the upstream (high-conversion) side of the heterology. The number of such crossovers seen is many times the frequency of events which would have been seen within the *b2* locus if the heterology had not been included. These crossovers did not have associated conversion events and

(a) **If the crossover is at the high conversion end of the event (left):**

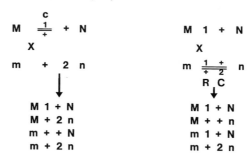

(b.) **If the crossover is at the variable end of the event within the gene:**

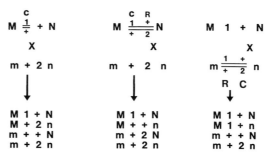

FIGURE 4. Origin of prototroph-containing tetrads with a crossover between flanking markers M/m and N/n from a heteroallelic cross, 1 +/+ 2, based on two hypotheses of the position of the crossover. It is assumed that the locus is highly polarized, with all events entering from the left. The double line indicates heteroduplex covering one or both allelic markers asymmetrically. C indicates resolution by conversion, and R, by restoration. Data for *his1* in *S. cerevisiae* do not conform to hypothesis a and give a very close fit to the expectation from hypothesis b (Savage and Hastings, 1981).

may constitute evidence that the recombination process is quite different from what is currently believed.

3. Crossover position interference

Most genetically studied fungi show crossover position interference, a phenomenon whereby the occurrence of a crossover is associated with a reduced chance of another crossover occurring nearby. Exceptions are seen in *Aspergillus nidulans* (Strickland, 1958) and *Ascobolus immersus* (A. Nicolas, Thèse Doctorat, Université Paris-Sud, Orsay, France, 1978), both of which show random distribution of crossovers.

The interference properties of recombination events can be seen in marked regions beyond the markers flanking the locus where recombination events are

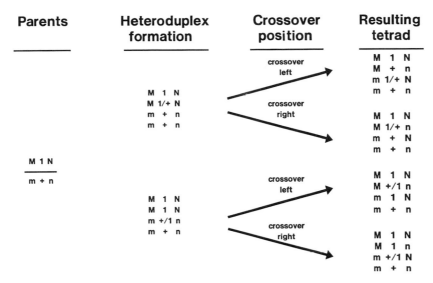

Parents	Heteroduplex formation	Crossover position	Resulting tetrad

FIGURE 5. Determination of crossover position by PMS. Asymmetrical heteroduplex formation without mismatch correction leads to 5:3 and 3:5 tetrads. In those events in which a crossover is associated with PMS, the crossover chromatid which does not show PMS reveals the position of the crossover relative to the site at which PMS is seen. The method is applicable only where all heteroduplex is asymmetrical.

detected. Recombination events which exert crossover position interference will reduce a genetic length of the outside region relative to its length in an unselected population. Using this method, Stadler (1959) found that in *N. crassa* crossover recombination events exert crossover position interference, but noncrossover recombination events do not. Mortimer and Fogel (1974) reported the same finding with data from *S. cerevisiae*. In Fig. 6, data from Savage (Ph.D. thesis) are used to illustrate the relative map length for crossover and noncrossover events at *his1* in *S. cerevisiae*. It can be seen that these data, too, show that crossovers in the outside region are reduced by 40% in crossover *his1* recombinant tetrads, but that the frequency is unaffected by noncrossover events or *his1*.

4. Complex events

A new trend in the genetic analysis of recombination is evident in two recent major works. The novelty is that the genetic markers used are free from the restrictions inherent in the use of phenotype of the offspring for following inheritance of parts of the genome. Now the markers are restriction site heterozygosities, which have been deliberately introduced by in vitro mutagenesis. Thus, the markers represent known changes in known positions and are situated between as well as within the genes. The result is that a more extensive tract of the genome is more intensively marked than has previously been possible.

Symington and Petes (1988) placed 12 restriction site heterozygosities in the 23-kilobase, 5-centimorgan interval between *LEU2* and the centromere of chro-

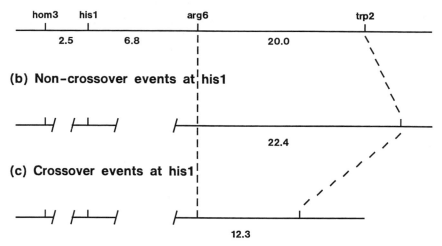

FIGURE 6. Changes in the map length of the *arg6* to *trp2* interval in *S. cerevisiae* in different samples of tetrads. Distances are given in percent recombination. The figure illustrates the principle that crossover recombination events show positive crossover position interference, while noncrossover recombination events do not. The map distances between *hom* and *arg* cannot be illustrated in maps b and c since these are selected samples. The sample size is 8,679 tetrads in map a, 579 tetrads in map b, and 531 tetrads in map c. (Data from Savage, Ph.D. thesis.)

mosome III of *S. cerevisiae*. The purpose was to investigate the relationship of physical to genetic distance, especially in the region of the centromere. Tetrads showing second-division segregation for *leu2* were scored for the presence or absence of restriction sites to determine the position of the crossover and its association with conversion. It was found that about half the crossovers were associated with a detectable conversion. Most conversion tracts were continuous and more than 4 kilobases in length. About 25% of the events studied were complex in the sense that they involved more than two chromatids, required more than one crossover or conversion tract to explain them, or had a conversion tract which was not contiguous with the crossover. Such a high frequency of complex events is not seen by conventional genetic techniques.

Borts and Haber (1987) also set out to test the hypothesis that crossovers are associated with conversion. These authors placed a *URA3* gene and portions of pBR322 between duplicated *MAT* loci. Crossovers were detected by the presence of two nonmating (*MATa MATα*) colonies in tetrads. The crossover tetrads were then scored for the presence or absence of nine restriction sites in the 20-kilobase interval in which the crossover had occurred. The findings were very similar to those of Symington and Petes (1988), except that a more even distribution of crossovers and conversions was seen. The major finding was the complexity of a large proportion of the events. It was inferred that the complexity was caused by the multiple sites of heterozygosity.

Borts and Haber (1987) suggested that the complexity is caused by further rounds of recombination induced by repair of heteroduplex generated by the initial recombination events. The occurrence of such secondary recombination is a prediction from the idea (Hastings, 1984) that yeast conversion proceeds by double-strand cutting at mismatches, followed by double-strand gap repair which is a recombination repair process.

5. Requirement for homology

Although both conversion and crossing-over appear to require homology, the extent of the homology is not yet established. Several experiments point toward the conclusion that it may be of the order of 1 kilobase or less since the homology need not be in homologous positions. Recombination between homologous sequences in nonhomologous positions is being referred to as ectopic recombination (Lichten et al., 1987). Petes (1980) showed that meiotic recombination in *S. cerevisiae* could occur unequally in the tandemly repeated rDNA genes. Several other papers have extended the concept of intrachromosomal recombination between duplicated genetic sequences in meiosis (Klein and Petes, 1981; Roeder, 1983; Klein, 1984; Jackson and Fink, 1985). In *Schizosaccharomyces pombe*, it was found that when repeated tRNA genes are located on different chromosomes, recombination between them occurs at a low level (Kohli et al., 1984; Amstutz et al., 1985). More extensive work on genes placed in nonhomologous chromsomes has established that this interchromosomal recombination is a general phenomenon (Jinks-Robertson and Petes, 1985). Jinks-Robertson and Petes (1986) studied both mitotic and meiotic interchromosomal recombination in *S. cerevisiae*, comparing it with the same recombination event occurring in the correct homologous position. They found that ectopic conversion is only one order of magnitude less frequent than the correctly placed event. They also found that the correlation of the conversion with crossing-over was the same when the reaction was ectopic as when the reaction was between sequences in the correct position. The ectopic crossing-over generated translocations.

Lichten et al. (1987) reported a very similar experiment, also with *S. cerevisiae*. Meiotic recombination between the same two *leu2* genes placed in the correct homologous position was compared with recombination seen when one of the alleles was at one of four other positions in the genome. They found that the frequency of ectopic conversion was comparable to the control value, but that it had acquired a frequency which was characteristic of the region into which the ectopic copy of *leu2* had been placed. Lichten et al. (1987) also constructed strains in which correct and ectopic sequences could compete for the interaction. They found that the presence of the homologously positioned allele did not reduce the frequency of ectopic recombination and also that the association of conversion with crossing-over was the same for ectopic as for correctly positioned recombination. This contrasts with the work of Klein (1984) on intrastrand meiotic recombination. She found that conversion between inverted repeats was not correlated with crossing-over as measured by inversion of the intervening segment. Kohli et al. (1984) also found no crossing-over associated with ectopic conversion between tRNA genes in *Schizosaccharomyces pombe*. Klein (1984) and Kohli et al. (1984) both suggested that a more substantial length of homology is needed for the formation of an associated crossover than for a conversion event with no crossing-over.

These findings are leading to the concept that chromosome synapsis, in the sense of the formation of synaptonemal complex, is not the initial recombination reaction in meiosis. Rather, it seems that there is initially a search for homology which can leave a conversion event as evidence of its occurrence but which also leaves a crossover when the homology is extensive. These ideas are discussed in detail by Carpenter (1987; this volume).

III. MITOTIC RECOMBINATION

Much less is known about mitotic than meiotic recombination in fungi. The absence of a regularized diploid phase in most filamentous ascomycetes has the effect of concentrating the work on *S. cerevisiae*.

Mitotic recombination has recently been reviewed by Esposito and Wagstaff (1981), Whitehouse (1982), Orr-Weaver and Szostak (1985), and Roman (1985). The subject overlaps with the subject of DNA repair. See, for example, the reviews by Haynes and Kunz (1981), Kunz and Haynes (1981), and Game (1983).

A. Conversion and Crossing-Over in Mitosis

That conversion is responsible for mitotic intragenic recombination was first shown by Roman (1957). As in meiosis, intragenic recombination to prototrophy results from conversion of one allele to wild type, while the other allele on the same molecule is not converted, so that it gives a measure of the frequency with which conversion lengths end in an interval. Conversion also leads to local homozygosity. This is often studied by visual selection of red colonies expressing recessive adenine requirements (see, for example, Zimmermann et al., 1975). Mitotic crossing-over may also lead to homozygosity because of the mitotic separation of sister centromeres. In the case of mitotic crossing-over, the homozygosity is reciprocal and applies to all genes distal to the site of the crossover. These three expressions of mitotic recombination are illustrated in Fig. 7. It also seems certain that sister chromatid interactions are very important in mitosis, but they leave no genetic trace, except for unequal sister interactions between repeated genes which may be either conversion or crossing-over (Petes, 1980; Szostak and Wu, 1980). Wildenberg (1970) discovered that mitotic conversion can occur before replication, that is, in G1. This has been extensively confirmed by Esposito (1978) and Roman and Fabre (1983) in work discussed below. The evidence for conversion occurring before replication is that two of the four chromatids coming from a diploid mitosis show the same conversion. For example, one may see both sister cells become protrophic at the same time from a heteroallelic diploid. It is clear that, in most cases, this is not caused by correction to wild type on both chromatids in symmetrical heteroduplex, since that would cause one copy of each allele to be replaced by the wild-type allele.

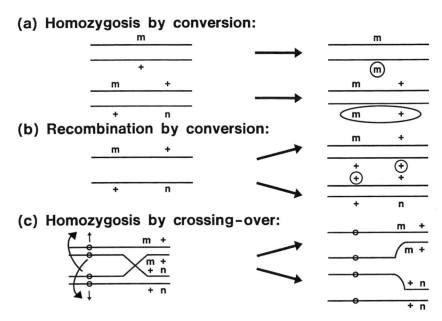

FIGURE 7. Three expressions of mitotic recombination. (a) Homozygosis by conversion, occurring at the two-strand stage, is shown. The cell may become homozygous for a recessive allele or for one of two distinguishable heteroalleles, as in the strain D7 of *S. cerevisiae* (Zimmermann et al., 1975). The region encircled is the part which has been converted. (b) Intragenic recombination, occurring by conversion of one of two heteroalleles to generate a wild-type chromosome, is shown. This occurs when a converted length ends between the alleles. (c) The consequences of a mitotic crossover are shown. The pattern of centromere separation shown here causes simultaneous homozygosis for both chromosomes for all material distal to the crossover. The alternative pattern of mitotic centromere separation would not lead to the reciprocal homozygosity but would still carry chromosomes with changed linkage relationships for markers flanking the crossover.

Rather, it stems from conversion to wild type of both copies of the same allele, since the same mutant allele is present in each cell, presumably because conversion was followed by replication. It has been shown by the use of synchronous cultures that, in fact, the event is induced in G1 and also that it actually occurs in G1. This last finding comes from an experiment by Fabre (1978), who used a diploid of *S. cerevisiae* carrying heteroalleles of a temperature-sensitive cell division cycle mutation. Cells held at restrictive temperature are unable to progress beyond G1. Fabre found that cells could escape the block to form colonies at restrictive temperatures. Thus, it is clear that recombination must have been completed in G1 to a point where recombinants could be expressed.

It has been established, by the use of cell division cycle mutations that stop the cell cycle at later stages, that G2 cells also show recombination, but at a lower level than cells in G1 (Fabre et al., 1983). The lower level seen in G2 is explained by the availability for recombination of sister chromatids, whereas in G1, only homologous chromosomes are available for interaction.

Esposito (1978) studied spontaneous mitotic recombination in a system in which he could select for prototrophy caused by conversion of an allele at *trp5* while detecting the associated crossover by red and white sectoring—caused by an adenine marker distal to *trp5*. The recombinants of interest amounted to about 10% of the colonies which grew without tryptophan. To be prototrophic for tryptophan and to be sectored, it is necessary that both daughter cells are prototrophic for tryptophan. For a colony to be sectored, it is necessary for the crossover to occur in G2. Analysis of meiotic products from the diploid cells in each sector confirmed that the conversion at *trp5* is occurring in G1 and that the associated crossover is occurring in G2. Esposito postulated that a cross-strand exchange formed during the conversion event was resolved by replication rather than by endonucleolytic cleavage, thereby appearing to occur in G2.

Further data of a similar type were obtained by Roman and Fabre (1983), using X rays to induce mitotic recombination. They found that in about one-third of the cases of G1 conversion occurring in association with G2 crossovers, there are other G2 crossovers involved in the event which cannot be explained by resolution of a cross-strand exchange near the conversion site by replication or by cleavage. They postulate that endonucleolytic cleavage at the conversion event in G1 leaves a substrate at which a crossover may occur in G2, possibly by a different mechanism.

B. Recombination Repair

1. Induction of mitotic recombination

All types of events of mitotic recombination occur spontaneously, although the rate is very much lower than that seen in meiosis. Mitotic recombination rates are strongly enhanced by exposure of cells to a wide variety of agents that cause damage to DNA. Examples of recombinogenic agents are X rays, which cause direct double-strand breakage (Ho and Mortimer, 1975; Resnick and Martin, 1976) and cross-linking agents such as psoralens, which are excised to form double-strand breaks before being repaired (Jachymczyk et al., 1981). Many agents which damage single DNA chains, such as the alkylating agent methyl methanesulfonate, cause double-strand breaks and become recombinogenic only at high doses. This is believed to stem from the formation and repair of close opposite lesions in the DNA (Chlebowicz and Jachymczyk, 1979). Some repair mutants of *S. cerevisiae* that have enhanced induction of recombination also show a high level of double-strand discontinuity appearing in DNA after treatment with DNA-damaging agents. This is true of *rad18* strains (Mowat et al., 1983) and *rad6* strains (DiCaprio and Cox, 1981).

2. Double-strand break repair

Resnick (1976) proposed that double-strand breaks are repaired by a recombination mechanism, whereby the information that has been damaged on both chains of a DNA molecule may be copied from a homologous sequence. Resnick offered a model for this process. The discovery of targeting of plasmid integration in *S. cerevisiae* supported and extended this point of view. It was reported by Hicks et al. (1979) that cutting a plasmid in a region of homology with the recipient genome greatly enhanced the efficiency with which the plasmid became integrated. This was extended by Orr-Weaver et al. (1981), who found that a plasmid which was gapped in the region of homology was integrated with the same efficiency as a plasmid which was merely cut and that the gap in the plasmid was filled with a copy of the chromosomal information. Further work involving a plasmid that could replicate autonomously showed that the gap in the plasmid could be filled by this conversion mechanism without becoming integrated—in other words, a noncrossover event. It was found that crossover and noncrossover gap-filling conversions were equally common (Orr-Weaver and Szostak, 1985). Extensive molecular evidence that yeast mating-type switch involves conversion into a double-strand gap is discussed by Strathern (this volume). Apart from Resnick (1976), further detailed models to explain double-strand gap filling by copying both nucleotide chains of a homologous molecule have been presented by Orr-Weaver et al. (1981) and Szostak et al. (1983).

It cannot be assumed that the primary double-strand cut or gap, or a double-strand discontinuity formed subsequent to the damage, forms the only recombinogenic substrates. Some recombinogenic processes are not expected to lead to double-strand discontinuity. Examples are thymidylate starvation (see review by Kunz and Haynes, 1982) and ligase deficiency (Game et al., 1979).

3. Recombination as a repair process

Repair of gapped plasmids by conversion seems to provide direct evidence for the proposition that mitotic recombination is a by-product of a DNA repair mechanism. Further substantial evidence for this proposal can be found in many aspects of the biology and genetics of mitotic recombination. X-ray-sensitive mutants of *S. cerevisiae* (the *rad52* epistasis group; see the review by Haynes and Kunz [1981]) are simultaneously defective in induced recombination and deficient in double-strand break repair (reviewed by Game, 1983; Resnick, 1986). While wild-type diploid strains have a substantial shoulder on the X-ray survival curve, this shoulder is absent in recombination-defective strains, implying that some X-ray-induced damage becomes irreparable when the recombination mechanism is impaired. The shoulder on the X-ray survival curve is also absent in diploids which are homozygous for mating type and in haploids (Mortimer, 1958; Ho and Mortimer, 1975; Game, 1983). Heterozygosity for mating type appears to provide the signal that homologous information is available for repair purposes. Haploid *S. cerevisiae* lacks the shoulder on the survival curve, but has a tail: about 10% of a population of growing cells shows much higher X-ray resistance than the general population. It has been shown (Brunborg and Williamson, 1978; Brunborg et al., 1980) that these X-ray-resistant cells are in the G2 phase, and hence the tail is attributed to the availability of sister chromatids for recombination repair (diploid cultures also show the tail).

Apart from the influence of the mating-type locus, there is other evidence that the recombination repair system is inducible. Fabre and Roman (1977) reported experiments in which a haploid strain was irradiated immediately before being mated to an unirradiated diploid strain. It was found that the presence of an irradiated nucleus induced recombination in unirradiated DNA even when the nuclei did not fuse. These experiments established not only that homologous recombination is inducible but also that a significant proportion of it is untargeted. That is, it is not necessary to induce damage in DNA to see enhancement of recombination. Apparently a substrate for recombination is present in undamaged DNA, but it does not normally lead to recombination because the homologous recombination system is not induced.

In *Ustilago maydis*, Holliday (1971) used a system in which nitrate reductase activity could be measured in a suspension of intact cells. When the cells carry heteroallelic mutations in the gene encoding nitrate reductase, the amount of enzyme activity gives a measure of the amount of recombination which has occurred. What is significant about this is that, unlike all normal ways of measuring recombination, it is unnecessary for the cell to live for its recombinant phenotype to be detected. The finding was that the recombination is occurring in the live cells, not in the cells which are going to die. The interpretation of this finding is that recombination both is inducible and bestows survival advantage. Thus, in this case, we can see that it is the recombination itself which enables the cell to survive, and therefore the recombination is itself the repair process.

One is left with the picture that recombination would occur without induced damage if the necessary gene products were present. Presumably, mitotic recombination is disadvantageous to the cell, causing, for example, homozygosis. Under some conditions of DNA damage, recombination becomes essential for survival, and it is then induced. It is interesting to note here that the tails on the survival curves for yeast cells, which have been attributed above to sister chromatid recombination, occur under conditions in which homologous recombination cannot be induced—in haploids and in homozygous mating-type diploids. Thus, it is homologous recombination repair, not all recombination repair, which requires induction.

IV. CONCLUSION

Although we see that there are two mechanisms of conversion, heteroduplex repair and double-strand gap filling, we are not free to conclude that there are two mechanisms of recombination in a broad sense, since it seems likely that gap filling, mismatched repair, and crossing-over are mixed in all the recombination processes we look at. It has been suggested that the mitotic process of double-strand gap filling may be involved in meiotic recombination as the initiating mechanism (Szostak et al., 1983) or as the process of mismatch repair following double-strand scission (Hastings, 1984). Conversely, it was pointed out above that the marker effects on recombination between close alleles in *S. cerevisiae*, which appear to be attributable to the properties of the mismatched repair mechanism, apply equally to meiotic, spontaneous mitotic, and induced mitotic recombination

(Moore et al., 1988). In *S. cerevisiae* meiotic and mitotic recombination seem to be very different, while meiotic recombination in *A. immersus* seems to occur by a third mechanism; yet, the properties of *A. immersus* meiotic recombination merge with those of *S. cerevisiae* when the conversion of heterologies is studied (Hamza et al., 1986).

The variation in the degree of association between conversion and crossing-over defies a unifying interpretation at present. When Kolodkin et al. (1986) introduced double-strand cuts into DNA during the early part of meiosis in *S. cerevisiae*, they saw mitotic conversion; it occurred at the two-strand stage. Yet, this conversion showed a 50% association with a G2 crossover, characteristic of meiotic rather than mitotic recombination. However, the hypothesis that the association might be strong in meiosis and weak in mitosis is contradicted by the absence of crossing-over between inverted repeats seen in meiosis by Klein (1984), and it is also contradicted by the high level of crossing-over seen in mitosis during repair of gapped plasmids (Orr-Weaver and Szostak, 1983). The concept of a minimum length of homology to establish a crossover is useful for explaining the meiotic data of Klein (1984) and of Kohli et al. (1984) but does not fit with the generally low association of spontaneous or induced mitotic conversion with crossing-over (see Esposito and Wagstaff, 1981).

The idea that the crossover is not a part of the conversion event, but something separate which is stimulated by some characteristic of a conversion event, has been around for a long time (see, for example, Sherman and Roman, 1963). This view is encouraged by the breakdown of the correlation between conversion and crossing-over and has been widely discussed in recent years (e.g., Whitehouse, 1982; Roman and Fabre, 1983; Fink and Petes, 1984; Rossignol et al., 1984; Hastings, 1986b; Carpenter, 1987). The variability in the association of crossovers with conversion events was discussed above. The converse situation, crossing-over without conversion, was described by Rossignol et al. (1984). It is also suggested by the *RAD52* independence of plasmid integration in yeast (Orr-Weaver et al., 1981). Temporal separation of conversion and crossing-over in mitosis was described above. Spatial separation of the events in meiosis in *Sordaria brevicollis* was described by Sang and Whitehouse (1979; 1983). Crossover and noncrossover chromosomal structures in *Drosophila* spp. look different when visualized in an electron microscope (see Carpenter, 1984 and this volume). However, although it is becoming harder to believe that conversion and crossing-over are parts of the same mechanism, it is still possible that the conflicting arguments will be resolved.

In summary, the evidence from fungi presented here is consistent with the following point of view: the function of mitotic recombination is repair by conversion, while the function of meiotic recombination is the production of crossovers. Thus, in mitosis, conversion is the result of a repair process whereby damaged information is replaced by copying a homologous molecule: crossing-over is an accidental and disadvantageous by-product of the repair process; meiotic conversion is a by-product of the initial reaction of chromosome synapsis and is also the result of the event of chromatid interaction which ensures the occurrence of sufficient crossing-over to produce orderly chromosomal segregation (see Hawley, this volume). Although we seem to understand many details of the mechanisms, we are still a long way from describing a single event, let alone understanding the differences and relationships between different types of event.

ACKNOWLEDGMENTS. I am very grateful to Wee-Shian Chan for help in the preparation of the manuscript and to Ann Childs for typing.

This work is supported by a grant from the Natural Sciences and Engineering Research Council of Canada.

LITERATURE CITED

Amstutz, H., P. Munz, W.-D. Heyer, U. Leupold, and J. Kohli. 1985. Concerted evolution of tRNA genes: intergenic conversion among three unlinked serine tRNA genes in *S. pombe*. *Cell* **40**:879–886.

Borts, R. H., and J. E. Haber. 1987. Meiotic recombination in yeast: alteration by multiple heterozygosities. *Science* **237**:1459–1465.

Brunborg, G., M. A. Resnick, and D. H. Williamson. 1980. Cell-cycle-specific repair of DNA double-strand breaks in *Saccharomyces cerevisiae*. *Radiat. Res.* **82**:547–548.

Brunborg, G., and D. H. Williamson. 1978. The relevance of the nuclear division cycle to radiosensitivity in yeast. *Mol. Gen. Genet.* **162**:277–286.

Carpenter, A. T. C. 1984. The meiotic roles of crossing-over and gene conversion. *Cold Spring Harbor Symp. Quant. Biol.* **49**:23–29.

Carpenter, A. T. C. 1987. Gene conversion, recombination nodules, and the initiation of meiotic synapsis. *BioEssays* **6**:232–236.

Catcheside, D. E. A. 1967. Regulation of the *am-1* locus in Neurospora: evidence of independent control of allelic recombination and gene expression. *Genetics* **59**:443–452.

Catcheside, D. E. A. 1986. A restriction and modification model for the initiation and control of recombination in *Neurospora*. *Genet. Res.* **47**:157–165.

Catcheside, D. G., and T. Angel. 1974. A *histidine-3* mutant, in *Neurospora crassa*, due to an interchange. *Aust. J. Biol. Sci.* **27**:219–229.

Chlebowicz, E., and W. J. Jachymczyk. 1979. Repair of MMS-induced DNA double-strand breaks in haploid cells of *Saccharomyces cerevisiae*, which requires the presence of a duplicate genome. *Mol. Gen. Genet.* **167**:279–286.

DiCaprio, L., and B. S. Cox. 1981. The effect of UV irradiation on the molecular weight of pre-existing and newly-synthesized DNA of yeast. *Mutat. Res.* **82**:69–85.

DiCaprio, L., and P. J. Hastings. 1976. Gene conversion and intragenic recombination at the *SUP6* locus and the surrounding region in *Saccharomyces cerevisiae*. *Genetics* **84**:697–721.

Emerson, S. 1966. Quantitative implications of the DNA-repair model of gene conversion. *Genetics* **53**:475–485.

Esposito, M. S. 1978. Evidence that spontaneous mitotic recombination occurs at the two-strand stage. *Proc. Natl. Acad. Sci. USA* **75**:4435–4440.

Esposito, M. S., and J. E. Wagstaff. 1981. Mechanisms of mitotic recombination, p. 341–370. *In* J. N. Strathern, E. W. Jones, and J. R. Broach (ed.), *The Molecular Biology of the Yeast Saccharomyces: Life Cycle and Inheritance*. Cold Spring Harbor Laboratory, Cold Spring Harbor, N.Y.

Fabre, F. 1978. Induced intragenic recombination in yeast can occur during the G1 mitotic phase. *Nature* (London) **272**:795–798.

Fabre, F., A. Boulet, and H. Roman. 1983. Gene conversion at different points in the mitotic cycle of *Saccharomyces cerevisiae*. *Mol. Gen. Genet.* **195**:139–143.

Fabre, F., and H. Roman. 1977. Genetic evidence for the inducibility of recombination competence in yeast. *Proc. Natl. Acad. Sci. USA* **74**:1667–1671.

Fincham, J. R. S., and R. Holliday. 1970. An explanation of fine structure map expansion in terms of excision repair. *Mol. Gen. Genet.* **109**:309–322.

Fink, G. R., and T. D. Petes. 1984. Gene conversion in the absence of reciprocal recombination. *Nature* (London) **310:**728–729.

Fink, G. R., and C. A. Styles. 1974. Gene conversion of deletions in the *HIS4* region of yeast. *Genetics* **77:**231–244.

Fogel, S., R. K. Mortimer, and K. Lusnak. 1981. Mechanisms of meiotic gene conversion, or "wanderings on a foreign strand," p. 289–339. *In* J. N. Strathern, E. W. Jones, and J. R. Broach (ed.), *The Molecular Biology of the Yeast Saccharomyces: Life Cycle and Inheritance*. Cold Spring Harbor Laboratory, Cold Spring Harbor, N.Y.

Fogel, S., R. Mortimer, K. Lusnak, and F. Tavares. 1979. Meiotic gene conversion: a signal of the basic recombination event in yeast. *Cold Spring Harbor Symp. Quant. Biol.* **43:**1325–1341.

Game, J. C. 1983. Radiation sensitive mutants and repair in yeast, p. 109–137. *In* J. F. T. Spencer, D. M. Spencer, and A. R. W. Smith (ed.), *Yeast Genetics: Fundamental and Applied Aspects*. Springer-Verlag, Berlin.

Game, J., L. Johnston, and R. von Borstel. 1979. Enhanced mitotic recombination in a ligase-defective mutant of the yeast *Saccharomyces cerevisiae*. *Proc. Natl. Acad. Sci. USA* **76:**4589–4592.

Gutz, H. 1971. Gene conversion: remarks on the quantitative implications of hybrid DNA models. *Genet. Res.* **17:**45–52.

Hamza, H., V. Haedens, A. Mekki-Berrada, and J.-L. Rossignol. 1981. Hybrid DNA formation during meiotic recombination. *Proc. Natl. Acad. Sci. USA* **78:**7648–7651.

Hamza, H., A. Kalogeropoulos, A. Nicolas, and J.-L. Rossignol. 1986. Two mechanisms for directional gene conversion. *Proc. Natl. Acad. Sci. USA* **83:**7386–7390.

Hamza, H., A. Nicolas, and J.-L. Rossignol. 1987. Large heterologies impose their gene conversion pattern onto closely linked point mutations. *Genetics* **116:**45–53.

Hastings, P. J. 1975. Some aspects of recombination in eukaryotic organisms. *Annu. Rev. Genet.* **9:**129–144.

Hastings, P. J. 1984. Measurement of restoration and conversion: its meaning for the mismatch repair hypothesis of conversion. *Cold Spring Harbor Symp. Quant. Biol.* **49:**49–53.

Hastings, P. J. 1986a. Meiotic recombination interpreted as heteroduplex correction, p. 107–137. *In* P. B. Moens (ed.), *Meiosis*. Academic Press, Inc., New York.

Hastings, P. J. 1986b. Models of heteroduplex formation, p. 139–156. *In* P. B. Moens (ed.), *Meiosis*. Academic Press, Inc., New York.

Hastings, P. J., A. Kalogeropoulos, and J.-L. Rossignol. 1980. Restoration to the parental genotype of mismatches formed in recombinant DNA heteroduplex. *Curr. Genet.* **2:**169–174.

Hastings, P. J., and E. A. Savage. 1984. Further evidence of a disparity between conversion and restoration in the *his1* locus of *Saccharomyces cerevisiae*. *Curr. Genet.* **8:**23–28.

Hastings, P. J., and H. L. K. Whitehouse. 1964. A polaron model of genetic recombination by the formation of hybrid deoxyribonucleic acid. *Nature* (London) **201:**1052–1054.

Haynes, R. H., and B. A. Kunz. 1981. DNA repair and mutagenesis in yeast, p. 371–414. *In* J. Strathern, E. W. Jones, and J. R. Broach (ed.), *The Molecular Biology of the Yeast Saccharomyces: Life Cycle and Inheritance*. Cold Spring Harbor Laboratory, Cold Spring Harbor, N.Y.

Hicks, J. B., A. Hinnen, and G. R. Fink. 1979. Properties of yeast transformation. *Cold Spring Harbor Symp. Quant. Biol.* **43:**1305–1313.

Ho, K. S., and R. K. Mortimer. 1975. X-ray-induced dominant lethality in a radiosensitive strain of yeast, p. 545–548. *In* P. C. Hanawalt and R. B. Setlow (ed.), *Molecular Mechanisms for Repair of DNA*. Plenum Publishing Corp., New York.

Holliday, R. 1962. Mutation and replication in *Ustilago maydis*. *Genet. Res.* **3:**472–486.

Holliday, R. 1964. A mechanism for gene conversion in fungi. *Genet. Res.* **5:**282–304.

Holliday, R. 1968. Genetic recombination in fungi, p. 157–174. *In* W. J. Peacock and R. D. Brock (ed.), *Replication and Recombination of Genetic Material*. Australian Academy of Science, Canberra, Australia.

Holliday, R. 1971. Biochemical measure of the time and frequency of radiation-induced recombination in *Ustilago*. *Nature* (London) **232:**233–236.

Holliday, R. 1984. The biological significance of meiosis. *Symp. Soc. Exp. Biol.* **38:**381–395.

Holliday, R., R. E. Halliwell, M. W. Evans, and V. Rowell. 1976. Genetic characterization of *rec1*, a mutant of *Ustilago maydis* defective in repair and recombination. *Genet. Res.* **27:**413–453.

Jachymczyk, W. J., R. C. von Borstel, M. R. A. Mowat, and P. J. Hastings. 1981. Repair of interstrand cross-links in DNA of *Saccharomyces cerevisiae* requires two systems for DNA repair: the *RAD3* system and the *RAD51* system. *Mol. Gen. Genet.* **182**:196–205.

Jackson, J. A., and G. R. Fink. 1985. Meiotic recombination between duplicated genetic elements in *Saccharomyces cerevisiae*. *Genetics* **109**:303–332.

Jinks-Robertson, S., and T. D. Petes. 1985. High-frequency meiotic gene conversion between repeated genes on nonhomologous chromosomes in yeast. *Proc. Natl. Acad. Sci. USA* **82**:3350–3354.

Jinks-Robertson, S., and T. D. Petes. 1986. Chromosomal translocations generated by high-frequency meiotic recombination between repeated yeast genes. *Genetics* **114**:731–752.

Kalogeropoulos, A., and J.-L. Rossignol. 1988. Hybrid DNA tracts may start at different sites during meiotic recombination in gene *b2* of Ascobolus. *EMBO J.* **7**:253–259.

Kitani, Y., and H. L. K. Whitehouse. 1974. Aberrant ascus genotype from crosses involving mutants at the *g* locus in *Sordaria fimicola*. *Genet. Res.* **24**:229–250.

Klein, H. L. 1984. Lack of association of intrachromosomal gene conversion and reciprocal exchange. *Nature* (London) **310**:748–753.

Klein, H. L., and T. D. Petes. 1981. Intrachromosomal gene conversion in yeast. *Nature* (London) **289**:144–148.

Kohli, J., P. Munz, R. Aebi, H. Amstutz, C. Gysler, W.-D. Heyer, L. Lehmann, P. Schuchert, P. Szankasi, P. Thuriaux, U. Leupold, J. Bell, V. Gamulin, H. Hottinger, D. Pearson, and D. Soll. 1984. Interallelic and intergenic conversion in three serine tRNA genes of *Schizosaccharomyces pombe*. *Cold Spring Harbor Symp. Quant. Biol.* **49**:31–40.

Kolodkin, A. L., A. J. S. Klar, and F. W. Stahl. 1986. Double-strand breaks can initiate meiotic recombination in *S. cerevisiae*. *Cell* **46**:733–740.

Kunz, B. A., and R. H. Haynes. 1981. Phenomenology and genetic control of mitotic recombination in yeast. *Annu. Rev. Genet.* **15**:57–89.

Kunz, B. A., and R. H. Haynes. 1982. DNA repair and the genetic effects of thymidylate stress in yeast. *Mutat. Res.* **93**:353–375.

Leblon, G. 1972a. Mechanism of gene conversion in *Ascobolus immersus*. I. Existence of a correlation between the origin of mutants induced by different mutagens and their conversion spectrum. *Mol. Gen. Genet.* **115**:36–48.

Leblon, G. 1972b. Mechanism of gene conversion in *Ascobolus immersus*. II. The relationships between the genetic alterations in *b1* or *b2* mutants and their conversion spectrum. *Mol. Gen. Genet.* **116**:322–335.

Leblon, G., and J.-L. Rossignol. 1973. Mechanism of gene conversion in *Ascobolus immersus*. III. The interaction of heteroalleles in the conversion process. *Mol. Gen. Genet.* **122**:165–182.

Lichten, M., R. Borts, and J. E. Harber. 1987. Meiotic gene conversion and crossing-over between dispersed homologous sequences occurs frequently in *Saccharomyces cerevisiae*. *Genetics* **115**:233–246.

Lissouba, P., J. Mousseau, G. Rizet, and J.-L. Rossignol. 1962. Fine structure of genes in the ascomycete *Ascobolus immersus*. *Adv. Genet.* **11**:343–380.

MacDonald, M. V., and H. L. K. Whitehouse. 1979. A *buff* spore colour mutant in *Sordaria brevicollis* showing high-frequency conversion. I. Characteristics of the mutant. *Genet. Res.* **34**:87–119.

Meselson, M. S. 1965. The duplication and recombination of genes, p. 3–16. *In* J. A. Moore (ed.), *New Ideas in Biology*. Natural History Press, New York.

Meselson, M. S., and C. M. Radding. 1975. A general model for genetic recombination. *Proc. Natl. Acad. Sci. USA* **72**:358–361.

Moore, C. W., D. M. Hampsey, J. F. Ernst, and F. Sherman. 1988. Differential mismatch repair can explain the disproportionalities between physical distances and recombination frequencies of *cyc1* mutations in yeast. *Genetics* **119**:21–34.

Mortimer, R. K. 1958. Radiobiological and genetic studies on a polyploid series (haploid to hexaploid) of *Saccharomyces cerevisiae*. *Radiat. Res.* **9**:312–326.

Mortimer, R. K., and S. Fogel. 1974. Genetic interference and gene conversion, p. 263–275. *In* R. Grell (ed.), *Mechanisms of Recombination*. Plenum Publishing Corp., New York.

Mowat, M. R. A., W. J. Jackymczyk, P. J. Hastings, and R. C. von Borstel. 1983. Repair of gamma-ray induced DNA strand breaks in radiation-sensitive mutant *rad18-2* of *Saccharomyces cerevisiae*. *Mol. Gen. Genet.* **189**:256–262.

Murray, N. E. 1963. Polarized recombination and fine structure within the *me-2* gene of *Neurospora crassa*. *Genetics* **48**:1163–1183.

Murray, N. E. 1970. Recombination events that span sites within neighbouring gene loci of *Neurospora*. *Genet. Res.* **15**:109–121.

Muster-Nassal, C., and R. Kolodner. 1986. Mismatch correction catalyzed by cell-free extracts of *Saccharomyces cerevisiae*. *Proc. Natl. Acad. Sci. USA* **83**:7618–7622.

Nicolas, A. 1982. Variation of crossover association frequencies with various aberrant segregation classes in *Ascobolus*. *Curr. Genet.* **6**:137–146.

Nicolas, A., and J.-L. Rossignol. 1983. Gene conversion: point mutation heterozygosities lower heteroduplex formation. *EMBO J.* **2**:2265–2270.

Orr-Weaver, T. L., and J. W. Szostak. 1983. Yeast recombination: the association between double strand gap repair and crossing over. *Proc. Natl. Acad. Sci. USA* **80**:4417–4421.

Orr-Weaver, T. L., and J. W. Szostak. 1985. Fungal recombination. *Microbiol. Rev.* **49**:33–58.

Orr-Weaver, T. L., J. W. Szostak, and R. J. Rothstein. 1981. Yeast transformation: a model system for the study of recombination. *Proc. Natl. Acad. Sci. USA* **78**:6354–6358.

Paquette, N., and J.-L. Rossignol. 1978. Gene conversion spectrum of 15 mutants giving post-meiotic segregation in the *b2* locus of *Ascobolus immersus*. *Mol. Gen. Genet.* **163**:313–326.

Petes, T. D. 1980. Unequal meiotic recombination within tandem arrays of yeast ribosomal DNA genes. *Cell* **19**:765–774.

Pukkila, P. J., M. D. Stephens, D. M. Binninger, and B. Errede. 1986. Frequency and directionality of gene conversion events involving the *CYC7-H3* mutation in *Saccharomyces cerevisiae*. *Genetics* **114**:347–361.

Resnick, M. A. 1976. The repair of double-strand breaks in DNA: a model involving recombination. *J. Theor. Biol.* **59**:97–106.

Resnick, M. A. 1986. Investigating the genetic control of biochemical events in meiotic recombination, p. 157–210. *In* P. B. Moens (ed.), *Meiosis*. Academic Press, Inc., New York.

Resnick, M. A., and P. Martin. 1976. The repair of double-stranded breaks in the nuclear DNA of *Saccharomyces cerevisiae* and its genetic control. *Mol. Gen. Genet.* **143**:119–129.

Roeder, G. S. 1983. Unequal crossing-over between yeast transposable elements. *Mol. Gen. Genet.* **190**:117–121.

Roman, H. 1957. Studies in gene mutation in *Saccharomyces*. *Cold Spring Harbor Symp. Quant. Biol.* **21**:175–185.

Roman, H. 1985. Gene conversion and crossing-over. *Environ. Mutagen.* **7**:923–932.

Roman, H., and F. Fabre. 1983. Gene conversion and associated reciprocal recombination are separable events in vegetative cells of *Saccharomyces cerevisiae*. *Proc. Natl. Acad. Sci. USA* **80**:6912–6916.

Rossignol, J.-L., A. Nicolas, H. Hamza, and T. Langin. 1984. Origins of gene conversion and reciprocal exchange in *Ascobolus*. *Cold Spring Harbor Symp. Quant. Biol.* **49**:639–649.

Rossignol, J.-L., N. Paquette, and A. Nicolas. 1979. Aberrant 4:4 asci, disparity in the direction of conversion and frequencies of conversion in *Ascobolus immersus*. *Cold Spring Harbor Symp. Quant. Biol.* **43**:1343–1352.

Sang, H., and H. L. K. Whitehouse. 1979. Genetic recombination at the *buff* spore color locus in *Sordaria brevicollis*. I. Analysis of flanking marker behaviour in crosses between *buff* mutants and wild type. *Mol. Gen. Genet.* **174**:327–334.

Sang, H., and H. L. K. Whitehouse. 1983. Genetic recombination at the *buff* spore color locus in *Sordaria brevicollis*. II. Analysis of flanking marker behaviour in crosses between *buff* mutants. *Genetics* **103**:161–178.

Savage, E. A., and P. J. Hastings. 1981. Marker effects and the nature of the recombination event at the *his1* locus of *Saccharomyces cerevisiae*. *Curr. Genet.* **3**:37–47.

Sherman, F., and H. Roman. 1963. Evidence of two types of allelic recombination in yeast. *Genetics* **48**:255–261.

Sigal, N., and B. Alberts. 1982. Genetic recombination: the nature of a crossed strand-exchange between two homologous DNA molecules. *J. Mol. Biol.* **71**:789–793.

Stadler, D. R. 1959. The relationship of gene conversion to crossing-over in *Neurospora*. *Proc. Natl. Acad. Sci. USA* **45**:1625–1629.

Stadler, D. R., and B. Kariya. 1969. Intragenic recombination at the *mtr* locus of *Neurospora* with segregation at an unselected site. *Genetics* **63**:291–316.

Stadler, D. R., and A. M. Towe. 1971. Evidence for meiotic recombination in *Ascobolus* involving only one member of a tetrad. *Genetics* **68**:401–413.

Stahl, F. W. 1969. One way to think about gene conversion. *Genetics* **61**(Suppl.):1–13.

Stahl, F. W. 1979. *Genetic Recombination: Thinking about It in Phage and Fungi.* W. H. Freeman and Co., San Francisco.

Strickland, W. N. 1958. An analysis of interference in *Aspergillus nidulans*. *Proc. R. Soc. Lond. B Biol. Sci.* **149**:82–101.

Symington, L. S., and T. D. Petes. 1988. Expansions and contractions of the genetic map relative to the physical map of yeast chromosome III. *Mol. Cell. Biol.* **8**:595–604.

Szostak, J. W., T. L. Orr-Weaver, R. J. Rothstein, and F. W. Stahl. 1983. The double-strand-break repair model of recombination. *Cell* **33**:25–35.

Szostak, J. W., and R. Wu. 1980. Unequal crossing-over in the ribosomal DNA of *Saccharomyces cerevisiae*. *Nature* (London) **284**:426–430.

Theivendirarajah, K., and H. L. K. Whitehouse. 1983. Further evidence that aberrant segregation and crossing-over in *Sordaria brevicollis* may be discrete, though associated, events. *Mol. Gen. Genet.* **190**:432–437.

Thuriaux, P. 1977. Is recombination confined to structural genes of the eukaryotic chromosome? *Nature* (London) **268**:460–462.

Thuriaux, P., M. Minet, P. Munz, A. Ahmad, D. Zbaeren, and U. Leupold. 1980. Gene conversion in nonsense suppressors of *Schizosaccharomyces pombe*. II. Specific marker effects. *Curr. Genet.* **1**: 89–95.

Wagner, R., C. Dohet, M. Jones, M.-P. Doutriaux, F. Hutchinson, and M. Radman. 1984. Involvement of *Escherichia coli* mismatch repair in DNA replication and recombination. *Cold Spring Harbor Symp. Quant. Biol.* **49**:611–615.

White, J. H., K. Lusnak, and S. Fogel. 1985. Mismatch-specific post-meiotic segregation frequency in yeast suggests a heteroduplex recombination intermediate. *Nature* (London) **315**:350–352.

Whitehouse, H. L. K. 1963. A theory of crossing-over by means of hybrid deoxyribonucleic acid. *Nature* (London) **199**:1034–1040.

Whitehouse, H. L. K. 1974. Genetic analysis of recombination at the *g* locus in *Sordaria fimicola*. *Genet. Res.* **24**:251–259.

Whitehouse, H. L. K. 1982. *Genetic Recombination: Understanding the Mechanisms.* John Wiley & Sons, Inc., New York.

Whitehouse, H. L. K., and P. J. Hastings. 1965. The analysis of genetic recombination on the polaron hybrid DNA model. *Genet. Res.* **6**:27–92.

Wildenberg, J. 1970. The relation of mitotic recombination to DNA replication in yeast pedigrees. *Genetics* **66**:291–304.

Yu-Sun, C. C., M. R. T. Wickramaratne, and H. L. K. Whitehouse. 1977. Mutagen specificity of conversion pattern in *Sordaria brevicollis*. *Genet. Res.* **29**:65–81.

Zimmermann, F. K., R. Kern, and H. Rasenberger. 1975. A yeast strain for simultaneous detection of induced mitotic crossing-over, mitotic gene conversion and reverse mutation. *Mutat. Res.* **28**:381–388.

Chapter 13

FLP Site-Specific Recombination System of *Saccharomyces cerevisiae*

Michael M. Cox

I. 2μm PLASMID

The 2μm plasmid is an autonomously replicating, circular DNA present in most strains of the yeast *Saccharomyces cerevisiae* at 50 to 100 copies per diploid cell (Broach, 1982). It is located in the nucleus. The plasmid contains 6,318 base pairs (bp), and the complete sequence is known (Hartley and Donelson, 1980). General properties of the plasmid were recently reviewed (Volkert et al., 1987). The plasmid is a prototype of a widespread class of plasmids in a variety of yeast species (Toh-e et al., 1987; Araki et al., 1985). The presence of the plasmid confers no important identifiable selective advantage to its host.

Important features of the 2μm plasmid with respect to this chapter are diagrammed in Fig. 1. Replication of the plasmid has been shown to be under stringent cell cycle control (Livingston and Kupfer, 1977; Zakian et al., 1979). The plasmid nevertheless exhibits high mitotic and meiotic stability. Replication is carried out by the host replication apparatus. The plasmid encodes two systems

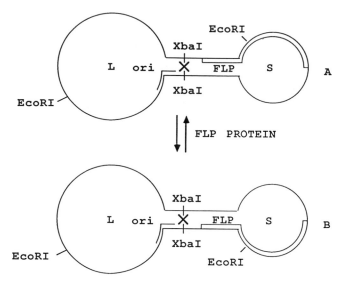

FIGURE 1. Yeast 2μm plasmid. The open reading frame encoding the FLP protein is shown. The FRT site includes base pairs spanning the *Xba*I restriction site. The interconversion between A and B forms involves site-specific recombination at the FRT site. L and S denote large and small unique sequence regions, respectively. The *Eco*RI sites serve as points of reference for the DNA inversion event. See text for details.

that enhance its stability. One of these is responsible for efficient partitioning of the plasmid during meiosis and mitosis. This system includes the products of two open reading frames designated REP1 and REP2 (not shown) and two *cis*-acting elements, i.e., the replication origin and a second site several hundred base pairs away from the origin, designated REP3 or STB (Jayaram et al., 1983; Kikuchi, 1983; Volkert et al., 1987). The REP3 locus appears to be the site of action of the REP1 and REP2 proteins (Jayaram et al., 1985; Murray and Cesareni, 1986). The second system is a plasmid-encoded capability to increase the plasmid copy number when it is low. This system consists, at least in part, of the site-specific recombination system described in this chapter.

A notable structural feature of the 2μm plasmid and related plasmids is the presence of a pair of long inverted repeats (599 bp in the case of the 2μm plasmid). The plasmid is generally found as a 1:1 mixture of two forms of equal size, designated A and B (Fig. 1). The A and B forms are related by a site-specific recombination event which occurs at a point localized to a short sequence within the 599-bp repeats (Broach et al., 1982). The recombination site has recently been designated FRT (FLP recombination target) (McLeod et al., 1986). Recombination is mediated by the product of an additional open reading frame on the plasmid designated FLP (Broach et al., 1982). No other yeast proteins are required for this reaction (Cox, 1983).

II. FLP SYSTEM

A. Historical Perspective

Efficient recombination between the 599-bp repeats in vivo, which produces an inversion of the unique regions separating the repeats, was first demonstrated by Beggs (1978). The demonstration that the 2μm plasmid encodes a protein required for this recombination event and that the protein acted in *trans* followed soon after (Gerbaud et al., 1979; Broach and Hicks, 1980). Broach et al. (1982) located the recombination site within a 65-bp region within each 599-bp repeat. Biochemical characterization was facilitated by the cloning and expression of the FLP gene in *Escherichia coli* (Cox, 1983; Vetter et al., 1983). This work established that no other yeast proteins are required for the reaction. The subsequent development of in vitro systems for the recombination event (Vetter et al., 1983; Meyer-Leon et al., 1984; Sadowski et al., 1984), by using extracts of *E. coli* strains expressing the FLP protein, has led to the extensive biochemical characterization described below. Complementary in vivo studies have also continued in *S. cerevisiae*.

B. Biological Function

Until recently, the function of the FLP site-specific recombination system was obscure. A number of similar plasmids have been found in several *Zygosaccharomyces* species, and each of them possesses a similar site-specific recombination system (Toh-e et al., 1987; Araki et al., 1985). These plasmids share little or no homology. Each possesses, however, a pair of long inverted repeats and encodes a site-specific recombination system which promotes inversion. The only detected homology occurs in the FLP genes at the amino acid level (Toh-e et al., 1987). The lack of homology at the nucleotide level, in spite of similarities in general structure, suggests that most components of these plasmids evolved independently. This further suggests that the general architecture of the plasmids, including the inversion system, has adaptive significance. Disruption of the FLP gene in the 2μm plasmid, however, has only a modest effect on plasmid stability (Broach and Hicks, 1980). Recently, Futcher (1986) proposed that the inversion mediated by the FLP system is required for plasmid copy-number amplification. His model explains how site-specific recombination can circumvent the cell cycle restriction on multiple replication initiation events. In this model (Fig. 2) the theta-form replicative intermediate of the plasmid is inverted by the FLP recombination system. The result is a double rolling circle in which the replication forks never converge, permitting the generation of many tandem copies of the plasmid from a single initiation event. This multimeric species can also be resolved into monomers by the FLP system. Experimental evidence for the central features of this model was quickly provided by Volkert and Broach (1986), who demonstrated that the FLP system is required for plasmid amplification in yeast cells grown under nonselective conditions.

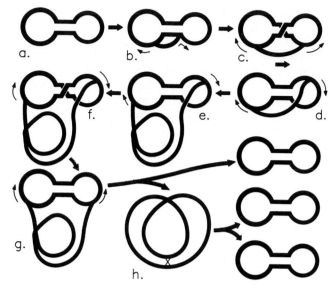

FIGURE 2. Double rolling circle model for amplification of 2μm plasmid copy number (from Futcher, 1986). Inversion (c) following replication of one 599-bp repeat (b → c) results in a double rolling circle (d). This leads to production of multiple copies of 2μm plasmid from a single replication initiation event. Resolution of this species into monomeric circles (h → k) can be achieved by FLP protein-promoted recombination between FRT sites in the same orientation on the multimer. Three copies of the plasmid are shown as products, although the number may be much larger. See text for details. (Adapted from Volkert and Broach, 1986.)

C. Fundamental Properties In Vivo and In Vitro

The properties of the FLP system place it in a class with the most straightforward of the known site-specific recombination systems. In many respects, it shares general properties with the *cre-lox* system of bacteriophage P1 (Hoess and Abremski, 1985). As described above, there is no evidence that any protein besides the FLP protein plays a specific role in this reaction, in contrast to a number of systems discussed elsewhere in this volume. As described below, the recombination site is also relatively simple. There is no evident restriction on the type of recombination reaction permitted. The FLP protein will promote inversions, deletions, or intermolecular recombination either in vitro (Meyer-Leon et al., 1984; Vetter et al., 1983) or in vivo (Cox, 1983; Royer and Hollenberg, 1977; Falco et al., 1982), if appropriate substrates are provided. The in vitro reaction requires only FLP protein, a substrate, and a simple set of buffer and ionic conditions. No high-energy cofactor or divalent cation is required. To a first approximation, there also appears to be no requirement for a supercoiled DNA substrate, with efficient recombination observed for relaxed substrates (this observation must now be amended as described below). The in vitro systems using FLP protein synthesized in *E. coli* have been successful in duplicating the key properties of the system observed in vivo in *S. cerevisiae*, especially with respect to the structure and properties of the recombination site.

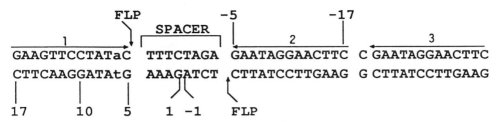

FIGURE 3. FRT site. Position numbers are as presented by Bruckner and Cox (1986). See text for details.

III. RECOMBINATION SITE

A. General Structure

Within the 65-bp region identified by Broach et al. (1982) as the region within which recombination occurs are several prominent structural features (Fig. 3). The most important are a set of three 13-bp repeats. The second and third repeats are separated by 1 bp and are in the same orientation. The first repeat is inverted with respect to the other two and is separated from the second repeat by an 8-bp spacer. The first repeat also has a 1-bp mismatch relative to the first two.

Deletion analysis demonstrated that the third repeat is unnecessary for recombination in vitro (Andrews et al., 1985; Senecoff et al., 1985). The function of this repeat is still unclear, but recent work suggests that it may have a slight effect on the reaction in vivo (Jayaram, 1985). Additional deletions revealed that most, but not all, of the first and second repeats (those flanking the spacer) are required. Deletion of 3 bp from the distal ends of one or both of these repeats has no detectable effect on the reaction. Further deletion leads to a gradual reduction in site function, with complete loss of site function occurring (in vitro) with deletions of 8 bp or more from either end (Andrews et al., 1985; Senecoff et al., 1985). The minimal site required for full function in vitro is therefore relatively small: 28 bp, including the spacer and the proximal 10 bp of each flanking repeat.

B. Spacer

The 8 bp of the spacer include the recognition sequence for the restriction endonuclease *Xba*I. Broach et al. (1982) observed that destruction of this restriction site prevented recombination in *S. cerevisiae*. Similar observations have been made in *E. coli* (Cox, 1983) and in vitro (Meyer-Leon et al., 1984), providing an experimental link between the latter systems and the recombination events in *S. cerevisiae*. The methods employed to destroy the restriction site either increased or decreased the size of the spacer, suggesting that spacer size is important for recombination site function. It has subsequently been demonstrated that an alteration of spacer size by ±1 bp is tolerated in the FLP system, although a reduction in site function is observed in reactions between these sites and sites with normal spacers. The site is inactivated by the addition of 2 bp in the spacer (Senecoff et al., 1985).

The reduction in site function observed when spacer size is altered by ±1 bp is restored in part when sites with identical alterations react with each other (Senecoff et al., 1985). This result suggested that homology between spacers of reacting sites is an important requirement for recombination. This interpretation has been confirmed with a variety of spacer mutations that do not alter spacer size (Senecoff and Cox, 1986; Andrews et al., 1987). The results of these experiments demonstrate that FLP protein does not recognize the central 6 bp of the spacer. Base substitutions can be made at any of these positions without affecting reaction efficiency as long as the reacting sites have homologous spacers.

The requirement for spacer homology implies that DNA-DNA pairing occurs in this region at some point during the recombination reaction. Similar homology requirements are observed in other site-specific recombination systems (Kikuchi and Nash, 1979; Bauer et al., 1985; Johnson and Simon, 1985). This pairing could occur before the cleavage event and reflect the formation of a four-stranded DNA intermediate such as that suggested by Kikuchi and Nash (1979). Alternatively, it could reflect events which occur subsequent to strand cleavage. In the case of the FLP system, the question has not yet been subjected to a clear experimental test.

The requirement for spacer homology serves an additional function in this system, that of proper alignment of two reaction sites which gives the reaction a directionality. In a normal intramolecular recombination reaction, a deletion will occur if the recombination sites are in the same orientation in the DNA molecule, while an inversion will occur if they are inverted with respect to each other. In some manner, the asymmetry of the sites is detected and utilized to align two sites in the same orientation during recombination. In the FRT site, the only asymmetric elements (once the third repeat is removed) are (i) a single-base-pair difference in the two remaining inverted repeats and (ii) the spacer sequence. The single mismatch in the repeats does not play a role in site alignment (Senecoff et al., 1985). The spacer, then, is the sole determinant of the directionality of the reaction. This property of the spacer is illustrated by results obtained with a recombination site in which 5 bp of the central 6 bp of the spacer were changed to produce a symmetrical spacer. This recombination site is fully functional, but the resulting reactions no longer exhibit directionality; i.e., inversion and deletion occur with equal frequency (Senecoff and Cox, 1986).

These results led to a hypothesis that the sequence of the central 6 bp of the spacer was irrelevant as long as homology was maintained in this region between reacting sites (Senecoff and Cox, 1986). Further mutational analysis, however, has revealed a number of spacer sequences which produce a site with reduced reaction efficiency. These studies (S. Umlauf and M. Cox, unpublished data) determined that no single-base-pair changes inactivated the recombination site; therefore, there is still no indication of protein-DNA interaction in this region. Instead, the effect of any given mutation depended on sequence context. Two patterns were evident. Mutations were generally deleterious if (i) they increased the G+C content of the spacer or (ii) they disrupted a polypurine tract which begins in the spacer and proceeds into the adjoining repeats. In most cases, two or more base substitutions with these properties were necessary before a significant deleterious effect was observed. In several cases in which the central

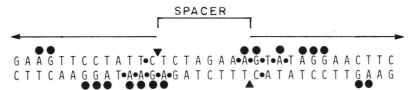

FIGURE 4. Identified contacts between FLP protein and purines and phosphate groups within the FRT site. Triangles, Points of cleavage; large circles, purine contacts; small circles, phosphate contacts.

6 bp of the spacer were all G+C, the site was inactivated in reactions employing nonsupercoiled DNA substrates. Interestingly, these mutations did not affect FLP protein binding. In fact, the binding affinity appeared to improve as the recombination efficiency decreased with these spacer mutants. In addition, the reactivity of each mutant site was improved when supercoiled rather than relaxed DNA substrates were used. These results suggest several things. The general sequence structure of the site is important for site function. This structure facilitates a recombination step subsequent to FLP protein binding. This step may involve unwinding of the DNA within the spacer region, as suggested by the effects of G+C content and supercoiling. The results permit a clear distinction to be drawn between the role of this sequence in site function as opposed to site recognition and binding by FLP protein.

C. Protein-DNA Interactions

The binding sites for FLP protein are the repeats that flank the spacer. Footprinting studies (Andrews et al., 1985) demonstrated that FLP protein protects the spacer and the two flanking repeats from DNase digestion. The binding sites were outlined in more detail by use of methylation protection and interference protocols (Bruckner and Cox, 1986). These studies mapped apparent purine and phosphate contacts within two 12-bp regions including each external base pair of the spacer and the first 11 bp of the adjacent repeats on each side of the recombination site. These contacts are summarized in Fig. 4. The phosphate contacts are clustered near the site at which FLP protein cleaves the DNA. The identified purine contacts are localized on one 180° face of a B-DNA helix. Both this work and the footprinting studies of Andrews et al. (1985) demonstrated that FLP protein binds to the third 13-bp repeat as well as the other two, even though this repeat is not required for recombination. Whether this binding has functional significance remains to be determined.

These results are consistent with and complementary to the results of the deletion analysis described above and the observed effects of mutations in the FLP binding sites. Mutations in this region have similar effects on recombination in vivo and in vitro. A few mutations in the FLP binding site have large effects on the recombination efficiency of the site (Prasad et al., 1986; Andrews et al., 1986; McLeod et al., 1986; Senecoff et al., 1988). Most of the mutations characterized in these studies, however, produce negligible effects on recombination. The

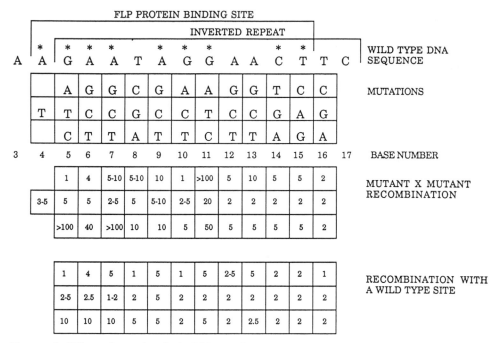

FIGURE 5. Effects of mutations in the FLP protein-DNA binding site. One FLP binding site is shown. Asterisks denote the purine contacts illustrated in Fig. 4. Numbers denote the factor by which the FLP protein concentration must be raised to produce a minimal reaction relative to a normal recombination site; i.e., higher numbers correspond to less reactive sites. See text for details.

results of a comprehensive mutational analysis carried out in vitro with pure FLP protein (Senecoff et al., 1988) are summarized in Fig. 5. Most of the base pair substitutions (31 of 37) within the FLP binding site have modest or negligible effects in vitro, although a change of at least 1 bp at each position produces a significant (fivefold or greater) effect. Several interesting patterns are evident. The results suggest at least three types of protein-DNA interaction. These are reflected by three different patterns of effects observed at different positions: (i) all three possible base pair changes produce small effects, (ii) all three changes produce large effects (position 11 only), and (iii) different changes produce different effects (positions 5 to 7). The patterns at positions 5 to 7 are particularly interesting. In each case one change, a G→C or A→T transversion, produces a large effect, while the effects of the other two changes are small. Finally, it is noteworthy that transition mutations generally have very small effects on recombination (with the exception of position 11). This reinforces the notion that the polypurine (polypyrimidine) tracts extending through the FRT site (positions −1 to 13, and −2 to −13) are important to site function.

As shown in Fig. 4, FRT sites with mutations in one of the two FLP binding sites react better with a normal site than with an identical mutant site. In some cases the improvement is greater than 10-fold. A wild-type site can therefore "rescue" a mutant site, suggesting protein-protein contacts in the complex within which recombination occurs (Prasad et al., 1986; Senecoff et al., 1988). Additional interactions between FLP monomers bound to the same FRT site are suggested by the much greater than additive effect observed when some mutations are present in both FLP binding sites rather than only one (Prasad et al., 1986; Senecoff et al., 1988).

D. Site of DNA Cleavage by FLP Protein

The FLP protein cleaves the two DNA strands at sites that are staggered by 8 bp. The cleavage points are at the spacer repeat junctions shown in Fig. 3. The protein becomes covalently attached to the DNA via a 3′ phosphate (Andrews et al., 1985; Senecoff et al., 1985) linked to a tyrosine (Gronastajski and Sadowski, 1985b). The eight-nucleotide single-stranded overhangs terminate in free 5′ hydroxyl groups.

E. Recombination Site Summary

The results described above thoroughly characterize the FLP recombination site. These studies reveal that the required site is relatively small, and a surprising degree of sequence flexibility is evident. In spite of the limitations described above, a wide variety of spacer sequences result in sites which are fully functional but do not cross-react. Sites with symmetrical spacers abolish reaction directionality. These properties of the site are useful experimentally (Bruckner and Cox, 1986; Senecoff et al., 1988; C. Gates and M. Cox, unpublished data).

IV. FLP PROTEIN

A. Expression in *E. coli* and Purification

Whereas expression of FLP protein in *E. coli* has been easy to detect, the levels of the protein obtained have been disappointing. Initial expression efforts employed either the *lac* or λ p_R promoters (Vetter et al., 1983; Cox, 1983). Upon induction, FLP protein is generally less than 0.1% of the total soluble protein. A wide variety of classical approaches to protein expression in *E. coli* have been utilized without success in attempts to improve the level of expression (Meyer-Leon et al., 1987; E. Wood and M. Cox, unpublished data). The reason for this limited expression remains obscure.

The levels of protein obtained to date have, however, been sufficient to permit purification of significant amounts of FLP protein. Purification to near homogeneity was accomplished with the aid of site-specific DNA affinity chromatography. The strategy employed the information outlined above with respect to the FRT site. Attempts to use the intact FRT site as an affinity ligand failed because of a tendency for FLP protein to form covalent adducts with the DNA. The observation that FLP protein bound specifically to the third 13-bp repeat, however, suggested an alternative strategy. FLP protein does not cleave DNA between the second and third 13-bp repeats. A polymer of 13-bp repeats in the same orientation, each linked to its neighbors as the second and third repeats in the FRT site are (Fig. 3), proved to be a successful affinity ligand (Meyer-Leon et al., 1987; C. A. Gates, L. Meyer-Leon, J. M. Attwood, E. A. Wood, and M. M. Cox, *in* R. Burgess, ed., *Protein Purification: Micro to Macro*, in press). Current protocols generate 1 mg of nearly homogeneous FLP protein from 100 g of *E. coli* host cells in 1 week.

B. Properties

The predicted molecular weight of the protein based on the DNA sequence of the gene is about 48,000. Migration of the protein in denaturing polyacrylamide gels and the amino-terminal sequence of the protein are consistent with the gene sequence. The purified protein is free from detectable nuclease activity and is stable for many months at −70°C. At 0°C the protein is stable for several weeks when stored in the presence of high (1 M) concentrations of salt. Under conditions previously determined to be optimal for activity (200 mM NaCl, 30°C, pH 7.5), the protein exhibits considerable instability, with complete loss of activity observed in the absence of DNA substrates in 20 to 30 min. Stability is improved significantly with the addition of bovine serum albumin and glycerol (Gates and Cox, unpublished data).

The pure protein promotes efficient recombination at levels which appear to be approximately stoichiometric with the number of recombination sites (one or two FLP monomers per site) in the reaction mixture under most conditions. When substrates with two recombination sites are employed, reactions are almost exclusively intramolecular regardless of the FLP protein concentration employed. Intermolecular recombination is enhanced in the presence of low levels of polyethylene glycol and in the presence of a number of basic DNA-binding proteins (see below).

Efforts have begun in several laboratories to generate mutations in the *flp* gene that affect the activity of the FLP protein. A strategy for isolating random mutations that inactivate FLP protein has been published (Govind and Jayaram, 1987). Mutations which change Tyr-343 to a serine or phenylalanine have also been described (Prasad et al., 1987). The inactivity of these mutants suggests that this tyrosine residue is the site of covalent attachment of the protein to the DNA.

V. RECOMBINATION REACTION

Site-specific recombination reactions provide an opportunity to examine many aspects of protein-DNA interactions. The enzyme must locate a specific binding site, juxtapose two such sites, cleave the DNA within each of the two sites, and then ligate the resulting ends with new partners in a precise way to yield recombinants. Each of the known enzymes promoting site-specific recombination is, in effect, a restriction enzyme and ligase in one package. As one of the simplest site-specific recombination systems, the FLP system recommends itself as an experimental vehicle to probe chemical details of this process. Until recently, however, progress was slowed by the lack of pure protein, although many properties of the system have been outlined with partially purified preparations. Unless noted, the information presented below was obtained with FLP protein fractions which were approximately 5% pure (Meyer-Leon et al., 1984; Andrews et al., 1987).

Nothing is yet known about the mechanism by which FLP protein locates its binding sites. A DNA-binding assay based on gel retardation has been used to resolve three binding complexes, which apparently correspond to the binding of one, two, or three FLP monomers to available 13-bp repeats (Andrews et al., 1987; Prasad et al., 1986; Senecoff et al., 1988). If a recombination site with two 13-bp repeats is employed, and a strong deleterious mutation is present in one repeat, FLP protein binds and cleaves the DNA adjacent to the wild-type repeat only (Senecoff et al., 1988). The efficiency of intramolecular recombination appears to decrease as the distance between the sites increases, suggesting that tracking of the enzyme along the DNA may be involved in site juxtaposition (Gronastajski and Sadowski, 1985a). This relationship is abolished, however, if KCl is omitted from the reaction, suggesting that under at least some conditions site juxtaposition may involve random collisions (Sadowski et al., 1987). The cleavage of the DNA and formation of a covalent adduct was described above. As with other site-specific recombination systems and topoisomerases, this covalent adduct preserves the high-energy bond and obviates the need for high-energy cofactors in the reaction.

A common observation with site-specific recombination systems is that levels of recombinase which are at least stoichiometric with the concentration of available recombination sites are generally required to produce an optimum reaction in vitro. This suggests that these proteins do not turn over and therefore are not true enzymes. Similar observations were made in the early in vitro experiments with FLP protein. Recent results, however, indicate that FLP protein turns over slowly when relaxed DNA substrates are employed (Gates and Cox, unpublished data). The observation of turnover depends in part on the use of conditions which enhance the stability of the protein. The turnover rate is increased 1.5- to 3-fold if supercoiled substrates are employed. The apparent turnover number is low, ~ 0.1 min^{-1} under one set of conditions, and the slow step in the process has not yet been identified.

A number of recombination sites that harbor spacer alterations affected the rate of recombination but not the rate of FLP protein binding (Umlauf and Cox, unpublished data). These mutations therefore do not affect recognition of the site by FLP protein, and their existence suggests that DNA structure plays a significant role in a kinetically significant step of this reaction. The reactivity of these mutant sites is increased if the DNA is supercoiled, indicating that the affected step involves DNA unwinding in the spacer region. These results suggest an intricate relationship between DNA structure or topology and the catalytic mechanism employed by the recombinase.

VI. EFFECTS OF OTHER PROTEINS ON FLP PROTEIN-PROMOTED RECOMBINATION

Purification of FLP protein was hampered to some degree by variations in the behavior of the protein from one fraction to the next. The variation involved the relative efficiencies of inter- and intramolecular recombination as well as increases and decreases in overall activity (Meyer-Leon et al., 1984, Gronastajski and Sadowski, 1985a). High levels of FLP protein led to a two- to threefold enhancement of intermolecular recombination, while even higher concentrations inhibited the reaction almost completely. These effects have been traced to *E. coli* proteins present as contaminants of the FLP protein preparations (Meyer-Leon et al., 1987). At least four *E. coli* proteins are capable of producing these effects, and each of the four has recently been purified to homogeneity (R. C. Bruckner and M. M. Cox, submitted for publication). One of these proteins (molecular weight, 27,000) has been identified as the protein designated the H protein (Hübscher et al., 1980), which is a histonelike protein that cross-reacts immunologically with histone H2A. Another is a prominent contaminant in many H protein preparations (molecular weight, 26,000). The other two proteins are smaller (molecular weights, 24,000 and 16,000) and presumably are DNA-binding proteins. Each of these proteins increases the rate of recombination two- to fivefold when present at low concentrations. Increasing the levels of these proteins leads first to an enhancement of intermolecular recombination relative to intramolecular recombination and ultimately to inhibition of both reactions. All four proteins are prominent contaminants at early stages of the FLP protein purification. N-terminal sequence analysis of each of these proteins has produced the surprising result that all four are ribosomal proteins. The *E. coli* H protein, in fact, is ribosomal protein S3. The others are L2, S4, and S5 in order of decreasing size (R. C. Bruckner and M. M. Cox, submitted for publication). The functional significance of this observation is not yet clear. Interestingly, histones produce similar effects on FLP protein-promoted recombination. These effects almost certainly do not reflect a specific interaction of these proteins with the FLP protein or FRT site. It is noteworthy, however, that the 2μm plasmid is packaged as chromatin in *S. cerevisiae*, and it is conceivable that histones play a role in regulating this recombination event in vivo.

VII. PROSPECTS

The FLP system has been thoroughly characterized in vivo and in vitro and can now be used to address more general questions related to the detailed mechanism of site-specific recombination. As a vehicle for the investigation of chemical steps in this process, it has the advantage of simplicity. No additional proteins or factors which could complicate experimental design or interpretation are required, and no restrictions on the type of reaction (inversion, deletion) are superimposed on the basic chemistry of site-specific recombination. The availability of pure protein, protein mutants, and a well-defined recombination site should ensure rapid progress. The properties of the system also recommend it as a potential reagent. The recombination site required is large enough that it is unlikely to occur at random in most DNA molecules but is small enough for convenient laboratory manipulation. The site exhibits considerable sequence flexibility, and suitable sites may exist in many chromosome-sized DNAs. Thus, it may be possible to use the FLP system to insert exogenous DNA into specific chromosomal sites.

LITERATURE CITED

Andrews, B. J., L. G. Beatty, and P. D. Sadowski. 1987. Isolation of intermediates in the binding of the FLP recombinase of the yeast plasmid 2-micron circle to its target sequence. *J. Mol. Biol.* **193:**345–358.

Andrews, B. J., M. McLeod, J. R. Broach, and P. D. Sadowski. 1986. Interaction of the FLP recombinase of the yeast 2-micron plasmid with mutated target sequences. *Mol. Cell. Biol.* **6:**2482–2489.

Andrews, B. J., G. A. Proteau, L. G. Beatty, and P. D. Sadowski. 1985. The FLP recombinase of the 2-micron circle DNA of yeast: interaction with its target sequences. *Cell* **40:**795–803.

Araki, H., H. Tatsumi, T. Sakurai, A. Jearnpipatkul, K. Ushio, T. Muta, and Y. Oshima. 1985. Molecular and functional organization of yeast plasmid pSR1. *J. Mol. Biol.* **182:**191–203.

Bauer, C. E., J. F. Gardner, and R. I. Gumport. 1985. Extent of sequence homology required for bacteriophage λ site-specific recombination. *J. Mol. Biol.* **181:**187–197.

Beggs, J. D. 1978. Transformation of yeast by a replicating hybrid plasmid. *Nature* (London) **275:**104–108.

Broach, J. R. 1982. The yeast plasmid 2μ circle, p. 445–470. *In* J. N. Strathern, E. W. Jones, and J. R. Broach (ed.), *The Molecular Biology of the Yeast Saccharomyces*. Cold Spring Harbor Laboratory, Cold Spring Harbor, N.Y.

Broach, J. R., V. R. Guarascio, and M. Jayaram. 1982. Recombination within the yeast plasmid 2 micron circle is site-specific. *Cell* **29:**227–234.

Broach, J. R., and J. B. Hicks. 1980. Replication and recombination functions associated with the yeast plasmid 2 micron circle. *Cell* **21:**501–508.

Bruckner, R. C., and M. M. Cox. 1986. Specific contacts between the FLP protein of the yeast 2 micron plasmid and its recombination site. *J. Biol. Chem.* **261:**11798–11807.

Cox, M. M. 1983. The FLP protein of the yeast 2 micron plasmid: expression of a eukaryotic genetic recombination system in *Escherichia coli. Proc. Natl. Acad. Sci. USA* **80:**4223–4227.

Falco, S. C., Y. Li, J. R. Broach, and D. Botstein. 1982. Genetic properties of chromosomally integrated 2μ plasmid in yeast. *Cell* **29:**573–584.

Futcher, A. B. 1986. Copy-number amplification of the 2 micron circle plasmid of *Saccharomyces cerevisiae. J. Theor. Biol.* **119:**197–204.

Gerbaud, C., P. Fournier, H. Blanc, M. Aigle, H. Heslot, and M. Guerineau. 1979. High frequency of yeast transformation by plasmids containing part or entire 2-μm yeast plasmid. *Gene* **5:**233–253.

Govind, N. S., and M. Jayaram. 1987. Rapid localization and characterization of random mutations within the 2μ circle site-specific recombinase: a general strategy for analysis of protein function. *Gene* **51:**31–41.

Gronastajski, R. M., and P. D. Sadowski. 1985a. The FLP protein of the 2 micron plasmid of yeast. Inter- and intramolecular reactions. *J. Biol. Chem.* **260:**12320–12327.

Gronastajski, R. M., and P. D. Sadowski. 1985b. The FLP recombinase of the yeast 2 micron plasmid attaches covalently to DNA via a phosphotyrosyl linkage. *Mol. Cell. Biol.* **5:**3274–3279.

Hartley, J. L., and J. E. Donelson. 1980. Nucleotide sequence of the yeast plasmid. *Nature* (London) **286:**860–865.

Hoess, R. H., and K. Abremski. 1985. Mechanism of strand cleavage and exchange in the cre-lox site-specific recombination system. *J. Mol. Biol.* **181:**351–362.

Hübscher, V., H. Lutz, and A. Kornberg. 1980. Novel histone H2A-like protein of *Escherichia coli.* *Proc. Natl. Acad. Sci. USA* **77:**5097–5101.

Jayaram, M. 1985. Two-micrometer circle site-specific recombination: the minimal substrate and possible role of flanking sequences. *Proc. Natl. Acad. Sci. USA* **82:**5875–5879.

Jayaram, M., Y.-Y. Li, and J. R. Broach. 1983. The yeast plasmid 2 micron circle encodes components required for its high copy propagation. *Cell* **34:**95–104.

Jayaram, M., A. Sutton, and J. R. Broach. 1985. Properties of REP3: a cis acting locus required for stable propagation of the yeast plasmid 2 micron circle. *Mol. Cell. Biol.* **5:**2466–2475.

Johnson, R. C., and M. I. Simon. 1985. Hin-mediated site-specific recombination requires two 26 bp recombination sites and a 60 bp recombinational enhancer. *Cell* **41:**781–791.

Kikuchi, Y. 1983. Yeast plasmid requires a *cis*-acting locus and two plasmid proteins for its stable maintenance. *Cell* **35:**487–493.

Kikuchi, Y., and H. A. Nash. 1979. Nicking-closing activity associated with bacteriophage λ int gene product. *Proc. Natl. Acad. Sci. USA* **76:**3760–3764.

Livingston, D. A., and D. A. Kupfer. 1977. Control of *Saccharomyces cerevisiae* 2 micron DNA replication by cell division cycle genes that control nuclear DNA replication. *J. Mol. Biol.* **116:**249–260.

McLeod, M., S. Craft, and J. R. Broach. 1986. Identification of the crossover site during FLP-mediated recombination in the yeast plasmid 2 micron circle. *Mol. Cell. Biol.* **6:**3357–3367.

Meyer-Leon, L., C. A. Gates, J. M. Attwood, E. A. Wood, and M. M. Cox. 1987. Purification of the FLP site-specific recombinase by affinity chromatography and re-examination of basic properties of the system. *Nucleic Acids Res.* **15:**6469–6488.

Meyer-Leon, L., J. F. Senecoff, R. C. Bruckner, and M. M. Cox. 1984. Site-specific recombination promoted by the FLP protein of the yeast 2 micron plasmid *in vitro. Cold Spring Harbor Symp. Quant. Biol.* **49:**797–804.

Murray, J. A. H., and G. Cesarini. 1986. Functional analysis of the yeast plasmid partition locus STB. *EMBO J.* **5:**3391–3399.

Prasad, P. V., D. Horensky, L.-J. Yound, and M. Jayaram. 1986. Substrate recognition by the 2 μm circle site-specific recombinase: effect of mutations within the symmetry elements of the minimal substrate. *Mol. Cell. Biol.* **5:**4329–4334.

Prasad, P. V., L.-J. Young, and M. Jayaram. 1987. Mutations in the 2-μm circle site-specific recombinase that abolish recombination without affecting substrate recognition. *Proc. Natl. Acad. Sci. USA* **84:**2189–2193.

Royer, H., and C. P. Hollenberg. 1977. *Saccharomyces cerevisiae* 2-μm DNA; an analysis of the monomer and its multimers by electron microscopy. *Mol. Gen. Genet.* **150:**271–284.

Sadowski, P. D., L. G. Beatty, D. Cleary, and S. Ollerhead. 1987. Mechanisms of action of the FLP recombinase of the 2 micron plasmid of yeast, p. 691–701. *In* R. McMacken and T. Kelly (ed.), *DNA Replication and Recombination.* Alan R. Liss, Inc., New York.

Sadowski, P. D., D. D. Lee, B. J. Andrews, D. Babineau, L. Beatty, M. J. Morse, G. A. Proteau, and D. Vetter. 1984. *In vitro* systems for the genetic recombination of the DNAs of bacteriophage T7 and yeast 2-micron circle. *Cold Spring Harbor Symp. Quant. Biol.* **49:**789–796.

Senecoff, J. F., R. C. Bruckner, and M. M. Cox. 1985. The FLP recombinase of the yeast 2-micron plasmid: characterization of its recombination site. *Proc. Natl. Acad. Sci. USA* **82:**7270–7274.

Senecoff, J. F., and M. M. Cox. 1986. Directionality in FLP protein-promoted site-specific recombination is mediated by DNA-DNA pairing. *J. Biol. Chem.* **261:**7380–7386.

Senecoff, J. F., P. J. Rossmeissl, and M. M. Cox. 1988. DNA recognition by the FLP recombinase of the yeast 2μ plasmid. A mutational analysis of the FLP binding site. *J. Mol. Biol.* **201:**405–421.

Toh-e, A., I. Utatsu, A. Utsunomiya, S. Sakamoto, and T. Imura. 1987. Two micron DNA-like plasmids from non-*Saccharomyces* yeasts, p. 425–437. *In* R. B. Wickner, A. Hinnebusch, A. M. Lambowitz, I. C. Gunsalus, and A. Hollaender (ed.), *Extrachromosomal Elements in Lower Eukaryotes.* Plenum Publishing Corp., New York.

Vetter, D., B. J. Andrews, L. Roberts-Beatty, and P. D. Sadowski. 1983. Site-specific recombination of yeast 2 micron DNA *in vitro. Proc. Natl. Acad. Sci. USA* **80:**7284–7288.

Volkert, F. C., and J. R. Broach. 1986. Site-specific recombination promotes plasmid amplification in yeast. *Cell* **46:**541–550.

Volkert, F. C., L. C. Wu, P. A. Fisher, and J. R. Broach. 1987. Survival strategies of the yeast plasmid 2 micron circle, p. 375–396. *In* R. B. Wickner, A. Hinnebusch, A. M. Lambowitz, I. C. Gunsalus, and A. Hollaender (ed.), *Extrachromosomal Elements in Lower Eukaryotes*. Plenum Publishing Corp., New York.

Zakian, V. A., B. J. Brewer, and W. L. Fangman. 1979. Replication of each copy of the yeast 2 micron DNA plasmid occurs during the S phase. *Cell* **17:**923–934.

Chapter 14

Control and Execution of Homothallic Switching in *Saccharomyces cerevisiae*

Jeffrey N. Strathern

I. INTRODUCTION

The homothallic yeast *Saccharomyces cerevisiae* has evolved a mechanism of developmental regulation involving a programmed rearrangement of the genome. The genes controlling cell type are activated by nonreciprocal recombination from a position in the genome where they are repressed, to a unique locus where they are expressed. This process resembles a gene conversion event and utilizes some of the functions involved in the repair of DNA damage. A very similar system is used in the homothallic yeast *Schizosaccharomyces pombe* (Egel and Gutz, 1981; Beach and Klar, 1984). Developmental regulatory mechanisms similar to homo- thallic switching have been demonstrated in trypanosomes (Bernards et al., 1981), *Neisseria gonorrhoeae* (Segal et al., 1985), and *Borrelia* spp. (Meier et al., 1985). For a review of these systems, see Borst and Greaves (1987). The detailed analysis of how *S. cerevisiae* uses a specific DNA rearrangement to change cell type provides not only a model for some classes of recombination and DNA damage repair but also a model for a novel regulatory mechanism utilized by several major disease-causing organisms. Recently, evidence suggesting a related mechanism in the activation of immunoglobin light chain genes in chickens has been presented (Reynaud et al., 1987; Thompson and Neiman, 1987).

Extensive genetic and physical analysis in the *Saccharomyces* mating-type system allows a complete description of the states of the DNA before and after the genome rearrangements (Strathern et al., 1980; Astell et al., 1981). Cell type in *S. cerevisiae* is controlled by the *MAT*a and *MAT*α alleles of the mating-type locus

445

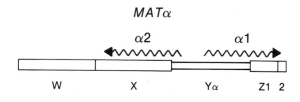

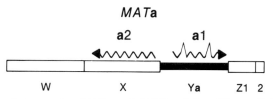

FIGURE 1. Difference between the *MATa* and *MATα* alleles. These alleles differ by a DNA substitution. The 642-base-pair **a**-specific region, Ya, includes a bidirectional promoter. The leftward transcript, **a2**, includes only sequences from the X region found in both *MATa* and *MATα* and has had no function ascribed to it. The rightward transcript, **a1**, is entirely contained within Ya. The *MATa1* gene has two intervening sequences, which may give rise to multiple functional messages. The 747-base-pair α-specific region, Yα, also has a bidirectional promoter. The open reading frame of *MATα2* begins in Yα and extends into the X region. The open reading frame of *MATα1* begins in Yα and extends into the Z region.

(MacKay and Manney, 1974; Strathern et al., 1981). These alleles differ by a DNA substitution, as diagrammed in Fig. 1 and discussed below. In heterothallic laboratory strains that are unable to switch mating type, these alleles are stably inherited. Thus, a/α diploids (*MATa/MATα*), which are themselves not capable of mating, can undergo meiosis and sporulation to generate 2**a** (*MATa*) and 2α (*MATα*) spores. The *MATa* spores grow into stable haploid **a** clones capable of mating to α cells to regenerate the a/α diploid part of the life cycle. Similarly, the *MATα* spores divide to form stable haploid α cultures.

The α1 and α2 proteins encoded by *MATα* and the **a1** protein encoded by *MATa* regulate the transcription of several genes involved in the sexual cycle. The α2 protein, encoded by *MATα2*, is a negative regulator. The α2 protein is particularly interesting because it is involved in the repression of two different sets of cell-type-specific genes (Strathern et al., 1981). In α cells, α2 has the role of a repressor of the **a**-specific genes: *MFa1* and *MFa2*, genes encoding a pheromone involved in coordinating mating (Brake et al., 1985); *STE2*, the receptor for the corresponding α pheromone α-factor (Jenness et al., 1983); *BAR1*, a protease that degrades α-factor (Sprague and Herskowitz, 1981); and *STE6*, an **a**-specific gene involved in **a**-factor biogenesis (Wilson and Herskowitz, 1984; Chan et al., 1983). A consensus DNA sequence required for regulation has been identified in the 5' noncoding region of these genes (Miller et al., 1985). Specific DNA-protein interactions between α2 protein and those sequences have been demonstrated (Johnson and Herskowitz, 1985). While such DNA-protein binding can even be demonstrated for α2 protein made in *Escherichia coli*, it is possible that proper regulation of these **a**-specific genes requires a second yeast protein in vivo.

In its second role, as a repressor of "haploid specific" genes, α2 protein requires the a1 protein, the product of the *MATa1* gene. Combined, the a1 and α2 proteins make a transcriptional repressor that acts on the following genes: *STE5*, a gene required for mating in **a** and α cells (Hartwell, 1980; Miller et al., 1985); *HO*, the gene encoding the endonuclease that initiates homothallic switching (Kostriken and Heffron, 1984; Jensen et al., 1983); *MATα1*, a positive regulator of α-specific genes (Strathern et al., 1981; Klar et al., 1981b); *SST2*, a gene involved in recovery from mating pheromone arrest (Chan and Otte, 1982; Dietzel and Kurjan, 1987); and *RME1*, a gene involved in the repression of functions involved in meiosis (Mitchell and Herskowitz, 1986). Although these are called haploid-specific genes, their expression reflects cell phenotype, not ploidy, and they are expressed in **a** and α cells but not **a**/α cells. The repression of these genes in **a**/α cells is mediated through a consensus DNA sequence found upstream of these genes (Miller et al., 1985; Siliciano and Tatchell, 1986; Dietzel and Kurjan, 1987). This sequence is related to, but different from, the sequence required for the repression of **a**-specific genes by α2 (Johnson and Herskowitz, 1985; Miller et al., 1985).

The α*1* protein is a positive regulator of α-specific gene expression (Strathern et al., 1981; Sprague et al., 1983). It appears to directly promote the expression of the α-specific gene *STE3*, the putative **a**-factor receptor (Hagen et al., 1986), and the genes encoding α-factor, *MFα1* and *MFα2* (Kurjan and Herskowitz, 1982; Emr et al., 1983). The α1 protein appears to act through binding to a consensus sequence found 5′ of the coding region of these genes (Jarvis et al., 1988) and may involve cooperative binding with another protein (Bender and Sprague, 1987).

Homothallic yeasts are able to differentiate from the **a** cell type to the α cell type and vice versa as a result of their ability to interconvert between the *MATa* and *MATα* alleles. This switching process in turn changes the expression of the cell-type-specific genes, as described above. In homothallic yeasts, **a**/α diploids again yield two *MATa* and two *MATα* spores; however, these spores grow into colonies that include *MATa*, *MATα*, and *MATa*/*MATα* cells (D. C. Hawthorne, *Abstr. Proc. 11th Int. Congr. Genet.*, 7:34–35, 1963). Homothallic *MATa* cells are able to make the DNA substitution necessary to become *MATα* cells because they have an unexpressed copy of the *MATα* sequence stored at a donor locus (*HMLα*; see Fig. 2). Similarly, *MATα* cells can switch to *MATa* by substituting a copy of the Y**a** region stored in *HMRa* for the Yα sequence at *MAT*. The details of how this process occurs, what functions are involved, and how it is regulated are the subjects of this review.

II. MATING-TYPE CASSETTES

The cloning of the *MATa* and *MATα* alleles (Hicks et al., 1979; Nasmyth and Tatchell, 1980) led to the demonstration that they differ by a DNA substitution defining an **a**-specific sequence, designated Y**a**, of 642 bases and an α-specific Yα sequence of 747 base pairs (Astell et al., 1981). While this difference made even more problematic the observation that cells could interconvert between *MATa*

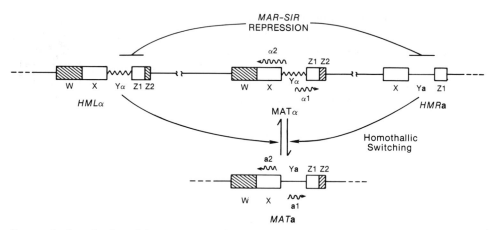

FIGURE 2. Organization of the cassettes on chromosome III. *HML* is on the left arm of chromosome III. *MAT* and *HMR* are on the right arm. W, X, Y, Z1, and Z2 represent regions defined by homologies between *HML*, *HMR*, and *MAT*. The W region is 723 base pairs in length and is homologous between *HML* and *MAT*. The X region (704 base pairs) is found at *HML*, *HMR*, and *MAT*. *HML*, *HMR*, and *MAT* can have either the **a**-specific sequence Y**a** (642 base pairs) or the α-specific sequence Yα (747 base pairs). Z1 (239 base pairs) is found at all three cassettes, whereas Z2 (89 base pairs) is found only at *HML* and *MAT*. The three cassettes are oriented in the same direction. The distance from *HML* to *MAT* is about 200 kilobases, and the distance from *MAT* to *HMR* is about 100 kilobases.

and *MAT*α, the use of these DNAs as hybridization probes to total digests of yeast DNA made it clear how this was accomplished. This approach revealed that *MAT*-related sequences, designated "cassettes," exist at two additional places in the yeast genome. These additional loci were identified as *HML* and *HMR* (Hicks et al., 1979), genes previously defined as having roles in homothallic switching (Harashima et al., 1974; Naumov and Tolstorukov, 1973). A series of genetic studies had suggested that the *HML* and *HMR* loci were the repositories of genetic information activated at *MAT* by homothallic switching (Harashima et al., 1974; Hicks et al., 1977; Blair et al., 1979; Haber and George, 1979; Hicks and Herskowitz, 1977; Klar et al., 1979; Rine et al., 1979; Strathern et al., 1979). Most laboratory strains have the *HMR***a** allele, which has the Y**a** sequence and flanking sequences X (703 bases) and Z1 (238 bases). Similarly, most strains have the *HML*α allele, which includes not only regions X, Yα, and Z1, but also additional regions of homology to *MAT*, W (723 bases) and Z2 (88 bases). Thus, homothallic *MAT***a** cells can switch to *MAT*α by substituting a copy of the Yα sequence stored in *HML*α for the Y**a** sequence at *MAT*. The substitution resembles a gene conversion in that it is not reciprocal. The *HML* Yα sequence is duplicated, and that *MAT* Y**a** sequence is destroyed. A similar process allows *MAT*α to use *HMR***a** as a donor of Y**a** to generate *MAT***a**. Isolates that have the same Y regions at *HML HMR* and *MAT* (*HML*α *HMR*α *MAT*α and *HML***a** *HMR***a** *MAT***a**) cannot change mating type, although they do undergo homologous cassette replacement. As described below, the W, X, Z1, and Z2 regions of flanking homology regions are important as sites for the initiation and resolution of the recombination intermediates in the switching pathway.

The *HML* and *HMR* loci encode complete and potentially functional copies of the *MAT* regulatory genes. They are kept silent by a set of at least four *trans*-acting negative regulators: *SIR1* (Rine et al., 1979), *MAR1* (*SIR2*) (Klar et al., 1979), *SIR3* (*CMT1*, *STE8*) (Haber and George, 1979; Rine and Herskowitz, 1987), and *SIR4* (*STE9*) (Rine and Herskowitz, 1987). Mutations in any of these genes leads to the expression of the cassettes at *HML* and *HMR*. For example, a *mar1⁻ HMLα MATa HMRa* strain will express **a** and α sequences and have the **a**/α phenotype. A description of the characterization of the *MAR/SIR* genes and the sites at which they act is beyond the scope of this review. However, it is important to note that the *MAR/SIR* gene products prevent the *HM* loci from being cleaved by the HO endonuclease and hence define the *HM* loci as donors and not recipients in the switching process (Klar et al., 1981c; Nasmyth, 1982b).

III. REGULATION OF THE *HO* GENE

Homothallic laboratory strains of yeasts differ from nonswitching (heterothallic) strains at a single locus designated *HO* (*D*) (Winge and Roberts 1949; Hawthorne, *Abstr. Proc. 11th Int. Congr. Genet.* 7:34–35, 1963). The *HO* gene encodes an endonuclease that makes a specific cleavage at the *MAT* locus as an initiating event in switching (see below).

The ability to undergo homothallic switching is strictly regulated relative to the sexual cycle and the cell division cycle. This appears to be due to multiple layers of control over expression of the *HO* gene. Switching is cell-type regulated: only **a** and α cells can switch. Cells with the **a**/α phenotype do not express the *HO* gene and hence are genetically stable (Jensen et al., 1983). The ability to distinguish the original cell from its daughter cell in budding yeast cells by microscopy, combined with the ability to determine the mating type of individual cells by their response to mating pheromones, made it possible to determine the distribution of mating switches of homothallic cells (Hicks and Herskowitz, 1976; Strathern and Herskowitz, 1979). Two major rules emerged from these studies (Fig. 3). (i) Switches were always observed in pairs of cells. For example, an α cell would divide to yield two cells, both of which were **a**. (ii) Only cells that had already divided at least once, so-called "mother cells" or "experienced cells," were capable of switching. Therefore, pedigrees started with spores or buds (inexperienced cells) would exhibit no switches until the four-cell stage. At that point the switches were always in the original cell and its second daughter. Both the "pair rule" and the "mother rule" reflect the control of transcription of the *HO* gene. The pair rule results from the fact that the *HO* gene is expressed in a brief portion of the cell cycle between commitment to exit from the G1 arrested stage of the cell cycle and the onset of DNA replication (Nasmyth, 1983). Hence, changes in the *MAT* allele occur prior to the duplication of *MAT* but in cells committed to DNA replication. Therefore, after the switch is complete, the *MAT* locus is replicated along with the rest of the genome. This timing results in the production of two cells with changed genotypes. The mother rule reflects the inability of inexperienced cells to express *HO*. Thus, mRNA from the *HO* gene is

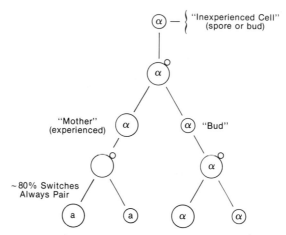

FIGURE 3. Pedigree analysis of homothallic switching. The mating type of a cell can be determined by its sensitivity to mating pheromones. Pedigrees initiated with an inexperienced cell, a bud or a spore, show no mating-type changes after the first division. After the second division, the two cells derived from the experienced cell are frequently opposite to the original cell in mating type. Cell-type changes are always found in pairs and always in cells derived from an experienced cell, one that has divided previously.

not detected in a population of new bud cells until the second time they pass through G1 (Nasmyth, 1983).

Recent emphasis has been placed on characterizing the *trans*-acting regulators of *HO* expression and the sites at which they work in order to understand the molecular basis of the cell cycle regulation of *HO* and the restriction of *HO* expression to mother cells. Functional analyses of the 5′ flanking region of the *HO* gene have been undertaken either by detecting expression of variants on plasmids (Jensen and Herskowitz, 1984) or by transplacing deletion variants back into the *HO* locus (Nasmyth, 1985). The DNA sequences upstream of *HO* have been subjected to deletion analysis in an effort to define the sites required for the cell type, cell cycle, and mother/daughter regulation of expression (Fig. 4). Sequences extending over 1,400 bases upstream of the coding sequences have been implicated in this regulation. The upstream regulatory sequence farthest from the start of transcription (URS1) is required for efficient expression and appears to contain sequences involved in the mother rule and the turnoff of the *HO* gene in a/α diploids (Nasmyth, 1985). A region designated URS2 has been shown to be involved in confining the expression to the late G1 stage. A more detailed analysis of the region upstream of *HO* revealed two reiterated sequence motifs. One sequence is present four times in URS1 and five times in URS2. Single copies of

FIGURE 4. Regulation of the *HO* gene. The upstream regulatory region of the *HO* gene includes sequences required for repression in a/α cells (solid arrows) and for restriction of expression to late G1 stage of the cell cycle (open arrows). Specific recognition sequences within URS1 for repression in inexperienced cells and promotion in experienced cells have not yet been identified.

A. a/α Cells

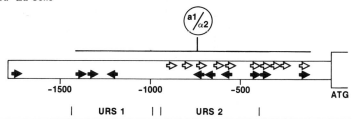

B. Haploid Daughter Cells

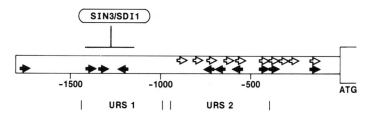

C. Haploid Mother Cells Not in Late G1

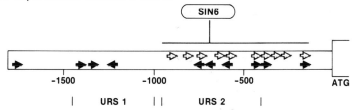

D. Haploid Mother Cells in Late G1

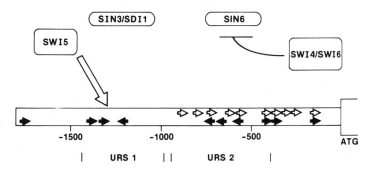

⇨	CACGAAAA sequences	Cell cycle regulation
➡	YCUTGTNN(A/T)NANNTACATCA sequences	Cell type regulation

this sequence, (T/C)C(A/G)TGTNN(A/T)NANNTACATCA, have been demonstrated to be sufficient to bring a heterologous gene under cell-type control (Miller et al., 1985). Thus, this sequence functions as an operator site through which the α2 and a1 proteins repress the expression of *HO*. The second reiterated motif (CACGAAAA) is found 10 times upstream of *HO* (Nasmyth, 1985). Fragments from *HO* containing this sequence are sufficient to make a heterologous gene subject to the same cell cycle restrictions on its expression as the *HO* gene (Breeden and Nasmyth, 1987).

Six genes (*SWI1*, *SWI2*, *SWI3*, *SWI4*, *SWI5*, and *SWI6*) have been identified as required for the expression of the *HO* gene (Haber and Garvik, 1977; Stern et al., 1984; Breeden and Nasmyth, 1987). *SWI4* and *SWI6* have been implicated in the cell cycle regulation of *HO* by demonstrating that the ability of the CAC GAAAA motif to act as an upstream activator sequence to a heterologous gene is cell cycle dependent and requires both *SWI4* and *SWI6* but not *SWI1*, *SWI2*, or *SWI5*. The interpretation of the role of *SWI4* and *SWI6* in the regulation of the *HO* gene is complicated by the observation that *HO* expression is independent of *SWI4* and *SWI6* when the CACGAAAA region is deleted. However, deletion of this region relieves the cell cycle regulation. Thus, while these sequences can be made to function as a UAS in the absence of URS1, they are not necessary for *HO* expression in the presence of URS1. The *SWI3* gene, like *SWI4* and *SWI6*, is necessary for expression of the CACGAAAA-driven heterologous gene but differs from *SWI4* and *SWI6* in that it is also necessary for expression of *HO* when the CACGAAAA region is deleted. These observations led to the proposal that the pair rule reflects the action of *SWI4* and *SWI6* in constraining the expression of *HO* to a period just after commitment to enter S phase "START" (Breeden and Nasmyth, 1987).

SWI1, *SWI2*, and *SWI5* are required for the expression of *HO* even when the CACGAAAA region is deleted (Breeden and Nasmyth, 1987). Thus, *SWI1*, *SWI2*, and *SWI5* exert their influence independent of the CACGAAAA motif and therefore act quite differently than *SWI4* and *SWI6*.

The above observations address the timing of the expression of *HO* but not the distinction between mother cells and daughter cells. This asymmetric distribution of developmental potential reflects a basic problem in development. The *SWI5* gene product plays a direct role in this mechanism. Nasmyth (1987) replaced the URS1 region of *HO* with a UAS of *GAL*. The UAS substitution still exhibited the pair rule (cell cycle regulation) and was still dependent on *SWI1*, *SWI2*, *SWI3*, *SWI4*, and *SWI6*. However, the promoter did not show mother/daughter asymmetry, and its expression was *SWI5* independent. It was argued that the dependence of this *GAL::HO* construct on the other *SWI* genes suggests that these products must be present in daughter cells and that only the *SWI5* gene product is missing in daughter cells. *SWI5* expression is cell cycle regulated, being expressed well after G1 at a distinctly different time than *HO*. *SWI5* may act directly on *HO* by binding to DNA in URS1 (D. Stillman et al. and K. Nagui and D. Rhodes, quoted in Nasmyth and Shore, 1987). Although these observations do not explain the mother/daughter asymmetry, they suggest that the *SWI5* product is sequestered into the mother or that its ability to act in daughters is specifically inhibited. The former view is supported by the observation that *SWI5* expressed from a *GAL* promoter allows daughters to switch.

The specific mechanisms by which the *SWI* gene products act as activators of *HO* is unknown. One possibility is that they inhibit repression of *HO* transcription. This model suggests that mutation in the repressor(s) would suppress the defect in expression of *HO* by *swi⁻* cells. Attempts to isolate such suppressors have identified several additional genes involved in the regulation of *HO*: *SIN1, 2, 3, 4, 5,* and *6* (Sternberg et al., 1987) and *SDI1* and *2* (Nasmyth et al., 1987b). While the analyses of these mutations are in their early stages, it is apparent that *sin3* and *sdi1* (suppressors of *swi5*) have roles in the mother/daughter asymmetry. Daughter cells switch in *sin3* and *sdi1* mutants. A DNA-binding activity that recognizes a site in URS1 was found in extracts of *SDI1* cells but not *sdi1* mutants (Nasmyth et al., 1987b). Thus, it is a reasonable model that mother/daughter regulation depends on the inhibition of *SIN3/SDI1* repression by *SWI5*. Again, this does not answer the question of how such asymmetry is generated. The genetics of *sin6* and *swi4* suggests a similar inhibition of *SWI4/SWI6* repression for the cell cycle-dependent expression of *HO*.

Together, these observations suggest a model for the regulation of *HO* in which at least three criteria must be met for expression to occur. First, repression of *HO* by *MATα2* and *MATa1* products (characteristic of a/α cells) must not be in effect. Second, the CACGAAAA region must be activated as a UAS, a process constrained to late G1 after the *cdc28* block, dependent on *SWI4* and *SWI6*, and perhaps brought about by the inhibition of a repressor, *SIN6*. Whether *SWI4* or *SWI6*, or both, show cell cycle regulation is not clear. Finally, expression of *HO* requires activation of URS1 in mother cells by a process that is dependent on *SWI5* and may reflect release of repression by *SIN3*.

IV. STRAND MECHANICS DURING SWITCHING

As discussed above, the timing and cell type specificity of the initiation of homothallic switching is largely determined by the regulation of transcription of the *HO* gene. The *HO* gene is known to encode an endonuclease that makes a cut at a unique site in *MAT*. Examination of the DNA from *S. cerevisiae* cultures undergoing homothallic switches revealed that some of the DNA from the *MAT* locus had been cleaved near the junction of the Y and Z1 regions (Strathern et al., 1982). An endonuclease capable of making this cleavage was identified in extracts of homothallic cells (Kostriken et al., 1983) and subsequently in extracts of *E. coli* carrying a clone of the *HO* gene (Kostriken and Heffron, 1984). The site of cleavage has been determined precisely in vitro (Kostriken et al., 1983) and in vivo (D. Raveh and J. Strathern, unpublished data). Mutations that render the *MAT* locus insensitive to *HO*-endonuclease cutting, and hence incapable of switching, have been identified and are designated *MATa-inc* and *MATα-inc* (Takano et al., 1973; Takano and Arima, 1979; Haber et al., 1980; Klar et al., 1984). Together with deletion analysis and site-directed mutagenesis (Weiffenbach

Definition of the HO cutting site

FIGURE 5. Recognition site for *HO* endonuclease. The *HO* endonuclease cleaves the *MAT* DNA with a four-base staggered cut and a 3′ extension. The cleavage is between bases seven and eight of the Z region (top strand). Essential bases defined by *inc* mutants or site-directed mutagenesis are indicated. Deletion analysis implicates sequences up to 12 bases into Z and 12 bases into Y.

et al., 1983; F. Heffron, personal communication), these studies suggest a rather large recognition site of up to 24 bases (Fig. 5).

Genetic evidence suggests that the *HO* endonuclease cleaves only at the *MAT* locus. Cells that have a *rad52* mutation and express *HO* die (Malone and Esposito, 1980; Weiffenbach and Haber, 1981). This reflects the defect in repair of double-strand DNA breaks in *rad52* mutants (Resnick and Martin, 1976). A single mutation to *MAT-inc* is sufficient to allow them to survive, indicating that *MAT* has the only site accessible to *HO* (Weiffenbach et al., 1983; Klar et al., 1984). The term accessible is germane because it is clear that *HML* and *HMR* have the recognition sequence for the *HO* endonuclease yet are not cleaved. DNase digestion experiments on chromatin indicate that the YZ junction is more protected at the *HM* loci that at *MAT* (Nasmyth, 1982b). In *mar1⁻* strains in which the *HM* loci are expressed, the DNase protection profiles of *HM* loci and *MAT* become equivalent, and the *HM* loci become both a target for the *HO*-endonuclease cleavage and recipients in the switching mechanism (Klar et al., 1981c).

However, it would be a mistake to view the switching process itself as merely the consequence of a double-strand break at *MAT*. Several features demonstrate that the process is highly ordered even after the endonuclease cleavage step.

A. Directionality

The frequency and distribution of cell-type switches in pedigrees of homo-thallic cells suggest that the donor *HM* locus used is not random. In *HO HMLα HMRa* cells there is a clear bias toward successful cell-type switching. In other words, a *MATα* allele after being cleaved by *HO* does not choose randomly which donor *HM* cassette to use. Mother α cells switch to *MATa* more than 80% of the time (Hicks and Herskowitz, 1976). Thus, there is a bias in *MATα* cells to use

*HMR*a as the donor. This bias, or directionality, has been shown to reflect the preference of α cells for *HMR* as a donor, not a preference for a donor with a **Y**a sequence (Klar et al., 1982; Jensen and Herskowitz, 1984). Similarly, *MAT*a cells have a preference for using *HML* as the donor. It is not yet clear how this directionality operates. These results are consistent with a model in which synapsis between *MAT* and one of the donor cassettes occurs in a cell type-specific fashion. Experiments moving the donor cassettes to other chromosomes or placing the *HML* cassette at *HMR* have not revealed sites involved in this process (Haber et al., 1981; A. Klar, personal communication).

B. Resolution

The second aspect of the outcome of homothallic switching that indicates that it is nonrandom is the observation that switches occur without the fusion of the donor *HM* locus to *MAT*. In contrast, gene conversion events in meiosis are associated with fusions (exchange of outside markers) about one-half of the time. The absence of such events in homothallic switching does not seem to be a direct consequence of initiating the event with *HO*. In experiments with diploids in which *MAT* is used as a donor to switch an allele on a sister chromosome, whether mitotically (Klar and Strathern, 1984) or meiotically (Kolodkin et al., 1986), recombination of outside markers was common.

Most models for homothallic switching suggest that *MAT* and the donor *HM* locus directly interact (no diffusible element). These models invoke the Z side of the cut as a primer for the synthesis of a duplicate of the donor. To understand the absence of fusions, one must postulate either that the intermediates in the switching pathway are different from gene conversion events or that a specific function exists to ensure that the resolution of the switching intermediate occurs without fusions. The structure shown in Fig. 6 has been proposed as an intermediate in both meiotic gene conversions initiated by double-strand breaks (Szostak et al., 1983) and homothallic switching (Strathern et al., 1982). Cleavage of the two Holliday junctions in opposite senses (the crossing strands of one and the noncrossing strands of the other) would result in a fusion. Three mechanisms have been proposed whereby this intermediate could be constrained during homothallic switching to avoid fusions. First, it could reflect the fact that the donors share only a short region of homology. If, for example, the Holliday junction-cleaving enzyme could be capable of cleaving only one sense, the second junction would have to isomerize prior to cleavage for the fusion to result. Second, homothallic switching normally occurs just prior to the replication of *MAT*, suggesting that replication might have a role in the resolution of switching. Third, there may exist specific functions and sites at which they act to resolve this intermediate. Results relevant to each of these proposals are described below.

The short regions of homology flanking the switched DNA may contribute to the concerted cleavage of the two junctions in the same sense. In strains in which the *HMR* allele has had most of the Z region deleted (*hmr1-Δ6*), the *hmr1-Δ6* locus can still serve as a donor (Weiffenbach and Haber, 1985). However, 10-fold more of the switches resulted in aberrant events than when the wild-type *HMR* donor

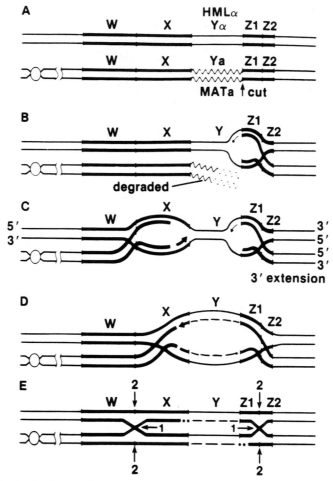

FIGURE 6. Switching by degradation of Y, 3' extension of the Z and X ends of *MAT* onto the donor locus, and resolution of a symmetric intermediate.

was used (Haber et al., 1980a). Similarly, in strains in which the length of the flanking homologies had been increased about 1 kilobase either by placing some of *HML* at *MAT* (Haber et al., 1982) or by placing some of *MAT* at *HMR* (Haber and Rogers, 1982), switching was efficient and generally did not result in the fusion of the donor and recipient loci. This experiment should be expanded using longer homologies and DNAs not related to the switching mechanism.

Homothallic switching normally occurs in cells committed to exit from G1 but prior to the replication of the *MAT* locus. Further, ARS elements flanking both *HML* and *HMR* have been identified (Abraham et al., 1984; Feldman et al., 1984). It has been postulated that the replication of the genome after the switch has a role in forcing the intermediate proposed in Fig. 6 to resolve without fusion of the donor and recipient loci. The passage of a replication fork through this intermediate could resolve it without cleavage of the Holliday junctions if a topoisomerase untangles the interwound strands. Constructions in which the *HO* gene has been put under the control of the galactose-regulatable *GAL1* promoter have made it possible to address this proposal. For example, *MATa* cells carrying the pGALHO plasmids can be induced for the *HO* endonuclease by the addition of galactose even when they have been previously arrested in G1 by α-factor. As assayed by the production of a characteristic *MATα* DNA fragment, these cells will complete the switching process without the formation of *MAT-HM* fusion fragments while still arrested in G1 (Raveh and Strathern, unpublished data). Hence, replication of the rest of the genome is not required for resolution. These results offer no support for the idea that the ARS elements found at *HML* and *HMR* have a role in the replication or resolution of the switching *MAT* cassettes.

There do not appear to be unique resolution sites for switching to the right or left of *MAT*. Tanaka et al. (1984) demonstrated that an *hmlα* mutation in the $\alpha2$ coding region (a single base deletion nine bases to the left of the X Y boundary) was transferred along with the Y2 region in over 99% of the switches. However, the observation that some switches did not transfer the mutation suggested that the extent of the DNA transferred was variable. Similarly, Klar et al. (1980) demonstrated that **a** alleles in the *MATa* Y**a** region were sometimes retained after homologous switches using *HMRa* as a donor. A more extensive analysis of the fates of restriction site polymorphisms throughout the X region of *MATα* has demonstrated that switches to *MATa* are sometimes, but not always, accompanied by loss of the restriction site. In other words, the restriction site polymorphism is sometimes coconverted when the Y region is switched. There is a coconversion gradient so that sites in X close to Y are switched at higher frequencies than sites nearer to the W region (C. McGill, B. Shafer, and J. Strathern, unpublished data). In these experiments the loss or retention of the restriction site difference could be independently scored in the two cells produced by one switch. It was frequently observed that the two cells did not have the same result. This is interpreted as the formation of a heteroduplex caused by the replacement of only one of the DNA strands at the restriction site followed by the replication of the DNA to give two different switching products. Even sequences in the *MAT* Z region to the right of the *HO*-endonuclease cleavage site are sometimes replaced during switching. These results demonstrate that a variable amount of *HM* sequence is copied into *MAT* during switching and suggest that resolution of the switching intermediate does not occur at a unique site.

C. In Search of Molecular Details

The result of switching is the replacement of the Y region resident at *MAT* by a duplicate of the Y region at the donor *HM* locus. Hence, it is a gene conversion event. Genetic analysis suggests that the Y region of *MAT* is discarded (Rine et al. 1981). The physical analysis indicates that the site-specific DNA cleavage of *MAT* by *HO* endonuclease is an initiating event. The end-product analysis described above suggests that the extent of the DNA replaced is variable and includes a mechanism that can generate heteroduplexes on the recipient *MAT* locus.

By placing the *HO* endonuclease under the control of a galactose-regulatable promoter, it has been possible to express *HO* independent of the cell type, cell cycle, and age of the cell (Jensen et al., 1983; Nasmyth, 1983). Thus, it is possible to induce homothallic switching in the majority of cells in a culture within a short period of time. This may make it possible to detect additional intermediates in the switching pathway. Analysis of DNA from such cultures has demonstrated that the cleavage seen in vivo (Raveh and Strathern, unpublished data) is the same as that reported for the in vitro reaction (Kostriken et al., 1983). The observation from these studies that the mobility of the *HO* endonuclease-cut DNA is independent of treatment by protease and runs at the position expected for naked DNA suggests that ends of the DNA are not covalently attached to a protein.

The use of the *GAL:HO* promoter fusion approach makes it possible to look at the fate of the *MAT* DNA in conditional mutants at nonpermissive conditions. For example, *rad52* cells that try to switch die. Hence, such cultures cannot be maintained. However, pGALHO *rad52* cultures are viable when grown on glucose, and analysis of the DNA from galactose-induced cultures indicates that the *HO* cut is made, but no completed switches occur. Thus, *rad52* mutants are blocked after the *HO* cleavage step (as expected) and prior to any detectable synthesis of the *HM* donor. In contrast, DNA ligase mutants (*cdc9^{ts}*) complete homothallic switching at the nonpermissive temperature. This may reflect the leaky nature of the mutation or the presence of another enzyme involved in ligating the DNAs together during homothallic switching.

The loss of the *MAT* Y region and its replacement by a duplicate of the donor *HM* Y region suggested that degradation of the *MAT* Y region and replication of the donor might be observed as intermediates in switching.

D. Working Model

Homothallic switching is initiated by a double-stranded cut and results in a unidirectional gene conversion-like transfer of genetic information from *HML* or *HMR* to *MAT*. Those features are reflected in the model in Fig. 6. However, there is considerable flexibility in the details of how switching can occur which can be illustrated by developing an extreme variant of this model, still compatible with the data.

The observation that fusions of *MAT* to the *HM* donors can rarely occur may reflect either the aberrant resolution of a normal switching intermediate (Haber et al., 1980b) or the formation of an aberrant switching intermediate, for example, one initiated by a cleavage at the *HM* locus. While it remains possible that homothallic switching proceeds via a diffusible intermediate, attempts to identify intermediates in this pathway genetically (Tanaka et al., 1984) or physically (Strathern et al., 1982) have not been successful. Therefore, it will be assumed in this working model that the transfer of DNA is accomplished by a copying mechanism using the cleaved *MAT* sequence as the primer. Both sides of *MAT* may provide primers, as shown in Fig. 6; however, the desired result can be achieved if only one side of *MAT* is used to copy sequences from the donor. It is most convenient to assign the role of primer to the Z region because it is

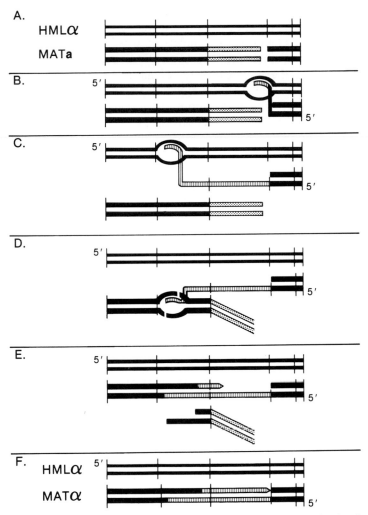

FIGURE 7. Switching by 3' extension of Z, invasion of the X region of *MAT* by the single-stranded duplicate of the donor, cleavage of *MAT* X, and second strand synthesis off the duplicate.

immediately homologous to the donor. In contrast, the W or X regions can serve as primers for a heterologous cassette switch only after the Y region has been discarded. Duplication of the donor may proceed via a replication fork synthesizing two new strands or may involve the synthesis of only a single strand of DNA made by elongating the primer, as shown in Fig. 7. The newly synthesized strand(s) may then be displaced from the donor and connected to the left side of *MAT* in a reaction independent of the donor. The old *MAT* Y region may be degraded from the cleaved end as a precursor to the ligation (or priming of the second strand synthesis), or it may be released as a unit as a consequence of the ligation of the new cassette to the left side of *MAT*.

The model in Fig. 7 is designed to be different from that implied by Fig. 6, to point out the molecular details that are yet to be resolved. For example, this model results in the production of two new strands of DNA at *MAT* while the *HM* locus retains its original atoms (conservative DNA replication). This would give a simple explanation for the lack of transfer of mutations from *MAT* to the *HM* loci during most switches. In contrast, the model in Fig. 6 can be resolved by cleavage to give one old and one new strand at donor and recipient. Alternatively, it can be resolved by topoisomerases to yield apparent conservative DNA replication (Nasmyth, 1982a). It should be noted that no specific role for the product of the *RAD52* gene is proposed and no sequences involved in the synapsis of *HM* loci to *MAT* are indicated. These details, as well as those implied by the differences between Fig. 6 and 7, are still to be defined. It is hoped that a combination of biochemical and genetic approaches will determine the process by which homothallic switching occurs and the enzymatic functions involved.

V. CONCLUSION

In both the regulation of homothallic switching and the mechanism of homothallic switching, research is focused on determining intermediates. The expression of the *HO* gene changes with cell age, cell cycle, and cell type. None of these features is static. Thus, *HO* gene regulation must be thought of as the constant interplay of positive and negative regulators. The challenge is to understand their interactions in a dynamic sense as *S. cerevisiae* passes through its cell and sexual cycles. End-product analysis of homothallic switching events may yield further details about sites involved in pairing, but the majority of progress in understanding how switching occurs will come from physical analysis of switching intermediates and the fates of such intermediates in strains defective in switching. Thus, the challenge is to develop systems to visualize intermediates reflecting the interaction of the donor and recipient loci, the replication of the donor, the loss of sequences from *MAT*, and the reformation of the intact *MAT* allele.

ACKNOWLEDGMENTS. Comments by David Garfinkel, Joan Curcio, and Carolyn McGill were very helpful in the preparation of this review. Thanks go to Fred Heffron for sharing results prior to publication and to Sylvia Lucas for her efforts on the manuscript.

Research was sponsored by the National Cancer Institute under contract no. NO1-CO-74101 with Bionetics Research, Inc.

LITERATURE CITED

Abraham, J., K. A. Nasmyth, J. N. Strathern, A. J. Klar, and J. B. Hicks. 1984. Regulation of mating-type information in yeast. Negative control requiring sequences both 5' and 3' to the regulated region. *J. Mol. Biol.* **176**:307–331.

Astell, C. R., L. Ahlstrom-Jonasson, M. Smith, K. Tatchell, K. A. Nasmyth, and B. D. Hall. 1981. The sequence of the DNAs coding for the mating-type loci of *Saccharomyces cerevisiae*. *Cell* **27**:15–23.

Beach, D. H., and A. J. S. Klar. 1984. Rearrangement of the transposable mating-type cassettes of fission yeast. *EMBO J.* **3**:603–610.

Bender, A., and G. F. Sprague, Jr. 1986. Yeast peptide pheromones, **a**-factor and α-factor, activate a common response mechanism in their target cells. *Cell* **47**:929–937.

Bender, A., and G. F. Sprague, Jr. 1987. *MATα1* protein, a yeast transcription activator, binds synergistically with a second protein to a set of cell-type-specific genes. *Cell* **50**:681–689.

Bernards, A., L. H. T. Vander Ploeg, A. C. C. Frasch, P. Borst, J. C. Boothroyd, S. Coleman, and G. A. M. Cross. 1981. Activation of trypanosome surface glycoprotein genes involves a duplication-transposition leading to an altered 3' end. *Cell* **27**:497–505.

Blair, L. C., P. J. Kushner, and I. Herskowitz. 1979. Mutations of the *HMa* and *HMα* loci and their bearing on the cassette model of mating type interconversion in yeast. *ICN-UCLA Symp. Mol. Cell. Biol.* **14**:13–26.

Borst, P., and D. R. Greaves. 1987. Programmed gene rearrangements altering gene expression. *Science* **235**:658–667.

Brake, A. J., C. Brenner, R. Najarian, P. Laybourn, and J. Merryweather. 1985. Structure of genes encoding precursors of the yeast peptide mating pheromone **a**-factor, p. 103–108. *In* M. J. Gething (ed.), *Protein Transport and Secretion*. Cold Spring Harbor Laboratory, Cold Spring Harbor, N.Y.

Breeden, L., and K. Nasmyth. 1987. Cell cycle regulation of *HO*: cis- and trans-acting regulators. *Cell* **48**:389–397.

Chan, R. K., L. M. Melnick, L. C. Blair, and J. Thorner. 1983. Extracellular suppression allows mating by pheromone-deficient sterile mutants of *Saccharomyces cerevisiae*. *J. Bacteriol.* **155**:903–906.

Chan, R. K., and C. A. Otte. 1982. Isolation and genetic analysis of *Saccharomyces cerevisiae* mutants supersensitive to G1 arrest by **a**-factor and α-factor pheromones. *Mol. Cell. Biol.* **2**:11–20.

Dietzel, C., and J. Kurjan. 1987. Pheromonal regulation and sequence of the *Saccharomyces cerevisiae SST2* gene: a model for desensitization to pheromone. *Mol. Cell. Biol.* **7**:4167–4177.

Egel, R., and H. Gutz. 1981. Gene activation by copy transposition in mating-type switching of a homothallic fission yeast. *Curr. Genet.* **3**:5–12.

Emr, S. D., R. Schekman, M. C. Flessel, and J. Thorner. 1983. An *MFα1-SUC2* (α-factor-invertase) gene fusion for study of protein localization and gene expression in yeast. *Proc. Natl. Acad. Sci. USA* **80**:7080–7084.

Feldman, J. B., J. B. Hicks, and J. R. Broach. 1984. Identification of sites required for repression of a silent mating type locus in yeast. *J. Mol. Biol.* **178**:815–834.

Haber, J., A. Comeau, P. Lie, D. Rogers, S. Stewart, M. Resnick, and B. Weiffenbach. 1982. Mechanism of homothallic switching in yeast mating type genes. *Recent Adv. Yeast Mol. Biol.* **1**: 332–347.

Haber, J., and B. Garvik. 1977. A new gene affecting the efficiency of mating type interconversions in homothallic strains of *Saccharomyces cerevisiae*. *Genetics* **87**:33–50.

Haber, J. E., and J. P. George. 1979. A mutation that permits the expression of normally silent copies of mating type information in *Saccharomyces cerevisiae*. *Genetics* **93**:13–35.

Haber, J., and D. Rogers. 1982. Transposition of a tandem duplication of yeast mating-type genes. *Nature* (London) **296**:768–773.

Haber, J. E., D. T. Rogers, and J. McCusker. 1980a. Homothallic conversions of yeast mating type genes occur by intrachromosomal recombination. *Cell* **22**:277–289.

Haber, J. E., L. Rowe, and D. T. Rogers. 1981. Transposition of yeast mating type genes from two translocations of the left arm of chromosome III. *Mol. Cell. Biol.* **1**:1106–1119.

Haber, J. E., W. T. Savage, S. M. Raposa, B. Weiffenbach, and L. B. Rowe. 1980b. Mutations preventing transpositions of yeast mating type alleles. *Proc. Natl. Acad. Sci. USA* **77**:2824–2828.

Hagen, D. C., G. McCaffrey, and Sprague, G. F., Jr. 1986. Evidence the yeast *STE3* gene encodes a receptor for the peptide pheromone **a** factor: gene sequence and implications for the structure of the presumed receptor. *Proc. Natl. Acad. Sci. USA* **83**:1418–1422.

Hagen, D. C., and G. F. Sprague, Jr. 1984. Induction of the yeast α-specific *STE3* gene by the peptide pheromone **a**-factor. *J. Mol. Biol.* **178**:835–852.

Harashima, S., Y. Nogi, and Y. Oshima. 1974. The genetic system controlling homothallism in *Saccharomyces* yeasts. *Genetics* **77**:639–650.

Hartwell, L. H. 1980. Mutants of *S. cerevisiae* unresponsive to cell division control by polypeptide mating pheromone. *J. Cell Biol.* **85**:811–823.

Hicks, J. B., and I. Herskowitz. 1976. Inter-conversion of yeast mating types. I. Direct observations of the action of the homothallism (*HO*) gene. *Genetics* **83**:245–258.

Hicks, J. B., and I. Herskowitz. 1977. Inter-conversion of yeast mating types. II. Restoration of mating ability to sterile mutants in homothallic and heterothallic strains. *Genetics* **85**:373–393.

Hicks, J. B., J. N. Strathern, and I. Herskowitz. 1977. The cassette model of mating type interconversion, p. 457–462. *In* A. Bukhari, S. Shapiro, and S. Adhya (ed.), *DNA Insertion Elements, Plasmids and Episomes.* Cold Spring Harbor Laboratory, Cold Spring Harbor, N.Y.

Hicks, J., J. N. Strathern, and A. J. Klar. 1979. Transposable mating type genes in *Saccharomyces cerevisiae. Nature* (London) **282**:478–483.

Jarvis, E. E., D. C. Hagen, and G. F. Sprague, Jr. 1988. Identification of a DNA segment that is necessary and sufficient for α-specific gene control in *Saccharomyces cerevisiae*: implications for regulation of α-specific and **a**-specific genes. *Mol. Cell. Biol.* **8**:309–320.

Jenness, D. D., A. C. Burkholder, and L. H. Hartwell. 1983. Binding of α-factor pheromone to yeast **a** cells: chemical and genetic evidence for an α-factor receptor. *Cell* **35**:521–529.

Jensen, R., and I. Herskowitz. 1984. Directionality and regulation of cassette substitution in yeast. *Cold Spring Harbor Symp. Quant. Biol.* **49**:97–104.

Jensen, R., F. Sprague, and I. Herskowitz. 1983. Regulation of yeast mating type interconversion: feedback control of *HO* gene expression by the mating-type locus. *Proc. Natl. Acad. Sci. USA* **80**: 3035–3039.

Johnson, A. D., and I. Herskowitz. 1985. A repressor (*MAT*α2 product) and its operator control expression of a set of cell-type-specific genes in yeast. *Cell* **42**:237–247.

Klar, A. J. S., S. Fogel, and K. Macleod. 1979. *Mar1*—a regulator of the *HM***a** *HM*α loci in *Saccharomyces cerevisiae. Genetics* **93**:37–50.

Klar, A. J. S., J. B. Hicks, and J. N. Strathern. 1981a. Irregular transpositions of mating-type genes in yeast. *Cold Spring Harbor Symp. Quant. Biol.* **45**:983–990.

Klar, A. J. S., J. B. Hicks, and J. N. Strathern. 1982. Directionality of yeast mating-type interconversion. *Cell* **28**:551–561.

Klar, A. J. S., J. McIndoo, J. N. Strathern, and J. B. Hicks. 1980. Evidence for a physical interaction between the transposed and the substituted sequences during mating-type gene transposition in yeast. *Cell* **22**:291–298.

Klar, A. J. S., and J. N. Strathern. 1984. Resolution of recombination intermediates generated during yeast mating-type switching. *Nature* (London) **310**:744–748.

Klar, A. J. S., J. N. Strathern, and J. A. Abraham. 1984. Involvement of double-strand chromosomal breaks for mating-type switching in *Saccharomyces cerevisiae. Cold Spring Harbor Symp. Quant. Biol.* **49**:77–88.

Klar, A. J. S., J. N. Strathern, J. R. Broach, and J. B. Hicks. 1981b. Regulation of transcription in expressed and unexpressed mating-type cassettes of yeast. *Nature* (London) **289**:239–244.

Klar, A. J. S., J. N. Strathern, and J. B. Hicks. 1981c. A position-effect control for gene transposition: state of expression of yeast mating-type genes affects their ability to switch. *Cell* **25**:517–524.

Kolodkin, A. L., A. J. S. Klar, and F. W. Stahl. 1986. Double-strand breaks can initiate meiotic recombination in *S. cerevisiae. Cell* **46**:733–740.

Kostriken, R., and F. Heffron. 1984. The product of the *HO* gene is a nuclease: purification and characterization of the enzyme. *Cold Spring Harbor Symp. Quant. Biol.* **49**:89–104.

Kostriken, R., J. N. Strathern, A. J. Klar, J. B. Hicks, and F. Heffron. 1983. A site-specific endonuclease essential for mating-type switching in *Saccharomyces cerevisiae. Cell* **35**:167–174.

Kurjan, J., and I. Herskowitz. 1982. Structure of a yeast pheromone gene (*MF*α): a putative α-factor precursor contains four tandem copies of mature α-factor. *Cell* **30**:933–943.

MacKay, V., and T. R. Manney. 1974. Mutations affecting sexual conjugation and related processes in *Saccharomyces cerevisiae.* I. Isolation and phenotypic characterization of nonmating mutants. *Genetics* **76**:255–271.

Malone, R., and R. E. Esposito. 1980. The *RAD52* gene is required for homothallic interconversion of mating-types and spontaneous recombination in yeast. *Proc. Natl. Acad. Sci. USA* **77**:503–507.

Meier, J. T., M. I. Simon, and A. G. Barbour. 1985. Antigenic variation is associated with DNA rearrangements in a relapsing fever borrelia. *Cell* **41**:403–409.

Miller, A. M., V. L. MacKay, and K. A. Nasmyth. 1985. Identification and comparison of two sequence elements that confer cell-type specific transcription in yeast. *Nature* (London) **314**:598–603.

Mitchell, A. P., and I. Herskowitz. 1986. Activation of meiosis and sporulation by repression of the *RME1* product in yeast. *Nature* (London) **319**:738–742.

Nasmyth, K. A. 1982a. Molecular genetics of yeast mating type. *Annu. Rev. Genet.* **16**:439–500.

Nasmyth, K. A. 1982b. The regulation of yeast mating type chromatin structure by *SIR*: an action at a distance affecting both transcription and transposition. *Cell* **30**:567–578.

Nasmyth, K. A. 1983. Molecular analysis of a cell lineage. *Nature* (London) **302**:670–676.

Nasmyth, K. A. 1985. At least 1400 base pairs of 5'-flanking DNA is required for the correct expression of the *HO* gene in yeast. *Cell* **42**:213–223.

Nasmyth, K. A. 1987. The determination of mother-cell-specific mating-type-switching in yeast by a specific regulator of *HO* transcription. *EMBO J.* **6**:243–248.

Nasmyth, K., A. Seddon, and G. Ammerer. 1987a. Cell cycle regulation of *SWI5* is required for mother-cell-specific *HO* transcription in yeast. *Cell* **49**:549–559.

Nasmyth, K., and D. Shore. 1987. Transcriptional regulation in the yeast life cycle. *Science* **237**:1162–1170.

Nasmyth, K., D. Stillman, and P. Kipling. 1987b. Both positive and negative regulators of *HO* are required for mother-cell-specific mating-type switching in yeast. *Cell* **48**:579–587.

Nasmyth, K. A., and K. Tatchell. 1980. The structure of transposeable yeast mating-type loci. *Cell* **19**:753–764.

Naumov, G. I., and I. I. Tolstorukov. 1973. Comparative genetics of yeast. X. Reidentification of mutators of mating-types in *Saccharomyces*. *Genetika* **9**:82–91.

Resnick, M. A., and P. Martin. 1976. The repair of double-strand breaks in the nuclear DNA of *Saccharomyces cerevisiae* and its genetic control. *Mol. Gen. Genet.* **143**:119–129.

Reynaud, C. A., V. Anquez, H. Grimal, and J. C. Weill. 1987. A hyperconversion mechanism generates the chicken light chain preimmune repertoire. *Cell* **48**:379–388.

Rine, J., and I. Herskowitz. 1987. Four genes responsible for a position effect on expression from *HML* and *HMR* in *Saccharomyces cerevisiae*. *Genetics* **116**:9–22.

Rine, J., R. Jensen, D. Hagen, L. Blair, and I. Herskowitz. 1981. Pattern of switching and fate of the replaced cassette in yeast mating-type interconversion. *Cold Spring Harbor Symp. Quant. Biol.* **45**:951–960.

Rine, J., J. N. Strathern, J. B. Hicks, and I. Herskowitz. 1979. A suppressor of mating-type locus mutations in *Saccharomyces cerevisiae*: evidence for and identification of cryptic mating-type loci. *Genetics* **93**:877–901.

Segal, E., E. Billyard, M. So, S. Storzbach, and T. F. Meyer. 1985. Role of chromosomal rearrangement in *N. gonorrhoeae* pilus phase variation. *Cell* **40**:293–300.

Siliciano, P. G., and K. Tatchell. 1986. Identification of the DNA sequences controlling the expression of the *MATα* locus of yeast. *Proc. Natl. Acad. Sci. USA* **83**:2320–2324.

Sprague, G. F., Jr., and I. Herskowitz. 1981. Control of yeast cell type by the mating-type locus. I. Identification and control of expression of the **a**-specific gene *BAR1*. *J. Mol. Biol.* **153**:305–321.

Sprague, G. F. Jr., R. Jensen, and I. Herskowitz. 1983. Control of yeast cell type of the mating-type locus: positive regulation of the α-specific *STE3* gene by the *MATα1* product. *Cell* **32**:409–415.

Stern, M., R. Jensen, and I. Herskowitz. 1984. Five *SWI* genes are required for expression of the *HO* gene in yeast. *J. Mol. Biol.* **178**:853–868.

Sternberg, P. W., M. J. Stern, I. Clark, and I. Herskowitz. 1987. Activation of the yeast *HO* gene by release from multiple negative controls. *Cell* **48**:567–577.

Strathern, J. N., and I. Herskowitz. 1979. Asymmetry and directionality in production of new cell types during clonal growth: the switching pattern of homothallic yeast. *Cell* **17**:371–381.

Strathern, J., J. Hicks, and I. Herskowitz. 1981. Control of cell type in yeast by the mating-type locus. The α1-α2 hypothesis. *J. Mol. Biol.* **147**:357–372.

Strathern, J. N., A. J. Klar, J. B. Hicks, J. A. Abraham, J. M. Ivy, K. A. Nasmyth, and C. McGill.

1982. Homothallic switching of yeast mating type cassettes is initiated by a double-stranded cut in the *MAT* locus. *Cell* **31:**183–192.

Strathern, J. N., C. S. Newlon, I. Herskowitz, and J. B. Hicks. 1979. Isolation of a circular derivative of yeast chromosome III: implications for the mechanism of mating type interconversion. *Cell* **18:** 309–319.

Strathern, J. N., E. Spatola, C. McGill, and J. B. Hicks. 1980. Structure and organization of transposable mating type cassettes in *Saccharomyces* yeasts. *Proc. Natl. Acad. Sci. USA* **77:**2839–2843.

Szostak, J. W., T. L. Orr-Weaver, R. J. Rothstein, and F. W. Stahl. 1983. The double-strand-break repair model for recombination. *Cell* **33:**25–35.

Takano, I., and K. Arima. 1979. Evidence of insensitivity of the α-*inc* allele to the function of the homothallic gene in *Saccharomyces* yeasts. *Genetics* **91:**245–254.

Takano, I., T. Kumasi, and Y. Oshima. 1973. An α mating-type allele insensitive to the mutagenic action of the homothallism genes system in *Saccharomyces cerevisiae*. *Mol. Gen. Genet.* **126:**19–28.

Tanaka, K., T. Oshima, H. Araki, S. Harashima, and Y. Oshima. 1984. Mating type control in *Saccharomyces cerevisiae*: a frameshift mutation at the common DNA sequence, X, of the *HMLα* locus. *Mol. Cell. Biol.* **4:**203–211.

Thompson, C. B., and P. E. Neiman. 1987. Somatic diversification of the chicken immunoglobulin light chain gene is limited to the rearranged variable gene segment. *Cell* **48:**369–378.

Weiffenbach, B., and J. Haber. 1981. Homothallic mating type switching generates lethal breaks in *rad52* strains of *Saccharomyces cerevisiae*. *Mol. Cell. Biol.* **6:**522–534.

Weiffenbach, B., and J. E. Haber. 1985. Homothallic switching of *Saccharomyces cerevisiae* mating type genes by using a donor containing a large internal deletion. *Mol. Cell. Biol.* **5:**2154–2158.

Weiffenbach, B., D. T. Rogers, J. E. Haber, M. Zoller, D. W. Russell, and M. Smith. 1983. Deletions and single base pair changes in the yeast mating type locus that prevent homothallic mating type conversions. *Proc. Natl. Acad. Sci. USA* **80:**3401–3405.

Wilson, K. L., and I. Herskowitz. 1984. Negative regulation of *STE6* gene expression by the α2 product of *Saccharomyces cerevisiae*. *Mol. Cell. Biol.* **4:**2420–2427.

Winge, O., and C. Roberts. 1949. A gene for diploidization of yeast. *C.R. Trav. Lab. Carlsberg Ser. Physiol.* **21:**77.

Chapter 15

Chromosome Synapsis and Meiotic Recombination

Craig N. Giroux

I. INTRODUCTION

At the molecular level, eucaryotes have evolved two distinct but overlapping sets of functions that ensure the faithful transmission of genetic information to the next generation. One set of functions is utilized by somatic cells to maintain genetic fidelity during the development, maturation, and growth of an organism. A second set of functions is used specifically by germ line cells to maintain genetic fidelity during gametogenesis and sexual reproduction. The fundamental genetic process operating in germ line cells is meiosis, whereby a diploid cell differentiates into a haploid cell. Reduction to the haploid state combines a cellular gametic differentiation, which exhibits great diversity among eucaryotes, with a highly conserved nuclear chromosomal differentiation. In addition to chromosome reduction, the major genetic consequence of meiosis is recombination resulting from the independent assortment of parental chromosomes during segregation accompanied by changes in the linkage relationships of genes residing·on homologs (crossing-over and gene conversion). Low levels of somatic recombination are important both for repair of chromosomal damage and for specific cellular differentiation events (often site specific) (see Engler and Storb, this volume). However, among eucaryotes elevated levels of general recombination are generally restricted to and specifically regulated in germ line cells undergoing meiotic differentiation.

Four general features distinguish germ line cells undergoing elevated levels of meiotic recombination from somatic cells, which exhibit the less frequent mitotic recombination events. (i) Meiotic recombination is regulated at the whole genome level as well as at the level of individual chromosomes. At least one recombination event (crossover) occurs between each pair of parental autosomal homologs (bivalent) in a meiotic nucleus; recombinant chromatids subsequently segregate from each other at the first meiotic division. Mitotic recombination usually occurs between only one pair of parental homologs and rarely, if ever, involves the entire chromosome complement of a somatic nucleus; recombinant homologous chromatids segregate independently of each other at the next mitotic division. (ii) Meiotic recombination occurs at the four-strand stage, whereas mitotic recombination may occur at the two- or four-strand stage. (iii) All parental homologs exhibit substantial physical interactions (homologous pairing) which are highly regulated during meiosis, whereas in most organisms somatic pairing is infrequent, if detectable at all, and often occurs between largely nonhomologous chromosomes. (iv) A unique chromosome-associated structure, the synaptonemal complex (SC), is elaborated in meiotic prophase nuclei and forms the axis of the paired homologs during close physical alignment (synapsis). Synapsis is not observed between homologs in somatic nuclei.

It is instructive to focus on these unique aspects of meiotic recombination in terms of the nuclear machinery that is observed or presumed to act on meiotic chromosomes. Examination of cell cycle mutants (Simchen, 1974, 1978) demonstrates that gene products are recruited from the mitotic cell cycle to serve an analogous role in meiosis. Genes required for normal DNA metabolism or repair in the cell cycle are also required for DNA or chromosomal metabolism in meiosis (Baker et al., 1978, 1980; Game, 1983). In addition, there are meiosis-specific gene products required for meiotic recombination and proper chromosome behavior (for example, *spo11* of *Saccharomyces cerevisiae*, discussed later). The SC is the prominent meiosis-specific nuclear organelle associated with paired chromosomes during prophase, suggesting that its composition, its morphogenesis, or both require these meiosis-specific gene products. Since molecular recombination between parental chromatids occurs at the DNA strand level during prophase (Borts et al., 1986), the SC is implicated by correlation, although not causally, in the mechanics of meiotic recombination. This poses the fundamental problem. How does the machinery that operates on chromosomes during prophase I relate to the final genetic consequences of meiosis: recombination and reduction?

This review focuses on the relationships among the SC, chromosome pairing and synapsis, and meiotic recombination. Two other chapters in this volume, those by Carpenter and by Hawley, address related aspects of structure and function in meiotic recombination. An excellent description of the classical cytogenetic features of meiosis can be found in the review by Rhoades (1961), and recent work is summarized in *Controlling Events in Meiosis* (Evans and Dickinson, 1984) and in *Meiosis* (Moens, 1987). Additional relevant reviews are of the structural biology of meiotic prophase (von Wettstein et al., 1984; Zickler, 1984; Holm, 1985), the structure of the SC (Moses, 1968; Wettstein and Sotelo, 1971; Westergaard and von Wettstein, 1972; Gillies, 1975a, 1984), the cytogenetics of crossing-over (Henderson, 1970; Peacock, 1971; Comings and Okada, 1972; Moens, 1973b, 1978), the genetic control of meiosis and chromosome pairing (Sears, 1976; Baker and Hall, 1976; Baker et al., 1976; Lindsley and Sandler, 1977; Golubovskaya, 1979; Gottschalk and Kaul, 1980a, b; Koduru and Rao, 1981; Esposito and Klapholz, 1981; Moens, 1982; Dawes, 1983; Carpenter, 1984), and the physiology and biochemistry of meiosis (Pukkila, 1977; Stern and Hotta, 1980; Stern, 1986; Dickinson, 1987).

II. CHROMOSOME CONJUNCTION: PAIRING AND SYNAPSIS

Any consideration of the mechanics of recombination must focus on an initial recognition of homology between the two participants in the event. At the DNA duplex level, the complementarity of opposite strands of a duplex provides an inherent mechanism of homologous recognition. At this first level of homology recognition, "DNA synapsis" may be naively considered as a dissociation and reassociation of double-stranded DNA molecules analogous to an in vitro hybridization reaction. Although necessary, it is not obvious that DNA strand complementarity is sufficient to explain meiotic chromosome conjunction and bivalent formation. At this second level of homology recognition, chromosome pairing initially colocalizes and orients homologs, and a progressively closer association follows, leading to their intimate alignment and the assembly of the SC (synapsis). Chromosome conjunction, homology recognition at the cytological level, is the subject of this review. A third level of homology recognition is evident in the meiosis of eucaryotes with complex genomes, such as certain hybrid polyploid plants, in which sets of chromosomes preferentially form metaphase bivalents with each other (homologous chromosomes) and not with partially homologous chromosomes of a presumed divergent evolutionary background (homocologous chromosomes) (Riley and Law, 1965; Sears, 1976). These three hierarchical levels of homology recognition are separable by genetic analysis: there are mutants which reduce or abolish all recombination (presumed defects at the DNA duplex level, for example, *rec1* of *Ustilago maydis*), mutants which abolish only meiotic synapsis and recombination (for example, *spo11* of *S. cerevisiae* and mutants which abolish only homoeology barriers (for example, Ph of wheat).

Conjunction, the precise side-by-side parallel alignment of homologs to produce a bivalent during meiotic prophase, includes both pairing and synapsis. The progress of chromosome pairing and synapsis has been well documented cytologically by light microscopy in a large number of diverse eucaryotes and serves to define the sequential stages of meiotic prophase (see Fig. 1) (for examples, see Wilson, 1925; Rhoades, 1950, 1961; Klásterská and Ramel, 1979). The discovery of the SC by electron microscope examination of sectioned meiotic nuclei added a new dimension to our understanding of chromosome behavior (Moses, 1956, 1958; Fawcett, 1956) and raised the question of whether homolog recognition is mediated or facilitated by this meiosis-specific organelle associated with intimately paired homologs (bivalents).

A representation of SC structure and morphogenesis is presented in Fig. 2. In

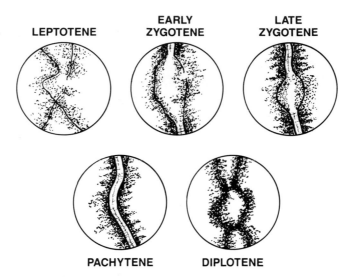

FIGURE 1. Generalized view of chromosome pairing and synapsis during the sequential stages of meiotic prophase I as observed by microscopy. A single pair of homologs is depicted as it progresses from presynapsis (chromosome condensation and pairing; leptotene and early zygotene) to synapsis (formation and association of the lateral elements with the central element to form the SC; zygotene and pachytene) and desynapsis (dissolution of the SC with the observation of chiasmata joining the two homologs together; diplotene). Chromatin condensation and formation of the proteinaceous axial element render duplicated chromosomes evident early in prophase, thus allowing the progress of synapsis to be observed. Chiasmata provide a physical demonstration of crossing-over between the homologous partners of a bivalent, but the precise stage of prophase at which exchange occurs is not known. (After a diagram provided by M. Moses.)

most organisms, chromosome condensation renders the homologs evident by light or electron microscopy just prior to or coincident with formation of the axial element which forms along each duplicated chromosome (leptotene stage, Fig. 2A and B). Chromosomal associations and pairing between homologs become obvious as synapsis initiates along the closely aligned axial elements, which now form the lateral elements of the SC (zygotene stage, Fig. 2B and C). The axial elements and lateral elements provide a reliable and faithful indication of the disposition of homologs and their interactions during the early events of meiosis (for examples, see Moses et al., 1984; Rasmussen and Holm, 1980; Gillies, 1975b; Moens, 1973a). Complete synapsis results in a continuous tripartite SC with a diameter of approximately 200 nm, composed of two lateral elements flanking and cross-connected by transverse filaments to a central element (pachytene stage; Fig. 2D and 3). Subsequently, desynapsis of the bivalents occurs, and chiasmata, presumed sites of crossing-over between the homologs, become apparent (diplotene stage, Fig. 2E and F). In some organisms a transient "diffuse" stage of chromatin decondensation occurs followed by a recondensation (diakinesis stage) and subsequent metaphase I segregation. Chromosome conjunction thus reflects the interaction of at least three discernible processes: condensation, pairing, and synapsis.

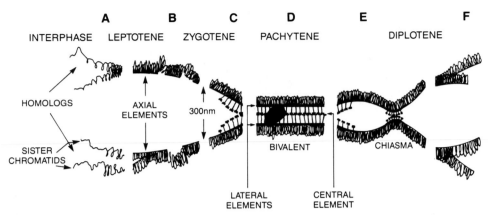

FIGURE 2. Representation of SC structure and morphogenesis derived from cytological data obtained in many species. (A) Duplicated sister chromatids begin condensation, which continues through most or all of meiotic prophase. (B) Axial elements form along each homolog, and portions of homologs initiate pairing at a limited number of sites. The distance separating homologs progressively diminishes, and the length of their regions of continuous pairing increases. (C) As homolog pairing becomes closer than approximately 300 nm, a central element forms between the two parallel aligned axial elements, which now form the lateral elements of the SC. (D) At complete synapsis, the SC has a ribbonlike appearance. Thin transverse filaments connect the central element to the lateral elements. During zygotene and pachytene, globular or rod-shaped nodules are frequently associated with the central region of the SC. (E) Desynapsis proceeds by an apparent reversal of synapsis. Dissolution of the central region is followed by separation of the lateral elements except in regions of chiasmata, where remnants of SC may be observed. (F) The individual axes of the sister chromatids can now be distinguished, but their relationship to the lateral elements of the SC or to the mitotic chromosome scaffold is unclear. Homologs remain joined at chiasmata, and sister chromatids remain associated except in the immediate vicinity of chiasmata.

III. CHROMOSOME PAIRING

Most considerations of homolog recognition have distinguished two phases of chromosome conjunction: a long-distance (>300 nm) interaction of two parental homologs that colocalizes or orients them nonrandomly with regard to each other in the meiotic nucleus (leptotene to zygotene stages) and a subsequent close or intimate (<300 nm) interaction that precisely aligns homologs with each other to form a bivalent (zygotene to pachytene stages). The intimate interaction is typically inferred to occur when the axes of homologs are similarly oriented and within 300 nm of each other; this interaction is coincident with assembly of the SC. For convenience, I shall refer both to the long-distance interaction and to the close interaction in association with SC assembly as chromosome pairing. Chromosome synapsis will refer only to assembly of the SC.

Direct cytological observation of the initiation of chromosome pairing and synapsis is not possible until meiotic condensation has proceeded sufficiently to spatially limit (define) the individual chromosomes. These events coincide or overlap during the early prophase stages of leptotene and zygotene, making their relationships difficult to decipher. At present, cytologically observed chromosome

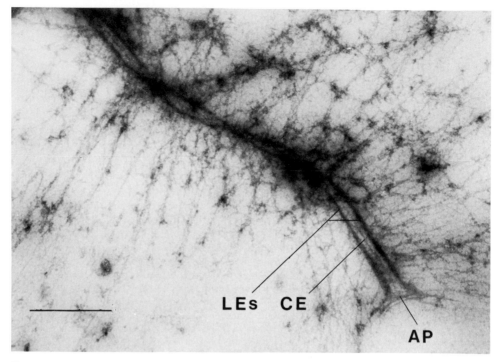

FIGURE 3. Electron micrograph of a whole-mount preparation from a Syrian hamster spermatocyte illustrating one end of a pachytene autosomal bivalent. The surface-spread preparation was stained with uranyl acetate. The parallel lateral elements (LEs) are sharply defined and allow the twists of the ribbonlike SC to be traced. The linear central element (CE) of the SC is also evident and is connected to the lateral elements by thin transverse filaments. A thickened attachment plaque (AP) anchors the bivalent to the nuclear envelope. Chromatin fibers are observed as flattened loops which radiate from the lateral elements of both homologs. Bar, 1 μm. (Reproduced from Moses and Solari, 1976.)

interactions are difficult to correlate with the presumed physical interactions of homologous DNA sequences. Thus, it is uncertain whether molecular recombination at the DNA level precedes or follows chromosome pairing and/or synapsis. This uncertainty reflects a longstanding fundamental question. What are the physical forces responsible for the attraction and subsequent interactions of homologous chromosomes? An extreme reductionist viewpoint would be that the random diffusion of DNA within a meiotic nucleus is both necessary and sufficient to provide the recognition of homology which results in chromosome pairing and synapsis at pachytene. This view ignores the possible constraint of progressive chromosome condensation on the free diffusion of homologous DNA sequences during meiotic prophase. An alternative extreme view would hold that the size and mass of meiotic chromosomes require that new long-range forces of attraction be invoked to account for pairing and synapsis (for example, see Fabergé, 1942). This concern has led to many models for somatic pairing or prealignment of meiotic homologs which essentially reduce the physical problem of homology recognition at the chromosome level to a one- or two-dimensional search instead of a three-dimensional search (telomere alignment, nuclear envelope attachment) or push the problem back one step by suggesting that they are already paired (for discussion, see Gillies, 1984; Zickler, 1984). Clearly, neither of these two extremes is appealing.

To address the mechanism(s) of homolog recognition, one needs comprehensive quantitative data on chromosome pairing and the early events of prophase, the leptotene and zygotene stages. Unfortunately, few organisms allow all stages of meiotic prophase to be observed with high resolution. To some extent this is a limitation of the preparative and cytological techniques which have been used. Smear or squash techniques and chromogenic dye staining of chromosomes (orcein or carmine, for example) allowed many of the original critical observations of meiotic chromosome behavior to be made by light microscopy, particularly on plant material. Light microscopy using these methods gives poor resolution of individual chromosome axes from the collapsed mass of chromatin during the initiation of pairing and synapsis, however. Serial section reconstruction of meiotic nuclei observed by electron microscopy has been profitably applied to analysis of SC behavior, but this is time-consuming and thus limits the number of cells which can be examined.

Recent development of surface-spreading methods that display the entire contents of a nucleus have greatly facilitated the rapid examination of large numbers of nuclei at sequential stages of meiotic prophase I (Counce and Meyer, 1973; Moses, 1977; Stack, 1982; Gillies, 1983a; Albini and Jones, 1984; Albini et al., 1984; Stack and Anderson, 1987). Spreading methods have been particularly effective in combination with improvements in fixation and silver staining which allow the SC and meiotic chromosome axes (axial element or lateral element) to be traced in spite of their association with large and relatively disperse amounts of chromatin (Dresser and Moses, 1979, 1980; Pathak and Hsu, 1979; Fletcher, 1979; Forejt and Goetz, 1979). It is anticipated that these improved methods of cytological examination will be particularly useful when applied to the analysis of meiotic mutants which change parameters of genetic recombination (Zickler and Simonet, 1981; Moens, 1982; Carpenter, 1984). In spite of these new cytological methods, several of the organisms most amenable to analysis by classical genetics and by cytogenetics at pachytene, such as maize and *D. melanogaster*, have pairing stages that are difficult to resolve in early meiotic prophase. In these cases, details of the behavior of individual chromosomes are obscured by their association into a dense clump, a synizetic knot and chromocenter, respectively (Rhoades, 1950; Gillies, 1975b; Nokkala and Puro, 1976; Dävring and Sunner, 1979).

Synapsis is first observable as a limited segment of assembled SC which intimately links the two homologs forming a meiotic bivalent. This initial segment of assembled SC may be telomeric or interstitial, the pattern and approximate number of segments being specific for each organism. Although a temporal progression of synapsis is difficult to determine for most organisms, increasingly more extensive segments of assembled SC are observed throughout zygotene until complete synapsis at pachytene yields continuous SCs which define the axes of the paired homologs (see, for example, Hasenkampf, 1984a; Jones and Croft,

1986; Holm, 1986). This typical and general pattern of SC assembly has been inferred to result from initiation at a limited number of nucleation sites, which subsequently elongate and join together to form a continuous SC. For interstitial nucleation of SC assembly, there is as yet no unambiguous evidence for specific sites of SC initiation. (Such postulated sites are frequently referred to as "pairing sites" in the literature. Unfortunately, the term "pairing site" has also been used in several other contexts.)

Regardless of whether SC assembly sites are specific or randomly located, the pertinent question is how they relate to the distribution of recombination events in the genome. This question has rarely been addressed, as it requires a detailed observation both of zygotene nuclei and subsequently of diplotene nuclei, where dissolution of the SC renders the distribution of chiasmata evident. In a recent study of two *Allium* species, using silver-stained, surface-spread preparations, no correspondence was found between the sequence of SC initiation and the distribution of chiasmata (Albini and Jones, 1987). This result may be misleading, however, since the number of intersititial SC initiations can far exceed the number of chiasmata (Holm, 1977; Gillies, 1985; Jones and Croft, 1986). If only a fraction of these SC initiations is effective in promoting crossing-over, as has been proposed, then this subclass of initiation sites could still be a significant determinant of the distribution of reciprocal recombination events (chiasmata).

IV. CHROMOSOME SYNAPSIS

The majority of the chromatin content of the two parental chromosomes composing a meiotic bivalent is located exterior to the SC, where it has often been observed to be arrayed in lateral loops anchored along the lateral elements (see Fig. 3; for reviews, see Luykx, 1974; Henderson, 1971; Weith and Traut, 1980; Rattner et al., 1980). Thus, a relatively small amount of chromosomal DNA is localized in direct association with the proteinaceous components of the SC, presumably at the anchorage sites of the chromatin loops. Evidence for the presence of DNA in the central region of the SC has been equivocal (Coleman and Moses, 1964; for discussion, see Moses, 1968). However, if there is DNA associated with the linear structure of the SC which is colinear with the bivalent axis, then it may be calculated that it composes a subset of between 0.006 (for lily) and 0.3% (for *S. cerevisiae*) of the DNA content of the genome (total length of SC divided by total DNA content expressed as length of free DNA [for discussion, see von Wettstein et al., 1984; Anderson et al., 1985]). This calculation should be considered approximate and underestimated, as it is likely that a putative chromatin thread colinear with the SC would have an organization or packing ratio of greater condensation than that of naked DNA in solution (for discussion, see Risley, 1986).

Virtually nothing is known about the disposition or organization of the small amount of chromosomal DNA directly associated with the SC. It has often been suggested that this small fraction of DNA consists of a specific set of pairing or synaptic initiation sites which promote synapsis through direct interaction with the protein components of the SC or by serving as foci of nucleation for homologous DNA interactions (for examples, see Sybenga, 1966; Holliday, 1977). Alternatively, random and not unique DNA sequences may be directly associated with the protein components of the SC at these proposed sites of synaptic initiation. Since early synapsing regions of homologs are observed at a stage when there is little apparent linear differentiation of the chromosomes, it has not been possible to determine whether sites of early synapsis are specific; this remains a critical unanswered question. In spite of the absence of data demonstrating the existence of specific sites of synaptic initiation, they are implicit in most contemporary models which attempt to unify chromosome synapsis and molecular recombination (for some examples, see Comings and Riggs, 1971; Holliday, 1977; Maguire, 1984; Chandley, 1986).

Local differences in structure along the length of the SC have been observed in many organisms (for review, see Dresser, 1987). In the extreme case, regions of consistent asynapsis such as the centromeres in *Allium ursinum* (Loidl, 1987) or the nucleolus organizer in *S. cerevisiae* (Dresser and Giroux, 1988) correlate with the observed local deficiency in meiotic recombination for these chromosomal loci. Other local differentiations such as relative thickenings of the lateral elements (Hasenkampf, 1984b) exhibit stage dependence, illustrating the dynamic nature of the SC, but a precise correlation with localized recombination behavior has not been determined.

The most promising correlation of a stage-dependent structural feature with recombination behavior is that of the "recombination nodule," a labile particle associated with the central element during SC morphogenesis (for a review, see Carpenter, this volume). Following the initial suggestion of the recombination nodule associated with synapsed homologs at pachytene in *D. melanogaster* females (Carpenter, 1975), its presence and distribution have been well correlated with the behavior of chiasmata in several organisms. This indirect evidence has led to the proposal that the recombination nodule is a large multienzyme "recombination complex" which operates on the intimately associated homologous chromosomes to effect molecular recombination at the DNA level. Neither the structural composition nor the function of the recombination nodule is known, however. A typical nodule has an approximate diameter of 90 nm (ranging from 30 to 200 nm depending on the stage of prophase and the species) and presumably is proteinaceous, as determined by its efficient staining with phosphotungstic acid and uranyl acetate-lead citrate. In contrast, silver nitrate staining rarely reveals nodules, although lateral elements are well resolved. This result suggests that this harsher staining procedure disrupts the association of the nodules with the assembled SC while preserving the more stable association of the synapsed lateral elements.

Recently, surface-spreading and compatible-staining methods have been developed which allow preservation and visualization of nodules associated with early stages of SC assembly prior to complete synapsis at pachytene (Solari, 1980; Stack, 1982; Gillies, 1983a; Albini and Jones, 1984). Examination of the early stages of SC morphogenesis in spread preparations of higher plants by these methods forces a reconsideration of the simple interpretation of a nodule as a recombination complex (for examples, see Stack and Anderson, 1986b; Albini and Jones, 1987). Prior to pachytene, nodules are observed in association with unpaired axial elements (leptotene stage) and in far greater abundance than the number of chiasmata later determined at diplotene. Presumptive regions of synaptic initiation are observed in the partially assembled SC at the subsequent zygotene stage, and these early synapsing regions are frequently associated with a "zygotene nodule." These observations raise the alternative possibilities that nodules have a role in the initiation of chromosome synapsis and the alignment of homologous chromosomes which could be either distinct from or directly related to their previously proposed involvement in the enzymology of recombination at the DNA duplex level (for further discussion, see Albini and Jones, 1987; Carpenter, 1987 and this volume).

Many of the studies which have examined chromatin structure in meiosis have focused on the transition from the physiologically active somatic state to the highly condensed dormant state present in a gamete (for review, see Meistrich and Brock, 1987). In mammalian spermatogenesis, the prominent feature of this process is a sequential substitution first of meiosis-specific histones and subsequently of protamines for the initial set of somatic histones (Brock et al., 1980; Rao et al., 1983). The consequence of this substitution is that during pachytene the nucleosome structure is more open (i.e., accessible to nuclease). Additionally, there is evidence for a change in overall DNA topology which serves to unfold nucleosomal DNA as spermatogenesis proceeds (Risley et al., 1986). Whole-mount electron microscopy has revealed a lampbrush-type organization of chromatin loops in late meiotic prophase and metaphase I for amphibian oocytes as well as for male meiosis in some insects, plants, and animals; this lampbrush organization is involved in active transcription required for subsequent development (Callan, 1982; Rattner et al., 1980; Henderson, 1971). These changes are consistent with gametogenic differentiation; there is no evidence at present that they reflect or are required for the activities of homologous chromosomes undergoing synapsis. Although synapsing chromosomes are condensed, it is unknown whether this state of chromosome organization contributes to the process of synapsis.

V. GENETIC INFERENCES ON CHROMOSOME SYNAPSIS

It is reasonable to assume that the genetic recombination events observed in meiotic progeny are a consequence both of the mechanics of chromosome pairing and synapsis as well as of the enzymology of molecular recombination operating at the DNA level. If this is true, then the distribution and extent of molecular recombination events along a chromosome should reflect the "effective pairing" of the two homologs bound by the SC to form a bivalent (Pritchard, 1960; Holliday, 1964). The underlying assumption of this argument is that chromosome synapsis initially juxtaposes homologous DNA sequences, providing the opportunity for their subsequent molecular recombination. By this assumption, the distribution and physical extent of recombination events are determined by the effectively paired regions of synapsis. Thus, a fine-structure genetic analysis of meiotic recombination events should reflect the fine structure of chromosome synapsis. Since it is uncertain whether synapsis precedes or determines molecular recombination, the concept of effective pairing remains abstract.

This concept has been developed as a theoretical approach for using gene conversion data from fungi to infer the close interactions of chromatids held together by the SC (Lamb and Wickramaratne, 1973; Lamb, 1977; Howell and Lamb, 1984). Since gene conversion frequencies at some loci are as high as 20%, this implies that the two parental homologs were effectively paired at this site in at least 20% of all meioses, a value much greater than would be expected from any hypothesis that assumes a limited size and random choice of effective pairing sites. To accommodate gene conversion frequencies of 20% with an estimated SC-associated DNA fraction of at most 0.3% (calculated value for *S. cerevisiae*; see previous discussion), one must conclude either that there are unique sites of synaptic initiation or that the DNA strand interactions leading to recombination occur predominantly in the DNA not directly associated with the SC. Since gene conversion tracts can routinely extend over distances of several kilobases, at least in fungi (Fogel et al., 1983; Symington and Petes, 1988), it is possible that recombination events or "effective pairing lengths" initiate on the DNA directly associated with the SC but extend processively to the DNA located external to the SC. As illustrated from this argument, although gene conversion data may provide tantalizing hints of synaptic behavior, it is not possible to interpret these data in terms of chromosome pairing and synapsis without making several untested simplifying assumptions (for other examples, see Welch et al., 1987; Maloney and Fogel, 1987). Such arguments can serve to identify critical questions, however. In this specific example, it is critical to know whether or not synapsis precedes molecular recombination.

An additional complication of the genetic approach to synapsis is provided by the observation that meiotic recombination readily occurs between directly repeated homologous sequences located 30 to 60 kilobases apart in the parental chromosomes of *D. melanogaster* (Goldberg et al., 1983; Davis et al., 1987). In these studies, duplications and deficiencies resulting from reciprocal exchange between transposon insertion alleles of the *white* locus are most readily interpreted as having arisen from unequal crossing-over associated with asymmetrical pairing of the homologous transposons. Since these events occur at a significant frequency relative to normal homologous exchange at the *white* locus, these data have been interpreted as suggesting that effective pairing during meiosis is "flexible" within an extended region of homology on the two synapsed homologs.

A similar result was observed for directly repeated Ty elements of *S. cerevisiae*, separated by an interval of 21 kilobases on chromosome III (Roeder, 1983). In this study, unequal reciprocal exchanges occurred at a frequency of 1% in meiosis, 100,000 times more frequently than in vegetative growth. It is not clear whether these phenomena reflect an intrinsic "flexibility" of homolog pairing and synapsis or whether they represent an additional contribution of a recombination behavior specific to transposons or repeated sequences. In this regard, "foldback synapsis" of the same lateral element with itself has been observed in zygotene plant meioses prior to complete homologous synapsis at pachytene (Holm, 1977; Hasenkampf, 1984a) and in late pachytene of triploid meioses for several organisms (Rasmussen, 1977; Loidl and Jones, 1986). This transient foldback, also observed in situations in which normal homologous pairing is disrupted in plants or rodents (Gillies, 1974, 1983b; Allen et al., 1987), may reflect the recognition of imperfect or local homology between related sequences located on the same homolog. Asymmetrical pairing of transposons and foldback synapsis illustrate the complexity of the homology search which two parental homologs must undergo in the process of faithful synapsis.

VI. CHROMOSOME SYNAPSIS AND MEIOTIC RECOMBINATION

The relationship between chromosome synapsis and meiotic recombination is further complicated by the phenomena described as "ectopic recombination" or "intergenic conversion," which refer to recombination events between closely related or allelic DNA sequences located on nonhomolog chromosomes (Slightom et al., 1980; Munz et al., 1982; Jinks-Robertson and Petes, 1985; Lichten et al., 1987). Although these events have been inferred by sequence comparisons of related gene families in mammals (for discussion, see Powers and Smithies, 1986), it is only in fungi that they have been directly observed. In the yeasts *S. cerevisiae* and *Schizosaccharomyces pombe*, ectopic recombination is detected both as nonreciprocal gene conversion events and as gene conversion events in association with reciprocal exchange which generates translocation chromosomes (Szankasi et al., 1986; Jinks-Robertson and Petes, 1986; Lichten et al., 1987). Ectopic recombination and homolog recombination are both stimulated 100- to 1,000-fold during meiosis, and both occur at the four-strand stage after premeiotic DNA synthesis (Jinks-Robertson and Petes, 1986; Junker et al., 1987).

Surprisingly, ectopic recombination between artificially constructed substrates in *S. cerevisiae* is comparable to homolog recombination with respect both to meiotic frequency and to the association of reciprocal exchanges with conversion events. In contrast, ectopic recombination during meiosis between endogenous dispersed repeated genes in these yeasts (Ty transposon in *S. cerevisiae* [Kupiec and Petes, 1988] and tRNA genes in *S. pombe* [Junker et al., 1987]) is rare in comparison with homolog recombination and is predominantly represented by gene conversion events without an associated reciprocal exchange. Since translocations are rarely observed during meiosis, these artificially constructed ectopic recombination substrates clearly behave differently than the endogenous dispersed yeast genes.

One possible resolution of these conflicting data is that dispersed repeated genes are subject to a suppression of normal meiotic recombination which serves to maintain the structural fidelity of the genome, whereas the artificial constructions are not subject to this hypothetical regulation. The precise location in the yeast genome of artificially constructed allelic sequences can vary their frequency of meiotic homolog gene conversion by about 40-fold, suggesting a second explanation for the differences in ectopic recombination data. The artificial constructions which exhibit high frequencies of ectopic recombination may be located in regions of the yeast genome which determine a recombination behavior distinct from those regions harboring endogenous repeated sequences (Lichten et al., 1987). In this regard, the tandemly repeated ribosomal cistrons in *S. cerevisiae* also exhibit a reduced level of meiotic homolog recombination and are the only region of the yeast genome exhibiting consistent asynapsis during meiosis (Petes et al., 1982; Dresser and Giroux, 1988).

If, however, the artificially constructed yeast recombination substrates accurately reflect the ability of their local "neighborhood" to engage in ectopic recombination and nonhomolog interactions, this raises the speculation that these events are a normal aspect of the homology search which must occur during meiotic recombination. Other lines of reasoning have similarly suggested that an early step in homolog recognition may be a testing for DNA sequence relatedness, the consequence of which would be chromosome synapsis and a subsequent high probability of meiotic crossing-over if the two interacting DNA sequences are located in the same relative position on homologous chromosomes. If the two related sequences are located in different regions of the genome, either on different homologs or in different positions on the same homolog, then their failed interaction may result in a gene conversion event without an associated reciprocal exchange (Powers and Smithies,1986; Smithies and Powers, 1986; Carpenter, 1987). By this interpretation, much or all of the elevated levels of meiotic recombination observed between related sequences in different chromosomal locations would be dependent on the same homology recognition mechanism which results in homolog pairing and synapsis.

Most molecular models for recombination account for gene conversion and reciprocal exchange as coupled events resulting from a common intermediate; a single initiating event can be resolved to yield either genetic outcome (see Hastings, this volume). However, if gene conversion events reflect a search for homology as a prelude to pairing and synapsis (leptotene-zygotene stages) whereas crossing-over events are observed as chiasmata at diplotene, then the relationship between "cytologically" defined conversion and reciprocal exchange is not obvious. Two lines of genetic inference from fungi suggest that there may be a mechanistic and/or temporal separation of gene conversion from crossing-over. First, reciprocal crossing-over events show interference with each other, but nonreciprocal gene conversion events do not show interference with crossovers (for discussion, see Mortimer and Fogel, 1974; Holliday, 1977). Second, detailed analyses of the fine structure of gene conversion events and associated reciprocal exchanges of flanking markers identify a class of crossover which is difficult to explain (by current models) as resulting from the same event that generated the adjacent gene conversion (for discussion, see Theivendirarajah and Whitehouse, 1983; Rossignol et al., 1984; Hastings, 1987). The simplest interpretation of these data is that a reciprocal exchange has occurred preferentially near a gene conversion event. These genetic inferences are consistent with, but do not prove, the view that gene conversion precedes associated crossing-over during fungal meiosis. Alternatively, these data are also consistent with two classes of gene conversion, one of which precedes and one of which is coincident with crossing-over.

These difficult-to-interpret genetic inferences from fungi become even more intriguing in the light of recent observations on synaptic initiation in higher plants. Observations of leptotene-zygotene pairing suggest that there are two distinct physical types of "SC-associated nodule" (see previous discussion of chromosome synapsis; Stack and Anderson, 1986a, b; Albini and Jones, 1987; Carpenter, 1987). The first type is abundant (in excess of the number of chiasmata present later at diplotene), is observed during leptotene-early zygotene, and is associated with partially assembled bivalents at the sites of synaptic initiation. The second type is less abundant (consistent with the number of expected chiasmata), is observed during late zygotene-pachytene, and is associated with more completely assembled SCs. It has been suggested that the first type of nodule is a precursor to the second; the relationship between these two types of nodule remains to be demonstrated, however (reviewed by Carpenter, this volume). The combination of these recent cytological observations with genetic inferences raises the speculation that the "early nodule" promotes (or is associated with) gene conversion and that the "late nodule" promotes (or is associated with) reciprocal exchange leading to the formation of a chiasma (Carpenter, 1987). At present, it is not possible to directly test this or similar suggestions, as there is no means of assigning a function to the SC structures observed at different stages of synapsis. Potentially, the correlation of recombination behavior with SC morphogenesis in meiotic mutants can address this problem.

This type of gene-conversion-as-homology-search model can account for the abundant "nonhomologous" interactions observed cytologically during the early stages of meiotic prophase of numerous organisms (McClintock, 1933; Douglas, 1966; Driscoll et al., 1979; Gillies, 1983b; Bojko, 1983). These nonhomologous interactions could reflect a search for homology that has found a localized region of DNA sequence similarity between two nonhomologs. Such early prophase interactions could also provide a basis for ectopic recombination if they can be demonstrated during fungal meiosis.

VII. ASSEMBLY AND MORPHOGENESIS OF THE SC

In spite of the nonhomologous interactions observed in early prophase, at the completion of synapsis in pachytene it is rare to observe SC formation between nonhomologs in a normal diploid organism; the fidelity of SC formation is very high. However, in genomes with multiple homologs (triploids, autotetraploids) or in haploids without homologs, chromosome pairing and even synapsis regularly occur between nonhomologs (for review, see von Wettstein et al., 1984; Gillies, 1984; Zickler, 1984). In triploid *Bombyx mori* (silk moth) oocytes (which form SCs but lack meiotic recombination and are achiasmate), trivalents with switches of pairing partners are observed frequently during the earlier stages of synapsis, followed by a lower proportion of trivalents (and correspondingly a greater number of bivalents) during late pachytene (Rasmussen, 1977). Surprisingly, the univalents (presumably released from earlier trivalent associations) observed during late pachytene undergo a high proportion of "nonhomologous" interactions, such as foldback synapsis and complex multipartner interactions. These results have led to the hypothesis that there is "two-phase pairing" during synapsis (for review, see Rasmussen and Holm, 1980).

In this model, the first phase of synapsis occurs during the leptotene-zygotene stages of prophase and is homologous; in a triploid, interactions among more than two homologs can form trivalent associations. These trivalent associations, as well as physical interlockings between synapsing homologs, are subsequently resolved to give completely synapsed bivalents and unpaired single chromosomes (univalents). The second phase of synapsis is postulated to occur during early to midpachytene and is nonhomologous; the free univalents in a triploid are now able to synapse in the absence of an available homolog. Subsequent examination of the stages of meiotic prophase in polyploids of chiasmate organisms supports the existence of a late phase of apparently nonhomologous synapsis; however, in at least one case (*Allium sphaerocephalon*) this nonhomologous synapsis appears restricted to previously unsynapsed regions (Loidl and Jones, 1986). The existence of a correction mechanism that converts multivalent synapsis into bivalent synapsis has been further challenged by quantitative analysis of polyploids in several organisms where multivalents are still observed (albeit reduced in frequency) during later stages of meiotic prophase or metaphase I (Rasmussen et al., 1981; Loidl and Jones, 1986; Loidl, 1986; Gillies et al., 1987). A suggested explanation of these polyploid data from chiasmate meioses is that crossing-over can inhibit the proposed correction process by stabilizing the multivalent synapses (Rasmussen, 1987). This interpretation remains contentious, however, and at present the action of a bivalent correction process in a normal diploid chiasmate meiosis is unsubstantiated (for discussion, see Gillies et al., 1987).

Independent support for two distinct aspects of chromosome synapsis, an initial homosynaptic (homology-dependent) and a subsequent heterosynaptic (homology-independent) phase, is provided by the phenomenon of synaptic adjustment (for review, see Moses et al., 1984). At the zygotene stage of male meiosis in mice heterozygous for chromosome rearrangements (tandem duplications or inversions), the SC is initially "displaced" from a continuous linear structure by the region of heterozygosity and forms a classical unsynapsed loop (Moses, 1977; Poorman et al., 1981; Moses and Poorman, 1981). Subsequent stages of meiotic prophase exhibit a quantitative and progressive decrease in the size of the displaced region of heterozygosity until at late pachytene there is no further evidence of any disturbance in the continuous linear SC of the heterozygous bivalent. This second phase of heterosynapsis has been interpreted as a release from the requirement for homology, since it appears to occur by desynapsis of homosynapsed SC adjacent to the heterozygosity and coincident resynapsis of this chromosomal region into the heterosynapsed SC encompassing the heterozygosity (Moses and Poorman, 1981; Sharp, 1986). This process of desynapsis and resynapsis presumably involves dissolution of the central region of the SC since lateral elements appear unbroken during synaptic adjustment.

Considerable variability has been observed in the extent and kinetics of synaptic adjustment in different species (for examples, see Moses et al., 1984; Maguire, 1981; Gillies, 1983b; Hale, 1986). In a study of male meiosis in the Sitka deer mouse, bivalents of two heterozygous pericentric inversions exhibited immediate heterosynapsis without an initial homosynaptic phase; intriguingly, crossing-over (chiasmata) was not observed in these inversions (Hale, 1986). Although synaptic adjustment clearly demonstrates the dynamic nature of synapsis, whether it reflects a normal aspect of SC morphogenesis or an aspect specific to the processing of structural heterozygosities is unclear.

Assembly and morphogenesis of the SC remain problematic in spite of an increasing body of ultrastructural data describing the leptotene and zygotene stages of meiotic prophase (see Fig. 2). It has been repeatedly observed that the axial elements are assembled (or evident) prior to observation of the central element. One interpretation of these observations is that the intimate parallel alignment of the two axial elements (now termed lateral elements) allows their overlap to be visualized as an apparently discrete fiber, the central element (for discussion, see Comings and Okada, 1972). By this and similar one-step models for SC assembly, the central region would be defined by the interaction of the lateral elements associated with the two synapsed homologs. In contrast, an alternative class of two-step assembly models assumes that the central element is an entity distinct from the lateral elements and is therefore assembled on the paired lateral elements as a subsequent stage of SC morphogenesis (for discussion, see von Wettstein, 1977; Lu, 1984). Unambiguous data discriminating between these alternative classes of models are lacking. Until SC components are identified biochemically, allowing their time of synthesis and their nuclear localization to be examined, a myriad of models for SC assembly may be entertained. By analogy to the analysis of bacteriophage morphogenesis, it is anticipated that meiotic mutants defective in specific components or steps of SC assembly would greatly facilitate the analysis of SC morphogenesis.

VIII. COMPOSITION OF THE SC

The conclusions from numerous examinations of SC preparations subjected to cytochemical stains or enzymatic digestions is that the SC (not including the externally associated chromatin) is largely proteinaceous, has some directly associated DNA, and has little or no directly associated RNA (for discussion, see Moses, 1968; Wettstein and Sotelo, 1971; Westergaard and von Wettstein, 1972; Moses, 1977; Gillies, 1985; Holm, 1985). Since the SC is attached to the homologous chromosomes, clearly both DNA and RNA are associated with the general structure of a bivalent. The density and compactness of the proteinaceous components of the SC have made it difficult to determine unambiguously whether nucleic acid is specifically associated with the longitudinal axes of lateral elements or the central region of the SC; neither of these substructures is sensitive to nuclease treatment in cytological preparations. There is limited evidence for a continuous DNA fiber associated with the longitudinal axes of the lateral elements from DNase treatment of spread SC preparations of rat spermatocytes (Solari,

1972), but this observation is dependent on a specific set of spreading conditions and has not been repeated. Electron microscopy autoradiographic evidence for DNA synthesis associated with the SC is consistent with the direct presence of DNA in the lateral elements or central region but does not have sufficient resolution to allow its precise localization (see, for example, Carpenter, 1981; Moses et al., 1984). Thus, there is no direct evidence of DNA associated with the central region of the SC. Consequently, models for the SC which assume that there is molecular interaction of homologous DNA strands located within the central region are subject to this caveat (for example, models which assume that "recombination nodules" act as macromolecular machines to recombine DNA).

The composition and function of the lateral elements is almost as enigmatic as that of the central region of the SC. When observed by electron microscopy, the lateral elements frequently exhibit a repetitive pattern of striations which is suggestive of the reiterative assembly of a common subunit, implying the presence of a limited number of abundant structural proteins (for examples, see Coleman and Moses, 1964; Zickler, 1973). Cytochemical staining is consistent with the presence of basic nonhistone proteins in the meiotic axes. The lateral elements are stained selectively with silver ions (see previous discussion) and are not disrupted either by nuclease digestion or by extraction with 2 M NaCl (Solari, 1972). These properties are similar to those observed for the mitotic chromosome scaffold and raise the question of the relationship between the two structures. Both structures serve as axes to which chromatin is attached, but there is no evidence that this functional similarity is reflected in structural similarities or in a similar protein composition (for discussion, see Moses, 1977; Risley, 1986). It is unclear whether a single lateral element is composed of or is organized from the mitotic axes of the two sister chromatids (see, for example, Tres, 1977), or whether it has a more complex multistranded organization (see, for example, del Mazo and Gil-Alberdi, 1986).

SC material is shed from meiotic chromosomes at diplotene in some organisms and may be observed as partial aggregates (polycomplexes) either in the nucleus or in the cytoplasm of germ cells (Roth, 1966; Fiil and Moens, 1973; for discussion, see Moses, 1968; Moens, 1973a; Gillies, 1984). The assembly and subsequent disassembly (or shedding) of the lateral elements are consistent with a role for these substructures in the initiation or maintenance, or both, of chromosome synapsis. There is no direct evidence, however, that the lateral elements themselves are the elements which bring the homologs into close alignment (pairing). It has been suggested that contractile proteins such as actin or myosin may act on the lateral elements to drive chromosome synapsis, but immunocytology with antibodies specific for these proteins has failed to support this hypothesis (for discussion, see Spyropoulos and Moens, 1984; Dresser, 1987).

The composition and structure of the SC would be more amenable to biochemical analysis if it were possible to isolate this structure intact and relatively pure from the meiotic nucleus. This has proved to be a difficult task, and at present only partial success has been achieved (for discussion, see Holm, 1985; Risley, 1986; Meistrich and Brock, 1987). Since SCs are found only in meiotic cells at a limited time in their differentiation, it is difficult to prepare a starting cell population which is not heavily contaminated with the germ line tissue surrounding pachytene meiocytes and with meiocytes in other stages of meiosis or gametogenesis. To minimize this problem, several investigators have chosen to use rodent spermatocytes, with which it is possible to achieve a relatively synchronous meiosis by physiological methods and with which tissue disruption and stage-specific cell enrichment are practical (Walmsley and Moses, 1981; Li et al., 1983; Ierardi et al., 1983; Heyting et al., 1985).

A second significant problem is that the SC is a large macromolecular suborganelle with physical associations to other nuclear components such as the nuclear envelope/pore-lamina complex (Comings and Okada, 1970; Engelhardt and Pusa, 1972; Fiil and Moens, 1973). Several investigators have exploited this feature by using nuclear matrix preparations of meiocytes to coisolate SCs (Comings and Okada, 1976; Gambino et al., 1981; Walmsley and Moses, 1981; Li et al., 1983; Ierardi et al., 1983; Raveh and Ben-Ze'ev, 1984). These preparations are heavily contaminated with nuclear lamina-pore complex, "nuclear matrix" residue, nucleoli, and other large insoluble components (acrosomes in testis preparations, for example), and this limits their utility as sources of material for protein analysis by sodium dodecyl sulfate-polyacrylamide gel electrophoresis or for subsequent biochemical analysis. SC preservation in these preparations is inferior to that found in unfractionated samples prepared directly for microscopy, but lateral elements, central elements, transverse filaments of the central region (less frequently), and nuclear envelope attachment plaques are present (Walmsley and Moses, 1981; Ierardi et al., 1983; Raveh and Ben-Ze'ev, 1984). These preparations have been used as evidence for an in vivo association between the SC and the meiotic "nuclear matrix," but it is equally likely that their copurification is a reflection of their similar masses, insolubility, and other physical properties.

Nuclear matrix/SC extractions of mouse or rat spermatocytes have been used as the starting material in studies that attempted to identify polypeptide components of the SC (Ierardi et al., 1983; Heyting et al., 1985, 1987; Behal et al., 1987; L. A. Ierardi, S. B. Moss, and A. R. Bellvé, submitted for publication). After denaturation by detergent or chaotropic agents, soluble polypeptides in these extractions were displayed by sodium dodecyl sulfate-polyacrylamide gel electrophoresis. By comparison with similar extractions of somatic tissue or of cell fractions enriched in other stages of spermatogenesis, candidate "pachytene or SC-specific" polypeptides were identified. Unfortunately, there is little if any agreement among the results of these studies, and reproducibility of the specific preparations appears to be difficult. Fractionation and characterization of the extracted polypeptides have been inadequate to allow more than tentative and heavily qualified interpretations of the results. These "SC preparations" are clearly enriched in a subset of polypeptides, but they are also clearly contaminated with non-SC components, in several cases with histones or lamina proteins. Some components present in the preparations may be due to artifactual redistribution of proteins during cell lysis or extraction (for discussion, see Chaly et al., 1985); mixing experiments with labeled cytoplasmic proteins indicate that they can constitute as much as 9% of the final SC preparation (Ierardi et al., 1983). In spite of the significant caveats, these preparations can serve as immunogens for the development of antibody probes against potential SC components (Heyting et al., 1987; Moens et al., 1987; Behal et al., 1987; M. E. Dresser and C. N. Giroux, unpublished data).

Recently, methods have been developed which allow antibody labeling of well-preserved SCs in spread preparations of mammalian meiotic nuclei (Moses et al., 1984; Dresser et al., 1987; Moens et al., 1987). These methods support both the screening of uncharacterized sera for anti-SC activity and the ultrastructural localization of nuclear proteins by using specific antibodies (for review, see Dresser, 1987). A 110-kilodalton DNA-binding protein identified in the nuclear matrix fraction of rat pachytene spermatocytes has been used to raise a polyclonal antibody with an apparent specificity for nuclear matrices of germ cells (Behal et al., 1987). In immunofluorescence microscopy of testicular preparations, this antibody labels pachytene nuclei, but its precise ultrastructural localization has not yet been determined. Two monoclonal antibodies raised against partially purified SCs from rat spermatocytes have been localized to the lateral elements of rat or mouse SCs by immunoelectron microscopy of spread meiotic nuclei (Heyting et al., 1987; Moens et al., 1987). This promising result is difficult to interpret for one of these antibodies (mAbIII15B8), since its specificity for meiotic cells (or the SC) is ambiguous. This first antibody was not tested on Western blots (immunoblots) against somatic cell protein (or nuclear matrix preparation), nor has specific recognition of the SC preparation which was used as immunogen been demonstrated. The second antibody (mAbII52F10) recognizes 30- and 33-kilodalton polypeptides present in a Western blot of the initial immunogen. These polypeptides may be enriched in or specific to spermatocyte nuclei (Heyting et al., 1988). It is critical to characterize the meiosis specificity of potential anti-SC antibodies in order to examine the possibility that an epitope common to a general chromatin- or nuclear matrix-associated protein is being recognized. Both of these studies illustrate the difficulty, even using antibodies for immunolocalization, of identifying a specific polypeptide as an SC component. Potentially, molecular genetic analysis of the corresponding structural gene could be used to unambiguously establish the connection between an SC component and a specific polypeptide.

IX. COMBINING FUNCTIONAL AND STRUCTURAL ANALYSIS OF CHROMOSOME SYNAPSIS

Identification of candidate SC components is the beginning of a biochemical approach to SC structure and function. The presumed biochemical complexity of the SC and the difficulty of determining the in vivo function of a nuclear component are limitations to a strictly biochemical approach. In the absence of an in vitro assay for SC function or morphogenesis, a complementary genetic approach can provide a means both to identify the gene products required for meiosis-specific chromosome behavior and to analyze their functions (for discussion, see Baker et al., 1976; Zickler and Simonet, 1981). Unfortunately, few meiotic systems support both detailed cytological and genetic investigation; thus, conclusions have often been derived from comparisons of data obtained in different organisms (see, for example, Wilson, 1925; Darlington, 1937; Rhoades, 1961). In order to rigorously establish and test specific hypotheses, however, experiments must be performed in a single meiotic system. Ideally, a combined approach in a single organism would allow a molecular determination of the gene products required for SC morphogenesis and function, a cytological determination of the in vivo role of these gene products in the structure of the meiotic nucleus, and a genetic analysis of the function of the gene products and structures in meiotic recombination and reductional segregation.

For a combined approach, meiotic systems with a well-developed cytology, such as lily or the mouse, are severely limited by the lack of meiotic mutants and of a practical means for genetic analysis of such mutants. Well-developed genetic systems such as maize or *D. melanogaster* have allowed identification of meiotic mutants and their subsequent cytological analysis, but this analysis has been laborious, and the number of mutants examined in detail has thus been limited. Recent advances in the preparation of meiotic squashes for light microscopy are extending routine cytological analysis to the early stages of prophase (leptotene and zygotene) in these organisms and should lead to increased ease of mutant characterization (for maize, see Maguire [1983]; for higher plants in general, see Stelly et al. [1984], Jongedijk [1987a, b]; for *D. melanogaster*, see Puro and Nokkala [1977]). The application of surface-spreading methods for the preparation of meiotic nuclei to these genetically favorable systems would greatly facilitate detailed ultrastructural analysis of mutant phenotypes (Gillies, 1983a).

Molecular cloning of the *D. melanogaster* meiotic mutants *mei-9* and *mei-41* by P-element transposon tagging and subsequent gene isolation is in progress (Boyd et al., 1987; A. H. Yamamoto and J. M. Mason, personal communication), as is deficiency mapping and chromosomal walking to isolate the *c(3)G* locus (P. Szauter, personal communication). Mutants at all three of these loci reduce meiotic recombination in *Drosophila* oocytes; *mei-9* appears to form normal SCs, *mei-41* has an altered distribution of SC-associated nodules, and *c(3)G* fails to assemble SCs (Meyer, 1964; Smith and King, 1968; Carpenter, 1979 and this volume). The wild-type function of these three genes is not known. Mutants *mei-9* and *mei-41* also reduce mitotic chromosome stability and confer mutagen sensitivity, suggesting that they have a role in general chromosome metabolism (for discussion, see Boyd et al., 1987). Mutant *c(3)G* may be primarily defective in meiosis, since it is reported to have normal levels of mitotic recombination and to exhibit somatic chromosome pairing in salivary glands (LeClerc, 1946); thus, its gene product is a candidate for an SC component or assembly function. Molecular cloning of these genes will enhance the prospect of identifying the proteins required for meiotic chromosome behavior in *D. melanogaster*.

Fungal systems have provided detailed genetic data on meiotic chromosome behavior. These studies have been facilitated by tetrad analysis, which allows the recovery of all four interacting chromatids from a single meiosis. Coupled with the ease of identifying large numbers of gene conversion events, this analysis provides a strong tool for testing mechanistic hypotheses of meiotic recombination (for discussion, see Hastings, 1987). The best developed of these fungal genetic systems is the yeast *S. cerevisiae*, which has the additional advantage of an efficient transformation system, allowing targeted gene manipulation and replacement. The life cycle of *S. cerevisiae* offers two major advantages over other systems for the isolation and analysis of meiotic mutants (for discussion, see Esposito and Klapholz, 1981; Moens, 1982; Dawes, 1983). First, there are interconvertible haploid and diploid vegetative forms such that meiosis can be made a nonessential part of the life cycle, allowing meiotic mutants to be stably propagated. Second, there are heterothallic (obligate outcrossing) and homothallic (self-mating and diploidizing) forms which allow mutagenesis of a haploid strain and subsequent homozygosis of recessive mutations to allow their expression in an isogenic diploid strain capable of meiosis. Unlike most meiotic systems, whose meiosis-specific protein components are difficult to identify (see previous discussion), *S. cerevisiae* can be synchronously induced to undergo meiosis and provide biochemical quantities of a relatively homogeneous unicellular population (for discussion, see Dawes, 1983).

In an attempt to identify genes required for meiotic differentiation, mutants which fail to yield viable sporulation products (or exhibit reduced sporulation frequencies) have been isolated in several fungi (for reviews, see Baker et al., 1976; Esposito and Klapholz, 1981). Subsequent genetic and cytological examination of these sporulation mutants has revealed a limited number of mutants with apparent defects in SC morphogenesis and/or proper meiotic chromosome behavior (*spo11* of *S. cerevisiae* [see discussion below], *mei4* of *S. pombe* [Shimoda et al., 1985], and several mutants of *Sordaria macrospora* [Zickler et al., 1985; Moreau et al., 1985; Huynh et al., 1986]). There is, however, considerable difficulty in determining the precise defect in these meiotic prophase mutants and therefore in determining which corresponding genes are specifically required for the mechanics of SC morphogenesis or meiotic recombination, or both. Mutants in regulatory genes required for the sequential developmental program of meiosis or in general DNA metabolism genes required for both somatic and meiotic chromosome behavior might be anticipated to have phenotypes which mimic specific defects in SC components or SC morphogenesis. Therefore, characterization of potential pairing or synapsis mutants requires a combination of cytological examination of the stages of meiosis, genetic determination of recombination behavior, and physiological analysis of DNA metabolism in vegetative versus meiotic cells. Fungal systems such as *Sordaria macrospora* (Zickler et al., 1984) or *Coprinus cinereus* (Lu, 1982; Pukkila et al., 1984) provide an efficient system for studying meiotic physiology but have not yet developed a genetic analysis of recombination or a molecular biology which can support a combined approach to meiotic chromosome behavior and recombination comparable to that possible in *S. cerevisiae* and potentially in *D. melanogaster*.

Exploiting the well-developed meiotic genetics and recombinant DNA technology of *S. cerevisiae* the *SPO11* gene, which is required specifically for meiotic prophase and not for the vegetative cell cycle, has been cloned by complementation of a mutant allele (Giroux et al., 1986; Atcheson et al., 1987). The *SPO11* gene is required both for meiotic recombination and for the subsequent proper

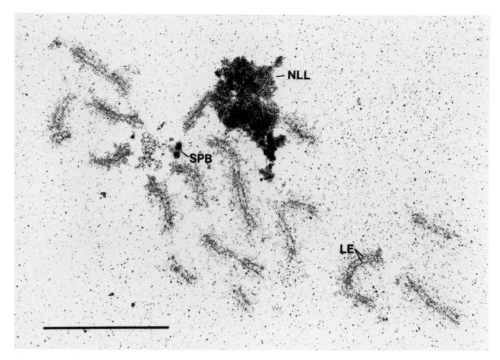

FIGURE 4. Electron micrograph of a surface-spread yeast pachytene nucleus stained with silver nitrate. The lateral elements (LE) of each of the 16 synaptonemal complexes, the nucleolus (NLL), and the duplicated spindle pole body (SPB) are well contrasted. Chromatin is partially extracted by the staining procedure but is visible as a less dense "fuzz" along each SC. Bar, 3 μm. (Reproduced from Dresser and Giroux, 1988.)

reductional chromosome segregation, but its precise function in meiotic chromosome behavior is undetermined. A serious limitation to the analysis of meiotic mutants in *S. cerevisiae*, as well as in other fungi, has been the small size of yeast chromosomes, which has required sample preparation by embedding and sectioning methods for cytological examination in an electron microscope. Recently, surface-spreading methods developed for larger organisms have been adapted for the cytological analysis of yeast meiotic nuclei with both light and electron microscopes, greatly facilitating the characterization of meiotic mutants (Dresser and Giroux, 1988; see Fig. 4). Using this new method, Giroux et al. (1986; C. N. Giroux, H. F. Tiano, and M. E. Dresser, unpublished data) showed that an isogenic mutant *S. cerevisiae* strain with a precise deletion of the *SPO11* gene is defective in chromosome synapsis. This result correlates well with the lack of meiotic recombination in *spo11* mutants and with the specific transcription of the wild-type gene during meiotic prophase (Giroux et al., 1986, unpublished data). Identification of the *SPO11* gene product and determination of its role in SC morphogenesis should now be possible.

X. SUMMARY

Chromosome conjunction results in the precise side-by-side parallel alignment of homologs to produce a bivalent during meiotic prophase. Initially, pairing serves to colocalize and orient the homologs within the meiotic nucleus. As pairing brings the aligned axes of homologs to within approximately 300 nm of each other, synapsis completes their intimate association, binding them together to form the SC. Although the SC may stabilize the pairing interaction of homologs, the relationship between synapsis and meiotic recombination is unclear.

Pairing precedes synapsis, and in the absence of a homologous pairing partner, subsequent nonhomologous or self- (foldback) synapsis often occurs. Thus, pairing may provide an initial check for chromosome homology. In the extreme case, pairing would be the sole determinant of homology recognition, providing the substrate for indiscriminate SC assembly. Synapsis does not occur without regard for homology, however. Close examination of the progress of synapsis during zygotene and pachytene in genomes with chromosome rearrangements or in polyploids reveals the phenomena of synaptic adjustment and of multivalent correction (to bivalents), respectively. If these processes reflect the synaptic behavior of normal as well as abnormal pairing configurations, then they suggest that synapsis is biphasic. Furthermore, they suggest that a homosynaptic (homology-dependent) phase in zygotene precedes a heterosynaptic (homology-independent) phase in pachytene. Formally, homosynapsis could result from residual homology-dependent pairing interactions (such as recombination) superimposed on the initiation of homology-independent synapsis. Thus, there is no conclusive evidence that chromosome synapsis is homology dependent.

However, since homologs preferentially synapse with each other, it has often been proposed that synapsis promotes or ensures homolog recombination. This recombination could be facilitated passively by stabilization of the DNA substrate (synapsed bivalent), allowing soluble recombination enzymes to act preferentially on homologs. Alternatively, components of the SC could participate directly in the enzymatic catalysis of molecular recombination. At present, there is no direct test of this proposition, and the two alternatives cannot be distinguished.

Although the order in which they occur is unknown, there is considerable indirect evidence from many organisms that chromosome synapsis is involved with meiotic recombination. The time during which crossing-over is estimated to occur corresponds well with the time course of synapsis (for reviews, see Peacock, 1968, 1971; Henderson, 1970). It is clear that chiasmata, visualized at diplotene, are the physical manifestation of crossing-over events (for review, see Jones, 1987). Genetic mutations or physiological manipulations of meiosis that disrupt SC formation similarly disrupt the formation of chiasmata and meiotic levels of recombination. The most convincing data derive from the meiosis-specific asynaptic mutant of *S. cerevisiae*, *spo11*, and the similar *c(3)G* mutant of *D. melanogaster* in which failure to assemble the SC correlates with the specific loss of meiotic (but not mitotic) levels of recombination. In contrast, there are mutants of both *S. cerevisiae* (*rad52* [M. E. Dresser, H. F. Tiano, and C. N.

Giroux, unpublished data]) and *D. melanogaster* (*mei-9* [Carpenter, 1979]) which are deficient in meiotic recombination, yet assemble an apparently normal SC. These mutants are also defective in mitotic DNA metabolism, suggesting that gene products required in the mitotic cell cycle are recruited for similar roles in meiotic DNA metabolism. In meiosis of interspecific hybrids, in which homologous pairing partners are absent, SC is formed between nonhomologous chromosomes but chiasmata are not observed (Menzel and Price, 1966). Although synapsis may be necessary for the regulated levels of meiotic recombination between homologs, it is clearly not sufficient.

Chromosome synapsis supports the intimate association of paired homologs during pachytene. Indirect and correlative cytological and genetic data support the involvement of the SC in meiotic recombination and subsequent reductional chromosome segregation at the first meiotic division. These data are inadequate, however, to directly elucidate the role of the SC in meiotic chromosome behavior or to reveal the underlying molecular mechanisms responsible for this behavior. To address these questions, a combined approach of cytology, genetics, and gene product identification in a single experimental system is required.

Recent advances in preparative and staining methods for meiotic cytology allow the rapid examination of large numbers of nuclei, overcoming the limitations of sectioning methods for the analysis of genetic mutants. These surface-spreading methods allow a quantitative evaluation of the progress of chromosome pairing and SC morphogenesis, revealing detailed features which are difficult to deduce from the serial reconstruction of a small number of nuclei. In addition, spread preparations support the immunocytological localization of specific gene products to nuclear structures. The identification of antibodies specific for bivalents (synapsed chromosomes) would complement the genetic analysis of meiotic mutants as an approach to establish the relationship between specific nuclear structures and meiotic chromosome behavior. Antibody screening of cDNA expression libraries, gene isolation, and subsequent "reverse genetics" in a meiotic system amenable to molecular genetic analysis would allow the functional role of synapsis-specific protein components to be determined. *D. melanogaster* and *Caenorhabditis elegans* are attractive metazoans for the molecular genetic analysis of meiosis, although cytological analysis is currently limited in these organisms. The most promising system for such a combined approach to meiotic chromosome behavior is the yeast *S. cerevisiae*, in which the molecular cloning, cytological, and genetic analysis of meiosis-specific genes has already been demonstrated.

ACKNOWLEDGMENTS. I thank the colleagues who provided me with preprints and reprints of their work, not all of which could be discussed in this limited review. I am grateful to Mike Dresser and Bob Malone for critical comments on the manuscript and for many helpful discussions.

LITERATURE CITED

Albini, S. M., and G. H. Jones. 1984. Synaptonemal complex-associated centromeres and recombination nodules in plant meiocytes prepared by an improved surface-spreading technique. *Exp. Cell Res.* **155**:588–592.

Albini, S. M., and G. H. Jones. 1987. Synaptonemal complex spreading in *Allium cepa* and *A. fistulosum*. I. The initiation and sequence of pairing. *Chromosoma* (Berlin) **95**:324–338.

Albini, S. M., G. H. Jones, and B. M. N. Wallace. 1984. A method for preparing two-dimensional surface-spreads of synaptonemal complexes from plant meiocytes for light and electron microscopy. *Exp. Cell Res.* **152**:280–285.

Allen, J. W., G. K. DeWeese, J. B. Gibson, P. A. Poorman, and M. J. Moses. 1987. Synaptonemal complex damage as a measure of chemical mutagen effects on mammalian germ cells. *Mutat. Res.* **190**:19–24.

Anderson, L. K., S. M. Stack, M. H. Fox, and Z. Chuanshan. 1985. The relationship between genome size and synaptonemal complex length in higher plants. *Exp. Cell Res.* **156**:367–378.

Atcheson, C. L., B. DiDomenico, S. Frackman, R. E. Esposito, and R. T. Elder. 1987. Isolation, DNA sequence, and regulation of a meiosis-specific eukaryotic recombination gene. *Proc. Natl. Acad. Sci. USA* **84**:8035–8039.

Baker, B. S., A. T. C. Carpenter, M. S. Esposito, R. E. Esposito, and L. Sandler. 1976. The genetic control of meiosis. *Annu. Rev. Genet.* **10**:53–134.

Baker, B. S., A. T. C. Carpenter, and P. Ripoll. 1978. The utilization during miotic cell division of loci controlling meiotic recombination and disjunction in *Drosophila melanogaster*. *Genetics* **90**:531–578.

Baker, B. S., M. Gatti, A. T. C. Carpenter, S. Pimpinelli, and D. A. Smith. 1980. Effects of recombination-deficient and repair-deficient loci on meiotic and mitotic chromosome behavior in *Drosophila melanogaster*, p. 188–208. *In* W. M. Generoso, M. D. Shelby, and F. J. deSerres (ed.), *DNA Repair and Mutagenesis in Eukaryotes*. Plenum Publishing Corp., New York.

Baker, B. S., and J. C. Hall. 1976. Meiotic mutants: genetic control of meiotic recombination and chromosome segregation, p. 351–434. *In* M. Ashburner and E. Novitski (ed.), *The Genetics and Biology of Drosophila*, vol. 1A. Academic Press, Inc., New York.

Behal, A., K. Prakash, and M. R. S. Rao. 1987. Identification of a meiotic prophase-specific nuclear matrix protein in the rat. *J. Biol. Chem.* **262**:10898–10902.

Bojko, M. 1983. Human meiosis. VIII. Chromosome pairing and formation of the synaptonemal complex in oocytes. *Carlsberg Res. Commun.* **48**:457–483.

Borts, R. H., M. Lichten, and J. E. Haber. 1986. Analysis of meiosis-defective mutations in yeast by physical monitoring of recombination. *Genetics* **113**:551–567.

Boyd, J. B., J. M. Mason, A. H. Yamamoto, R. K. Brodberg, S. S. Banga, and K. Sakaguchi. 1987. A genetic and molecular analysis of DNA repair in *Drosophila*, p. 39–60. *In* A. Collins, R. T. Johnson, and J. M. Boyle (ed.), *Molecular Biology of DNA Repair* (*J. Cell Sci.*, Suppl. 6). The Company of Biologists Ltd., Cambridge.

Brock, W. A., P. K. Trostle, and M. L. Meistrich. 1980. Meiotic synthesis of testis histones in the rat. *Proc. Natl. Acad. Sci. USA* **77**:371–375.

Callan, H. G. 1982. Lampbrush chromosomes. *Proc. R. Soc. Lond. B Biol. Sci.* **214**:417–448.

Carpenter, A. T. C. 1975. Electron microscopy of meiosis in *Drosophila melanogaster* females. II. The recombination nodule—a recombination-associated structure at pachytene? *Proc. Natl. Acad. Sci. USA* **72**:3186–3189.

Carpenter, A. T. C. 1979. Recombination nodules and synaptonemal complex in recombination-defective females of *Drosophila melanogaster*. *Chromosoma* (Berlin) **75**:259–292.

Carpenter, A. T. C. 1981. EM autoradiographic evidence that DNA synthesis occurs at recombination nodules during meiosis in *Drosophila melanogaster* females. *Chromosoma* (Berlin) **83**:59–80.

Carpenter, A. T. C. 1984. Genic control of meiosis. *Chromosomes Today* **8**:70–79.

Carpenter, A. T. C. 1987. Gene conversion, recombination nodules, and the initiation of meiotic synapsis. *BioEssays* **6**:232–236.

Chaly, N., J. E. Little, and D. L. Brown. 1985. Localization of nuclear antigens during preparation of nuclear matrices *in situ*. *Can. J. Biochem. Cell Biol.* **63:**644–653.

Chandley, A. C. 1986. A model for effective pairing and recombination at meiosis based on early replicating sites (R-bands) along chromosomes. *Hum. Genet.* **72:**50–57.

Coleman, J. R., and M. J. Moses. 1964. DNA and the fine structure of synaptic chromosomes in the domestic rooster (*Gallus domesticus*). *J. Cell Biol.* **23:**63–78.

Comings, D. E., and T. A. Okada. 1970. Association of chromatin fibers with the annuli of the nuclear membrane. *Exp. Cell Res.* **62:**293–302.

Comings, D. E., and T. A. Okada. 1972. Architecture of meiotic cells and mechanisms of chromosome pairing. *Adv. Cell Mol. Biol.* **2:**310–384.

Comings, D. E., and T. A. Okada. 1976. Nuclear proteins. III. The fibrillar nature of the nuclear matrix. *Exp. Cell Res.* **103:**341–360.

Comings, D. E., and A. D. Riggs. 1971. Molecular mechanisms of chromosome pairing, folding and function. *Nature* (London) **233:**48–50.

Counce, S. J., and G. F. Meyer. 1973. Differentiation of the synaptonemal complex and the kinetochore in Locusta spermatocytes studied by whole mount electron microscopy. *Chromosoma* (Berlin) **44:**231–253.

Darlington, C. D. 1937. *Recent Advances in Cytology*, 2nd ed. The Blakiston Co., Philadelphia.

Davis, P. S., M. W. Shen, and B. H. Judd. 1987. Asymmetrical pairings of transposons in and proximal to the white locus of Drosophila account for four classes or regularly occurring exchange products. *Proc. Natl. Acad. Sci. USA* **84:**174–178.

Dävring, L., and M. Sunner. 1979. Cytological evidence for procentric synapsis of meiotic chromosomes in female *Drosophila melanogaster*: the behavior of an extra Y chromosome. *Hereditas* **91:**53–64.

Dawes, I. W. 1983. Genetic control and gene expression during meiosis and sporulation in *Saccharomyces cerevisiae*, p. 29–64. *In* J. F. T. Spencer, D. M. Spencer, and A. R. W. Smith (ed.), *Yeast Genetics, Fundamental and Applied Aspects*. Springer-Verlag, New York.

del Mazo, J., and L. Gil-Alberdi. 1986. Multistranded organization of the lateral elements of the synaptonemal complex in the rat and mouse. *Cytogenet. Cell Genet.* **41:**219–224.

Dickinson, H. G. 1987. The physiology and biochemistry of meiosis in the anther. *Int. Rev. Cytol.* **107:**79–109.

Douglas, L. T. 1966. Meiosis. I. Association of non-homologous bivalents during spermatogenesis in white mice. *Genetica* (The Hague) **37:**466–480.

Dresser, M. E. 1987. The synaptonemal complex of meiosis: an immunocytochemical approach, p. 245–274. *In* P. B. Moens (ed.), *Meiosis*. Academic Press, Inc., New York.

Dresser, M. E., and C. N. Giroux. 1988. Meiotic chromosome behavior in spread preparations of yeast. *J. Cell Biol.* **106:**567–573.

Dresser, M. E., and M. J. Moses. 1979. Silver staining of synaptonemal complexes in surface spreads for light and electron microscopy. *Exp. Cell Res.* **121:**416–419.

Dresser, M. E., and M. J. Moses. 1980. Synaptonemal complex karyotyping in spermatocytes of the chinese hamster (*Cricetulus griseus*). *Chromosoma* (Berlin) **76:**1–22.

Dresser, M. E., D. Pisetsky, R. Warren, G. McCarty, and M. Moses. 1987. A new method for the cytological analysis of autoantibody specificities using whole mount, surface-spread meiotic nuclei. *J. Immunol. Methods* **104:**111–121.

Driscoll, D. J., C. G. Palmer, and A. Melman. 1979. Nonhomologous associations of C-heterochromatin at human male meiotic prophase. *Cytogenet. Cell Genet.* **23:**23–32.

Engelhardt, P., and K. Pusa. 1972. Nuclear pore complexes: "press-stud" elements of chromosomes in pairing and control. *Nature* (London) *New Biol.* **240:**163–166.

Esposito, R. E., and S. Klapholz. 1981. Meiosis and ascospore development, p. 211–287. *In* J. N. Strathern, E. W. Jones, and J. R. Broach (ed.), *The Molecular Biology of the Yeast Saccharomyces: Life Cycle and Inheritance*. Cold Spring Harbor Laboratory, Cold Spring Harbor, N.Y.

Evans, C. W., and H. G. Dickinson (ed.). 1984. *Controlling Events in Meiosis. Symposia of the Society for Experimental Biology*, vol. 38. Cambridge University Press, London.

Fabergé, A. C. 1942. Homologous chromosome pairing: the physical problem. *J. Genet.* **34:**121–145.

Fawcett, D. W. 1956. The fine structure of chromosomes in the meiotic prophase of vertebrate spermatocytes. *J. Biophys. Biochem. Cytol.* **2**:403–406.

Fiil, A., and P. B. Moens. 1973. The development, structure and function of modified synaptonemal complexes in mosquito oocytes. *Chromosoma* (Berlin) **41**:37–62.

Fletcher, J. M. 1979. Light microscope analysis of meiotic prophase chromosomes by silver staining. *Chromosoma* (Berlin) **72**:241–248.

Fogel, S., R. K. Mortimer, and K. Lusnak. 1983. Meiotic gene conversion in yeast: molecular and experimental perspectives, p. 65–107. *In* J. F. T. Spencer, D. M. Spencer, and A. R. W. Smith (ed.), *Yeast Genetics, Fundamental and Applied Aspects.* Springer-Verlag, New York.

Forejt, J., and P. Goetz. 1979. Synaptonemal complexes of mouse and human pachytene chromosomes visualized by silver staining in air-dried preparations. *Chromosoma* (Berlin) **73**:255–261.

Gambino, J., R. A. Eckhardt, and M. S. Risley. 1981. Nuclear matrices containing synaptonemal complexes from *Xenopus laevis. J. Cell Biol.* **91**:63a.

Game, J. C. 1983. Radiation-sensitive mutants and repair in yeast, p. 109–137. *In* J. F. T. Spencer, D. M. Spencer, and A. R. W. Smith (ed.), *Yeast Genetics, Fundamental and Applied Aspects.* Springer-Verlag, New York.

Gillies, C. B. 1974. The nature and extent of synaptonemal complex formation in haploid barley. *Chromosoma* (Berlin) **48**:441–453.

Gillies, C. B. 1975a. Synaptonemal complex and chromosome structure. *Annu. Rev. Genet.* **9**:91–109.

Gillies, C. B. 1975b. An ultrastructural analysis of chromosomal pairing in maize. *C.R. Trav. Lab. Carlsberg* **40**:135–161.

Gillies, C. B. 1983a. Spreading plant synaptonemal complexes for electron microscopy, p. 115–122. *In Kew Chromosome Conference II.* George Allen & Unwin, London.

Gillies, C. B. 1983b. Ultrastructural studies of the association of homologous and non-homologous parts of chromosomes in the mid-prophase of meiosis in *Zea mays. Maydica* **28**:265–287.

Gillies, C. B. 1984. The synaptonemal complex in higher plants. *Crit. Rev. Plant Sci.* **2**:81–116.

Gillies, C. B. 1985. An electron microscopic study of synaptonemal complex formation at zygotene in rye. *Chromosoma* (Berlin) **92**:165–175.

Gillies, C. B., J. Kuspira, and R. N. Bhambhani. 1987. Genetic and cytogenetic analyses of the A genome of *Triticum monococcum.* IV. Synaptonemal complex formation in autotetraploids. *Genome* **29**:309–318.

Giroux, C. N., H. F. Tiano, and M. E. Dresser. 1986a. Analysis of *SPO11*, a gene required for meiotic recombination in yeast. *Yeast* **2**:S133.

Giroux, C. N., H. F. Tiano, M. E. Dresser, and M. Moses. 1986b. Molecular cloning and analysis of genes required for meiotic recombination and DNA metabolism in yeast. *J. Cell. Biochem. Suppl.* **10B**:215.

Goldberg, M. L., J.-Y. Sheen, W. J. Gehring, and M. M. Green. 1983. Unequal crossing-over associated with asymmetrical synapsis between nomadic elements in the *Drosophila melanogaster* genome. *Proc. Natl. Acad. Sci. USA* **80**:5017–5021.

Golubovskaya, I. N. 1979. Genetic control of meiosis. *Int. Rev. Cytol.* **58**:247–290.

Gottschalk, W., and M. L. H. Kaul. 1980a. Asynapsis and desynapsis in flowering plants. I. Asynapsis. *Nucleus* (Paris) **23**:1–15.

Gottschalk, W., and M. L. H. Kaul. 1980b. Asynapsis and desynapsis in flowering plants. II. Desynapsis. *Nucleus* (Paris) **23**:97–120.

Hale, D. W. 1986. Heterosynapsis and suppression of chiasmata within heterozygous pericentric inversions of the Sitka deer mouse. *Chromosoma* (Berlin) **94**:425–432.

Hasenkampf, C. A. 1984a. Synaptonemal complex formation in pollen mother cells of *Tradescantia. Chromosoma* (Berlin) **90**:275–284.

Hasenkampf, C. A. 1984b. Longitudinal axis thickenings in whole-mount spreads of synaptonemal complexes from *Tradescantia. Chromosoma* (Berlin) **90**:285–288.

Hastings, P. J. 1987. Models of heteroduplex formation, p. 139–156. *In* P. B. Moens (ed.), *Meiosis.* Academic Press, Inc., New York.

Henderson, S. A. 1970. The time and place of meiotic crossing-over. *Annu. Rev. Genet.* **4**:295–324.

Henderson, S. A. 1971. Grades of chromatid organisation in mitotic and meiotic chromosomes. I. The morphological features. *Chromosoma* (Berlin) **35**:28–40.

Heyting, C., R. J. Dettmers, A. J. J. Dietrich, E. J. W. Redeker, and A. C. G. Vink. 1988. Two major components of synaptonemal complexes are specific for meiotic prophase nuclei. *Chromosoma* (Berlin) **96**:325–332.

Heyting, C., A. J. J. Dietrich, E. J. W. Redeker, and A. C. G. Vink. 1985. Structure and composition of synaptonemal complexes, isolated from rat spermatocytes. *Eur. J. Cell Biol.* **36**:307–314.

Heyting, C., P. B. Moens, W. van Raamsdonk, A. J. J. Dietrich, A. C. G. Vink, and E. J. W. Redeker. 1987. Identification of two major components of the lateral elements of synaptonemal complexes of the rat. *Eur. J. Cell Biol.* **43**:148–154.

Holliday, R. 1964. A mechanism for gene conversion in fungi. *Genet. Res.* **5**:282–304.

Holliday, R. 1977. Recombination and meiosis. *Philos. Trans. R. Soc. Lond. B Biol. Sci.* **277**:259–370.

Holm, P. B. 1977. Three-dimensional reconstruction of chromosome pairing during the zygotene stage of meiosis in *Lilium longiflorum* (Thunb.). *Carlsberg Res. Commun.* **42**:103–151.

Holm, P. B. 1985. Ultrastructural characterization of meiosis, p. 39–90. *In Six Papers in the Biological Sciences*. Royal Danish Academy of Sciences and Letters, Copenhagen.

Holm, P. B. 1986. Chromosome pairing and chiasma formation in allohexaploid wheat, *Triticum aestivum* analyzed by spreading of meiotic nuclei. *Carlsberg Res. Commun.* **51**:239–294.

Howell, W. M., and B. C. Lamb. 1984. Two locally acting genetic controls of gene conversion, *ccf-5* and *ccf-6*, in *Ascobolus immersus*. *Genet. Res.* **43**:107–121.

Huynh, A. D., G. Leblon, and D. Zickler. 1986. Indirect intergenic suppression of a radiosensitive mutant of *Sordaria macrospora* defective in sister-chromatid cohesiveness. *Curr. Genet.* **10**:545–555.

Ierardi, L. A., S. B. Moss, and A. R. Bellvé. 1983. Synaptonemal complexes are integral components of the isolated mouse spermatocyte nuclear matrix. *J. Cell Biol.* **96**:1717–1726.

Jinks-Robertson, S., and T. D. Petes. 1985. High-frequency meiotic gene conversion between repeated genes on nonhomologous chromosomes in yeast. *Proc. Natl. Acad. Sci. USA* **82**:3350–3354.

Jinks-Robertson, S., and T. D. Petes. 1986. Chromosomal translocations generated by high-frequency meiotic recombination between repeated yeast genes. *Genetics* **114**:731–752.

Jones, G. H. 1987. Chiasmata, p. 213–244. *In* P. B. Moens (ed.), *Meiosis*. Academic Press, Inc., New York.

Jones, G. H., and J. A. Croft. 1986. Surface spreading of synaptonemal complexes in locusts. II. Zygotene pairing behavior. *Chromosoma* (Berlin) **93**:489–495.

Jongedijk, E. 1987a. A quick enzyme squash technique for detailed studies on female meiosis in *Solanum*. *Stain Technol.* **62**:135–141.

Jongedijk, E. 1987b. A rapid methyl salicylate clearing technique for routine phase-contrast observations on female meiosis in *Solanum*. *J. Microsc.* **146**:157–162.

Junker, A., E. Lehmann, and P. Munz. 1987. Genetic analysis of particular aspects of intergenic conversion in *Schizosaccharomyces pombe*. *Curr. Genet.* **12**:119–125.

Klásterská, I., and C. Ramel. 1979. Prophase of plants meiosis: sequences and interpretation of stages. *Genetica* (The Hague) **51**:15–20.

Koduru, P. R. K., and M. K. Rao. 1981. Cytogenetics of synaptic mutants in higher plants. *Theor. Appl. Genet.* **59**:197–214.

Kupiec, M., and T. D. Petes. 1988. Meiotic recombination between repeated transposable elements in *Saccharomyces cerevisiae*. *Mol. Cell. Biol.* **8**:2942–2954.

Lamb, B. C. 1977. The use of gene conversion to study synaptinemal complex structure and molecular details of chromatid pairing in meiosis. *Mol. Gen. Genet.* **157**:31–37.

Lamb, B. C., and M. R. T. Wickramaratne. 1973. Corresponding-site interference, synaptinemal complex structure, and 8+:0m and 7+:1m octads from wild-type mutant crosses of *Ascobolus immersus*. *Genet. Res.* **22**:113–124.

LeClerc, G. 1946. Occurrence of mitotic crossing over without meiotic crossing over. *Science* **102**:553–554.

Li, S., M. L. Meistrich, W. A. Brock, T. C. Hsu, and M. T. Kuo. 1983. Isolation and preliminary characterization of the synaptonemal complex from rat pachytene spermatocytes. *Exp. Cell Res.* **144**:63–74.

Lichten, M., R. H. Borts, and J. E. Haber. 1987. Meiotic gene conversion and crossing over between dispersed homologous sequences occurs frequently in *Saccharomyces cerevisiae*. *Genetics* **115**:233–246.

Lindsley, D. L., and L. Sandler. 1977. The genetic analysis of meiosis in female *Drosophila melanogaster*. *Philos. Trans. R. Soc. Lond. B Biol. Sci.* **277**:295–312.

Loidl, J. 1986. Synaptonemal complex spreading in *Allium*. II. Tetraploid *A. vineale*. *Can. J. Genet. Cytol.* **28**:754–761.

Loidl, J. 1987. Synaptonemal complex spreading in *Allium ursinum*: pericentric asynapsis and axial thickenings. *J. Cell Sci.* **87**:439–448.

Loidl, J., and G. H. Jones. 1986. Synaptonemal complex spreading in *Allium*. I. Triploid *A. sphaerocephalon*. *Chromosoma* (Berlin) **93**:420–428.

Lu, B. C. 1982. Replication of deoxyribonucleic acid and crossing over in *Coprinus*, p. 93–112. *In* K. Wells and E. K. Wells (ed.), *Basidium and Basidiocarp: Evolution, Cytology, Function, and Development*. Springer-Verlag, New York.

Lu, B. C. 1984. The cellular program of the formation and dissolution of the synaptonemal complex in *Coprinus*. *J. Cell Sci.* **67**:25–43.

Luykx, P. 1974. The organization of meiotic chromosomes, p. 163–207. *In* H. Busch (ed.), *The Cell Nucleus*, vol. 2. Academic Press, Inc., New York.

Maguire, M. P. 1981. A search for the synaptic adjustment phenomenon in maize. *Chromosoma* (Berlin) **81**:717–725.

Maguire, M. P. 1983. Homologue pairing and synaptic behavior at zygotene in maize. *Cytologia* **48**:811–818.

Maguire, M. P. 1984. The mechanism of meiotic homologue pairing. *J. Theor. Biol.* **106**:605–615.

Maloney, D. H., and S. Fogel. 1987. Gene conversion, unequal crossing-over and mispairing at a non-tandem duplication during meiosis of *Saccharomyces cerevisiae*. *Curr. Genet.* **12**:1–7.

McClintock, B. 1933. The association of non-homologous parts of chromosomes in the mid-prophase of meiosis in *Zea mays*. *Z. Zellforsch. Mikrosk. Anat.* **19**:191–237.

Meistrich, M. L., and W. A. Brock. 1987. Proteins of the meiotic cell nucleus, p. 333–353. *In* P. B. Moens (ed.), *Meiosis*. Academic Press, Inc., New York.

Menzel, M. Y., and J. M. Price. 1966. Fine structure of synapsed chromosomes in F1 *Lycopersicon esculentum-Solanum lycopersicoides* and its parents. *Am. J. Bot.* **53**:1979–1086.

Meyer, G. F. 1964. A possible correlation between the submicroscopic structure of meiotic chromosomes and crossing over, p. 461–462. *In* M. Titlbach (ed.), *Proceedings of the 3rd European Regional Conference on Electron Microscopy*. Czechoslovak Academy of Sciences, Prague.

Moens, P. 1973a. Quantitative electron microscopy of chromosome organization at meiotic prophase. *Cold Spring Harbor Symp. Quant. Biol.* **38**:99–107.

Moens, P. 1973b. Mechanisms of chromosome synapsis at meiotic prophase. *Int. Rev. Cytol.* **35**:117–134.

Moens, P. 1978. Ultrastructural studies of chiasma distribution. *Annu. Rev. Genet.* **12**:433–450.

Moens, P. 1982. Mutants of yeast meiosis (*Saccharomyces cerevisiae*). *Can. J. Genet. Cytol.* **24**:243–256.

Moens, P. (ed.). 1987. *Meiosis*. Academic Press, Inc., New York.

Moens, P., C. Heyting, A. J. J. Dietrich, W. van Raamsdonk, and Q. Chen. 1987. Synaptonemal complex antigen location and conservation. *J. Cell Biol.* **105**:93–103.

Moreau, P. J. F., D. Zickler, and G. Leblon. 1985. One class of mutants with disturbed centromere cleavage and chromosome pairing in *Sordaria macrospora*. *Mol. Gen. Genet.* **198**:189–197.

Mortimer, R. K., and S. Fogel. 1974. Genetical interference and gene conversion, p. 263–275. *In* R. F. Grell (ed.), *Mechanisms in Recombination*. Plenum Publishing Corp., New York.

Moses, M. J. 1956. Chromosomal structures in crayfish spermatocytes. *J. Biophys. Biochem. Cytol.* **2**:215–218.

Moses, M. J. 1958. The relation between the axial complex of meiotic prophase chromosomes and chromosome pairing in a Salamander (*Plethodon cinereus*). *J. Biophys. Biochem. Cytol.* **4**:633–638.

Moses, M. J. 1968. Synaptinemal complex. *Annu. Rev. Genet.* **2**:363–412.

Moses, M. J. 1977. Microspreading and the synaptonemal complex in cytogenetic studies. *Chromosomes Today* **6**:71–82.

Moses, M. J., M. E. Dresser, and P. A. Poorman. 1984. Composition and role of the synaptonemal complex. *Symp. Soc. Exp. Biol.* **38**:245–270.

Moses, M. J., and P. A. Poorman. 1981. Synaptonemal complex analysis of mouse chromosomal rearrangements. II. Synaptic adjustment in a tandem duplication. *Chromosoma* (Berlin) **81**:519–535.

Moses, M. J., and A. Solari. 1976. Positive contrast staining and protected drying of surface spreads: electron microscopy of the synaptonemal complex by a new method. *J. Ultrastruct. Res.* **54**:109–114.

Munz, P., H. Amstutz, J. Kohli, and U. Leupold. 1982. Recombination between dispersed serine tRNA genes in *Schizosaccharomyces pombe*. *Nature* (London) **300**:225–231.

Nokkala, S., and J. Puro. 1976. Cytological evidence for a chromocenter in *Drosophila melanogaster* oocytes. *Hereditas* **83**:265–268.

Pathak, S., and T. C. Hsu. 1979. Silver-stained structures in mammalian meiotic prophase. *Chromosoma* (Berlin) **70**:195–203.

Peacock, W. J. 1968. Chiasmata and crossing-over, p. 242–252. *In* W. J. Peacock and R. D. Brock (ed.), *Replication and Recombination of the Genetic Material.* Australian Academy of Science, Canberra, Australia.

Peacock, W. J. 1971. Cytogenetic aspects of the mechanism of recombination in higher organisms. *Stadler Genet. Symp.* **1** and **2**:123–152.

Petes, T. D., S. Smolik-Utlaut, and M. McMahon. 1982. Recombination in yeast ribosomal DNA. *Recent Adv. Yeast Mol. Biol.* **1**:69–75.

Poorman, P. A., M. J. Moses, L. B. Russell, and N. L. A. Cacheiro. 1981. Synaptonemal complex analysis of mouse chromosomal rearrangements. I. Cytogenetic observation on a tandom duplication. *Chromosoma* (Berlin) **81**:507–518.

Powers, P. A., and O. Smithies. 1986. Short gene conversions in the human fetal globin gene region: a by-product of chromosome pairing during meiosis? *Genetics* **112**:343–358.

Pritchard, R. H. 1960. Localized negative interference and its bearing on models of gene recombination. *Genet. Res.* **1**:1–24.

Pukkila, P. J. 1977. Biochemical analysis of genetic recombination in eukaryotes. *Heredity* **39**:193–217.

Pukkila, P. J., B. M. Yashar, and D. M. Binninger. 1984. Analysis of meiotic development in *Coprinus cinereus*. *Symp. Soc. Exp. Biol.* **38**:177–194.

Puro, J., and S. Nokkala. 1977. Meiotic segregation of chromosomes in *Drosophila melanogaster* oocytes. *Chromosoma* (Berlin) **63**:273–286.

Rao, B. J., S. K. Brahmachari, and M. R. Satyanarayana. 1983. Structural organization of the meiotic prophase chromatin in the rat testis. *J. Biol. Chem.* **258**:13478–13485.

Rasmussen, S. W. 1977. Chromosome pairing in triploid females of *Bombyx mori* analyzed by three dimensional reconstructions of synaptonemal complexes. *Carlsberg Res. Commun.* **42**:163–197.

Rasmussen, S. W. 1987. Chromosome pairing in autotetraploid Bombyx males. Inhibition of multivalent correction by crossing over. *Carlsberg Res. Commun.* **52**:211–242.

Rasmussen, S. W., and P. B. Holm. 1980. Mechanics of meiosis. *Hereditas* **93**:187–216.

Rasmussen, S. W., P. B. Holm, B. C. Lu, D. Zickler, and J. Sage. 1981. Synaptonemal complex formation and distribution of recombination nodules in pachytene trivalents of triploid *Coprinus cinereus*. *Carlsberg Res. Commun.* **46**:347–360.

Rattner, J. B., M. Goldsmith, and B. A. Hamkalo. 1980. Chromatin organization during meiotic prophase of *Bombyx mori*. *Chromosoma* (Berlin) **79**:215–224.

Raveh, D., and A. Ben-Ze'ev. 1984. The synaptonemal complex as part of the nuclear matrix of the flour moth, *Ephestia kuehniella*. *Exp. Cell Res.* **153**:99–108.

Rhoades, M. M. 1950. Meiosis in maize. *J. Hered.* **41**:59–67.

Rhoades, M. M. 1961. Meiosis, p. 1–75. *In* J. Brachet and A. E. Mirsky (ed.), *The Cell*, vol. 3. Academic Press, Inc., New York.

Riley, R., and C. N. Law. 1965. Genetic variation in chromosome pairing. *Adv. Genet.* **13**:57–107.

Risley, M. S. 1986. The organization of meiotic chromosomes and synaptonemal complexes, p. 126–151. *In* M. S. Risley (ed.), *Chromosome Structure and Function.* Van Nostrand Rheinhold Co., New York.

Risley, M. S., S. Einheber, and D. A. Bumcrot. 1986. Changes in DNA topology during spermatogenesis. *Chromosoma* (Berlin) **94**:217–227.

Roeder, G. S. 1983. Unequal crossing-over between yeast transposable elements. *Mol. Gen. Genet.* **190**:117–121.

Rossignol, J. L., A. Nicolas, H. Hamza, and T. Langin. 1984. Origins of gene conversion and reciprocal exchange in *Ascobolus. Cold Spring Harbor Symp. Quant. Biol.* **49**:13–21.

Roth, T. F. 1966. Changes in the synaptinemal complex during meiotic prophase in mosquito oocytes. *Protoplasma* **61**:346–386.

Sears, E. R. 1976. Genetic control of chromosome pairing in wheat. *Annu. Rev. Genet.* **10**:31–51.

Sharp, P. J. 1986. Synaptic adjustment at a C-band heterozygosity. *Cytogenet. Cell Genet.* **41**:56–57.

Shimoda, C., A. Hirata, M. Kishida, T. Hashida, and K. Tanaka. 1985. Characterization of meiosis-deficient mutants by electron microscopy and mapping of four essential genes in the fission yeast *Schizosaccharomyces pombe. Mol. Gen. Genet.* **200**:252–257.

Simchen, G. 1974. Are mitotic functions required in meiosis? *Genetics* **76**:745–753.

Simchen, G. 1978. Cell cycle mutants. *Annu. Rev. Genet.* **12**:161–191.

Slightom, J. L., A. E. Blechl, and O. Smithies. 1980. Human fetal $^G\gamma$- and $^A\gamma$-globin genes: complete nucleotide sequences suggest that DNA can be exchanged between these duplicated genes. *Cell* **21**:627–638.

Smith, P. A., and R. C. King. 1968. Genetic control of synaptonemal complexes in *Drosophila melanogaster. Genetics* **60**:335–351.

Smithies, O., and P. A. Powers. 1986. Gene conversions and their relation to homologous chromosome pairing. *Philos. Trans. R. Soc. London B Biol. Sci.* **312**:291–302.

Solari, A. J. 1972. Ultrastructure and composition of the synaptonemal complex in spread and negatively stained spermatocytes of the golden hamster and the albino rat. *Chromosoma* (Berlin) **39**:237–263.

Solari, A. J. 1980. Synaptonemal complexes and associated structures in microspread human spermatocytes. *Chromosoma* (Berlin) **81**:315–337.

Spyropoulos, B., and P. B. Moens. 1984. The synaptonemal complex: does it have contractile proteins? *Can. J. Genet. Cytol.* **26**:776–781.

Stack, S. 1982. Two-dimensional spreads of synaptonemal complexes from solanaceous plants. I. The technique. *Stain Technol.* **57**:265–272.

Stack, S., and L. K. Anderson. 1986a. Two-dimensional spreads of synaptonemal complexes from solanaceous plants. II. Synapsis in *Lycopersicon esculentum* (tomato). *Am. J. Bot.* **73**:264–281.

Stack, S., and L. K. Anderson. 1986b. Two-dimensional spreads of synaptonemal complexes from solanaceous plants. III. Recombination nodules in crossing over in *Lycopersicon esculentum* (tomato). *Chromosoma* (Berlin) **94**:253–258.

Stack, S., and L. Anderson. 1987. Hypotonic bursting method for spreading synaptonemal complexes of *Zea mays. Heredity* **78**:178–182.

Stelly, D. M., S. J. Peloquin, R. G. Palmer, and C. F. Crane. 1984. Mayer's hemalum-methyl salicylate: a stain-clearing technique for observations within whole ovules. *Stain Technol.* **59**:155–161.

Stern, H. 1986. Meiosis: some considerations, p. 29–43. *In* A. V. Grimstone, H. Harris, and R. T. Johnson (ed.), *Prospects in Cell Biology* (*J. Cell Sci.*, Suppl. 4). The Company of Biologists Ltd., Cambridge.

Stern, H., and Y. Hotta. 1980. The organization of DNA metabolism during the recombinational phase of meiosis with special reference to humans. *Mol. Cell. Biochem.* **29**:145–158.

Sybenga, J. 1966. The zygomere as hypothetical unit of chromosome pairing initiation. *Genetica* (The Hague) **37**:186–198.

Symington, L. S., and T. D. Petes. 1988. Expansions and contractions of the genetic map relative to the physical map of yeast chromosome III. *Mol. Cell. Biol.* **8**:595–604.

Szankasi, P., C. Gysler, U. Zehntner, U. Leupold, J. Kohli, and P. Munz. 1986. Mitotic recombination between dispersed but related tRNA genes of *Schizosaccharomyces pombe* generates a reciprocal translocation. *Mol. Gen. Genet.* **202**:394–402.

Theivendirarajah, K., and H. L. K. Whitehouse. 1983. Further evidence that aberrant segregation and crossing over in *Sordaria brevicollis* may be discrete, though associated, events. *Mol. Gen. Genet.* **190**:432–437.

Tres, L. L. 1977. Extensive pairing of the XY bivalent in mouse spermatocytes as visualized by whole-mount electron microscopy. *J. Cell Sci.* **25:**1–15.

von Wettstein, D. 1977. The assembly of the synaptinemal complex. *Philos. Trans. R. Soc. London Biol. Sci.* **277:**235–243.

von Wettstein, D., S. W. Rasmussen, and P. B. Holm. 1984. The synaptonemal complex in genetic segregation. *Annu. Rev. Genet.* **18:**331–413.

Walmsley, M., and M. J. Moses. 1981. Isolation of synaptonemal complexes from hamster spermatocytes. *Exp. Cell Res.* **133:**405–411.

Weith, A., and W. Traut. 1980. Synaptonemal complexes with associated chromatin in a moth, *Ephestia kuehniella* Z. *Chromosoma* (Berlin) **78:**275–291.

Welch, J. W., D. H. Maloney, and S. Fogel. 1987. Synaptic relations in meiotic gene conversion at the iterated *cup1^r* locus of *S. cerevisiae. Exper. Suppl.* **52:**431–437.

Westergaard, M., and D. von Wettstein. 1972. The synaptonemal complex. *Annu. Rev. Genet.* **6:**71–110.

Wettstein, R., and J. R. Sotelo. 1971. The molecular architecture of synaptonemal complexes. *Adv. Cell Mol. Biol.* **1:**109–152.

Wilson, E. B. 1925. *The Cell in Development and Heredity*, 3rd ed. The Macmillan Co., New York.

Zickler, D. 1973. Fine structure of chromosome pairing in ten Ascomycetes. Meiotic and premeiotic (mitotic) synaptonemal complexes. *Chromosoma* (Berlin) **40:**401–416.

Zickler, D. 1984. Données récentes sur la prophase I de méiose (revue). *Ann. Sci. Nat. Bot. Biol. Veg.* **13:**177–197.

Zickler, D., L. de Lares, P. J. F. Moreau, and G. Leblon. 1985. Defective pairing and synaptonemal complex formation in a *Sordaria* mutant (spo44) with a translocated segment of the nuclear organizer. *Chromosoma* (Berlin) **92:**37–47.

Zickler, D., G. Leblon, V. Haedens, A. Collard, and P. Thuriaux. 1984. Linkage group-chromosome correlations in *Sordaria macrospora*: chromosome identification by three dimensional reconstruction of their synaptonemal complex. *Curr. Genet.* **8:**57–67.

Zickler, D., and J. M. Simonet. 1981. The use of mutants in the analysis of meiosis, p. 168–177. *In* H. G. Schweiger (ed.), *International Cell Biology 1980–1981*. Springer-Verlag, New York.

Chapter 16

Exchange and Chromosomal Segregation in Eucaryotes

R. Scott Hawley

I. INTRODUCTION

The laws of Mendel are the genetic consequences of the mechanisms by which eucaryotic chromosomes are segregated at meiosis. If we are to understand the mechanisms by which eucaryotic heredity operates, we must possess a thorough understanding of the mechanisms which underlie the behavior of chromosomes during meiosis. Specifically, we wish to understand how chromosomes pair, how

497

they recombine, how disjunctional partners are chosen, and how segregation occurs.

A. Relationship between Exchange and Segregation

The basic pathway of meiosis is pairing → exchange → disjunction. The mechanisms of pairing at the DNA and chromosomal levels are dealt with elsewhere in this volume (see especially Radding, Stasiak, and Egelman, and Giroux), as are the meiotic mechanisms by which the sites of exchange are chosen and the exchange events executed (see Carpenter). In this chapter I consider the relationship of pairing and exchange to the ultimate meiotic event, the proper segregation of homologous chromosomes at the first meiotic division.

As shown by Nicklas (1977), the ability of a bivalent to segregate reductionally at anaphase I is an intrinsic property of that bivalent, rather than of the meiosis I spindle. This is to say that a meiosis I bivalent maintains its ability to segregate reductionally even after it has been moved onto a meiosis II spindle. Nearly a century of genetic analysis has supported a model of chromosome segregation popularized by Darlington (1932), in which chiasmata are the distinctive features of meiotic bivalents that ensure segregation. Chiasmata are visible at diplotene-diakinesis as a region where two nonsister chromatids appear to precisely break and rejoin, and thus crossover between the two paired homologs.

We now accept that chiasmata are the cytological manifestation of genetic exchanges (Janssens, 1909; Beadle, 1932a; Brown and Zohary, 1955; Jones, 1971). However, this was a vigorously debated issue during the first half of this century. A discussion of the controversy and the evidence that eventually settled it is presented by Whitehouse (1969).

B. Chiasma Formation

Exchange occurs between paired homologs in early prophase and results in the physical exchange of genetic material between two nonsister chromatids (Creighton and McClintock, 1931; Stern, 1931). The ultrastructural manifestations of exchange are the so-called late recombination nodules. These nodules appear during pachytene with both a frequency and a distribution paralleling that of normal meiotic exchange. More extensive evidence that late nodules mark the site of reciprocal meiotic exchanges (and are in fact involved in executing such events) is reviewed by Carpenter (this volume).

Holm and Rasmussen (1980) were able to follow the formation of chiasmata from late recombination nodules in the spermatocytes of the silk moth, *Bombyx mori*. By mid- to late pachytene the nodules, which are attached laterally to the synaptonemal complex, appear larger and more electron dense. By late pachytene the nodules are directly associated with the chromatin and have become still larger and irregular in shape. These presumed derivatives of the late nodules are known as chromatin nodules. Most of the synaptonemal complex is stripped from the bivalent at the beginning of diplotene. However, the chromatin nodules are in most cases still associated with short remnants of the synaptonemal complex. By late diplotene the site of the chiasma is marked by a circular structure which bridges the two homologs. By the beginning of metaphase the circular structure is surrounded by a chromatin bridge. The evidence that chromatin nodules, circular structures, and chromatin bridges represent sequential modifications of the late recombination nodules during chiasma formation has been reviewed by von Wettstein et al. (1984). The basis of this argument is that the distributions and frequencies of these elements are similar and that the various structures appear sequentially. Moreover, as pointed out by von Wettstein et al. (1984), "reconstructions of mid-diakinesis nuclei revealed that these chromatin bridges with their circular component constitute the chiasmata." Finally, it should be noted that the disappearance of the structures at metaphase correlated well with the timing of chiasma resolution.

Similar observations regarding the meiotic ultrastructure of chiasma formation have been made by Holm et al. (1981) in the mushroom *Coprinus cinereus*. Observations of fragments of synaptonemal complex associated with recombination nodules, or their derivatives, in a number of other organisms were reviewed by von Wettstein et al. (1984). Although sequential transformations of nodules into chiasmata have not been observed in all organisms, the data presented here indicate that in at least some organisms the formation of a mature chiasma requires functions that occur subsequent to the recombinational event, such as modifications of the recombination nodule and/or the synaptonemal complex. The resulting chiasmata hold the bivalent together during homolog repulsion at diplotene-diakinesis and continue to link the bivalents together until anaphase.

C. Mechanical Basis of Chiasma Function at Metaphase

To quote Maguire (1974), "persistent chiasmata normally serve the vitally important function of assuring the regular disjunction of homologous centromeres." The manner by which chiasmata ensure disjunction is explained by a simple physical model described below. During prometaphase, each centromere (or kinetochore) of the bivalent attaches to the spindle and begins to move toward one of the two poles.

Most bivalents achieve a bipolar orientation immediately. For example, in *Tipula oleracea* 90% of the bivalents achieve a bipolar orientation immediately (Bauer et al., 1961). Similar observations have also been made in *Melanoplus differentialis* (Nicklas, 1967). The high frequency of initial proper orientation is due to the fact that chromosomal spindle fibers connect to the pole to which a given centromere most nearly points (Nicklas, 1967). Moreover, at the start of prometaphase the two homologous centromeres are usually oriented in opposite directions, such that if one centromere is pointed at one pole, the other centromere is pointed at the opposite pole (Ostergren, 1951, as modified by Nicklas, 1967). In the words of Nicklas (1974), "while the upper half-bivalent's kinetochores most nearly face the upper pole, those of its partner most nearly face the lower pole, simply because bivalents are so constructed." Moreover, as might be expected, in long flexible bivalents the frequency of successful orientation is decreased (Nicklas, 1971). Thus, the tendency for proper orientation may be

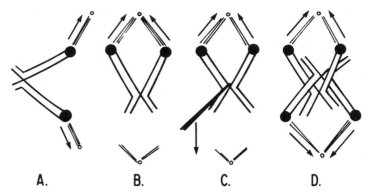

FIGURE 1. Bivalent orientation and stability. (A) A properly oriented and stable bivalent. (B) A maloriented bivalent. (C) A maloriented bivalent stabilized by pulling on the chiasma with a micromanipulator needle. (D) Two interlocked and maloriented bivalents. See text for explanation. (Modified from Nicklas, 1974.)

simply thought of as the ability of chiasmata to constrain the kinetochores of two homologs such that they are oriented in opposite directions (Nicklas, 1974).

Once the two centromeres have oriented toward opposite poles, the progression of the two centromeres toward the poles is halted at the metaphase plate by the chiasma. This represents a stable position in which the bivalent will remain until anaphase I (Fig. 1A). However, some bivalents do become maloriented so that the two centromeres are attached to the same pole. In this case the entire bivalent moves to that pole during prometaphase (Fig. 1B), at which time the centromeres are able to reorient (Hughes-Schrader, 1943; Bauer et al., 1961). If the centromeres reorient in opposite directions, the bivalent will move toward the center of the spindle and then be stabilized at the plate by a chiasma. This process of poleward movement and reorientation continues until a stable bipolar orientation is achieved.

Hughes-Schrader (1943) suggested that reorientation might involve the loss of spindle fiber (microtubule) attachments to one pole followed by the establishment of a new attachment to the same or opposite pole. That this is the case has been demonstrated by experiments in which a micromanipulator was used to physically detach a bivalent from the spindle in living grasshopper spermatocytes (Nicklas, 1967; Nicklas and Staehly, 1967). Following detachment, the bivalent was placed in various positions within the cell. After a pause of variable duration, the detached chromosomes moved, centromeres first, back to the spindle and centromere-pole attachments reformed. Moreover, a given centromere, attached to one pole prior to detachment, often reorients toward the opposite pole upon reattachment. Thus, centromeres are capable of breaking and reforming spindle attachments until a stable position at the metaphase plate is achieved.

Stability at the metaphase plate is achieved by balancing the tension forces between two oppositely oriented centromeres and the two poles of the spindle. Indeed, Nicklas and Koch (1969) and Henderson and Koch (1970) have demonstrated that the application of tension by artificial means can stabilize maloriented bivalents (Fig. 1C and D). In the first case stabilization is achieved by tugging on the chiasma of a maloriented bivalent with a micromanipulator needle. In this case the two centromeres do not reorient, but rather the malorientation is stable indefinitely. If this tension is not released before anaphase I, the two homologs will proceed to the same pole (nondisjunction). If, however, the tension is released prior to anaphase I, reorientation almost always follows, until bipolar orientation is achieved. In the second case a maloriented bivalent is stabilized by interlocking it with a second maloriented bivalent. This event results in nondisjunction of both of the bivalents involved. As shown by Buss and Henderson (1971), interlocked bivalents can also be produced by heat shock during meiotic prophase (i.e., without micromanipulation). These interlocked bivalents also achieve a stable orientation and undergo simultaneous nondisjunction.

Thus, the ability of bivalents to orient their centromeres toward opposite poles is achieved by balancing the tension between the bivalent and the two spindle poles. This balancing of tension requires chiasmata to hold the two homologs together.

D. Behavior of Achiasmate Bivalents

In many organisms achiasmate bivalents fall apart at diplotene-diakinesis, resulting in the premature dissociation of the bivalent into two univalents prior to prometaphase I. When this occurs, the two homologs attach to the spindle separately, and stability at the metaphase plate is never achieved. Rather, both chromosomes move up and down the spindle, frequently reorienting. Their disjunctional fate at anaphase I is determined solely by their orientation on the spindle at the time anaphase begins. In this instance normal disjunctional events and nondisjunctional events (in which both homologs proceed to the same pole) occur with equal frequency. Workers using a variety of organisms have also shown that such univalents are frequently unstable; chromosome loss and breakage are not uncommon events at both meiotic divisions and in subsequent mitotic divisions (Sears, 1952; Prakken, 1943; see below). Thus, chiasmata provide an essential device for stabilizing bivalents at the metaphase plate during prometaphase. Indeed, in most organisms, prometaphase continues until all the bivalents are stably balanced on the metaphase plate. The cell then waits briefly in metaphase until anaphase begins.

The distinguishing event that marks the end of metaphase and the beginning of anaphase is the abolition of sister chromatid cohesion except at the centromere. The release of sister chromatid cohesion dissolves the chiasma linking the two homologs, which then proceed to opposite poles of the spindle. Whether chiasmata undergo terminalization (i.e., move or wind to the telomeres) prior to sister chromatid separation at the end of metaphase remains a hotly debated issue. Although some observations would seem to require at least some movement of the chiasma as the bivalent moves from prophase to metaphase (Maguire, 1979), it is clear that, in at least one case, the observed terminality of chiasmata at metaphase I actually reflects differences in the extent of condensation along the bivalent (Jones, 1978). In other organisms, there does not appear to be any change in chiasma distribution prior to metaphase (see Jones, 1977). This issue is fully discussed by Carpenter (this volume).

In summary, the vast number of data argue that chiasmata are the physical manifestation of exchange and support a model in which chiasmata are necessary and sufficient to ensure disjunction. If chiasmata are in fact the primary means of ensuring disjunction, then there are two testable corollaries: (i) genetic exchange should be necessary and sufficient for normal disjunction, and (ii) the failure of exchange should lead to failed disjunction. Moreover, several alternative segregation systems do exist that bypass the requirement for exchange; these are themselves illuminating.

II. EXCHANGE IS USUALLY NECESSARY AND SUFFICIENT FOR DISJUNCTION

I will begin by inquiring whether there is enough exchange to ensure disjunction. I will first consider the disjunctional consequences of reducing exchange and then ask whether an exchange event is really entirely sufficient to ensure disjunction.

A. How Much Exchange Is There?

If chiasmata are the major means of ensuring disjunction, there must be enough exchange events to ensure at least one exchange per bivalent in the vast majority of meioses. Indeed in most organisms the average number of exchanges per cell is greater than n, the haploid number of chromosomes (i.e., there are more chiasmata per cell than there are bivalents). Moreover, the distribution of the number of chiasma per bivalent is much narrower than that predicted by a Poisson distribution (for review, see Jones, 1984, 1987), so that there are almost no achiasmate bivalents and few with high numbers of chiasmata.

For example in rye (*Secale cereale*) only 1.4% of the bivalents are achiasmate. The remaining 98% of the bivalents possess one (26.2%), two (70.5%), or three (1.7%) exchanges (Jones, 1967, 1974). These observations differ markedly from those predicted by a Poisson distribution of chiasmata per bivalent, in that there are large excesses of bivalents with one or two chiasmata and corresponding deficiencies of bivalents with zero or two chiasmata. Thus, in rye not only is there a sufficient number of chiasmata to ensure disjunction, but also those chiasmata are distributed so that the vast majority of bivalents have at least one chiasma. That this distribution is under tight genetic control is indicated by a genetic variant in which the chromosomal distribution of chiasmata is randomized, which is to say that the frequencies of bivalents with zero, one, two, or three chiasmata correspond to a Poisson distribution (Jones, 1967). In this variant achiasmate bivalents are frequent, as are the resulting univalents, despite the fact that the total number of chiasmata remains unchanged (Jones, 1967).

Although virtually all organisms have at least n chiasmata per cell, the number of chiasmata per cell is not sufficient to guarantee one exchange per bivalent if chiasmata are distributed randomly. Rather, there exist regulatory systems that ensure the formation of at least one chiasma per bivalent at each meiosis. There are, in addition, subchromosomal controls, including regional differences in the frequency of exchange, a universal absence of exchange in heterochromatin, and chiasma interference (for review, see Jones, 1987; Carpenter, this volume). As discussed below, both interference and regional differences in exchange frequency can be altered by mutations that reduce the overall frequency of exchange. This suggests a commonality to the control mechanisms responsible for chiasma distribution and for chiasma interference.

Although the nature of these controls of chiasma distribution remains a mystery, it is tempting to speculate that such constraints on chiasma distribution have functional significance for the actual mechanics of bivalent orientation and disjunction. For example, it has been clearly shown in *Drosophila melanogaster* that exchange is strongly suppressed in the vicinity of the centromere (Mather, 1939; Beadle, 1932b). Perhaps exchanges occurring too close to the centromere would prevent the proper orientation or separation of the bivalent.

B. Reducing Exchange Increases Nondisjunction

If exchange provides the basis for disjunction, then mutations that reduce the frequency of crossing-over are predicted to increase the frequency of nondisjunction. There are a host of organisms in which recombination-deficient mutations have been studied; these include *D. melanogaster*, corn (*Zea mays*), yeast (*Saccharomyces cerevisiae*), rye (*Secale cereale*), tomato (*Lycopersicon esculentum*), and wheat (*Triticum aestivum*). Such mutations in these organisms were reviewed by Baker et al. (1976).

The study of the mutations that reduce exchange is hampered by the fact that in most organisms achiasmate bivalents fail to segregate properly. This leads to nondisjunction at meiosis I and therefore to aneuploid gametes. Since aneuploidy is generally lethal, special techniques or circumstances are required for a systematic study of the fate of achiasmate bivalents. Three types of approaches have been successful in this regard. First, in organisms with good meiotic cytology the meiotic fate of nonexchange bivalents can be followed directly. The difficulty in this approach lies in having cytological markers that allow for the unambiguous determination of the exchange status of chromosomes at anaphase. As discussed below, the existence of easily visible chromosomal polymorphisms in corn allows one to circumvent this problem. Second, although aneuploid spores in *S. cerevisiae* are usually inviable, certain mutations allow meiotic cells to finish prophase and then proceed directly to meiosis II without a reductional division and thus to produce two diploid gametes (Malone and Esposito, 1981; Klapholtz et al., 1985). Since the failure of disjunction of achiasmate bivalents occurs at anaphase I, these mutations bypass the step at which nondisjunction occurs and thus prevent the lethality due to the production of aneuploid spores. The extent of aneuploidy resulting from an exchange reduction can then be measured by comparing spore viability in the presence or absence of the first meiotic division. Finally, in *D. melanogaster* the combination of a small number of chromosomes,

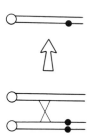

FIGURE 2. Exchange in a bivalent heterozygous for a chromosomal knob. The open circles represent centromeres; the knob is indicated as a filled circle.

viable aneuploids, and a disjunctional backup system allows one to easily assess the disjunctional consequences of reducing exchange. Each of these approaches is discussed briefly below.

1. Meiotic mutations in corn

The literature on mutations that suppress exchange in plants is voluminous and cannot be thoroughly reviewed here. The interested reader is referred to Baker et al. (1976). I will consider the asynaptic mutation in maize as a typical example.

The asynaptic mutation was discovered by Beadle and McClintock (1928). In meiocytes bearing this mutation, synapsis at zygotene and pachytene is frequently incomplete, with short arms and the intercalary regions of long arms having the greatest tendency for asynapsis. This pattern of asynapsis correlates with the observed distribution of chiasmata at diplotene-diakinesis. This suggests that the primary defect in asynaptic meiocytes is in synapsis and that exchange occurs only in chromosomal regions in which synapsis is achieved.

The amount of asynapsis induced by the asynaptic mutation varies greatly from plant to plant. The extent of asynapsis in a given plant is highly correlated with the fraction of bivalents in that plant which precociously separate to form univalents (Miller, 1963). That those chromosomes which nondisjoin are in fact nonexchange was demonstrated by examining plants in which one or more of the homolog pairs were heterozygous for a distally located chromosomal knob. Knobs are dispensable bits of heterochromatin which can be heterozygous and can be followed cytologically. Thus, they provide useful cytogenetic markers for the presence or absence of exchange on a given bivalent. As shown in Fig. 2, in a knob heterozygote exchange between the centromere and a knob will result in the formation of two homologs in which the sister chromatids are differentially marked (one chromatid has a knob while the other is knobless). These are called heteromorphic dyads, in contrast to homomorphic dyads (the nonexchange state) in which both chromatids are alike (either knobbed or knobless).

As shown by Miller (1963), those chromosomes which have become univalents at diakinesis are never observed to be heteromorphic for the knob. On the other hand, all bivalents that segregate normally at anaphase I are pairs of heteromorphic dyads. Thus, the failure of proper disjunction at anaphase I can be directly attributed to a failure of exchange. In the presence of exchange, the bivalent is maintained during the homolog repulsion that occurs at diplotene-diakinesis, and the bivalent can thus orient as a bivalent of the metaphase plate. In the absence of exchange the bivalent separates precociously at diplotene-diakinesis to form two univalents which attach to the spindle independently of each other.

These data argue that in the absence of exchange, nondisjunction is a frequent event. Moreover, the resulting univalents are extremely unstable. For example, univalents frequently divide equationally at meiosis I (i.e., sister chromatid disjunction at anaphase I). The resulting monads are frequently lost at meiosis II. The centromeres of univalents are also prone to undergo misdivision. Misdivision describes the transverse, as opposed to equatorial, division of a univalent centromere. The result is two truly telocentric chromosomes, each with two chromatids. Rehealing of the two broken ends at each half-centromere results in the formation of isochromosomes. Indeed, in plants homozygous for the asynaptic mutation, univalents were shown to undergo equational division, meiotic loss, and misdivision at high frequencies (Miller, 1963).

The meiotic instability of univalents is not unique to this mutation or to corn as a species. Sears (1952) showed that in wheat univalents are prone to misdivision as well as to loss or misbehavior at meiosis II. Similar observations regarding the loss or instability of univalents have also been made in *D. melanogaster* (Sturtevant, 1929; Sandler and Braver, 1954; Baker, 1975; Carpenter, 1973). Thus, the requirement for exchange goes beyond simply ensuring disjunction. Rather, proper maintenance of the bivalent is essential to maintain chromosome integrity.

2. Recombination-deficient mutations in *S. cerevisiae*

Because of the enormous genetic and molecular flexibility available to the yeast geneticist, organisms such as *S. cerevisiae* are excellent tools for the study of meiotic phenomena. Indeed, studies in this organism strongly support the argument that normal levels of genetic exchange are required for the proper segregation of homologous chromosomes at meiosis I.

Since *S. cerevisiae* has 17 chromosomes, it has, until recently, been difficult to analyze the relationship between recombination and proper meiotic segregation in this organism. This is to say that events that increase the frequency of nondisjunction reduce the probability that each of the four products of meiosis (ascospores) will receive at least one copy of every chromosome, resulting in a high frequency of inviable ascospores. Indeed, "mutations creating deficiencies in meiotic recombination would therefore be expected to lead almost exclusively to the formation of inviable ascospores" (Malone and Esposito, 1981).

Given this problem of failed disjunction leading to virtual sterility, it initially proved difficult to identify and analyze recombination- and/or disjunction-deficient mutants in *S. cerevisiae*. The first available tool for this purpose involved exposing cells only briefly to sporulation medium. Under these conditions genetically normal cells undergo high frequencies of recombination and then return to mitotic growth. Thus, cells can undergo normal levels of both gene conversion and meiotic crossing-over without being irrevocably committed to a reductional division at meiosis I. Using this method, Game et al. (1980) and Prakash et al. (1980) suggested that a number of radiation-sensitive (*rad*) mutations, which also caused ascospore inviability, were indeed recombination defective. The problem with this technique lies in the fact that substantial losses in viability after exposure to sporulation medium were observed for most of the mutations tested. This raises the possibility that a postmeiotic block might also prevent a return to meiosis, and thus, the only cells which would recover are those which have not yet undergone recombination. Such an artifact would result in the erroneous identification of a deficiency in recombination.

The identification of the *spo13-1* mutation allowed the normal meiotic requirement for segregation at meiosis I to be circumvented (Malone and Esposito, 1981; Klapholtz et al., 1985). Diploids homozygous for *spo13-1* (but otherwise genetically normal) undergo normal levels of meiotic recombination, bypass reductional segregation, and then undergo a single equational (i.e., mitotic-like) division. These events produce asci bearing two diploid spores. Thus, if exchange is necessary for normal disjunction, and if failed disjunction is the cause of sterility in recombination-deficient mutations, then this sterility should be rescued by *spo13-1*. Indeed, although the recombination-deficient mutation *rad50-1* produces few or no viable ascospores after meiosis, the *rad50-1 spo13-1* double mutation produces mainly viable but nonrecombinant two-spored asci after meiosis and sporulation (Malone and Esposito, 1981). These studies demonstrated that the *rad50-1* mutation reduced exchange by 150-fold and that this exchange reduction was lethal only when followed by a reductional division.

Similar studies have demonstrated that the defects in segregation and spore viability observed in diploids homozygous for another mutation, *spo11-1*, result from a defect in meiotic recombination (Klapholtz et al., 1985). Although synaptonemal complexes are formed when *spo11-1* diploids enter meiosis, recombination is reduced by more than 100-fold. After the two meiotic divisions, fewer than 1% of the asci are viable. However, *spo11-1 spo13-1* diploids have high spore viability. If the presence of synaptonemal complex in *spo11-1* diploids may be taken as an indicator that homologous pairing is not greatly affected by this mutation, then the failed segregation is best explained by failed recombination.

3. Recombination-deficient mutations in *D. melanogaster*

There have been several large-scale searches for mutations that disrupt female meiosis in *D. melanogaster* (Sandler et al., 1968; Baker and Carpenter, 1972). Each of these screens has been based on detecting mutations which increased nondisjunction (for review, see Baker and Hall, 1976). Perhaps not surprisingly, for most of these mutants increased nondisjunction is accompanied by decreased exchange. I will consider four of these mutations, *c(3)G*, *mei-41*, *mei-9*, and *mei-218*, in some detail to demonstrate the relationship of the exchange and disjunctional defects.

Recombinational effects. Each of these mutations reduces exchange on the X chromosome as well as on the major autosomes (Baker and Carpenter, 1972; Carpenter, 1979) and thus is defective in general cell functions rather than chromosome-specific functions. The degree of exchange reduction ranges from absolute in the presence of *c(3)G* to just over a twofold decrease in the presence of *mei-41*. Both *mei-9* and *mei-218* reduce exchange by approximately 13-fold.

Females homozygous for *c(3)G* display an almost complete absence of meiotic exchange (Gowen and Gowen, 1922; Hall, 1972). There is insufficient residual exchange to allow any inference as to whether or not *c(3)G* affects the distribution of the few remaining exchanges.

However, in the cases of *mei-41* and *mei-218* exchange is reduced less severely and in a polar manner, such that distal euchromatic exchanges are more strongly reduced than are proximal euchromatic exchanges. This is the converse of the exchange distribution observed in wild-type meioses. Chiasma interference (as measured by the frequency of double exchanges) is also altered by *mei-41* (Baker and Carpenter, 1972; but see Baker and Hall, 1976). Following a convention established by Bridges (1916), subsequent workers have suggested that mutants like *mei-41* and *mei-218*, which alter the constraints on chiasma distribution, are defective in an aspect of the recombination process which determines the positions at which an exchange can occur (Sandler et al., 1968; Lindsley et al., 1968; Carpenter and Sandler, 1974). These mutations are thus said to be precondition mutations.

As pointed out by Baker and Carpenter (1972), these precondition mutations act both to reduce the total frequency of exchange and to make the residual exchange more uniformly distributed with respect to the physical length of euchromatic regions (i.e., to void the positional constraints on chiasma localization). Indeed, even the fourth chromosomal euchromatin becomes available for exchange in the presence of such mutations (Sandler and Szauter, 1978). These mutations do not, however, obviate the block to heterochromatic exchange (Carpenter and Baker, 1982). This is most likely a consequence of the fact that the block to heterochromatic exchange is the abnormal morphology of synaptonemal complex in heterochromatic intervals (Carpenter and Baker, 1974), rather than the result of the more global and *trans*-acting functions which regulate exchange in the euchromatin.

The five known alleles of *mei-9* differ in the frequency of residual exchange (ranging from residual exchange frequencies of 8 to 16% of the exchange observed in controls). Unlike the precondition mutations described above, *mei-9* reduces the frequency of exchange events along the chromosomes in a manner proportional to exchange in wild type and does not appear to affect chiasma interference. Following the criteria described above, *mei-9* is considered to be defective in the exchange process itself, rather than in the establishment of preconditions for exchange.

No normal synaptonemal complex is visible in meiocytes of females homozygous for *c(3)G* (Meyer, 1964; Smith and King, 1968; Rasmussen, 1975). This suggests that *c(3)G* may encode either some function necessary for normal synapsis or some component of the synaptonemal complex itself. On the other hand, Carpenter (1979) demonstrated that *mei-41*, *mei-9*, and *mei-218* do not affect synaptonemal complex structure, continuity, or temporal behavior. This suggests that, at least in a gross sense, chromosome pairing proceeds normally in these mutants. However, two of these mutations, *mei-41* and *mei-218*, affect both the number and morphology of spherical recombination nodules in a manner which parallels their effects on recombination (Carpenter, 1979 and this volume).

Disjunctional effects. The four recombination-defective mutations described above increase nondisjunction of all chromosomes at meiosis I dramatically. There is no effect on disjunction at meiosis II or on either meiotic division in males, which have one of the rare exchange-independent meiotic systems (see below). Several lines of evidence argue that the increase in nondisjunction is not the consequence of a defect in the segregational system, but rather is the result of an indirect effect of the suppression of exchange caused by these mutations.

First, a comparison of the various exchange-deficient mutants studied reveals that the increase in nondisjunction is proportional to the extent of the exchange defect (for review, see Baker and Hall, 1976). Second, nondisjunction is limited to those bivalents in which exchange did not occur, while exchange bivalents disjoin normally (Parry, 1973; Carpenter and Sandler, 1974). Therefore, the residual exchanges that do occur are sufficient to ensure the disjunction of the bivalents involved, and the chromosomes that fail to disjoin normally are the chromosomes that have failed to undergo exchange. Third, the observed patterns of nondisjunctional events (such as nonhomologous segregations) follow the rules for achiasmate segregations via the distributive system in *D. melanogaster* (see below).

The simple expectation was that the relationship between the amount of nondisjunction and the fraction of nonexchange tetrads would be linear. In actuality, the frequency of X nondisjunction is proportional to the cube of the frequency of nonexchange X tetrads (see Baker and Hall, 1976). This nonlinear relationship is due to the presence of a second disjunctional system in *D. melanogaster*. This system, known as distributive segregation, regularizes the segregation of nonexchange chromosomes. Although the distributive system is described in detail below, to be able to understand the fate of achiasmate bivalents in *D. melanogaster*, we must first consider it briefly here.

Wild-type *Drosophila* females have fairly high frequencies of nonexchange bivalents; the frequency of nonexchange X chromosomal bivalents (E_0) is approximately 5%, and E_0 frequencies for the autosomes are only slightly lower (2 to 4%). Nevertheless, the normal frequency of nondisjunction for these bivalents is on the order of 1 to 5 nondisjunctional events per 1,000 meioses, suggesting that the homologs composing these bivalents normally segregate to opposite poles at anaphase I. Moreover, the tiny fourth chromosomes are always achiasmate; yet, fourth chromosome segregation is highly precise (the frequency of spontaneous nondisjunction is approximately 5 nondisjunctional events per 1,000 meioses). These data do not contradict the thesis that chiasmata are necessary for normal disjunction; rather, they reveal the existence of an achiasmate backup system (distributive segregation) which can, to some extent, provide for normal disjunction when exchange does not occur (Grell, 1976; but see below).

The distributive system also protects the meiotic system from the effects of regional depressions in the frequency of exchange. For example, in females heterozygous for an X chromosome aberration that strongly suppresses recombination, such as the multiply inverted X chromosomal balancer *FM7*, X chromosome disjunction remains essentially normal as a consequence of the distributive system.

Unfortunately, the distributive system segregates chromosomes on the basis of size and shape, rather than on the basis of homology (see below). Thus, the presence of one pair of large chromosomes appears to represent a natural limit of the system's capacity to ensure regular disjunction. Since the larger chromosomes are similar in size, in meioses in which both the X chromosomes and a pair of major autosomes are nonexchange, the distributive system becomes overloaded, and XX ↔ AA disjunctions (i.e., cases where two X chromosomes segregate from two autosomes) become commonplace (Carpenter, 1973).

That these XX ↔ AA segregations are mediated by the distributive system was shown by Baker and Carpenter (1972). They examined *mei-9*–induced nondisjunction in females homozygous for the *nod* mutation, which oblates the distributive system (Carpenter, 1973). Although simultaneous nondisjunctional events in *mei-9* females are almost entirely nonhomologous (XX ↔ AA), in *mei-9 nod* females the simultaneous nondisjunctional events are now randomized (i.e., XX ↔ AA segregations are no more frequent than 0000 ↔ XXAA).

Thus, the most reasonable explanation for the nonlinear relationship between the frequency of nonexchange X chromosomes and nondisjunction is that a failure of exchange on two or more bivalents is necessary for nondisjunction. In meiocytes with only a single nonexchange bivalent, regular segregation can be guaranteed by the distributive system. Only when both the X chromosomes and one pair of major autosomes are nonexchange does X nondisjunction result. Accordingly, the frequency of X chromosome nondisjunction would not be expected to rise linearly with the fraction of X chromosome E_0s. Instead, the frequency of X chromosome nondisjunction would be expected to be linearly related to a metric that assays the fraction of meiocytes in which the X chromosome and both arms of one major autosome are simultaneously nonexchange.

Given that the X chromosome is similar in size to the arms of the two large metacentric autosomes, and that failures of exchange on different chromosomal arms are independent events (Parry, 1973), the probability that the X chromosome and both arms of one major autosome will be simultaneously nonexchange may be estimated to be E_0^3. This in fact is the observed result. Numerous other results also support this model (Baker and Hall, 1976).

Cytology. While cytological data in support of this model are scanty, some data are available for wild type and for the *c(3)G* mutation (Puro and Nokkala, 1977). Although metaphase and anaphase cytology in *D. melanogaster* is difficult, some observations have been made in both *D. melanogaster* and *D. subobscura*. For example, in normal female meiosis, distributively segregating chromosomes, such as the fourth chromosome, appear to segregate to the poles precociously. This suggests that distributive segregation may be similar to the system of

high frequency (the segregation pattern here is complex because all four chromosomes enter the distributive system). Thus, we can say that exchange is both necessary and sufficient to ensure the regular disjunction of these two chromosomes despite the fact that they possess nonhomologous centromeres (i.e., the disjunction of two centromeres at anaphase I is not a consequence of centromere homology, but rather of the linkage of those two centromeres by exchange). Exchange and disjunction in females heterozygous for translocations involving the X and Y chromosomes are similar (Hawley, 1980, unpublished data).

Another example of the ability of homologous exchanges to ensure the segregation of nonidentical centromeres is X-Y disjunction in mammalian males. In mammalian males X-Y homology is limited to a small region at the tip of the short arms of these two chromosomes, the so-called pseudo-autosomal region (Cooke et al., 1985; Simmler et al., 1985). On the basis of studies using both light and electron microscopy, X-Y pairing also appears to be limited to this chromosomal region (Pearson and Bobrow, 1970; Moses et al., 1975). Moreover, in males exchange occurs in this interval at a frequency of approximately one exchange per meiosis (Singh and Jones, 1982; Rouyer et al., 1986). These observations support a half-century-old model of Koller and Darlington (1934), which postulated a requirement for X-Y pairing and recombination to ensure sex chromosome segregation.

B. Nonhomologous Exchanges Ensure Nonhomologous Disjunctions

As noted above, breakage and repair of two nonhomologous chromosomes can result in a translocation. Early studies of radiation genetics in *D. melanogaster* showed that when females were irradiated during meiosis but prior to anaphase I, only one element of the newly induced translocation (so-called half-translocations) could be recovered, although both elements of the translocation could be recovered from irradiated sperm. In an elegant series of papers, Parker and his students solved this mystery by demonstrating that when females are irradiated prior to anaphase I, half-translocations result from breakage and repair of nonhomologous chromatids at the four-strand stage (i.e., just as reciprocal exchanges involve any two nonsister chromatids), and that this radiation-induced exchange between nonhomologs is sufficient to ensure the meiosis I disjunction of the two chromosomes involved (for review, see Parker and Williamson, 1976).

C. When Exchange Does Not Ensure Disjunction

Two exceptions to the rule that exchange ensures disjunction are worth noting. First, when Merriam and Frost (1964) examined the exchange status of spontaneous diplo-X ova in *D. melanogaster*, they observed that 49.5% of these nondisjunctional events had occurred in meiocytes with double and triple exchanges between the X chromosomes. Since double and triple exchanges occurred in only 29.7% of the meiocytes in which regular disjunction occurred, they suggested that too many exchanges may occasionally present mechanical difficulties in separating homologs. That this is not a common problem is shown by the fact that the frequency of spontaneous X chromosomal nondisjunction in meiocytes with two or three exchanges was only 0.0016.

The second exception arose during the analysis of the meiotic behavior of artificial chromosomes in *S. cerevisiae*. Dawson et al. (1986) recently uncovered what seems to be a yeast analog of the distributive system in *D. melanogaster*. This system is "capable of segregating pairs of nonrecombinant artificial chromosomes, regardless of the extent of their sequence homology." Even more relevant to our concerns is their observation that the fidelity of such segregations is not improved by genetic exchange (in one-third of the observed cases of recombination, the homologs proceeded to the same pole). Perhaps these artificial chromosomes are simply too small for chiasmata to function, or perhaps specific chromosomal sites are required to hold sister chromatids together at meiosis and thus ensure chiasma function.

IV. ARE THERE MUTATIONS THAT AFFECT CHIASMA FUNCTION?

In vacuo, it is not obvious why an exchange should hold two homologs together. Why doesn't the chiasma just slip off during homolog repulsion? However, although homologs repel each other at diplotene, sister chromatids do not separate until anaphase I. Darlington (1932) suggested that this sister chromatid cohesion generates resistance to the free separation of the bivalents and serves as a chiasma binder. In other words, the mechanical model that underlies discussions of chiasma function is one in which homologs are held together as long as sister chromatid cohesion is maintained on both sides of the chiasma. The release of this cohesion, which occurs at anaphase I, allows homologs to separate at the appropriate time (once centromere orientations have been established).

If there is a chiasma binder, whether or not it is simply sister chromatid cohesion, then we should expect to find mutations which impair chiasma function without impairing exchange. Indeed, several such mutations have been identified in corn, *D. melanogaster*, and *S. cerevisiae*. At least in two cases the defect in chiasma maintenance is also associated with a failure of normal sister chromatid cohesion. This correlation is consistent with Darlington's model and also with a more modern model in which retention of the lateral elements of the synaptonemal complex or some remnant of these structures, through the end of metaphase I, is postulated to be the mechanism for sister chromatid cohesion (Maguire, 1978a).

A. Desynaptic Mutation in Corn

The desynaptic mutation (*dy*) provides the most elegant evidence that exchange and chiasma function are separate events under separate genetic control (Maguire, 1978a). Although levels of exchange are normal, chiasmata frequently dissolve at diplotene-diakinesis in maize microsporocytes. These attributes were deduced by examining the disjunctional behavior of bivalents which were heterozygous for a distally located chromosomal knob. Pachytene synapsis appears to occur normally in maize microsporocytes homozygous for *dy*. However, following pachytene synapsis precocious desynapsis of paired homologs is frequently observed. Indeed, by diakinesis bivalents have often separated to pairs of univalents, each with a knob-carrying and a knobless chromatid, which now disjoin independently (see Fig. 2).

As noted above, exchange between heterozygous knobs and the centromere gives rise to heteromorphic dyads when sister chromatid cohesion lapses (normally at anaphase I). Heteromorphic dyads are seen at diakinesis in maize microsporocytes homozygous for *dy*, and the frequency is that expected from wild-type levels of exchange between the knob and the centromere. The *dy* mutation does not appear to reduce the frequency of exchange; rather, the precocious separation of such exchange bivalents prior to anaphase I demonstrates a failure of the chiasma to be maintained.

A second aspect of the *dy* mutation also deserves consideration. At prophase II the chromosomes show precocious sister chromatid separation, suggesting a defect in sister centromere cohesion. Maguire (1974, 1978b, 1979, 1982) has argued that both properties depend on the normal functioning of the synaptonemal complex. A great deal of evidence suggests that at least short stretches of the synaptonemal complex are essential and sufficient for exchange (for reviews, see von Wettstein et al., 1984; Carpenter, 1987). However, in maize, and in most organisms, the synaptonemal complex extends the full length of each bivalent and thus connects the chiasma with the two centromeres. Rather than assuming that the extension of the synaptonemal complex along the entire bivalent plays some further role in exchange, Maguire (1979) proposed that the synaptonemal complex also functions by providing for "the sister chromatid cohesiveness required for maintenance of chiasmata."

If the synaptonemal complex does provide a means of ensuring sister chromatid cohesion, then we can imagine a simple model of chiasma function in which the lateral elements of the synaptonemal complex hold sister chromatids together. The resulting cohesions on either side of the exchange then serve to link the two centromeres together. The balancing of spindle forces on these two linked centromeres (see section I.C.) will then guarantee proper orientation at metaphase I. The major difficulty with this model is that, although in some organisms remnants of the synaptonemal complex persist between sister chromatids until metaphase (Moens and Church, 1979), in most organisms the synaptonemal complex is stripped from the bivalent at diplotene. Thus, one has to propose either that the ability of the synaptonemal complex to ensure sister chromatid cohesion is not generally dependent on the continued presence of the complex after pachytene or that some remnant(s) of the complex can serve this function.

Maguire (1974) also speculated that the sticky mutation in maize, in which chromosomes fail to separate after zygotene and are ripped apart by the spindle at anaphase I (Beadle, 1932c, 1937), may conversely represent a defect which enhances sister chromatid cohesiveness to the degree that chiasmata cannot be resolved.

B. *ald*, *ord*, and *ca*nd Mutations in *D. melanogaster*

To my knowledge, there are no mutations in *D. melanogaster* with chiasma-defective phenotypes as clear-cut as that possessed by the desynaptic mutation. As noted above, most of the meiotic mutants examined in *D. melanogaster* increase nondisjunction by decreasing exchange. In such mutations, exchange chromosomes disjoin normally, demonstrating that when exchange does occur, chiasma function is normal. The *ald*, *ord*, and *ca*nd mutations provide interesting exceptions to this generality.

During the discussion of these mutations, it should be kept in mind that the phenotype of chiasma-defective mutations in *D. melanogaster* might be complicated by the distributive system. If, for example, exchanges simply failed to hold homologs together, the resulting univalents would be expected to disjoin via the distributive system as nonexchange bivalents routinely do. Thus, one expected phenotype of chiasma maintenance mutations in *D. melanogaster* is that exchange chromosomes should routinely segregate by the distributive system. The *ald* mutation may be a case in point.

The *ald* mutation was isolated and characterized by O'Tousa (1982). This mutation is recessive and maps to the third chromosome. The phenotype of *ald* is complex and includes effects on exchange-mediated disjunction and on size recognition within the distributive system. We are concerned here only with the effect of *ald* on X chromosomal exchange and disjunction. In *ald* females X chromosomal exchange occurs with a normal frequency and distribution. However, in a high percentage of cases, these exchange X chromosomes enter the distributive system (which as was noted above is normally reserved for the disjunction of nonexchange bivalents). Indeed, in XXY females bearing *ald* the two exchange X chromosomes frequently segregate from the Y chromosome rather than from each other. For example, in one experiment XXY; *ald/ald* females exhibited 17% XX ↔ Y segregation, and 76% of these nondisjunctional events involved X chromosomal bivalents which had undergone exchange (O'Tousa, 1982). Given that X chromosomes undergo exchange in approximately 90% of the meioses executed by these females, one can calculate that exchange in *ald* females results in the X chromosomes entering the distributive system 15% of the time. By contrast, exchange X chromosomes in otherwise wild-type XXY females engage in XX ↔ Y segregation with a frequency of approximately 0.5% (data of O'Tousa, 1982; but see below).

That exchange X chromosomes correctly disjoin from the Y chromosome (i.e., XXY ↔ 0 segregations are rare) argues that, at least in this respect, distributive disjunctions function normally. This suggests that *ald* fails to preclude exchange chromosomes from entering the distributive system, but does not thereafter impair their disjunctional ability. These data suggest that one component of the *ald* defect may be in chiasma maintenance. (Under this hypothesis normal chiasma function, rather than exchange per se, is required to keep exchange chromosomes out of the distributive system, a point for which there is independent evidence; see below.)

However, two lines of evidence suggest that such an interpretation of the *ald* mutation should be viewed with caution. First, *ald* has no effect on the disjunction of exchange autosomes. If *ald*$^+$ does in fact play a role in chiasma maintenance, such a chromosomal specificity is hard to understand. Second, *ald* also interferes with proper partner recognition in the distributive system. This is most easily observed for the small fourth chromosomes whose choice of disjunction partners is highly abnormal in *ald* females. Nonetheless, the effect of this mutation on the disjunction of exchange X chromosomes is intriguing and deserves further study.

The second mutation that may have a defect in chiasma maintenance is *ord* (Mason, 1976). In females homozygous for *ord*, exchange is strongly reduced. However, unlike the exchange-deficient mutations considered above, the probability of nondisjunction in *ord* females is the same for both exchange and nonexchange bivalents (except that centromere-proximal exchanges do tend to regularize disjunction in *ord* females). Thus, *ord* appears to affect not only the exchange process itself but also the ability of those residual exchanges that do occur to ensure segregation.

In both males and females *ord* causes a dramatic increase in both reductional and equational nondisjunction, as detected genetically. In male meiosis, Goldstein (1980) cytologically observed abnormal sister chromatid association during prophase as well as precocious sister chromatid separation during anaphase I. "The bulk of the first division misbehavior [in *ord*] consists of sister chromatids disjoining from one another, a process which usually occurs only during the second meiotic division" (Goldstein, 1980). If, as seems likely, the same defect in sister chromatid cohesion is present in females, then *ord* may indeed be a *Drosophila* analog of the desynaptic mutation described in corn. This interpretation of the *ord* mutation would further strengthen the argument that exchanges cannot ensure disjunction in the absence of proper sister chromatid cohesion. Finally, the defect in sister chromatid cohesion may also explain the recombination defect, if this proper sister chromatid association is an aspect of the assembly and/or function of the synaptonemal complex.

Finally, I shall consider briefly the claret nondisjunctional mutation (*ca*nd) (E. B. Lewis and W. Gencarella, *Genetics* **37**:600–601, 1952). An analogous meiotic mutation known as claret simulans (*ca simulans*) had been previously isolated by Sturtevant (1929) in *D. simulans*. Both these mutations are pleiotropic in that they result both in meiotic anomalies and in an abnormal eye color (hence the name claret). However, two lines of evidence suggest that in each case the same mutational event has damaged two tightly linked genes, one necessary for proper eye pigment and the other required for normal meiotic chromosome behavior. First, other claret mutations affect eye color without affecting meiosis, and second, O'Tousa and Szauter (1985) have isolated an allele of *ca*nd that affects meiosis but does not alter eye pigmentation. This mutation is known as nonclaret-disjunctional (*ncd*).

The *ca*nd, *ncd*, and *ca simulans* mutations all increase nondisjunction at meiosis I in females. (There is no effect on meiotic chromosome behavior in males.) It is clear that exchange chromosomes nondisjoin at high frequency in females bearing these mutations. Indeed, Davis (1969) concluded that *ca*nd does not influence either the frequency or the distribution of exchange and that chromosomes nondisjoin independently of whether or not an exchange has occurred. Although it is tempting to suggest that these mutations define a locus involved in chiasma maintenance, at least in the cases of ca^{nd} and ncd, distributive disjunction is even more severely affected than is exchange-mediated disjunction. These observations have led Carpenter (1973) and Baker and Hall (1976) to suggest that exchange may even to some extent ameliorate the disjunctional defects caused by mutations at this locus. Moreover, the ca^{nd} and $ca\ simulans$ mutations also cause a high frequency of chromosome loss and a high frequency of loss of maternally derived chromosomes during early embryonic mitoses. This loss appears to be independent of a chromosome's previous disjunctional behavior at meiosis and suggests a general defect in chromosome structure or behavior, or both. Thus, it seems more reasonable to suggest that these mutations cause a global defect in segregation at meiosis I, rather than a specific failure of chiasma maintenance. Indeed, in a cytological analysis of the claret simulans mutation, Wald (1936) observed an abnormally broad spindle at meiosis I and a wide separation of chromosomes at the poles of the first meiotic division. Similar observations have been made by P. Roberts (cited by Davis, 1969) and by Kimble and Church (1983) in $D.\ melanogaster$.

In summary, while mutations at the locus defined by ncd and ca^{nd} appear to define a global meiotic function, mutations such as ald and ord may indeed define functions involved in the maintenance of chiasmata.

C. $red1-1$ Mutation in S. cerevisiae

I am not aware of a yeast mutation that is known to be defective in chiasma maintenance. However, preliminary data obtained by B. Rockmill and S. Roeder (personal communication) suggest that the $red1-1$ mutation in S. cerevisiae may define such a function. Diploids homozygous for $red1-1$ complete both meiotic divisions; however, only a few viable spores are produced, and they are highly aneuploid as a consequence of nondisjunction at meiosis I. On the other hand $red1-1$ $spo13-1$ diploids produce viable two-spored asci (for a description of the phenotype of $spo13-1$, see above). Analysis of these asci demonstrates that $red1-1$ does not decrease the frequency of meiotic exchange. Rather, $red1-1$ appears to define a disjunctional defect for chromosomes which have previously undergone exchange.

Whether $red1-1$ defines a defect in chiasma function or a more general defect, such as that proposed for ca^{nd} above, remains to be seen. Nonetheless, the intriguing phenotype of this mutation indicates the power of this system for the genetic analysis of meiosis. This is especially true in that the ability to rapidly clone genes in S. cerevisiae by complementing mutant defects has allowed the rapid cloning and molecular analysis of this locus (Rockmill, Thompson, and Roeder, personal communication). The $red1-1$ mutation has been shown to define a locus which is expressed only in cells entering meiosis.

V. ACHIASMATE MECHANISMS OF DISJUNCTION

In the previous sections I detailed the evidence that exchange is usually both necessary and sufficient to ensure disjunction. I also discussed some possible models of chiasma function in both mechanical and ultrastructural terms. However, there do exist disjunctional systems that do not rely on chiasmata. Such systems are instructive and will be considered below.

A. Achiasmate Disjunction in *B. mori* Females

Mutations, like the *dy* mutation in maize described above, that uncouple exchange and disjunction demonstrate that a second event (chiasma maintenance) is required in order for exchange to ensure disjunction. Indeed, the ultrastructural studies described above indicate that in at least some organisms the formation of a mature chiasma requires functions which occur subsequent to the exchange event, such as modifications of the recombination nodule or the synaptonemal complex, or both. It is interesting in this light to consider a case in which such modifications alone, in the absence of exchange, are sufficient to ensure disjunction.

Meiosis in the oocytes of *B. mori* is entirely achiasmate despite the presence of synaptonemal complex (Rasmussen, 1976, 1977). As might be expected from the lack of genetically detectable exchange, recombination nodules are not observed. However, in this instance the synaptonemal complex is not disassembled at diplotene, but rather is modified and elaborated as prophase continues (for excellent micrographs, see Rasmussen, 1977). This modified synaptonemal complex continues to link the two homologs until bivalent separation occurs at anaphase. After homolog separation, the material forming the synaptonemal complex is left behind at the metaphase plate. As reviewed by von Wettstein et al. (1984), similar cases of disjunction of achiasmate meioses mediated by synaptonemal complex modification have been described in a number of other organisms.

We now consider a second system, used in the achiasmate male meiosis of *D. melanogaster*, which does not depend on the synaptonemal complex at all, but rather on which specific pairing sites function to produce pseudochiasmata.

B. Pairing-Site-Mediated Disjunction

Like *B. mori* females, *D. melanogaster* males exhibit an achiasmate meiosis. Moreover, *Drosophila* males do not even produce a synaptonemal complex. At least for the sex chromosomes, disjunction appears to be mediated by specific chromosomal sites known as collochores (Cooper, 1964).

1. Pairing sites in *D. melanogaster*

In *Drosophila* males X chromosomes bearing complete or near complete deficiencies for the basal heterochromatin exhibit high levels of nondisjunction at meiosis I. The first evidence that this failed disjunction is a consequence of the deletion of specific sex chromosome pairing sites came from genetic analysis of X-ray-induced heterochromatic deficiencies and duplications (Muller and Painter, 1932; Gershenson, 1940; Lindsley and Sandler, 1958; McKee and Lindsley, 1987). These studies demonstrated that the basal heterochromatin of the X chromosome (Xh) is nonuniform in terms of its ability to promote sex chromosome disjunction in males; this result implies that specific regions of the X chromosome, when deleted or present as free duplications, have a much greater effect on sex chromosome disjunction than do other regions. To explain these data, Gershenson (1940) postulated the existence of three major X chromosomal pairing sites.

Cytological verification of these sites was provided by Cooper (1964). He observed that X-Y pairing was limited to the X heterochromatin, and that not all regions of the X chromosome could serve as sites for X-Y pairing. To quote Cooper (1964), these data ''may all be accounted for by supposing that there are particular, localized, cohesive elements, or 'collochores,' in Xh and Y. They may be conceived as chromosomal organelles analogous to a kinetochore or a nucleolus organizer, and like them, perhaps divisible into functionable fractions.'' Cytologically, these paired sites are visible as a thin thread connecting specific sites on the X and Y chromosomes.

Recently, evidence from several investigators showed that at least one of these X chromosomal pairing sites may lie near to or within the nucleolus organizer, which is to say the ribosomal DNA (Appels and Hilliker, 1982; McKee and Lindsley, 1987). Indeed, the insertion of a single ribosomal DNA cistron into Df(1)X1, a grossly deleted X chromosome, which by itself lacks detectable pairing activity, substantially restores the disjunctional capacity of this chromosome (B. McKee, personal communication). This observation opens the door to a molecular analysis of these sites.

It should also be noted that collochorelike structures are not apparently involved in facilitating autosomal disjunction in *Drosophila* males (Yamamoto, 1979). This suggests the existence of at least one other system of achiasmate disjunction operating in the same meiotic process. A third system of achiasmate disjunction in *D. melanogaster* is considered below.

C. Distributive Segregation System in *D. melanogaster*

The distributive system in *D. melanogaster* functions during female meiosis to ensure the segregation of nonexchange chromosomes. As noted above, distributive segregation uses various aspects of chromosome structure (such as size and shape), rather than homology, to determine disjunctional partners (for review, see Grell, 1976).

The normally achiasmate chromosome four is an obligate denizen of the distributive system, while in wild-type females the two X chromosomes enter the distributive system only in the 5% of the meiocytes in which no exchange occurs. In those 5% of the meioses, both the X and the fourth chromosomes are present in the distributive system; however, disjunction is nonetheless X ↔ X and 4 ↔ 4. That this regularity is a function of size recognition and not homology has been clearly shown by elegant experiments using size and homology variant chromosomes, such as grossly deleted X chromosomes (for a review, see Grell, 1976). This system is sufficiently accurate to ensure normal disjunction even in cases where X chromosomal recombination is severely depressed by chromosome aberrations, such as multiply inverted balancer chromosomes.

The first indication of such a system was noted by Bridges (1916). During his analysis of spontaneous nondisjunction in *Drosophila* females, Bridges obtained and characterized XXY females. These females arose by the fertilization of an XX egg by a Y-bearing sperm. Bridges examined sex chromosome disjunction in these XXY females and observed that when the X chromosomes underwent exchange they segregated from each other faithfully, with the Y going at random. However, when the X chromosomes failed to undergo exchange, the two X chromosomes segregated from the Y chromosome. Bridges (1916) argued that this nondisjunction resulted from pairing of an X and Y chromosome, with the remaining X segregating at random. However, Cooper (1948) showed that these segregations are best explained by the formation of an XXY trivalent in which the Y directs the two X chromosomes (i.e., XX ↔ Y segregation). This process of XX ↔ Y segregation is referred to as secondary nondisjunction. In a series of elegant experiments, Grell demonstrated that secondary nondisjunction is but one aspect of a disjunctional mechanism, known as the distributive system, which has the following properties (for review, see Grell, 1976).

First, the distributive system acts almost exclusively on nonexchange chromosomes and guarantees their segregation at anaphase I. Second, when there are more than two nonexchange chromosomes, disjunctional patterns are determined by the sizes of the chromosomes involved. For example, in most wild-type meioses, only the fourth chromosomes fall under the purview of the distributive system, and these are routinely segregated to opposite poles at anaphase I. In the 5% of the cases where nonexchange X chromosomes are also present in the cell, the large disparity in size between the X and the fourth chromosomes allows the two pairs of chromosomes to segregate regularly from their homologs. However, if there are three nonexchange chromosomes of similar size, as in XXY females with nonexchange X chromosomes, trivalents are formed at high frequency. Like size recognition, this ability of a metacentric chromosome, such as the Y chromosome, to segregate from two acrocentric chromosomes (the X chromosomes) is a hallmark of distributive disjunction. Indeed, homology plays no role at all in determining distributive disjunctions; rather, disjunctional patterns are determined solely by the sizes and shapes of the chromosomes in the distributive system (for review of the arguments that led to this conclusion, see Grell, 1976). Thus, the distributive system represents a true backup system for exchange-mediated disjunction. This system is also under strict genetic control, as demonstrated by the properties of meiotic mutants that are specifically defective in distributive disjunction.

At least three mutations that affect distributive pairing have been identified in *D. melanogaster*. The first mutation to be characterized, *nod* (no distributive disjunction), apparently randomizes disjunctions within the distributive system, but has no effect either on the frequency of exchange or on the proper disjunction of exchange chromosomes (Carpenter, 1973). For example, we have recently observed frequencies of X chromosome nondisjunction of 50% in females of the genotype *nod/FM7*; *nod*, where *FM7* is an X chromosome balancer that effectively prohibits exchange (P. Zhang and R. S. Hawley, unpublished data). The *nod* mutation also causes a high frequency of chromosome loss, nondisjunction at meiosis II, and somatic chromosome loss in early embryonic mitoses. In all three cases, this misbehavior is restricted to maternal chromosomes that failed to undergo exchange (Carpenter, 1973; Zhang and Hawley, unpublished data). In XXY females homozygous for *nod* the two nonexchange X chromosomes still comigrate to the same pole but now segregate independently of the Y chromosome (Carpenter, 1973). This observation has led Carpenter (1973) to suggest that trivalent formation occurs normally in *nod* females and that *nod* affects the disjunctional process per se.

A second mutation, *Axs* (Aberrant X segregation), that affects only the distributive system has been recently isolated and characterized in my laboratory (A. Zitron and R. S. Hawley, unpublished data). Like *nod*, *Axs* has no effect either on the frequency of exchange or on the proper disjunction of exchange chromosomes. Rather, *Axs* appears to affect only the choice of partners within the distributive system, such that both X chromosomes frequently segregate from one of the fourth chromosomes, rather than from each other. Unlike *nod*, *Axs* does not cause chromosomes to misbehave after meiosis I. This is consistent with the suggestion that *Axs* allows incorrect partner choice, rather than disrupting segregation and thus forming univalents.

The third mutation affecting distributive disjunction is *ald*, which has been thoroughly described above. The existence of meiotic mutations that are specific to the distributive system argues that this process does indeed represent a separate and discrete mechanism of chromosome disjunction which can ensure the regular segregation of chromosomes that have failed to undergo exchange.

It should, however, be noted that exchange chromosomes do occasionally participate in distributive disjunctions. For example, in an otherwise genetically normal XXY female, 10% of the XX ↔ Y segregations will involve X chromosomes that have undergone exchange (data from O'Tousa, 1982; but see also the appendix of Carpenter, 1973). Although these are minor effects (involving fewer than 1.0% of the X chromosomes that have undergone exchange; see above), they do suggest that exchange between two chromosomes per se is usually, but not always, sufficient to exclude those chromosomes from the distributive system. (Intrachromosomal exchanges, such as those which occur between the arms of compound chromosomes, such as attached Xs, are never sufficient to exclude that chromosome from the distributive system [for review, see Grell, 1976, and Holm, 1976].) Moreover, as a consequence of her analysis of exchange and autosomal nondisjunction in *nod* females, Carpenter (1973) has argued that distal exchanges may be somewhat less likely to exclude a bivalent from the distributive system than are more proximal exchanges. These data suggest that, in addition to exchange, exclusion from the distributive system may require other events such as proper chiasma function and maintenance.

Although the mechanism of distributive disjunction remains obscure, the cytological observations of Puro and Nokkala (1977) are suggestive. During metaphase I achiasmate bivalents (such as the fourth chromosomes) coorient on the same arc of the metaphase spindle and move precociously toward opposite poles. Similar phenomena have been observed cytologically in a variety of insect meioses. For example, in mantispid males, Hughes-Schrader (1969) documented a segregational mechanism in which the sex chromosomes and autosomal univalents (when present) segregate precociously along the same arc of the spindle without prior synapsis or physical connection; she has termed this phenomenon "distance segregation." A similar phenomenon has been documented in *Calocoris quandripuctatus* (Nokkala, 1986a; but see also Nokkala, 1986b).

It should also be noted that there exist two major schools of thought in terms of the timing of partner choice within the distributive system and the timing of pairing for exchange. According to the first model, distributive associations are not made until after homologs pair and exchange; indeed, this model argues for a second round of chromosome pairing (for review of this hypothesis, see Grell, 1976). According to the second model, known as the chromocentral hypothesis, distributive associations occur in the meiotic chromocenter prior to homolog pairing, but can be subsequently voided or overridden by exchange and chiasmata (Novitski, 1964, 1975, 1978). In support of the chromocentral hypothesis, Novitski (1978) has analyzed disjunction and exchange in a female heterozygous for a *T(Y;A)* (a translocation involving the Y chromosome and an autosome). Confining his attention to cases in which the X chromosomes have nondisjoined, and in which exchange has occurred on the autosome on both sides of the break point, he observed that the X chromosomes segregated faithfully from the *T(Y;A)*. In other words, the chiasmata on both sides of the translocation break point did not preclude the arms of the Y chromosome from orienting nonexchange X chromosomes within the distributive system. Although it is difficult to explain this observation in terms of models which invoke a second, independent round of pairing from which chiasmate bivalents are entirely excluded, it seems to me that such arguments are premature. A mechanistic understanding of this process must await a detailed characterization at the molecular and ultrastructural levels.

Finally, it should also be noted that evidence exists for an *S. cerevisiae* analog of the distributive segregation system (Dawson et al., 1986; see also section III.C.). This system allows for the segregation of nonrecombinant artificial chromosomes which are not homologous.

Taken together, the various systems of achiasmate conjunction reveal a remarkable diversity of means to compensate for a lack of chiasmata, and in that sense also reveal the importance of chiasmata in more normal meiotic processes. The existence of multiple systems for ensuring segregation, and the fact that, at least in *D. melanogaster* and *S. cerevisiae*, multiple systems operate in the same meiosis, points to the essential problem of ensuring proper segregation. Given the importance of proper segregation, it is not surprising that evolution has selected for the development of redundant or "fail-safe" systems to ensure faithful segregation.

VI. SUMMARY

I have detailed the argument that chiasmata, and thus exchanges, are the primary means of ensuring disjunction in eucaryotes. I have also reviewed our understanding of chiasma function at the genetic, ultrastructural, and mechanical levels. It should, however, be clear that a molecular understanding of this process is not currently available. The exact mechanism(s) by which an exchange event develops into a chiasma capable of ensuring disjunction can only be elucidated by identifying the genes that encode this process and by isolating and characterizing their products. Such genetic, molecular, and structural studies will provide the grist for the next generation of inquiry into the physical basis of the meiotic process.

ACKNOWLEDGMENTS. I am greatly indebted to A. T. C. Carpenter for her guidance, wisdom, and friendship during the writing of this manuscript. The willingness of B. Rockmill, E. Thompson, S. Roeder, B. McKee, P. Zhang, and A. Zitron to share unpublished data is also gratefully acknowledged, as is the kindness of M. P. Maguire and A. T. C. Carpenter in reading and criticizing this manuscript. Finally, I am grateful to Jeanne Mosca for her limitless patience and encouragement during the course of this effort.

This chapter is dedicated to the memory of Larry Sandler. Many of us began our study of meiosis under his careful tutelage, and his energy and intellect were seminal to this discipline. His untimely death in February 1987 leaves us all saddened.

LITERATURE CITED

Appels, R., and A. J. Hilliker. 1982. The cytogenetic boundaries of the rDNA region within heterochromatin of the X chromosome of *Drosophila melanogaster* and their relation to meiotic pairing sites. *Genet. Res.* **39**:149–156.

Baker, B. S. 1975. Paternal loss (*pal*): a meiotic mutant in *D. melanogaster* causing loss of paternal chromosomes. *Genetics* **80**:267–296.

Baker, B. S., and A. T. C. Carpenter. 1972. Genetic analysis of sex chromosomal meiotic mutants in *Drosophila melanogaster*. *Genetics* **71**:255–286.

Baker, B. S., A. T. C. Carpenter, M. S. Esposito, R. E. Esposito, and L. Sandler. 1976. The genetic control of meiosis. *Annu. Rev. Genet.* **10**:53–134.

Baker, B. S., and J. C. Hall. 1976. Meiotic mutants: genetic control of meiotic recombination and chromosome segregation, p. 351–434. *In* M. A. Ashburner and E. Novitski (ed.), *Genetics and Biology of Drosophila*, vol. 1A. Academic Press, Inc., New York.

Bauer, H., R. Dietz, and C. Robbelen. 1961. Die Spermatocytenteilungen der Tipuliden. III. Das Bewegung-sverhalten der Chromosomen in Translokations-heterozygoten von *Tipula oleracea*. *Chromosoma* (Berlin) **12**:116–189.

Beadle, G. W. 1932a. The relationship of crossing over to chromosome association in *Zea-Enchlaena* hybrids. *Genetics* **17**:481–501.

Beadle, G. W. 1932b. A possible influence of the spindle fiber on crossing over in Drosophila. *Proc. Natl. Acad. Sci. USA* **18**:160–165.

Beadle, G. W. 1932c. A gene for sticky chromosomes in *Zea mays*. *Z. Indukt. Abstammungs. Vererbungsl.* **63**:195–217.

Beadle, G. W. 1937. Chromosome aberrations and gene mutation in sticky chromosome plants of *Zea mays*. *Cytologia* (Tokyo) **5**:110–121.

Beadle, G. W., and B. McClintock. 1928. A genic disturbance in *Zea mays*. *Science* **68**:433.

Bridges, C. B. 1916. Nondisjunction as the proof of the chromosome theory of heredity. *Genetics* **1**: 1–51, 107–163.

Brown, S. W., and D. Zohary. 1955. The relationship of chiasmata and crossing over in *Lilium formosanum*. *Genetics* **40**:850–873.

Buss, M. E., and S. A. Henderson. 1971. The induction of orientational instability and bivalent inter-locking at meiosis. *Chromosoma* (Berlin) **35**:153–183.

Carpenter, A. T. C. 1973. A mutant defective in distributive disjunction in *Drosophila melanogaster*. *Genetics* **73**:393–428.

Carpenter, A. T. C. 1979. Recombination nodules in recombination-defective females of *Drosophila melanogaster*. *Chromosoma* (Berlin) **83**:59–80.

Carpenter, A. T. C. 1987. Gene conversion, recombination nodules, and the initiation of meiotic synapsis. *Bioessays* **6**:232–236.

Carpenter, A. T. C., and B. S. Baker. 1974. Genic control of meiosis and some observations on the synaptonemal complex in *Drosophila melanogaster*, p. 365–375. *In* R. F. Grell (ed.), *Mechanisms in Recombination*. Plenum Publishing Corp., New York.

Carpenter, A. T. C., and B. S. Baker. 1982. On the control of the distribution of meiotic exchange in *Drosophila melanogaster*. *Genetics* **101**:81–89.

Carpenter, A. T. C., and L. Sandler. 1974. On recombination-defective meiotic mutants in *Drosophila melanogaster*. *Genetics* **76**:453–475.

Cooke, H. J., W. R. A. Brown, and G. A. Rappold. 1985. Hypervariable telomeric sequences from the human sex chromosomes are pseudoautosomal. *Nature* (London) **317**:687–692

Cooper, K. W. 1948. A new theory of secondary nondisjunction in female *Drosophila melanogaster*. *Proc. Natl. Acad. Sci. USA* **34**:179–187.

Cooper, K., W. 1964. Meiotic conjunctive elements not involving chiasmata. *Proc. Natl. Acad. Sci. USA* **52**:1248–1255.

Creighton, H. B., and B. McClintock. 1931. A correlation of cytological and genetical crossing over in *Zea mays*. *Proc. Natl. Acad. Sci. USA* **17**:492–497.

Darlington, C. D. 1932. *Recent Advances in Cytology*. The Blakiston Co., Philadelphia.

Davis, D. G. 1969. Chromosome behavior under the influence of claret-nondisjunctional in *Drosophila melanogaster*. *Genetics* **61**:577–594.

Dawson, A., A. Murray, and J. W. Szostak. 1986. An alternative pathway for chromosome segregation in yeast. *Science* **234**:713–717.

Fahmy, O. G. 1952. The cytology and genetics of *Drosophila subobscura*. VI. Maturation, fertilization and cleavage in normal eggs and in the presence of the *crossover-suppressor* gene. *J. Genet.* **50**:486–506.

Game, J. C., T. S. Zamb, R. Braun, M. Resnick, and R. Roth. 1980. The role of radiation (*rad*) genes in meiotic recombination in yeast. *Genetics* **94**:51–68.

Gershenson, S. 1940. The nature of the so-called genetically inert parts of chromosomes. *Vidensk. Akad. Nauk URRS* **3**:116.

Goldstein, L. S. B. 1980. Mechanisms of chromosome orientation revealed by two meiotic mutants in *Drosophila melanogaster*. *Chromosoma* (Berlin) **78**:79–111.

Gowen, M. S., and J. W. Gowen. 1922. Complete linkage in *Drosophila melanogaster*. *Am. Nat.* **56**: 286–288.

Grell, R. F. 1976. Distributive pairing, p. 435–486. *In* M. A. Ashburner and E. Novitski (ed.), *Genetics and Biology of Drosophila melanogaster*, vol. 1A. Academic Press, Inc., New York.

Hall, J. C. 1972. Chromosome segregation influenced by two alleles of the meiotic mutant *C(3)G* in *Drosophila melanogaster*. *Genetics* **71**:367–400.

Hawley, R. S. 1980. Chromosomal sites necessary for normal levels of meiotic recombination in *Drosophila melanogaster*. I. Evidence for and mapping of the sites. *Genetics* **94**:625–646.

Henderson, S. A., and C. A. Koch. 1970. Co-orientation stability by physical tension: a demonstration with experimentally interlocked bivalents. *Chromosoma* (Berlin) **29**:207–216.

Holm, D. 1976. Compound autosomes, p. 529–561. *In* M. A. Ashburner and E. Novitski (ed.), *Genetics and Biology of Drosophila*, vol. 1B. Academic Press, Inc., New York.

Holm, D. B., and S. W. Rasmussen. 1980. Chromosome pairing, recombination nodules and chiasma formation in diploid Bombyx males. *Carlsberg Res. Commun.* **43**:329–350.

Holm, D. B., S. W. Rasmussen, D. Zickler, B. C. Lu, and J. Sage. 1981. Chromosome pairing, recombination nodules and chiasma formation in the Basidiomycete *Coprinus cinereus*. *Carlsberg Res. Commun.* **46**:305–346.

Hughes-Schrader, S. 1943. Polarization, kinetochore movements, and bivalent structure in the meiosis of male mantids. *Biol Bull.* (Woods Hole) **85**:265–300.

Hughes-Schrader, S. 1969. Distance segregation and compound sex chromosomes in mantispids (Neuroptera: Mantispidae). *Chromosoma* (Berlin) **27**:109–129.

Janssens, F. A. 1909. Spermatogenese dans les. batracens. V. La theorie de las chiasma-typie, nouvelle interpretation des cineses de maturation. *Cellule* **25**:387–411.

Jones, G. H. 1967. The control of chiasma distribution in rye. *Chromosoma* (Berlin) **22**:69–70.

Jones, G. H. 1971. The analysis of exchanges in tritium-labelled meiotic chromosomes. II. *Stethopyna grossum. Chromosoma* (Berlin) **34**:367–382.

Jones, G. H. 1974. Correlated components of chiasma variation and the control of chiasma distribution in rye. *Heredity* **32**:375–387.

Jones, G. H. 1977. A test for early terminalization of chiasmata in diplotene spermatocytes of *Schistocerca gregoria. Chromosoma* (Berlin) **63**:287–294.

Jones, G. H. 1978. Giemsa C-banding of rye meiotic chromosomes and the nature of terminal chiasmata. *Chromosoma* (Berlin) **66**:45–57.

Jones, G. H. 1984. The control of chiasma distribution, p. 293–320. *In* C. W. Evans and H. G. Dickinson (ed.), *Controlling Events in Meiosis*. The Company of Biologists, Ltd., Cambridge.

Jones, G. H. 1987. Chiasmata, p. 213–244. *In* P. B. Moens (ed.), *Meiosis*. Academic Press, Inc. (London), Ltd., London.

Kimble, M., and K. Church. 1983. Meiosis and early cleavage in *Drosophila melanogaster* eggs: effect of the claret-non-disjunctional mutation. *J. Cell Sci.* **62**:301–318.

Klapholtz, S., C. S. Waddell, and R. E. Esposito. 1985. The role of the *spo11* gene in meiotic recombination in yeast. *Genetics* **110**:187–216.

Koller, P. C., and C. D. Darlington. 1934. The genetical and mechanical properties of sex chromosomes. I. *Rattus norvegicus. J. Genet.* **29**:159–173.

Lindsley, D. L., and L. Sandler. 1958. The meiotic behavior of grossly deleted *X* chromosomes in *Drosophila melanogaster. Genetics* **43**:547–563.

Lindsley, D. L., L. Sandler, B. Nicolleti, and G. Trippa. 1968. Genetic control of recombination in *Drosophila*, p. 253–276. *In* W. J. Peacock and R. D. Brock (ed.), *Replication and Recombination of Genetic Material*. Australian Academy of Science, Canberra, Australia.

Maguire, M. P. 1974. The need for a chiasma binder. *J. Theor. Biol.* **48**:485–487.

Maguire, M. P. 1978a. Evidence for separate genetic control of crossing over and chiasma maintenance in maize. *Chromosoma* (Berlin) **65**:173–183.

Maguire, M. P. 1978b. A possible role for the synaptonemal complex in chiasma maintenance. *Exp. Cell Res.* **112**:297–308.

Maguire, M. P. 1979. Direct cytological evidence for true terminalization of chiasmata in maize. *Chromosoma* (Berlin) **71**:283–287.

Maguire, M. P. 1982. Evidence for a role of the SC in meiosis II. *Chromosoma* (Berlin) **84**:675–686.

Malone, R. E., and R. Esposito. 1981. Recombination less meiosis in *Saccharomyces cerevisiae. Mol. Cell. Biol.* **1**:891–901.

Mason, J. M. 1976. Orientation disrupter (*ord*): a recombination-defective and disjunction-defective meiotic mutant in *Drosophila melanogaster. Genetics* **84**:545–572.

Mather, K. 1939. Crossover and heterochromatin in the *X* chromosome of *Drosophila melanogaster. J. Genet.* **33**:207–235.

McKee, B., and D. L. Lindsley. 1987. Inseparability of *X*-heterochromatic functions responsible for *X: Y* pairing, meiotic drive, and male fertility in *Drosophila melanogaster. Genetics* **116**:399–407.

Merriam, J. R., and J. N. Frost. 1964. Exchange and nondisjunction of the *X* chromosomes in female *Drosophila melanogaster. Genetics* **49**:109–122.

Meyer, G. 1964. A possible correlation between submicroscopic structure of meiotic chromosomes and crossing over, p. 461–462. *In Proceedings of the 3rd European Regional Conference on Electron Microscopy, Prague.* Czechoslovakia Academy of Science, Prague.

Miller, O. 1963. Cytological studies in asynaptic maize. *Genetics* **48**:1445–1466.

Moens, P. B., and K. Church. 1979. The distribution of synaptonemal complex material in metaphase I bivalents of *Locusta* and *Chloeatis. Chromosoma* (Berlin) **73**:247–254.

Moses, M. J., S. J. Counce, and D. F. Paulson. 1975. Synaptonemal complex complement of man in spreads of spermatocytes, with details of the sex chromosome pair. *Science* **187**:363–365.

Muller, H. J., and T. S. Painter. 1932. The differentiation of the sex chromosomes of Drosophila into genetically active and inert regions. *Z. Indukt. Abstammungs Vererbungsl.* **62**:316–365.

Nicklas, R. B. 1967. Chromosome micromanipulation. II. Induced reorientation and the experimental control of segregation in meiosis. *Chromosoma* (Berlin) **21**:17–50.

Nicklas, R. B. 1971. Mitosis, p. 225–297. *In* D. M. Prescott, L. Goldstein, and E. McConkey (ed.), *Advances in Cell Biology*, vol. 2. Appleton-Century Crofts, New York.

Nicklas, R. B. 1974. Chromosome segregation mechanisms. *Genetics* **78**:205–213.

Nicklas, R. B. 1977. Chromosome distribution: experiments on cell hybrids and *in vitro. Philos. Trans. R. Soc. London B Biol. Sci.* **277**:267–276.

Nicklas, R. B., and C. A. Koch. 1969. Chromosome micromanipulation. III. Spindle fiber tension and the reorientation of mal-oriented chromosomes. *J. Cell Biol.* **43**:40–50.

Nicklas, R. B., and C. A. Staehly. 1967. Chromosome micromanipulation. I. The mechanics of chromosome attachment to the spindle. *Chromosoma* (Berlin) **21**:1–16.

Nokkala, S. 1986a. The mechanisms behind the regular segregation of autosomal univalents in *Calocoris quandripunatatus. Hereditas* **105**:199–204.

Nokkala, S. 1986b. The meiotic behaviour of B-chromosomes and their effect on the segregation of sex chromosomes in males of *Hemerobius marginatus. Hereditas* **105**:221–227.

Novitski, E. 1964. An alternative to the distributive pairing hypothesis in Drosophila. *Genetics* **50**:1449–1451.

Novitski, E. 1975. Evidence for the single phase theory of meiosis. *Genetics* **79**:63–71.

Novitski, E. 1978. The relation of exchange to nondisjunction in heterologous chromosome pairs in the Drosophila female. *Genetics* **88**:499–503.

Ostergren, B. 1951. The mechanism of co-orientation in bivalents and multivalents. *Hereditas* **37**:85–156.

O'Tousa, J. 1982. Meiotic chromosome behaviour influenced by mutation-altered disjunction in *Drosophila melanogaster* females. *Genetics* **102**:503–524.

O'Tousa, J., and P. Szauter. 1985. The initial characterization of non-claret disjunctional (*ncd*): evidence that ca^{nd} is the double mutant, *ca ncd. Dros. Inform. Serv.* **55**:119.

Parker, R., and J. H. Williamson. 1976. Aberration induction and segregation in oocytes, p. 1252–1268. *In* M. A. Ashburner and E. Novitski (ed.), *Genetics and Biology of Drosophila*, vol. 1B. Academic Press, Inc., New York.

Parry, D. M. 1973. A meiotic mutant affecting recombination in female *Drosophila melanogaster. Genetics* **73**:465–486.

Pearson, P. L., and M. Bobrow. 1970. Definitive evidence for the short arm of the *Y* chromosome associating with the *X* chromosome during meiosis in the human male. *Nature* (London) **226**:959–961.

Prakash, S., L. Prakash, W. Burke, and B. Montelone. 1980. Effects of the *RAD52* gene on recombination in *Saccharomyces cerevisiae. Genetics* **90**:49–68.

Prakken, R. 1943. Studies of asynapsis in rye. *Hereditas* **29**:475–495.

Puro, J., and S. Nokkala. 1977. Meiotic segregation of chromosomes in *Drosophila melanogaster* oocytes. *Chromosoma* (Berlin) **63**:273–286.

Rasmussen, S. W. 1975. Ultra-structural studies of meiosis in males and females of the $c(3)G^{17}$ mutant of *Drosophila melanogaster* Meigen. C. R. Trav. Lab. Carlsberg **40**:163–173.

Rasmussen, S. W. 1976. The meiotic prophase in *Bombyx mori* females analyzed by three dimensional reconstructions of synaptonemal complexes. *Chromosoma* (Berlin) **54**:254–293.

Rasmussen, S. W. 1977. The transformation of the synaptonemal complex into the "elimination chromatin" of *Bombyx mori* oocytes. *Chromosoma* (Berlin) **60**:205–221.

Rouyer, F., M.-C. Simmler, C. Johnson, G. Vergnaud, H. J. Cooke, and J. Weissenbach. 1986. A gradient of sex linkage in the psuedo-autosomal region of the human sex chromosomes. *Nature* (London) **319**:291–295.

Sandler, L., and G. Braver. 1954. The meiotic loss of unpaired chromosomes in *Drosophila melanogaster*. *Genetics* **39**:365–377.

Sandler, L. M., D. L. Lindsley, B. Nicolleti, and G. Trippa. 1968. Mutants affecting meiosis in natural populations of *Drosophila melanogaster*. *Genetics* **60**:525–558.

Sandler, L., and P. Szauter. 1978. The effect of recombination defective mutations on fourth-chromosome recombination in *Drosophila melanogaster*. *Genetics* **90**:699–712.

Sears, E. R. 1952. Misdivision of univalents in common wheat. *Chromosoma* (Berlin) **4**:S.535–550.

Simmler, M-C., F. Rouyer, G. Vergnaud, M. Nystron-Lahti, K. Yan Ngo, A. de la Chapelle, and J. Weissenbach. 1985. Pseudoautosomal DNA sequences in the pairing region of the human sex chromosome. *Nature* (London) **317**:692–697.

Singh, L., and K. W. Jones. 1982. Sex reversal in the mouse (*Mus musculus*) is caused by a recurrent nonreciprocal crossover involving the *X* and an aberrant *Y* chromosome. *Cell* **28**:205–216.

Smith, P. A., and R. C. King. 1968. Genetic control of synaptonemal complexes in *Drosophila melanogaster*. *Genetics* **60**:335–351.

Spurway, H. 1946. A sex-linked recessive crossover suppressor in *D. subsobscura*. *Dros. Inform. Serv.* **20**:91.

Stern, C. 1931. Zytologish-gene-tische Untersuchungen als Beweise fur die Morgansche Theorie des Faktoren austauchs. *Biol. Zentralbl.* **51**:54–587.

Sturtevant, A. H. 1929. The claret mutant type of *Drosophila simulans*; a study of chromosome elimination and cell lineage. *Z. Wiss. Zool.* **135**:323–356.

von Wettstein, D., S. W. Rasmussen, and P. B. Holm. 1984. The synaptonemal complex in genetic recombination. *Annu. Rev. Genet.* **18**:331–413.

Wald, H. 1936. Cytological studies on the claret mutant type of *Drosophila simulans*. *Genetics* **21**:264–279.

Warren, A. C., A. Chakravati, C. Wong, S. A. Slaugenhaut, S. L. Halloran, P. C. Watkins, C. Metaxotou, and S. E. Antonarakis. 1987. Evidence for reduced recombination on the nondisjoined chromosomes 21 in Down syndrome. *Science* **237**:652–654.

Whitehouse, H. L. K. 1969. *Towards an Understanding of the Mechanism of Heredity*, 2nd ed. Edward Arnold Ltd., London.

Yamamoto, M. 1979. Cytological studies of heterochromatic function in *Drosophila melanogaster*: autosomal meiotic pairing. *Chromosoma* (Berlin) **72**:293–328.

Chapter 17

Thoughts on Recombination Nodules, Meiotic Recombination, and Chiasmata

Adelaide T. C. Carpenter

I. INTRODUCTION

In this chapter I explore several aspects of meiotic recombination in eucaryotes that distinguish it from recombination occurring in other settings. All organisms that utilize DNA as their genetic material have the enzymatic capacities not only to replicate it accurately but also to repair a variety of damages arising from mistakes, spontaneous instabilities, or exogenous insults. The functioning of some of these repair pathways can result in recombination events. However, both spontaneous and induced recombination events are relatively infrequent during mitotic (vegetative) growth; in contrast, during meiosis virtually all eucaryotes undergo general (non-site-specific) recombination at high levels, and meiotic recombination displays a number of characteristics (interference, for example) that are not observed in mitotic events. The properties of recombination also differ substantially between eucaryotes and procaryotes. When students of procaryotic genetics discovered recombination, they abandoned the principles of meiotic recombination for the pragmatic reason that the rules for eucaryotic meiosis do not fit procaryotes; conversely, the rules for procaryotic recombination do not fit eucaryotic meiosis. Because of the differences between these systems, I will digress briefly to review some terminology before focusing on the unique properties of eucaryotic meiotic recombination.

There are four terms in common usage that describe reassortment of genetic information: recombination, crossing-over, exchange, and gene conversion. These terms describe overlapping, but not identical, ranges of phenomena. Recombination (the process inferred from the observation of recombinant progeny) includes all processes except mutation that yield new genotypes; both crossing-over and gene conversion are recombination, but so is the independent assortment of nonhomologous chromosomes. Crossing-over (the process inferred from the observation of crossover progeny) is the process that yields physical exchange between homologous chromosomes. A single crossover event yields two reciprocal crossover chromosomes, one of which has only the markers of homolog A to the left of the event and those of homolog B to its right; the other has the markers of B to the left and those of A to the right. The term "exchange" also refers to the process that generates crossover chromosomes but emphasizes that meiotic crossing-over occurs after replication, that is, at the four-strand (= chromatid), or tetrad, stage. Thus, a single exchange event yields two reciprocal crossover chromatids; the other two chromatids are uninvolved and remain parental in genotype.

That all meiotic crossing-over occurs as exchange was demonstrated as soon as techniques became available for the high-frequency recovery of two chromatids per meiosis (Anderson, 1925), and Weinstein (1936) developed analytic techniques for calculating meiotic exchange parameters from single-chromatid data. Briefly, for a pair of fully marked homologs, in some meioses there will be no exchanges (E_0) anywhere along the full length of the chromosomes; all progeny from E_0 meioses will carry parental (noncrossover) chromosomes. In some meioses there will be a single exchange (E_1); half the progeny from E_1 meioses will carry single crossover chromosomes and the other half will carry noncrossovers. In some meioses there will be two exchanges (E_2); since there is no chromatid interference, one-fourth of E_2 events will have both the exchanges between the same two chromatids (two-strand events), one-fourth will have them between different chromatids (four-strand events), and one-half will have one chromatid involved in both events, two each involved in one, and one chromatid uninvolved (three-strand events), for an overall frequency of one-fourth parental, one-half single crossover, and one-fourth double crossover chromosomes among the progeny of E_2 meioses. Similarly, E_3 events yield 1:3:3:1 ratios of noncrossover, single crossover, double crossover, and triple crossover progeny. It is therefore simple to calculate the array of meioses that produced a given set of progeny. If, for example, triple crossovers are the highest multiple observed, then the number of E_3 meioses is eight times the observed number of triple crossover progeny, and the frequency of E_3 meioses equals the number of them divided by the total number of progeny observed. However, three-eighths of E_3s give double crossover chromatids, so that the number of E_2s = 4 (number of double crossovers observed − $\frac{3}{8}E_3$); similarly, E_1 = 2 (number of single crossovers observed − $\frac{3}{8}E_3$ − $\frac{1}{2}E_2$), and E_0 = number of noncrossovers observed − ($\frac{1}{8}E_3$ + $\frac{1}{4}E_2$ + $\frac{1}{2}E_1$). This analysis can be done on the chromosome as a whole, but it is even more informative when the locations of the exchanges are considered during calculations of the tetrad frequencies and compared with the cytological sizes of the intervals (see Fig. 2; also see Charles, 1938; Stephens, 1961). Since exchange occurs at the four-strand or tetrad stage of meiosis, Weinstein (1936) called this process "tetrad analysis"; the equivalent process in fungi, involving direct scoring of all four products of each meiosis, was also quite reasonably called "tetrad analysis." Finally, gene conversion is defined as the nonreciprocal transfer of genetic information from one region to a second homologous region; it is discussed in section II.

With respect to meiotic recombination two different levels of questions have been addressed repeatedly over the past 80 years and are still unsolved. (i) What molecular events are involved in going from two parental chromatids to two crossover chromatids? (ii) How are the numbers and locations of events regulated between and along chromosomes? The first question is considered elsewhere; the second is addressed in this chapter, but first we need to review how these events are observed.

Both genetic and cytological techniques have been used to identify the number and location of exchange events during meiosis. Janssens (1909) suggested that high frequencies of crossing-over occur during meiosis to explain the morphology of chiasmata several years before crossing-over was discovered genetically (Morgan, 1911). Although the relationship between crossing-over (detected genetically) and chiasmata (detected cytologically) was a burning issue in the early decades of this century (for excellent discussions, see Lewis and John, 1972; Whitehouse, 1969), all workers in the field have long been in agreement that, in general, a meiotic exchange event will become a chiasma during late prophase-metaphase I and that all chiasmata have resulted from exchanges that occurred earlier in meiosis, probably during pachytene, the stage of maximum synapsis. The two techniques, genetics and cytology, therefore detect the occurrence of the same phenomenon, and in the few organisms where the process can be effectively studied by both techniques, equivalent answers result in terms of numbers and average locations of the events. Moreover, the equivalence indicates the function of the overall process of general meiotic recombination, since chiasmata are, in virtually all organisms, essential for the regular segregation of homologs at anaphase I (see Hawley, this volume). However, the two techniques have different strengths, weaknesses, and resolving power. Chiasma counts yield information on overall cellular numbers and distributions of events, but since chromosomes are relatively condensed by the stages when chiasmata can be detected, and since chromosome morphology is generally not finely detailed at this time, specific locations are poorly resolved; moreover, there are always some (the number depending on the organism) instances where it is impossible to distinguish a chiasma from a twist of the bivalent, and some organisms have such small chromosomes that detailed chiasma cytology is impossible. Genetics, on the other hand, can yield information on any interval that can be spanned by marker genes, but the number of intervals that can be assayed simultaneously is limited by the tolerance of the specific organism for the mutations those marker genes cause and, realistically, the range of phenotypes that can be assayed simultaneously. For *Drosophila melanogaster* and corn, for example, it is possible to get detailed information on all events along any one chromosome, but only parts of chromosomes can be monitored if different chromosomes are assayed simultaneously; in *Saccharomyces cerevisiae*, it is generally difficult to assay genetically all events along even a single chromosome.

Since crossovers become chiasmata which are in turn essential for regular segregation of homologs, the minimum number of exchange events per pair of homologs per meiosis is one. In theory, an organism could evolve any one of an infinite number of strategies to guarantee this minimum number: I will consider first the two extremes. On the one hand, there could be a large number of events per cell, randomly distributed along and across chromosomes (= constant probability per unit physical length of DNA), large enough so that the shortest chromosome would, to some (evolutionarily selectable) degree of certainty, have at least one exchange per meiosis. Taking "random" to mean "Poisson," and assuming that having 5% of meioses (those with no exchange on the shortest chromosome) at risk for nondisjunction is tolerable, then the total number of events per cell must be high enough so that the shortest chromosome has an average of three exchanges (= 150 map units)—the longer chromosomes, of course, would have higher average levels of exchange. Not even *S. cerevisiae* has this high a level of exchange. On the other hand, one event per chromosome could be guaranteed by evolving special sites for exchange and a mechanism for performing one and only one exchange each meiosis at each such site. Although organisms that exhibit "localized chiasmata" exist, either as observed directly by cytology (for a review, see John and Lewis, 1965) or as deduced from the pattern of gene distribution along the recombination map (e.g., *Caenorhabditis elegans*), it is likely that even these (rare) cases are not examples of special exchange sites but rather are extreme cases of the more common, but conceptually more complex, method of chromosomal control described below (see Jones, 1967, 1987).

Most organisms have evolved systems of distributions of numbers and locations of exchanges that are intermediate between these two extremes. Typically, the mean number of exchanges (chiasmata) observed per meiosis is much lower than that predicted by using the Poisson expectation as described above, although it is at least equal to n, the haploid number of chromosomes. Moreover, there are a number of departures from expectations predicted from various kinds of probability distributions. (i) The observed number of total chiasmata per cell, of course, is not a constant; there is between-cell variance. However, there is less variance than predicted (Jones, 1984). (ii) Similarly, the observed numbers of chiasmata per chromosome, although variable, are also less variable than expected (Weinstein, 1936). (For example, for the *D. melanogaster* X chromosome, the observed frequencies of the different tetrad types for the entire chromosome are E_0 [no exchange] = 5%, E_1 [one exchange] = 60%, E_2 [two exchanges] = 30%, and E_3 [three exchanges] = 5%, for a mean of 1.35 exchanges per meiosis [67 map units]; if this were the mean of a Poisson variable, then the expected frequencies of the various tetrad types would be E_0 = 26%, E_1 = 35%, E_2 = 24%, E_3 = 11%, E_4 = 4%, and E_5 = 1%. Many fewer E_0 and high-order tetrads are observed than expected, with concomitant increases in the frequencies of E_1 and E_2 tetrads.) (iii) In chromosomes that experience two or more exchanges (chiasmata), the events are nonrandomly far apart; that is, there is positive chiasma interference (Muller, 1916). (iv) Although chromosome length does appear to affect chiasma frequency (in that long chromosomes generally have higher means than short ones of the same species), the effect is not a simple one because normalizing for length does not yield a constant even within a species (generally, the mean per unit length is higher for short chromosomes than for long ones [Jones, 1984]). (v) Although in most organisms an exchange (chiasma) can occur anywhere along the euchromatin of a chromosome arm, the probability of such an event is not constant per unit length; rather, there are species-specific patterns (Bridges, 1935).

These cellular properties of exchange have been documented in many species, and it has long been known that, although the basic species characteristics are remarkably consistent, they are subject to environmental and genetic perturbations. Two levels of deduction come from these studies. The first is that, not surprisingly, these cellular and chromosomal patterns of exchange are under genetic control. The second is that these different types of patterns are different aspects of a common regulatory pathway, since manipulations that affect one tend to affect several in the same way. For example, recombination-defective mutations in *D. melanogaster* that relax the constraints on intrachromosomal distribution (giving recombination maps that are more nearly proportional to physical distance than is observed in wild type) also both relax interference and give interchromosomal distributions that are more random (Baker and Hall, 1976). Similarly, a mutant genotype in rye that abolishes the normal subterminal localization of chiasmata in that species simultaneously greatly increases the variance in chiasma number between chromosomes and between cells (Jones, 1967). That the most common type of control of chiasmata (exchange) is a complex cellular process has several implications. Firstly, of course, is that we would like to understand how this process operates. Although we can set some limits (models that are likely to be too simple; see below), we do not yet have workable mechanistic models for the overall process. Secondly, it is important to remember that the various aspects of the process are interconnected when the experiments at hand focus on one specific aspect; manipulations that affect one aspect may well affect others too, thereby yielding results that are misleading when taken in vacuo.

In addition to the analyses of the cellular properties of exchange by the light microscopic examination of chiasmata and by the genetics of crossing-over, the identification of a subcellular structure—the recombination nodule—that appears to mediate exchange means that it has become possible to analyze recombination parameters cytologically during the meiotic stages when it is occurring. I will first explore the evidence that links these structures with recombination and then explore possible applications.

II. LATE RECOMBINATION NODULES

A. Properties

Recombination nodules are small, dense structures that are present during specific stages of meiosis in positions that are suggestive of roles in recombination (Fig. 1). They are of two types, those that are present earlier (generally during

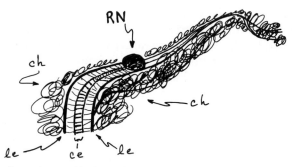

FIGURE 1. Diagram of a pachytene bivalent with synaptonemal complex and a recombination nodule. RN, recombination nodule; ch, chromatin; le, lateral element; ce, central element. The synaptonemal complex is typically 0.1 μm wide (the distance between lateral elements), and recombination nodules are generally somewhat smaller than this in diameter.

zygotene, the stage of meiotic prophase during which synapsis occurs and the synaptonemal complex is laid down) and those that are present later (during pachytene, the stage during which synapsis and synaptonemal complex extension are as complete as is typical for the species). Different sets of names have been used to distinguish between these two types of recombination nodules (based on morphology differences, stage differences, etc.), but no set of names turn out to apply to all species; the only consistent difference is that of relative temporal order, so I will refer to the two types as "early" and "late" here. Hypotheses regarding possible functions of early nodules are still quite speculative, and I will delay discussing them; however, hypotheses for the function of late recombination nodules have become quite specific, and there are extensive data that are consistent with these hypotheses (for compilation, see von Wettstein et al., 1984). Late recombination nodules are hypothesized to mark the sites of ongoing exchange events; specific hypotheses differ slightly with respect to details, but the most global hypothesis is that they form at or just before the initiation of molecular recombination events and mediate those events (Carpenter, 1975). In some (but not all) organisms late nodules, morphologically modified, remain associated with the bivalent until diplotene, so here they have been postulated to function also in the chromosomal reorganization that yields the visible chiasma (Holm and Rasmussen, 1980).

The evidence that initially led to the nodule = site of exchange hypothesis is that in all respects (maximum number per cell, numbers per chromosome, locations along chromosome arms, and interference) late nodules exactly parallel exchange events in *D. melanogaster* females (Carpenter, 1975). Equivalent parallels between late nodules and chiasmata (and/or exchanges) have been shown for a large number of species; these species span a wide range of phenotypes for numbers and distributions of chiasmata (for reviews, see Carpenter, 1979b; von Wettstein et al., 1984; Holm, 1986), and additional examples continue to be reported. Since late recombination nodules are present much closer in time to the actual physical exchange between homologs than either chiasmata or crossovers, they are expected to give more insightful data on the process that regulates the number and distribution of exchanges than either of these other two methods. It is this issue—the advantages and disadvantages of using nodule distributions to deduce the parameters of the cellular process for regulation of exchange distribution and what kinds of information we can expect to recover—that I wish to address here.

The first problem, both technical and physiological, is that of detection. Two main techniques are in general use for the detection of recombination nodules: serial-section reconstruction, which is expected to preserve all nodules present but is relatively laborious, and spreading the synaptonemal complex from swollen meiotic prophase nuclei, which is for many organisms much faster than sectioning but which may possibly be less reliable. Firstly, the ammoniacal silver spreading procedure rarely retains and/or visualizes nodules (Stack and Anderson, 1986a); the phosphotungstic acid spreading procedure does preserve and visualize nodules. Secondly, even with phosphotungstic acid some nuclei that should have nodules from apparent substage of meiosis do not (see, e.g., Holm, 1986); it is assumed that these represent nuclei that have been particularly severely affected by the rigors of the technique. Finally, in at least some organisms all nuclei that have nodules have fewer than expected from chiasma counts (e.g., Gillies, 1983). It may be that nodules are ephemeral in these species (see below); however, it is also possible that the nodule-bivalent association is fragile and that a proportion of nodules are broken loose by the spreading regimen. Moreover, spreading and sectioning also differ in the kinds of problems that affect nodule recognition. In sectioned material there is generally a lot of chromatin, and it can be difficult to recognize each nodule, especially when they are small (see, e.g., Byers and Goetsch, 1975); on the other hand, surface dirt is distinctive and easy to discount. In spread preparations the chromatin is dispersed, but it can be difficult to distinguish between nodules and dirt in specific instances. All workers in the field are aware of these problems and are constantly working to refine the techniques to generate ever-more-reliable detection of nodules, and as long as nodule underrepresentation is unbiased this does not complicate analysis unduly; however, the possibility that susceptibility to underrepresentation may be preferential must always be borne in mind.

Although in a wide variety of organisms "late" nodules consistently parallel the exchange characteristics of the organism, the timing of nodule presence varies widely. Some organisms attain their full number of "late" nodules early in pachytene and retain them, with or without morphological modifications, through diplotene, when the modified nodule clearly is at the site of its chiasma (Zickler, 1977; for discussion, see Holm, 1986); in these organisms, the physiology is not expected to cause difficulties in determining nodule number. However, in other organisms late nodules are ephemeral (e.g., Holm and Rasmussen, 1983; Gillies, 1983; Bernelot-Moens and Moens, 1986; Glamann, 1986; G. H. Jones and S. M. Albini, *in* P. E. Brandham and M. D. Bennett, ed., *Kew Chromosome Conference III*, in press), and here the physiology may complicate the analysis in the same way that detachment of nodules during spreading may: nuclei may never have all of the nodules they will experience present at the same time. If the underrepresentation is random, then the errors introduced are analogous to those that accrue when genetic mapping experiments involve distant markers: some events will not be detected, so maps will be somewhat too short, and the apparent tetrad distributions will overall be shifted down in ranks. Various kinds of nonrandomness could, of course, complicate analysis. Interestingly enough, in one organism that exhibits ephemeral late ("spherical") nodules, *D. melanogaster* females, nuclei that have close to the maximum number of late nodules also have distributions of numbers per arm that closely approximate those of exchanges, whereas nuclei with lower numbers have a paucity of arms with more than one nodule. Nodules therefore appear to accumulate per nucleus, rather than per arm, until the maximum number is attained; then they disappear in the same manner (Carpenter, 1979a). Locations of late nodules along the arms in nuclei before and after the maximum indicate no obvious redistribution of nodule position during accumulation or disappearance, suggesting that, once at a site, the nodule stays there. These features of nodule behavior need to be addressed in additional organisms.

(Several aspects of the data from *D. melanogaster* have been reinterpreted by von Wettstein et al. [1984], who conclude that the deductions from *D. melanogaster* are not convincing, although later deductions in other organisms are. Their reanalysis involves summing the chromosomal positions of early and late nodules and finding that this summed distribution no longer matches the known exchange parameters very well. This is not surprising, since the distributions of early [ellipsoidal] nodules do not match exchange at all whereas those of late [spherical] nodules do, so the sum must match less well. They justify this manipulation because, for a subset of the sample of nuclei, the mean number of total nodules per nucleus matches the mean number of exchanges. For this mean to have significance it must be assumed that each of these 12 nuclei has present all nodules it will ever experience [no asynchrony in time of nodule presence], and this generates a between-cell variance in total numbers of nodules [ranging from three to nine] that is unreasonably large for total exchanges per cell [$\bar{x} = 5.6$]. Moreover, von Wettstein et al. then reluctantly suggest that "not all nodules are visible at a single stage" to account for the absence of a group of nuclei having a mean number of nodules twice that of exchanges. It is unreasonable to reject the hypothesis of ephemerality of both early and late nodules, mix the two, and then invoke ephemerality on the total. The interested reader is urged to read the relevant papers.)

A priori, late nodules could be present before, during, after, or throughout the DNA and chromatid interactions that culminate in physical exchange of homologous chromatids; the distributional data indicate only that nodules mark the site (Holliday, 1977). As far as I know, the only evidence on relative times of nodule presence and exchange comes from *D. melanogaster*, and it is as follows. (i) Late nodules are specifically labeled by tritiated thymidine; this observation is consistent with their being sites of repair DNA synthesis (Carpenter, 1981), an event that has been proposed to accompany recombination (e.g., Meselson and Radding, 1975). Late nodules therefore appear to mark sites of ongoing recombination events. (ii) Mutants at the *mei-9* locus, which reduce the frequency of exchange events without affecting any of the distributional parameters and are therefore postulated to be defective in a step after site choice but before final accomplishment of the physical exchange (Baker and Carpenter, 1972; see also Carpenter, 1984; Hawley, this volume), have no apparent effect whatsoever on late nodules; in two independent *mei-9* mutants, nodule numbers, distributions, etc., are indistinguishable from those of controls (Carpenter, 1979b). These observations are consistent with late nodules being present at sites where exchange events will occur (and with most late nodules in *mei-9* mutants being defective). (iii) Mutations at three additional loci (*mei-218*, *mei-41*, and *mei-S282*), all of which alter both frequency of exchange events and the distributional characteristics of the few events that occur and are therefore postulated to be defective in site choice itself (Baker and Carpenter, 1972; Parry, 1973), display late nodule numbers and distributions that parallel their effects on exchange (Carpenter, 1979b and unpublished data). As far as I know, this is the only direct experimental test of the hypothesis that late nodules mark exchange sites. Although as an experimenter I am, of course, satisfied by these observations, as a worker in the field I will be happier when tests of these points have been performed in additional organisms.

Additional examples of correlations between late recombination nodules and exchanges (chiasmata) continue to be reported, and surely whole-cell and chromosomal observations of late nodules both in wild type and in various kinds of experimental situations are likely to give insights into the ranges of general cellular controls of exchange distribution and perhaps even some clues as to the physical mechanism(s). The importance of seeking ways to test hypotheses cannot be overemphasized. I will give two examples from *D. melanogaster*; the point is not to advertise any particular organism but rather to illustrate the complexity of the problem.

In *D. melanogaster*, the five major chromosomal arms exhibit similar patterns of exchange distribution, illustrated in Fig. 2 for the X chromosome. One might postulate that this distribution is generated by some number of *cis*-acting sites, each of which regulates the probability of exchange in its vicinity (perhaps by influencing the binding of late nodules, as has been postulated for the human distribution [see Holm, 1986]). This simple hypothesis predicts that a chromosomal segment will retain its intrinsic exchange frequency wherever it is in the genome. However, numerous experiments that move proximal segments distally and vice versa show that the important factor for exchange frequency is the new chromosome position, not the original (see Fig. 2). (Of course, these experiments are done with homozygous aberrations, since aberration heterozygosity imparts distortions of its own.) Although there might still be *cis*-acting local modifiers, they are not a sufficient explanation for the general, chromosomal pattern, which appears to be generated with the whole arm used as substrate and is under general cellular genetic control (see, e.g., Baker and Carpenter, 1972; Hawley, this volume).

A second topic concerns the regulation of the total number of exchange events per cell. This number is a species constant (Darlington, 1932), and this fact has suggested to many people that regulation might involve some component or

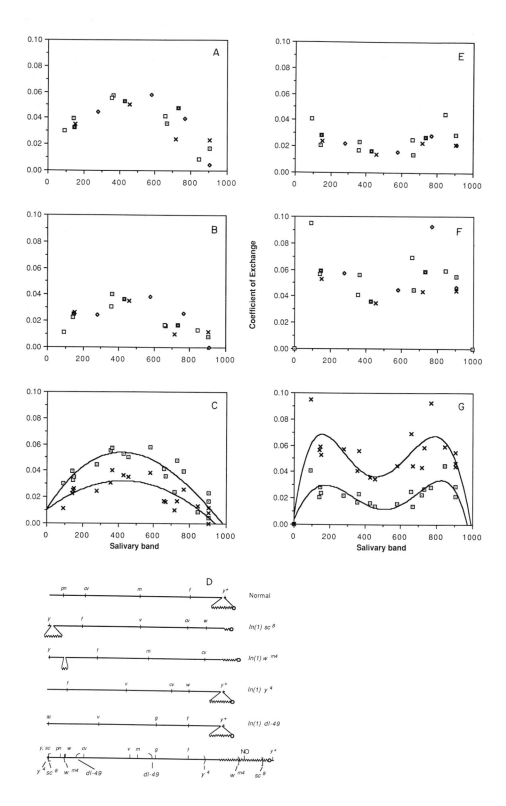

reaction step occurring in limiting quantity. Stochastic models would predict that the intercellular distribution of total numbers should follow a Poisson distribution; the observed distributions are less variable than predicted (see Jones, 1987). Moreover, observations on the effects of aberration heterozygosity in *D. melanogaster* seem to eliminate any simple "limiting component" model here. Aberration heterozygosity causes a decrease in exchange in the immediate vicinity of the breakpoints and, generally, an increase elsewhere; the early interpretations were that this reflected redistribution of a limited component, but more extensive data demonstrated that the increases could be much greater than the decreases—in other words, the cell total of exchanges goes up (Lucchesi and Suzuki, 1968; Hewitt, 1967). The limiting-component model might be rescuable by postulating that the limiting component can be reutilized under special conditions. However, the "new" high-order tetrads (E_2s and E_3s) have exactly the same exchange locations as do tetrads of that rank in isosequential genotypes; that is, interference distances remain constant when tetrads of the same rank are

---

FIGURE 2. Distribution of the events in E_1 (single exchange) and E_2 (double exchange) tetrads of the X chromosome of *D. melanogaster*. The figures compare normal homozygotes and four inversion homozygotes without (normal third chromosomes) and with ($TM2/+$, heterozygotes for multiply inverted third chromosomes) the imposition of the interchromosomal effect (data from Szauter, 1984). The data have been analyzed by calculating regional tetrads (see text and Charles, 1938; Stephens, 1961) from the original strand data, dividing the regional frequencies of E_1 and E_2 tetrads by the number of salivary bands in that region and then dividing by two. This gives a measure of the frequency of crossing-over per unit physical called the coefficient of exchange (see Lindsley and Sandler, 1977); here the calculations are by tetrads rather than by total map distance. The coefficient of exchange for an interval (ordinate) is plotted against the midpoint of the region monitored (as salivary-chromosome bands: abscissa). (A) E_1 tetrads for the five X chromosomes without additional aberrations. (B) E_1 tetrads for the five X chromosomes with the multiply inverted third chromosome *TM2* heterozygous. (C) Least-square curves for the data in A and B. The best fits are to the polynomials $y = 0.01 + 0.04x - 0.003x^2$ and $y = 0.01 + 0.02x - 0.001x^2$. The different X chromosome genotypes give equivalent patterns even though the inversions move substantial regions of the chromosome (see the physical maps in D); the frequency of crossing-over of a region is therefore not intrinsic (for example, determined by its base sequence) but rather is governed by its position along the chromosome. In the presence of inversion heterozygosity, the frequency of E_1 tetrads is reduced but the pattern is unchanged. (D) Physical maps of the markers and the breakpoints of the five X chromosomes. Straight lines symbolize euchromatin, wavy lines symbolize heterochromatin, and circles symbolize centromeres. Euchromatic markers and breakpoints are positioned here by their location on the salivary map; positions of heterochromatic breakpoints derive from observations of mitotic metaphases. Heterochromatin is looped out in the top five maps because, since there is no meiotic exchange within heterochromatin, it does not contribute to the coefficients of exchange plotted in A, B, C, E, F, and G. The bottom diagram summarizes the breakpoints of the four inversion chromosomes as superimposed on a chromosome of normal sequence. Note that these inversions do rearrange regions that might have harbored *cis*-acting regulators of exchange frequency had there been any. (E) E_2 tetrads for the five X chromosomes without additional aberrations. (F) E_2 tetrads for the five X chromosomes with the multiply inverted third chromosome *TM2* heterozygous. (G) Least-square curves for the data in E and F. The best fits are to the polynomials $y = 0.006 + 0.178x - 0.012x^2 + 0.001x^3$ and $y = 0.004 + 0.060x - 0.003x^2 + 0.0004x^3$. The different X chromosome genotypes continue to give equivalent patterns. In the presence of inversion heterozygosity on the third chromosome, the frequency of E_2 tetrads on the X is increased but the pattern is unchanged. Symbols for panels A, B, E, and F: □, $+/+$; □, $ln(1)sc^8/ln(1)sc^8$; ■, $ln(1)w^{m4}/ln(1)w^{m4}$; ◇, $ln(1)y^4/ln(1)y^4$; ×, $ln(1)dl-49/ln(1)dl-49$. Symbols for panels C and G: □, normal third chromosomes; ×, $TM2/+$.

compared (see Fig. 2). This implies that the temporal extents of occurrence of all exchanges overlap (Charles, 1938; Stephens, 1961). (Inclusive interference values decrease, of course, because each arm is experiencing a higher average number of exchanges.) The hypothesis that is most consistent with the observations is that of Lucchesi and Suzuki (1968), who proposed that synapsis and the stage during which exchange can occur are both temporally regulated but at least semi-independently; in genotypes with synaptic difficulties the exchange stage is prolonged, potentially yielding a higher total number of exchanges per cell than in wild type. As Lucchesi and Suzuki note, this hypothesis simultaneously accounts for the chromosomal genotypes (some translocation homozygotes) in which total exchange is lower than in wild type. Of course, it would be useful to have the cytological correlates for these genetic observations in *D. melanogaster*, but it would be even more useful to have synapsis as well as late nodule and chiasma data in a range of aberration heterozygotes in a range of species.

B. The Issue of Chiasma Terminalization

For any particular species, correspondence between the exchange phenotype of that species and its late nodule phenotype must be established before experimental interventions can be interpreted. For the majority of species studied in both the past and the future, it is nodules and chiasma numbers and positions that will be compared rather than nodules and exchange. Insofar as chiasmata change position along the bivalent as it progresses through prophase I to metaphase I of meiosis, chiasma positions, numbers, or both will correlate imperfectly with the distributions of exchanges and also, presumably, of late recombination nodules. The issue of whether, and if so how often and in which species, chiasmata move is still incompletely resolved. Darlington (1932) pro-posed that chiasmata are moved distally from their points of formation (i.e., undergo terminalization) to reconcile observations of apparently terminal chias-mata at metaphase I in *Campanula persicifolia* with the concept that chiasmata are the result of exchange, since reciprocal crossover events must be at least slightly subterminal. It is ironic that one part of this hypothesis (that chiasmata are the visible manifestation of genetic exchange) is universally accepted within the field but viewed with skepticism outside it, whereas the other part (whether chiasma terminalization ever occurs) is a topic of considerable controversy within the field whereas those outside assume it is a universal feature of meiotic chromosome behavior.

That most organisms do not exhibit complete terminalization has always been recognized (Darlington, 1932); many organisms reproducibly exhibit interstitial chiasmata at metaphase I. For many organisms there is no demonstrable change in chiasma distribution from early diplotene to metaphase I (for review, see Jones, 1977), and Tease and Jones (1978) have shown directly that chiasmata do not move away from the point of exchange in *Locusta migratoria*. Moreover, Jones (1978) has convincingly demonstrated that the touted terminality of chiasmata in rye at metaphase I is an artifact of chromatin condensation; the chiasmata are, in fact, still subterminal at metaphase I, but the short arms distal to each chiasma are packed in with the proximal portion of the bivalent. If this kind of compaction is general, and I agree with Jones that this is likely, then even in *C. persicifolia* distal chiasmata need not move (compare Fig. 32 of Darlington [1932, p. 104 in the 1964 edition] with Fig. 8 to 15 of Jones [1978], noting also that, especially in early work, it was not always possible to distinguish interstitial coils from true chiasmata). The philosophical necessity for chiasma movement is therefore removed. However, the basic problem for our present purposes remains: the apparent locations of chiasmata at metaphase I may not be a reliable indication of the locations of the exchanges, especially when exchanges are subterminal. When chromatin condensation, mechanical effects of bivalent-spindle interactions, or both affect apparent chiasma location, metaphase chiasma distributions and pachytene late nodule distributions will differ.

There are other reasons to be concerned about the general issue of chiasma terminalization. The chiasma closest to the kinetochore serves as the anchor point that holds the bivalent together during prometaphase I orientation and resists premature homolog separation at full metaphase (see Hawley, this volume); indeed, the poleward forces during metaphase frequently cause the chromatin proximal to this chiasma to be substantially stretched (for excellent photographs, see Lewis and John, 1972). Even interstitial chiasmata at metaphase I are therefore "bound" (see Maguire, 1974), and this is an active process, i.e., requires specific gene activity (Maguire, 1978). Distal movement of chiasmata, if it be an integral meiotic feature in any organism (see Maguire, 1979), requires that chiasmata not be irrevocably bound distally during the process of movement, presumably during diplotene-diakinesis. Moreover, movement would seem to require that sister chromatids dissociate downstream of the direction of movement whereas nonsister chromatids must then associate after the chiasma has passed. Both of these, if they occur, must be active processes, processes conceptually difficult in any organism but especially so in those reported to maintain synaptonemal complex segments at chiasmata until metaphase I (for review, see von Wettstein et al., 1984). Contemporary models for the mechanism of terminalization have been ably reviewed by Swanson (1957); Swanson's own model (that it is a passive by-product of homolog repulsion and chromatid coiling, both presumably active processes) appears now to be the only viable one. Some organisms retain a modified form of late nodules and/or synaptonemal complex segments past pachytene; Holm (1986) has postulated that these structures function to prohibit chiasma movement. This, of course, assumes that terminalization is a regular feature of meiosis but one that can be actively circumvented. Given the unambiguous demonstrations that chiasmata do not move in several species (for reviews, see Jones, 1987; Hawley, this volume), it seems to me premature to assert chiasma movement in the absence of unambiguous proof. Unfortunately, definitive tests for chiasma movement are difficult—meiotic cells are difficult to label with 5-bromodeoxyuridine or other agents that permit differentiation of sister chromatids, and most organisms do not have cytological markers equivalent to the dispensable knobs Maguire (1979) used in maize.

In fact, the possibility that late nodules themselves move has been raised; the available evidence suggests that they do not change position (no significant differences between locations of late nodules present early in the late-nodule period [mid-pachytene] versus in its middle and at its end [still mid-pachytene] in *D. melanogaster* females [Carpenter, 1979a]; equivalent observations have been made in human male material [see von Wettstein et al., 1984]). This is certainly an issue that needs additional checks in additional organisms.

C. Interference

The final issue with respect to exchange/nodule distributions is that of interference. Interference (I) is defined as $1 - C$, where C is the coefficient of coincidence, and C = (number of double crossovers observed)/(number of double crossovers expected if events in the two intervals are independent). Because factors of 50 cancel, the same formulas hold for pairs of exchanges (E_2s) and pairs of chiasmata (X_2s). Neither interference nor the coefficient of coincidence has meaning except as calculated for specified pairs of intervals, but those intervals need not be adjacent. It is therefore possible to hold one interval constant and examine interference with a variety of second intervals. This has been done both genetically and cytologically (see, e.g., Charles, 1938; Stephens, 1961; Jones, 1984), and the pattern seems to be universal; interference is complete (there are no double crossovers) within very small intervals (10 map units or so), decreases as the two intervals become farther apart, reaches a minimum (little or no interference) at about 45 map units of separation, and then increases again. This means not only that close simultaneous events are nonrandomly unlikely but also that when there are two events they tend to be a modal distance apart (Charles, 1938; Stephens, 1961).

Another way of seeing interference is to take just E_2 tetrads and plot where the two events are (Fig. 2); indeed, the more distal event has a distribution around a modal position, as does the proximal one, but in meioses in which the distal event is very distal the proximal one is likely to be on the distal side of its distribution, etc. This relationship can be treated numerically by calculating the average interexchange interval in these E_2 tetrads (in physical, not genetic, units) and comparing it with the mean interval expected from chance positioning of two events on a line of fixed length: $\int_0^1 [(1 - x)^2/2 + x^2/2]dx$ = one-third of the total physical length; the general formula is $1/(n + 1)$, where n is the number of events to be positioned. Consequently, if the two events are on average greater than one-third of the total length apart, there is interference, whereas if they average one-third there is none. E_3s show the same properties, except that, of course, there are three modal positions, their means are closer together than are the two of E_2s, and the expected mean (chance) separation distance is one-fourth. Both chiasma and recombination nodule data can be interpreted in the same two ways. It is important to realize that just as the coefficient of coincidence applies only to two specified intervals, having no meaning in an "average" sense, neither does an "average" internodule distance have meaning unless bivalents with different total numbers of nodules have been analyzed separately.

III. EARLY RECOMBINATION NODULES

A. Properties

As mentioned above, all organisms examined have two types of recombination nodules, based on size, shape, and time of appearance. The shape and size differences simplify sorting the two types when their temporal distributions overlap (for references, see von Wettstein et al., 1984). The earlier type of nodule usually is present during synapsis and the formation of the synaptonemal complex, that is, in zygotene, although there are exceptions (e.g., *D. melanogaster* females [Carpenter, 1979a]). Early nodules are always more frequent (2 to 20 times, depending on the organism) than late nodules, chiasmata, or exchanges, and they always have distributions that are different from those of exchanges, chiasmata, or late nodules; generally, their distributions are completely random between and along the euchromatin of chromosomes (that is, there is no interference), but in at least two organisms (*Allium cepa* and *A. fistulosum* [Albini and Jones, 1987]) their distribution is, instead, regular.

There are two unanswered questions with respect to early nodules. What is their meiotic function, and what is their relationship to the later nodules that correspond so closely to exchanges? Being synthesists we would prefer to think (at least as a first approximation) that all organisms that follow the standard meiotic sequence will give the same answer to these questions, but given the clearly wide variety of strategies across organisms for, for example, details of synaptic initiation, I think that at this point it will be more profitable to focus on broad generalities rather than getting sidetracked by specific differences.

As it happened, both of the two organisms in which (independently) early nodules were first noticed, *Homo sapiens* and *D. melanogaster* (Rasmussen and Holm, 1978; Carpenter 1979a), have twice as many early nodules as late nodules, exchanges, and chiasmata. Since there are two recognizably distinct meiotic recombinational outcomes, simple gene conversion (transferral of information from one homolog to the other without concomitant exchange of flanking markers) and exchange (reciprocal crossing-over between two homologous chromatids with or without a detected gene conversion event), and since late nodules correlate with exchange, it was obvious to hypothesize that early nodules correlate in some way with gene conversion events. There are two alternative scenarios for such a correlation. If the two genetic outcomes are physically derived from one common pool of initial events, then early nodules correspond to that pool and late nodules are derived from a subset of the early ones (Rasmussen and Holm, 1978). If, on the other hand, simple gene conversion and exchange, however similar biochemically, are derived from different sets of events, then early nodules correspond solely to simple gene conversion and late nodules arise afresh (Carpenter, 1979a). At this point we lack definitive evidence either from genetics or from cytology to distinguish between these two alternatives as universal mechanisms, and of course it is entirely possible that some organisms use one progression, others use the other, or that neither is close enough to reality to stand the test of time. Moreover, several plant species have recently been discovered (Stack and

Anderson, 1986a, Albini and Jones, 1987) to have their early nodules present very early (at or just before initiation of synaptonemal complex formation) in very high numbers, so that here these early nodules may function during synapsis (Albini and Jones, 1987; for a synthesis, see Carpenter, 1987; Giroux, this volume), especially since in *Allium* spp. these nodules equally dramatically disappear as synapsis is complete (Albini and Jones, 1987). Considerations of function of early nodules will be continued below, but first it is necessary to discuss briefly the problem of whether physical continuity exists between early and late nodules.

Early and late nodules clearly are related structures; they are similar in overall morphology, have identical staining characteristics, and are both located along the axis of the bivalent. The differences in shape and size, although useful as markers, are not dramatic in most species, and species differ with respect to which form corresponds to which nodule type (for example, early nodules in humans are spherical whereas late ones are ellipsoidal; in *D. melanogaster* the converse is true); consequently, the differences in form probably do not indicate complete differences in function. However, it is still not clear in any organism whether the two forms are independent structures or whether late nodules are derived from a subset of early ones.

This is still an open question because of limitations of the available techniques. Although it is possible to do real-time, dynamic studies by light microscope cinematography, recombination nodules are too small to be resolved by conventional light microscopy; the shorter wavelengths available by electron microscopy are needed to be able to resolve recombination nodules as discrete structures. Moreover, although recombination nodules are unequivocably real and specific structures, they are nevertheless small dense balls, and cellular context, to wit their association with the bivalent and synaptonemal complex, provides essential clues for their recognition. Consequently, recombination nodules have so far been studied only by electron microscopy, and electron microscopy is perforce a static observational technique. Of course, dynamic processes can be studied by electron microscopy and sequences can be inferred. For example, when the numbers and chromosomal distributions of structures that differ in morphology but appear sequentially are congruent, as is the case for late recombination nodules, chromatin nodules, circular structures, and chiasmata in *Bombyx mori* males (see von Wettstein et al., 1984), there is no reason to be particularly skeptical of the straightforward hypothesis that the different morphologies represent sequential modifications of one basic structure (until, of course, there is evidence to the contrary). However, when the numbers and chromosomal distributions are not congruent, as is the case for early and late nodules in all organisms, direct transformation of the earlier into the later is certainly not disproved but the lack of congruence can hardly be taken as evidence in favor of the transformation hypothesis. Even structures that have been interpreted as intermediates between the two forms (e.g., Gillies, 1979) are not unambiguous: the apparent intermediates could be immature, newly initiated late nodules rather than genuine intermediates between early and late forms. Additional static observations cannot discriminate between these two alternatives; physical pulse-marking of early nodules is needed to determine whether a subset of them chases into late nodules.

B. Function

A variety of functions have been hypothesized for early nodules: they may mediate gene conversion with and without exchange (Rasmussen and Holm, 1978), serve as excess potential sites for exchange (Stack and Anderson, 1986b), function in synapsis (Albini and Jones, 1987), and function in a homology-checking aspect of synapsis that can lead to simple gene conversion but not to exchange (Carpenter, 1987). The correlation with synapsis is temporal—in many organisms early nodules are present during the stage of synapsis (zygotene), and in at least three species of plants (Stack and Anderson, 1986a; Albini and Jones, 1987) they are present at or before the initiation of synapsis. In the one exception, *D. melanogaster* females (Carpenter, 1979a), homologs enter meiosis in the same very close juxtaposition that is typical of mitotic cells (Metz, 1916; Bridges, 1916; Grell, 1967), so here a synaptic function for early nodules may not be necessary.

Most of the hypotheses for early nodule function involve gene conversion as a logical extension of the hypotheses for late nodule function as mediators of exchange. However, testing these hypotheses or even assessing their reasonableness is difficult because a distressingly high proportion of what we "know" about meiotic gene conversion is derived from a family of models (e.g., Meselson and Radding, 1975; Szostak et al., 1983) that are primarily focused on the molecular details of DNA interactions during recombination events. It is very much more difficult to examine frequencies and chromosomal distributions of simple gene conversion events than it is of exchanges or chiasmata because simple gene conversion events, by definition, affect only a small segment of a chromosome—since they do not give rise to crossovers, they can be detected only when they chance to occur over a genetic marker. Granted, when a large number of loci have been examined for any one organism, patterns can be built up and reasonable extrapolations can be made for that organism. However, even if it be true for *S. cerevisiae* that, on average, the frequency of simple gene conversion events is equal to that of exchanges (Fogel et al., 1971), this equivalence need not hold for other organisms—unless the models that assume that this equivalence (in yeast meiosis) reflects an unregulated branch point in a unitary molecular pathway in fact reflect reality. Even if the generality of this assumption had not been challenged (Carpenter, 1982, 1984), it would still be shaky to push it too far. The arguments quickly become circular. For example, if one argues that the 2:1 ratio of early to late nodules in humans implies that early nodules correlate with total recombination events (simple gene conversion plus exchange) because the models predict a 2:1 ratio in all organisms, what is one to do with the observations of vast excesses of early to late nodules in tomatos and onions? Obviously, one cannot now extend the argument to say that because early nodules are equivalent to total events in humans they are here, too, and thus there must be a vast excess of simple gene conversion events; nor can one conclude that most of these early nodules are nonfunctional solely to bring the effective ratio down. The observations are disparate unless the "rule" of equivalence of simple gene conversion and exchange is not general—or else early nodules do not, after all, reflect any aspect of meiotic recombination in a predictable way. By no means do I intend to denigrate anyone's hypotheses; I do hope, however, to stimulate both thought and interventive experiments. It would be lovely if we could extrapolate with confidence from cell and chromosomal distributions of early nodules to cell and chromosomal distributions of gene conversion events across organisms: unlike the situation for exchange, chiasmata, and late nodules, we have no other hope of having access to this information. At this point we lack that confidence. Since I personally find it annoying to have to cite some armchair theoretician for listing obvious predictions when I have not only seen those predictions but also done the work of devising and carrying out experiments to test them, I am going to forbear making such a list in the hope that this will encourage experiments—which need not cite this review! The issues, both with respect to gene conversion and with respect to early nodules, are of sufficient concern and import that no one laboratory or organism can hope to settle them.

Acknowledgments. The financial supports of Public Health Service grant GM23338 from the National Institutes of Health and of National Science Foundation grant DCB-8712888 are gratefully acknowledged.

I thank Gareth Jones and Bruce Baker for their critically helpful comments on the manuscript and Gerald Smith for pointing out a number of phrases that could have been misinterpreted.

LITERATURE CITED

Albini, S. M., and G. H. Jones. 1987. Synaptonemal complex spreading in *Allium cepa* and *A. fistulosum*. I. The initiation and sequence of pairing. *Chromosoma* (Berlin) 95:324–338.

Anderson, E. G. 1925. Crossing over in a case of attached X chromosomes in *Drosophila melanogaster*. *Genetics* 10:403–417.

Baker, B. S., and A. T. C. Carpenter. 1972. Genetic analysis of sex chromosomal meiotic mutants in *Drosophila melanogaster*. *Genetics* 71:255–286.

Baker, B. S., and J. C. Hall. 1976. Meiotic mutants: genetic control of meiotic recombination and chromosome segregation, p. 351–434. *In* M. A. Ashburner and E. Novitski (ed.), *Genetics and Biology and Drosophila*, vol. 1A. Academic Press, Inc., New York.

Bernelot-Moens, C., and P. B. Moens. 1986. Recombination nodules and chiasma localization in two Orthoptera. *Chromosoma* (Berlin) 93:220–236.

Bridges, C. B. 1916. Nondisjunction as the proof of the chromosome theory of heredity (concluded). *Genetics* 1:107–163.

Bridges, C. B. 1935. Salivary chromosome maps. *J. Hered.* 26:60–64.

Byers, B., and L. Goetsch. 1975. Electron microscopic observations on the meiotic karyotype of diploid and tetraploid *Saccharomyces cerevisiae*. *Proc. Natl. Acad. Sci. USA* 72:5056–5060.

Carpenter, A. T. C. 1975. Electron microscopy of meiosis in *Drosophila melanogaster* females. II. The recombination nodule—a recombination-associated structure at pachytene? *Proc. Natl. Acad. Sci. USA* 72:3186–3189.

Carpenter, A. T. C. 1979a. Synaptonemal complex and recombination nodules in wild-type *Drosophila melanogaster* females. *Genetics* 92:511–541.

Carpenter, A. T. C. 1979b. Recombination nodules and synaptonemal complex in recombination-defective females of *Drosophila melanogaster*. *Chromosoma* (Berlin) 75:259–292.

Carpenter, A. T. C. 1981. EM autoradiographic evidence that DNA synthesis occurs at recombination nodules during meiosis in *Drosophila melanogaster* females. *Chromosoma* (Berlin) **83**:59–80.

Carpenter, A. T. C. 1982. Mismatch repair, gene conversion, and crossing-over in two recombination-defective mutants of *Drosophila melanogaster*. *Proc. Natl. Acad. Sci. USA* **79**:5961–5965.

Carpenter, A. T. C. 1984. Meiotic roles of crossing-over and of gene conversion. *Cold Spring Harbor Symp. Quant. Biol.* **49**:23–29.

Carpenter, A. T. C. 1987. Gene conversion, recombination nodules and the initiation of meiotic synapsis. *BioEssays* **6**:232–236.

Charles, D. R. 1938. The spatial distribution of crossovers in X-chromosome tetrads of *Drosophila melanogaster*. *J. Genet.* **36**:103–126.

Darlington, C. D. 1932. *Recent Advances in Cytology*. The Blakiston Co., Philadelphia (3rd ed., updated 1965, as *Cytology*, J & A Churchill Ltd., London).

Fogel, S., D. D. Hurst, and R. K. Mortimer. 1971. Gene conversion in unselected tetrads from multipoint crosses. *Stadler Genet. Symp.* **1** and **2**:89–110.

Gillies, C. B. 1979. The relationship between synaptonemal complexes, recombination nodules and crossing over in *Neurospora crassa* bivalents and translocation quadrivalents. *Genetics* **91**:1–17.

Gillies, C. B. 1983. Ultrastructural studies of the association of homologous and non-homologous parts of chromosomes in the mid-prophase of meiosis in *Zea mays*. *Maydica* **28**:265–287.

Glamann, J. 1986. Crossing over in the male mouse as analysed by recombination nodules and bars. *Carlsberg Res. Commun.* **51**:143–162.

Grell, R. F. 1967. Pairing at the chromosomal level. *J. Cell. Physiol.* **70**(Suppl. 1):119–145.

Hewitt, G. M. 1967. An interchange which raises chiasma frequency. *Chromosoma* (Berlin) **21**:285–295.

Holliday, R. 1977. Recombination and meiosis. *Philos. Trans. R. Soc. London B Biol. Sci.* **277**:359–370.

Holm, P. B. 1986. Ultrastructural analysis of meiotic recombination and chiasma formation. *Tokai J. Exp. Clin. Med.* **11**:415–436.

Holm, P. B., and S. W. Rasmussen. 1980. Chromosome pairing, recombination nodules and chiasma formation in diploid Bombyx males. *Carlsberg Res. Commun.* **43**:329–350.

Holm, P. B., and S. W. Rasmussen. 1983. Human meiosis. VI. Crossing over in human spermatocytes. *Carlsberg Res. Commun.* **48**:385–413.

Janssens, P. A. 1909. Spermatogénèse dans les Batraciens. V. La théorie de la chiasmatypie. Nouvelles interprétation des cinèses de maturation. *Cellule* **25**:387–411.

John, B., and K. R. Lewis. 1965. *The Meiotic System*, vol. 6, F1. *Protoplasmatologia*. Springer-Verlag, New York.

Jones, G. H. 1967. The control of chiasma distribution in rye. *Chromosoma* (Berlin) **22**:69–90.

Jones, G. H. 1977. A test for early terminalization of chiasmata in diplotene spermatocytes of *Schistocerca gregaria*. *Chromosoma* (Berlin) **63**:287–294.

Jones, G. H. 1978. Giemsa C-banding of rye meiotic chromosomes and the nature of "terminal" chiasmata. *Chromosoma* (Berlin) **66**:45–57.

Jones, G. H. 1984. The control of chiasma distribution, p. 293–320. *In* C. W. Evans and H. G. Dickinson (ed.), *Controlling Events in Meiosis*. The Company of Biologists Ltd., Cambridge.

Jones, G. H. 1987. Chiasmata, p. 213–244. In P. B. Moens (ed.), *Meiosis*. Academic Press, Inc. (London), Ltd., London.

Lewis, K. R., and B. John. 1972. *The Matter of Mendelian Heredity*, 2nd ed. Halsted Press, a Division of John Wiley & Sons, Inc., New York.

Lindsley, D. L., and L. Sandler. 1977. The genetic analysis of meiosis in female *Drosophila melanogaster*. *Philos. Trans. R. Soc. London B Biol. Sci.* **277**:295–312.

Lucchesi, J. C., and D. T. Suzuki. 1968. The interchromosomal control of recombination. *Annu. Rev. Genet.* **2**:53–86.

Maguire, M. P. 1974. The need for a chiasma binder. *J. Theor. Biol.* **48**:485–487.

Maguire, M. P. 1978. Evidence for separate genetic control of crossing over and chiasma maintenance in maize. *Chromosoma* (Berlin) **65**:173–183.

Maguire, M. P. 1979. Direct cytological evidence for true terminalization of chiasmata in maize. *Chromosoma* (Berlin) **71**:283–287.

Meselson, M. S., and C. R. Radding. 1975. A general model for genetic recombination. *Proc. Natl. Acad. Sci. USA* **72:**358–361.

Metz, C. W. 1916. Chromosome studies in Diptera. II. The paired association of chromosomes in the Diptera and its significance. *J. Exp. Zool.* **21:**213–279.

Morgan, T. H. 1911. The application of the conception of pure lines to sex-limited inheritance and to sexual dimorphism. *Am. Nat.* **45:**65–78.

Muller, H. J. 1916. The mechanism of crossing-over. *Am. Nat.* **50:**193–211, 284–305, 350–366, 421–434.

Parry, D. M. 1973. A meiotic mutant affecting recombination in *Drosophila melanogaster. Genetics* **73:**465–486.

Rasmussen, S. W., and P. B. Holm. 1978. Human meiosis. II. Chromosome pairing and recombination nodules in human spermatocytes. *Carlsberg Res. Commun.* **43:**423–438.

Stack, S. M., and L. K. Anderson. 1986a. Two-dimensional spreads of synaptonemal complexes from solanaceous plants. II. Synapsis in *Lycopersicon esculentum* (tomato). *Am. J. Bot.* **73:**264–281.

Stack, S., and L. Anderson. 1986b. Two-dimensional spreads of synaptonemal complexes from solanaceous plants. III. Recombination nodules and crossing over in *Lycopersicon esculentum* (tomato). *Chromosoma* (Berlin) **94:**253–258.

Stephens, S. G. 1961. A remote coincidence. *Am. Nat.* **95:**279–293.

Swanson, C. P. 1957. *Cytology and Cytogenetics*, p. 214–218. Prentice-Hall, Inc., Engelwood Cliffs, N.J.

Szauter, P. 1984. An analysis of regional constraints on exchange in *Drosophila melanogaster* using recombination-defective meiotic mutants. *Genetics* **106:**45–71.

Szostak, J. W., T. L. Orr-Weaver, R. J. Rothstein, and F. W. Stahl. 1983. The double-strand-break-repair model for recombination. *Cell* **33:**25–35.

Tease, C., and G. H. Jones. 1978. Analysis of exchanges in differentially stained meiotic chromosomes of *Locusta migratoria* after BrdU-substitution and FPG staining. *Chromosoma* (Berlin) **69:**163–178.

von Wettstein, D., S. W. Rasmussen, and P. B. Holm. 1984. The synaptonemal complex in genetic recombination. *Annu. Rev. Genet.* **18:**331–413.

Weinstein, A. 1936. The theory of multiple-strand crossing over. *Genetics* **21:**155–199.

Whitehouse, H. L. K. 1969. *Towards an Understanding of the Mechanism of Heredity*, 2nd ed. Edward Arnold Ltd., London.

Zickler, D. 1977. Development of the synaptonemal complexes and the "Recombination Nodules" during prophase in the seven bivalents of the fungus *Sordaria macrospora* Auersw. *Chromosoma* (Berlin) **61:**289–316.

Chapter 18

Homologous Recombination in Mitotically Dividing Mammalian Cells

Suresh Subramani and Brent L. Seaton

I. INTRODUCTION

In the past decade a large number of genes from a variety of organisms have been isolated and sequenced. The introduction of these cloned genes and their altered derivatives into various cells has provided valuable insights into the functions and mechanisms of regulation of these cellular genes. In mammalian cells, however, the full utility of this approach has been limited by the nonhomologous integration of vectors (carrying the genes of interest) into apparently random locations in the cellular genome (Kato et al., 1986; Robins et al., 1981), even when the vectors contain DNA sequences homologous to one or more sites in the genome.

Although it is possible, in principle, to correct genetic defects in cell lines or animals with a cloned functional copy of the affected gene, the nonspecific integration of the vector may yield unexpected results and may even be deleterious to the cell. The integration of the DNA into essential transcription units may generate recessive or dominant mutations (Copeland et al., 1983; Covarrubias et al., 1985; Covarrubias et al., 1987; Gordon, 1986; Hooper et al., 1987; Kuehn et al., 1987; Mark et al., 1985; Palmiter et al., 1984; Schnieke et al., 1983; Westaway et al., 1984; Woychik et al., 1985). Alternatively, the appropriate expression and/ or regulation of the introduced gene may be hampered by position effects (Maniatis, 1985; Palmiter and Brinster, 1985). The nonhomologous integration of vectors into chromosomes also renders it impossible to introduce derivatives of a chromosomal gene into the same chromosomal context for studies on gene regulation. This problem has been circumvented partially by the use of transient assay systems. However, these procedures suffer from the drawback that the introduction of a large number of DNA copies into a cell may titrate out limiting protein factors involved in the regulation of genes. Therefore, much of the impetus for studying homologous recombination in mammalian cells is derived from a desire to target genes to predetermined chromosomal locations and to rescue chromosomal alleles onto extrachromosomal vectors. In addition to this, many groups have been investigating a wide variety of phenomena in which somatic recombination events have been implicated. These include the excision of DNA from the genome (Jones and Potter, 1985; Krolewski et al., 1984), DNA rearrangements (Brack et al., 1978; Hochtl and Zachau, 1983; Lewis et al., 1982; Tonegawa, 1983), amplification of genes (Schimke, 1982; Stark and Wahl, 1984) including oncogenes (Alitalo et al., 1983; Collins and Groudine, 1982; Kohl et al., 1983), sister chromatid exchange (Latt, 1974, 1981), the maintenance of sequence homogeneity among members of multigene families (Baltimore, 1981), the activation of gene expression (Reth et al., 1985; Tonegawa, 1983), and chromosome translocations associated with certain forms of neoplasia (Croce and Klein, 1985; Yunis, 1983). With the advent of suitable model systems in the past 5 years, considerable progress has been achieved in unraveling some of the complexities and properties of the recombinational machinery in mitotically dividing mammalian cells. These advances are to be the focus of this review (written in summer 1987). Whenever possible, we have attempted to compare similar recombination events in mammalian cells and in the yeast *Saccharomyces cerevisiae*.

II. EXTRACHROMOSOMAL RECOMBINATION IN MAMMALIAN CELLS

Studies on extrachromosomal recombination have been undertaken, in part, to explore whether similar or different mechanisms operate in chromosomal and extrachromosomal recombination events. An additional impetus stems from the fact that the assays for extrachromosomal recombination are invariably faster, the substrates are easier to manipulate, and large numbers of independent products can be easily isolated and characterized.

Early attempts to study homologous recombination at the molecular level involved the use of mutant DNA viruses such as simian virus 40 (SV40) (Dubbs et al., 1974; Upcroft et al., 1980; Vogel, 1980; Vogel et al., 1977), adenovirus (Young and Silverstein, 1980; Volkert and Young, 1983; Wolgemuth and Hsu, 1980), polyoma virus (Miller et al., 1976), or herpes simplex virus (Dasgupta and Summers, 1980; Mocarski and Roizman, 1982). Subsequently, we and others used molecules from which wild-type viral genomes could be regenerated by recombination (Rubnitz and Subramani, 1984; Subramani and Berg, 1983; Wake and

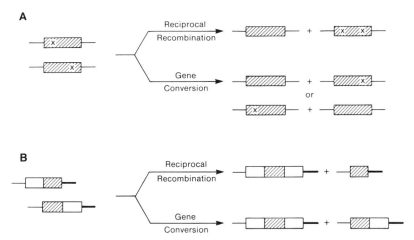

FIGURE 1. Recombination between duplicated sequences. The consequences of reciprocal or nonreciprocal recombination (gene conversion) are shown. The homologous regions in the selectable marker gene and in the adjacent vector sequences are indicated by the hatched areas and the solid bars, respectively. The thin lines are vector DNAs that do not share homology. (A) Recombination between two complete genes with different mutations. The symbol × denotes a point mutation, a deletion, or a linker insertion mutation. (B) Recombination between truncated but overlapping fragments of the marker gene. If homology exists in the vector sequences as shown, then gene conversion or reciprocal recombination can regenerate a functional marker gene.

Wilson, 1979). While the use of viruses permits the detection of rare recombinants, such studies are restricted by the host range of the virus, the lysis of cells infected by lytic viruses, and the possibility that some of the large viruses might encode or modulate recombinases.

Much of the recent progress in the field can be attributed to the demonstration and quantitation of homologous recombination by using the regeneration of a selectable marker gene (e.g., the neomycin resistance gene, *neo*, or the thymidine kinase gene, *tk*) from two mutant copies of the gene, after transfection of cells with plasmid- or virus-based vectors bearing the mutant copies. The mutants are the result of linker insertions, point mutations, or deletions that render the marker gene nonfunctional. Some of these markers (e.g., *tk*) can be used only in cell lines auxotrophic for that gene product, whereas other dominant selectable markers such as *neo* or the bacterial xanthine-guanine phosphoribosyltransferase gene (*gpt*) can be used in many cell types (Subramani and Southern, 1983). The partners in the recombination event have been constructed such that either intramolecular (Lin et al., 1984a; Rubnitz and Subramani, 1985, 1986, 1987; Subramani and Rubnitz, 1985; J. Rubnitz, Ph.D. thesis, University of California, San Diego, 1987) or intermolecular (Brenner et al., 1984, 1985, 1986; Folger et al., 1982, 1985a; Kucherlapati et al., 1984b; Robert de Saint Vincent and Wahl, 1983; Shapira et al., 1983; Smithies et al., 1985b; Wong and Capecchi, 1986) interactions can be studied. In every case, either reciprocal recombination or gene conversion can generate a functional marker gene which can be expressed in bacterial or mammalian cells, or both (Fig. 1).

TABLE 1
Direct selection for a recombined marker gene after transfection
of mammalian cells

Marker gene	Cell line	Recombination frequency (%)[a]	Reference
tk	LTk⁻	11	Small and Scangos (1983)
		20	Shapira et al. (1983)
neo	LTk	1	Kucherlapati et al. (1984b)
	LTk⁻	3–8	Folger et al. (1985a)
	EJ	7	Kucherlapati et al. (1984b)
	3T6	15	Subramani and Rubnitz (1985)
	COS	5–10	Subramani and Rubnitz (1985)

[a] Recombination frequency is the ratio of G418^r colonies obtained with the substrate to the number of colonies obtained with a plasmid carrying a functional *neo* gene.

Characterization of the recombination products and determination of the recombination frequencies have been accomplished by four methods. The first, based on direct selection for a recombined selectable marker gene, provides information regarding the number of cells in a transfected cell population that are capable of catalyzing at least one recombination event. Since the selection is applied a few days after transfection, intra- or intermolecular recombination is presumed to occur extrachromosomally. The recombined product is then integrated and expressed stably from the genome. However, it is difficult to rigorously rule out the integration of one substrate molecule into the chromosome, followed by the recombination of that integrated DNA with another substrate molecule, nor is it possible to exclude intramolecular recombination events in the newly integrated DNA. Whatever the mechanism, these experiments show that about 5 to 20% of the cells that are capable of expressing a marker gene stably are also competent to mediate at least one recombination event (Table 1). Replication of the substrates is unnecessary for recombination, and the frequencies of recombination are not dependent on the direct or inverted repeat configuration of homologous sequences when recombination is intramolecular (Subramani and Rubnitz, 1985). The recombination products have been characterized either by Southern blot analysis of the genomic DNAs from recombinant cell lines (Brenner et al., 1985; Folger et al., 1982; Lin et al., 1984b; Shapira et al., 1983; Subramani and Rubnitz, 1985) or by plasmid rescue (Brenner et al., 1985; Folger et al., 1985a; Kucherlapati et al., 1984b) of the integrated vector DNA in recombination-deficient bacteria. When detailed product analyses were undertaken, the products arose predominantly by nonreciprocal gene conversion (Folger et al., 1985a) rather than by reciprocal recombination, and as discussed later, the frequency of recombination was enhanced by double-strand breaks within but not outside the region of homology. Interestingly, most of the recombination events occur within 1 h after the microinjection of the plasmid DNA into the nuclei of cells (Folger et al., 1985a).

An alternative method of quantitating extrachromosomal recombination involves the measurement of the ratio of recombined to unrecombined molecules recovered from mammalian cells in which the substrates have undergone replication. For example, ampicillin-resistant plasmid recombination substrates containing two truncated, nonfunctional but overlapping fragments of the *neo* gene were allowed to replicate in monkey COS cells (Gluzman, 1981) for 1 to 4 days; the plasmid DNAs were recovered by the procedure of Hirt (1967) and then were used to transform recombination-deficient bacteria to either ampicillin or neomycin resistance. The recombination frequencies, as measured by the ratio of Neo^r to Amp^r colonies, were 1×10^{-2} to 4×10^{-2} for intramolecular recombination between directly repeated sequences sharing 420 base pairs (bp) of homology, whereas for inverted repeat plasmids the frequencies were 1×10^{-3} to 6×10^{-3}. The intermolecular recombination frequency between plasmids carrying the truncated fragments of the *neo* gene (see above) was 10^{-3} to 10^{-4} (Rubnitz and Subramani, 1985). Analogous studies in COS cells coinfected with two different deletions in the *neo* gene yielded recombination frequencies of 2×10^{-4} to 7×10^{-4} (Kucherlapati et al., 1984a).

This bacterial transformation assay for intra- or intermolecular, reciprocal or nonreciprocal recombination takes only a few days to perform and has the advantage that large numbers of truly independent recombination products can be isolated and characterized. Though replication of the DNA substrate and product are not essential for recombination, the act of replication does stimulate recombination more than can be accounted for by a simple increase in plasmid copy number (J. Rubnitz and S. Subramani, unpublished data). Thus, recombination assays involving the screening of recombinants in bacteria are best suited for mammalian cells in which the recombination substrates replicate. These include COS cells for plasmids carrying the SV40 origin of DNA replication (*ori*) and lymphoid cells for plasmids with the Epstein-Barr virus *ori* and EBNA-1 gene (Yates et al., 1985).

The high frequency of recombination suggested by the bacterial transformation assay raised concerns that the frequency might be aberrantly high because events initiated in the mammalian cells may have been completed in bacteria. This problem has been circumvented by the use of Southern blots of DNA which has been passaged through COS cells (Rubnitz and Subramani, 1985). The recombination substrate and product DNAs are recovered from the COS cells and used directly for Southern blot analysis. The use of suitable radioactive probes and restriction enzyme digests easily discriminates between the recombination substrate and the product. Densitometric scanning of the autoradiograms shows that the recombination frequency (10^{-2}) is similar to that obtained by using the transformation assays and is substantially higher than that obtained in control experiments involving direct transformation of bacteria with the substrates or Southern blot analysis of the recombination substrates not passaged through COS cells. Thus, these assays do indeed quantitate recombination events that occur in the animal cells.

The Southern blot procedure is also suitable for an estimate of the stimulation of the frequency of recombination by double-strand breaks. The increase in the absolute amount of recombination product can be used to determine the stimulation in experiments in which the circular and linear substrates are transfected into mammalian cells separately (Rubnitz and Subramani, 1985). Southern blot analysis, however, is not amenable for an analysis of independent recombination products.

A new procedure for measuring the population of recombined molecules in mammalian cells is based on gene expression from the regenerated copy of a highly sensitive reporter gene (e.g., firefly luciferase). Cells are infected transiently with a substrate containing two nonfunctional but overlapping segments of the reporter gene. Protein extracts of these cells made 2 days later show 10 to 17% enzymatic activity relative to that in cells transfected with an intact copy of the reporter gene (unpublished data). In addition, the extrachromosomal DNA recovered from the mammalian cells can be used to transform bacteria, and colonies expressing the recombined luciferase gene can be identified by a film assay that detects glowing colonies (Wood and Deluca, 1987).

For the extrachromosomal recombination assays in which replication of the substrates and products occurs, the inverse relationship between the ability of the plasmids to replicate and their size should be borne in mind (Calos, 1986). If the recombination product is smaller or bigger than the substrate, the frequency of recombination may be overestimated or underestimated, depending on the extent to which the product replicates with a higher or lower efficiency, respectively, than the substrate.

III. EXTRACHROMOSOMAL GENE CONVERSION

In the substrates described earlier, reciprocal or nonreciprocal recombination can restore the marker or reporter gene. Other substrates in which gene conversion alone restores the marker gene have also been developed. These substrates contain a 10-bp linker insertion or a deletion (22 or 167 bp) mutation in an otherwise complete *neo* gene and also an internal homologous fragment of the gene (Rubnitz and Subramani, 1986, 1987). Although nonreciprocal gene conversion or double-reciprocal recombination can restore a functional copy of the marker gene, no double-reciprocal recombination is detectable and gene conversion appears to predominate. As measured by the bacterial transformation assay, the frequency of gene conversion in COS cells is 1×10^{-4} to 6×10^{-4} for these substrates, and the frequencies are 14- to 750-fold higher than those obtained upon introduction of the substrates directly into bacteria without prior passage through COS cells. Irrespective of the type of mutation, linearization of the molecules at the site of the insertion/deletion prior to transfection stimulates the frequency 5- to 12-fold, when cells expressing G418[r] are selected directly after transfection of mouse NIH 3T6 cells with the substrates (Rubnitz and Subramani, 1987).

Single-stranded DNA can also serve as the donor of genetic information in extrachromosomal, intermolecular gene conversion events under conditions in which extensive conversion of the single-stranded DNA into its double-stranded counterpart is undetectable (Rauth et al., 1986). Similar homologous interactions between chromosomes in yeast cells and single-stranded DNAs transfected into these cells have been described (Simon and Moore, 1987).

IV. HOMOLOGY REQUIREMENTS FOR RECOMBINATION

All of the proposed mechanisms for homologous recombination invoke DNA sequence homology for the initial synapsis and subsequent branch migration events (Holliday, 1964, 1974; Meselson and Radding, 1975; Szostak et al., 1983). Because of its central role in recombination, several groups have addressed the homology requirements for recombination in mammalian cells. Using pBR322-SV40 hybrid plasmids in monkey CV1 cells, we showed that approximately 200 bp of homology was necessary for intramolecular, extrachromosomal excision of wild-type SV40 from the plasmids. The frequency of this process decreased linearly from 5,243 to 214 bp, dropped ninefold between 214 and 163 bp, and then fell linearly down to 14 bp of homology (Rubnitz and Subramani, 1984). Interestingly, a low level of recombination was observed with as little as 14 bp of homology.

For intermolecular, extrachromosomal gene conversion (or double-reciprocal recombination) between plasmids in human EJ bladder carcinoma cells, greater than 330 bp of homology was necessary for optimal recombination and low levels were observed with as little as 25 bp of homology (Ayares et al., 1986).

The homology requirement for efficient intrachromosomal gene conversion between duplicated chromosomal sequences has been investigated in mouse L cells (Liskay et al., 1987). Conversion was efficient and proportional to the extent of homology between 295 and 1,800 bp. In contrast, conversion rates were 7- and 100-fold lower for 200- and 95-bp regions of homology, respectively. Thus, these results are also consistent with the notion that greater than 200 bp of homology is required for efficient gene conversion between repeated chromosomal sequences in mammalian cells.

A comparison of the homology requirements for recombination in other organisms revealed that it is 50 bp for bacteriophage T4 (Singer et al., 1982) and 20 to 74 bp for *Escherichia coli* (Watt et al., 1985). It has been suggested that the minimum recognition length for homologous pairing may prevent the participation of the recombination machinery in deleterious recombination reactions, and consequently, organisms with larger genomes would have larger homology requirements (Thomas, 1966). The observation that greater than 200 bp of homology is required for efficient homologous recombination in mammalian cells is consistent with this suggestion. Mammalian cells contain highly reiterated sequences distributed throughout the genome, and it is possible that a minimum recognition length in conjunction with other mechanisms prevents recombination between these repeated sequences.

V. EFFECT OF CELL CYCLE POSITION ON
EXTRACHROMOSOMAL RECOMBINATION

Wong and Capecchi (1987) examined the ability of cells to mediate intermolecular homologous recombination between DNA sequences at various stages of the cell cycle. Two nonreplicating plasmids containing truncated but overlapping fragments of the adenine phosphoribosyltransferase (*aprt*) gene were coinjected into the nuclei of Rat-20 (adenine phosphoribosyltransferase-negative) cells synchronized by mitotic shake-off. The injections were done at 3-h intervals after reattachment of mitotic cells to glass cover slips. The cells were incubated with [³H]adenine for 24 h, and the *aprt* activity generated as a consequence of recombination was monitored by an autoradiographic assay involving the deposition of silver grains over the cells. Homologous recombination activity was initially low in early G1, rose 10- to 15-fold by early to mid-S phase, and then declined as cells proceeded through the S phase and reentered G1. Thus, the peak of recombination activity was in early to mid-S phase, but replication of the substrate was not necessary for recombination. It has been shown previously that mitotic conversion events in yeast chromosomes occur primarily in the G1 phase before DNA replication occurs (Esposito, 1978; Esposito and Wagstaff, 1982; Fabre, 1978; Golin and Esposito, 1981).

VI. EFFECT OF DOUBLE-STRAND BREAKS ON
EXTRACHROMOSOMAL RECOMBINATION

The work of Orr-Weaver et al. (1981) on *S. cerevisiae* demonstrated that recombination between incoming plasmids and homologous chromosomal regions could be enhanced by treatment of the plasmid DNA. Most remarkable was a 13- to 3,750-fold increase in recombination frequency when double-strand breaks or gaps were introduced within the region of homology in the plasmid DNA. Several experiments suggest similar effects in mammalian cells. Double-strand breaks stimulate extrachromosomal intramolecular crossover events 10- to 100-fold (Lin et al., 1984a, b; Rubnitz and Subramani, 1985) and extrachromosomal intramolecular gene conversion events 5- to 15-fold (Rubnitz and Subramani, 1987). Linear molecules are preferred over their supercoiled counterparts as substrates for intermolecular recombination. For intermolecular reactions, several groups have reported a 5- to 20-fold increase in recombination frequency when one of the two plasmid substrates was linearized and a 35- to 100-fold stimulation when both partners were in linear form (Brenner et al., 1985, 1986; Folger et al., 1982; Kucherlapati et al., 1984b; Song et al., 1985; Wake et al., 1985). The degree of stimulation differs with the location of the double-strand break, but the general observation is that a greater degree of stimulation is apparent when the break is made near or within the region of homology, whereas breaks made more distal to the region of homology result in little or no stimulation. Linear molecules that lack terminal phosphates, have blunt ends, or have dideoxynucleotides at their 3′ ends do not stimulate recombination more than restriction enzyme-generated double-strand breaks. Also, short stretches of nonhomology (30 to 325 bp) at the site of the double-strand break do not diminish recombination (Wake et al., 1985).

VII. EFFECT OF INSERTIONS AND DELETIONS ON EXTRACHROMOSOMAL RECOMBINATION

Deletion or insertion heterologies could affect the frequencies and products of gene conversion in multiple ways. The size of the heterology could determine the frequency of its incorporation into a region of heteroduplex DNA (Bianchi and Radding, 1983). Heterologies of different sizes may also hinder the branch migration of DNA to different extents (Dasgupta and Radding, 1982). Finally, the ability of deletions and insertions to be gene converted depends on the way in which the mismatch-repair system deals with such heterologies (Hastings, 1984; Hastings et al., 1980). The effect of insertions on intermolecular recombination between two mutant *tk* genes was evaluated by placing insertions between 598- and 99-bp regions of homology. Reconstruction of the intact *tk* gene required a recombination event within the 99-bp region of homology. For uncut plasmids, the recombination frequency showed a progressive decline as the heterologous insert increased in size from 8 to 2,172 bp. When the plasmids carrying the insertions were linearized within the 598-bp region of homology, recombination in LTk⁻ cells was stimulated two- to fourfold (relative to values for uncut DNA), but only for inserts smaller than 24 bp. These results suggest either that heterologies larger than 24 nucleotides cannot be incorporated into heteroduplex DNA or that such heteroduplexes cannot be repaired efficiently (Brenner et al., 1985). The latter possibility seems unlikely because, as described later, single-strand loops between 25 and 247 nucleotides are repaired extremely efficiently and the repair is biased 2:1 in favor of the strand without the loop (Weiss and Wilson, 1987). Deletions of at least 167 nucleotides in the initiating DNA can be repaired in COS cells without substantial drops in the frequency of extrachromosomal gene conversion (Rubnitz and Subramani, 1987).

VIII. NONCONSERVATIVE NATURE OF EXTRACHROMOSOMAL RECOMBINATION

Most models of homologous recombination depict the intra- or intermolecular exchange of genetic information as being conservative; that is, all of the substrate sequences are recovered in the products. Many of the recombination experiments in mammalian cells suffer from the drawback that not all of the products are recoverable for analysis. A careful analysis reveals, however, that nonconservative mechanisms (in which part of the substrate DNA is either degraded or lost) may play a dominant role in both intra- and intermolecular extrachromosomal recombination. Chakrabarti and Seidman (1986) used a vector containing two overlapping, yet individually defective, portions of the SV40 T-antigen gene arranged as direct repeats. Between the two repeats, an SV40 origin and other markers were inserted. The vector could not replicate in CV1 monkey cells unless recombination reconstructed the T-antigen gene. Conservative recombination was expected to yield two smaller plasmids, whereas only one of the smaller plasmids would have been recovered if nonconservative recombination prevailed. Under conditions in which a single copy of the supercoiled substrate was introduced into CV1 cells, only one of the two expected products was found. Control experiments in COS cells showed that there was no inherent difficulty in forming either plasmid product and that both products would have been recovered in the presence of T antigen. When the substrate plasmid was cut within the region of T-antigen homology, again only one of the two plasmid products was recovered. Similar conclusions were reached for intermolecular extrachromosomal reactions.

It should be noted that not all recombination reactions in mammalian cells are nonconservative. Reports concerning chromosome inversions (Malissen et al., 1986) and reciprocal translocation events involved in Burkitt's lymphoma (Moulding et al., 1985) suggest the possibility of conservative intra- or interchromosomal recombination. More recently, Okazaki et al. (1987) reported the isolation of small circular DNAs which are representative of site-specific recombination events during thymocyte differentiation. These extrachromosomal DNAs were shown to be the excision products of a V-D or D-J joining event in the T-cell-receptor variable region. The reports of inversions in an intrachromosomal recombination substrate (Subramani and Rubnitz, 1985) and the site-specific integration of a plasmid into the β-globin locus (Smithies et al., 1985a) also suggest the conservation of DNA sequences during recombination.

IX. EXTRACHROMOSOMAL REPAIR OF HETERODUPLEX DNA

All models of recombination invoke the formation of heteroduplex DNA as an important intermediate. Several of these suggest that the repair of mismatched heteroduplexes can contribute to gene conversion, marker effects, and high negative interference (Norkin, 1970; White and Fox, 1974; White et al., 1985). Investigations of the mechanism of mismatch repair in procaryotes have suggested two distinct heteroduplex repair systems. One of these senses the methylation status of the parental strands and may function in the rectification of mistakes made during replication (Wagner and Meselson, 1976), while the other is methylation independent and may play a role in the correction of mismatches that arise during recombination (Fishel et al., 1986; Kolodner and Fishel, 1984). Because mismatched duplexes play a role in recombination and repair, artificially generated heteroduplexes introduced into mammalian cells have been exploited to enhance our understanding of these processes.

Ayares et al. (1987) introduced heteroduplexes prepared from derivatives of the shuttle vector pSV2neo into monkey COS cells. After replication, the recovered plasmids were introduced into *E. coli* DH1, and plasmid DNA from ampicillin-resistant colonies was characterized to determine the efficiency and strand bias of repair for different types of mismatches. Single-strand gaps and free single-strand nicks were repaired with almost 100% efficiency. Small single-strand loops (8 to 10 nucleotides) were excised (10% of the time) less efficiently than large loops (248 to 283 nucleotides), which were excised 43% of the time. Markers that were 58 nucleotides apart were corepaired almost 100% of the time, while those that were 1,000 or more nucleotides apart were almost never corepaired.

In an analogous study, Weiss and Wilson (1987) transfected heteroduplexes constructed from SV40 wild-type and deletion mutant DNAs into monkey CV1 cells. Each heteroduplex contained one or more single-strand loops in the intron for the large T antigen, which is nonessential for lytic infection. Analysis of 1,123 individual viral progeny indicated that single-strand loops (25 to 247 nucleotides) were corrected prior to replication with an efficiency of almost 100%. This is in qualitative but not quantitative agreement with the work of Ayares et al. (1987) described above. The repair was accurate 98% of the time, and it was biased 2:1 in favor of the strand without the loop. The efficiency, accuracy, and strand bias of repair were unaffected by loop sizes in the range of 25 to 247 nucleotides. Control experiments strongly suggested that the repair was genuine (i.e., not due to the loss of one strand) and that it was not due to recombination between progeny. The excision tract associated with repair of single-stranded loops rarely exceeded 200 to 400 nucleotides. This value is somewhat smaller than the average repair tract length of 500 to 600 nucleotides observed for plasmids in *S. cerevisiae* (Ahn and Livingston, 1986; Struhl, 1987) and substantially below the 3-kilobase tract lengths described for *E. coli* (Wagner and Meselson, 1976). It is interesting to note in this context that the coconversion tract length for chromosomal gene conversion in L cells is less than 358 nucleotides (Liskay and Stachelek, 1986).

Earlier work by Folger et al. (1985b) involving the microinjection of heteroduplex DNAs containing different amber mutations in the *neo* gene suggested that under nonreplicating conditions (in LTk⁻ cells), repair of single-base-pair mismatches was also extremely efficient (almost 100%). Furthermore, the efficiency of generation of G418ʳ clones was 10-fold higher than could be accounted for by intermolecular recombination between DNA duplexes containing the mutations. These results suggest that the regeneration of the functional *neo* gene occurs by mismatch repair. The recent development of an in vitro system to study mismatch repair in mammalian cell extracts (Glazer et al., 1987) will undoubtedly shed more light on the efficiency of repair of different mismatches and the enzymology of the process.

X. MODELS FOR EXTRACHROMOSOMAL RECOMBINATION

Two essential features of the extrachromosomal recombination process in mammalian cells are that the process is nonconservative and that double-strand breaks in the region of sequence homology enhance recombination. Two models have been proposed to explain these results. The first, proposed by Lin et al. (1984b, 1987), suggests that the ends created by restriction enzyme digestion of the recombination substrate are acted upon by a 3′ exonuclease that degrades DNA in opposite directions from the two ends, exposing in the process complementary sequences. This is followed by pairing of the complementary strands, producing a substrate for an endonuclease that cuts DNA at the junction between the paired and unpaired regions. The marker gene is then reconstructed by a replicative gap-filling process followed by ligation.

The alternative model, proposed by Wake et al. (1985), is similar except for the suggestion that a helicase (rather than an exonuclease) initially unwinds and exposes the complementary strands which pair together. Repair of the unpaired DNA then occurs as a result of a nucleolytic (exo or endo) step followed by gap filling and ligation.

Both of these models explain the stimulation of recombination by the double-strand breaks, as does the model of double-strand break repair (Szostak et al., 1983). However, the models described above are distinct from the model of double-strand break repair in that they incorporate features to explain the nonconservative nature of the extrachromosomal recombination process. The extension of these models to chromosomal recombination is premature at present because events that appear to be conservative have been observed in several instances (see above).

XI. CHROMOSOMAL RECOMBINATION IN MAMMALIAN CELLS

The experimental strategy exploited to examine chromosomal recombination involves the initial placement of a suitable recombination substrate (preferably as a single copy) in the genome of an animal cell line. Since transformation of mammalian cells with marker genes occurs at a low efficiency (10^{-3} to 10^{-4} per cell), the plasmid containing the recombination substrate invariably has a second selectable marker gene (e.g., bacterial *gpt* or *neo*) which is used to identify clones of cells that have potentially taken up the recombination substrate and integrated it into the genome. These lines are then screened by Southern blot analysis of DNA from independent clones to ensure that the recombination substrate, consisting of two mutant copies of a different selectable marker gene, is in the unrecombined configuration and preferably as a single, integrated copy. The cells are then shifted into the appropriate selective medium to isolate cell lines in which an intrachromosomal recombination event (either intrachromatid or between sister chromatids) has occurred. The rates of chromosomal recombination events are measured by fluctuation tests (Luria and Delbruck, 1943), without which the recombination rates can be aberrantly high (by 10- to 100-fold) because sibling clones are counted as independent products. The products of recombination are analyzed by Southern blot analysis of genomic DNAs from independent recombinant cell lines followed by a comparison of the integrated DNAs with those of the parental cell lines from which these recombinants were derived. Plasmid rescue has also been used to characterize the products in detail. As was the case for extrachromosomal recombination, only cells with the DNAs (chromatids) containing the recombined marker gene are selected out of the population (Fig. 2). Consequently, other partners or products (chromatids) in the recombination event

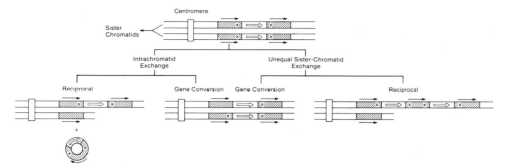

FIGURE 2. Chromosomal recombination between directly repeated sequences. Sister chromatids, each containing two different mutant alleles (×) of the same selectable marker gene (hatched boxes), are shown. The open arrows between the mutant genes indicate another marker gene. The thin arrows show the orientation of the marker genes. In an intrachromatid reciprocal exchange, the chromatid undergoing recombination suffers a deletion. In a reciprocal unequal sister chromatid exchange, one chromatid shows a deletion while the other exhibits an amplification event. The donor of genetic information for gene conversion can be the same or the sister chromatid. Note that in mammalian cells, selection for cells expressing the marker gene results in the loss of the cell containing the chromatid with the mutant genes.

are often lost, resulting in some ambiguity as to whether the event being analyzed occurred between mutant genes within a chromatid or between sister chromatids.

Plasmids containing truncated, nonfunctional copies of *neo* and *tk* genes have been used to examine chromosomal recombination in 3T6 and LTk⁻ cells, respectively (Lin et al., 1984a; Subramani and Rubnitz, 1985). These studies were done with multiple copies of the substrates in the chromosomes. Similar experiments have been undertaken with two mutant alleles of the *neo* gene and the *tk* gene (each as a single copy) in 3T6 and LTk⁻ cells, respectively (Liskay and Stachelek, 1983, 1986; Liskay et al., 1984; Smith and Berg, 1984). All of these substrates could undergo gene conversion or reciprocal recombination either within a chromatid or between sister chromatids (Fig. 2).

The rates of recombination were 10^{-6} to 10^{-8} per cell generation in 3T6 cells (Subramani and Rubnitz, 1985) and 10^{-6} in LTk⁻ cells (Liskay and Stachelek, 1986; Liskay et al., 1984). About 80% of the events observed in the L cells were due to gene conversion; the remaining 20% were accounted for by reciprocal recombination. Interestingly, while position effects on recombination were obvious in 3T6 cells when the *neo* gene was used (Smith and Berg, 1984; Subramani and Rubnitz, 1985), no position effects were observed in LTk⁻ cells when the *tk* gene substrates were used (Liskay and Stachelek, 1986; Liskay et al., 1984). Chromosomal deletion and inversion events were detected as predicted for substrates with either direct or inverted repeat configurations of the homologous sequences (Subramani and Rubnitz, 1985).

Substrates containing a linker insertion mutation in the *neo* gene and an internal homologous fragment (overlapping the site of linker insertion) of the gene appear to undergo only gene conversion in the chromosome of 3T6 cells. The rates of gene conversion were 10^{-6} to 10^{-8} per cell generation. Double-reciprocal recombination could also have generated a functional *neo* gene in these experiments. However, no evidence of intrachromatid double-reciprocal recombination was found in many independent recombinants (Rubnitz and Subramani, 1986, 1987). Though double-reciprocal recombination between sister chromatids cannot be rigorously excluded, it seems unlikely that double-reciprocal recombination could occur at a rate of 10^{-8} per cell generation unless a high degree of negative interference was also involved.

Members of multigene families are often remarkably homologous even though they accumulate base changes as a group over time. Two mechanisms have been proposed for the maintenance of DNA sequence homogeneity among members of such families. The first involves reciprocal recombination in the form of unequal crossovers and has the drawback that it alters gene dosage as a consequence of gene duplication or deletion events. The second mechanism, involving nonreciprocal or gene conversion events, has been invoked for a number of genes such as the alpha- (Michelson and Orkin, 1983), beta- (Erhart et al., 1985), gamma- (Shen et al., 1981), and zeta-globin genes (Hill et al., 1985); murine serum amyloid A genes (Lowell et al., 1986); bovine vasopressin and oxytocin genes (Ruppert et al., 1984); rodent cytochrome P-450 genes (Atchison and Adesnik, 1986); and immunoglobulin genes (Akimenko et al., 1986; Bentley and Rabbits, 1983; Clark et al., 1982). The 5- to 10-fold predominance of gene conversion over reciprocal recombination (Liskay and Stachelek, 1983; Liskay et al., 1984; Rubnitz and Subramani, 1986) lends credence to the proposed role of gene conversion as the primary force in the concerted evolution of multigene families. Furthermore, this mechanism is also attractive because it does not change gene dosage or the overall organization of genes.

Rates for interchromosomal recombination events have not been reported, but they are probably lower than 10^{-8} per cell generation. However, several recent reports suggest that they can occur between homologous chromosomes during mitotic growth (Cavenee et al., 1983; Potter et al., 1987; Wasmuth and Hall, 1984). Recombination between sequences on different chromosomes occurs at rates of 1.3×10^{-8} to 2.4×10^{-8} per cell generation in *S. cerevisiae*. As is the case for intrachromosomal recombination, the majority of these events can be accounted for by gene conversion, and approximately 10% are the result of reciprocal translocations (Ernst et al., 1982; Mikus and Petes, 1982; Sugawara and Szostak, 1983).

XII. CONVERSION TRACT LENGTHS FOR CHROMOSOMAL RECOMBINATION

The length of DNA sequence information transferred during intrachromosomal gene conversion events was investigated by Liskay and Stachelek (1986). Plasmids harboring different combinations of two defective *tk* genes were placed in the chromosome of LTk⁻ cells. One of the *tk* sequences consisted of an internal fragment of the gene and acted as the donor of genetic information, while the other was a linker insertion mutant and served as the recipient in gene conversion events. The coconversion of another restriction site which served as a silent polymorphic marker was also analyzed. The conversion tract length was found to be less than 358 bp, and it involved contiguous blocks of DNA. In *S. cerevisiae*, tract lengths larger than 10 kilobases have been observed during both mitosis (S. R. Judd and T. Petes, personal communication) and meiosis (L. Symington and T. Petes, personal communication).

XIII. HOMOLOGOUS INTERACTIONS BETWEEN PLASMIDS AND CHROMOSOMES

The nonspecific homology-independent integration of DNAs into the mammalian genome raised questions as to whether homologous interactions between incoming DNAs and the chromosome were at all possible. Reports by several groups make it quite certain that homologous interactions between plasmids and chromosomal genes are indeed detectable, albeit at a low frequency. Smith and Berg (1984) and Lin et al. (1985) constructed mouse cell lines with integrated copies of nonfunctional *neo* or *tk* genes. Calcium phosphate transfection of these lines with plasmids carrying nonfunctional but overlapping fragments of the marker gene led to restoration of the selectable marker gene. Both groups found the event to be extremely rare (10^{-5} that of normal *tk* transformation). Linearization of the input plasmid (i.e., double-strand breaks) appeared to be essential for the recombination event. Somewhat surprisingly, only 1 of 10 Tk$^-$ cell lines containing defective *tk* genes could be transformed to Tk$^+$ by homologous insertion of the complementary defective *tk* gene. Furthermore, relatively little illegitimate insertion of the introduced *tk* DNA into cellular DNA was detected in those cells that were transformed to the Tk$^+$ phenotype by homologous recombination (Lin et al., 1985).

Thomas et al. (1986) corrected mutant *neo* genes in the chromosome of L cells via homologous recombination using an injected plasmid DNA that carried a different mutation in the *neo* gene. The frequency of appearance of G418^r clones was as high as 1 in 1,000 injected cells, and two classes of G418^r cell lines were obtained. In the first class a wild-type *neo* gene was generated by gene conversion either of the chromosomal target or of the incoming plasmid, which then integrated elsewhere into the genome. The second class became G418^r by "heteroduplex-induced mutagenesis." These mutations appear to result from the incorrect repair of a heteroduplex formed between the introduced and chromosomal sequences (Thomas and Capecchi, 1986).

The only well-documented example of gene targeting into an endogenous, nonselectable chromosomal locus is the work of Smithies et al. (1985a), who introduced a DNA sequence into the chromosomal β-globin locus in a human-mouse hybrid cell line (Hu11). The planned modification was achieved in 1 of 1,000 transformed cells whether or not the target gene was expressed. Several significant elements contributed to the success of this experiment. The presence of specific restriction endonuclease recognition sites on the donor plasmid made it easy to distinguish between the normal chromosomal copy of the globin gene and the one containing the integrated plasmid. A *neo* gene incorporated into the incoming plasmid facilitated the identification of cells that took up the donor DNA, and a cell line with an active globin gene was used as the recipient because of concerns that the insertion of the plasmid into a silent locus could turn off the expression of the *neo* gene and thus eliminate true recombinants from the population of transformed cells. The input DNA also contained a bacterial suppressor gene (*supF*), so that the DNA fragment diagnostic for the recombination event could be rescued in bacteriophage lambda from a mixture of total genomic DNA. Finally, the plasmid was linearized within the region of homology to stimulate recombination. While the studies mentioned above are significant and encouraging, the frequencies of targeting are still too low for many practical applications. For every 100 to 1,000 cells in which nonspecific integration of incoming DNAs is evident, only 1 exhibits proper gene targeting by homologous recombination (Smithies et al., 1985a; Thomas et al., 1986). This is in direct contrast to organisms such as *S. cerevisiae* (Rothstein, 1983), *Aspergillus nidulans* (Miller et al., 1985), and *Dictyostelium* spp. (DeLozanne and Spudich, 1987), for which gene replacement and disruption techniques have been available for some time.

Besides the emphasis on the site-specific integration of DNAs into predetermined chromosomal loci, the rescue of chromosomal sequences onto extrachromosomal molecules is also of practical importance, as illustrated by the application of gapped plasmids in *S. cerevisiae* for the cloning of chromosomal alleles of genes for which a cloned fragment and restriction map are available (Orr-Weaver et al., 1983). Three groups have succeeded in rescuing chromosomal sequences onto extrachromosomally replicating SV40 molecules by homologous recombination in mammalian cells. Shaul et al. (1985) found that wild-type SV40 was generated upon passage of an early region replacement mutant of SV40 in monkey COS cells which contain an integrated T-antigen gene. Jasin et al. (1985) showed that SV40 molecules lacking an enhancer (for the early promoter) were able to acquire this sequence from the chromosomal T-antigen gene in COS cells. Subramani (1986) characterized and quantitated the rescue of a 1,018-bp sequence from the T-antigen gene of COS cells onto extrachromosomally replicating defective SV40 molecules.

In all of these rescue experiments, the recombination could have occurred either by a double-reciprocal event (resulting in the loss or substitution of chromosomal sequences with those on the incoming DNA) or by gene conversion. Unfortunately, it is difficult to address the mechanism because cells in which recombination occurs are killed by the wild-type virus.

We have determined that the rescue occurs predominantly by gene conversion, using a COS cell line with a single integrated copy of the 3′ two-thirds of the *neo* gene (unpublished data). Introduction of a linearized plasmid carrying the 5′ two-thirds of the *neo* gene resulted in G418^r colonies which arose by gene conversion of the plasmid from the chromosome followed by the integration of the repaired plasmid elsewhere in the genome (see Fig. 3). This fits the prediction of the double-strand break repair model (Szostak et al., 1983). It will be interesting

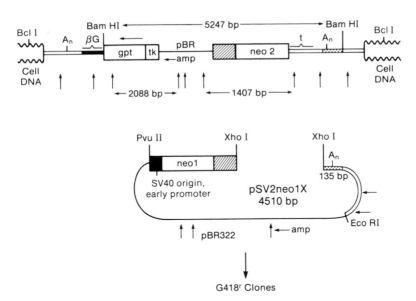

FIGURE 3. Rescue of chromosomal sequences onto extrachromosomal molecules. Rescue occurs primarily by gene conversion and not double-reciprocal recombination. The COS/neo2 cell line contains a single integrated copy of the neo2 segment (3' two-thirds of the *neo* gene) as shown at the top. The linearized plasmid pSV2neo1X contains the 5' two-thirds of the *neo* gene and has two regions of homology with the chromosomal sequences (hatched and stippled areas). Transfection of linear pSV2neo1X into the COS/neo2 line results in recombinants (G418^r colonies). All of nine recombinants analyzed arose by filling in the gap in the incoming plasmid followed by integration of the recombined plasmid elsewhere in the genome (Seaton and Subramani, unpublished data). Vertical arrows show the location of *Dra*I sites. A$_n$ denotes a polyadenylation signal.

to test whether the integration of the incoming plasmid will occur at the homologous site if the plasmid is incapable of replication.

Cell lines derived from the rare cells in which gene targeting occurs do not exhibit a higher frequency of recombination when they are retested using extrachromosomal recombination assays (unpublished data). Thus, these cells are not aberrant in any detectable way.

XIV. UNANSWERED QUESTIONS AND FUTURE DIRECTIONS

A major paradox is the high frequency of extrachromosomal recombination events relative to chromosomal recombination. These frequencies differ by several orders of magnitude. One clue to this puzzling observation is that extrachromosomal recombination occurs at this high frequency only for about 90 min after the substrates are injected into the nucleus (Folger et al., 1985a). This suggests that naked DNA may be a better substrate for recombination enzymes. However, the possibility that separate enzymatic machinery catalyzes extrachromosomal events is not ruled out.

The ubiquity of the extrachromosomal events also raises questions regarding the biological necessity for such high levels of recombination. The nonconservative nature of these events suggests that perhaps these processes are catalyzed by some machinery whose primary role might involve repair of damaged DNA. In contrast, chromosomal recombination appears to occur in a conservative manner.

A comparison of the chromosomal mitotic recombination rates in mammalian cells (10^{-6} per cell generation or lower) and in yeast cells (10^{-4} per cell generation [Jackson and Fink, 1981]) suggests that homologous recombination events in these organisms occur at vastly different rates. However, in both organisms gene conversion occurs about 5- to 10-fold more frequently than reciprocal recombination. For substrates that could undergo both homologous and nonhomologous events extrachromosomally, it was found that the frequency of nonhomologous recombination in mammalian cells was higher than the frequency of homologous recombination between 2.3-kilobase (Brouillette and Chartrand, 1987) or even 5.2-kilobase (Subramani and Berg, 1983) regions of homology. The higher ratio of nonhomologous to homologous events and the larger size of the genome in mammalian cells may contribute to the nonspecific integration of vector DNAs into the mammalian genome. This implies that the suppression of other nonhomologous events in the same cell may be an important, and perhaps independent, avenue of pursuit that will complement efforts devoted to the optimization of homologous interactions between plasmids and the chromosome. Recent results (Chang and Wilson, 1987) suggesting that the addition of dideoxynucleotides to the 3′ ends of linearized DNAs decreases the frequency of nonhomologous end joining by a factor of five or six relative to homologous recombination may prove useful in altering the ratio of random to homology-dependent targeting events in mammalian cells. In addition, the limited data suggesting that few nonhomologous events occur in cell lines that exhibit targeting need to be extended and substantiated further.

Previous studies with yeast cells and mammalian cells have indicated that transcriptional activity and/or chromatin structure affect recombination. For example, in the mating type locus (*MAT*) of *S. cerevisiae*, only the actively transcribed *MAT* locus is cleaved by the *HO* endonuclease (and is the recipient of genetic information); the nontranscribed *HML* and *HMR* loci are not (Klar et al., 1981; Strathern et al., 1982). Transcription by RNA polymerase I through the ribosomal DNA genes in yeast cells also enhances genetic exchange (Voelkel-Meiman et al., 1987). In mammalian cells, transcription is involved in the regulation of recombination in immunoglobulin rearrangements (Blackwell et al., 1986), and circumstantial evidence exists for the integration of retroviral or other foreign DNAs into transcriptionally active sites (Rohdewohld et al., 1987; Schulz et al., 1987). Furthermore, the high frequency (15 to 20%) of generation of insertional mutations (including nonlethal ones) in transgenic mice (Covarrubias et al., 1987) suggests that the integration of vector DNA into transcription units must occur often. In view of these observations, further experiments are necessary to explore the connection, if any, between transcription and/or chromatin structure of chromosomal genes and the efficiency of targeting.

The targeting experiments done until now should be considered as the first important step toward the development of gene reconstruction, replacement, or disruption techniques. Improvements will be necessary to make these procedures work for nonselected or autosomal loci, or both. The parameters that govern the insertion of a plasmid at a homologous site or the directionality of gene conversion are also poorly understood. An understanding of the mechanisms of these events will be necessary before the system can be manipulated, for example, to correct a defect in a resident chromosomal gene without insertion of vector sequences elsewhere in the genome.

The importance of recombination and repair make it quite likely that multiple pathways exist for these processes in mammalian cells, as they do in bacteria (Clark, 1973). The paucity of genetic and biochemical information regarding the recombinational machinery has been a hurdle. However, the development of in vitro systems to analyze recombination (Hsieh et al., 1986; Kucherlapati et al., 1985; Lopez et al., 1987; Ganea et al., 1987) and mismatch repair (Glazer et al., 1987) appears to be very promising in this regard.

Many specific features of chromosomal recombination such as the association of reciprocal exchange with conversion, the question of parity in gene conversion, and the effect of double-strand breaks on the directionality of conversion remain unanswered. Recombination events in gene amplification units are also of great interest, as are sequences or drugs that might stimulate extrachromosomal or chromosomal recombination. If progress in the field continues to be as rapid as it has been in the past few years, the answers to these and other questions should be forthcoming soon.

ACKNOWLEDGMENTS. This work was supported by Public Health Service grants GM31253 and CA01062 to S.S. from the National Institutes of Health.

LITERATURE CITED

Ahn, B.-Y., and D. M. Livingston. 1986. Mitotic gene conversion lengths, coconversion patterns, and the incidence of reciprocal recombination in a *Saccharomyces cerevisiae* plasmid system. *Mol. Cell. Biol.* **6**:3685–3693.

Akimenko, M. A., B. Mariame, and F. Rougeon. 1986. Evolution of the immunoglobulin kappa light chain locus in the rabbit: evidence for differential gene conversion events. *Proc. Natl. Acad. Sci. USA* **83**:5180–5183.

Alitalo, K., M. Schwab, C. C. Lin, H. E. Varmus, and J. M. Bishop. 1983. Homogeneously staining chromosomal regions contain amplified copies of an abundantly expressed cellular oncogene (c-myc) in malignant neuroendocrine cells from a human colon carcinoma. *Proc. Natl. Acad. Sci. USA* **80**:1707–1711.

Atchison, M., and M. Adesnik. 1986. Gene conversion in a cytochrome P-450 gene family. *Proc. Natl. Acad. Sci. USA* **83**:2300–2304.

Ayares, D., L. Chekuri, K.-Y. Song, and R. Kucherlapati. 1986. Sequence homology requirements for intermolecular recombination in mammalian cells. *Proc. Natl. Acad. Sci. USA* **83**:5199–5203.

Ayares, D., D. Ganea, L. Chekuri, C. R. Campbell, and R. Kucherlapati. 1987. Repair of single-stranded DNA nicks, gaps, and loops in mammalian cells. *Mol. Cell. Biol.* **7**:1656–1662.

Baltimore, D. 1981. Gene conversion: some implications for immunoglobulin genes. *Cell* **24**:592–594.

Bentley, D. L., and T. H. Rabbits. 1983. Evolution of immunoglobulin V genes: evidence indicating that recently duplicated human V kappa sequences have diverged by gene conversion. *Cell* **32**:181–189.

Bianchi, M. E., and C. M. Radding. 1983. Insertions, deletions and mismatches in heteroduplex DNA made by recA protein. *Cell* **35**:511–520.

Blackwell, T. K., M. W. Moore, G. D. Yancopoulos, H. Suh, S. Lutzker, E. Selsing, and F. W. Alt. 1986. Recombination between immunoglobulin variable region gene sequences is enhanced by transcription. *Nature* (London) **324**:585–589.

Brack, C., M. Hirama, R. Lenhard-Schuller, and S. Tonegawa. 1978. A complete immunoglobulin gene is created by somatic recombination. *Cell* **15**:1–14.

Brenner, D. A., S. Kato, R. A. Anderson, A. C. Smigocki, and R. D. Camerini-Otero. 1984. The recombination and integration of DNAs introduced into mouse L cells. *Cold Spring Harbor Symp. Quant. Biol.* **4**:123–138.

Brenner, D. A., A. C. Smigocki, and R. D. Camerini-Otero. 1985. Effect of insertions, deletions, and double-strand breaks on homologous recombination in mouse L cells. *Mol. Cell. Biol.* **5**:684–691.

Brenner, D. A., A. C. Smigocki, and R. D. Camerini-Otero. 1986. Double-strand gap repair results in homologous recombination in mouse L cells. *Proc. Natl. Acad. Sci. USA* **83**:1762–1766.

Brouillette, S., and P. Chartrand. 1987. Intermolecular recombination assay for mammalian cells that produces recombinants carrying both homologous and nonhomologous junctions. *Mol. Cell. Biol.* **7**:2248–2255.

Calos, M. P. 1986. Mutation of autonomously replicating plasmids, p. 243–262. *In* R. Kucherlapati (ed.), *Gene Transfer.* Plenum Publishing Corp., New York.

Cavenee, W. K., T. P. Dryja, R. A. Phillips, W. F. Benedict, R. Godbout, B. L. Gallie, A. L. Murphree, L. C. Strong, and R. L. White. 1983. Expression of recessive alleles by chromosomal mechanisms in retinoblastoma. *Nature* (London) **305**:779–784.

Chakrabarti, S., and M. M. Seidman. 1986. Intramolecular recombination between transfected repeated sequences in mammalian cells is nonconservative. *Mol. Cell. Biol.* **6**:2520–2526.

Chang, X.-B., and J. H. Wilson. 1987. Modification of DNA ends can decrease end joining relative to homologous recombination in mammalian cells. *Proc. Natl. Acad. Sci. USA* **84**:4959–4963.

Clark, A. J. 1973. Recombination-deficient mutants of *Escherichia coli* and other bacteria. *Annu. Rev. Genet.* **7**:67–86.

Clark, S. H., J. L. Claflin, and S. Rudikoff. 1982. Polymorphism in immunoglobulin heavy chains suggesting gene conversion. *Proc. Natl. Acad. Sci. USA* **79**:3280–3284.

Collins, S., and M. Groudine. 1982. Amplification of endogenous myc-related DNA sequences in a human myeloid leukemia cell line. *Nature* (London) **298**:679–681.

Copeland, N. G., N. A. Jenkins, and B. K. Lee. 1983. Association of the lethal yellow (A^y) coat color mutation with an ecotropic murine leukemia virus genome. *Proc. Natl. Acad. Sci. USA* **80**:247–249.

Covarrubias, L., Y. Nishida, and B. Mintz. 1985. Early developmental mutations due to DNA rearrangements in transgenic mouse embryos. *Cold Spring Harbor Symp. Quant. Biol.* **50**:447–452.

Covarrubias, L., Y. Nishida, M. Terao, P. D'Eustachio, and B. Mintz. 1987. Cellular DNA rearrangements and early developmental arrest caused by DNA insertion in transgenic mouse embryos. *Mol. Cell. Biol.* **7**:2243–2247.

Croce, C. M., and G. Klein. 1985. Chromosome translocations and human cancer. *Sci. Am.* **252**:54–60.

Dasgupta, C., and C. M. Radding. 1982. Polar branch migration promoted by recA protein: effect of mismatched base pairs. *Proc. Natl. Acad. Sci. USA* **79**:762–766.

Dasgupta, U. B., and W. C. Summers. 1980. Genetic recombination of Herpes-simplex virus, the role of the host cell and uv-irradiation of the virus. *Mol. Gen. Genet.* **178**:617–623.

DeLozanne, A., and J. A. Spudich. 1987. Disruption of the *Dictyostelium* myosin heavy chain gene by homologous recombination. *Science* **236**:1086–1091.

Dubbs, D. R., M. Rachmeler, and S. Kit. 1974. Recombination between temperature-sensitive mutants of simian virus 40. *Virology* **57**:161–174.

Erhart, M. A., K. S. Simons, and S. Weaver. 1985. Evolution of the mouse β-globin genes: a recent gene conversion in the Hbb^s haplotype. *Mol. Biol. Evol.* **2**:304–320.

Ernst, J. F., J. W. Stewart, and F. Sherman. 1982. Formation of composite iso-cytochromes *c* by recombination between non-allelic genes of yeast. *J. Mol. Biol.* **161**:373–394.

Esposito, M. 1978. Evidence that spontaneous mitotic recombination occurs at the two-strand stage. *Proc. Natl. Acad. Sci. USA* **75**:4436–4440.

Esposito, M. E., and J. S. Wagstaff. 1982. Mechanisms of mitotic recombination, p. 341–370. *In* J. N.

Strathern, E. W. Jones, and J. R. Broach (ed.), *The Molecular Biology of the Yeast Saccharomyces*, vol. 1. Cold Spring Harbor Laboratory, Cold Spring Harbor, N.Y.

Fabre, F. 1978. Induced intragenic recombination in yeast can occur during the G_1 mitotic phase. *Nature* (London) **272:**795–798.

Fishel, R. A., E. C. Siegel, and R. Kolodner. 1986. Gene conversion in *Escherichia coli*: resolution of heteroallelic mismatched nucleotides by co-repair. *J. Mol. Biol.* **188:**147–157.

Folger, K. R., K. Thomas, and M. R. Capecchi. 1985a. Nonreciprocal exchanges of information between DNA duplexes coinjected into mammalian cell nucleic. *Mol. Cell. Biol.* **5:**59–69.

Folger, K. R., K. Thomas, and M. R. Capecchi. 1985b. Efficient correction of mismatched bases in plasmid heteroduplexes injected into cultured mammalian cell nuclei. *Mol. Cell. Biol.* **5:**70–74.

Folger, K. R., E. A. Wong, G. Wahl, and M. R. Capecchi. 1982. Patterns of integration of DNA microinjected into cultured mammalian cells: evidence for homologous recombination between injected plasmid DNA molecules. *Mol. Cell. Biol.* **2:**1372–1387.

Ganea, D., P. Moore, L. Chekuri, and R. Kucherlapati. 1987. Characterization of an ATP-dependent DNA strand transferase from human cells. *Mol. Cell. Biol.* **7:**3124–3130.

Glazer, P. M., S. N. Sarkar, G. E. Chisholm, and W. C. Summers. 1987. DNA mismatch repair detected in human cell extracts. *Mol. Cell. Biol.* **7:**218–224.

Gluzman, Y. 1981. SV40-transformed simian cells support the replication of early SV40 mutants. *Cell* **23:**175–182.

Golin, J., and M. Esposito. 1981. Mitotic recombination: mismatch correction and replication resolution of Holliday structures formed at the two-strand stage in *Saccharomyces. Mol. Gen. Genet.* **183:**252–263.

Gordon, J. W. 1986. A foreign dihydrofolate reductase gene in transgenic mice acts as a dominant mutation. *Mol. Cell. Biol.* **6:**2158–2167.

Hastings, P. J. 1984. Measurement of restoration and conversion: its meaning for the mismatch repair hypothesis of conversion. *Cold Spring Harbor Symp. Quant. Biol.* **49:**49–53.

Hastings, P. J., A. Kalogeropoulos, and J.-L. Rossignol. 1980. Restoration to the parental genotype of mismatches formed in recombinant DNA heteroduplex. *Curr. Genet.* **2:**169–174.

Hill, A. V., R. D. Nicholls, S. L. Thein, and D. R. Higgs. 1985. Recombination within the human embryonic zeta-globin locus: a common zeta-zeta chromosome produced by gene conversion of the psi zeta gene. *Cell* **42:**809–819.

Hirt, B. 1967. Selective extraction of polyoma DNA from infected mouse cell cultures. *J. Mol. Biol.* **26:**365–369.

Hochtl, J., and H. G. Zachau. 1983. A novel type of aberrant recombination in immunoglobulin genes and its implications for V-J joining mechanisms. *Nature* (London) **302:**260–263.

Holliday, R. 1964. A mechanism for gene conversion in fungi. *Genet. Res.* **5:**282–304.

Holliday, R. 1974. Molecular aspects of genetic exchange and gene conversion. *Genetics* **78:**273–285.

Hooper, M., K. Hardy, A. Handyside, S. Hunter, and M. Monk. 1987. HPRT-deficient (Lesch-Nyhan) mouse embryos derived from the germline colonization by cultured cells. *Nature* (London) **326:**292–295.

Hsieh, P., M. S. Meyn, and R. D. Camerini-Otero. 1986. Partial purification and characterization of a recombinase from human cells. *Cell* **44:**885–894.

Jackson, J. A., and G. R. Fink. 1981. Gene conversion between duplicated genetic elements in yeast. *Nature* (London) **292:**306–311.

Jasin, M., J. De Villiers, F. Weber, and W. Schaffner. 1985. High frequency of homologous recombination in mammalian cells between endogenous and introduced SV40 genomes. *Cell* **43:**695–703.

Jones, R. S., and S. S. Potter. 1985. *L1* sequences in HeLa extrachromosomal circular DNA: evidence for circularization by homologous recombination. *Proc. Natl. Acad. Sci. USA* **82:**1989–1993.

Kato, S., R. A. Anderson, and R. D. Camerini-Otero. 1986. Foreign DNA introduced by calcium phosphate is integrated into repetitive DNA elements of the mouse L cell genome. *Mol. Cell. Biol.* **6:**1787–1795.

Klar, A. J. S., J. N. Strathern, and J. B. Hicks. 1981. A position-effect control for gene transposition: state of expression of yeast mating-type genes affects their ability to switch. *Cell* **25:**517–524.

Kohl, N. E., N. Kanda, R. R. Schreck, G. Bruns, S. A. Latt, F. Gilbert, and F. W. Alt. 1983.

Transposition and amplification of oncogene-related sequences in human neuroblastomas. *Cell* **35:** 359–367.

Kolodner, R., and R. Fishel. 1984. An *Escherichia coli* cell-free system that catalyzes the repair of symmetrically methylated heteroduplex DNA. *Cold Spring Harbor Symp. Quant. Biol.* **49:**603–609.

Krolewski, J. J., C. W. Schindler, and M. G. Rush. 1984. Structure of extrachromosomal circular DNAs containing both the Alu family of dispersed repetitive sequences and other regions of chromosomal DNA. *J. Mol. Biol.* **174:**41–54.

Kucherlapati, R. S., D. Ayares, A. Hanneken, K. Noonan, S. Rauth, J. M. Spencer, L. Wallace, and P. D. Moore. 1984a. Homologous recombination in monkey cells and human cell free extracts. *Cold Spring Harbor Symp. Quant. Biol.* **49:**191–197.

Kucherlapati, R. S., E. M. Eves, K.-Y. Song, B. S. Morse, and O. Smithies. 1984b. Homologous recombination between plasmids in mammalian cells can be enhanced by treatment of input DNA. *Proc. Natl. Acad. Sci. USA* **81:**3153–3157.

Kucherlapati, R. S., J. Spencer, and P. D. Moore. 1985. Homologous recombination catalyzed by human cell extracts. *Mol. Cell. Biol.* **5:**714–720.

Kuehn, M. R., A. Bradley, E. J. Robertson, and M. J. Evans. 1987. A potential animal model for Lesch-Nyhan syndrome through introduction of HPRT mutations into mice. *Nature* (London) **326:** 295–298.

Latt, S. A. 1974. Localization of sister-chromatid exchanges in human chromosomes. *Science* **185:**74–76.

Latt, S. A. 1981. Sister-chromatid exchange formation. *Annu. Rev. Genet.* **15:**11–55.

Lewis, S., N. Rosenberg, F. Alt, and D. Baltimore. 1982. Continuing kappa-gene rearrangement in a cell line transformed by Abelson murine leukemia virus. *Cell* **30:**807–816.

Lin, F.-L., K. Sperle, and N. Sternberg. 1984a. Homologous recombination in mouse L cells. *Cold Spring Harbor Symp. Quant. Biol.* **49:**139–149.

Lin, F.-L., K. Sperle, and N. Sternberg. 1984b. Model for homologous recombination during transfer of DNA into mouse L cells: role for DNA ends in the recombination process. *Mol. Cell. Biol.* **4:** 1020–1034.

Lin, F.-L., K. Sperle, and N. Sternberg. 1985. Recombination in mouse L cells between DNA introduced into cells and homologous chromosomal sequences. *Proc. Natl. Acad. Sci. USA* **82:** 1391–1395.

Lin, F.-L., K. Sperle, and N. Sternberg. 1987. Extrachromosomal recombination in mammalian cells as studied with single- and double-stranded DNA substrates. *Mol. Cell. Biol.* **7:**129–140.

Liskay, R. M., A. Letsou, and J. L. Stachelek. 1987. Homology requirement for efficient gene conversion between duplicated chromosomal sequences in mammalian cells. *Genetics* **115:**161–167.

Liskay, R. M., and J. L. Stachelek. 1983. Evidence for intrachromosomal gene conversion in cultured mouse cells. *Cell* **35:**157–165.

Liskay, R. M., and J. L. Stachelek. 1986. Information transfer between duplicated chromosomal sequences in mammalian cells involves contiguous regions of DNA. *Proc. Natl. Acad. Sci. USA* **83:** 1802–1806.

Liskay, R. M., J. L. Stachelek, and A. Letsou. 1984. Homologous recombination between repeated chromosomal sequences in mouse cells. *Cold Spring Harbor Symp. Quant. Biol.* **49:**183–189.

Lopez, B., S. Rousset, and J. Coppey. 1987. Homologous recombination intermediates between two duplex DNAs catalyzed by human cell extracts. *Nucleic Acids Res.* **15:**5643–5655.

Lowell, C. A., D. A. Potter, R. S. Stearman, and J. F. Morrow. 1986. Structure of the murine serum amyloid A family. Gene conversion. *J. Biol. Chem.* **261:**8442–8452.

Luria, S. E., and M. Delbruck. 1943. Mutations of bacteria from virus sensitivity to virus resistance. *Genetics* **28:**491–511.

Malissen, M., C. McCoy, D. Blanc, J. Trucy, C. DeVaux, A.-M. Schmitt-Verhulst, F. Fitch, L. Hood, and B. Malissen. 1986. Direct evidence for chromosomal inversion during T-cell receptor β-gene rearrangements. *Nature* (London) **319:**28–33.

Maniatis, T. 1985. Targeting in mammalian cells. *Nature* (London) **317:**205–206.

Mark, W. H., K. Signorelli, and E. Lacy. 1985. An insertion mutation in a transgenic mouse line results in developmental arrest at day 5 of gestation. *Cold Spring Harbor Symp. Quant. Biol.* **50:**453–463.

Meselson, M. S., and C. M. Radding. 1975. A general model for genetic recombination. *Proc. Natl. Acad. Sci. USA* **72:**358–361.

Michelson, A. M., and S. H. Orkin. 1983. Boundaries of gene conversion within the duplicated human alpha-globin genes. Concerted evolution by segmental recombination. *J. Biol. Chem.* **258:**15245–15254.

Mikus, M. D., and T. Petes. 1982. Recombination between genes located on nonhomologous chromosomes in *Saccharomyces cerevisiae. Genetics* **101:**369–404.

Miller, B. L., K. Y. Miller, and W. E. Timberlake. 1985. Direct and indirect gene replacements in *Aspergillus nidulans. Mol. Cell. Biol.* **5:**1714–1721.

Miller, L. K., B. E. Cooke, and M. Fried. 1976. Fate of mismatched base-pair regions in polyoma heteroduplex DNA during infection of mouse cells. *Proc. Natl. Acad. Sci. USA* **73:**3073–3077.

Mocarski, E. S., and B. Roizman. 1982. Herpes virus-dependent amplification and inversion of cell-associated viral thymidine kinase gene flanked by viral *a* sequences and linked to an origin of viral DNA replication. *Proc. Natl. Acad. Sci. USA* **79:**5626–5630.

Moulding, C., A. Rapoport, P. Goldman, J. Battey, G. M. Lenoir, and P. Leder. 1985. Structural analysis of both products of a reciprocal translocation between c-myc and immunoglobulin loci in Burkitt's lymphoma. *Nucleic Acids Res.* **13:**2141–2152.

Norkin, L. C. 1970. Marker-specific effects in genetic recombination. *J. Mol. Biol.* **51:**633–655.

Okazaki, K., D. D. Davis, and H. Sakano. 1987. T cell receptor β-gene sequences in the circular DNA of thymocyte nuclei: direct evidence for intramolecular DNA deletion in V-D-J joining. *Cell* **49:**477–485.

Orr-Weaver, T. L., J. W. Szostak, and R. J. Rothstein. 1981. Yeast transformation: a model for the study of recombination. *Proc. Natl. Acad. Sci. USA* **78:**6354–6358.

Orr-Weaver, T. L., J. W. Szostak, and R. J. Rothstein. 1983. Genetic applications of yeast transformation with linear and gapped plasmids. *Methods Enzymol.* **101:**228–245.

Palmiter, R. D., and R. L. Brinster. 1985. Transgenic mice. *Cell* **41:**343–345.

Palmiter, R. D., T. M. Wilkie, H. Y. Chen, and R. L. Brinster. 1984. Transmission distortion and mosaicism in an unusual transgenic mouse pedigree. *Cell* **36:**869–877.

Potter, T. A., R. A. Zeff, W. Frankel, and T. V. Rajan. 1987. Mitotic recombination between homologous chromosomes generates H-2 somatic cell variants in vitro. *Proc. Natl. Acad. Sci. USA* **84:**1634–1637.

Rauth, S., K.-Y. Song, D. Ayares, L. Wallace, P. D. Moore, and R. Kucherlapati. 1986. Transfection and homologous recombination involving single stranded DNA substrates in mammalian cells and nuclear extracts. *Proc. Natl. Acad. Sci. USA* **83:**5587–5591.

Reth, M. G., P. Ammirati, S. Jackson, and F. W. Alt. 1985. Regulated progression of a cultured pre-B-cell line to the B-cell stage. *Nature* (London) **317:**353–355.

Robert de Saint Vincent, B., and G. M. Wahl. 1983. Homologous recombination in mammalian cells mediates formation of a functional gene from two overlapping gene fragments. *Proc. Natl. Acad. Sci. USA* **80:**2002–2006.

Robins, D. M., S. Ripley, A. S. Henderson, and R. Axel. 1981. Transforming DNA integrates into the host chromosome. *Cell* **23:**29–39.

Rohdewohld, H., H. Weiher, W. Reik, R. Jaenisch, and M. Breindl. 1987. Retrovirus integration and chromatin structure: Moloney murine leukemia proviral integration sites map near DNase I-hypersensitive sites. *J. Virol.* **61:**336–343.

Rothstein, R. J. 1983. One-step gene disruption in yeast. *Methods Enzymol.* **101:**202–211.

Rubnitz, J., and S. Subramani. 1984. The minimum amount of homology required for homologous recombination in mammalian cells. *Mol. Cell. Biol.* **4:**2253–2258.

Rubnitz, J., and S. Subramani. 1985. Rapid assay for extrachromosomal homologous recombination in monkey cells. *Mol. Cell. Biol.* **5:**529–537.

Rubnitz, J., and S. Subramani. 1986. Extrachromosomal and chromosomal gene conversion in mammalian cells. *Mol. Cell. Biol.* **6:**1608–1614.

Rubnitz, J., and S. Subramani. 1987. Correction of deletions in mammalian cells by gene conversion. *Somatic Cell Mol. Genet.* **13:**183–190.

Ruppert, S., G. Scherer, and G. Schutz. 1984. Recent gene conversion involving bovine vasopressin and oxytocin precursor genes suggested by nucleotide sequence. *Nature* (London) **308:**554–557.

Schimke, R. T. 1982. *Gene Amplification*. Cold Spring Harbor Laboratory, Cold Spring Harbor, N.Y.

Schnieke, A., K. Harbers, and R. Jaenisch. 1983. Embryonic lethal mutations in mice induced by retrovirus insertion into the alpha1(1) collagen gene. *Nature* (London) **304**:315–320.

Schulz, M., U. Freisem-Rabien, R. Jessberger, and W. Doerfler. 1987. Transcriptional activities of mammalian genomes at sites of recombination with foreign DNA. *J. Virol.* **61**:344–353.

Shapira, G., J. L. Stachelek, A. Letsou, L. K. Soodak, and R. M. Liskay. 1983. Novel use of synthetic oligonucleotide insertion mutants for the study of homologous recombination in mammalian cells. *Proc. Natl. Acad. Sci. USA* **80**:4827–4831.

Shaul, Y., O. Laub, M. D. Walker, and W. J. Rutter. 1985. Homologous recombination between a defective virus and a chromosomal sequence in mammalian cells. *Proc. Natl. Acad. Sci. USA* **82**: 3781–3784.

Shen, S. H., J. L. Slightom, and O. Smithies. 1981. A history of the human fetal globin gene duplication. *Cell* **26**:191–203.

Simon, J. R., and P. Moore. 1987. Homologous recombination between single-stranded DNA and chromosomal genes in *Saccharomyces cerevisiae*. *Mol. Cell. Biol.* **7**:2329–2334.

Singer, B. S., L. Gold, P. Gauss, and D. H. Doherty. 1982. Determination of the amount of homology required for recombination in bacteriophage T4. *Cell* **31**:25–33.

Small, J., and G. Scangos. 1983. Recombination during gene transfer into mouse cells can restore the function of deleted genes. *Science* **219**:174–176.

Smith, A. J. H., and P. Berg. 1984. Homologous recombination between defective neo genes in mouse 3T6 cells. *Cold Spring Harbor Symp. Quant. Biol.* **49**:171–181.

Smithies, O., R. G. Gregg, S. S. Boggs, M. A. Koralewski, and R. S. Kucherlapati. 1985a. Insertion of DNA sequences into the human chromosomal β-globin locus by homologous recombination. *Nature* (London) **317**:1230–1234.

Smithies, O., M. A. Koralewski, K.-Y. Song, and R. S. Kucherlapati. 1985b. Homologous recombination with DNA introduced into mammalian cells. *Cold Spring Harbor Symp. Quant. Biol.* **49**:161–170.

Song, K.-Y., L. Chekuri, S. Rauth, S. Ehrlich, and R. Kucherlapati. 1985. Effect of double-strand breaks on homologous recombination in mammalian cells and extracts. *Mol. Cell. Biol.* **5**:3331–3336.

Stark, G. R., and G. M. Wahl. 1984. Gene amplification. *Annu. Rev. Biochem.* **53**:447–491.

Strathern, J. N., A. Klar, J. Hicks, J. Abraham, J. Ivy, K. Nasmyth, and C. McGill. 1982. Homothallic switching of yeast mating type cassettes is initiated by a double-stranded cut in the MAT locus. *Cell* **31**:183–192.

Struhl, K. 1987. Effect of deletion and insertion on double-strand break repair in *Saccharomyces cerevisiae*. *Mol. Cell. Biol.* **7**:1300–1303.

Subramani, S. 1986. Rescue of chromosomal T-antigen sequences onto extrachromosomally replicating defective simian virus 40 DNA by homologous recombination. *Mol. Cell. Biol.* **6**:1320–1325.

Subramani, S., and P. Berg. 1983. Homologous and nonhomologous recombination in mammalian cells. *Mol. Cell. Biol.* **3**:1040–1052.

Subramani, S., and J. Rubnitz. 1985. Recombination events after transient infection and stable integration of DNA into mouse cells. *Mol. Cell. Biol.* **5**:659–666.

Subramani, S., and P. J. Southern. 1983. Analysis of gene expression using simian virus 40 vectors. *Anal. Biochem.* **135**:1–15.

Sugawara, N., and J. Szostak. 1983. Recombination between sequences in nonhomologous positions. *Proc. Natl. Acad. Sci. USA* **80**:5675–5679.

Szostak, J. W., T. L. Orr-Weaver, R. J. Rothstein, and F. W. Stahl. 1983. The double-strand-break repair model for recombination. *Cell* **33**:25–35.

Thomas, C. A. 1966. Recombination of DNA molecules. *Prog. Nucleic Acid Res. Mol. Biol.* **5**:315–348.

Thomas, K. R., and M. R. Capecchi. 1986. Introduction of homologous DNA sequences into mammalian cells induces mutations in the cognate gene. *Nature* (London) **324**:34–38.

Thomas, K. R., K. R. Folger, and M. R. Capecchi. 1986. High frequency targeting of genes to specific sites in the mammalian genome. *Cell* **44**:419–428.

Tonegawa, S. 1983. Somatic generation of antibody diversity. *Nature* (London) **302**:575–581.

Upcroft, P., B. Carter, and C. Kidson. 1980. Analysis of recombination in mammalian cells using SV40 genome segments having homologous overlapping termini. *Nucleic Acids Res.* **8**:2725–2736.

Voelkel-Meiman, K., R. L. Keil, and G. S. Roeder. 1987. Recombination-stimulating sequences in yeast ribosomal DNA correspond to sequences regulating transcription by RNA polymerase I. *Cell* **48:** 1071–1079.

Vogel, T. 1980. Recombination between endogenous and exogenous simian virus 40 genes. *Virology* **104:**73–83.

Vogel, T., Y. Gluzman, and E. Winocour. 1977. Recombination between endogenous and exogenous simian virus 40 genes: biochemical evidence for genetic exchange. *J. Virol.* **24:**541–550.

Volkert, F. C., and C. S. H. Young. 1983. The genetic analysis of recombination using adenovirus overlapping terminal DNA fragments. *Virology* **125:**175–193.

Wagner, R., Jr., and M. Meselson. 1976. Repair tracts in mismatched DNA heteroduplexes. *Proc. Natl. Acad. Sci. USA* **73:**4135–4139.

Wake, C. T., F. Vernaleone, and J. H. Wilson. 1985. Topological requirements for homologous recombination among DNA molecules transfected into mammalian cells. *Mol. Cell. Biol.* **5:**2080–2089.

Wake, C. T., and J. H. Wilson. 1979. Simian virus 40 recombinants are produced at high frequency during infection with genetically mixed oligomeric DNA. *Proc. Natl. Acad. Sci. USA* **76:**2876–2880.

Wasmuth, J. J., and L. V. Hall. 1984. Genetic demonstration of mitotic recombination in cultured Chinese hamster cell hybrids. *Cell* **36:**697–707.

Watt, V. M., C. J. Ingles, M. S. Urden, and W. J. Rutter. 1985. Homology requirements for recombination in *Escherichia coli. Proc. Natl. Acad. Sci. USA* **82:**4768–4772.

Weiss, U., and J. H. Wilson. 1987. Repair of single-stranded loops in heteroduplex DNA transfected into mammalian cells. *Proc. Natl. Acad. Sci. USA* **84:**1619–1623.

Westaway, D., G. Payne, and H. E. Varmus. 1984. Proviral deletions and oncogene base-substitutions in insertionally mutagenized c-myc alleles may contribute to the progression of avian bursal tumors. *Proc. Natl. Acad. Sci. USA* **81:**843–847.

White, J. H., K. Lusnak, and S. Fogel. 1985. Mismatch-specific post-meiotic segregation frequency in yeast suggests a heteroduplex recombination intermediate. *Nature* (London) **315:**350–352.

White, R. L., and M. S. Fox. 1974. On the molecular basis of high negative interference. *Proc. Natl. Acad. Sci. USA* **71:**1544–1548.

Wolgemuth, D. J., and M.-T. Hsu. 1980. Visualization of genetic recombination intermediates of human adenovirus type 2 DNA from infected HeLa cells. *Nature* (London) **287:**168–171.

Wong, E. A., and M. R. Capecchi. 1986. Analysis of homologous recombination in cultured mammalian cells in transient expression and stable transformation assays. *Somatic Cell Mol. Genet.* **12:**63–72.

Wong, E. A., and M. R. Capecchi. 1987. Homologous recombination between coinjected DNA sequences peaks in early to mid-S phase. *Mol. Cell. Biol.* **7:**2294–2295.

Wood, K., and M. Deluca. 1987. Photographic detection of luminescence in *E. coli* containing the gene for firefly luciferase. *Anal. Biochem.* **161:**501–507.

Woychik, R. P., T. A. Stewart, L. G. Davis, P. D'Eustachio, and P. Leder. 1985. An inherited limb deformity created by insertional mutagenesis in a transgenic mouse. *Nature* (London) **318:**36–40.

Yates, J. L., N. Warren, and B. Sugden. 1985. Stable replication of plasmids derived from Epstein-Barr virus in various mammalian cells. *Nature* (London) **313:**812–815.

Young, C. S. H., and S. J. Silverstein. 1980. The kinetics of adenovirus recombination in homotypic and heterotypic genetics crosses. *Virology* **101:**503–515.

Yunis, J. J. 1983. The chromosomal basis of human neoplasia. *Science* **221:**227–236.

somal gene modification (gene targeting) by homologous recombination, substantial efforts are directed toward this goal.

One approach to achieve high-frequency gene targeting is empirical. In this approach, information gathered from studies of recombination in procaryotes and some eucaryotes, especially *Saccharomyces cerevisiae*, is used to design vectors and targeting strategies to improve the efficiency of homologous recombination or reduce the frequency of nonhomologous recombination. It is almost certain that this approach will yield important information in the near future.

The second approach is to understand the biochemical aspects of both homologous and nonhomologous recombination and to purify the enzymes responsible for the various steps. This approach may yield fundamental information about recombination and may eventually help us devise novel and efficient ways of gene targeting. In this chapter we summarize the current status of the efforts to understand the biochemical aspects of homologous recombination in mammalian cells.

Two types of experiments have helped characterize the biochemistry of mammalian somatic cell recombination. In one class of experiments, appropriate recombination substrates in the form of bacterial plasmids or bacteriophage DNA are incubated with cell-free extracts, and the reaction products are used to transform recombination-deficient (*recA*) *Escherichia coli* mutants. In this case the bacteria are used as a tool to isolate the end products of recombination. In this line of investigation the nature of the substrates and the nature of the end products allow deduction of the processes by which recombinant molecules are generated. In the second approach, extracts of mammalian cells or fractions derived from them are used to catalyze the formation of intermediates in recombination. The availability of appropriate substrates and assays to detect specific intermediates would help in purification of specific proteins or protein complexes which are capable of catalyzing a particular reaction. The two approaches are complementary to each other because ultimately it is necessary to use purified proteins to reconstitute a system capable of yielding finished products of recombination.

The first sections of this chapter deal with our current knowledge of the total recombination reaction catalyzed by extracts, and the later sections deal with attempts to identify and isolate an activity that is considered to be crucial for recombination, a strand exchange protein. Much of the search is directed toward isolation of a protein(s) with properties similar to those of the *E. coli* RecA protein and the *Ustilago maydis* Rec1 protein (see Radding, this volume).

II. BIOLOGICAL ASSAY FOR RECOMBINATION

The first step toward understanding the biochemical basis of recombination is to develop a reliable assay which can measure recombination (or a portion of it) catalyzed by cell extracts. One such in vitro recombination assay developed by us (Kucherlapati et al., 1985) is based on the principle that even a small number of recombinant molecules formed from the substrate DNAs can be detected in a biological assay by transfection into *recA* mutant bacteria which are essentially

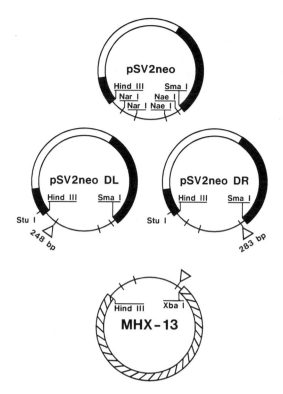

FIGURE 1. Substrates used in the recombination reactions. Plasmids pSV2neo DL and pSV2neo DR were derived by removal of appropriate restriction fragments from pSV2neo (Kucherlapati et al., 1984). Phage MHX-13 was constructed by transfering the *neo* gene fragment from pSV2neo DR into M13mp11 (Rauth et al., 1986). Symbols: ☐, pBR322; ■, SV40; ▨, M13; ——, *neo* gene.

defective in homologous recombination (see Mahajan, this volume). The substrates consist of two different DNA molecules, each containing a defective or incomplete gene but sharing a region of homology. Recombination in the region of homology produces an intact gene that can be selected in bacterial transformants under appropriate conditions. We have used substrates derived from the plasmid pSV2neo (Fig. 1), which was constructed by Southern and Berg (1982). This plasmid contains the plasmid pBR322 replication origin, the ampicillin resistance gene, and an aminoglycoside phosphorylase gene (*neo*) which confers on bacteria resistance to neomycin or kanamycin. Two deletion derivatives, DL and DR, lack 248 base pairs from the 5' end and 283 base pairs from the 3' end of the *neo* gene, respectively (Kucherlapati et al., 1984). Each deletion completely abolishes the function of the *neo* gene, so that bacteria containing DL or DR plasmids are ampicillin resistant but kanamycin sensitive. The two deletions are separated by 501 base pairs of DNA so that a recombination event between DL and DR in this region can generate an intact *neo* gene which is detected by the resulting kanamycin-resistant bacteria. The ability of human cell-free extracts or fractions derived from them to catalyze recombination can therefore be determined by incubating mixtures of DL and DR plasmids with protein fractions under appropriate conditions and determining the ratio of kanamycin-resistant to ampicillin-resistant bacteria after transformation of *recA E. coli* mutants (Kucher-lapati et al., 1985). Control experiments in which each of the two plasmids is incubated with the protein fractions separately, or not at all, can be performed to confirm that the recombination (or some aspect of it) results from the in vitro incubation.

Similar assays have been utilized by other laboratories. Symington et al. (1983) have used plasmids containing linker insertion mutations in a tetracycline resistance gene to detect homologous recombination catalyzed by extracts from the yeast *S. cerevisiae*. This assay is directly analogous to the pSV2neo system, and recombination is measured by the ratio of tetracycline-resistant to ampicillin-resistant transformants obtained in *recA* mutants. Bacteriophage lambda DNA containing various amber mutations was used by Darby and Blattner (1984) to assay recombination activity in extracts of several murine cell lines. One substrate consisted of a lambda molecule with an amber mutation in the lysis gene *S*, and the other carried amber mutations in two head protein genes (either *A* and *B* or *W* and *E*). After reaction with extracts, the lambda DNA was packaged in vitro and plated on *E. coli* Su$^+$ and Su$^-$ (amber suppressing and nonsuppressing, respectively) strains to determine the frequency of wild-type phage DNA produced by recombination. This assay detects exchange in an interval spanning about 40,000 base pairs, whereas the tetracycline and neomycin gene assays detect intragenic events over the range of several hundred base pairs. Lopez et al. (1987) utilized a bacteriophage M13mp8 containing a linker insertion mutation in the *E. coli lacZ* gene to detect recombination catalyzed by nuclear extracts from HeLa cells. The wild-type *lacZ* sequence was present on a nonreplicating, gel-purified restriction fragment, and recombination was measured as the frequency of β-galactosidase–positive phage plaques by using a color assay.

Each of these assays utilizes *recA* mutant bacteria, and thus it can be reasoned that a minimal requirement for the extracts to catalyze recombination would be a DNA strand transferase activity that would complement the *recA* deficiency (see Radding, this volume). However, the requirement of the lambda system for a product molecule capable of being packaged into phage particles and the observations that with each of the plasmid systems dimeric recombinant molecules can be detected prior to transformation imply that the crude extracts used in these experiments can accomplish a more or less complete recombination reaction that is presumed to require several enzyme activities in addition to DNA strand transferase (see below).

III. PREPARATION OF NUCLEAR EXTRACTS AND ASSAYS FOR RECOMBINATION

The enzyme preparations initially assayed for recombination activity were unfractionated, nucleic acid-free protein extracts from nuclei of mammalian cells grown in culture (Kucherlapati et al., 1985). Nuclei from approximately 2×10^8

cells are prepared by Dounce homogenization in hypotonic buffer. The source of the extract is usually the human bladder carcinoma cell line EJ. The isolated nuclei are adjusted to 0.5 M with respect to NaCl and disrupted by brief sonication. After centrifugation at $100,000 \times g$ for 1 h to remove debris, the resulting supernatant is passed through a small DEAE-Sepharose column equilibrated with 0.5 M NaCl. Contaminating nucleic acids are retained by the column at this salt concentration, while the vast majority of the proteins pass through. The protein-containing fractions are concentrated by ammonium sulfate precipitation and are redissolved in a buffer containing 50 mM Tris hydrochloride (pH 7.5), 1 mM EDTA, 1 mM dithiothreitol, 0.1 mM phenylmethylsulfonyl fluoride, and 10% glycerol; they are then quickly frozen in liquid nitrogen. Usually, 1 ml of extract containing 1 to 2 mg of protein is obtained.

The nuclear extracts used by Lopez et al. (1987) were prepared by a procedure analogous to that described above. Darby and Blattner (1984) also used a similar procedure except that the DEAE column and ammonium sulfate precipitation were omitted. In contrast, the yeast extracts of Symington et al. (1983) were not prepared from nuclei but consisted of whole cell extracts obtained by lysis of cells with zymolase followed by removal of debris by high-speed centrifugation and concentration of proteins by ammonium sulfate precipitation.

For the in vitro recombination reactions conducted by us (Kucherlapati et al., 1985), 0.5 µg of each substrate molecule, DL and DR, was incubated with 10 to 50 µg of protein fraction for 60 min at 37°C in a reaction mixture of 100 µl containing 20 mM Tris (pH 7.4), 10 mM MgSO$_4$, 120 mM NaCl, 1 mM ATP, 0.1 mM each dATP, dCTP, dGTP, and TTP, and 0.001% gelatin. The reactions were stopped by addition of EDTA, and the DNA was reisolated by treatment with pronase, phenol extraction, and ethanol precipitation. The DNA was redissolved in transformation buffer, and 10 to 25% of the recovered DNA was used to transform competent E. coli DH1 by the method of Mandel and Higa (1970) as modified by Maniatis et al. (1982). The fraction of recombined molecules is determined by comparison of the numbers of kanamycin-resistant and ampicillin-resistant bacteria obtained. Control experiments using mixtures of DL and DR that have not been exposed to extract give rise to kanamycin resistance at low frequencies (below 10^{-5}), presumably as a result of some recA-independent recombinational activity in E. coli DH1. Since incubation with the extracts might produce alterations in the substrates, such as nicks or single-strand regions, that might enhance the background recombination in DH1, a routinely performed control is to incubate each substrate separately with extract and then to mix them prior to transformation. These reactions result in even lower frequencies of recombination than the unreacted substrate mixture, and in fact for certain DNA substrate conformations, e.g., double-stranded linear × single-stranded circular DNA (see below), no kanamycin-resistant transformants are obtained.

To confirm that the kanamycin-resistant transformants arise from substrate plasmids that have undergone homologous recombination, we isolate the plasmids present in the transformants and analyze them by suitable restriction enzyme digestion. In all cases examined (Kucherlapati et al., 1985) the kanamycin-resistant transformants contain plasmids with a complete pSV2neo genome including an intact wild-type *neo* gene that could arise only through a homologous exchange of sequences present in the two substrates.

IV. RECOMBINATION ACTIVITY IN MAMMALIAN NUCLEAR EXTRACTS

We have examined recombination activity using substrates with a number of different structural conformations. Since it was known that double-strand breaks in the region of homology stimulate recombination in vivo (for a list of references, see Szostak et al., 1983), one of the standard substrate combinations utilized has been pSV2neo DL linearized by restriction cutting at the site of the deletion together with intact pSV2neo DR. When these substrates are incubated with 25 to 50 μg of protein from a crude nuclear extract of EJ cells, recombinant plasmids are recovered at frequencies ranging from 10^{-4} to $>10^{-3}$, compared with frequencies of $<10^{-5}$ for the unincubated or separately incubated controls (Kucherlapati et al., 1985). Reactions incubated with cytoplasmic extracts also produced recombinants, although at lower frequencies than were obtained with the nuclear extracts. Since the nuclei and cytoplasm were not extensively purified, it is uncertain whether the cytoplasmic activity was due to a genuine localization of recombination enzymes in the cytoplasm or simply resulted from nuclear contamination. Boiled extracts produced no detectable recombinants. The standard reaction mixture contains $MgCl_2$, ATP, and all four of the deoxynucleoside triphosphates. Removal of any one of these components reduced the recovery of recombinants by a factor of 10 or more (Kucherlapati et al., 1985). A similar requirement for Mg^{2+} and deoxynucleoside triphosphates was observed for murine cell extract-catalyzed recombination by Darby and Blattner (1984), who used the lambda DNA assay. However, these authors did not examine recombination in the absence of ATP. Lopez et al. (1987) did not detect a requirement for ATP in their assay, although the requirement was tested only for a DNA fraction enriched for recombinant molecules. Since this fraction was measured by use of radioisotopes and contained 99% nonrecombinant molecules, the rest probably being concatemers, the involvement of ATP in the homologous recombination reaction detected by this assay must remain in question.

Hotta et al. (1985) described experiments in which they compared the in vitro recombination catalyzed by nuclear extracts from lily plants, yeast cells, and mice. Using an assay similar to that used by Symington et al. (1983), they showed that the extracts can catalyze a recombination reaction requiring Mg^{2+} and ATP. As with the human cell extracts of Kucherlapati et al. (1985), Hotta et al. (1985) observed that double-strand breaks in the region of homology greatly enhance the frequency of recombination.

The ability of nuclear extracts prepared from EJ cells to catalyze recombination between substrate molecules in various structural conformations is summarized in Fig. 2. When both DL and DR substrates are covalently closed circular, double-stranded plasmids, the frequency of recombination is very low, being close to the minimum level detectable with this assay. However, when one or both of the substrates contain a double-strand break introduced at the site of

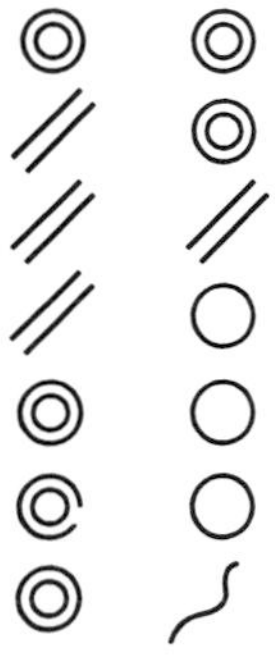

SUBSTRATES		FREQUENCY	PRODUCTS				
DL	DR	kan^R/amp^R(10^6)	monomers			dimers	
			wt	wt+DR	wt+DL	wt+DR	wt+DL
◎	◎	≤ 5	ND	ND	ND	ND	ND
∥	◎	200–2000	33	15	3	14	4
∥	∥	>1000	31	0	0	0	0
∥	○	200–700	79	0	9	0	0
◎	○	20	0	0	30	0	0
◎	○	90	16	0	6	0	0
◎	⌒	40	ND	ND	ND	ND	ND

FIGURE 2. Frequencies and products of in vitro recombination with substrates of different conformations. All products were pSV2neo derivatives. In pSV2neo DL, double- or single-strand cuts were introduced by restriction endonuclease at the site of DL. Double-strand substrates were pSV2neo DR with breaks introduced by restriction endonuclease at the site of DR. Single-strand substrates were the M13 derivative MHX or MSX; one or two random single-strand breaks were introduced with DNase I where indicated. No dimers containing two wild-type genomes or a wild-type genome plus the double deletion (DL, DR) *neo* gene were recovered. The values 3 and 9 for monomers containing the wild-type genome plus the double-deletion *neo* gene probably resulted from recombination events involving intact pSV2neo DL, due to inefficient restriction endonuclease digestion. (Data from Kucherlapati et al., 1985.)

the deletion by cutting with an appropriate restriction endonuclease, much higher levels of recombination are obtained. No increase in recombination is observed if the double-strand break is introduced elsewhere in the plasmid outside of the *neo* gene. Stimulation of recombination by double-strand breaks has also been observed in vivo (Kucherlapati et al., 1984; Song et al., 1985), although the magnitude of the effect is only 5- to 100-fold, less than that seen in vitro. Although part of this discrepancy is due to the difference in the way the frequencies are calculated (the in vitro frequencies are inversely related to the number of ampicillin-resistant colonies recovered, which is decreased by the introduction of double-strand breaks), it is apparent that the in vitro reaction is much more dependent on the presence of breaks than is in vivo recombination. Since all current models of recombination predict the introduction of DNA strand breaks prior to the commencement of strand exchange, it is likely that the nucleases responsible for these breaks in vivo are less active in the extracts relative to the enzyme(s) involved in strand exchange.

Purified DNA strand exchange enzymes such as the *E. coli* RecA protein and the *U. maydis* protein Rec1 have a strong affinity for single-stranded DNA and efficiently catalyze exchange between homologous single- and double-stranded substrates (Cox and Lehman, 1981a, b, 1982; Dasgupta et al., 1980; Kmiec and

Holloman, 1982; Tsang et al., 1985). To determine whether the enzymes responsible for the in vitro recombination reaction were able to utilize single-stranded DNA, reactions were carried out with DNA substrates containing the *neo* DR mutant gene cloned into the phage M13. The results (Fig. 2) indicate that the recombination reaction can proceed efficiently when one of the substrate molecules is present in single-stranded form. The participation of single-stranded molecules in recombination is consistent with current models of homologous recombination and has also been shown to occur in vivo (Rauth et al., 1986). When one of the substrates is single stranded, significant levels of recombination in vitro can be detected without the explicit introduction of breaks, although in this situation too there is a large increase when the double-stranded substrate has a double-strand break. With these substrates there is an indication that single-strand breaks in the double-stranded partner may also stimulate recombination, although the effect is considerably smaller than that with double-strand breaks.

We have assayed in vitro recombination activity with extracts prepared from a variety of transformed cell lines. In addition to the human bladder carcinoma line EJ, other lines in which we have detected activity are mouse L cells (S. Rauth, P. Moore, and R. Kucherlapati, unpublished data), the hamster cell line CHO K1 (Moore et al., 1985), and a mouse neuroblastoma line NEI (P. Moore, unpublished data). Using the lambda DNA assay, Darby and Blattner (1984) found recombination in extracts from several transformed murine cell lines in the pre-B- and pre-T-cell lineages.

V. RECOMBINATION IN A REPAIR-DEFICIENT CELL LINE

Since there is a strong association between recombination and the repair of double-strand breaks (Szostak et al., 1983; Resnick and Martin 1976), mutant cell lines defective in double-strand break repair are strong candidates for deficiencies in homologous recombination. The mutant *xrs-5*, an X ray-sensitive Chinese hamster cell line (Jeggo and Kemp, 1983), is defective in the repair of double-strand breaks (Kemp et al., 1984). This mutant has a fourfold reduction of plasmid-plasmid recombination assayed in vivo by transfection with pSV2neo DL and DR (Moore et al., 1985). The deficiency in homologous recombination in vivo is similar in magnitude to the defect in double-strand break repair (Kemp et al., 1984), and a comparable reduction in the nonhomologous integration of transfected DNA has been observed (Moore et al., 1985; Hamilton and Thacker, 1987; also see Thompson, this volume). It is therefore likely that the primary defect in this mutant is in a process common to both homologous and nonhomologous recombination pathways. Tests of nuclear extracts from the *xrs-5* mutant in the in vitro recombination assay revealed no significant difference in activity compared with extracts from the wild-type parental cell line (Moore et al., 1985). Several explanations can be offered for the differences between the in vivo and in vitro recombination results. The defect in *xrs-5*, which in vivo is simply quantitative, might be in an activity that is not rate limiting in the extracts. Alternatively, it may be replaced in vitro by a different but related activity; e.g., cells may possess

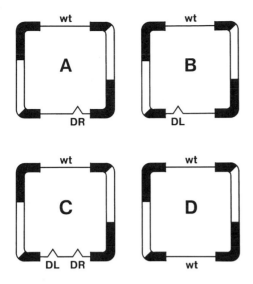

FIGURE 3. Nature of dimeric products with at least one wild-type *neo* gene that could result from homologous recombination between pSV2neo DR and pSV2neo DL. Symbols as in Fig. 1.

several exonucleases with particular cellular functions but equivalent activity. Alternatively, the defect may be in a function carried out by the *recA* mutant bacteria used to recover and assay the recombinant molecules. Taken together with the deficiency in nonhomologous recombination, it can be concluded that the *xrs-5* mutant is not defective in the strand exchange reactions central to the process of homologous recombination.

VI. MECHANISM OF THE RECOMBINATION REACTION

Information as to the nature of the reactions taking place in the extracts can be gained from an analysis of the plasmids present in the kanamycin-resistant transformants. In our experiments, the recovery in all cases of plasmids containing sequences identical to the wild-type pSV2neo genome indicates that the kanamycin resistance arose through homologous recombination rather than some other mechanism, such as repair or mutation. The production of an intact *neo* gene by homologous recombination could occur either through a gene conversion event at one of the mutant sites or by a reciprocal exchange event between the two sites (see Hastings, this volume). Unambiguous distinction between these events is possible only when both of the participating substrate molecules can be recovered and analyzed. In crosses between certain pairs of substrates, dimeric recombinant plasmids produced by reciprocal recombination are recovered (Fig. 2). There are four possible classes of dimeric molecules that could give rise to kanamycin resistance, depending on the nature of the recombination event (Fig. 3). For the cross between linear and circular double-stranded substrates, all the dimeric plasmids recovered fall into classes A and B, which are indicative of gene conversion events accompanying a reciprocal exchange. The substrate containing

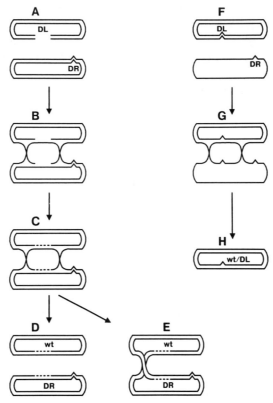

FIGURE 4. Proposed mechanism for recombination catalyzed by extracts. (A–E) Recombination involving DL substrate with a double-strand break at the site of the deletion. (B) Strand exchange at broken ends forms two Holliday junctions flanking the gap. (C) Repair synthesis restores the wild-type sequence at the site of DL. If the DR substrate is double stranded, repair synthesis (dotted line) may occur on both strands. (D) Resolution of both Holliday junctions in the same orientation produces two monomeric plasmids containing wild-type and DR *neo* genes, respectively. (E) Resolution of the two Holliday junctions in opposite directions produces a dimeric plasmid containing one wild-type and one DR *neo* gene. The crossover could occur at either junction, the products being identical. If the DR substrate either has a double-strand break or is single stranded, resolution in the crossover orientation would produce a nonviable gapped molecule. (F–H) Recombination involving an intact DL substrate and a single-stranded DR substrate. (G) Strand exchange forms a D-loop structure with heteroduplex spanning the site of DL flanked by two exchange points. (H) Resolution of this structure produces a monomeric plasmid containing a region of heteroduplex DNA with a deletion/wild-type mismatch at the site of DL. Resolution in the crossover orientation would not form a viable product if DR remains single stranded.

the double-strand break preferentially acts as the recipient of genetic information in the conversion event.

The observations on the nature of the recombinant products can be readily explained by the double-strand break repair model of recombination (Szostak et al., 1983), and a likely mechanism is shown in Fig. 4A–E. Strand exchange between each of the broken ends and the homologous sequences present in the other substrate forms two Holliday junctions flanking the cut site. Repair synthesis primed by the broken ends and use of the intact molecule as template restores a copy of the wild-type sequence at the site of the gap. When plasmids containing multiple restriction site heterozygosities are used as substrates, coconversion of markers adjacent to the cut site is frequently observed (Song et al., 1985). This probably results from exonucleolytic action at the double-strand break to produce an extended gap. Coconversion of markers up to 3,000 base pairs from the deletion has been detected. If, as illustrated in Fig. 4C, the substrate acting as donor of genetic information is an intact double-stranded circle, resolution of the two Holliday junctions can result in either two monomeric plasmids or a dimeric molecule containing one wild-type and one mutant *neo* gene with a single deletion originating from the donor substrate. If the donor sequence were either a single-stranded or a linear double-stranded molecule, successful resolution in the crossover configuration would not be possible by this model; as expected, only monomeric products are obtained (Fig. 2).

In this same cross between linear and circular double-stranded substrates, several classes of kanamycin-resistant transformants containing monomeric plasmids resolved in the noncrossover configuration were recovered (Fig. 2). The majority class simply contained a monomeric wild-type plasmid. However, a second major class contained mixtures of both wild-type and DR monomeric plasmids. These monomeric mixtures could arise either by branch migration of the structure in Fig. 4C, to form a heteroduplex at the site of the DR deletion followed by resolution of the Holliday junctions in vitro and segregation of the heteroduplex in *E. coli*, or by transformation with the intermediate shown in Fig. 4C, which becomes resolved in *E. coli*. If branch migration did extend through the DR deletion, a similar class of products would be expected from the comparable reaction with single-stranded DR. The recovery of this class only with the double-stranded DR suggests that the second explanation may be correct. In this particular experiment a small number of dimers and monomeric mixtures were recovered containing a wild-type gene together with a DL *neo* gene. Since precise rejoining of a double-strand break at the DL site could not occur by recombination with DR sequences, it is possible that these molecules resulted from incomplete cutting at the DL site by the restriction endonuclease *Nar*I in this experiment. In the reciprocal cross, where DR was cut efficiently with the enzyme *Sal*I, the equivalent class of product was not observed (Song et al., 1985). It is therefore believed that these plasmids result from recombination between intact circular DL and DR substrates. If this interpretation is correct, then recombination between intact plasmids may also proceed via gene conversion rather than reciprocal exchange between the two deletion sites.

In a recombination reaction between two intact substrates, e.g., double-stranded circular DL and the single-stranded circular derivative of DR phage MHX-13, strand exchange can lead to the formation of heteroduplex DNA containing mismatched sequences at the site of the DL deletion (Fig. 4G). The kanamycin-resistant transformants from this reaction all contained mixtures of monomeric wild-type and DL plasmids. The exclusive recovery of monomeric mixtures suggests that the product of the in vitro reaction was the heteroduplex intermediate which segregated into the two component genomes upon replication following the transformation of the *recA E. coli* mutant.

VII. NATURE OF ENZYMATIC ACTIVITIES REQUIRED FOR RECOMBINATION

An important activity likely to be involved in the in vitro recombination reaction is an enzyme capable of catalyzing strand exchange between the substrate molecules. Numerous models of recombination (e.g., Holliday, 1964; Meselson and Radding, 1975; Szostak et al., 1983) predict that both gene conversion and reciprocal exchange are dependent on such strand transfer. Since the assay system involves transformation into *recA* mutant bacteria, it would appear that a minimum requirement of the in vitro reaction is an activity that complements the deficiency of the RecA protein, which is responsible for strand exchange in *E. coli* (see Radding, this volume; Mahajan, this volume). Purified recombination enzymes such as the RecA and Rec1 proteins efficiently catalyze strand exchange between double-stranded linear and single-stranded circular substrates (see Radding, this volume). The ability of the human nuclear extracts to promote recombination with these same substrates suggests an equivalent strand exchange mechanism. Both RecA and Rec1 proteins require ATP for strand exchange (Weinstock et al., 1981; Kmiec and Holloman, 1982), and the requirement for ATP in the recombination reaction could be due to an analogous activity in the extracts. Furthermore, we have identified such an ATP-dependent strand exchange activity after fractionation of the extracts (see below). In vitro recombination catalyzed by yeast extracts also requires ATP (Symington et al., 1983), perhaps because of this type of activity. One of the properties of purified strand exchange enzymes is their stochiometric binding to single-stranded DNA such that heterologous single-stranded DNA will competitively inhibit the exchange reaction. Surprisingly, although heterologous DNA does inhibit recombination catalyzed by the human nuclear extracts, it is double-stranded rather than single-stranded DNA that is most effective (Rauth et al., 1986).

The dependence of the reaction on the presence of the four deoxynucleoside triphosphates indicates a requirement for DNA synthesis and implicates a DNA polymerase in the extracts. However, experiments using aphidicolin, ddTTP or *N*-ethylmaleimide, which are specific inhibitors of the various mammalian DNA polymerases, failed to show a consistent inhibiting effect equivalent to the omission of deoxynucleoside triphosphates (D. Clark and R. Kucherlapati, unpublished data). Perhaps the reaction has a nonspecific polymerase requirement, so that the inhibition of one species of enzyme can be compensated by the activity of the others. In contrast, recombination catalyzed by yeast extracts does not appear to require DNA synthesis and occurs efficiently in the absence of added deoxynucleoside triphosphates (Symington et al., 1983).

The initiation of recombination is believed to involve an action of endonucleases that introduces single- or double-strand breaks into the DNA. The strong dependence of the recombination reaction on experimentally introduced breaks suggests that the human nuclear extracts have only a limited capacity for such initiation events. The yeast extracts, however, catalyze recombination between intact molecules very efficiently and are stimulated only about 4-fold by double-strand breaks, as opposed to 100-fold or greater stimulation in the human extracts, suggesting a difference between the two extracts in their capabilities for generating the recombination-initiating strand breaks. The action of exonucleases is not essential for recombination to occur in the human nuclear extract, but the coconversion of markers adjacent to the double-strand break sites indicates that an exonuclease may play a role in the reaction. With the human extracts, conversion of a linker insertion heterozygosity about 250 base pairs from the double-strand break was observed in over 90% of the recombinants (Song et al., 1985). In the yeast system, however, conversion of a similar mutant site at a comparable distance from the double-strand break occurred in only a quarter of the events (Symington et al., 1983).

The exclusive recovery of mixed monomers from the cross between double-stranded circles and single-stranded circles would appear to indicate that mismatch correction, at least for deletions or insertions of this size, is not very efficient in vitro. Another enzyme activity specific to recombination is one capable of resolving Holliday junctions into the two component duplexes. The recovery of simple wild-type monomeric products from recombination between linear and circular double-stranded substrates suggests that resolution does occur in the extracts, although the mixed monomer products could result from a proportion of recombination intermediates remaining unresolved before transformation into *E. coli.*

VIII. PURIFICATION OF RECOMBINATION ENZYMES FROM *U. MAYDIS* AND MAMMALIAN CELLS

The successful development of cell-free systems that are capable of catalyzing homologous recombination between plasmid molecules has led to efforts to purify proteins that are capable of mediating specific steps in recombination. Major efforts in this area are directed toward identification and isolation of proteins capable of promoting DNA strand exchange.

Molecular models of recombination attempt to explain the extensive genetic observations of recombination and provide a framework for thinking about recombination (Holliday, 1964; Meselson and Radding, 1975; Szostak et al., 1983). One feature common to all of these models is the formation of heteroduplex DNA, which might result from the invasion of single-stranded DNA into its homologous duplex. This process is referred to as strand transfer or strand exchange. The bacterial RecA protein and the fungal Rec1 protein are capable of efficiently catalyzing this reaction (see Radding, this volume), and their behavior forms the basic model in the search for other eucaryotic strand exchange enzymes. Many assays devised to study the RecA and Rec1 proteins are used directly or in a modified form to study the mammalian recombination enzymes. Some of these assays are described below.

IX. ASSAYS FOR STRAND EXCHANGE ACTIVITY

As described earlier, the in vitro reaction to study recombination involved incubating the recombination substrates with crude nuclear extracts and using bacteria to detect recombination events. As a result, it is not possible to clearly establish how many steps of the reaction are catalyzed by the extract and what the contribution of the bacterial recombination system is in the generation of fully completed recombination products. It was necessary to develop individual assays for the strand transfer reactions.

A. D-Loop Assay

The D-loop assay is based on a method developed by Shibata et al. (1979). Typically, a uniformly labeled linear double-stranded DNA is mixed with its homologous single-stranded DNA in a circular form and incubated with the extract or fractions derived from it. The products of the reaction are then passed through a nitrocellulose filter. The double-stranded DNA does not bind to the filter, and any joint molecules that resulted from homologous interaction between the substrates would generate D-loops, whose single-stranded regions bind to the filter. The amount of labeled DNA bound to the filter provides a measure of the enzymatic activity. As a control, the reaction is conducted in the absence of single-stranded DNA or in the presence of heterologous single-stranded DNA.

B. Gel Assay

In the gel assay single-stranded circular DNA and its homologous linear double-stranded DNA are incubated with the extracts or fractions, and the products of the reaction are separated on an agarose gel (West et al., 1981). Of the substrates, the single-stranded DNA migrates rapidly and the linear double-stranded DNA migrates less rapidly. A joint molecule derived from interaction of the two substrates migrates slower than either parental molecule. If the reaction goes to completion, the products will be a nicked double-stranded circle and a linear single-stranded DNA molecule. The former product can be readily distinguished from the parental molecules.

C. S1 Nuclease Protection Assays

Assays 1 and 2 measure the number of molecules that have participated in the strand transfer reaction. They do not accurately measure the extent of the heteroduplex formation. This aspect can be measured by using a uniformly radiolabeled single-stranded DNA and unlabeled double-stranded DNA as the substrates. The products of the reaction are treated with single-strand-specific S1 nuclease, and the amount of radioactivity that is insensitive to the nuclease is measured (Kmiec and Holloman, 1982). This assay measures the extent of heteroduplex formation.

D. Visualization of Joint Molecules

Perhaps the most convincing assay for recombination is the direct visualization of intermediates with an electron microscope. Substrates of the type described above are quite useful in this assay. This assay could provide substantial information about the nature of the reactions. For example, the joint molecule formation described in assays 1 and 2 above could result from exonucleolytic digestion of part of a strand in the double-stranded DNA followed by annealing of the single-stranded DNA regions. Such a product can be readily distinguished by direct visualization of products.

E. Other Assays

Other methods which may or may not be able to directly identify the strand transfer activity have been used. In one assay, the ability of the fractions to catalyze annealing of homologous single-stranded DNA molecules is measured (Bryant and Lehman, 1985; Weinstock et al., 1979). This can be achieved by measuring change in hyperchromicity or by resistance to single-strand-specific nucleases. Yet another assay was borrowed from the study of properties of the bacterial RecA protein. This protein is capable of forming nucleoprotein networks, which are easily precipitable from the reaction mixture (Chow and Radding, 1985). It has been proposed that such networks are presynaptic intermediates catalyzed by RecA protein. This reaction is not homology dependent and thus cannot be used as a direct measure of the strand transfer activity.

X. PURIFICATION OF ENZYMATIC ACTIVITIES

The successes in purifying and characterizing the bacterial RecA protein (Ogawa et al., 1978; Roberts et al., 1978) led to efforts to isolate similar proteins from eucaryotic organisms. The most successful of these attempts was the isolation of the Rec1 protein from the fungus *U. maydis*.

Mitotic cell extracts of *U. maydis* were monitored for an enzyme that is capable of promoting DNA strand exchange. The assays used are (i) the D-loop assay, (ii) a filter binding assay of a protein-DNA complex which can be generated by strand exchange, and (iii) a single-strand reannealing assay. After fractionation of the extract by conventional chromatographic procedures, Kmiec and Holloman (1982) reported identification of a 70,000-kilodalton polypeptide which was capable of catalyzing strand exchange in a homology- and ATP-dependent fashion. They have also shown that this enzymatic activity is absent in a mutant strain of *U. maydis*, *rec1*, which has pleiotropic effects on recombination. Other recombination mutants such as *rec2* and *uvs3* contained high levels of this activity. Holloman and colleagues examined the properties of this protein extensively.

DNA strand exchange catalyzed by Rec1 protein proceeds in two steps, a synapsis step followed by strand transfer. During the initial phases of the reaction there does not seem to be a stringent requirement for ATP, but in the later stages of the reaction ATP and ATP hydrolysis seem to be essential. In several assays, the substrates used for the reactions are single-stranded circular molecules and their homologous linear double-stranded molecules. In this reaction, one of the single-stranded molecules in the duplex pairs with the circular molecule to form the heteroduplex. Kmiec and Holloman (1983) showed that for Rec1 action, homologous DNA ends are necessary. They have also shown that the strand transfer reaction is polarized and the 5'-complementary end of the double-stranded molecule initiates the joint molecule formation. These observations reveal that the action of Rec1 protein can be distinguished from that of the purified bacterial RecA protein, which starts the reaction with a free 3' end (Kahn et al., 1981; Cox and Lehman, 1981a, b; West et al., 1981).

Kmiec and Holloman (1984) examined the structure of the DNA that is involved in homologous pairing. On the basis of observations about the types of joint molecules that could be formed, they concluded that a left-handed DNA molecule might be a reaction intermediate. They tested this feature by use of anti-Z-DNA antibodies and a plasmid which contained a stretch of alternating purines and pyrimidines which, when brominated and when the superhelical density of the plasmid is appropriately altered, is known to acquire a left-handed structure. They have shown that anti-Z-DNA antibodies bind to the joint molecules (Kmiec and Holloman, 1984). They have also shown that Rec1 protein binds much more efficiently with Z-DNA and that the strand exchange reaction can be inhibited by competing Z-DNA but not by B-DNA (Kmiec et al., 1985; Kmiec and Holloman, 1986). These observations indicate that the strand exchange proteins from procaryotic and eucaryotic organisms may share some interesting properties and may act by similar mechanisms.

Several investigators have partially purified enzymatic activities from mammalian cells capable of promoting DNA strand exchange. Hsieh et al. (1986) have purified a "recombinase" from human lymphoblastoid cells (RPMI 1788). Nuclear extracts were prepared by isolating and lysing the nuclei. The proteins were precipitated with ammonium sulfate, dialyzed, passed through a cation exchange column, and fractionated by high-pressure liquid chromatography. The fraction which had enriched recombinase activity was reprecipitated with ammonium sulfate and passed through the cation exchange column a second time. The active fraction was referred to as fraction V.

Fraction V catalyzed a strand exchange reaction between a linear double-stranded DNA and its homologous single-stranded circular DNA as measured by a gel assay (assay 2 described above). Hsieh et al. (1986) showed that the formation of joint molecules required a protein(s) present in fraction V and required Mg^{2+}. The fraction V does not contain any significant exonucleolytic activity. The product is not the result of exonucleolytic action and reannealing, since the three-stranded structure expected from the strand displacement is observable by electron microscopy. In these experiments the strand transfer was not extensive, involving only about 150 nucleotides of DNA. The strand displacement by this enzyme proceeds exclusively in a $3' \rightarrow 5'$ direction. A surprising observation that stemmed from these studies is that addition or deletion of ATP from the reaction did not seem to have a significant effect on the reaction. Kolodner et al. (1987) reported that an enzyme purified from *S. cerevisiae* promotes strand exchange in an ATP-independent manner.

A second report of partial purification of a DNA strand transfer activity has appeared recently from our laboratory (Ganea et al., 1987). In this case, the source of the activity is a human bladder carcinoma cell line, EJ. On the basis of observations of in vitro-catalyzed recombination by Kucherlapati et al. (1985), Song et al. (1985), and Rauth et al. (1986), Ganea et al. (1987) fractionated human EJ cell nuclear extracts in search of a strand exchange activity. These investigators initially assayed their fractions for DNA-dependent ATPase activity, and fractions that exhibited this activity were tested for DNA strand transfer activity by the D-loop assay (described in section IV.A.).

The nuclear extracts prepared by Ganea et al. (1987) contained a biologically relevant activity that was capable of catalyzing genetically detectable recombination events between homologous pSV2neo deletion plasmids (see section IV above). The nuclear extract was fractionated on a 30 to 50% glycerol gradient, and individual fractions were assayed for biological activity and DNA-dependent ATPase activity. A peak of each activity was detected at a region corresponding to proteins of 70,000 to 100,000 molecular weight, and a second peak was detected at a region corresponding to proteins of greater than 250,000 molecular weight. The fact that the biological activity can be detected in a region corresponding to a molecular weight of 70,000 indicated that the activity is not restricted to a complex but may involve individual enzyme activities. Passage of the extracts sequentially through phosphocellulose, DNA cellulose, and an anion exchange column yielded a fraction that had a number of biochemical characteristics of strand exchange enzymes.

The highly purified fraction had DNA-dependent ATPase activity as well as strand exchange activity, as measured by the D-loop assay. Joint molecule formation was complete in 10 min, and it was homology dependent. The reaction required Mg^{2+}, and the removal of ATP or the presence of a nonhydrolyzable analog of ATP prevented the reaction from proceeding. The S1 nuclease assay indicated that as much as 3,000 base pairs of DNA could be in heteroduplex. The fraction also contained DNA reannealing activity and was capable of catalyzing nucleoprotein network formation.

Ganea et al. (1987) also examined the biological activity of fractions obtained from the DNA cellulose column. They observed that the strand exchange activity alone is not sufficient for the biologically assayed recombination reaction, since this DNA cellulose fraction alone is inactive in this assay. However, mixing the strand exchange fraction with a separate fraction from the column reconstituted biological activity, implicating the requirement of the strand exchange activity together with one or more additional enzyme activities in this reaction.

Cassuto et al. (1987) have also reported on the partial purification of an enzymatic activity which is capable of promoting strand exchange. This activity was fractionated from extracts derived from human diploid fibroblasts. This fraction promoted complete strand exchange in a homology-dependent manner and was ATP dependent. The properties of the enzymatic activity described by Cassuto et al. (1987) are shared by the enzyme(s) described by Ganea et al. (1987).

Fishel et al. (1988) fractionated a nuclear extract prepared from a human T-lymphoblast cell line. Using a Z-DNA affinity column, they identified a fraction that promoted DNA strand exchange in a homology-dependent and ATP-dependent fashion. This fraction contained a 72-kilodalton peptide that bound ATP. Fishel et al. (1988) consider this peptide to be the prime candidate to play a key role in the strand transfer reaction.

Keene and Ljungquist (1984) reported the fractionation of extracts from normal human cells and cells from patients with Bloom's syndrome. They obtained a fraction that catalyzed D-loop formation in an ATP- and homology-dependent manner. They also reported that cells from patients with Bloom's syndrome had a higher level of activity than normal cells. This is somewhat surprising in the light of recent observations that the lesion in these patients involves a DNA ligase (Chan et al., 1987; Willis and Lindahl, 1987).

Hotta et al. (1985) purified proteins from mitotic and meiotic mouse cells that have a DNA-dependent ATPase activity and that form joint molecules detected by the D-loop assay. The subunit size of the mouse meiotic Rec protein is 45,000, and the size of that from mitotic cells is 75,000. Examination of the proteins on nondenaturing gels indicated that the active forms are dimers in each case. These proteins bind to single-stranded DNA and exhibit additional properties analogous to those of RecA and Rec1 proteins.

Taken together, results from several laboratories suggest that there are at least two different proteins in mammalian somatic cells which are capable of promoting strand exchange, of which one is ATP dependent and the other is ATP independent. The bacterial RecA and the fungal Rec1 proteins, which play important roles in recombination, are ATP dependent (see Radding, this volume). A recent report by Kolodner et al. (1987) indicates that a protein isolated from yeast cells catalyzes strand transfer in the absence of ATP. Since a completely pure protein isolated from mammalian cells is not available, it is difficult to draw definitive conclusions about the nature of the proteins involved in the reaction or the processes and intermediates of recombination in these cells.

The goal of many of the above-described experiments is the biochemical characterization of the processes of recombination and purification of the relevant proteins to homogeneity. Such purification will permit cloning of the corresponding genes with the attendant possibility of preparing large quantities of the protein. Efforts in this direction are under way in several laboratories.

ACKNOWLEDGMENTS. Original work reported in this article was supported by grants from the National Institutes of Health.

The technical help of S. Frey and C. Boudos is appreciated. The manuscript was expertly prepared by V. Cummins.

LITERATURE CITED

Bryant, S. R., and I. R. Lehman. 1985. On the mechanism of renaturation of complementary DNA strands by the recA protein of E. coli. *Proc. Natl. Acad. Sci. USA* **82:**279–301.

Campbell, C. E., and R. G. Worton. 1981. Segregation of recessive phenotypes in somatic cell hybrids: role of mitotic recombination, gene inactivation, and chromosome disjunction. *Mol. Cell. Biol.* **1:** 336–346.

Cassuto, E., L. Lightfoot, and P. Howard-Flanders. 1987. Partial purification of an activity from human cells that promotes homologous pairing and the formation of heteroduplex DNA in the presence of ATP. *Mol. Gen. Genet.* **208:**10–14.

Chan, J. Y. H., F. F. Becker, J. German, and J. H. Ray. 1987. Altered DNA ligase I activity in Bloom's syndrome cells. *Nature* (London) **325:**357–359.

Chow, S. A., and C. M. Radding. 1985. Ionic inhibition of formation of RecA nucleoprotein networks blocks homologous pairing. *Proc. Natl. Acad. Sci. USA* **82:**5646–5650.

Cox, M. M., and I. R. Lehman. 1981a. Directionality and polarity in recA protein-promoted branch migration. *Proc. Natl. Acad. Sci. USA* **78:**6018–6022.

Cox, M. M., and I. R. Lehman. 1981b. RecA protein of *Escherichia coli* promotes branch migration, a kinetically distinct phase of DNA strand exchange. *Proc. Natl. Acad. Sci. USA* **78:**3433–3437.

Cox, M. M., and I. R. Lehman. 1982. RecA protein-promoted DNA strand exchange: stable complexes of recA protein and single-stranded DNA formed in the presence of ATP and single-stranded DNA binding protein. *J. Biol. Chem.* **257:**8523–8532.

Darby, V., and F. Blattner. 1984. Homologous recombination catalyzed by mammalian cell extracts in vitro. *Science* **226:**1213–1215.

Dasgupta, C., T. Shibata, R. P. Cunningham, and C. M. Radding. 1980. The topology of homologous pairing promoted by recA protein. *Cell* **22:**437–446.

Doetschman, T., R. G. Gregg, N. Maeda, M. L. Hooper, D. W. Melton, S. Thompson, and O. Smithies. 1987. Targetted correction of a mutant HPRT gene in mouse embryonic stem cells. *Nature* (London) **330:**576–578.

Fishel, R. A., K. Detmer, and A. Rich. 1988. Identification of homologous pairing and strand exchange activity from a human tumor cell line based on Z-DNA affinity chromatography. *Proc. Natl. Acad. Sci. USA* **85:**36–40.

Ganea, D., P. Moore, L. Chekuri, and R. Kucherlapati. 1987. Characterization of an ATP-dependent DNA strand transferase from human cells. *Mol. Cell. Biol.* **7:**3124–3130.

Hamilton, A. A., and J. Thacker. 1987. Gene recombination in X-ray-sensitive hamster cells. *Mol. Cell. Biol.* **7:**1409–1414.

Holliday, R. 1964. A mechanism for gene conversion in fungi. *Genet. Res.* **5:**282–304.

Hotta, Y., S. Tabata, R. A. Bouchard, R. Pinon, and H. Stern. 1985. General recombination mechanisms in extracts of meiotic cells. *Chromosoma* (Berlin) **93:**140–151.

Hsieh, P., M. S. Meyn, and R. D. Camerini-Otero. 1986. Partial purification and characterization of a recombinase from human cells. *Cell* **44:**885–894.

Jeggo, P. A., and L. M. Kemp. 1983. X-ray sensitive mutants of CHO cell line, isolation and cross sensitivity to other DNA damaging agents. *Mutat. Res.* **112:**312–327.

Kahn, R., R. P. Cunningham, C. Dasgupta, and C. M. Radding. 1981. Polarity of heteroduplex formation promoted by *Escherichia coli* recA protein. *Proc. Natl. Acad. Sci. USA* **78:**4786–4790.

Keene, K., and S. Ljungquist. 1984. A DNA-recombinogenic activity in human cells. *Nucleic Acids Res.* **2:**3057–3068.

Kemp, L. M., S. G. Sedgewick, and P. A. Jeggo. 1984. X-ray sensitive mutants of CHO cells defective in double strand break rejoining. *Mutat. Res.* **132:**189–196.

Kmiec, E. B., K. J. Angelides, and W. H. Holloman. 1985. Left handed DNA and the synaptic pairing reaction promoted by Ustilago Rec1 protein. *Cell* **40:**139–145.

Kmiec, E., and W. K. Holloman. 1982. Homologous pairing of DNA molecules promoted by a protein from Ustilago. *Cell* **29:**367–374.

Kmiec, E., and W. K. Holloman. 1983. Heteroduplex formation and polarity during strand transfer promoted by Ustilago Rec 1 protein. *Cell* **33:**857–864.

Kmiec, E., and W. K. Holloman. 1984. Synapsis promoted by Ustilago Rec1 protein. *Cell* **36:**593–598.

Kmiec, E. B., and W. K. Holloman. 1986. Homologous pairing of DNA molecules by Ustilago Rec1 protein is promoted by sequences of Z-DNA. *Cell* **44:**545–554.

Kolodner, R., D. H. Evans, and P. T. Morrison. 1987. Purification and characterization of an activity from *Saccharomyces cerevisiae* that catalyzes homologous pairing and strand exchange. *Proc. Natl. Acad. Sci. USA* **84:**5560–5564.

Kucherlapati, R. K., E. M. Eves, K. Y. Song, B. S. Morse, and O. Smithies. 1984. Homologous recombination between plasmids in mammalian cells can be enhanced by treatment of input DNA. *Proc. Natl. Acad. Sci. USA* **81:**3153–3157.

Kucherlapati, R. S., J. Spencer, and P. D. Moore. 1985. Homologous recombination catalyzed by human cell extracts. *Mol. Cell. Biol.* **5:**714–720.

Lin, F. L., K. Sperle, and N. Sternberg. 1985. Recombination in mouse cells between DNA introduced into cells and homologous chromosomal sequences. *Proc. Natl. Acad. Sci. USA* **82:**1391–1395.

Lopez, B., S. Rousseet, and J. Coppey. 1987. Homologous recombination intermediates between two duplex DNAs catalyzed by human cell extract. *Nucleic Acids Res.* **15:**5643–5655.

Mandel, M., and A. Higa. 1970. Calcium dependent bacteriophage DNA infection. *J. Mol. Biol.* **53:**159–162.

Maniatis, T., E. F. Fritsch, and J. Sambrook. 1982. Molecular cloning: a laboratory manual. Cold Spring Harbor Press, Cold Spring Harbor, N.Y.

Meselson, M. S., and C. M. Radding. 1975. A general model for genetic recombination. *Proc. Natl. Acad. Sci. USA* **72:**358–361.

Moore, P. D., K. Y. Song, L. Chekuri, L. Wallace, and R. S. Kucherlapati. 1985. Homologous recombination in a chinese hamster X-ray sensitive mutant. *Mutat. Res.* **160:**149–155.

Ogawa, T., H. Wakibo, T. Tsurimoto, T. Horil, H. Masukata, and H. Ogawa. 1978. Characteristics of purified RecA protein and the regulation of its synthesis in vivo. *Cold Spring Harbor Symp. Quant. Biol.* **43:**909–915.

Rauth, S., K.-Y. Song, D. Ayares, L. Wallace, P. D. Moore, and R. Kucherlapati. 1986. Transfection and homologous recombination involving single-stranded DNA substrates in mammalian cells and nuclear extracts. *Proc. Natl. Acad. Sci. USA* **83:**5587–5591.

Resnick, M. A., and P. Martin. 1976. The repair of double strand breaks in the nuclear DNA of *Saccharomyces cerevisiae* and its genetic control. *Mol. Gen. Genet.* **143:**119–219.

Roberts, J. W., C. W. Roberts, N. L. Craig, and E. M. Phizicky. 1978. Activity of E. coli recA gene product. *Cold Spring Harbor Symp. Quant. Biol.* **43:**917–921.

Rosenstrauss, M. J., and L. A. Chasin. 1978. Separation of linked markers in chinese hamster cell hybrids—mitotic recombination is not involved. *Genetics* **90:**735–760.

Shibata, T., C. DasGupta, R. P. Cunningham, and C. M. Radding. 1979. Purified E. coli rec A protein catalyzes homologous pairing of superhelical DNA and single-stranded fragments. *Proc. Natl. Acad. Sci. USA* **76:**1642–1648.

Smith, A. J. H., and P. Berg. 1984. Homologous recombination between defective neo genes in mouse 3T6 cells *Cold Spring Harbor Symp. Quant. Biol.* **49:**171–181.

Smithies, O., R. G. Gregg, S. S. Boggs, M. A. Korelewski, and R. S. Kucherlapati. 1985. Insertion of DNA sequences into the human chromosomal beta-globin locus by homologous recombination. *Nature* (London) **317:**230–234.

Song, K.-Y., L. Chekuri, S. Rauth, S. Ehrlich, and R. Kucherlapati. 1985. Effect of double-strand breaks on homologous recombination in mammalian cells and extracts. *Mol. Cell. Biol.* **5:**3331–3336.

Song, K. Y., F. Schwartz, N. Maeda, O. Smithies, and R. Kucherlapati. 1987. Accurate modification of a chromosomal plasmid by homologous recombination in human cells. *Proc. Natl. Acad. Sci. USA* **84:**6820–6824.

Southern, P. J., and P. Berg. 1982. Transformation of mammalian cells to antibiotic resistance with a bacterial gene under control of the SV40 early region promoter. *J. Mol. Appl. Genet.* **1:**327–341.

Symington, L. S., L. M. Fogarty, and R. Kolodner. 1983. Genetic recombination of homologous plasmids catalyzed by cell-free extracts of S. cerevisiae. *Cell* **35:**805–813.

Szostak, J. W., T. L. Orr-Weaver, R. J. Rothstein, and F. W. Stahl. 1983. The double-strand-break repair model for recombination. *Cell* **33:**25–35.

Tarrant, G. M., and R. Holliday. 1977. A search for allelic recombination in chinese hamster cell hybrids. *Mol. Gen. Genet.* **156:**273–279.

Thomas, K., and M. R. Capecchi. 1987. Site-directed mutagenesis by gene targeting in mouse embryo-derived stem cells. *Cell* **51:**503–512.

Thomas, K. R., K. R. Folger, and M. R. Capecchi. 1986. High frequency targetting of genes to specific sites in the mammalian genome. *Cell* **44:**419–428.

Tsang, S. S., K. Muniyappa, E. Azhderian, D. K. Gonda, C. M. Radding, and J. W. Chase. 1985. Intermediates in homologous pairing promoted by RecA protein. Isolation and characterization of active presynaptic complexes. *J. Mol. Biol.* **185:**295–309.

Weinstock, G. M., K. McEntee, and I. R. Lehman. 1979. ATP catalyzed renaturation of DNA catalyzed by the RecA protein of E. coli. *Proc. Natl. Acad. Sci. USA* **76:**126–130.

Weinstock, G. M., K. McEntee, and I. R. Lehman. 1981. Hydrolysis of nucleoside triphosphates, catalyzed by the Rec A protein of E. coli. *J. Biol. Chem.* **256:**8829–8834.

West, S. C., E. Cassuto, and P. Howard-Flanders. 1981. RecA protein promotes homologous pairing and strand exchange reactions between duplex DNA molecules. *Proc. Natl. Acad. Sci. USA* **78:** 2100–2104.

Willis, A. E., and T. Lindahl. 1987. DNA ligase I deficiency in Bloom's syndrome. *Nature* (London) **325:**355–357.

Chapter 20

Mammalian Cell Mutations Affecting Recombination

Larry Thompson

I. INTRODUCTION

In human and other mammalian cells, the fundamental processes of recombination are now being studied at the molecular level. By using DNA precipitates of calcium phosphate, or microinjection or electroporation of DNA in solution, defined DNA substrates for recombination can be readily transferred into cultured cells and examined with respect to the resulting recombination products. The proteins that mediate these processes may well be those involved in genetic exchange during meiosis or those that participate in DNA replication, repair, or recombination processes during the mitotic cell cycle, such as sister chromatid exchange (SCE).

To analyze the recombination machinery genetically, mutations in the many loci involved are needed. Such mutant lines can be derived from humans with genetic diseases or produced in culture by mutagenesis and selection for phenotypes likely to be coupled to recombinational defects. For example, mutant cell lines isolated on the basis of hypersensitivity to various DNA-damaging agents often prove to be defective in DNA repair processes. Such mutants are also candidates for having altered recombination properties, as demonstrated by mutations in other eucaryotes such as *Saccharomyces cerevisiae* (Haynes and Kunz, 1981; Game, 1983) and *Drosophila melanogaster* (Boyd et al., 1983).

Pleiotrophic expression of repair mutants for different facets of DNA metabolism is a well-known occurrence.

In this article mutant mammalian cell lines altered in several different recombination events are surveyed. These events can be divided into two classes: (i) SCE, which can be observed in all dividing cells, and (ii) events that occur following transfection of foreign DNA. The latter events include both homologous and nonhomologous recombination, as well as a specific kind of break rejoining.

II. MUTATIONS AFFECTING SCE

A. Nature of SCE

SCE, or reciprocal chromatid interchange, was first described in plant cells by Taylor et al. (1957), who used tritiated thymidine and autoradiography to distinguish the daughter chromatids. Later, when bromodeoxyuridine (BrdUrd) incorporation for two cycles was followed by fluorescence microscopy and staining with either the Hoechst dye 33258 (Latt, 1973) or acridine orange (Kato, 1974), sister chromatids could be much more clearly differentiated. Subsequently, the fluorescence plus Giemsa method was devised to obtain permanent, higher resolution preparations of differentially stained chromatids (Perry and Wolff, 1974; Wolff and Perry, 1974). This procedure involves BrdUrd incorporation followed by staining with Hoechst 33258, exposure to light, and then Giemsa staining. When these methods are used, the chromatids containing bifilarly substituted DNA at the second metaphase stain more lightly than those having unifilar substitution. Although they must be observed at mitosis, SCEs probably occur during the S phase of the cell cycle (Wolff et al., 1974). SCEs are readily induced by UV radiation (254 nm). However, no increase is seen when cells are irradiated in G_2 and examined at the first mitosis (Wolff et al., 1974). Some SCEs are truly spontaneous (Brewen and Peacock, 1969; Mazrimas and Stetka, 1978; O'Neill, 1984; Stetka and Spahn, 1984; Pinkel et al., 1985), but the incorporated BrdUrd normally used to obtain chromatid differentiation contributes to the base-line SCE frequency, as discussed below.

SCEs are a more sensitive cytogenetic indicator of DNA damage than are chromosomal aberrations (Perry and Evans, 1975; Kato, 1977b; Wolff, 1982; Sandberg, 1982), and they are induced by a wide variety of chemical mutagens as well as radiation. These increases can be of the order of 10-fold or more of the base-line frequency. SCE as an assay for genotoxicity is used extensively to detect environmental mutagens (Tice and Hollaender, 1984). Despite intensive study, the molecular basis of SCE remains unclear. A variety of models have been described, but experimental systems for rigorously testing these models have not been available.

A major unanswered question is what DNA structures are the initiating substrates for SCE. As discussed by Kato (1977a), SCEs might occur between the two daughter DNA duplexes after replication, or they might arise directly at the DNA replication fork. The latter type of model appears to be more plausible (see discussion by Ishii and Bender, 1980) and has been further developed by several authors. Ishii and Bender (1980) presented a "replication-detour" model in the context of SCEs induced by UV radiation (Fig. 1). These authors first proposed that spontaneous SCEs occur at DNA replication forks. They postulated that occasionally both parental strands are nicked and that sometimes the resulting ends rejoin with the nascent daughter strands of the same polarity (Fig. 1A). To explain the origin of a UV-induced SCE, the authors proposed that during replication a block occurs in a daughter strand opposite a lesion (i.e., pyrimidine dimer) (Fig. 1B). If the parental strand containing the dimer becomes nicked at the replication fork, it could rejoin with the daughter strand of the same polarity. The other parental strand would then become nicked near the end of the daughter strand that is complementary to the dimer-containing parental strand. Denaturation of this nicked parental strand followed by annealing to the dimer-containing parental strand would overcome the block and allow the daughter strands to be completed. The end result is that error-free replication proceeds past the lesion, and an SCE is produced. Ishii and Bender (1980) discussed the possible involvement of a topoisomerase II activity in mediating the SCE.

Cleaver (1981) also presented an SCE model involving the action of topoisomerases to explain the correlation observed between SCE frequency and replicon size in a variety of cell lines. In this context, it is interesting to note that intercalating agents [such as 4'-(9-acridinylamino)methanesulfon-m-anisidide and 5-iminodaunorubicin], which trap topoisomerase II-DNA complexes, are relatively efficient inducers of SCE (Pommier et al., 1985).

Painter (1980) presented a provocative model, which is a variation of the model of Ishii and Bender (1980). Painter proposed that SCEs occur at the junctions of replicon clusters, which appear to be the functional units for the control of DNA replication (Hand, 1978; Painter, 1978). At a junction between a completely replicated cluster and a partially replicated cluster, exchange would occur by a double-strand break in the unreplicated duplex, followed by rejoining of the parental strands to the adjacent replicated daughter strands of correct polarity. The likelihood of an SCE occurring might depend on the length of time between the completion of replication in the two replicon clusters. With exposure to an SCE-inducing agent (such as UV radiation), the role of a lesion blocking chain elongation would be to delay the completion of replication in a cluster that is adjacent to one that has replicated, thereby creating a long-lived junction structure. Thus, in this model exchange would not occur at the immediate site of a lesion, but rather would be limited to specific structurally specialized regions of the genome. This model helps explain why only a very small fraction of DNA lesions are converted to SCEs (e.g., $\sim 10^{-3}$ for UV-induced pyrimidine dimers as discussed by Ishii and Bender [1980]). Evidence from both CHO (Chinese hamster ovary) cells (Painter and Mathur, 1984) and frog cells (Chao and Rosenstein, 1984) supports Painter's model.

More recently, Saffhill and Ockey (1985) obtained insight into molecular mechanisms based on the properties of incorporated BrdUrd. BrdUrd-dependent SCEs occur primarily during the second cycle as replication proceeds on a BrdUrd-containing template; BrdUrd incorporated into the nascent strand has

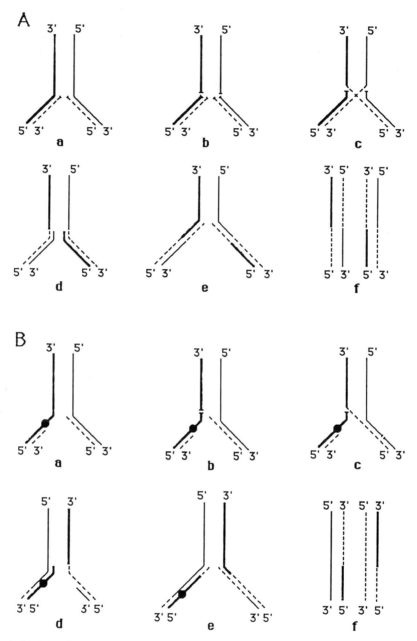

FIGURE 1. Models of Ishii and Bender (1980) to account for spontaneous and UV-induced SCEs. Parental strands are represented by solid lines, and daughter strands, by dashed lines. The two parental strands are distinguished by thick and thin lines; the thick line represents the BrdUrd-substituted strand resulting from one cycle of labeling. The daughter strands may or may not contain BrdUrd. (A) Spontaneous SCEs: a, a DNA replication fork; b, a double-strand break occurs in the parental duplex adjacent to the site of replication; c, parental strands rejoin with daughter strands of the same polarity; d, same structure as in c, redrawn; e, replication continues; f, an SCE is observed

little effect (Dillehay et al., 1983; O'Neill et al., 1983; Suzuki and Yosida, 1983). (At very high BrdUrd concentrations, 12 to 300 μM, there is another pathway for SCE induction that involves nucleotide pool perturbations and is independent of the level of previous incorporation of BrdUrd into DNA [Davidson et al., 1980].) Incorporated BrdUrd is rapidly converted to uracil (Saffhill and Ockey, 1985), which in turn is rapidly removed from DNA, presumably by uracil glycosylase (Makino and Munakata, 1979). The resulting apyrimidinic sites would then be acted on by apyrimidinic endonuclease, producing strand breaks and initiating a sequence of reactions resulting in repair of the site. Incorporation of additional BrdUrd from repair (if present in the medium during the second cycle) would further promote a high level of strand breakage in the template strand during replication. These breaks (or gaps) encountered by replication forks would be the initiating events for SCEs, as indicated by the model shown in Fig. 2. Subsequent molecular events completing the SCE (see Fig. 2 legend for details) can be considered a repair process to ensure that genetic information is faithfully preserved during replication. The Saffhill-Ockey model is consistent with several observations.

(i) 3-Aminobenzamide (3-AB), a strong inhibitor of ADP-ribosyltransferase, is also a potent inducer of SCE in protocols utilizing BrdUrd (Oikawa et al., 1980). This induction occurs most efficiently when 3-AB is present during replication of a BrdUrd-substituted template (Natarajan et al., 1981; Zwanenburg and Natarajan, 1984). Therefore, after removal of uracil residues derived from BrdUrd, 3-AB may retard the ligation step of repair (Lehmann and Broughton, 1984) by inhibiting the addition of poly(ADP-ribose) to DNA ligase II, which appears to be activated by the polymer (Creissen and Shall, 1982). Thus, 3-AB would produce SCEs by prolonging the lifetime of breaks or gaps resulting from BrdUrd substitution DNA, thereby increasing the likelihood that replication forks will encounter these lesions.

(ii) 3-AB interacts synergistically with the lesions from alkylation damage (methyl methanesulfonate), but only when both treatments occur in the same cell cycle (Morgan and Cleaver, 1982; Saffhill and Ockey, 1985).

(iii) In plant cells (*Allium cepa*) transient treatment with inhibitors of uracil glycosylase enhanced SCEs when given after exposure to visible light, which causes debromination of the BrdUrd used to measure SCEs (Maldonado et al., 1985). This paradoxical effect might be explained by the short half-life of uracil residues in the DNA of these cells. Cells known to be in G_1 phase when treated with inhibitor should have more removal of uracil during S phase, allowing repair intermediates to interact with replication forks.

---

in the sister chromatids at mitosis. (B) SCEs induced by UV radiation or bulky adducts as a consequence of "replication-detour": a, polymerization is blocked in the leading strand by a pyrimidine dimer; b, a nick is introduced into the BrdUrd-substituted parental strand downstream of the dimer; c, the 5' end at the break joins with the daughter strand of the same polarity, and a nick is introduced into the other parental strand at a position near the end of the blocked daughter strand; d, denaturation of the nicked parental strand occurs, followed by reannealling to the region of the parental strand containing the dimer; e, replication continues; f, an SCE is observed at mitosis.

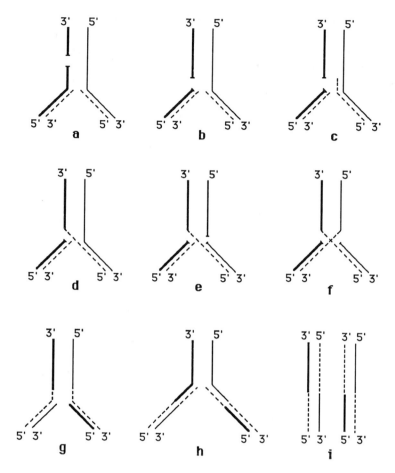

FIGURE 2. Model of Saffhill and Ockey (1985) for SCE induction initiated by single-strand breaks or gaps in front of the replication fork. (See Fig. 1 legend for description of symbols.) a, A replication fork in the vicinity of a break or gap; b, replication continues until the lesion is reached; c, synthesis continues (in the case of a gap) on the intact strand (in the case of a break without a gap, no further synthesis would occur); d, ligation of the newly synthesized strand occurs to the end of the break with correct polarity; e, an endonuclease cut occurs in the parental strand without the break; f, the 5′ end of the newly replicated strand made on the other template ligates with the 3′ end of the break; g, equivalent figure to f, redrawn for clarity; h, normal replication proceeds; i, an SCE is observed when the cell enters mitosis.

(iv) Cells from Bloom's syndrome patients, which are defective in DNA ligase I (Willis et al., 1987), have very high SCE levels (see section II.C.).

These relationships support the idea that breaks in the template strand can play an important role in SCE formation, though not necessarily by the mechanism shown in Fig. 2. However, some data seem inconsistent with models invoking strand breaks as initiating events in SCE. For example, Speit et al. (1984) argue against the importance of breaks based on the failure of bleomycin (a strand-breaking agent) to induce SCE when given during S phase. Evidence against the idea that inhibitors of ADP-ribosyltransferase (such as 3-AB) induce SCEs through the pathway that involves incorporated BrdUrd was presented by Lindahl-Kiessling and Shall (1987).

Several studies involving simple alkylating (methylating or ethylating) agents have attempted to correlate the induction of SCEs with the presence of specific DNA adducts (e.g., O^6-methylguanine, 3-methyladenine, or 7-methylguanine). No particular adduct formed by methylating or ethylating agents showed a strong correlation with SCE (Swenson et al., 1980; Heflich et al., 1982; Morris et al., 1983). In a related study, the frequency of SCEs induced by methylation or ethylation damage correlated well with the extent of cell killing, whereas SCEs did not correlate with induced mutations (Morris et al., 1982). This last study suggests that some common lesion, e.g., strand breaks resulting from repair processes, may be an important intermediate in the production of SCEs by alkylating agents.

Several studies have clearly shown that some DNA-damaging agents can induce SCEs while not producing detectable mutations. These agents include *trans*-diamminedichloroplatinum(II) hydrogen peroxide, and radiations from sunlamps and fluorescent lamps filtered to remove wavelengths greater than 345 nm (Bradley et al., 1979; Speit, 1986). All of these agents, except for the *trans*-Pt compound, produced detectable strand breaks or alkali-labile lesions, or both (Bradley et al., 1979). *trans*-Pt produces DNA intrastrand cross-links and DNA-protein cross-links. Thus, some types of lesions appear to be highly efficient in producing SCEs, and because of the differing types of lesions produced by the aforementioned compounds, no single type of lesion appears to be uniquely responsible for their induction.

Another unresolved issue is whether an SCE is itself a mutagenic event. According to the models discussed above, one can imagine both error-free and error-prone pathways of strand exchange. Base slippage occurring during the rejoining steps would introduce frameshift mutations. Answering questions such as this will require analysis of SCE at the molecular level. So far, efforts to detect hybrid DNA structures that would be candidates for SCE junctions have not been successful, partly because of the very low fraction of DNA present in such structures (see discussion by Latt, 1982).

B. Mutations Isolated in Culture

Several mutants of rodent cell lines isolated as abnormally sensitive to killing by simple alkylating agents or X rays show perturbations in their rate of SCE formation, either base-line or mutagen induced (see Table 1). Changes in base-line frequency may be due either to alteration in the genuinely spontaneous rate of SCE occurrence or to a modified response to BrdUrd, which is used to differentiate sister chromatids (Perry and Wolff, 1974).

Among the mutants in Table 1, the CHO line EM9 clearly has the most pronounced elevation in base-line rate of SCE, resulting in ~100 SCEs per cell. (The parental line, clone AA8, has ~9 SCEs per cell.) The frequency in the mutant is similar to that in many cell lines from humans with Bloom's syndrome (see

TABLE 1
Hypersensitive mutants isolated in culture with altered SCE

Mutant line	Parental line	Species	SCE frequency[a]		Basis of iso-lation[b]	Enhance-ment in chromo-somal aber-rations (ratio)[c]	Reference
			Fold increase in base-line	Fold increase in induced (agent)			
EM9	CHO AA8	Hamster	12	7 (EMS)	EMS[s]	6.5[d]	Thompson et al. (1980, 1982b)
M10	L5178Y	Mouse	0.65	1 (MMS)	MMS[s]	9.1	Tsuji et al. (1987a, b)
ES2	L5178Y	Mouse	1.4	ND	EMS[s]	3.2	Tsuji et al. (1987a)
MS1	L5178Y	Mouse	1.8	1 (MMS)	MMS[s]	4.0	Tsuji et al. (1987a, b)
ES4	L5178Y	Mouse	2.0	ND	EMS[s]	2–3	Tsuji et al. (1984, 1987a)
AC12	L5178Y	Mouse	4.0	ND	Aph[r]	4–7	Tsuji et al. (1986b, 1987a)
xrs5	CHO-K1	Hamster	1.0	1.5 (X ray)[e]	X ray[s]	1.4	Darroudi and Natarajan (1987a, b)
xrs6	CHO-K1	Hamster	1.0	2.2 (X ray)[e]	X ray[s]	1.3	Darroudi and Natarajan (1987a, b)

[a] Results are given for conditions of two cycles of labeling at a BrdUrd concentration of ~1 μg/ml (3.3 μM). EMS, ethyl methane sulfonate; MMS, methyl methanesulfonate; ND, not determined.
[b] EMS[s], sensitive to EMS; MMS[s], sensitive to MMS; Aph[r], resistant to aphidicolin; X ray[s], sensitive to X rays.
[c] Ratio of chromosomal aberrations per cell in mutant line divided by the value of parental cells, both in the absence of BrdUrd; chromosomal gaps were excluded.
[d] Unpublished result of A. V. Carrano.
[e] Increase in frequency at 150 rads; parental line showed no X-ray induction up to this dose.

section II.C. below). In EM9 cells the enhanced exchanges can be attributed predominantly, if not completely, to the presence of BrdUrd. The efficiency of BrdUrd incorporation in EM9 is normal (Thompson et al., 1982a). The use of a monoclonal antibody to BrdUrd-containing DNA allowed SCE detection at low levels of BrdUrd substitution and showed a large reduction in the SCE frequency in EM9 as the BrdUrd level was reduced (Pinkel et al., 1985). At the lowest level of substitution (0.6%) at which SCEs could be visualized, the frequency was ~25 SCEs per cell (and still declining), compared with more than 100 SCEs per cell at 60% substitution. Moreover, in other experiments on EM9 using standard BrdUrd procedures (at 60% substitution), SCE frequencies were very similar whether one or two cycles of labeling with BrdUrd occurred (Dillehay et al., 1983). This result implies that the SCEs arise predominantly during the second cycle when DNA replication occurs on a BrdUrd-substituted template.

Among five mutants of the mouse lymphoma line L5178Y, four show an elevation in SCE frequency, but only one (line AC12) shows more than a twofold increase (Table 1). This line was isolated as being resistant to aphidicolin (a potent inhibitor of DNA polymerase α) (Huberman, 1981), whereas the others were isolated as being sensitive to alkylating agents. The fifth line (M10) has a slight reduction in SCE frequency. All five mutants seem to have alterations in their spontaneous rates. The curves for frequency of SCE versus BrdUrd concentration (or percentage substitution) are flat below 1 μg of BrdUrd per ml down to the reported limit of detection at 3 to 5% substitution (Tsuji et al., 1984, 1986b, 1987a). Thus, the SCE values at these concentrations of BrdUrd should be equivalent to those at zero BrdUrd. Genetic tests using hybrid cells indicate that each of the five mutants has a defect in a different locus. Moreover, M10 is unusual because of its dominant phenotype in hybrids, and the phenotypes of MS1 and ES4 are semidominant. These results suggest that many genes may influence SCE rates.

Two mutants derived from CHO cells as being X-ray sensitive, xrs5 and xrs6 (Jeggo and Kemp, 1983), do not have an altered base-line SCE but do show modestly enhanced frequencies of X-ray-induced SCEs (Darroudi and Natarajan, 1987a, b). EM9 is the only other mutant shown in Table 1 that is reported to have a hypersensitive response to induced SCEs, and again the effect is comparatively large. The mutants M10 and MS1 did not show enhanced SCE induction from the agent (methyl methanesulfonate) that was used to isolate them.

Not all mutants with enhanced sensitivity to killing by DNA damage (Collins and Johnson, 1987) have been tested for altered SCE frequencies. However, the V79 Chinese hamster mutants irs1 and irs2 (Jones et al., 1987), which have enhanced sensitivity to killing by X rays and chromosomal instability (irs1 only), have normal base-line SCE frequencies (J. D. Tucker, N. J. Jones, A. V. Carrano, and L. H. Thompson, unpublished data). CHO mutants UV5 and UV20, which are deficient in nucleotide excision repair (bulky adduct repair), have essentially normal base-line levels but approximately sevenfold elevated frequencies of UV-induced SCEs (Thompson et al., 1982b). In this case the elevation is most likely due to a dose modification effect, i.e., higher levels of unrepaired photoproducts present during DNA replication, rather than any change in the enzymatic machinery that produces SCEs. In contrast, a virally transformed muntjac line (SVM) is very sensitive to killing and extremely sensitive to SCE induction by UV radiation, although nucleotide excision repair appears normal (Pillidge et al., 1986).

Insight concerning the factors (gene products) that determine SCE frequencies will come with an understanding of the specific defects of mutant lines such as those listed in Table 1. Among these mutants, some understanding as to the biochemical defects exists only for the CHO lines. Mutant EM9, which should be especially useful for studying SCE because of its high frequency, has a reduced rate of DNA strand break rejoining after exposure to either X rays or methylating and ethylating agents (Thompson et al., 1982a; Schwartz et al., 1987). This phenotype suggests a defect in some component of repair for a broad spectrum of DNA lesions. Many of the properties of EM9 cells mimic the phenotype of normal CHO cells treated with inhibitors of ADP-ribosyltransferase (Thompson and Carrano, 1983). One possible mode of action of these inhibitors is to delay the rejoining of DNA strand breaks during repair (Shall, 1984). Thus, the possibility of a defect in ADP-ribosyltransferase or associated functions was examined in EM9 cells, but none was seen (Ikejima et al., 1984). DNA ligases I and II (Chan et al., 1984) and apurinic/apyrimidinic endonucleases (La Belle et al., 1984) also appeared normal. In EM9 cells, higher levels of strand breaks arising from

incorporated BrdUrd may promote SCEs, and the reduced rate of DNA maturation associated with replication on a BrdUrd-substituted template may also play a
role (Dillehay et al., 1983).

The CHO cell mutants *xrs*5 and *xrs*6 have a reduced rate of rejoining
double-strand breaks after X irradiation (Kemp et al., 1984). The enhanced
induction of SCEs by X rays in these mutants (Table 1) suggests that double-
strand breaks can promote SCEs. For example, if such breaks occur immediately
downstream of a replication fork, they could lead to exchanges when rejoining of
the broken parental strands with daughter strands occurs (see section II.A.).

Besides the putative repair mutants that were isolated on the basis of
hypersensitivity to DNA damage, temperature-sensitive mutants showing altered
SCE have been described by Tsuji et al. (1986a). These mutants, obtained in
mouse FM3A cells, fall into three classes in terms of their responses to a transient
shift to nonpermissive temperature: (i) increased SCEs, (ii) increased SCEs and
chromosomal aberrations, or (iii) increased aberrations only. The mutants have
induced SCE frequencies ranging up to nine times that of the parental cells.
Interestingly, in several mutants ~20% of the chromosomal breaks result from
incomplete SCEs, a pattern also seen in mutant EM9 (Carrano et al., 1986). These
results obtained with conditional mutants reinforce the idea that the functions
essential for DNA replication are intimately associated with the processes that
govern SCE rates and chromosomal integrity.

C. Bloom's Syndrome and Other Human Genetic Disorders

One cancer-prone human disease, Bloom's syndrome, is characterized by the
distinctive features of chromosome instability and a greatly elevated (often
>10-fold) frequency of SCE in blood lymphocytes or skin fibroblasts (Chaganti et
al., 1974; German and Schonberg, 1980). In some individuals, however, unexplained phenotypic heterogeneity is seen. In one study (German et al., 1977), 5 of
21 individuals had a bimodal distribution of SCEs per lymphocyte, with a
percentage of cells (between 3 and 47%) in the normal range. Moreover, some
individuals show only slightly increased SCE frequency (Willis et al., 1987). An
interesting feature of the chromosome instability of Bloom's syndrome cells is the
tendency for exchanges to occur between the chromatids of homologous chromosomes at homologous sites, as inferred from the presence of symmetrical
quadriradial figures at metaphase (Chaganti et al., 1974).

The elevation of SCE frequency in Bloom's syndrome cells is partly due to a
greater enhancement of SCE by the incorporated BrdUrd than in normal cells
(Shiraishi et al., 1983; Shiraishi and Ohtsuki, 1987; Heartlein et al., 1987).
However, the evidence obtained by using sensitive antibodies to BrdUrd-
substituted DNA points toward a significant proportion of the SCE elevation as
being truly spontaneous (Shiraishi and Ohtsuki, 1987; Heartlein et al., 1987).
Topoisomerase II, which may be involved in the SCE process, was studied in
BrdUrd-substituted cells. The amount of salt-extractable activity of this enzyme
was distinctly lower in several Bloom's syndrome fibroblast lines than in
fibroblast lines from normal individuals (Heartlein et al., 1987). The authors

speculated that an alteration in topoisomerase II may have increased the affinity of the enzyme for BrdUrd-substituted DNA.

Cells from individuals with Bloom's syndrome also show enhanced sensitivity to induced SCEs after exposure to ethyl methanesulfonate (Krepinsky et al., 1979), ethyl nitrosourea (Kurihara et al., 1987a), and UV radiation (Kurihara et al., 1987b). Hypersensitivity of these cells to killing by ethyl nitrosourea (Kurihara et al., 1987b) and, sometimes, to UV radiation has also been reported (Giannelli et al., 1977; Krepinsky et al., 1980; Kurihara et al., 1987b).

The primary biochemical defect in Bloom's syndrome cells appears to be a structural defect in DNA ligase I, which is the predominant ligase activity in dividing cells (Soderhall and Lindahl, 1976). In cell lines from various individuals with Bloom's syndrome, ligase I activity was reduced or showed altered thermolability or aggregation properties (Willis and Lindahl, 1987; Chan et al., 1987; Willis et al., 1987). A defect in ligase I is consistent with the retarded rates of replication fork progression (Hand and German, 1977; Kapp, 1982) and the delayed maturation of nascent DNA to high-molecular-weight forms (Giannelli et al., 1977). Thus, during DNA replication in Bloom's syndrome cells, long-lived strand interruptions from insufficient ligation could help account for the high frequency of SCEs, according to models such as that described in Fig. 2. This biochemical defect is also consistent with evidence for a single complementation group of genetic defects in Bloom's syndrome cells (R. Weksberg, personal communication), since DNA ligase I contains a single polypeptide (Teraoka and Tsukada, 1985). Because the CHO mutant line EM9 has unaltered DNA ligase I, the recessive mutations in Bloom's syndrome cells would be expected to be complemented by fusion with EM9 cells. This prediction was recently confirmed: EM9-Bloom's syndrome cell hybrids had normalized SCE frequencies (Ray et al., 1987).

Base-line frequencies of SCE are not significantly elevated in the fibroblasts of other cancer-prone human disorders such as xeroderma pigmentosum (Wolff et al., 1975), ataxia telangiectasia (Galloway and Evans, 1975), and Fanconi's anemia (Kato and Stich, 1976). These syndromes have been extensively studied for DNA repair defects and for chromosome instability, which is characteristic of ataxia telangiectasia and Fanconi's anemia. Modest increases in base-line SCE frequency have been documented for dyskeratosis congenita (twofold increase [Burgdorf et al., 1977]) and for cells from an immunodeficient patient (50% increase), which are defective in rejoining Okazaki-type fragments during DNA replication (Henderson et al., 1985).

In mutant human cell lines in which DNA damage is poorly repaired, one would expect hypersensitivity to induced SCEs in parallel with their hypersensitivity to cell killing, chromosomal aberrations, and mutation induction. This property has been seen. For example, xeroderma pigmentosum cell lines from various complementation groups are much more sensitive to induction of SCE by UV irradiation (de Weerd-Kastelein et al., 1977; Cheng et al., 1978). Fibroblasts from patients with Fanconi's anemia, although very sensitive to killing by mitomycin C (~10-fold, but variable), are only ~2-fold more sensitive to the induction of SCEs by this agent (Kano and Fujiwara, 1981, 1982). Whether there is a repair defect in the unhooking step (cleavage) of interstrand cross-links common to Fanconi's anemia cell lines remains unclear, as the repair deficiency reported by Fujiwara (1982) was not confirmed by Poll et al. (1984).

III. MUTATIONS AFFECTING RECOMBINATION
DURING DNA TRANSFECTION

The development of reliable procedures for transfecting cells with foreign DNAs (see reviews by Kucherlapati and Skoultchi, 1984; Abraham, 1985) allows one to examine mutant cell lines for possible defects in various parameters associated with rearrangement and integration into chromosomal DNA. The development of plasmids carrying selectable marker genes that are efficiently expressed in a variety of cell types has made this approach practical. The studies summarized here used the *Escherichia coli gpt* (xanthine-guanine phosphoribosyl-transferase) gene, as in the plasmid pSV2*gpt* (Mulligan and Berg, 1981), or the transposon-derived *neo* gene, which confers resistance to the aminoglycoside G418 (Geneticin, GIBCO Laboratories) by its encoded 3' phosphotransferase activity (Colbere-Garapin et al., 1981; Southern and Berg, 1982). In Table 2 the results obtained with mutant cell lines are listed according to the type of recombination activity described below that is detected in the assay. Often, not all of the three different assays were performed on a given mutant. In their present form, DNA transfection procedures have limited quantitative reproducibility, making precise measurement difficult. Nevertheless, several mutations show pronounced defects in particular assays. Since in most instances the cell lines being compared are not isogenic, an altered recombination property in a mutant line may not derive from the primary mutation. Rodent mutant lines are usually obtained after potent mutagenesis (Thompson, 1979), and lines of human origin carry the intrinsic genetic polymorphism of the species.

A. Nonhomologous Recombination

When DNA is introduced ("transfected") by calcium phosphate precipitates or other procedures, it becomes stably integrated into the chromosomal DNA of relatively few cells (frequency $\leq 10^{-3}$). (The term "stably" is used in a relative sense. Transfected DNA seems consistently to have higher mutation rates than native sequences [Thacker et al., 1983; Gebara et al., 1987].) Such integration does not depend on regions of extensive homology and appears to occur at random sites throughout the genome. Nonhomologous integration can be stimulated by DNA damage present in either the foreign DNA or the recipient cells (see section III.D.). Different cell lines vary greatly in their ability to generate stable transformants (Hoeijmakers et al., 1987). Since integration of foreign DNA into chromosomal DNA is most likely a multistep process, numerous mutational defects could presumably manifest an altered efficiency of integration.

Among the mutants in Table 2, only the *xrs* mutants isolated by Jeggo and Kemp (1983) show evidence of impairment in overall transfection efficiency. A

TABLE 2
Mutations affecting recombinational events associated with DNA transfection

Mutant line	Parental or control line[b]	Efficiency[a]			Reference
		Nonhomologous recombination[c]	Homologous recombination	Faithful rejoining of restriction enzyme cut	
EM9	CHO AA8	1.03 ± 0.08 (l)	0.37 ± 0.04[d]	ND[e]	Hoy et al. (1987)
xrs1	CHO K1	0.2–1.0 (l)	~0.4	~0.8[f]	Hamilton and Thacker (1987) Debenham et al. (1987)
xrs7	CHO K1	0.60 ± 0.15 (l)	~0.5	~0.8[f]	Hamilton and Thacker (1987)
xrs5	CHO K1	0.25	~0.2	ND	Moore et al. (1986)
irs1	V79-4	0.65 ± 0.24 1.53 ± 0.58 (l)	ND	0.24 ± 0.04, 0.39[g]	Debenham et al. (1988)
irs2	V79-4	0.73 ± 0.26 1.20 ± 0.41 (l)	ND	0.92 ± 0.11, 1.13[g]	Debenham et al. (1988)
AT5BIVA	MRC5CV	0.92 (l)	0.80	0.05[f]	Cox et al. (1984)
AT5BIVA	MRC5CV	0.95 ± 0.63 (l)	ND	0.04 ± 0.01,[f] 0.13[h]	Cox et al. (1986)
XP12ROSV	MRC5CV	0.85 (l)	ND	1.0[f]	Cox et al. (1986)
AT5BIVA	MRC5CV	ND	ND	~0.1	Debenham et al. (1987)
XP12ROSV	MRC5CV	ND	ND	~0.6	Debenham et al. (1987)

[a] For each recombinational assay the efficiency is the ratio of the frequency in mutant cells divided by the frequency in parental cells. Values for homologous recombination and faithful rejoining are normalized for the transformation efficiency of controls with intact marker DNA.
[b] In studies with transformed human cells, control cells are genetically unrelated transformed cells.
[c] Values marked (l) indicate that linearized plasmid DNA was used; other studies used circular DNA.
[d] Standard error of the mean, calculated by compounding (Wilson, 1952), is given when sufficient data were presented.
[e] Not determined because of insufficient data.
[f] Assays done using single-gene vector; other rejoin assays utilized double-gene vectors.
[g] Two different restriction enzymes (KpnI and EcoRV) were used.
[h] The two values were obtained using two different plasmids and assays (see text).

convincing reduction was described by Hamilton and Thacker (1987) for xrs1. While at 5 µg of DNA per flask this mutant had a normal frequency, at higher DNA concentrations the frequency relative to parental cells declined by almost fivefold. This result was confirmed qualitatively, in a sense, by the study of Moore et al. (1986) using the line xrs5, which belongs to the same complementation group (Jeggo, 1985). This defect in the xrs mutants is consistent with the biochemical phenotype, which involves a deficiency in rejoining double-strand breaks after ionizing radiation (Kemp et al., 1984).

B. Homologous Recombination

A variety of studies have shown that mammalian cells are able to perform homologous recombination quite efficiently (Folger et al., 1982; Small and

Scangos, 1983; Shapira et al., 1983; de Saint Vincent and Wahl, 1983; Miller and Temin, 1983). In this type of assay cells are transfected with either a mixture of two overlapping gene fragments or a pair of genes that contain mutations in different regions. Intracellular recombination produces an intact, functional gene, which may be detected by scoring for the dominant phenotype associated with expression of the gene. In Table 2, mutant EM9 (introduced above as having high SCE) shows a definite reduction compared with the parental CHO cells in its ability to carry out homologous recombination; there was no evidence that EM9 is altered in its ability to integrate transfected DNA (Hoy et al., 1987). The *xrs* mutants may also have a defect in homologous recombination. Hamilton and Thacker (1987) found a reduction of ~50% in *xrs*1 and *xrs*7. However, the experimental interpretation is complicated by the reduced transfection efficiency. Moore et al. (1986) reported a sixfold reduction for *xrs*5, but this result was not fully convincing since cell viability determinations after DNA treatment were not made and the experimental variation was considerable.

C. Rejoin Fidelity of Restriction Enzyme Cleavage

A third type of assay that has been applied to mutant cell lines determines the ability of cells to faithfully rejoin double-strand breaks generated in plasmids by digestion with restriction enzymes. Initial experiments were done simply by cleaving the *gpt* gene within a plasmid and using it as the selectable marker for transfection (Cox et al., 1984, 1986). For example, a *Kpn*I cut in the coding sequence of the *gpt* gene can lead to loss of gene function if there is degradation of strand termini before rejoining occurs (Cox et al., 1984). However, more suitable plasmids were developed that contain a second selectable gene acting as a control for DNA transfer and gene expression (Cox et al., 1986; Debenham et al., 1987, 1988). In the more recent studies an uncut *neo* gene was used to select cells that have received the DNA, and a cut *gpt* gene was used to assay rejoining. Therefore, after transfection, colonies selected for G418 resistance (determined by a functional *neo* gene) are tested for resistance to mycophenolic acid (determined by a functional *gpt* gene). The percentage of colonies that continue growth gives a measure of faithful repair of the double-strand break.

This type of assay has been applied to most of the mutants listed in Table 2, with the notable exception of EM9. The *xrs* mutants did not show a defect in rejoining as determined by this assay, with the qualification that most of the results were obtained using single-marker plasmids. Two complementing *irs* mutants from V79 hamster cells (Jones et al., 1988) were examined by Debenham et al. (1988). A substantial reduction in rejoining in *irs*1 was seen, the magnitude depending on the restriction enzyme used. The fidelity of rejoining implied by resistance to mycophenolic acid was confirmed by Southern blotting analysis of transformant DNAs. Resistant lines contained a restriction fragment, of the appropriate size, spanning the *gpt* coding sequence. Unlike the *xrs* mutants isolated by Jeggo (1985), *irs*1 does not show a deficiency in the rejoining of double-strand breaks in cellular DNA after exposure to ionizing radiation (N. J. Jones and L. H. Thompson, unpublished data), but it is very sensitive to radiation-induced chromosomal aberrations (Tucker et al., unpublished data). It is unclear why the plasmid rejoining assay gives results different from those obtained in assays on cellular DNA with the mutants. The plasmid assay may reflect a particular class of genomic repair, such as repair that is preferential for transcribing genes. This type of repair has recently been described for the repair of UV radiation damage (Bohr et al., 1985, 1986; Madhani et al., 1986; Mellon et al., 1986).

The other mutant line with reduced rejoining activity listed in Table 2, AT5BIVA, is a virally transformed human line from a patient with the cancer-prone disorder ataxia telangiectasia. This line, while having normal transfection efficiency, shows a severe defect in faithful break rejoining. Southern blotting analysis confirmed that misrepair occurring in mycophenolic acid-sensitive clones involved deletion or rearrangement of the sequences spanning the target cleavage site. Ataxia telangiectasia cell lines are characteristically hypersensitive to ionizing radiations, but previous studies have indicated a normal efficiency in break rejoining (both single- and double-strand breaks) in assays on cellular DNA (see review by Lehmann, 1982). If these new results can be extended and confirmed in other ataxia telangiectasia lines, the hypothesis (Cox et al., 1986) that the primary defect in this disorder involves excessive endonucleolytic degradation or dissociation of strand termini will warrant further exploration.

Mutations that confer UV sensitivity and affect nucleotide excision repair do not seem to influence DNA transformation efficiency. CHO mutants in complementation groups 1 and 2 for UV sensitivity give frequencies of plasmid transfer similar to those of parental lines (Abraham et al., 1982; Thompson et al., 1987). The human disorder xeroderma pigmentosum does not show any abnormality with respect to DNA transformation efficiency or fidelity of double-strand break rejoining (Table 2).

D. Effects with Damaged Molecules or Damaged Cells

When plasmid DNA carrying a selectable marker gene is exposed to a DNA-damaging agent before transfection, the repair of damage can be studied in normal versus repair-deficient mutants. Repair within the target gene can lead to its rescue and normal expression. For example, when CHO mutants UV5 and UV4 (UV complementation groups 1 and 2, respectively) are transfected with UV-irradiated pSV2*gpt* DNA, the dose-dependent inhibition of transfection efficiency occurs at UV doses that are about four times lower than those that produce equivalent inhibition with parental CHO cells (Thompson et al., 1987). These results suggest that the parental cells are able to repair much of the UV damage present in the transfected DNA. In another study CHO mutants UVL1 and UVL10, which belong to the same two complementation groups, respectively, were transfected with UV-irradiated plasmids in the presence of carrier DNA (10 μg per dish) (Nairn et al., 1988). With pSV2*aprt* as the target DNA, a more sensitive response was seen in mutant cells, but with the *gpt* gene no difference was seen between normal and mutant lines. The reason for the differences between genes is unclear.

Other studies have reported that UV or *N*-acetoxy-2-acetylaminofluorene treatment of plasmid DNA can stimulate transformation. Spivak et al. (1984) and van Duin et al. (1985) found that UV damage could enhance the transformation efficiency severalfold in both normal and repair-deficient xeroderma pigmentosum cells; this effect was attributed to cyclobutane dimers (van Duin et al., 1985). At none of the UV doses tested did xeroderma pigmentosum cells show a more sensitive response than normal cells. Enhanced transformation was also seen in normal CHO cells at low UV fluences, but not in the mutant lines UV4 and UV5, which are deficient in nucleotide excision repair (Thompson et al., 1987). Treatment of the recipient cells with either ionizing or UV radiation enhances transformation efficiency (Debenham and Webb, 1984). However, a possible role of nucleotide excision repair in this enhancement is unclear because, of the CHO mutants UV5 and UV20, only UV20 showed an enhanced response (Perez and Skarsgard, 1986). Both mutants appear to be totally deficient in the incision step of nucleotide excision repair (Thompson et al., 1982b, 1984).

IV. CONCLUSIONS AND FUTURE OUTLOOK

Mutant lines of mammalian cells have been used only in the past 3 years for studying recombination at the molecular level. Likewise, most of the handful of mutant lines with altered SCE levels induced in culture were reported quite recently. Since there are probably more genes governing recombination than these few mutants reveal, more effort is needed to get a larger spectrum of mutations. Most of the alterations in recombination events listed in Table 2 are relatively modest in magnitude. Only the ataxia telangiectasia cell line was altered by more than 10-fold. Similarly, in Table 1 the EM9 line is the only mutant having at least a 10-fold increase in SCE frequency. Some recombination enzymes, such as topoisomerase II, are most likely essential for chromosome stability and cell viability. Fully deficient mutants of this type will be obtainable only as temperature-conditional mutations, like those isolated in *Saccharomyces pombe* (Uemura and Yanagida, 1986; Uemura et al., 1987).

Also needed is a detailed molecular characterization of those mutations that show pronounced alterations in phenotype. Among the mutants discussed here, the primary biochemical change has not been identified in any of those induced in culture. The recent identification of a DNA ligase deficiency in Bloom's syndrome cells (Willis et al., 1987), which is most likely due to point mutation, lends encouragement for determining the changes underlying other mutations. Identifying the protein alterations in mutants is a much greater task than isolating the mutants. Many of these mutations in DNA recombination or repair functions may lie in genes whose products are presently unknown.

The study of recombinational defects is currently limited by the inefficiency and inherent variability associated with the calcium phosphate DNA transfection procedures. Human mutant lines have been particularly difficult to stably transform with present methods, and the rodent lines from which mutants have been obtained vary widely in their transfection efficiency. Improvements in the sensitivity and reproducibility of DNA transfer look promising with the newer technique of electroporation (Potter et al., 1984; Toneguzzo and Keating, 1986; Chu et al., 1987).

A long-range, potentially highly productive approach is the isolation of the normal gene that corrects (complements) a mutation. This can be done by using DNA transfection procedures (for gene isolation strategies, see discussion by Friedberg et al., 1987; Thompson, 1988). Three genes involved in repair or recombination, or both, have recently been isolated by using CHO cell mutants: genes correcting complementation groups 1 and 2 of UV excision repair deficiency in rodent cells (Westerveld et al., 1984; van Duin et al., 1986; Weber et al., 1988) and a third gene that complements the CHO mutant EM9 (K. W. Brookman, L. H. Thompson, C. C. Collins, S. A. Stewart, J. L. Minkler, and A. V. Carrano, *Environ. Mutagen.* **9**(Suppl. 8):20). In each case the hamster cell mutation was complemented by a gene derived from normal human cells. When a functional gene has been isolated, the corresponding cDNA can be obtained by screening a cDNA library made with an expression vector (Okayama and Berg, 1983, 1985; Chen and Okayama, 1987). A fraction of recombinants (roughly 10%) that contain sequences for the gene of interest in these libraries have full-length cDNAs (Alkhatib et al., 1987; L. H. Thompson, unpublished data), but sometimes the complete sequence may have to be reconstructed from several incomplete clones (van Duin et al., 1986). Once a complete cDNA gene sequence is available in an appropriate expression vector, making the protein becomes possible, and a function for the protein can be sought.

ACKNOWLEDGMENTS. This work was performed under U.S. Department of Energy contract W-7405-ENG-48 with the Lawrence Livermore National Laboratory.

I thank Nigel Jones and Christine Weber for reading and discussing the manuscript.

LITERATURE CITED

Abraham, I. 1985. DNA-mediated gene transfer, p. 181–210. *In* M. M. Gottesman (ed.), *Molecular Cell Genetics.* John Wiley & Sons, Inc., New York.

Abraham, I., J. S. Tyagi, and M. M. Gottesman. 1982. Transfer of genes to Chinese hamster ovary cells by DNA-mediated transformation. *Somatic Cell Genet.* **8**:23–39.

Alkhatib, H. M., D. Chen, B. Cherney, K. Bhatia, V. Notario, C. Giri, G. Stein, E. Slattery, R. G. Roeder, and M. E. Smulson. 1987. Cloning and expression of cDNA for human poly(ADP-ribose) polymerase. *Proc. Natl. Acad. Sci. USA* **84**:1224–1228.

Bohr, V. A., D. S. Okumoto, and P. C. Hanawalt. 1986. Survival of UV-irradiated mammalian cells correlates with efficient DNA repair in an essential gene. *Proc. Natl. Acad. Sci. USA* **83**:3830–3833.

Bohr, V. A., C. A. Smith, D. S. Okumoto, and P. C. Hanawalt. 1985. DNA repair in an active gene: removal of pyrimidine dimers from the DHFR gene of CHO cells is much more efficient than in the genome overall. *Cell* **40**:359–369.

Boyd, J. B., P. V. Harris, J. M. Presley, and M. Narachi. 1983. *Drosophila melanogaster:* a model eukaryote for the study of DNA repair. *UCLA Symp. Mol. Cell. Biol. New Ser.* **11**:107–123.

Bradley, M. O., I. C. Hsu, and C. C. Harris. 1979. Relationships between sister chromatid exchange and mutagenicity, toxicity, and DNA damage. *Nature* (London) **182**:318–320.

Brewen, J. G., and W. J. Peacock. 1969. The effect of tritiated thymidine on sister-chromatid exchange in ring chromosome. *Mutat. Res.* **7:**433–440.

Burgdorf, W., K. Kurvink, and J. Cervenka. 1977. Sister chromatid exchange in dyskeratosis congenita lymphocytes. *J. Med. Genet.* **14:**256–257.

Carrano, A. V., J. L. Minkler, L. E. Dillehay, and L. H. Thompson. 1986. Incorporated bromodeoxyuridine enhances the sister-chromatid exchange and chromosomal aberration frequencies in an EMS-sensitive Chinese hamster cell line. *Mutat. Res.* **162:**233–239.

Chaganti, R. S. K., S. Schonberg, and J. German. 1974. A manyfold increase in sister chromatid exchanges in Bloom's syndrome lymphocytes. *Proc. Natl. Acad. Sci. USA* **71:**4508–4512.

Chan, J. Y. H., F. F. Becker, J. German, and J. H. Ray. 1987. Altered DNA ligase I activity in Bloom's syndrome cells. *Nature* (London) **325:**357–359.

Chan, J. Y. II., L. II. Thompson, and F. F. Becker. 1984. DNA-ligase activities appear normal in the CHO mutant EM9. *Mutat. Res.* **131:**209–214.

Chao, C. C.-K., and B. S. Rosenstein. 1984. Inhibition of the UV induction of sister-chromatid exchanges in ICR-2A frog cells by pretreatment with γ-rays. *Mutat. Res.* **139:**35–39.

Chen, C., and H. Okayama. 1987. High-efficiency transformation of mammalian cells by plasmid DNA. *Mol. Cell. Biol.* **7:**2745–2752.

Cheng, W.-S., R. E. Tarone, A. D. Andrews, J. S. Whang-Peng, and J. H. Robbins. 1978. Ultraviolet light-induced sister chromatid exchanges in xeroderma pigmentosum and in Cockayne's syndrome lymphocyte cell lines. *Cancer Res.* **38:**1601–1609.

Chu, G., H. Hayakawa, and P. Berg. 1987. Electroporation for the efficient transfection of mammalian cells with DNA. *Nucleic Acids Res.* **15:**1311–1326.

Cleaver, J. E. 1981. Correlations between sister chromatid exchange frequencies and replicon sizes. *Exp. Cell Res.* **136:**27–30.

Colbere-Garapin, F., F. Horodinceanu, P. Kourilsky, and A.-C. Garapin. 1981. A new dominant hybrid selective marker for higher eukaryotic cells. *J. Mol. Biol.* **150:**1–14.

Collins, A., and R. T. Johnson. 1987. DNA repair mutants of higher eukaryotes, p. 61–82. *In* A. Collins, R. T. Johnson, and J. M. Boyle (ed.), *Molecular Biology of DNA Repair* (*J. Cell Sci.*, Suppl. 6). The Company of Biologists Ltd., Cambridge.

Cox, R., P. G. Debenham, W. K. Masson, and M. B. T. Webb. 1986. Ataxia-telangiectasia: a human mutation giving high-frequency misrepair of DNA double-stranded scissions. *Mol. Biol. Med.* **3:**229–244.

Cox, R., W. K. Masson, P. G. Debenham, and M. B. T. Webb. 1984. The use of recombinant DNA plasmids for the determination of DNA-repair and recombination in cultured mammalian cells. *Br. J. Cancer* **49**(Suppl. 6):67–72.

Creissen, D., and S. Shall. 1982. Regulation of DNA ligase activity by poly(ADP-ribose). *Nature* (London) **296:**271–272.

Darroudi, F., and A. T. Natarajan. 1987a. Cytological characterization of Chinese hamster ovary X-ray-sensitive mutant cells, xrs5 and xrs6. I. Induction of chromosomal aberrations by X-irradiation and its modulation with 3-aminobenzamide and caffeine. *Mutat. Res.* **177:**133–148.

Darroudi, F., and A. T. Natarajan. 1987b. Cytological characterization of Chinese hamster ovary X-ray-sensitive mutant cells, xrs5 and xrs6. II. Induction of sister-chromatid exchanges and chromosomal aberrations by X-rays and UV-irradiation and their modulation by inhibitors of poly(ADP-ribose) synthetase and α polymerase. *Mutat. Res.* **177:**149–160.

Davidson, R. L., E. R. Kaufman, C. P. Dougherty, A. M. Ouellette, C. M. DiFolco, and S. A. Latt. 1980. Induction of sister chromatid exchanges by BUdR is largely independent of the BUdR content of DNA. *Nature* (London) **284:**74–76.

Debenham, P. G., N. J. Jones, and M. B. T. Webb. 1988. Vector-mediated DNA double strand break repair analysis in normal, and radiation sensitive, Chinese hamster V79 cells. *Mutat. Res.* **199:**1–9.

Debenham, P. G., and M. B. T. Webb. 1984. The effect of X-rays and ultraviolet light on DNA-mediated gene transfer in mammalian cells. *Int. J. Radiat. Biol.* **46:**555–568.

Debenham, P. G., M. B. T. Webb, N. J. Jones, and R. Cox. 1987. Molecular studies on the nature of the repair defect in ataxia-telangiectasia and their implications for cellular radiobiology, p. 177–189. *In* A. Collins, R. T. Johnson, and J. M. Boyle (ed.), *Molecular Biology of DNA Repair* (*J. Cell Sci.*, Suppl. 6). The Company of Biologists Ltd., Cambridge.

de Saint Vincent, B. R., and G. M. Wahl. 1983. Homologous recombination in mammalian cells mediates formation of a functional gene from overlapping gene fragments. *Proc. Natl. Acad. Sci. USA* **180**:2002–2006.

de Weerd-Kastelein, E. A., W. Keijzer, G. Rainaldi, and D. Bootsma. 1977. Induction of sister chromatid exchanges in xeroderma pigmentosum cells after exposure to ultraviolet light. *Mutat. Res.* **45**:253–261.

Dillehay, L. E., L. H. Thompson, J. L. Minkler, and A. V. Carrano. 1983. The relationship between sister-chromatid exchange and perturbations in DNA replication in mutant EM9 and normal CHO cells. *Mutat. Res.* **109**:283–296.

Folger, K. R., E. A. Wong, G. Wahl, and M. R. Capecchi. 1982. Patterns of integration of DNA microinjected into cultured mammalian cells: evidence for homologous recombination between injected plasmid DNA molecules. *Mol. Cell. Biol.* **2**:1372–1387.

Friedberg, E. C., C. Backendorf, J. Burke, A. Collins, L. Grossman, J. H. J. Hoeijmakers, A. R. Lehmann, E. Seeberg, G. P. van der Shans, and A. A. van Zeeland. 1987. Molecular aspects of DNA repair. *Mutat. Res.* **184**:67–86.

Fujiwara, Y. 1982. Defective repair of mitomycin C crosslinks in Fanconi's anemia and loss in confluent normal human and xeroderma pigmentosum cells. *Biochim. Biophys. Acta* **699**:217–225.

Galloway, S. M., and H. J. Evans. 1975. Sister chromatid exchange in human chromosomes from normal individuals and patients with ataxia telangiectasia. *Cytogenet. Cell Genet.* **15**:17–29.

Game, J. C. 1983. Radiation-sensitive mutants and repair in yeast, p. 109–137. *In* J. F. T. Spencer, D. M. Spencer, and A. R. W. Smith (ed.), *Yeast Genetics, Fundamental and Applied Aspects.* Springer-Verlag, New York.

Gebara, M. M., C. Drevon, S. A. Harcourt, H. Steingrimsdottir, M. R. James, J. F. Burke, C. F. Arlett, and A. R. Lehmann. 1987. Inactivation of a transfected gene in human fibroblasts can occur by deletion, amplification, phenotypic switching, or methylation. *Mol. Cell. Biol.* **7**:1459–1464.

German, J., and S. Schonberg. 1980. Bloom's syndrome. IX. Review of cytological and biochemical aspects, p. 175–186. *In* H. V. Gelboin et al. (ed.), *Genetic and Environmental Factors in Experimental and Human Cancer.* Japan Scientific Societies Press, Tokyo.

German, J., S. Schonberg, E. Louie, and R. S. K. Chaganti. 1977. Bloom's syndrome. IV. Sister-chromatid exchanges in lymphocytes. *Am. J. Hum. Genet.* **29**:248–255.

Giannelli, F., P. F. Benson, S. A. Pawsey, and P. E. Polani. 1977. Ultraviolet light sensitivity and delayed DNA-chain maturation in Bloom's syndrome fibroblasts. *Nature* (London) **265**:466–469.

Hamilton, A. A., and J. Thacker. 1987. Gene recombination in X-ray-sensitive hamster cells. *Mol. Cell. Biol.* **7**:1409–1414.

Hand, R. 1978. Eucaryotic DNA: organization of the genome for replication. *Cell* **15**:317–325.

Hand, R., and J. German. 1977. Bloom's syndrome: DNA replication in cultured fibroblasts and lymphocytes. *Hum. Genet.* **38**:297–306.

Haynes, R. H., and B. A. Kunz. 1981. DNA repair and mutagenesis in yeast, p. 371–414. *In* J. N. Strathern, E. W. Jones, and J. R. Broach (ed.), *The Molecular Biology of the Yeast Saccharomyces, Life Cycle and Inheritance.* Cold Spring Harbor Laboratory, Cold Spring Harbor, N.Y.

Heartlein, M. W., H. Tsuji, and S. A. Latt. 1987. 5-Bromodeoxyuridine-dependent increase in sister chromatid exchange formation in Bloom's syndrome is associated with reduction in topoisomerase II activity. *Exp. Cell Res.* **169**:245–254.

Heflich, R. H., D. T. Beranek, R. L. Kodell, and S. M. Morris. 1982. Induction of mutations and sister chromatid exchanges in Chinese hamster ovary cells by ethylating agents. Relationship to specific DNA adducts. *Mutat. Res.* **106**:147–161.

Henderson, L. M., C. F. Arlett, S. A. Harcourt, A. R. Lehmann, and B. C. Broughton. 1985. Cells from an immunodeficient patient (46BR) with a defect in DNA ligase are hypomutable but hypersensitive to the induction of sister chromatid exchanges. *Proc. Natl. Acad. Sci. USA* **82**:2044–2048.

Hoeijmakers, J. H. J., H. Odijk, and A. Westerveld. 1987. Differences between rodent and human cell lines in the amount of integrated DNA after transfection. *Exp. Cell Res.* **169**:111–119.

Hoy, C. A., J. C. Fuscoe, and L. H. Thompson. 1987. Recombination and ligation of transfected DNA in CHO mutant EM9, which has high levels of sister chromatid exchange. *Mol. Cell. Biol.* **7**:2007–2011.

Huberman, J. A. 1981. New views of the biochemistry of eucaryotic DNA replication revealed by aphidicolin, an unusual inhibitor of DNA polymerase α. *Cell* **23:**647–648.

Ikejima, M., D. Bohannon, D. M. Gill, and L. H. Thompson. 1984. Poly(ADP-ribose) metabolism appears normal in EM9, a mutagen-sensitive mutant of CHO cells. *Mutat. Res.* **128:**213–220.

Ishii, Y., and M. A. Bender. 1980. Effects of inhibitors of DNA synthesis on spontaneous and ultraviolet light-induced sister-chromatid exchanges in Chinese hamster cells. *Mutat. Res.* **79:**19–32.

Jeggo, P. A. 1985. Genetic analysis of X-ray-sensitive mutants of the CHO cell line. *Mutat. Res.* **146:**265–270.

Jeggo, P. A., and L. M. Kemp. 1983. X-ray sensitive mutants of Chinese hamster ovary cell line: isolation and cross-sensitivity to other DNA-damaging agents. *Mutat. Res.* **112:**313–327.

Jones N. J., R. Cox, and J. Thacker. 1987. Isolation and cross-sensitivity of X-ray-sensitive mutants of V79-4 hamster cells. *Mutat. Res.* **183:**279–286.

Jones, N. J., R. Cox, and J. Thacker. 1988. Six complementation groups for ionizing radiation sensitivity in Chinese hamster cells. *Mutat. Res.* **193:**139–144.

Kano, Y., and Y. Fujiwara. 1981. Roles of DNA interstrand crosslinking and its repair in the induction of sister-chromatid exchange and a higher induction in Fanconi's anemia cells. *Mutat. Res.* **81:**365–375.

Kano, Y., and Y. Fujiwara. 1982. Higher inductions of twin and single sister chromatid exchanges by cross-linking agents in Fanconi's anemia cells. *Hum. Genet.* **60:**233–238.

Kapp, L. N. 1982. DNA fork displacement rates in Bloom's syndrome fibroblasts. *Biochim. Biophys. Acta* **696:**226–227.

Kato, H. 1974. Spontaneous sister chromatid exchanges detected by a BUdR-labelling method. *Nature* (London) **251:**70–72.

Kato, H. 1977a. Mechanisms for sister chromatid exchanges and their relation to the production of chromosomal aberrations. *Chromosoma* (Berlin) **59:**179–191.

Kato, H. 1977b. Spontaneous and induced sister chromatid exchanges as revealed by the BUdR-labeling method. *Int. Rev. Cytol.* **49:**55–97.

Kato, H., and H. F. Stich. 1976. Sister chromatid exchanges in aging and repair-deficient human fibroblasts. *Nature* (London) **260:**447–448.

Kemp, L. M., S. G. Sedgwick, and P. A. Jeggo. 1984. X-ray sensitive mutants of Chinese hamster ovary cells defective in double-strand break rejoining. *Mutat. Res.* **132:**189–196.

Krepinsky, A. B., J. A. Heddle, and J. German. 1979. Sensitivity of Bloom's syndrome lymphocytes to ethyl methanesulfonate. *Hum. Genet.* **50:**151–156.

Krepinsky, A. B., A. J. Rainbow, and J. A. Heddle. 1980. Studies on the ultraviolet light sensitivity of Bloom's syndrome fibroblasts. *Mutat. Res.* **69:**357–368.

Kucherlapati, R., and A. I. Skoultchi. 1984. Introduction of purified genes into mammalian cells. *Crit. Rev. Biochem.* **16:**349–379.

Kurihara, T., M. Inoue, and K. Tatsumi. 1987a. Hypersensitivity of Bloom's syndrome fibroblasts to N-ethyl-N-nitrosourea. *Mutat. Res.* **184:**147–151.

Kurihara, T., K. Tatsumi, H. Takahashi, and M. Inoue. 1987b. Sister-chromatid exchanges induced by ultraviolet light in Bloom's syndrome fibroblasts. *Mutat. Res.* **183:**197–202.

La Belle, M., S. Linn, and L. H. Thompson. 1984. Apurinic/apyrimidinic endonuclease activities appear normal in the CHO-cell ethyl methanesulfonate sensitive mutant, EM9. *Mutat. Res.* **141:**41–44.

Latt, S. A. 1973. Microfluorometric detection of deoxyribonucleic acid replication in human metaphase chromosomes. *Proc. Natl. Acad. Sci. USA* **70:**3395–3399.

Latt, S. A. 1982. Sister chromatid exchange: new methods for detection, p. 17–40. *In* S. Wolff (ed.), *Sister Chromatid Exchange.* John Wiley & Sons, Inc., New York.

Lehmann, A. R. 1982. The cellular and molecular responses of ataxia-telangiectasia cells to DNA damage, p. 83–101. *In* B. A. Bridges and D. G. Harnden (ed.), *Ataxia-telangiectasia.* John Wiley & Sons, Inc., New York.

Lehmann, A. R., and B. C. Broughton. 1984. Poly(ADP-ribosylation) reduces the steady-state level of breaks in DNA following treatment of human cells with alkylating agents. *Carcinogenesis* (London) **5:**117–119.

Lindahl-Kiessling, K., and S. Shall. 1987. Nicotinamide deficiency and benzamide-induced sister chromatid exchanges. *Carcinogenesis* (London) **8**:1185–1188.

Madhani, H. D., V. A. Bohr, and P. C. Hanawalt. 1986. Differential DNA repair in transcriptionally active and inactive proto-oncogenes: c-abl and c-mos. *Cell* **45**:417–423.

Makino, F., and N. Munakata. 1979. Excision of uracil from bromodeoxyuridine-substituted and U.V.-irradiated DNA in cultured mouse lymphoma cells. *Int. J. Radiat. Biol.* **36**:349–357.

Maldonado, A., P. Hernandez, and C. Gutierrez. 1985. Inhibition of uracil-DNA glycosylase increases SCEs in BrdU-treated and visible light-irradiated cells. *Exp. Cell Res.* **161**:172–180.

Mazrimas, J. A., and D. G. Stetka. 1978. Direct evidence for the role of incorporated BUdR in the induction of sister chromatid exchanges. *Exp. Cell Res.* **117**:23–30.

Mellon, I., V. A. Bohr, C. A. Smith, and P. C. Hanawalt. 1986. Preferential DNA repair of an active gene in human cells. *Proc. Natl. Acad. Sci. USA* **83**:8878–8882.

Miller, C. K., and H. M. Temin. 1983. High efficiency ligation and recombination of DNA fragments by vertebrate cells. *Science* **220**:606–609.

Moore, P. D., K.-Y. Song, L. Chekuri, L. Wallace, and R. S. Kucherlapati. 1986. Homologous recombination in a Chinese hamster X-ray-sensitive mutant. *Mutat. Res.* **160**:149–155.

Morgan, W. F., and J. E. Cleaver. 1982. 3-Aminobenzamide synergistically increases sister-chromatid exchanges in cells exposed to methyl methanesulfonate but not to ultraviolet light. *Mutat. Res.* **104**:361–366.

Morris, S. M., D. T. Beranek, and R. H. Heflich. 1983. The relationship between sister-chromatid exchange induction and the formation of specific methylated DNA adducts in Chinese hamster ovary cells. *Mutat. Res.* **121**:261–266.

Morris, S. M., R. H. Heflich, D. T. Beranek, and R. L. Kodell. 1982. Alkylation-induced sister-chromatid exchanges correlate with reduced cell survival, not mutations. *Mutat. Res.* **105**:163–168.

Mulligan, R. C., and P. Berg. 1981. Selection for animal cells that express the *Escherichia coli* gene coding for xanthine-guanine phosphoribosyltransferase. *Proc. Natl. Acad. Sci. USA* **78**:2072–2076.

Nairn, R. S., R. M. Humphrey, and G. M. Adair. 1988. Transformation of U.V. hypersensitive Chinese hamster ovary cell mutants with U.V. irradiated plasmids. *Int. J. Radiat. Biol.* **53**:249–260.

Natarajan, A. T., I. Csukas, and A. A. van Zeeland. 1981. Contribution of incorporated 5-bromodeoxyuridine in DNA to the frequency of sister chromatid exchanges induced by inhibitors of poly-(ADP-ribose)-polymerase. *Mutat. Res.* **84**:125–132.

Oikawa, A., H. Tohda, M. Kanai, M. Miwa, and T. Sugimura. 1980. Inhibitors of poly(adenosine diphosphate ribose) polymerase induce sister chromatid exchanges. *Biochem. Biophys. Res. Commun.* **97**:1311–1316.

Okayama, H., and P. Berg. 1983. A cDNA cloning vector that permits expression of cDNA inserts in mammalian cells. *Mol. Cell. Biol.* **3**:280–289.

Okayama, H., and P. Berg. 1985. Bacteriophage lambda vector for transducing a cDNA clone library into mammalian cells. *Mol. Cell. Biol.* **5**:1136–1142.

O'Neill, J. P. 1984. Quantification of the induction of SCE due to the replication of unsubstituted and BrdU- or CldU-substituted DNA in CHO cells. *Mutat. Res.* **140**:21–25.

O'Neill, J. P., M. W. Heartlein, and R. J. Preston. 1983. Sister-chromatid exchanges and gene mutations are induced by the replication of 5-brom- and 5-chloro-deoxyuridine substituted DNA. *Mutat. Res.* **109**:259–270.

Painter, R. B. 1978. Inhibition of DNA replicon initiation by 4-nitroquinoline 1-oxide, adriamycin, and ethyleneimine. *Cancer Res.* **38**:4445–4449.

Painter, R. B. 1980. A replication model for sister-chromatid exchange. *Mutat. Res.* **70**:337–341.

Painter, R. B., and V. Mathur. 1984. Effect of prior X-irradiation on ultraviolet light-induced sister-chromatid exchange. *Mutat. Res.* **139**:123–126.

Perez, C. F., and L. D. Skarsgard. 1986. Radiation enhancement of the efficiency of DNA-mediated gene transfer in CHO UV-sensitive mutants. *Radiat. Res.* **106**:401–407.

Perry, P., and H. J. Evans. 1975. Cytological detection of mutagen-carcinogen exposure by sister chromatid exchange. *Nature* (London) **258**:121–125.

Perry, P., and S. Wolff. 1974. New Giemsa method for the differential staining of sister chromatids. *Nature* (London) **251**:156–158.

Pillidge, L., S. R. R. Musk, R. T. Johnson, and C. A. Waldren. 1986. Excessive chromosome fragility and abundance of sister-chromatid exchanges by UV in an Indian muntjac cell line defective in postreplication (daughter strand) repair. *Mutat. Res.* **166**:265–273.

Pinkel, D., L. H. Thompson, J. W. Gray, and M. Vanderlaan. 1985. Measurement of sister chromatid exchanges at very low bromodeoxyuridine substitution levels using a monoclonal antibody in Chinese hamster ovary cells. *Cancer Res.* **45**:5795–5798.

Poll, E. H. A., F. Arwert, H. T. Kortbeek, and A. W. Eriksson. 1984. Fanconi anaemia cells are not uniformly deficient in unhooking of DNA interstrand crosslinks, induced by mitomycin C or 8-methoxypsoralen plus UVA. *Hum. Genet.* **68**:228–234.

Pommier, Y., L. A. Zwelling, C.-S. Kao-Shan, J. Whang-Peng, and M. O. Bradley. 1985. Correlations between intercalator-induced DNA strand breaks and sister chromatid exchanges, mutations, and cytotoxicity in Chinese hamster cells. *Cancer Res.* **45**:3143–3149.

Potter, H., L. Weir, and P. Leder. 1984. Enhancer-dependent expression of human κ immunoglobulin genes introduced into mouse pre-B lymphocytes by electroporation. *Proc. Natl. Acad. Sci. USA* **81**: 7161–7165.

Ray, J. H., E. Louie, and J. German. 1987. Different mutations are responsible for the elevated sister-chromatid exchange frequencies characteristic of Bloom's syndrome and hamster EM9 cells. *Proc. Natl. Acad. Sci. USA* **84**:2368–2371.

Saffhill, R., and C. H. Ockey. 1985. Strand breaks arising from the repair of the 5-bromodeoxyuridine-substituted template and methyl methanesulphonate-induced lesions can explain the formation of sister chromatid exchanges. *Chromosoma* (Berlin) **92**:218–224.

Sandberg, A. A. (ed.). 1982. *Sister Chromatid Exchange.* Alan R. Liss, New York.

Schwartz, J. L., S. Giovanazzi, and R. R. Weichselbaum. 1987. Recovery from sublethal and potentially lethal damage in an X ray sensitive CHO cell. *Radiat. Res.* **111**:58–67.

Shall, S. 1984. ADP-ribose in DNA repair: a new component of DNA excision repair. *Adv. Radiat. Biol.* **11**:1–69.

Shapira, G., J. L. Stachelek, A. Letsou, L. Soodak, and R. M. Liskay. 1983. Novel use of synthetic oligonucleotide insertion mutants for the study of homologous recombination in mammalian cells. *Proc. Natl. Acad. Sci. USA* **80**:4827–4831.

Shiraishi, Y., and Y. Ohtsuki. 1987. SCE levels in Bloom-syndrome cells at very low bromodeoxyuridine (BrdU) concentrations: monoclonal anti-BrdU antibody. *Mutat. Res.* **176**:157–164.

Shiraishi, Y., T. H. Yosida, and A. A. Sandberg. 1983. Analyses of bromodeoxyuridine-associated sister chromatid exchanges (SCEs) in Bloom syndrome based on cell fusion: single and twin SCEs in endoreduplication. *Proc. Natl. Acad. Sci. USA* **80**:4369–4373.

Small, J., and G. Scangos. 1983. Recombination during gene transfer into mouse cells can restore the function of deleted genes. *Science* **219**:174–176.

Soderhall, S., and T. Lindahl. 1976. DNA ligases of eukaryotes. *FEBS Lett.* **67**:1–8.

Southern, P. J., and P. Berg. 1982. Transformation of mammalian cells to antibiotic resistance with a bacterial gene under control of the SV40 early region promoter. *J. Mol. Appl. Genet.* **1**:327–341.

Speit, G. 1986. The relationship between the induction of SCEs and mutations in Chinese hamster cells. I. Experiments with hydrogen peroxide and caffeine. *Mutat. Res.* **174**:21–26.

Speit, G., R. Hochsattel, and W. Vogel. 1984. The contribution of DNA strand-breaks to the formation of chromosome aberrations and SCEs, p. 229–244. *In* R. R. Tice and A. Hollaender (ed.), *Sister Chromatid Exchange.* Plenum Publishing Corp., New York.

Spivak, G., A. K. Ganesan, and P. C. Hanawalt. 1984. Enhanced transformation of human cells by UV-irradiated pSV2 plasmids. *Mol. Cell. Biol.* **4**:1169–1171.

Stetka, D. G., Jr., and M. C. Spahn. 1984. SCEs are induced by replication of BrdU-substituted DNA templates, but not by incorporation of BrdU into nascent DNA. *Mutat. Res.* **140**:33–42.

Suzuki, H., and T. H. Yosida. 1983. Frequency of sister-chromatid exchanges depending on the amount of 5-bromodeoxyuridine incorporated into parental DNA. *Mutat. Res.* **111**:277–282.

Swenson, D. H., P. R. Harbach, and R. J. Trzos. 1980. The relationship between alkylation of specific DNA bases and induction of sister chromatid exchange. *Carcinogenesis* (London) **1**:931–936.

Taylor, J. H., P. S. Woods, and W. L. Hughes. 1957. The organization and duplication of chromosomes as revealed by autoradiographic studies using tritium-labeled thymidine. *Proc. Natl. Acad. Sci. USA* **43**:122–128.

Teraoka, H., and K. Tsukada. 1985. Biosynthesis of mammalian DNA ligase. *J. Biol. Chem.* **260:**2937–2940.

Thacker, J., P. G. Debenham, A. Stretch, and M. B. T. Webb. 1983. The use of a cloned bacterial gene to study mutation in mammalian cells. *Mutat. Res.* **111:**9–23.

Thompson, L. H. 1979. Mutant isolation. *Methods Enzymol.* **58:**308–322.

Thompson, L. H. 1988. Use of Chinese hamster ovary cell mutants to study human DNA repair genes, p. 115–132. *In* E. Friedberg and P. Hanawalt (ed.), *DNA Repair*, vol. 3. Marcel Dekker, New York.

Thompson, L. H., K. W. Brookman, L. E. Dillehay, A. V. Carrano, J. A. Mazrimas, C. L. Mooney, and J. L. Minkler. 1982a. A CHO-cell strain having hypersensitivity to mutagens, a defect in DNA strand-break repair, and an extraordinary baseline frequency of sister-chromatid exchange. *Mutat. Res.* **95:**427–440.

Thompson, L. H., K. W. Brookman, L. E. Dillehay, C. L. Mooney, and A. V. Carrano. 1982b. Hypersensitivity to mutation and sister-chromatid-exchange induction in CHO cell mutants defective in incising DNA containing UV lesions. *Somatic Cell Genet.* **8:**759–773.

Thompson, L. H., K. W. Brookman, and C. L. Mooney. 1984. Repair of DNA adducts in asynchronous CHO cells and the role of repair in cell killing and mutation induction in synchronous cells treated with 7-bromomethylbenz[a]anthracene. *Somatic Cell Genet.* **10:**183–194.

Thompson, L. H., and A. V. Carrano. 1983. Analysis of mammalian cell mutagenesis and DNA repair using in vitro selected CHO cell mutants. *UCLA Symp. Mol. Cell. Biol. New Ser.* **11:**125–143.

Thompson, L. H., J. S. Rubin, J. E. Cleaver, G. F. Whitmore, and K. Brookman. 1980. A screening method for isolating DNA repair-deficient mutants of CHO cells. *Somatic Cell Genet.* **6:**391–405.

Thompson, L. H., E. P. Salazar, K. W. Brookman, C. C. Collins, S. A. Stewart, D. B. Busch, and C. A. Weber. 1987. Recent progress with the DNA repair mutants of Chinese hamster ovary cells, p. 97–110. *In* A. Collins, R. T. Johnson, and J. M. Boyle (ed.), *Molecular Biology of DNA Repair* (*J. Cell Sci.*, Suppl. 6). The Company of Biologists Ltd., Cambridge.

Tice, R. R., and A. Hollaender (ed.). 1984. *Sister Chromatid Exchange. 25 Years of Experimental Research.* Plenum Publishing Corp., New York.

Toneguzzo, F., and A. Keating. 1986. Stable expression of selectable genes introduced into human hematopoietic stem cells by electric field-mediated DNA transfer. *Proc. Natl. Acad. Sci. USA* **83:**3496–3499.

Tsuji, H., M. Hyodo, S. Tsuji, T.-A. Hori, I. Tobari, and K. Sato. 1986a. Isolation of temperature-sensitive mouse FM3A cell mutants exhibiting conditional chromosomal instability. *Somatic Cell Mol. Genet.* **12:**595–610.

Tsuji, H., T. Shiomi, and I. Tobari. 1984. High induction of sister chromatid exchange and chromosome aberration by 5-bromodeoxyuridine in an ethylmethane- sulfonate-sensitive mouse lymphoma cell mutant (ES4), p. 109–125. *In* R. R. Tice and A. Hollaender (ed.), *Sister Chromatid Exchange. 25 Years of Experimental Research.* Plenum Publishing Corp., New York.

Tsuji, H., T. Shiomi, S. Tsuji, I. Tobari, D. Ayusawa, K. Shimizu, and T. Seno. 1986b. Aphidicolin-resistant mutants of mouse lymphoma L5178Y cells with a high incidence of spontaneous sister chromatid exchanges. *Genetics* **113:**433–447.

Tsuji, H., E.-I. Takahashi, S. Tsuji, I. Tobari, T. Shiomi, H. Hama-Inaba, and K. Sato. 1987a. Chromosomal instability in mutagen-sensitive mutants isolated from mouse lymphoma L5178Y cells. I. Five different genes participate in the formation of baseline sister-chromatid exchanges and spontaneous chromosomal aberrations. *Mutat. Res.* **178:**99–106.

Tsuji, H., E.-I. Takahashi, S. Tsuji, I. Tobari, T. Shiomi, and K. Sato. 1987b. Chromosomal instability in mutagen-sensitive mutants isolated from mouse lymphoma L5178Y cells. II. Abnormal induction of sister-chromatid exchanges and chromosomal aberrations by mutagens in an ionizing radiation-sensitive mutant (M10) and an alkylating agent-sensitive mutant (MS1). *Mutat. Res.* **178:**107–116.

Uemura, T., H. Ohkura, Y. Adachi, K. Morino, K. Shiozaki, and M. Yanagida. 1987. DNA topoisomerase II is required for condensation and separation of mitotic chromosomes in S. pombe. *Cell* **50:**917–925.

Uemura, T., and M. Yanagida. 1986. Mitotic spindle pulls but fails to separate chromosomes in type II DNA topoisomerase mutants: uncoordinated mitosis. *EMBO J.* **5:**1003–1010.

van Duin, M., J. de Wit, H. Odijk, A. Westerveld, A. Yasui, M. H. M. Koken, J. H. J. Hoeijmakers,

and D. Bootsma. 1986. Molecular characterization of the human excision repair gene *ERCC1*: cDNA cloning and amino acid homology with the yeast DNA repair gene *RAD10*. *Cell* **44**:913–923.

van Duin, M., A. Westerveld, and J. H. J. Hoeijmakers. 1985. UV stimulation of DNA-mediated transformation of human cells. *Mol. Cell. Biol.* **5**:734–741.

Weber, C. A., E. P. Salazar, S. A. Stewart, and L. H. Thompson. 1988. Molecular cloning and biological characterization of a human gene, *ERCC2*, that corrects the nucleotide excision repair defect in CHO UV5 cells. *Mol. Cell. Biol.* **8**:1137–1146.

Westerveld, A., J. H. J. Hoeijmakers, M. van Duin, J. de Wit, H. Odijk, A. Pastink, R. D. Wood, and D. Bootsma. 1984. Molecular cloning of a human DNA repair gene. *Nature* (London) **310**:425–428.

Willis, A. E., and T. Lindahl. 1987. DNA ligase I deficiency in Bloom's syndrome. *Nature* (London) **325**:355–357.

Willis, A. E., R. Weksberg, S. Tomlinson, and T. Lindahl. 1987. Structural alterations of DNA ligase I in Bloom syndrome. *Proc. Natl. Acad. Sci. USA* **84**:8016–8020.

Wilson, E. B., Jr. 1952. *An Introduction to Scientific Research*, p. 272. McGraw-Hill Book Co., New York.

Wolff, S. (ed.). 1982. *Sister Chromatid Exchange*. John Wiley & Sons, Inc., New York.

Wolff, S., J. Bodycote, and R. B. Painter. 1974. Sister chromatid exchanges induced in Chinese hamster cells by UV irradiation of different stages of the cell cycle: the necessity for cells to pass through S. *Mutat. Res.* **25**:73–81.

Wolff, S., J. Bodycote, G. H. Thomas, and J. E. Cleaver. 1975. Sister chromatid exchange in xeroderma pigmentosum cells that are defective in DNA excision repair or post-replication repair. *Genetics* **81**:349–355.

Wolff, S., and P. Perry. 1974. Differential Giemsa staining of sister chromatids and the study of sister chromatid exchanges without autoradiography. *Chromosoma* (Berlin) **48**:341–353.

Zwanenburg, T. S. B., and A. T. Natarajan. 1984. 3-Aminobenzamide-induced sister chromatid exchanges are dependent on incorporated bromodeoxyuridine in DNA. *Cytogenet. Cell Genet.* **38**:278–281.

Chapter 21

Illegitimate Recombination in Mammalian Cells

David Roth and John Wilson

I. INTRODUCTION

Mammalian cells readily integrate foreign DNA into their chromosomes. Regardless of how the DNA is introduced, the principal mode of integration deposits input DNA randomly—or at least widely—around the genome. This process, which we will refer to as random integration, requires little, if any, nucleotide sequence homology at the joints between foreign DNA and chromosomal DNA. The absence of significant homology indicates that random integration arises from illegitimate (or nonhomologous) recombination. Random integration is one of many kinds of illegitimate recombination. Operationally, illegitimate events are the grab bag of genetic rearrangements left over when homologous and site-specific events are set aside. In addition to random integration, other illegitimate recombination events occur during chromosome translocation, exon shuffling, and gene amplification, to name a few. Although these rearrangements of genetic information threaten the organism with genetic disease (for review, see M. Meuth, *in* D. E. Berg and M. M. Howe, ed., *Mobile DNA*, in press) and cancer (Yunis, 1983; Duesberg, 1987; Stenman et al., 1987), they also hold evolutionary promise for the species in the form of hybrid genes (Gilbert et al., 1986) and remodeled genomes (Schimke et al., 1986).

By examining random integration in some detail, we hope to illuminate general features of illegitimate recombination. Not that random integration is clearly understood—quite the contrary—but it serves to illustrate common principles and highlight some puzzles. The theme we will develop is that illegitimate recombination is basically a two-step process: DNA ends are generated in the first step and joined in the second. We present the idea that DNA ends arise primarily from errors of DNA metabolism, that is, as mistakes in replication, repair, recombination, or transcription (Franklin, 1971; Michel and Ehrlich, 1986). And we suggest that DNA ends are subsequently eliminated by sticking them together, regardless of the terminal sequences (Roth and Wilson, 1986; Chang and Wilson, 1987). From this perspective, end joining is the general defense mechanism in mammalian cells for dealing with broken chromosomes (Wake et al., 1985). This sequence-independent mechanism for break repair contrasts sharply with the mechanism in bacterial and yeast cells, which is more dependent on homologous recombination (Szostak et al., 1983; Stahl, 1986).

Random integration is interesting from another perspective as well: it provides a nuisance factor in any experiment designed to add foreign DNA to mammalian chromosomes by targeted (or homologous) recombination. In contrast to random integration, targeted recombination depends on extensive nucleotide sequence homology to align the exchange. In mammalian cells, targeted recombination is 100- to 10,000-fold less frequent than random integration (Fig. 1) (Folger et al., 1984; Lin et al., 1984, 1985; Smith and Berg, 1984; Smithies et al., 1984, 1985; Jasin et al., 1985; Thomas et al., 1986; Thomas and Capecchi, 1986, 1987; Doetschman et al., 1987; G. M. Adair, R. S. Nairn, M. M. Seidman, and J. H. Wilson, manuscript in preparation; see Subramani and Seaton, this volume; Kucherlapati and Moore, this volume). As a consequence, in the absence of selection for the homologous event, gene replacement and gene disruption experiments, which depend on homologous recombination, are difficult in mammalian cells. In essence, random integration creates a "haystack" of transformants that must be sorted through to find the rare targeted events. Although we will focus primarily on the haystack in this chapter, to set the stage, we will begin by comparing random integration with targeted recombination.

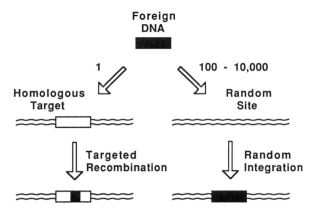

FIGURE 1. Integration of foreign DNA into the chromosome by either targeted recombination or random integration. Chromosomal DNA is represented by the pairs of wavy lines. A foreign linear DNA duplex is represented by the filled bar. The numbers beside the arrows represent the relative frequencies of targeted recombination and random integration.

II. BALANCE BETWEEN RANDOM INTEGRATION AND TARGETED RECOMBINATION VARIES ENORMOUSLY FROM *SACCHAROMYCES CEREVISIAE* TO MAMMALS

In contrast to mammalian cells, which usually incorporate DNA randomly, yeast cells (*S. cerevisiae*) target DNA to homologous chromosomal sites, with targeted events commonly outnumbering random events by more than 10 to 1 (Hinnen et al., 1978; Scherer and Davis, 1979; Rothstein, 1983). Thus, the ratio of targeted to random events goes from greater than 10:1 in yeast cells to less than 1:100 in mammalian cells. The >1,000-fold difference in these ratios is one reason that yeast genetics is such a pleasure and mammalian genetics is such a "challenge." Although unfortunate for the mammalian geneticists, the ratios raise a deeper question: why does targeting efficiency differ so dramatically from *S. cerevisiae* to mammals?

One question should be addressed at the outset: does targeted integration into mammalian chromosomes actually occur? Or do the rare, "apparently" targeted events really result from the accidental random integration of a sequence into its homologous counterpart in exactly the right way to reconstruct a functional gene? Calculations suggest that the observed frequency of targeted events is probably at least 1,000-fold above chance expectation. (For a 2-kilobase [kb] stretch of input DNA, the chance of an accidental targeted integration in a mammalian genome [2×10^6 kb] is 1 in 10^6, if integration is equally likely throughout the genome and the input DNA, or 1 in 10^9, if integration occurs only at the ends of the input DNA. Since about 1 in 1,000 integrations is targeted, the observed frequency of targeted events is 10^3- to 10^6-fold above chance expectation.) Thus, true targeted recombination does occur in mammalian cells. This conclusion is really no surprise since homologous recombination, upon which targeted events depend, has been amply demonstrated between foreign DNA molecules in cells, between homologous chromosomal sequences, and between homologous plasmids in cell-free extracts (see Subramani and Seaton, this volume; Kucherlapati and Moore, this volume).

Targeted recombination could be less efficient in mammalian cells than in *S. cerevisiae* simply because much more DNA must be searched to find a single homologous counterpart. In mammalian cells, a target segment of unique DNA is 200 times more dilute than in yeast cells, given the difference in genome size. Thus, the change in the ratio of targeted to random events could simply reflect the difference in genome size. In concert with this thinking, the slime mold (*Dictyostelium discoideum*), which has an intermediate genome size, incorporates foreign DNA with more equal numbers of targeted and random events (DeLozanne and Spudich, 1987). However, other fungi, such as *Schizosaccharomyces pombe* and *Neurospora crassa*, which have genomes of about the same size as that of *S. cerevisiae*, also generate more equal numbers of targeted and random events (Wright, Maundrell, and Shall, 1986; Case, 1986). Thus, there seems to be no simple correlation between genome size and frequency of random integration.

If targeted recombination in mammalian cells depended on the abundance of the target, one might expect that targeting efficiency would vary in a way that reflects the number of copies of the target. In *S. cerevisiae*, this expectation apparently is met since attachment of a segment of rDNA to a plasmid increases the frequency of transformation by 100- to 200-fold (Szostak and Wu, 1979) and there are about 140 ribosomal genes in *S. cerevisiae* (Schweizer et al., 1969). However, in mammalian cells no effect of target number has been demonstrated in studies using, as targets, "chromosomal plasmids" in the range of 1 to 5 copies (Thomas et al., 1986), ribosomal genes at about 200 copies (Steele et al., 1984), and repetitive sequences at a copy number around 100,000 (Wallenburg et al., 1987). Individually, none of these experiments is conclusive; collectively, they suggest that targeted recombination may be independent of target number and, by implication, that genome size is not solely responsible for the rarity of targeted recombination in mammalian cells.

If genome size is not the basis for the species difference in targeting efficiency, what is? The difference may lie in the enzymology of targeted recombination and random integration in yeast and mammalian cells. For example, one or more mammalian enzymes involved in homologous recombination (that is, in synapsis, heteroduplex formation, or crossover resolution) might be less effective than its yeast counterpart. Not enough is known about these enzymes in yeast and mammalian cells to evaluate this possibility. Alternatively, the enzymology for random integration might be much more effective in mammalian cells. As we shall discuss, mammalian cells have a remarkable capacity to join DNA ends together, and end joining is a likely step in random integration. By contrast, the direct joining of ends is rarer in yeast cells (Orr-Weaver and Szostak, 1983; Suzuki et al., 1983; Kunes et al., 1984).

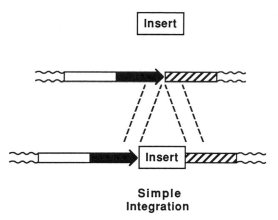

**Simple
Integration**

FIGURE 2. Generation of a simple insert by random integration of foreign linear DNA by end-to-end ligation at a chromosome break. A linear DNA duplex (marked "insert") is shown integrating by ligation to two chromosome fragments generated by a double-strand chromosome break. The two flanking sequences generated by breakage are designated by the filled arrow and the hatched box.

III. RANDOM INTEGRATION OCCURS THROUGH DNA ENDS

One important source of information about the mechanism of random integration is the structure of the chromosomal site before and after integration. To characterize an integration site fully, one must clone chromosomal DNA from both sides of the integrated DNA and then use the cloned flanking DNA to probe the organization of the undisturbed parental chromosome. If the flanks are not adjacent, one must walk along the chromosome to determine the linkage relationship. In the case of retroviral integration, which is under control of a viral protein (see Skalka, this volume), the flanking DNA is arranged as expected for a simple insertion into a chromosome (Fig. 2). However, this straightforward method, when applied to the random integration of foreign DNA, has met with two difficulties. First, it has sometimes proved difficult to clone a unique chromosomal sequence flanking the integration site, because of the structure of the junction or the presence of a repetitive element, and, therefore, impossible to identify the parental site of integration (Doerfler, 1982; Hasson et al., 1984; Covarrubias et al., 1986; Kato et al., 1986). Second, and more important, substantial chromosome rearrangements usually accompany integration of foreign DNA, making it difficult to decipher the arrangement of flanking sequences (Brinster et al., 1981; Wagner et al., 1981; Hayday et al., 1982; Stringer, 1982; Doerfler et al., 1983; Koshy et al., 1983; Covarrubias et al., 1986; Wallenburg et al., 1987; Wilkie and Palmiter, 1987). Because of these problems, only two random integration sites are fully characterized in mammalian cells (Gahlmann and Doerfler, 1983; Wilkie and Palmiter, 1987). Even so, the information from these and the many partially characterized sites offers several clues to the mechanism of integration.

A. Random Integration Sites Are Widely Distributed in the Genome

Cells that have integrated foreign DNA into their chromosomes commonly contain anywhere from 1 to more than 100 copies of the foreign DNA (Robins et al., 1981; Doerfler et al., 1983; Gordon and Ruddle, 1985). In general, the foreign DNA is inserted without internal deletion (Doerfler et al., 1983), although there are some notable exceptions involving extensive rearrangement of the integrated DNA (Doerfler et al., 1983; Ruley and Fried, 1983). Integration can occur at one or many different sites within the same cell (Perucho et al., 1980; Robins et al., 1981; Thomas et al., 1986). In general, high concentrations of DNA tend to give transformants carrying several copies of the foreign DNA, often linked together at one or a few sites; low concentrations of DNA tend to give transformants carrying fewer copies of the foreign DNA, which are often integrated as single copies at independent sites (Perucho et al., 1980; Robins et al., 1981; Boggs et al., 1986; Covarrubias et al., 1986; Thomas et al., 1986). There are many potential sites for random integration in a mammalian genome and no obvious preferred sites (Botchan et al., 1976; Ketner and Kelly, 1976; Robins et al., 1981; Scangos and Ruddle, 1981; Stringer, 1981, 1982; Mendelsohn et al., 1982; Doerfler et al., 1983; Wallenburg et al., 1987).

The distribution of integration sites is unlikely to be truly random, however, because the input DNA probably does not have equal access to all parts of the genome. The overrepresentation of repetitive sequences at integration sites argues for some unevenness in random integration (Kato et al., 1986; Wallenburg et al., 1984, 1987). Random integration may be influenced by normal metabolic activities, such as replication and transcription, which transiently expose chromosomal sequences, perhaps making them more accessible (Doerfler et al., 1983). The possible importance of chromosomal access has precedent in retroviral integration (Rohdewohld et al., 1987), immune system rearrangements (Blackwell et al., 1986), and homologous recombination in *S. cerevisiae* (Voelkel-Meiman et al., 1987), which have all been linked to transcriptional activity. Random integration could be favored by transcription (Schulz et al., 1987), by replication (Botchan et al., 1980; Chia and Rigby, 1981; Wilkie and Palmiter, 1987), or by other DNA metabolic processes.

B. Random Integration Is Stimulated by Free DNA Ends

Microinjected linear DNA molecules transform cultured cells 40-fold more efficiently than circles (Folger et al., 1982), and linear molecules are more efficiently integrated into mouse eggs (Palmiter et al., 1984). The difference between linear and circular molecules is not so obvious during transfection because the input DNA is broken on the way in (Calos et al., 1983; Razzaque et al., 1983; Wake et al., 1984). The estimate for DEAE-mediated transfection is one double-strand break per 5 to 15 kb (Wake et al., 1984), which would break a high fraction of input circular molecules. The structures of integration sites at which linear input molecules have joined to chromosomes suggest that DNA ends stimulate random integration directly. Most commonly, linear molecules are joined to chromosomal sequences at or near the original ends of the input DNA (Folger et al., 1982; Doerfler et al., 1983; Thomas et al., 1986). These results and others strongly suggest that ends, at least in the input DNA, play a crucial role in the integration reaction (Gusew et al., 1987).

The importance of ends is further supported by another observation. Occasionally, a few extra nucleotides of unknown origin are found exactly at the joint between the input and chromosomal sequences (Botchan et al., 1980; Gahlmann and Doerfler, 1983; Williams and Fried, 1986; Wilkie and Palmiter, 1987). Filler DNA, as it is sometimes called, is relatively common at illegitimate junctions in mammalian cells (Wilson et al., 1982; Neuberger and Calabi, 1983; Tsujimoto et al., 1985; Bakhashi et al., 1987; D. B. Roth and J. H. Wilson, submitted for publication). Filler DNA (called N regions) at VDJ joints in immune system rearrangements supports the idea that VDJ joining passes through an intermediate with free DNA ends (Alt and Baltimore, 1982; Desiderio et al., 1984; Landau et al., 1987). The presence of filler DNA at integration junctions suggests, likewise, that random integration involves free DNA ends, to which extra nucleotides could be added by polymerization or ligation (Roth and Wilson, submitted).

C. Random Integration Requires Little or No Sequence Homology at Junctions between Foreign DNA and Chromosomes

When foreign DNA integrates into a chromosome, it generates a rather distinctive junction, usually with an abrupt boundary between foreign and chromosomal DNA; that is, the sequences are not scrambled at the junction (Botchan et al., 1980; Deuring et al., 1981; Stringer, 1981, 1982; Stabel and Doerfler, 1982; Doerfler et al., 1983; Hasson et al., 1984). As is typical of illegitimate junctions in mammalian cells, the homology exactly at a junction is generally less than five nucleotides (Roth et al., 1985; Meuth, in press). Outside the junction, foreign DNA and chromosomal DNA show very little similarity to each other, except for short patches of complementary nucleotides (Stringer, 1982; Doerfler et al., 1983). This so-called patchy homology, which flanks these and other illegitimate junctions, is within statistical expectations; therefore, its presence in the neighborhood of a junction does not necessarily indicate an involvement in the recombination process (Gutai and Nathans, 1978; Doerfler et al., 1983; Savageau et al., 1983). Since DNA ends are implicated in random integration, it is particularly noteworthy that end joining in transfected DNA molecules yields junctions that are indistinguishable from those generated by random integration (Roth et al., 1985; Roth and Wilson, 1986).

D. Random Integration Involves Rearrangement of Chromosomal Sequences at the Site of Integration

With one exception (Gahlmann and Doerfler, 1983), all the chromosomal sites of integration that have been examined (more than 100) have undergone a rearrangement of chromosomal sequences (Brinster et al., 1981; Wagner et al., 1981; Hayday et al., 1982; Stringer, 1982; Doerfler et al., 1983; Koshy et al., 1983;

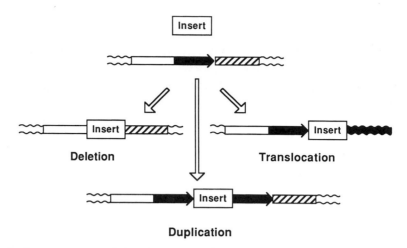

Duplication

FIGURE 3. Rearrangements of the chromosomal insertion site that usually accompany random integration of foreign DNA. Three types of chromosomal rearrangement are diagrammed: deletion (the filled arrow represents the segment deleted from the parental chromosome), duplication (the filled arrow in the parental DNA is duplicated in the recombinant), and translocation. Symbols are described in the Fig. 2 legend.

Covarrubias et al., 1986; Wallenburg et al., 1987; Wilkie and Palmiter, 1987). Many of these sites involve viral DNAs—simian virus 40 (SV40), polyomavirus, or adenovirus—which are unstable under certain conditions and can generate rearrangements after integration (Clayton and Rigby, 1981; Hiscott et al., 1981; Doerfler et al., 1983). However, the same patterns of rearrangements are seen with transfected and microinjected DNA as well (Brinster et al., 1981; Wagner et al., 1981; Covarrubias et al., 1986; Wallenburg et al., 1987; Wilkie and Palmiter, 1987). Thus, chromosomal rearrangements are probably generated during the integrative process.

Several simple rearrangements—deletion, duplication, inversion, and translocation—could be generated during random integration by the inserted segment (Fig. 3 and 4). The rearrangement at one characterized site carries a 5-kb duplication of the chromosome at the site of integration (Wilkie and Palmiter, 1987). In other cases, chromosomal DNA may be deleted during integration, but the "gaps" have not been defined by chromosome walking in the parental genome (Botchan et al., 1980; Hayday et al., 1982; Stringer, 1981, 1982; Wallenburg et al., 1987). Minimum estimates of the size of the gap range from 3 kb to well over 100 kb (Stringer, 1982; Wallenburg et al., 1987). Because the strategy for chromosome walking (toward or away from the insert) varies with the type of rearrangement (Fig. 4), it is prudent to walk in both directions along the parental chromosome when trying to define a rearrangement. Of course, if the foreign DNA becomes integrated by linking broken chromosomes in the act of translocation, chromosome walks can never meet. When random integration was analyzed by cytological techniques, visible chromosomal rearrangements were detected (Robins et al., 1981), but no random integration site has yet been assigned definitely to a

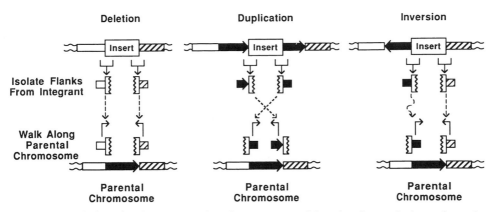

FIGURE 4. Variations in the strategy for chromosome walking that is required to clone the unrearranged insertion site from the parental chromosome. For each rearrangement, the first step is to isolate the chromosomal DNA that flanks the insert, by using the insert as a probe. Once the flanks are isolated, as shown, they are used as probes to walk along the parental chromosome in the direction of the arrows.

translocation (Croce and Koprowski, 1975; Robins et al., 1981; Woychik et al., 1985; Covarrubias et al., 1987).

To summarize, foreign DNA integrates through its ends at many sites in chromosomes, leaving simple junctions with limited homology and the occasional presence of extra nucleotides. These junctions are indistinguishable from those formed by the joining of foreign DNA ends outside the chromosome. Most peculiarly, random integration is associated with substantial rearrangements of DNA at the site of integration. Chromosome rearrangement is the singular puzzle of random integration. Its solution may hold the key to the mechanism of random integration. We will return to this puzzle, but we will first consider the variety of ways that DNA ends can be generated and joined in mammalian cells.

IV. MANY ERRORS OF DNA METABOLISM GENERATE ENDS

DNA is in constant metabolic turmoil; it is periodically replicated, continually transcribed, frequently repaired, and often recombined. These processes are extremely accurate, but none is perfect. Many potential errors of DNA metabolism can produce free DNA ends, as summarized below.

A. Replication

There are several ways in which errors during DNA replication can generate ends. As shown in Fig. 5, reannealing of the parental strands, reinitiation at an origin, or breakage of the same parental strand at diverging replication forks each could release a linear chromosomal fragment, yet leave the chromosome intact. Reannealing of parental strands accompanied by the melting out of daughter

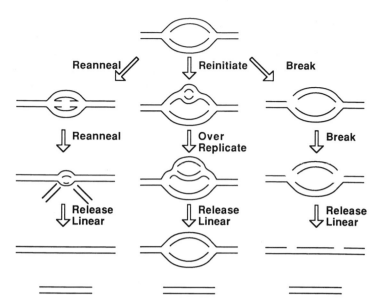

FIGURE 5. Generation of free linear DNA molecules by errors during DNA replication. A replication fork is shown at the top of the figure. In the scheme shown on the left, the newly synthesized strands anneal, resulting in release of a linear duplex. In the middle pathway, overreplication occurs on one of the daughter duplexes; when the two replication forks meet, a linear duplex is released. In the pathway shown on the right, a nick occurs on one side of the replication fork. A subsequent nick on the other side of the replication bubble results in release of a linear molecule.

strands is a process known to occur during some DNA isolation procedures but is as yet unproven in vivo (Zannis-Hadjopoulos et al., 1981). Overreplication (reinitiation) at an origin has been observed as a programmed developmental step in the amplification of chorion genes in *Drosophila melanogaster* (Spradling, 1981; de Cicco and Spradling, 1984) and during excision of integrated SV40 sequences: the latter is the original observation of "onion skin" replication (Botchan et al., 1979). Overreplication has also been proposed as an initial step in gene amplification (Stark and Wahl, 1984). Breakage at replication forks may be much more frequent than breakage elsewhere because single strands are exposed at the fork. Thus, movement of the fork across a single-strand nick, introduction of a nick into the single-stranded region, or the failure of topoisomerase I to close a transient nick could yield double-strand breaks.

B. Repair

Mammalian cells repair abnormal structures in duplex DNA very efficiently. Potential repair events (many of which have been demonstrated) include: (i) repair of mismatched base pairs, which could be generated by recombination or by misincorporation of a base at a replication fork (Miller et al., 1976; Wake and Wilson, 1980; Abastado et al., 1984; Folger et al., 1985; Hare and Taylor, 1985); (ii) repair of the many different kinds of base modification that result from environmental insult (Cornforth and Bedford, 1983; Brash et al., 1987); and (iii) repair of single-strand loops, which can occur by slipped mispairing at a replication fork or as a heteroduplex intermediate in homologous recombination (Abastado et al., 1984; Hasson et al., 1984; Thomas and Capecchi, 1986; Ayares et al., 1987; Weiss and Wilson, 1987). These repair events involve breakage of phosphodiester bonds in DNA. If a repair-introduced break is close to a preexisting nick on the opposite strand, a chromosome break can result. Alternatively, if repair is initiated on both strands at the same time, breaks can also occur. Simultaneous repair on both strands, accompanied by formation of double-strand breaks, has been proposed to underlie the low viability of multiply mismatched heteroduplexes introduced into bacteria (Doutriaux et al., 1986; see Radman, this volume). It has also been suggested as the basis for formation of deletions between mismatch sites that are generated when multiply mismatched heteroduplexes are introduced into mammalian cells (Weiss and Wilson, 1988).

Other abnormal or unusual duplex structures may trigger repair systems inappropriately, leading to strand breakage. For example, palindromes, which can form hairpin structures by intrastrand base pairing, have been identified adjacent to sites of homologous and illegitimate recombination (Lehrman et al., 1985; Krawinkel et al., 1986; Nicholls et al., 1987; Hyrien et al., 1987). Similarly, the boundary regions between B- and Z-DNA may also be sensitive to breakage, which could initiate homologous or illegitimate recombination events (Slightom et al., 1980; Nordheim and Rich, 1983; Stringer, 1985; Bullock et al., 1986; Kato et al., 1986).

C. Recombination

Both homologous and site-specific recombination occur in mammalian cells (see Subramani, this volume; Engler and Storb, this volume; Thompson, this volume; Kucherlapati and Moore, this volume). (i) Recombination between homologous sequences is thought to occur by an orderly process involving pairing (to sense homology), breakage of four single strands, and their reunion to form recombinant duplexes (Meselson and Radding, 1975; Szostak et al., 1983). The exact order of breakage, pairing, and reunion is unclear and may not be uniform from species to species. If any of the single-strand breaks were near preexisting nicks on the opposite strand, a chromosome break could be generated. (ii) Developmental rearrangements in the immune system are thought to proceed through double-strand breaks (Alt and Baltimore, 1982; Desiderio et al., 1984; Landau et al., 1987; see Engler and Storb, this volume). If the broken ends escape the recombination machinery, they could participate in abnormal rearrangements. Indeed, many well-characterized tumor cell translocations involve just such rearrangements.

D. Transcription

In principle, transcription could also lead to an increased frequency of double-strand breaks, since single strands are exposed during the transit of RNA polymerase. However, the length of exposed single strand is much less extensive than during replication, and breaks in both strands would be needed to cause a chromosome break. Transcription may generate DNA ends more commonly via the indirect route of reverse transcription of RNA molecules into DNA. Formation of processed pseudogenes and the dispersal of certain repetitive sequences are thought to involve reverse transcripts (Lemischka and Sharp, 1982; Katzir et al., 1985; Weiner et al., 1986).

This list of metabolic errors is not all inclusive. There are undoubtedly other end-generating errors associated with these or other aspects of chromosome function. Nor are metabolic errors the sole source of DNA ends in mammalian cells: certain kinds of DNA damage (X rays, for example) also generate DNA breaks, and shear forces inside cells may occasionally snap the duplex (Cornforth and Bedford, 1983; McClintock, 1984). The important point is that ends, regardless of how they are generated, become substrates for a very active repair system, which sticks ends together with remarkable facility. End joining seems to be the dominant strategy used by mammalian cells to repair broken chromosomes. In combination, the generation and joining of ends—a simple, if nonspecific, two-step pathway—can produce new arrangements of genetic information, and may account for much of the illegitimate recombination in mammalian cells.

V. MAMMALIAN CELLS CAN JOIN VIRTUALLY ANY TWO DNA ENDS TOGETHER

Half a century ago, it was noted that broken chromosomes rejoin efficiently, that is, that broken ends are, in some sense, sticky (McClintock, 1938, 1984). Related observations at the molecular level initially showed that transfected DNA molecules often are joined together prior to chromosome integration (Perucho et al., 1980). More recently, in vivo studies of end joining in mammalian cells have revealed an unexpected versatility in joining mechanisms (Roth and Wilson, 1986). Cells not only can ligate blunt or complementary restriction enzyme-generated ends, but also they can join mismatched ends by single-strand ligation, which uses no homology, or after terminal pairing, which uses one to five nucleotides of homology (Roth and Wilson, 1986). These two reaction pathways are novel and may be catalyzed by enzymes that are present at much reduced levels in bacterial and yeast cells (Orr-Weaver and Szostak, 1983; Suzuki et al., 1983; Conley and Saunders, 1984; Kunes et al., 1984). As if this were not enough, topoisomerase I, which is implicated in chromosome breakage, may also be involved in the joining of ends (Bullock et al., 1985; see Champoux and Bullock, this volume). Thus, mammalian cells have a remarkable ability to join ends.

Are results that are obtained in established cell lines valid for cells in their native state? This question is especially pertinent, because most established cell lines are aneuploid and contain a highly rearranged chromosome complement. Is high efficiency end joining induced (or selected for) by the requirements of life in culture, perhaps as a prerequisite for chromosome rearrangement? This potential complication has been dealt with experimentally by comparing end joining in an established line of monkey cells (CV1) to end joining in primary AGMK cells, from which the CV1 cell line was derived. The results show that these two cell lines have identical end-joining capabilities (H. Zheng, X.-B. Chang, and J. H. Wilson, submitted for publication). Thus, the ability of mammalian cells to join ends appears to be a native property and not a peculiar adaptation to culture conditions.

Although many systems have contributed to our knowledge of end joining in mammalian cells (Miller et al., 1976; Folger et al., 1982; Miller and Temin, 1983; Kopchik and Stacey, 1984; Brinster et al., 1985; Munz and Young, 1987), end joining has been studied most intensively with SV40 genomes transfected into cultured monkey cells (Mertz and Davis, 1972; Wilson, 1977; Subramanian, 1979; Upcroft et al., 1980; Johnson et al., 1982; Wilson et al., 1982; Wake et al., 1984; Roth and Wilson, 1985, 1986; Roth et al., 1985; Chang and Wilson, 1987; Zheng et al., submitted). Since these studies are in agreement, we will describe only the SV40 results. The genome of SV40, which is a small (5.2-kb) double-stranded DNA circle (Tooze, 1980), offers several advantages for studying cell-mediated extrachromosomal end joining. (i) When SV40 is introduced as a linear molecule, viral gene expression and the viral life cycle cannot be initiated until the ends are joined. Thus, the end-joining capabilities of the cells, not the virus, are being assessed. (ii) Since SV40 forms plaques, end joining can be assessed by plaque assay, which allows a statistical analysis. The plaques also serve as a ready source of clones of individual recombinants, which can be examined in detail. (iii) The intron in the SV40 T-antigen gene provides a convenient stretch of DNA, about 350 nucleotides long, in which to assess end joining (Thimmappaya and Shenk, 1979; Volckaert et al., 1979). Since the intron is nonessential for lytic infection, no particular sequence must be reconstructed by end joining. Thus, a broad spectrum of end-joining events within the intron can be recovered as viable virus.

SV40 has not only been valuable for analyzing end-joining mechanisms, but it has also long been a favorite experimental tool for studying integration and excision from chromosomes. Elegant early studies demonstrated widespread chromosomal integration (Croce and Koprowski, 1975; Botchan et al., 1976; Ketner and Kelly, 1976), described the peculiar chromosomal rearrangements that accompany random integration (Botchan et al., 1980; Stringer, 1982), defined the phenomenon of onion skin replication (Botchan et al., 1979), and showed that targeted recombination into homologous chromosomal sequences was rare (Botchan et al., 1979). More recent studies suggest that excision of SV40-linked DNA from chromosomes may involve topoisomerase I at the breakage step (Bullock et al., 1984, 1985; see Champoux and Bullock, this volume). Thus, a broad base of observations made with the SV40 system exists.

VI. MAMMALIAN CELLS JOIN ENDS BY DIRECT LIGATION AND BY USING SHORT SEQUENCE HOMOLOGIES

The mechanism of end joining in monkey cells was studied in detail by using the experimental strategy outlined in Fig. 6 (Roth and Wilson, 1986). The goal was

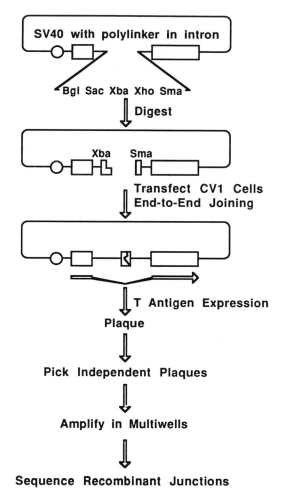

FIGURE 6. Substrates for end joining produced by cleaving SV40 DNA in the intron of the T-antigen gene. The large rectangles represent the exons encoding T antigen, and the circle represents the origin of replication. The small boxes on the ends of the linear molecule represent the structures of the ends produced by restriction digestion; the *Sma*I end is blunt, and the *Xba*I end has a four-nucleotide 5' extension.

to examine a large number of products of end joining and to deduce features of the end-joining mechanism by comparing recombinant junctions with the structures of the input linear molecules. By placing a polylinker in the intron of the T-antigen gene, a variety of linear molecules bearing mismatched ends could be readily created by digestion with different pairs of restriction enzymes. The resulting linear SV40 genomes were transfected into CV1 monkey cells, which circularized the molecules by joining the ends together. Because end joining reconstructed the T-antigen gene, viable genomes were generated, thereby allowing plaques to form. Independent plaques (from separate plates) were picked and amplified, and

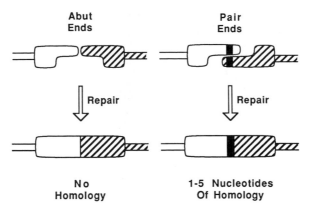

FIGURE 7. Two mechanisms by which free DNA ends are joined: single-strand ligation and pairing of short homologies. The cross-hatched and open boxes denote the two "parental" ends involved in joining. On the left, the two single-strand extensions can be abutted and ligated directly by single-strand ligation, followed by repair. This process generates a junction with zero nucleotides of homology. On the right, the single-strand extensions are of opposite polarity, so they cannot be ligated directly. These ends can be joined by a mechanism involving formation of 1 to 5 base pairs (shaded area), followed by repair. The resulting junction exhibits one to five nucleotides of homology, corresponding to the nucleotides used in pairing.

the nucleotide sequences across the end-join junctions were determined (Roth and Wilson, 1986).

The nucleotide sequences of 199 junctions, which were derived from seven linear substrates with different pairs of mismatched ends, revealed that 97% of the end-joining events are within 15 nucleotides and 83% are within 5 nucleotides of the ends of the input DNA molecules (Roth and Wilson, 1986). More than 60% of the junctions involve either the terminal nucleotide of a blunt end or the single-stranded portion of ends with a 5' or 3' overhang (Roth and Wilson, 1986). Thus, end joining does not require extensive modification of the ends. The retention of the intron sequences at the input ends, despite their dispensability for viral growth, emphasizes the role of the ends in the joining reaction.

Given the capacity of mammalian cells to join blunt ends (Subramanian, 1979; Upcroft et al., 1980; Miller and Temin, 1983; Wake et al., 1984), one simple, general strategy for joining any pair of mismatched ends would be to make the ends flush—either by extending the short strand or exonucleolytically shortening the long strand—and then to join the ends of blunt-end ligation (Subramanian, 1979). However, in monkey cells this mechanism, although seemingly reasonable, is rarely used and accounts for less than 2% of the analyzed junctions (Roth and Wilson, 1986). Instead, the predominant mechanism for joining any particular pair of ends depends on the structural features of the termini (Roth and Wilson, 1986). If single strands from the two ends can be abutted (a 5' extension opposite a 3' extension or either extension opposite a blunt end) the major species of junction (50 to 70% of all junctions) appears to arise by direct single-strand ligation (Roth and Wilson, 1986). This mechanism is illustrated in Fig. 7 for a DNA molecule with a 5' extension and a 3' extension. Since 3' extensions cannot be filled in by any known DNA polymerase, yet the sequence of the extension is retained in the product, the joining reaction presumably involves at least one single strand. The 3' extension joins directly either to the 5' extension or to the blunt end formed by filling in the 5' extension. Although single-strand ligation to another single strand or to a blunt end has not been demonstrated for any characterized DNA ligase (Lehman, 1974; Zimmerman and Pheiffer, 1983), bacteriophage T4 RNA ligase is capable of ligating single strands of DNA (McCoy and Gumport, 1980).

If the ends cannot be abutted—that is, if the linear molecule has two 3' extensions or two 5' extensions—they are joined by mechanisms that appear to involve pairing of short, one- to five-nucleotide homologies (Roth and Wilson, 1986). This mechanism is illustrated for two 3' extensions in Fig. 7, which shows the most common form of joining, that is, joining mediated by pairing between complementary nucleotides in the single-strand extensions. Short sequence pairing provides an explanation for the frequent appearance of one to five nucleotides of homology at illegitimate recombination junctions in mammalian cells (Roth et al., 1985; Roth and Wilson, 1986). The role that these homologies play in the joining reaction is unclear, since pairing between such short homologies (one- and two-nucleotide homologies are the most common) provides insufficient stability to hold the duplexes together. Thus, it seems likely that proteins in some way enhance the stability of such end associations until they can be joined covalently. Activities that apparently juxtapose DNA ends have been observed in frog oocytes (Bayne et al., 1984) and in HeLa cell extracts (Roth and Wilson, unpublished data). Whether either of these activities participates in joining DNA ends is not yet known.

Another mechanism of end joining has also been proposed, based on studies of excision of SV40 DNA from chromosomes (see Champoux and Bullock, this volume). When SV40-transformed cells are fused with permissive cells, molecules containing the SV40 origin of replication are released from chromosomes and exist extrachromosomally as circles (Gluzman, 1981; Conrad et al., 1982; Miller et al., 1984). The sites of intracellular circularization are enriched for the most common topoisomerase I cleavage sites, CTT and GTT (Been et al., 1984; Edwards et al., 1982; Bullock et al., 1984). The abortive action of topoisomerase I during replication—to cleave but not rejoin—could liberate linear DNA from the chromosome (Bullock et al., 1984, 1985). If topoisomerase I is attached to one end of the linear molecule, it could then circularize the molecule by carrying out an intramolecular rejoining of the ends, as has been demonstrated in vitro (Been and Champoux, 1981; Halligan et al., 1982). This mechanism has experimental support in bacteria, where it is thought to underlie one class of illegitimate recombination events (Ikeda, 1986; see Allgood and Silhavy, this volume). Whether topoisomerase I joins ends in mammalian cells is less certain. The presence of two- to three-nucleotide homologies at the six characterized junctions is unexpected on the basis of a direct end-to-end joining by topoisomerase I and suggests that short homologies may stabilize the joining reaction (Bullock et al., 1984, 1985). If short homologies are used (Bullock et al., 1985), then topoisomerase I could be bypassed by the end-joining activity discussed above (Fig. 7). (Because topoisomerase I adds to DNA by cleavage in the interior of a strand rather than at the ends, it is not likely to be responsible for joining transfected DNA ends, which are usually preserved in the joining reaction [Halligan et al., 1982].)

The capacity for eucaryotic cells to join mismatched ends efficiently is not limited to mammalian cells. Recently, in vitro studies using extracts from frog eggs demonstrated that efficient end joining also occurs in amphibians (Pfeiffer and Vielmetter, 1988). Curiously, this end-joining activity is present only in fully developed eggs and is not detected in vitro or in vivo in frog oocytes (Grzesiuk and Carroll, 1987; Pfeiffer and Vielmetter, 1988). Not only is end joining efficient in eggs, but the mechanism, as deduced from sequence analysis of junctions (Pfeiffer and Vielmetter, 1988), is virtually indistinguishable from the mechanism in monkey cells (Roth and Wilson, 1986). In both frog eggs and monkey cells, the mechanism of joining depends on the structure of the ends: ends that can be abutted are joined by single-strand ligation, and ends that cannot be abutted are joined by pairing of short homologies (Fig. 7) (Roth and Wilson, 1986; Pfeiffer and Vielmetter, 1988). Such a similar pattern of end joining in amphibians and mammals suggests that these end-joining mechanisms are very widespread in multicellular eucaryotes. This comparison also emphasizes the general biological relevance and evolutionary importance of direct end joining.

VII. NOVEL LIGATION MECHANISMS MAY BE RESPONSIBLE FOR FILLER DNA

The sequences of more than 500 illegitimate recombination junctions from a variety of mammalian cells have been determined. Approximately 10% of the junctions generated in nonlymphoid cells contain extra nucleotides of uncertain origin, termed filler DNA (Roth et al., 1985; Roth and Wilson, 1986; Meuth, in press; Roth and Wilson, submitted). Filler DNA is also present at about half of the reciprocal chromosome translocations associated with lymphoid cancers, many of which apparently result from mistakes accompanying immune rearrangements (Neuberger and Calabi, 1983; Tsujimoto et al., 1985; Bakhashi et al., 1987; Croce, 1987). Thus, filler DNAs are common at illegitimate junctions in mammalian cells.

Filler DNAs, known as N regions, are also present at 5 to 90% of the junctions between V, D, and J gene segments in immune system rearrangements (Landau et al., 1987; Concannon et al., 1986). N regions are thought to be formed by the template-independent addition of nucleotides to broken 3' ends by the enzyme terminal transferase (Landau et al., 1987; Alt and Baltimore, 1982). Filler DNAs at illegitimate junctions, as well, are probably generated by addition of nucleotides to broken ends. This hypothesis is supported by the frequent "trapping" of extra nucleotides precisely between the ends of linear DNA molecules transfected into monkey cells (Roth et al., 1985; Roth and Wilson, 1986, submitted). However, filler DNAs produced in nonlymphoid cells are unlikely to be due to terminal transferase, since its expression appears to be confined to cells of the lymphoid lineage (Chang, 1971).

A comparison of filler DNAs at 40 junctions from nonlymphoid cells with 97 N regions reveals several differences that are consistent with different mecha-

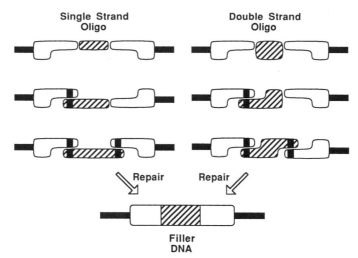

FIGURE 8. Insertion of filler DNA by ligation of oligonucleotides onto broken ends. The hatched bars represent oligonucleotides, the open bars denote broken DNA ends with short single-strand extensions, and the filled bars indicate the continuation of the chromosomal DNA. The scheme on the left depicts incorporation of single-stranded oligonucleotides by single-strand ligation (top), by a combination of single-strand ligation and pairing (shaded areas) of short sequence homologies (middle), and by pairing alone (bottom). The scheme on the right illustrates the same three types of ligation using double-stranded oligonucleotides. Repair events complete the process of filler DNA incorporation.

nisms for addition (Roth and Wilson, submitted). Filler DNAs from nonlymphoid cells show a distinct peak of 1-nucleotide additions (40% of the junctions), but a relatively flat distribution from 2 to 40 nucleotides, with 25% longer than 14 nucleotides. By contrast, N regions decrease progressively in number with increasing length, with none longer than 13 nucleotides. Furthermore, the overall G+C content of N regions is 57%, which is consistent with the in vitro preference of terminal transferase for addition of G nucleotides (Alt and Baltimore, 1982), whereas the overall G+C content of filler DNAs from nonlymphoid cells is only 37%. Filler DNAs from 18 oncogenic translocations in lymphoid cells have intermediate characteristics and, therefore, may represent a mixture of the different mechanisms responsible for addition of extra nucleotides (Roth and Wilson, submitted).

In filler DNAs from nonlymphoid cells, the peak of 1-nucleotide additions and the flat distribution from 2 to 40 nucleotides may represent different mechanisms (Roth and Wilson, submitted). One likely mechanism that could account for the flat distribution is the addition of preformed blocks of single- or double-stranded oligonucleotides to DNA ends by the highly efficient ligation mechanisms present in mammalian cells. As shown in Fig. 8, oligonucleotides could be joined to ends by direct ligation or by pairing of short homologies followed by repair (see Fig. 7). The recent observation that mammalian cells contain a pool of oligonucleotides with a length distribution similar to that of filler DNA (Plesner et al., 1987) suggests that endogenous oligonucleotides may provide one source of extra nucleotides. Indeed, half of the filler DNAs longer than 10 nucleotides seem to be derived from other DNA sequences in the cell (Anderson et al., 1984; Mager et al., 1985; Williams and Fried, 1986).

VIII. HOW MIGHT END JOINING ACCOUNT FOR RANDOM INTEGRATION?

Although foreign DNA can integrate at a simple break (Doerfler et al., 1983), it is far more common, almost universal, for integration to cause a chromosomal rearrangement (Brinster et al., 1981; Wagner et al., 1981; Hayday et al., 1982; Stringer, 1982; Doerfler et al., 1983; Koshy et al., 1983; Covarrubias et al., 1986; Wallenburg et al., 1987; Wilkie and Palmiter, 1987). Since there are so few characterized integration sites, it is difficult to know exactly how rearranged a typical integration site is. The simple kinds of rearrangement that might accompany integration are deletions, duplications, inversions, and translocations (Fig. 3 and 4). There is good evidence for integrations at deletions and duplications (Wallenburg et al., 1987; Wilkie and Palmiter, 1987) and no evidence to rule out integrations involving inversions or translocations (Robins et al., 1981; Stringer, 1982). The one fully characterized rearrangement is a duplication of 5 kb, which was interpreted as an integration in a replication bubble (Wilkie and Palmiter, 1987). Integration into a replication bubble to produce a duplication or a deletion is shown in Fig. 9. Because replicating loops are packed together along the chromosome axis (Razin et al., 1986), the same types of rearrangement (only much larger) would occur if the breaks were in different replication bubbles (Fig. 10).

Replicating forms of SV40 DNA have been suggested as active species in integration (Botchan et al., 1980; Chia and Rigby, 1981), but chromosomal replication may not be a universal requirement for random integration. If it were, one might expect that random integration would occur most frequently during S phase. For circular input DNA, random integration does peak during S phase (Giulotto and Israel, 1984; Wong and Capecchi, 1985); however, for linear input DNA the frequency of random integration does not vary substantially during the cell cycle (Wong and Capecchi, 1985). These observations are difficult to interpret, but they may indicate that there are mechanisms other than replication for generating chromosome ends to which foreign DNA can attach. For example, all of four adenovirus integration events occurred at active sites of transcription in the parental chromosome (Schulz et al., 1987), raising the possibility that transcription (or the associated changes in chromatin structure) may also increase chromosome breakage (Fig. 10).

Although rearrangements at the site of integration can be understood as shown in Fig. 9 and 10, the diagrams do not account for the very low frequency (one example) of simple integration events, in which the foreign DNA bridges the two ends of a single break (Gahlmann and Doerfler, 1983). This is the real puzzle of random integration: why are there so many rearrangements and so few simple insertions? One possibility is that integration usually is simple, but the site of

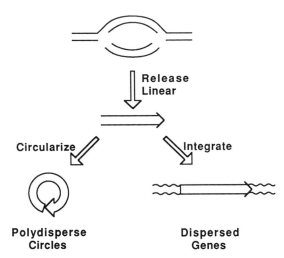

FIGURE 11. Replication bubbles as a potential source of polydisperse circles and dispersed genes. A replication bubble can generate a linear duplex by any of several mechanisms (see Fig. 5). The linear molecules may then circularize by intramolecular ligation, thereby giving rise to circular forms (as shown on the left). Alternatively, the linear species could integrate elsewhere in the genome by illegitimate recombination, giving rise to dispersed genes, as shown on the right.

could result from the imprecise rejoining of a chromosome after a single break—a situation analogous to the intramolecular joining of the ends of transfected DNA molecules. And the large deletions could result from rejoining the wrong partners in a pair of chromosome breaks. In both classes of deletion the junctions are the same as those formed by end joining in transfected molecules: they have zero to seven nucleotides of homology at the junction, there is very little homology outside the junction, and filler DNA is present at similar frequencies (Roth et al., 1985; Nalbantoglu et al., 1986; Roth and Wilson, 1986; Ashman and Davidson, 1987; Meuth, in press; Roth and Wilson, submitted).

Illegitimate recombination plays a key role in gene amplification, the dispersal of genes and exons, and the formation of extrachromosomal circular DNA. These examples may all be linked to replication. Replication bubbles are a potential source of linear chromosome fragments (Fig. 5), which are very powerful agents for promoting genetic rearrangements. It seems likely that the diverse set of circular DNA molecules that exists in virtually all mammalian cell populations (Rush et al., 1971; DeLap et al., 1978; Yamagishi et al., 1982; Schindler and Rush, 1985; Wiberg et al., 1986; Maurer et al., 1987) is generated by intramolecular end joining of linear molecules that are released from chromosomes, probably at replication bubbles (Fig. 11). These circles, which are derived from many places in the genome, are present at concentrations up to a few thousand per cell (DeLap et al., 1978). Their function, if any, is unknown. However, the same processes that generate these circles may underlie gene amplification (Stark and Wahl, 1984; Schimke et al., 1986; Maurer et al., 1987). Linear chromosome segments formed by onion skin replication (Fig. 5) could join

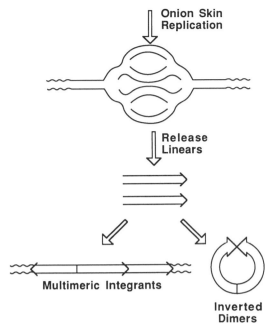

FIGURE 12. Effects of onion skin replication. Onion skin replication can give rise to free linears by the mechanism diagrammed in Fig. 5. These linear molecules can then join end-to-end to form multimers, which could integrate elsewhere in the genome by illegitimate recombination, as shown on the left. Linears could also join end-to-end to give rise to inverted dimers, as shown on the right.

together end-to-end (Fig. 12) to give the initial burst of amplification that seems to be an early step in the process of gene amplification (Varshavsky, 1981; Zieg et al., 1983; Stark and Wahl, 1984; Schimke et al., 1986). However, linear molecules might also join end-to-end to create an inverted dimer (Fig. 12), another attractive intermediate in gene amplification (Ford and Fried, 1986; Nalbantoglu and Meuth, 1986; Saito and Stark, 1986; Looney and Hamlin, 1987; Hyrien et al., 1988). By analogy with the *S. cerevisiae* 2μm circle, an inverted dimer could amplify even while remaining under cell cycle control (Futcher, 1986; Giulotto et al., 1986; see Cox, this volume). Finally, the linear molecules generated by replication (Fig. 5) could also participate directly in the dispersal of genes and exons during evolution, by integrating into the chromosome in much the same way foreign DNA molecules do (Fig. 12) (Schimke et al., 1986). If such integrations also cause chromosomal rearrangements, then dispersal might also be evolutionarily important as a random rearrangement generator.

Illegitimate recombination may be a fundamental step in the genesis of cancer. Many types of cancer have now been identified with particular chromosome translocations (Yunis, 1983; Schimke, 1984; Croce, 1987; Duesberg, 1987). The majority of the characterized translocations are linked to errors in the programmed rearrangements that occur in the immune system. These translocations are proposed to occur by chromosome breakage followed by end joining

ACKNOWLEDGMENTS. We are especially grateful to Susan Berget and Sharon Y. Roth for many valuable suggestions on the structure and organization of this chapter. We also thank Sandra Pennington, Tom Porter, and Neal Proctor for their helpful comments and criticisms. Finally, we thank Jeffrey Cossman, Walter Doerfler, Mark Meuth, Michael Seidman, and James Stringer for taking the time to read and comment on the chapter. We are indebted to Mark Meuth for sharing his excellent chapter on illegitimate recombination with us prior to publication.

This work was supported by Public Health Service grants CA15743 and GM38219 from the National Institutes of Health and by grant Q-977 from the Robert A. Welch Foundation. D.R. was in the Medical Scientist Training Program and was supported by the Edward J. and Josephine Hudson Scholars Fund.

LITERATURE CITED

Abastado, J.-P., B. Cami, T. H. Dinh, J. Igolen, and P. Kourilsky. 1984. Processing of complex heteroduplexes in E. coli and COS-1 monkey cells. *Proc. Natl. Acad. Sci. USA* **81**:5792–5796.

Alt, F. W., and D. Baltimore. 1982. Joining of immunoglobulin heavy chain gene segments: implications from a chromosome with evidence of three D-J_H fusions. *Proc. Natl. Acad. Sci. USA* **79**:4118–4122.

Anderson, R. A., S. Kato, and R. D. Camerini-Otero. 1984. A pattern of partially homologous recombination in mouse L cells. *Proc. Natl. Acad. Sci. USA* **81**:206–210.

Ashman, C. R., and R. L. Davidson. 1987. Sequence analysis of spontaneous mutations in a shuttle vector gene integrated into mammalian chromosomal DNA. *Proc. Natl. Acad. Sci. USA* **84**:3354–3358.

Ayares, D., D. Ganea, L. Chekuri, C. R. Campbell, and R. Kucherlapati. 1987. Repair of single-stranded DNA nicks, gaps, and loops in mammalian cells. *Mol. Cell. Biol.* **7**:1656–1662.

Bakhashi, A., J. J. Wright, W. Graninger, M. Seto, J. Owens, J. Cossman, J. P. Jensen, P. Goldman, and S. J. Korsmeyer. 1987. Mechanism of the t(14;18) chromosomal translocation: structural analysis of both derivative 14 and 18 reciprocal partners. *Proc. Natl. Acad. Sci. USA* **84**:2396–2400.

Bayne, M. L., R. F. Alexander, and R. M. Benbow. 1984. DNA binding protein from ovaries of the frog, Xenopus laevis which promotes concatenation of linear DNA. *J. Mol. Biol.* **172**:87–108.

Been, M. D., R. R. Burgess, and J. J. Champoux. 1984. Nucleotide sequence preference at rat liver and wheat germ type I DNA topoisomerase breakage sites in duplex SV40 DNA. *Nucleic Acids Res.* **12**:3097–3114.

Been, M. D., and J. J. Champoux. 1981. DNA breakage and closure by rat liver type I topoisomerase: separation of the half-reactions by using a single-stranded DNA substrate. *Proc. Natl. Acad. Sci. USA* **78**:2883–2887.

Blackwell, T. K., M. W. Moore, G. D. Yancopoulos, H. Suh, S. Lutzker, E. Selsing, and F. W. Alt. 1986. Recombination between immunoglobulin variable region gene segments is enhanced by transcription. *Nature* (London) **324**:585–589.

Boggs, S. S., R. G. Gregg, N. Borenstein, and O. Smithies. 1986. Efficient transformation and frequent single-site, single-copy insertion of DNA can be obtained in mouse erythroleukemia cells transformed by electroporation. *Exp. Hematol.* (New York) **14**:988–994.

Borst, P., and D. R. Greaves. 1987. Programmed gene rearrangements altering gene expression. *Science* **235**:658–667.

Botchan, M., J. Stringer, T. Mitchison, and J. Sambrook. 1980. Integration and excision of SV40 DNA from the chromosome of a transformed cell. *Cell* **20**:143–152.

Botchan, M., W. Topp, and J. Sambrook. 1976. The arrangement of simian virus 40 sequences in the DNA of transformed cells. *Cell* **9**:269–287.

Botchan, M., W. Topp, and J. Sambrook. 1979. Studies on simian virus 40 excision from cellular chromosomes. *Cold Spring Harbor Symp. Quant. Biol.* **43**:709–719.

Brash, D. E., S. Seethram, K. H. Kraemer, M. M. Seidman, and A. Bredberg. 1987. Photoproduct frequency is not the major determinant of UV base substitution hot spots or cold spots in human cells. *Proc. Natl. Acad. Sci. USA* **84:**3782–3786.

Breimer, L. H., J. Nalbantoglu, and M. Meuth. 1986. Structure and sequence of mutations induced by ionizing radiation of selectable loci in Chinese hamster ovary cells. *J. Mol. Biol.* **192:**669–674.

Brinster, R. L., H. Y. Chen, M. Trumbauer, A. W. Senear, R. Warren, and R. D. Palmiter. 1981. Somatic expression of herpes thymidine kinase in mice following injection of a fusion gene into eggs. *Cell* **27:**223–231.

Brinster, R. L., H. Y. Chen, M. E. Trumbauer, M. K. Yagle, and R. D. Palmiter. 1985. Factors affecting the efficiency of introducing foreign DNA into mice by microinjecting eggs. *Proc. Natl. Acad. Sci. USA* **82:**4438–4442.

Bullock, P., J. J. Champoux, and M. R. Botchan. 1985. Association of crossover points with topoisomerase I cleavage sites: a model for nonhomologous recombination. *Science* **230:**954–958.

Bullock, P., W. Forrester, and M. Botchan. 1984. DNA sequence studies of simian virus 40 chromosomal excision and integration in rat cells. *J. Mol. Biol.* **174:**55–84.

Bullock, P., J. Miller, and M. Botchan. 1986. Effects of poly[d(pGpT · d(pApC)] and poly[d(pCpG) · d(pCpG)] repeats on homologous recombination in somatic cells. *Mol. Cell. Biol.* **6:**3948–3953.

Calos, M. P., J. S. Lebkowski, and M. R. Botchan. 1983. High mutation frequency in DNA transfected into mammalian cells. *Proc. Natl. Acad. Sci. USA* **80:**3015–3019.

Capecchi, M. R. 1980. High efficiency transformation by direct microinjection of DNA into cultured mammalian cells. *Cell* **22:**479–488.

Case, M. E. 1986. Genetical and molecular analysis of QA-2 transformants in Neurospora crassa. *Genetics* **113:**569–587.

Chang, L. M. S. 1971. Development of terminal deoxynucleotidyl transferase activity in embryonic calf thymus gland. *Biochem. Biophys. Res. Commun.* **44:**124–131.

Chang, X.-B., and J. H. Wilson. 1987. Modification of DNA ends can decrease end joining relative to homologous recombination in mammalian cells. *Proc. Natl. Acad. Sci. USA* **84:**4959–4963.

Chia, W., and P. W. J. Rigby. 1981. Fate of viral DNA in nonpermissive cells infected with simian virus 40. *Proc. Natl. Acad. Sci. USA* **78:**6638–6642.

Clayton, C. E., and P. W. J. Rigby. 1981. Cloning and characterization of the integrated viral DNA from three lines of SV40-transformed mouse cells. *Cell* **25:**547–559.

Concannon, P., L. A. Pickering, P. Kung, and L. Hood. 1986. Diversity and structure of human T-cell receptor V_{beta} genes. *Proc. Natl. Acad. Sci. USA* **83:**6598–6602.

Conley, E. C., and J. R. Saunders. 1984. Recombination-dependent recircularization of linearized pBR322 plasmid DNA following transformation of Escherichia coli. *Mol. Gen. Genet.* **194:**211–218.

Conrad, S. E., C.-P. Liu, and M. Botchan. 1982. Fragment spanning the SV40 replication origin is the only DNA sequence required in cis for viral excision. *Science* **218:**1223–1225.

Cornforth, M. N., and J. S. Bedford. 1983. X-ray-induced breakage and rejoining of human interphase chromosomes. *Science* **222:**1141–1143.

Covarrubias, L., Y. Nishida, and B. Mintz. 1986. Early postimplantation embryo lethality due to DNA rearrangements in a transgenic mouse strain. *Proc. Natl. Acad. Sci. USA* **83:**6020–6024.

Covarrubias, L., Y. Nishida, M. Terao, P. D'Eustachio, and B. Mintz. 1987. Cellular DNA rearrangements and early developmental arrest caused by DNA insertion in transgenic mouse embryos. *Mol. Cell. Biol.* **7:**2243–2247.

Croce, C. M. 1987. Role of chromosome translocations in human neoplasia. *Cell* **49:**155–156.

Croce, C. M., and H. Koprowski. 1975. Assignment of gene(s) for cell transformation to human chromosome 7 carrying the SV40 genome. *Proc. Natl. Acad. Sci. USA* **72:**1658–1662.

de Cicco, D. V., and A. C. Spradling. 1984. Localization of a cis-acting element responsible for the developmentally regulated amplification of Drosophila chorion genes. *Cell* **38:**45–54.

DeLap, R. J., M. G. Rush, D. Zouzias, and S. Khan. 1978. Isolation and preliminary characterization of the small circular DNA present in African green monkey kidney (BSC-1) cells. *Plasmid* **1:**508–521.

DeLozanne, A., and J. A. Spudich. 1987. Disruption of the Dictyostelium myosine heavy chain gene by homologous recombination. *Science* **236:**1086–1091.

Desiderio, S. V., G. D. Yancopoulos, M. Paskind, E. Thomas, M. A. Boss, N. Landau, F. W. Alt, and

D. Baltimore. 1984. Insertion of N regions into heavy-chain genes is correlated with expression of terminal deoxytransferase in B cells. *Nature* (London) **311**:752–755.

Deuring, R., U. Winterhoff, F. Pamanoi, S. Stabel, and W. Doerfler. 1981. Site of linkage between adenovirus type 12 and cell DNAs in hamster tumor line CLAC3. *Nature* (London) **293**:81–84.

Doerfler, W. 1982. Uptake, fixation and expression of foreign DNA in mammalian cells: the organization of integrated adenovirus sequences. *Curr. Top. Microbiol. Immunol.* **101**:127–194.

Doerfler, W., R. Gahlmann, S. Stabel, R. Deuring, U. Lichtenberg, M. Schulz, D. Eick, and R. Leisten. 1983. On the mechanism of recombination between adenoviral and cellular DNAs: the structure of junction sites. *Curr. Top. Microbiol. Immunol.* **109**:194–228.

Doetschman, T., R. G. Gregg, N. Maeda, M. L. Hooper, D. W. Melton, S. Thompson, and O. Smithies. 1987. Targetted correction of a mutant HPRT gene in mouse embryonic stem cells. *Nature* (London) **330**:576–578.

Doutriaux, M. P., R. Wagner, and M. Radman. 1986. Mismatch-stimulated killing. *Proc. Natl. Acad. Sci. USA* **83**:2576–2578.

Duesberg, P. H. 1987. Cancer genes: rare recombinants instead of activated oncogenes. *Proc. Natl. Acad. Sci. USA* **84**:2117–2124.

Edwards, K. A., B. D. Halligan, J. L. Davis, N. L. Nivera, and L. F. Liu. 1982. Recognition sites of eukaryotic DNA topoisomerase I: DNA nucleotide sequencing analysis of topo I cleavage sites on SV40 DNA. *Nucleic Acids Res.* **10**:2565–2576.

Efstratiadis, A., J. W. Posakony, T. Maniatis, R. M. Lawn, C. O'Connell, R. A. Spritz, J. K. DeRiel, B. G. Forget, S. M. Weissman, J. L. Slightom, A. E. Blechl, O. Smithies, F. E. Baralle, C. C. Shoulders, and N. J. Proudfoot. 1980. The structure and evolution of the human beta-globin gene family. *Cell* **21**:653–668.

Folger, K. R., K. R. Thomas, and M. R. Capecchi. 1984. Analysis of homologous recombination in cultured mammalian cells. *Cold Spring Harbor Symp. Quant. Biol.* **49**:123–138.

Folger, K. R., K. R. Thomas, and M. R. Capecchi. 1985. Efficient correction of mismatched bases in plasmid heteroduplexes injected into cultured mammalian cell nuclei. *Mol. Cell. Biol.* **5**:70–74.

Folger, K. R., E. A. Wong, G. Wahl, and M. R. Capecchi. 1982. Patterns of integration of DNA microinjected into cultured mammalian cells: evidence for homologous recombination between injected plasmid DNA molecules. *Mol. Cell. Biol.* **2**:1372–1387.

Ford, M., and M. Fried. 1986. Large inverted duplications are associated with gene amplification. *Cell* **45**:425–430.

Franklin, N. 1971. Illegitimate recombination, p. 175–194. *In* A. D. Hershey (ed.), *The Bacteriophage Lambda*. Cold Spring Harbor Laboratory, Cold Spring Harbor, N.Y.

Futcher, A. B. 1986. Copy number amplification of the 2 micron circle plasmid of Saccharomyces cerevisiae. *J. Theor. Biol.* **119**:197–204.

Gahlmann, R., and W. Doerfler. 1983. Integration of viral DNA into the genome of the adenovirus type 2-transformed hamster cell line HE5 without loss or alteration of cellular nucleotides. *Nucleic Acids Res.* **11**:7347–7361.

Gerondakis, S., S. Cory, and J. Adams. 1984. Translocation of the myc cellular oncogene to the immunoglobulin heavy chain locus in murine plasmacytomas is an imprecise reciprocal exchange. *Cell* **36**:973–982.

Gilbert, W., M. Marchionni, and G. McKnight. 1986. On the antiquity of introns. *Cell* **46**:151–153.

Giulotto, E., and N. Israel. 1984. DNA-mediated gene transfer is more efficient during S-phase of the cell cycle. *Biochem. Biophys. Res. Commun.* **118**:310–316.

Giulotto, E., I. Saito, and G. R. Stark. 1986. Structure of DNA formed in the first step of CAD gene amplification. *EMBO J.* **5**:2115–2121.

Gluzman, Y. 1981. SV40-transformed simian cells support the replication of early SV40 mutants. *Cell* **23**:175–182.

Gordon, J. W., and F. H. Ruddle. 1985. DNA-mediated genetic transformation of mouse embryos and bone marrow—a review. *Gene* **33**:121–136.

Grzesiuk, E., and D. Carroll. 1987. Recombination of DNAs in Xenopus oocytes based on short homologous overlaps. *Nucleic Acids Res.* **15**:971–985.

Gusew, N., A. Nepveu, and P. Chartrand. 1987. Linear DNA must have free ends to transform rat cells efficiently. *Mol. Gen. Genet.* **206**:121–125.

Gutai, M. W., and D. Nathans. 1978. Evolutionary variants of simian virus 40: cellular DNA sequences and sequences at recombinant joints of substituted variants. *J. Mol. Biol.* **126:**275–288.

Halligan, B. D., J. L. Davis, K. A. Edwards, and L. F. Liu. 1982. Intra- and intermolecular strand transfer by HeLa DNA topoisomerase 1. *J. Biol. Chem.* **257:**3995–4000.

Hare, J. T., and H. J. Taylor. 1985. One role for DNA methylation in vertebrate cells is strand discrimination in mismatch repair. *Proc. Natl. Acad. Sci. USA* **82:**7350–7354.

Hasson, J.-F., E. Mougneau, F. Cuzin, and M. Yaniv. 1984. Simian virus 40 illegitimate recombination occurs near short direct repeats. *J. Mol. Biol.* **177:**53–68.

Hayday, A., H. E. Ruley, and M. Fried. 1982. Structural and biological analysis of integrated polyoma virus DNA and its adjacent host sequences cloned from transformed rat cells. *J. Virol.* **44:**67–77.

Hinnen, A., J. B. Hicks, and G. R. Fink. 1978. Transformation of yeast. *Proc. Natl. Acad. Sci. USA* **75:**1929–1933.

Hiscott, J. B., D. Murphy, and V. Defendi. 1981. Instability of integrated viral DNA in mouse cells transformed by simian virus 40. *Proc. Natl. Acad. Sci. USA* **78:**1736–1740.

Hope, T. J., R. J. Aguilera, M. E. Minie, and H. Sakano. 1986. Endonucleolytic activity that cleaves immunoglobulin recombination sequences. *Science* **231:**1141–1145.

Hyrien, O., M. DeBatisse, G. Buttin, and B. Robert de Saint Vincent. 1987. A hotspot for novel amplification joints in a mosaic of *Alu*-like repeats and palindromic A+T-rich DNA. *EMBO J.* **6:**2401–2408.

Hyrien, O., M. DeBatisse, G. Buttin, and B. Robert de Saint Vincent. 1988. The multicopy appearance of a large inverted duplication and the sequence at the inversion joint suggest a new model for gene amplification. *EMBO J.* **7:**407–417.

Ikeda, H. 1986. Bacteriophage T4 DNA topoisomerase mediates illegitimate recombination in vitro. *Proc. Natl. Acad. Sci. USA* **83:**922–926.

Jasin, M., J. de Villiers, F. Weber, and W. Schaffner. 1985. High frequency of homologous recombination in mammalian cells between endogenous and introduced SV40 genomes. *Cell* **43:**695–703.

Johnson, A. D., A. Barkan, and J. E. Mertz. 1982. Nucleotide sequence analysis of the recombinant joints in 16 naturally arising deletion mutants of simian virus 40. *Virology* **12:**464–469.

Kato, S., R. A. Anderson, and R. Camerini-Otero. 1986. Foreign DNA introduced by calcium phosphate is integrated into repetitive DNA elements of the mouse L cell genome. *Mol. Cell. Biol.* **6:**1787–1795.

Katzir, N., G. Rechavi, J. B. Cohen, T. Unger, F. Simoni, S. Segal, D. Cohen, and D. Givol. 1985. "Retroposon" insertion into the cellular oncogene c-myc in canine transmissible venereal tumor. *Proc. Natl. Acad. Sci. USA* **82:**1054–1058.

Ketner, G., and T. J. Kelly. 1976. Integrated simian virus 40 sequences in transformed cell DNA: analysis using restriction endonucleases. *Proc. Natl. Acad. Sci. USA* **73:**1102–1106.

Kopchick, J. J., and D. W. Stacey. 1984. Differences in intracellular DNA ligation after microinjection and transfection. *Mol. Cell. Biol.* **4:**240–246.

Koshy, R., S. Kock, A. Freytag von Loringhoven, R. Kahmann, K. Murray, and P. H. Hoschneider. 1983. Integration of hepatitis B virus DNA: evidence for integration in the single-stranded gap. *Cell* **34:**215–223.

Krawinkel, U., G. Zoebelein, and A. L. M. Bothwell. 1986. Palindromic sequences are associated with sites of DNA breakage during gene expression. *Nucleic Acids Res.* **14:**3871–3882.

Kunes, S., D. Botstein, and M. S. Fox. 1984. Formation of inverted dimer plasmids after transformation of yeast with linearized plasmid DNA. *Cold Spring Harbor Symp. Quant. Biol.* **49:**617–628.

Kurosawa, Y., and S. Tonegawa. 1982. Organization, structure, and assembly of immunoglobulin heavy chain diversity segments. *J. Exp. Med.* **155:**201–218.

Landau, N. R., D. G. Schatz, M. Rosa, and D. Baltimore. 1987. Increased frequency of N-region insertion in a murine pre-B-cell line infected with a terminal deoxynucleotidyl transferase retroviral expression vector. *Mol. Cell. Biol.* **7:**3237–3243.

Lehman, I. R. 1974. DNA ligase: structure, mechanism, and function. *Science* **186:**790–797.

Lehrman, M. A., W. J. Schneider, T. C. Sudhof, M. S. Brown, J. L. Goldstein, and D. W. Russell. 1985. Mutation in LDL receptor: Alu-Alu recombination deletes exons encoding transmembrane and cytoplasmic domains. *Science* **227:**140–146.

Lemischka, I., and P. A. Sharp. 1982. The sequences of an expressed rat alpha-tubulin gene and a pseudogene with an inserted repetitive element. *Nature* (London) **300**:330–335.

Lewis, S., A. Gifford, and D. Baltimore. 1985. DNA elements are asymmetrically joined during the site-specific recombination of kappa immunoglobulin genes. *Science* **228**:677–685.

Lin, F. L., K. Sperle, and N. Sternberg. 1984. Homologous recombination in mouse L cells. *Cold Spring Harbor Symp. Quant. Biol.* **49**:139–149.

Lin, F. L., K. Sperle, and N. Sternberg. 1985. Recombination in mouse L cells between DNA introduced into cells and homologous chromosomal sequences. *Proc. Natl. Acad. Sci. USA* **82**:1391–1395.

Looney, J. E., and J. L. Hamlin. 1987. Isolation of the amplified dihydrofolate reductase domain from methotrexate-resistant Chinese hamster ovary cells. *Mol. Cell. Biol.* **7**:569–577.

Mager, D. L., P. S. Henthorn, and O. Smithies. 1985. A Chinese $G_{gamma} + A_{(gamma\ delta\ beta)}^0$ thalassemia deletion: comparison of other deletions in the human beta-globin gene cluster and sequence analysis of the breakpoints. *Nucleic Acids Res.* **13**:6559–6575.

Maurer, B. J., E. Lai, B. A. Hamkalo, L. Hood, and G. Attardi. 1987. Novel submicroscopic extrachromosomal elements containing amplified genes in human cells. *Nature* (London) **327**:434–437.

McClintock, B. 1938. The production of homozygous deficient tissues with mutant characteristics by means of the aberrant mitotic behavior of ring-shaped chromosomes. *Genetics* **23**:315–376.

McClintock, B. 1984. The significance of responses of the genome to challenge. *Science* **226**:792–801.

McCoy, M. I. M., and R. I. Gumport. 1980. T4 ribonucleic acid ligase joins single-strand oligo(deoxyribonucleotides). *Biochemistry* **19**:635–642.

Mendelsohn, E., N. Baran, A. Neer, and H. Manor. 1982. Integration site of polyoma virus DNA in the inducible LPT line of polyoma-transformed rat cells. *J. Virol.* **41**:192–209.

Mertz, J. E., and R. W. Davis. 1972. Cleavage of DNA by R1 restriction endonuclease generates cohesive ends. *Proc. Natl. Acad. Sci. USA* **69**:3370–3374.

Meselson, M. S., and C. M. Radding. 1975. A general model for genetic recombination. *Proc. Natl. Acad. Sci. USA* **72**:358–361.

Michel, B., and S. D. Ehrlich. 1986. Illegitimate recombination occurs between the replication origin of the plasmid pC194 and a progressing replication fork. *EMBO J.* **5**:3691–3696.

Miller, C. K., and H. M. Temin. 1983. High efficiency ligation and recombination of DNA fragments by vertebrate cells. *Science* **220**:606–609.

Miller, J., P. Bullock, and M. Botchan. 1984. Simian virus 40 T antigen is required for viral excision from chromosomes. *Proc. Natl. Acad. Sci. USA* **81**:7534–7538.

Miller, L., B. Cooke, and M. Fried. 1976. Fate of mismatched base-pair regions in polyoma heteroduplex DNA during infection of mouse cells. *Proc. Natl. Acad. Sci. USA* **73**:3073–3077.

Munz, P. L., and C. S. H. Young. 1987. The creation of adenovirus genomes with viable, stable, internal redundancies centered about the E2b region. *Virology* **158**:52–60.

Nalbantoglu, J., O. Goncalves, and M. Meuth. 1983. Structure of mutant alleles at the aprt locus of Chinese hamster ovary cells. *J. Mol. Biol.* **167**:575–594.

Nalbantoglu, J., D. Hartley, G. Phear, B. Tear, and M. Meuth. 1986. Spontaneous deletion formation at the aprt locus of hamster cells: the presence of short sequence homologies and dyad symmetries at deletion termini. *EMBO J.* **5**:1199–1204.

Nalbantoglu, J., and M. Meuth. 1986. DNA amplification–deletion in a spontaneous mutation of the hamster aprt locus: structure and sequence of the novel joint. *Nucleic Acids Res.* **14**:8361–8371.

Nalbantoglu, J., G. Phear, and M. Meuth. 1987. DNA sequence analysis of spontaneous mutations at the *aprt* locus of hamster cells. *Mol. Cell. Biol.* **7**:1445–1449.

Neuberger, M. S., and F. Calabi. 1983. Reciprocal chromosome translocation between c-myc and immunoglobulin gamma 2b genes. *Nature* (London) **305**:240–243.

Nicholls, R. D., N. Fischel-Ghodsian, and D. R. Higgs. 1987. Recombination at the human alpha-globin gene cluster: sequence features and topological constraints. *Cell* **49**:369–378.

Nordheim, A., and A. Rich. 1983. The sequence $(dC-dA)_n$-$(dG-dT)_n$ forms left-handed Z-DNA in negatively supercoiled plasmids. *Proc. Natl. Acad. Sci. USA* **80**:1821–1825.

Okazaki, K., D. D. Davis, and H. Sakano. 1987. T cell receptor beta gene sequences in the circular

DNA of thymocyte nuclei: direct evidence for intramolecular DNA deletion in V-D-J joining. *Cell* **49:**477–485.

Orr-Weaver, T. L., and J. W. Szostak. 1983. Yeast recombination: the association between double-strand gap repair and crossing-over. *Proc. Natl. Acad. Sci. USA* **80:**4417–4421.

Palmiter, R. D., T. M. Wilkie, H. Y. Chen, and R. L. Brinster. 1984. Transmission distortion and mosaicism in an unusual transgenic mouse pedigree. *Cell* **36:**869–877.

Perucho, M. D., D. Hanahan, and M. Wigler. 1980. Genetic and physical linkage of exogenous sequences in transformed cells. *Cell* **22:**309–317.

Pfeiffer, P., and W. Vielmetter. 1988. Joining of nonhomologous DNA double strand breaks in vitro. *Nucleic Acids Res.* **16:**907–924.

Piccoli, S. P., P. G. Caimi, and M. D. Cole. 1984. A conserved sequence at c-myc oncogene chromosomal translocation breakpoints in plasmacytomas. *Nature* (London) **310:**327–330.

Plesner, P., J. Goodchild, H. M. Kalckar, and P. C. Zamecnik. 1987. Oligonucleotides with rapid turnover of the phosphate groups occur endogenously in eukaryotic cells. *Proc. Natl. Acad. Sci. USA* **84:**1936–1939.

Razin, S. V., M. G. Kekelidze, E. M. Lukanidin, K. Scherrer, and G. P. Georgiev. 1986. Replication origins are attached to the nuclear skeleton. *Nucleic Acids Res.* **14:**8189–8207.

Razzaque, A., H. Mizusawa, and M. M. Seidman. 1983. Rearrangement and mutagenesis of a shuttle vector plasmid after passage in mammalian cells. *Proc. Natl. Acad. Sci. USA* **80:**3010–3014.

Robins, D. M., S. Ripley, A. S. Henderson, and R. Axel. 1981. Transforming DNA integrates into the host chromosome. *Cell* **23:**29–39.

Rohdewohld, H., H. Weiher, W. Reik, R. Jaenisch, and M. Breindl. 1987. Retrovirus integration and chromatin structure: Moloney murine leukemia proviral integration sites map near DNase I-hypersensitive sites. *J. Virol.* **61:**336–343.

Roth, D. B., T. N. Porter, and J. H. Wilson. 1985. Mechanisms of nonhomologous recombination in mammalian cells. *Mol. Cell. Biol.* **5:**2599–2607.

Roth, D. B., and J. H. Wilson. 1985. Relative rates of homologous and nonhomologous recombination in transfected DNA. *Proc. Natl. Acad. Sci. USA* **82:**3355–3359.

Roth, D. B., and J. H. Wilson. 1986. Nonhomologous recombination in mammalian cells: role for short sequence homologies in the joining reaction. *Mol. Cell. Biol.* **6:**4295–4304.

Rothstein, R. 1983. One-step gene disruption in yeast. *Methods Enzymol.* **101:**202–211.

Ruley, H. E., and M. Fried. 1983. Clustered illegitimate recombination events in mammalian cells involving very short sequence homologies. *Nature* (London) **304:**181–184.

Rush, M. G., R. Eason, and J. Vinograd. 1971. Identification and properties of complex forms of SV40 DNA isolated from SV40-infected African green monkey (BSC-1) cells. *Biochim. Biophys. Acta* **228:**585–594.

Saito, I., and G. R. Stark. 1986. Charomids: cosmid vectors for efficient cloning and mapping of large or small restriction fragments. *Proc. Natl. Acad. Sci. USA* **83:**8664–8668.

Savageau, M. A., R. Metter, and W. W. Brockman. 1983. Statistical significance of partial base-pairing potentials: implications for recombination of SV40 DNA in eukaryotic cells. *Nucleic Acids Res.* **11:**6559–6570.

Scangos, G., and F. H. Ruddle. 1981. Mechanisms and applications of DNA-mediated gene transfer in mammalian cells—a review. *Gene* **14:**1–10.

Scherer, S., and R. W. Davis. 1979. Replacement of chromosome segments with altered DNA sequences constructed in vitro. *Proc. Natl. Acad. Sci. USA* **76:**4951–4955.

Schimke, R. T. 1984. Gene amplification in cultured animal cells. *Cell* **37:**705–713.

Schimke, R. T., S. W. Sherwood, A. B. Hill, and R. N. Johnston. 1986. Overreplication and recombination of DNA in higher eukaryotes: potential consequences and biological implications. *Proc. Natl. Acad. Sci. USA* **83:**2157–2161.

Schindler, C. W., and M. G. Rush. 1985. Discrete size classes of monkey extrachromosomal circular DNA containing the L1 family of long interspersed nucleotide sequences are produced by a general non-sequence specific mechanism. *Nucleic Acids Res.* **13:**8247–8258.

Schulz, M., U. Freisem-Rabien, R. Jessberger, and W. Doerfler. 1987. Transcriptional activities of mammalian genomes at sites of recombination with foreign DNA. *J. Virol.* **61:**344–353.

Schweizer, E., C. MacKechnie, and H. O. Halvorson. 1969. The redundancy of ribosomal and transfer RNA genes in Saccharomyces cerevisiae. *J. Mol. Biol.* **40:**261–277.

Slightom, J. L., A. E. Blechl, and O. Smithies. 1980. Human fetal G_{gamma} and A_{gamma}-globin genes: complete nucleotide sequences suggest that DNA can be exchanged between these duplicated genes. *Cell* **21:**627–638.

Smith, A. J. H., and P. Berg. 1984. Homologous recombination between defective neo genes in mouse 3T6 cells. *Cold Spring Harbor Symp. Quant. Biol.* **49:**171–181.

Smithies, O., R. G. Gregg, S. S. Boggs, M. A. Koralewski, and R. S. Kucherlapati. 1985. Insertion of DNA sequences into the human beta-globin locus by homologous recombination. *Nature* (London) **317:**230–234.

Smithies, O., M. A. Koralewski, K.-Y. Song, and R. S. Kucherlapati. 1984. Homologous recombination with DNA introduced into mammalian cells. *Cold Spring Harbor Symp. Quant. Biol.* **49:**161–170.

Spradling, A. C. 1981. The organization and amplification of two chromosomal domains containing Drosophila chorion genes. *Cell* **27:**193–201.

Stabel, S., and W. Doerfler. 1982. Nucleotide sequence at the site of junction between adenovirus type 12 DNA and repetitive hamster cell DNA in transformed cell line CLAC1. *Nucleic Acids Res.* **10:** 8007–8023.

Stahl, F. W. 1986. Roles of double-strand breaks in generalized genetic recombination. *Prog. Nucleic Acid Res. Mol. Biol.* **33:**169–194.

Stark, G. R., and G. M. Wahl. 1984. Gene amplification. *Annu. Rev. Biochem.* **53:**447–491.

Steele, R. E., A. H. Bakken, and R. H. Reeder. 1984. Plasmids containing mouse rDNA do not recombine with cellular ribosomal genes when introduced into cultured mouse cells. *Mol. Cell. Biol.* **4:**576–582.

Stenman, G., E. O. Delorme, C. C. Lau, and R. Sager. 1987. Transfection with plasmid pSV2gptEJ chromosome rearrangements in CHEF cells. *Proc. Natl. Acad. Sci. USA* **84:**184–188.

Stringer, J. R. 1981. Integrated simian virus 40 DNA: nucleotide sequences at cell-virus recombinant junctions. *J. Virol.* **38:**671–679.

Stringer, J. R. 1982. DNA sequence homology and chromosomal deletion at a site of SV40 DNA integration. *Nature* (London) **296:**363–366.

Stringer, J. R. 1985. Recombination between poly[d(GT) · d(CA)] sequences in simian virus 40-infected cultured cells. *Mol. Cell. Biol.* **5:**1247–1259.

Subramanian, K. S. 1979. Segments of SV40 DNA spanning most of the leader sequence of the major late viral messenger RNA are dispensable. *Proc. Natl. Acad. Sci. USA* **76:**2556–2560.

Suzuki, K., Y. Imai, I. Yamashito, and S. Fukui. 1983. In vivo ligation of linear DNA molecules of circular forms in the yeast *Saccharomyces cerevisiae*. *J. Bacteriol.* **155:**747–754.

Szostak, J. W., T. L. Orr-Weaver, R. J. Rothstein, and F. S. Stahl. 1983. The double-strand-break repair model for recombination. *Cell* **33:**25–35.

Szostak, J. W., and R. Wu. 1979. Insertion of a genetic marker into the ribosomal DNA of yeast. *Plasmid* **2:**536–554.

Thimmappaya, B., and T. Shenk. 1979. Nucleotide sequence analysis of viable deletion mutants lacking segments of the simian virus 40 genome coding for small-t antigen. *J. Virol.* **30:**668–673.

Thomas, K. R., and M. R. Capecchi. 1986. Introduction of homologous DNA sequences into mammalian cells induces mutations in the cognate gene. *Nature* (London) **324:**34–38.

Thomas, K. R., and M. R. Capecchi. 1987. Site-directed mutagenesis by gene targeting in mouse embryo-derived stem cells. *Cell* **51:**503–512.

Thomas, K. R., K. R. Folger, and M. R. Capecchi. 1986. High frequency targeting of genes to specific sites in the mammalian genome. *Cell* **44:**419–428.

Tonegawa, S. 1983. Somatic generation of antibody diversity. *Nature* (London) **302:**575–581.

Tooze, J. 1980. *DNA Tumor Viruses.* Cold Spring Harbor Laboratory, Cold Spring Harbor, N.Y.

Tsujimoto, Y., J. Gorham, J. Cossman, E. Jaffe, and C. M. Croce. 1985. The t(14;18) chromosomal translocations involved in B-cell neoplasms result from mistakes in VDJ joining. *Science* **229:**1390–1393.

Upcroft, P., B. Carter, and C. Kidson. 1980. Mammalian cell functions mediating recombination of genetic elements. *Nucleic Acids Res.* **8:**5835–5844.

Varshavsky, A. 1981. On the possibility of metabolic control of replicon "misfiring": relationship to emergence of malignant phenotypes in mammalian cell lineages. *Proc. Natl. Acad. Sci. USA* **78:** 3673–3677.

Voelkel-Meiman, K., R. L. Keil, and G. S. Roeder. 1987. Recombination-stimulating sequences in yeast ribosomal DNA correspond to sequences regulating transcription by RNA polymerase I. *Cell* **48:** 1071–1079.

Volckaert, G., J. Feunteun, L. V. Crawford, P. Berg, and W. Fiers. 1979. Nucleotide sequence deletions within the coding region for small-t antigen of simian virus 40. *J. Virol.* **30:**674–682.

Wagner, E. F., T. A. Stewart, and B. Mintz. 1981. The human beta-globin gene and a functional viral thymidine kinase gene in developing mice. *Proc. Natl. Acad. Sci. USA* **78:**5016–5020.

Wake, C. T., T. Gudewicz, T. Porter, A. White, and J. H. Wilson. 1984. How damaged is the biologically active subpopulation of transfected DNA? *Mol. Cell. Biol.* **4:**387–398.

Wake, C. T., F. Vernaleone, and J. H. Wilson. 1985. Topological requirements for homologous recombination among DNA molecules transfected into mammalian cells. *Mol. Cell. Biol.* **5:**2080–2089.

Wake, C. T., and J. H. Wilson. 1980. Defined oligomeric SV40 DNA: a sensitive probe of general recombination in somatic cells. *Cell* **21:**141–148.

Wallenburg, J. C., A. Nepveu, and P. Chartrand. 1984. Random and nonrandom integration of a polyomavirus DNA molecule containing highly repetitive cellular sequences. *J. Virol.* **50:**678–683.

Wallenburg, J. C., A. Nepveu, and P. Chartrand. 1987. Integration of a vector containing rodent repetitive elements in the rat genome. *Nucleic Acids Res.* **15:**7849–7863.

Weiner, A. M., P. L. Deininger, and A. Efstratiadis. 1986. Nonviral retroposons: genes, pseudogenes, and transposable elements generated by the reverse flow of genetic information. *Annu. Rev. Biochem.* **55:**631–662.

Weiss, U., and J. H. Wilson. 1987. Repair of single-stranded loops in heteroduplex DNA transfected into mammalian cells. *Proc. Natl. Acad. Sci. USA* **84:**1619–1623.

Weiss, U., and J. H. Wilson. 1988. Heteroduplex-induced mutagenesis in mammalian cells. *Nucleic Acids Res.* **16:**2313–2322.

Wiberg, F. C., P. Sunnerhagen, and G. Bjursell. 1986. New, small circular DNA in transfected mammalian cells. *Mol. Cell. Biol.* **6:**653–662.

Wilkie, T. M., and R. D. Palmiter. 1987. Analysis of the integrant in MyK-103 transgenic mice in which males fail to transmit the integrant. *Mol. Cell. Biol.* **7:**1646–1655.

Williams, T. J., and M. Fried. 1986. Inverted duplication-transposition event in mammalian cells at an illegitimate recombination join. *Mol. Cell. Biol.* **6:**2179–2184.

Wilson, J. H. 1977. Genetic analysis of host range mutant viruses suggests an uncoating defect in simian virus 40-resistant monkey cells. *Proc. Natl. Acad. Sci. USA* **74:**3503–3507.

Wilson, J. H., P. Berget, and J. M. Pipas. 1982. Somatic cells efficiently join unrelated DNA segments end-to-end. *Mol. Cell. Biol.* **2:**1258–1269.

Wong, E. A., and M. R. Capecchi. 1985. Effect of cell cycle position on transformation by microinjection. *Somatic Cell Mol. Genet.* **12:**43–51.

Wong, E. A., and M. R. Capecchi. 1987. Homologous recombination between coinjected DNA sequences peaks in early to mid-S phase. *Mol. Cell. Biol.* **7:**2294–2295.

Woychik, R. P., T. A. Stewart, L. G. Davis, P. D'Eustachio, and P. Leder. 1985. An inherited limb deformity created by insertional mutagenesis in a transgenic mouse. *Nature* (London) **318:**36–40.

Wright, A. P. H., K. Maundrell, and S. Shall. 1986. Transformation of Schizosaccharomyces pombe by nonhomologous, unstable integration of plasmids in the genome. *Curr. Genet.* **10:**503–508.

Yamagishi, H., T. Kunisada, and T. Tsuda. 1982. Small circular DNA complexes in eucaryotic cells. *Plasmid* **8:**299–306.

Yunis, J. J. 1983. The chromosomal basis of human neoplasia. *Science* **221:**227–236.

Zannis-Hadjopoulos, M., M. Persico, and R. G. Martin. 1981. The remarkable instability of replication loops provides a general method for the isolation of origins of DNA replication. *Cell* **27:**155–163.

Zieg, J., C. E. Clayton, F. Ardeshir, E. Giulotto, E. A. Swyryd, and G. R. Stark. 1983. Properties of single-step mutants of Syrian hamster cell lines resistant to *N*-(phosphonacetyl)-L-aspartate. *Mol. Cell. Biol.* **3:**2089–2098.

Zimmerman, S. B., and B. H. Pheiffer. 1983. Macromolecular crowding allows blunt-end ligation by DNA ligases from rat liver or Escherichia coli. *Proc. Natl. Acad. Sci. USA* **80:**5852–5856.

Chapter 22

Possible Role for the Eucaryotic Type I Topoisomerase in Illegitimate Recombination

James J. Champoux and Peter A. Bullock

I. INTRODUCTION

The class of enzymes called topoisomerases catalyze the temporary interruption of phosphodiester bonds in duplex DNA (for review, see Wang, 1985). During the time the DNA strand is broken, one of the two ends generated by the break is covalently attached to the enzyme by way of a phosphoester bond to a tyrosine in the active site. Type I enzymes introduce a transient break in one of the two strands, while type II enzymes break both strands to generate a transient staggered double-strand cut. All of the type II enzymes appear to be multisubunit structures, two identical subunits of which are responsible for the breaking and joining of the two DNA strands.

These activities endow topoisomerases with the capacity to facilitate a variety of topological manipulations of duplex DNA in the cell. It is clear that the combined activity of the DNA gyrase and the *Escherichia coli* type I topoisomerase are responsible for maintaining and monitoring the negative superhelical state of the *E. coli* chromosome (DiNardo et al., 1982; Pruss et al., 1982). The type II topoisomerases in both procaryotic and eucaryotic cells segregate daughter chromosomes after DNA replication by the iterative passing of a region of duplex DNA from one molecule through a double-strand break in another DNA molecule (Steck and Drlica, 1984; DiNardo et al., 1984; Yang et al., 1987). Recent evidence suggests a role for the eucaryotic type I enzyme (topoisomerase I) in transcription (Fleischmann et al., 1984; Gilmour et al., 1986; Brill et al., 1987; Gilmour and Elgin, 1987) and DNA replication (Yang et al., 1987). In both cases, it is likely that torsional strain introduced into the DNA is relieved by nicking-closing cycles catalyzed by the enzyme. In all of these examples, the breakage and joining reactions catalyzed by the topoisomerases are coupled such that the two ends created in the breakage reaction are rejoined, restoring continuity to the DNA strand.

Evidence has accumulated that for some topoisomerases and topoisomerase-like enzymes the breakage and rejoining steps can be uncoupled. The result of such uncoupling is that different, unrelated ends can be joined together by the enzyme. For example, the *E. coli* DNA gyrase (Ikeda et al., 1982, 1984) and the bacteriophage T4 DNA topoisomerase (Ikeda, 1986), both type II enzymes, can promote illegitimate recombination in vitro. In these cases it is hypothesized that new DNA joints are formed after exchange of the DNA-bound enzyme subunits during the lifetime of the double-strand breaks (Ikeda et al., 1982). In addition, other types of enzymes are functionally very similar to topoisomerases but do not rejoin the two ends created by the breakage reaction. Instead, ends from different breakage events are joined together to generate a recombinant structure. These enzymes have been referred to as DNA-strand transferases (Wang, 1985). The bacteriophage λ Int protein (Craig and Nash, 1983), the Tn*3* (γδ) resolvase (Reed and Grindley, 1981), the bacteriophage P1 Cre protein (Abremski et al., 1986), and the *Saccharomyces cerevisiae* 2μm DNA FLP recombinase (Andrews et al., 1985) catalyze site-specific recombination reactions in which broken strands are exchanged between the breaking and rejoining reactions (for reviews, see Hatfull and Grindley, this volume; Cox, this volume). The phage φX174 A protein (Eisenberg et al., 1977; Brown et al., 1984) and the M13 and fd phage gene *II* proteins (Meyer and Geider, 1979, 1982) initiate rolling-circle replication by nicking the respective double-stranded circular DNAs and subsequently relink the displaced single strands to form the circular product. Most pertinent to this discussion, the formation of a class of deletions in M13 DNA requires the gene *II* protein. Moreover, one endpoint of the deletions is always located at the normal site of nicking by the gene *II* protein (Michel and Ehrlich, 1986), suggesting that this protein can occasionally rejoin the wrong ends and in the process promote illegitimate recombination. The fundamental difference between these reactions and the classical topoisomerase-catalyzed reactions is that, during the time the strand or strands are broken, an exchange occurs so that new DNA combinations are created.

Rare DNA rearrangements involving no homology were discovered in bacteria. Franklin (1967) first defined such rearrangements as illegitimate recombination and determined that they occur independently of the normal mechanisms of homologous recombination (see also Anderson, 1987; Allgood and Silhavy, this volume). Similarly, in eucaryotic cells DNA rearrangements often occur between sequences that share little or no homology (see Roth and Wilson, this volume). Genetic alterations such as inversions, deletions, transpositions, duplications, and amplifications of chromosomal fragments result from these nonhomologous or illegitimate forms of recombination. The purpose of this review is to examine the evidence suggesting a possible involvement of the eucaryotic type I topoisomerase in illegitimate recombination. The evidence takes two forms. (i) In vitro studies with the purified enzyme have revealed DNA substrates with unusual structures that are broken but not rejoined by the enzyme. In these uncoupled reactions the enzyme remains covalently attached to one end of the broken strand and can attach that end to another DNA molecule bearing a free 5'-hydroxyl. (ii) The nucleotide sequences surrounding exchange sites utilized for illegitimate recombination in vivo correlate with the preferred recognition sequences of topoisomerase I in vitro.

II. SEQUENCE PREFERENCE OF TOPOISOMERASE I

Addition of detergents or alkali to a topoisomerase I reaction mixture results in the covalent attachment of a small fraction of the enzyme molecules to the 3' end of broken strands of DNA (Champoux, 1977, 1978, 1981). These structures are interpreted to represent the nicked intermediates in the topoisomerase I-catalyzed reaction, but a direct demonstration of an intermediate status has not been possible because the bound enzyme is inactive. Nevertheless, the distribution of nucleotides in the vicinity of 223 break sites for the rat liver enzyme on duplex simian virus 40 (SV40) DNA was compiled (Been et al., 1984). Although there is no clear preference for base sequence 3' to the break site, the enzyme does exhibit a preference for the four nucleotides towards the 5' end of the broken strand (labeled -1 to -4). The consensus sequence for the strong break sites is

$$-4 \qquad -3 \qquad -2 \quad -1$$
$$\text{5'-(A or T)-(G or C)-(A or T)-T-3'}$$

where the 3'-terminal T is the nucleotide to which the enzyme is covalently bound. Some breakage was also observed with C residues in the -1 position and with A residues at the -3 position; furthermore, some weak break sites were found to deviate substantially from this consensus sequence. Using the same DNA substrate, virtually the same consensus sequence was found for the wheat germ topoisomerase I. The conservation of the recognition sequence between flowering plants and mammals makes it likely that the DNA-binding sites on the enzymes are sufficiently conserved that essentially the same consensus sequence holds for all multicellular eucaryotic type I topoisomerases. We assume this is true in the following discussion of the sequences at crossover sites in various species.

Since not every site that conforms to the consensus sequence is broken by the enzyme, it is useful to tabulate the frequency with which triplets of bases (positions -3 to -1) that are subsets of the consensus sequence are observed break sites. Different sequences are broken with very different frequencies (Table 1). This analysis reveals a hierarchy of sequence preference that is not obvious from the simple tabulation of preferred bases shown above. However, this analysis does not take into account that sites with the same sequence are broken with very different efficiencies. Clearly, some feature of the structure of DNA in addition to simple base sequence influences breakage by the enzyme as detected in these experiments.

These results must be qualified in two ways. First, they are derived from in vitro experiments carried out with purified enzyme. It is possible that the

TABLE 1
Frequency of breakage of triplet sequences by rat liver
topoisomerase I[a]

Sequence			No. of sites present	% of sites broken
−3	−2	−1		
C	T	T	35	86
C	T	C	24	50
G	T	T	39	87
G	T	C	13	54
C	A	T	32	63
C	A	C	31	39
G	A	T	23	74
G	A	C	20	45
A	A	T	35	69
A	A	C	31	45
A	T	T	25	40
A	T	C	13	0

[a] Data taken from Been et al. (1984). The substrate DNAs for this study were restriction fragments of SV40 DNA. The −1 position is the nucleotide to which the enzyme is covalently attached. The −2 and −3 positions are the two nucleotides towards the 5′ end of the strand from the point of attachment.

association of DNA with histones in the cell influences the specificity of the topoisomerase. Second, breakage sites as deduced from detergent-treated reactions may not reflect the strongest binding sites for the enzyme or the sites where the enzyme is most active in catalyzing cycles of nicking and closing. The ability to detect breakage may be a reflection of the lifetime of the intermediate in the reaction and therefore may correlate with the ability of the enzyme to close the nick rather than with the binding of the enzyme to a particular site or with its nicking activity at that site.

Westergaard and his colleagues have identified a repeated hexadecameric sequence (with minor variations) in the nontranscribed spacers of the rDNA of *Tetrahymena pyriformis* that is preferentially cleaved not only by the *Tetrahymena* topoisomerase I, but also by other eucaryotic type I topoisomerases (Andersen et al., 1985; Bonven et al., 1985; Christiansen et al., 1987). Breakage by topoisomerase I in vivo as well as in vitro always occurs between the sixth and seventh nucleotides in the 16-mer following the sequence ACTT; this tetranucleotide fits the consensus found earlier in the studies using SV40 DNA as a substrate. It is not known why this particular site has such a high affinity for topoisomerase I, but since the repeats occur in DNase I-hypersensitive regions, they may be important in the transcription of the rRNA genes. It would be interesting to know whether these high-affinity topoisomerase I sites are hot spots for illegitimate recombination.

III. DNA SUBSTRATES THAT UNCOUPLE THE TOPOISOMERASE I REACTION

The first indication that topoisomerase I might be able to catalyze strand breakage without concomitant strand closure came from studies using three

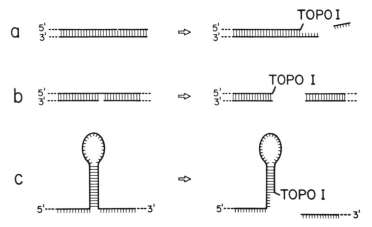

FIGURE 1. Diagrammatic representation of DNA substrates that are subject to breakage without closure. Action of topoisomerase I within a few base pairs of the end of double-stranded DNA (a) or in the intact strand opposite a single-strand break (b) can lead to permanent breakage of a DNA strand. Similar breakage can result from action of the enzyme within a region of secondary structure in what is otherwise single-stranded DNA (c). In each case, the enzyme (TOPO I) becomes covalently attached to the 3′ end of the broken strand and can attach the end to another DNA molecule bearing a free 5′-hydroxyl.

unusual DNA structures as substrates for the enzyme (Fig. 1). (i) The enzyme can break single-stranded DNAs with resultant attachment of the enzyme to the 3′ end of the broken strand (Been and Champoux, 1980). Mapping studies on the break points suggested that breaks occur only at sites that can base pair with a nearby region on the single strand. The requirement for regions of secondary structure for breakage of single-stranded DNA has been confirmed for several sites on single-stranded fragments of simian virus 40 (SV40) DNA by using a deletion analysis (Been and Champoux, 1984). Importantly, the covalently attached enzyme remains active and, under the appropriate conditions, can link the broken strand to another DNA molecule, providing the acceptor contains a free 5′-hydroxyl (Been and Champoux, 1981). This result provided the first evidence that a covalent complex between the enzyme and the end of the broken strand is likely to be an intermediate in the topoisomerase reaction. (ii) The enzyme breaks a short duplex oligonucleotide without reclosure if the substrate contains a topo-isomerase I recognition sequence near its end (Champoux et al., 1984). Since the bound enzyme can catalyze the joining of the oligonucleotide to which it is attached to another oligonucleotide, the bound enzyme remains active. (iii) Similarly, nicking by the topoisomerase in the vicinity of a single-strand break but in the unbroken strand can lead to double-strand cleavage (McCoubrey and Champoux, 1986). With a nicked circular substrate, phosphorylation by polynu-cleotide kinase of the 5′-hydroxyl end produced by the topoisomerase prevented reclosure and allowed trapping of the linear products. With a linear duplex substrate containing a single-strand break at a unique location, phosphorylation was not necessary to trap the products; moreover, the two linear products had the lengths predicted for breakage of the intact strand of the DNA in the vicinity of the preexisting nick.

In all three cases, topoisomerase I can introduce a permanent break in a DNA strand at a point near the end of the strand (i.e., near a double- or single-strand break) or near a point where the strand ceases to be base paired with a complementary strand (i.e., regions of secondary structure in single-stranded DNA). Apparently when topoisomerase I binds to the DNA at such locations, nicking can lead to unpairing of the DNA at the site of the nick such that the two ends are no longer juxtaposed and cannot be rejoined. The fact that the bound enzyme remains active and can, in some cases, join the end of the broken strand to another DNA molecule (Been and Champoux, 1981; Champoux et al., 1984) provides the biochemical rationale for suggesting that conditions might occur in the cell under which topoisomerase I catalyzes illegitimate recombination.

IV. ASSOCIATION OF TOPOISOMERASE I RECOGNITION SEQUENCES WITH EXCHANGE POINTS FOR ILLEGITIMATE RECOMBINATION

A. SV40 Excision

One approach to elucidating the mechanism of illegitimate recombination is to determine the nucleotide sequence at the two sites involved in the exchange. (Each of the parental sites involved in an exchange is defined here as an exchange site for recombination; the exact point of the exchange cannot always be assigned because of redundancies in the sequence or the presence of weak homologies at the sites.) Identification of the sequence requirements for these recombination events might allow one to infer an enzymatic mechanism.

One detailed study of illegitimate recombination involved an analysis of the sequences used during SV40 excision from a proviral locus in the transformed rat cell line 14B (Bullock et al., 1984). Six pairs of excision sites were examined, and a sequence feature common to all pairs of parental domains was the presence of several small regions of partial homology. However, with the exception of the homologies (2 or 3 base pairs) at the exchange sites, free energy and statistical arguments were raised against the involvement of the additional regions of partial homology in excision. Moreover, Savageau et al. (1983) raised similar arguments against the involvement of patchy homologies in other nonhomologous recombination systems.

Further analysis of the six pairs of excision sites revealed a high incidence of sequences in the vicinity of the exchange sites for excision that matched the topoisomerase I consensus sequence described above. This study was extended by carrying out in vitro breakage studies with the purified rat enzyme on cloned DNAs containing the exchange sites. The results demonstrated that four pairs of exchange sites used during SV40 excision from the 14B cell line were indeed associated with topoisomerase I break sites (Bullock et al., 1985). A statistical analysis of these results revealed that there was a very low probability that this association occurred simply as a result of chance. Interestingly, while topoisomerase I sites were found in one of the two strands at the appropriate position for both exchange sites in every case, in only one case were topoisomerase I sites associated with both strands at the two exchange sites. This observation suggests that in these cases, topoisomerase I-mediated illegitimate recombination may initiate in regions that contain single-strand gaps which can be converted to double-stranded breaks by topoisomerase I, as discussed in section III.

B. Other Examples

Several groups of investigators have interpreted their sequence data for exchange sites involved in illegitimate recombination as implicating topoisomerase I in the recombination process on the basis of the presence of the consensus sequence described above at the exchange sites. Hogan and Faust (1986) noted that deletions in the minute virus of mice chromosome could involve topoisomerase I, since preferred topoisomerase I cleavage sites were frequently found at recombination junctions in this virus. Wilkie and Palmiter (1987) established the MyK-103 transgenic mouse pedigree by microinjecting several hundred copies of a linearized plasmid into the pronucleus of an F2 hybrid egg. Analysis of the MyK-103 plasmid integrant and flanking mouse DNA revealed that the preferred topoisomerase I break sites, GTT and CTT, were located at several break points. Similarly, Winocour et al. (1987) presented the nucleotide sequences across 16 recombinant junctions derived from the products of recombination of various viral DNA molecules in monkey cells. They observed that many of the parental exchange points were associated with topoisomerase I sites and suggested that factors such as the conformation of the SV40 chromatin could play a crucial role in determining topoisomerase I access to the DNA. Hyrien et al. (1987) have identified a DNA region that is frequently involved in amplification-associated rearrangements in Chinese hamster fibroblasts. Analysis of a novel joint within the amplified domain indicated that the rearrangement occurred at a possible topoisomerase I cleavage site. Finally, Krawinkel et al. (1986) analyzed an apparent gene conversion event involving different immunoglobulin V_H genes, and they noted that one of the exchange sites occurs precisely at a potential topoisomerase I cleavage site.

It is noteworthy that in these five studies, topoisomerase I break sequences were usually found associated with only one of the two parental exchange sites. Thus, unlike SV40 excision, in which topoisomerase I break sites were identified in vitro on one of the two strands at both exchange sites, in these examples topoisomerase I may be directly responsible for breaking only one of the four strands of DNA participating in the illegitimate recombination event.

In the absence of a direct assay for topoisomerase I involvement in illegitimate recombination, the sequences surrounding 496 parental exchange sites (including those cases cited above) that have participated in illegitimate recombination (Konopka, 1988) have been subjected to a computer-aided analysis to search for sequence features that might be important for the recombination event (A. K. Konopka and P. A. Bullock, unpublished data). Whereas many sites have topoisomerase I sequences within or near the exchange site, many do not. This observation could have several explanations. First, as noted above, topoisomerase I-mediated illegitimate recombination may require breakage of only one of the four strands of DNA participating in the recombination event (see model below). The other three strand breaks may therefore not be caused by topoisomerase I. Thus, the presence of topoisomerase I sites at exchange sites could be obscured by the presence of what might be random sequences present at the majority of the break sites for illegitimate recombination. Second, it is possible that the sites where topoisomerase I acts in vivo are different from the sites identified in the in vitro breakage analysis.

Third, we suspect that there are multiple pathways for illegitimate recombination in mammalian cells, most of which may be independent of topoisomerase I (see Roth and Wilson, this volume). For example, Wake et al. (1984) showed that mammalian cells join blunt ends very efficiently, and this was confirmed by the sequence analysis of the recombination products of transfected linear DNA molecules (Roth et al., 1985). Thus, nuclease breakage plus trimming of transfected DNA followed by blunt end ligation may be responsible for many of the examples analyzed. It is possible that certain examples in the collection of 496 exchange sites reflect a mechanism that has been termed "slipped mispairing" (Efstratiadis et al., 1980; Albertini et al., 1982). According to this hypothesis, the passage of a replication fork through a region of DNA containing two short direct repeats permits the mispairing of the repeats and the subsequent deletion of the intervening DNA (see Allgood and Silhavy, this volume). In addition, it has been noted that approximately 10% of the sequences in the collection contain runs of five or more alternating purines and pyrimidines at the exchange sites (Konopka and Bullock, unpublished data). This is interesting because alternating purines and pyrimidines appear to stimulate homologous recombination in vertebrate somatic cells (Bullock et al., 1986). Finally, it is important to emphasize that the context of illegitimate recombination may determine the pathway employed by the process. Thus, while the context for SV40 excision is believed to be replicating DNA (Botchan et al., 1979, 1980; Bullock and Botchan, 1982), we do not know whether the other examples of illegitimate recombination in the collection are associated with replicating DNA. It is conceivable that the chromatin structure of nonreplicating DNA may preclude access of an enzyme such as topoisomerase I to the DNA, while the state of replicating DNA may allow such access. In fact, Winocour et al. (1987) have concluded that SV40 recombination can proceed by at least two distinct pathways: one independent of DNA replication and the other associated with replication.

V. MODEL FOR TOPOISOMERASE I-MEDIATED ILLEGITIMATE RECOMBINATION

Topoisomerase I might mediate illegitimate recombination as diagrammed in Fig. 2. This scheme is a variation of the model first proposed by Bullock et al. (1985). The recognition sites 5'-CTT-3' and 5'-GTT-3' are shown embedded in a single-stranded region of DNA for two reasons. First, as we have previously noted, the topoisomerase I cleavage sites associated with the SV40 excision

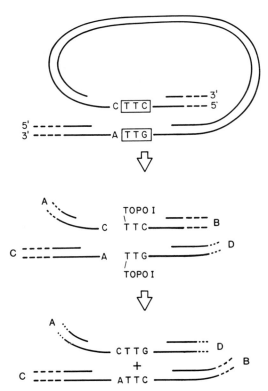

FIGURE 2. Deletion model for topoisomerase I-mediated illegitimate recombination. The topoisomerase I recognition sequences 5'-CTT-3' and 5'-GTT-3' are shown as part of a single-stranded region in the DNA (see text). Breakage and attachment of the enzyme to both of the 3'-T residues are shown in the first step. In the second step, each of the enzymes links the 3' end to which it is attached to the 5' end generated by the other enzyme to complete the exchange.

exchange sites were found in only one of the two strands. Second, on the basis of our in vitro studies, we expect that breakage without closure by topoisomerase I does not normally occur on a purely double-stranded DNA, but can occur on gapped or broken substrates. Although the model shows topoisomerase I producing the breaks at both crossover sites, this is not required by the most general form of the hypothesis. Given an end to which topoisomerase I is attached, the recombinant joint could be formed by topoisomerase I-mediated attachment of the enzyme-bound 3'-phosphate to any free end bearing a 5'-hydroxyl.

Illegitimate recombination is a rare process. Assuming topoisomerase I is responsible for at least a fraction of the observed illegitimate recombination events in eucaryotic cells, what prevents the enzyme from runaway scrambling of the cellular DNA? Two factors are likely to be important. First, on the basis of the in vitro results, it seems probable that some unusual DNA structure is a prerequisite for the uncoupled breakage of a DNA strand by topoisomerase I. Second, under normal circumstances, control mechanisms in the cell may prevent topoisomerase I from acting on such unusual structures when they occur during replication or in damaged DNA. One such mechanism could be the attachment of poly(ADP-ribose) to the enzyme. Poly(ADP-ribose) synthetase is stimulated by DNA that contains breaks or gaps, and poly(ADP-ribosylation) of topoisomerase I inhibits the enzyme (Ferro et al., 1983; Ferro and Olivera, 1984). Support for the notion that poly(ADP-ribosylation) of topoisomerase I may regulate the enzyme in vivo comes from a recent study showing that DNA damage in the presence of camptothecin, a drug that causes intracellular breakage of DNA by topoisomerase I (Hsiang et al., 1985), is enhanced when poly(ADP-ribose) synthetase is inhibited by 3-aminobenzamide (Mattern et al., 1987). Thus, the temporary attachment of poly(ADP-ribose) chains to topoisomerase I in the vicinity of single-stranded gaps or DNA breaks may prevent the enzyme from catalyzing the kind of strand breakage that could lead to illegitimate recombination. According to this hypothesis, illegitimate recombination mediated by topoisomerase I would occur in animal cells only when this control mechanism is disabled.

ACKNOWLEDGMENTS. Support was provided by National Science Foundation grants PCM-8300944 and DMB-8603208. P.A.B. is an American Cancer Society postdoctoral fellow (grant PF-2807) in the laboratory of J. Hurwitz (Public Health Service grant GM 34559 from the National Institutes of Health) and is grateful for his encouragement during the course of this work.

LITERATURE CITED

Abremski K., A. Wierzbicki, B. Frommer, and R. H. Hoess. 1986. Bacteriophage P1 cre-*loxP* site-specific recombination. *J. Biol. Chem.* **261**:391–396.

Albertini, A. M., M. Hofer, M. P. Calos, and J. H. Miller. 1982. On the formation of spontaneous deletions: the importance of short sequence homologies in the generation of large deletions. *Cell* **29**:319–328.

Andersen, A. H., E. Gocke, B. J. Bonven, O. F. Nielsen, and O. Westergaard. 1985. Topoisomerase I has a strong binding preference for a conserved hexadecameric sequence in the promoter region of the rRNA gene from *Tetrahymena pyriformis*. *Nucleic Acids Res.* **13**:1543–1557.

Anderson, P. 1987. Twenty years of illegitimate recombination. *Genetics* **115**:581–584.

Andrews, B. J., G. A. Proteau, L. G. Beatty, and P. D. Sadowski. 1985. The FLP recombinase of the 2μ circle DNA of yeast: interaction with its target sequences. *Cell* **40**:795–803.

Been, M. D., R. R. Burgess, and J. J. Champoux. 1984. Nucleotide sequence preference at rat liver and wheat germ type 1 DNA topoisomerase breakage sites in duplex SV40 DNA. *Nucleic Acids Res.* **12**:3097–3114.

Been, M. D., and J. J. Champoux. 1980. Breakage of single-stranded DNA by rat liver nicking-closing enzyme with the formation of a DNA-enzyme complex. *Nucleic Acids Res.* **8**:6129–6142.

Been, M. D., and J. J. Champoux. 1981. DNA breakage and closure by rat liver type 1 topoisomerase: separation of the half-reactions by using a single-stranded DNA substrate. *Proc. Natl. Acad. Sci. USA* **78**:2883–2887.

Been, M. D., and J. J. Champoux. 1984. Breakage of single-stranded DNA by eukaryotic type 1 topoisomerase occurs only at regions with the potential for base-pairing. *J. Mol. Biol.* **180**:515–531.

Bonven, B. J., E. Gocke, and O. Westergaard. 1985. A high affinity topoisomerase I binding sequence is clustered at DNAase I hypersensitive sites in *Tetrahymena* R-chromatin. *Cell* **41**:541–551.

Botchan, M., J. Stringer, T. Mitchison, and J. Sambrook. 1980. Integration and excision of SV40 DNA from the chromosome of a transformed cell. *Cell* **20**:143–152.

Botchan, M., W. Topp, and J. Sambrook. 1979. Studies on simian virus 40 excision from cellular chromosomes. *Cold Spring Harbor Symp. Quant. Biol.* **43**:709–719.

Brill, S. J., S. DiNardo, K. Voelkel-Meiman, and R. Sternglanz. 1987. Need for DNA topoisomerase activity as a swivel for DNA replication and for transcription of ribosomal RNA. *Nature* (London) **326**:414–416.

Brown, D. R., M. J. Roth, D. Reinberg, and J. Hurwitz. 1984. Analysis of Bacteriophage φX174 gene *A* protein-mediated termination and reinitiation of φX DNA synthesis. I. Characterization of the termination and reinitiation reactions. *J. Biol. Chem.* **259**:10545–10555.

Bullock, P., and M. Botchan. 1982. Molecular events in the excision of SV40 DNA from the chromosomes of cultured mammalian cells, p. 215–224. *In* R. T. Schimke (ed.), *Gene Amplification*. Cold Spring Harbor Laboratory Press, Cold Spring Harbor, N.Y.

Bullock, P., J. J. Champoux, and M. Botchan. 1985. Association of crossover points with topoisomerase I cleavage sites: a model for nonhomologous recombination. *Science* **230**:954–958.

Bullock, P., W. Forrester, and M. Botchan. 1984. DNA sequence studies of simian virus 40 chromosomal excision and integration in rat cells. *J. Mol. Biol.* **174**:55–84.

Bullock, P., J. Miller, and M. Botchan. 1986. Effects of poly[d(pGpT)·d(pApC)] and poly[d(pCpG·d(pCpG)] repeats on homologous recombination in somatic cells. *Mol. Cell. Biol.* **6**:3948–3953.

Champoux, J. J. 1977. Strand breakage by the DNA untwisting enzyme results in covalent attachment of the enzyme to DNA. *Proc. Natl. Acad. Sci. USA* **74**:3800–3804.

Champoux, J. J. 1978. Mechanism of the reaction catalyzed by the DNA untwisting enzyme: attachment of the enzyme to 3′-terminus of the nicked DNA. *J. Mol. Biol.* **118**:441–446.

Champoux, J. J. 1981. DNA is linked to the rat liver DNA nicking-closing enzyme by a phosphodiester bond to tyrosine. *J. Biol. Chem.* **256**:4805–4809.

Champoux, J. J., W. K. McCoubrey, Jr., and M. D. Been. 1984. DNA structural features that lead to strand breakage by eukaryotic type-I topoisomerase. *Cold Spring Harbor Symp. Quant. Biol.* **49**:435–442.

Christiansen, K., B. J. Bonven, and O. Westergaard. 1987. Mapping of sequence-specific chromatin proteins by a novel method: topoisomerase I on *Tetrahymena* ribosomal chromatin. *J. Mol. Biol.* **193**:517–525.

Craig, N. L., and H. A. Nash. 1983. The mechanism of phage λ site-specific recombination: site-specific breakage of DNA by Int topoisomerase. *Cell* **35**:795–803.

DiNardo, S., K. Voelkel, and R. Sternglanz. 1984. DNA topoisomerase II mutant of *Saccharomyces cerevisiae*: topoisomerase II is required for segregation of daughter molecules at the termination of DNA replication. *Proc. Natl. Acad. Sci. USA* **81**:2616–2620.

DiNardo, S., K. A. Voelkel, R. Sternglanz, A. E. Reynolds, and A. Wright. 1982. *Escherichia coli* DNA topoisomerase I mutants have compensatory mutations in DNA gyrase genes. *Cell* **31**:43–51.

Efstratiadis, A., J. W. Posakony, T. Maniatis, R. M. Lawn, C. O'Connell, R. A. Spritz, J. K. DeRiel, B. G. Forget, S. M. Weissman, J. L. Slightom, A. E. Blechl, O. Smithies, F. E. Baralle, C. C. Shoulders, and N. J. Proudfoot. 1980. The structure and evolution of the human beta-globin gene family. *Cell* **21**:653–668.

Eisenberg, S., J. Griffith, and A. Kornberg. 1977. φX174 cistron *A* protein is a multifunctional enzyme in DNA replication. *Proc. Natl. Acad. Sci. USA* **74**:3198–3202.

Ferro, A. M., N. P. Higgins, and B. M. Olivera. 1983. Poly(ADP-ribosylation) of a DNA topoisomerase. *J. Biol. Chem.* **258**:6000–6003.

Ferro, A. M., and B. M. Olivera. 1984. Poly(ADP-ribosylation) of DNA topoisomerase I from calf thymus. *J. Biol. Chem.* **259**:547–554.

Fleischmann, G., G. Pflugfelder, E. K. Steiner, K. Javaherian, G. C. Howard, J. C. Wang, and S. C. R. Elgin. 1984. *Drosophila* DNA topoisomerase I is associated with transcriptionally active regions of the genome. *Proc. Natl. Acad. Sci. USA* **81**:6958–6962.

Franklin, N. C. 1967. Extraordinary recombinational events in *Escherichia coli*. Their independence of the *rec*⁺ function. *Genetics* **55**:699–707.

Gilmour, D. S., and S. C. R. Elgin. 1987. Localization of specific topoisomerase I interactions within the transcribed region of active heat shock genes by using the inhibitor camptothecin. *Mol. Cell. Biol.* **7**:141–148.

Gilmour, D. S., G. Pflugfelder, J. C. Wang, and J. T. Lis. 1986. Topoisomerase I interacts with transcribed regions in *Drosophila* cells. *Cell* **44**:401–407.

Hogan, A., and E. A. Faust. 1986. Nonhomologous recombination in the parvovirus chromosome: role for a CT$_T^A$TT$_T^C$ motif. *Mol. Cell. Biol.* **6**:3005–3009.

Hsiang, Y.-H., R. Hertzberg, S. Hecht, and L. F. Liu. 1985. Camptothecin induces protein-linked DNA breaks via mammalian DNA topoisomerase I. *J. Biol. Chem.* **260:**14873–14878.

Hyrien, O., M. Debatisse, G. Buttin, and B. R. de Saint Vincent. 1987. A hotspot for novel amplification joints in a mosaic of *Alu*-like repeats and palindromic A+T-rich DNA. *EMBO J.* **6:**2401–2408.

Ikeda, H. 1986. Illegitimate recombination mediated by T4 DNA topoisomerase *in vitro*. Recombinants between phage and plasmid DNA molecules. *Mol. Gen. Genet.* **202:**518–520.

Ikeda, H., K. Aoki, A. Naito. 1982. Illegitimate recombination mediated *in vitro* by DNA gyrase of *Escherichia coli*. Structure of recombinant DNA molecules. *Proc. Natl. Acad. Sci. USA* **79:**3724–3728.

Ikeda, H., I. Kawasaki, and M. Gellert. 1984. Mechanism of illegitimate recombination: common sites for recombination and cleavage mediated by *E. coli* DNA gyrase. *Mol. Gen. Genet.* **196:**546–549.

Konopka, A. K. 1988. Compilation of DNA strand exchange sites for nonhomologous recombination in somatic cells. *Nucleic Acids Res.* **16:**1739–1758.

Krawinkel, U., G. Zoebelein, and A. L. M. Bothwell. 1986. Palindromic sequences are associated with sites of DNA breakage during gene conversion. *Nucleic Acids Res.* **14:**3871–3882.

Mattern, M. R., S.-M. Mong, H. F. Bartus, C. K. Mirabelli, S. T. Crooke, and R. K. Johnson. 1987. Relationship between the intracellular effects of camptothecin and the inhibition of DNA topoisomerase I in cultured L1210 cells. *Cancer Res.* **47:**1793–1798.

McCoubrey, W. K., Jr., and J. J. Champoux. 1986. The role of single-strand breaks in the catenation reaction catalyzed by the rat type I topoisomerase. *J. Biol. Chem.* **261:**5130–5137.

Meyer, T. F., and K. Geider. 1979. Bacteriophage fd gene II-protein. II. Specific cleavage and relaxation of supercoiled RF from filamentous phages. *J. Biol. Chem.* **254:**12642–12646.

Meyer, T. F., and K. Geider. 1982. Enzymatic synthesis of bacteriophage fd viral DNA. *Nature* (London) **296:**828–832.

Michel, B., and S. D. Ehrlich. 1986. Illegitimate recombination at the replication origin of bacteriophage M13. *Proc. Natl. Acad. Sci. USA* **83:**3386–3390.

Pruss, G. J., S. H. Manes, and K. Drlica. 1982. *Escherichia coli* DNA topoisomerase I mutants: increased supercoiling is corrected by mutations near gyrase genes. *Cell* **31:**35–42.

Reed, R. R., and N. D. F. Grindley. 1981. Transposon-mediated site-specific recombination *in vitro*: DNA cleavage and protein-DNA linkage at the recombination site. *Cell* **25:**721–728.

Roth, D. B., T. N. Porter, and J. H. Wilson. 1985. Mechanisms of nonhomologous recombination in mammalian cells. *Mol. Cell. Biol.* **5:**2599–2607.

Savageau, M. A., R. Metter, and W. W. Brockman. 1983. Statistical significance of partial base-pairing potential: implications for recombination of SV40 DNA in eukaryotic cells. *Nucleic Acids Res.* **11:**6559–6570.

Steck, T. R., and K. Drlica. 1984. Bacterial chromosome segregation: evidence for DNA gyrase involvement in decatenation. *Cell* **36:**1081–1088.

Wake, C. T., T. Gudewicz, T. Porter, W. White, and J. H. Wilson. 1984. How damaged is the biologically active subpopulation of transfected DNA? *Mol. Cell. Biol.* **4:**387–398.

Wang, J. C. 1985. DNA topoisomerases. *Annu. Rev. Biochem.* **54:**665–697.

Wilkie, T. M., and R. D. Palmiter. 1987. Analysis of the integrant in MyK-103 transgenic mice in which males fail to transmit the integrant. *Mol. Cell. Biol.* **7:**1646–1655.

Winocour, E., T. Chitlaru, K. Tsutsui, R. Ben-Levy, and Y. Shaul. 1987. Multiple pathways for simian virus 40 nonhomologous recombination. *Cancer Cells* (Cold Spring Harbor) **4:**509–516.

Yang, L., M. S. Wold, J. J. Li, T. J. Kelly, and L. F. Liu. 1987. Roles of DNA topoisomerases in simian virus 40 DNA replication *in vitro*. *Proc. Natl. Acad. Sci. USA* **84:**950–954.

Chapter 23

Immunoglobulin Gene Rearrangement

Peter Engler and Ursula Storb

I. INTRODUCTION

A. Antibodies and Immunoglobulin Genes

Antibodies are a class of proteins which are able to bind to an essentially limitless array of antigen molecules and yet are able to carry out a rather restricted set of biological functions such as activating the complement cascade and crossing the placental barrier. This functional dichotomy is reflected in the structure of the antibody molecule.

A typical mouse immunoglobulin molecule consists of four polypeptide chains: two identical heavy (H) chains which are disulfide linked to each other and two identical light (L) chains disulfide bonded to the H chains (Fig. 1). Each chain is folded into characteristic globular domains (two for L chains and four or five for H chains, depending on the class). Peptide-mapping and amino acid-sequencing

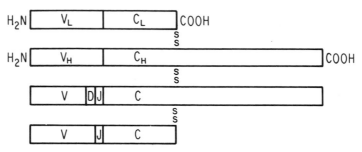

FIGURE 1. Schematic diagram of a typical immunoglobulin molecule. The two L chains are identical, as are the two H chains. The top half indicates the variable (V) and constant (C) regions of the protein chains, while the bottom half indicates the gene segments (V, D, J, and C) that code for these regions.

studies showed that the amino-terminal end has a high degree of variability while the carboxy-terminal end is essentially invariant. As expected, antigen-binding activity is associated with the variable portion (V) and the constant portion (C) is responsible for conserved functions.

For many years, it has been appreciated that a single polypeptide consisting of both variable and constant regions must be encoded by an unusual set of genes. One possible explanation was that each immunoglobulin chain is encoded by a unique germ line gene. The incredible variety of antibodies made it necessary to postulate that a very significant fraction of the genome is devoted to coding for immunoglobulin. At the other extreme were theories suggesting that antibodies are encoded by a small number of germ line genes subject to alteration by mutation in those cells destined to produce antibodies. For a thorough historical discussion of theories of antibody diversity, see Kindt and Capra (1984).

More than 20 years ago, Dreyer and Bennett (1965) made the radical proposal that V regions of light chains are encoded by multiple genetic elements that are separate from the single element encoding the C region (Fig. 2). They suggested that during the development of an antibody-producing cell, the V and C elements are brought into proximity, thus creating a functional light-chain gene. Furthermore, they suggested that this process might be analogous to the integration of the bacteriophage lambda into the bacterial chromosome. Implicit in this suggestion is the idea that a site-specific recombination event is responsible for the formation of an active immunoglobulin gene.

Today, as a result of a tremendous amount of work done in many laboratories, we know that the production of functional immunoglobulin genes is indeed a result of somatic rearrangement of coding elements. This chapter is concerned with the processes of rearrangement that give rise to L-chain and H-chain immunoglobulin genes. We will focus on mouse (more specifically, the BALB/c mouse strain) immunoglobulin genes, although occasional reference will be made to other species and to T-cell receptor (TCR) genes.

B. Organization of Immunoglobulin Loci

In all mammalian species for which information is available, immunoglobulin chains are encoded by three unlinked loci. H chains of all classes are encoded by

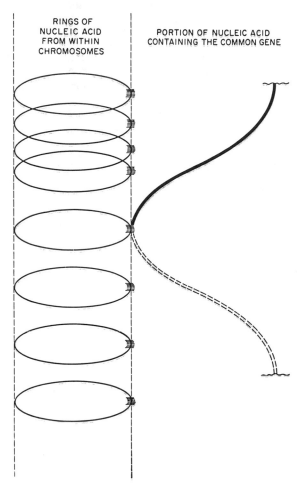

RINGS OF
NUCLEIC ACID
FROM WITHIN
CHROMOSOMES

PORTION OF NUCLEIC ACID
CONTAINING THE COMMON GENE

FIGURE 2. An early proposal of immunoglobulin gene rearrangement. The original caption reads: "Diagram of the proposed genetic mechanism which accounts for the amino acid sequence variations found in L chains. Genetic material which codes for the 'variable' portion of L-chain molecules is inserted into that which codes for the 'common' region of amino acid sequence by a mechanism similar to the insertion of the λ virus into a bacterial chromosome." (From Dreyer and Bennett [1965]; reproduced with permission of the authors.)

a gene complex found on chromosome 12 in the mouse. In contrast, the two classes of L chains are encoded by separate loci. The κ locus is found on mouse chromosome 6, while the λ locus is on chromosome 16. TCR receptor genes encoding the α, β, and γ chains are found on murine chromosomes 14, 6, and 13, respectively. The β locus maps very close to the immunoglobulin κ locus, and recombinants, presumably very rare, have been found between the κ and β loci (Denny et al., 1986).

The H-chain locus (Fig. 3) consists of a large number (~100) of V-region gene segments, perhaps 12 D segments, 4 J-region segments, and 8 C-region genes

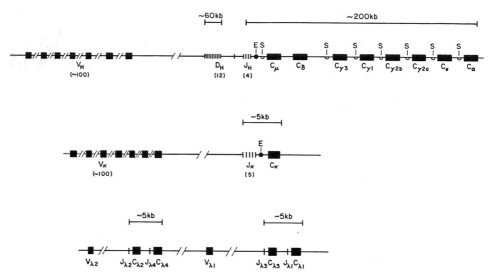

FIGURE 3. Physical organization of the three immunoglobulin loci of the mouse. Transcriptional enhancers are indicated by E; switch regions are indicated by S. These maps are not to scale, although certain selected distances are indicated. See text for explanation.

(Shimizu et al., 1982). The J_H cluster, covering ~1 kilobase (kb), is separated from C_μ (the most 5' C_H gene) by an intron of ~6 kb. This intron contains the transcriptional enhancer as well as sequences involved in the class switch recombination. The C_H genes span a region of nearly 200 kb. Although the structure of the D-J-C region is known in considerable detail, the V-region cluster is less well understood.

Primarily from Southern blot analysis, the V_H genes can be grouped into seven distinct families which vary in complexity from only 2 to over 50 members (Brodeur and Riblet, 1984). Although certain V_H genes have been linked to other members, the overall organization is not clear. Furthermore, no V has been physically linked to a D region, and even the orientation of V and D genes is not certain. The relative orientation of the recombining gene segments has important implications for the rearrangement process. The use of modern "megamapping" techniques such as pulse field electrophoresis and cloning in artificial *Saccharomyces cerevisiae* chromosomes should increase our understanding of the organization of this locus.

The κ locus is a bit simpler in that it consists of many Vs and five Js, but only a single C region. As with the H locus, there are a large number (perhaps 100 to 200) of V segments which can be grouped into a smaller number of families (Cory et al., 1981). The cluster of five J segments (the middle J segment cannot be used to produce a functional κ protein because of a 5' splice site alteration) is separated from the C_κ gene by an intron of 2.5 kb (Sakano et al., 1979; Max et al., 1979, 1981). Like the J_H-C_μ intron, it contains a transcriptional enhancer element, but in contrast to the H intron, it has no switch region. Again, V and J gene segments have not been physically linked, nor are their relative orientations known. As discussed below in section II.C., it is entirely possible that at least some V regions are in opposite transcriptional orientation with respect to J and C.

The λ locus is unusual in that it consists of four C segments, each with its own J region (Selsing et al., 1982). It has just been mapped (J. Miller, S. Ogden, M. McMullen, H. Andres, and U. Storb, *J. Immunol.*, in press; U. Storb, D. Haasch, B. Arp, P. Sanchez, P.-A. Cazenave, and J. Miller, submitted for publication). The λ gene order is V2-JC2-JC4-V1-JC3-JC1, with all V and JC genes in the same transcriptional orientation. Surprisingly, the distances between the V_λ and C_λ genes are inversely proportional to their rearrangement frequencies.

C. Overview of B-Cell Development

Immunoglobulin gene rearrangement is a process critical to the development of antibody-forming cells, and thus it is not surprising that this process is strictly controlled at a number of levels. Although we are very far from a complete understanding of the regulation of gene rearrangement, a few basic facts will be reviewed here (for a review, see Cooper et al., 1983); a more detailed discussion of the control of rearrangement is presented in section III.

Antibodies are produced exclusively by members of the B lymphoid cell lineage (so named because they were first identified as derived from the *b*ursa of Fabricius in the chicken). In the mouse, B-cell development begins in the fetal liver (the primary hematopoietic organ in the embryo) or in later development in the bone marrow. Pluripotent hematopoietic stem cells give rise to all blood cell types. From these multipotent stem cells are derived precursors of the myeloid lineage (erythrocytes, megakaryocytes, granulocytes, and macrophages) and of the lymphoid lineage. Differentiation of these lymphocyte precursors occurs after migration to the thymus (where T-cell development occurs) or to the various sites of B-cell development.

Gene rearrangements occur in an ordered fashion during B-cell development. H-chain loci rearrange first, with the initial step being the joining of a D segment to a J segment. Only after D-J_H rearrangements have occurred do V_H gene segments recombine with Ds. As discussed more fully in later sections, not all rearrangement events produce functional immunoglobulin polypeptide chains (because of frameshifts or other problems), but only after a proper series of rearrangements is the cell able to produce μ chains. Those cells making intracellular μ chains but no L chains are referred to as pre-B cells. Once a μ chain is made, pre-B cells proceed to rearrange L-chain loci.

After a productive κ or λ rearrangement, the cell is able to assemble a complete immunoglobulin M (IgM) molecule which can be either secreted or in a membrane-bound form. The membrane-bound form of immunoglobulin serves as a receptor permitting clonal expansion in response to antigenic stimulation. The progeny of this cell that has carried out productive rearrangements of both H and L genes will continue to produce antibody of the specificity determined by the V regions. Certain progeny, however, are able to produce antibody of different H-chain classes while retaining the same V regions. This process is termed class switching and appears to be mediated by homologous recombination between

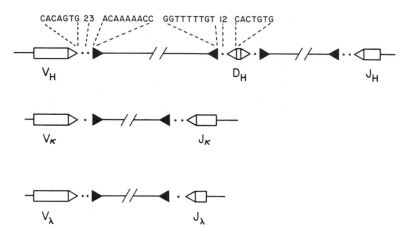

FIGURE 4. Arrangement of recombination signal sequences in the three immunoglobulin loci of the mouse. Heptamers are indicated by an open triangle; filled triangles indicate the nonamer sequences. Spacers are indicated by dots (two dots = two turns = 23 bp; one dot = one turn = 12 bp).

large blocks of repeated sequences found ~2 kb 5' of all C_H genes (with the exception of C_δ).

II. MECHANISM OF REARRANGEMENT

A. Recombination Signal Sequences

When germ line V and J gene segments were first sequenced and compared with their rearranged counterparts, it was noted that the point of recombination was close to a highly conserved sequence consisting of a palindromic heptamer, a spacer region of conserved length (but not sequence), and an AT-rich nonamer (Sakano et al., 1979; Max et al., 1979). These characteristic sequences of 7-mer/spacer/9-mer are found 3' of all V coding regions and 5' of all J coding regions; these recombination signal sequences (RSSs) also flank D regions in H-chain genes (Fig. 4). These sequences can, in principle, base pair with each other (i.e., V 7-mer with J 7-mer and V 9-mer with J 9-mer) to yield a stem loop structure, and it was hypothesized that this may be an intermediate in recombination. Although the existence of such a structure has not been absolutely ruled out, it is unlikely for a number of reasons. In all likelihood the 7-mer/spacer/9-mer sequences constitute the sites recognized by DNA-binding proteins involved in recombination.

Each set of 7-mer and 9-mer is separated by a spacer of either 12 or 23 base pairs (bp) (although deviations of 1 bp in length are sometimes seen). While the length of these spacer regions is highly conserved, the sequence is not; there are no convincing patterns of conserved residues. The spacers are commonly referred to as being "one turn" or "two turn," although of course 12 bp is significantly more than the helix periodicity of 10.4 nucleotides per turn.

Recombination seems always to occur between two gene segments having different spacer lengths: the so-called one-turn/two-turn rule (Early et al., 1980). All known V_κ segments have one-turn spacers, while the J_κ segments have two-turn spacers. The situation is the opposite in lambda segments: the V 7-mer and 9-mer are separated by 23 bp while 12 bp separates their J counterparts. Both V_H and J_H RSSs have two-turn spacers, while D regions are flanked by one-turn RSSs; indeed, the structure of D regions, including their associated RSSs, was predicted in advance of their isolation by application of the one-turn/two-turn rule.

The TCR loci all appear to abide by the one-turn/two-turn rule (Kronenberg et al., 1986), although there is a peculiarity: D_β segments are flanked on the 5′ side by a one-turn RSS and on the 3′ side by a two-turn signal. V_β regions, having a two-turn signal, could join directly to J_β (one-turn) segments, and multiple D_β segments could join to each other. The frequency of these events and their significance for the generation of the diversity of the TCR repertoire are not yet clear.

The structure of the RSSs is not restricted to mice or even to mammals. The chicken λ locus consists of a single functional V (although ~25 pseudo-V genes exist) and a single J segment (Reynaud et al., 1985). The J_λ RSS consists of a 9-mer with only a single deviation from the mammalian consensus sequence (C for T in the ninth position), while the 7-mer matches perfectly; these sequences are separated by a 12-bp spacer. The V_λ 7-mer differs at only the central nucleotide, while the 9-mer is perfect; these sequences are separated by a 23-bp spacer. Even a *Caiman* V_H gene has an RSS that differs from the mouse 7-mer/9-mer consensus sequence at only a single position (Litman et al., 1983).

The universality of the one-turn/two-turn rule for joining of immunoglobulin and TCR gene segments has implications for the mechanism of rearrangement. It suggests that the recombinase contains two separate binding domains (or distinct polypeptides), each specific for one type of RSS.

The sequences of 91 different RSSs are presented in Fig. 5. From this compilation it is clear that while most RSSs differ somewhat from the consensus, there is a high degree of sequence conservation. In particular the three nucleotides at the 3′ end of the 7-mer are essentially invariant; only in pseudogenes are deviations from the GTG sequence seen. Obviously, this suggests that these are critical bases for binding and/or cutting by the recombinase. Knowledge of the exact sequence requirements for gene rearrangement awaits a detailed mutational analysis (see Akira et al., 1987).

Although the conservation of the RSSs among the various rearranging loci and throughout evolution strongly suggests that they have an important role in recombination, more direct evidence has been provided by experiments using artificially constructed recombination substrates. These substrates, when introduced into certain pre-B-cell lines, undergo rearrangements similar to those of endogenous genes (Lewis et al., 1984, 1985; Blackwell and Alt, 1984; Yancopoulos et al., 1986; Blackwell et al., 1986; Hesse et al., 1987; Engler and Storb, 1987; Akira et al., 1987). Such experiments are of considerable importance because they prove that all sequences necessary for rearrangement are present on

Consensus		5' GGTTTTTGT	(bp)	CACTGTG 3'
Mouse	J1	---------	(23)	------- ↓
Ig κ	J2	A--------	(23)	--G----
	J3ψ	--G------	(21)	------A
	J4	---------	(24)	-------
	J5	---------	(23)	-------
Human	J1	----C---	(23)	-------
Ig κ	J2	A--------	(23)	--T----
	J3	---------	(23)	-------
	J4	---------	(23)	-------
	J5	---------	(23)	-------
Mouse	J1	-T------C	(12)	---A---
Ig λ	J2	------G-G	(12)	--T----
	J3	-----AG-G	(12)	-------
	J4ψ	-T------C	(12)	--G---T
Human	J1	------G--	(12)	-------
Ig λ	J2	---------	(12)	---A---
	J3	---------	(12)	---A---
	J4ψ	-----CAA-	(12)	--G--AC
	J5ψ	-----CAA-	(12)	--G--AA
Mouse	J1	A-----A--	(22)	G------
Ig H	J2	---------	(23)	T-G----
	J3	AT--A----	(23)	--A----
	J4	---------	(22)	T-T----
Human	J1	------C--	(21)	---C---
Ig H	J2	T--------	(23)	G------
	J3	------G--	(23)	-C-----
	J4	---------	(23)	---A---
	J5	----C--TG	(22)	---A---
	J6	---------	(22)	--T----
Mouse	DQ52	------GAC	(12)	---A---
Ig H	DSP22	-A-------	(12)	---A---
	DSP23	-A-------	(12)	T------
	DSP24	-A-------	(12)	T------
	DSP25	-A-------	(12)	T------
	DSP26	-A-------	(12)	T------
	DSP27	-A-------	(12)	T------
	DSP28	-A-------	(12)	T------
	DFL1	-C-------	(12)	T------
	DFL2	-C-------	(12)	T------
Human	D1	--A------	(12)	-------
Ig H	D2	--A------	(12)	-------
	D3	--A------	(12)	-------
	D4	--A------	(12)	-------
	DQ52	-------G	(13)	-------

Mouse	J1.1	TA-----C-	(13)	------- ↓
TCR β	J1.2	CCA-Λ--CG	(12)	TGA----
	J1.3	------GAA	(12)	GG-----
	J1.4	A-----ACC	(12)	TGT----
	J1.5	--AG-----	(13)	T------
	J1.6	------ACC	(12)	AG-----
	J1.7ψ	-C-CCA-T-	(12)	GG-----
	J2.1	-AA--C-TG	(12)	TG-----
	J2.2	-----G--C	(12)	G------
	J2.3	A--------	(12)	GG-----
	J2.4	A--------	(12)	GG-----
	J2.5	A--------	(12)	GG-----
	J2.6ψ	-----C-C-	(12)	GGT----
	J2.7	-----G---	(12)	-T-----
Human	J1.1	-A-----CAC	(12)	-------
TCR β	J1.2	CC----A-A	(12)	TTA----
	J1.3	------GAA	(12)	GG-----
	J1.4	------CC-	(12)	TGT----
	J1.5	--G---GCC	(12)	-------
	J1.6	--G---TA-	(12)	AG-----
	J2.1	-AA--C--G	(12)	-------
	J2.2	-----GC-C	(12)	GG-----
	J2.3	---------	(12)	GG-----
	J2.4	A----C---	(12)	GG-----
	J2.5	---------	(11)	GG-C---
	J2.6	--------C	(12)	GG-----
	J2.7ψ	-----GCA-	(12)	-T-C---
Mouse	JPHDS	A--------	(12)	-------
TCR α	JTT11	A---A----	(12)	GG-----
	J80	-----A---	(12)	--GG---
	J84	A--------	(12)	-------
	J19	-----A---	(12)	-------
	J65	---------	(12)	---A---
	J45	A-------C	(12)	--G----
Human	JA1	----A-CTC	(12)	---A---
TCR α	JB	T-------A	(12)	-------
	JC	T-G------	(12)	--TA---
	JD	CCA------	(12)	---A---
	JEψ?	--------A	(12)	----A--
Mouse	J10.3	-A-------	(12)	-------
TCR γ	J13.4	-A-------	(12)	-------
Human	J1	A------A	(12)	-------
TCR γ	J2	A------A	(12)	-------
Mouse	D1	CT-------	(12)	--T----
TCR β	D2	CT-------	(12)	--T----
Human	D1	T-------	(12)	--T----
TCR β	D2	CA-------	(12)	--T----

5' **GGTTTTTGT** 3' ----- 5' **CACTGTG** 3' --- ↓ [Coding Region]

71 73 86 98 92 82 79 79 71 (%)   63 75 74 86 99 98 97 (%)

FIGURE 5. Compilation of 91 recombination signal sequences. Hyphens indicate identity with the consensus sequences. Pseudogenes are marked with ψ. (Reproduced from Akira et al. [1987] with permission.)

the transfected DNA. Furthermore, they prove that rearrangement is not restricted to only a few "privileged" chromosomal sites. Finally, transfection studies as well as cell-free in vitro experiments promise to greatly expand our understanding of the process of immunoglobulin and TCR gene rearrangement.

The first artificial substrates used to demonstrate V_κ to J_κ joining (Lewis et al., 1984) or D_H to J_H rearrangement (Blackwell and Alt, 1984) included the coding regions of the gene segments as well as substantial flanking sequences totaling in excess of 5 kb. These pioneering studies clearly showed that introduced DNA could be recombined by pre-B cells in a manner resembling the rearrangement of endogenous genes. Since the constructs integrated into the host genome in a random fashion, it was clear that multiple chromosomal sites are compatible with rearrangement.

More recently, rearrangements of two other substrates, both based on V_κ to J_κ joining, have been described that contain far fewer immunoglobulin sequences, suggesting that only very minimal DNA sequences are required for efficient and accurate rearrangement.

One of these constructs contains a polyoma replicon and so is maintained as an episome (Hesse et al., 1987). It contains not a single nucleotide from the V_κ or J_κ coding region and only 59 nucleotides in addition to the V and J RSSs. After transfer into a pre-B cell line, reisolation of the plasmid, and assay by transfection of *Escherichia coli* (rearrangement is accompanied by the deletion of a transcriptional terminator that prevents expression of chloramphenicol resistance in the unrearranged configuration), V-J joining could be detected at a frequency of 0.4%. This value probably represents an underestimate of the true rearrangement frequency because presumably not all of the molecules recovered after transfection of the pre-B cell had entered the cell. In any case it is clear that chromosomal integration is not a prerequisite for rearrangement and that V_κ and J_κ coding sequences are not required for recombination.

Another recombination indicator has been described that depends upon a V_κ-J_κ rearrangement to delete an interposed translational block (Engler and Storb, 1987). A rearrangement allows expression of a selectable marker gene directly in the transfected pre-B cell. This construct also contains only very minimal immunoglobulin sequences: in addition to a bit of V_κ and J_κ coding sequences it has only 4 nucleotides 3' of the V_κ 9-mer and 43 nucleotides 5' of the $J_{\kappa 1}$ 9-mer. Even without selection for expression of the marker gene, over 80% of the pre-B cells transfected show evidence of rearrangement.

Recently, recombination substrates containing only the 7-mer/spacer/9-mer sequences were shown to rearrange by deletion or by inversion depending on the arrangement of the RSSs (Akira et al., 1987). Point mutations in the 7-mer (T for C at position 1 or G for C at position 3) greatly decreased the rearrangement efficiency, although transfected cells harboring a rearrangement could still be selected. A substrate containing two two-turn spacers appeared to be totally inactive.

Other sites, outside of the 7-mer/9-mer sequences, have been implicated as being involved in rearrangement (Weaver and Baltimore, 1987). Two 16-bp sites, each containing imperfect inverted repeats, have been shown to be bound by a protein specifically found in pre-B and B cells. One of these sites, termed K1, is immediately 5′ of the $J_{\kappa 1}$ 9-mer; the other, termed K2, is 38 bp 5′ of the K1 site. However, the K2 site is not present in two of the constructs previously described (Hesse et al., 1987; Engler and Storb, 1987), and neither K1 nor K2 is present in another construct (Akira et al., 1987); yet, all of these substrates rearrange efficiently. Thus, while we can say that K1 and K2 are not absolutely required for rearrangement, we cannot rule out the possibility that these sites may be principally involved in regulation of the rearrangement process.

B. Recombination Enzymes

Cells of the B lymphoid lineage rearrange immunoglobulin gene segments, while T cells rearrange TCR gene segments (although some T-cell lines have D-J_H immunoglobulin rearrangements [Kurosawa et al., 1981]). Furthermore, the rearrangement of the various loci within a lineage is an ordered process. Certain 7-mer/9-mer sequences of TCR segments show substantial deviation from the consensus (e.g., $J_{\beta 1.2}$ differs at five of nine and three of seven positions), suggesting the possibility that a different recombinase might carry out rearrangement of TCR genes. However, it seems likely that a single recombinase is responsible for rearrangement of all loci and that rearrangement is controlled at a different level (see section III.).

Many of the data concerning this point have come from studies using transfected substrates. Introduced κ gene segments were shown to rearrange in a pre-B cell line previously shown to carry out endogenous V_κ-J_κ joining in culture (Lewis et al., 1984). Likewise, rearrangement of a transfected D-J_H substrate was shown to occur in a line capable of rearranging endogenous H gene segments (Blackwell and Alt, 1984). When the analogous experiment was performed with a substrate containing TCR β gene segments, joining was frequently observed even though these pre-B cells did not rearrange their endogenous TCR genes (Yanco-poulos et al., 1986). These results supported the idea that a common recombinase may rearrange all immunoglobulin and TCR loci and led to the proposal that the accessibility of a given locus to the recombinase is the controlling factor in rearrangement.

Other experiments using a cell line (38B9) in the H-chain phase of endogenous gene assembly have indicated that this cell line is capable of rearranging transfected L-chain gene segments as well. When a rearrangement substrate containing the herpes simplex virus thymidine kinase (*tk*) gene flanked on one side by $V_{\lambda 1}$ and on the other side by $J_{\lambda 1}$ was introduced into 38B9, V to J rearrangements could be detected (Blackwell et al., 1986). Interestingly, only copies that were transcriptionally active showed evidence of rearrangement. Also, a different rearrangement test gene, containing RSS derived from V_κ and J_κ segments, transfected into the same cell line rearranged in virtually every cell that took up the DNA (Engler and Storb, 1987). This occurred even without selection for expression or rearrangement.

It is clear that the recombinase is not ubiquitously expressed; most cell types will not rearrange transfected recombination substrates, even though they are in an "accessible" state. No rearrangement of transfected substrates has been detected in fibroblasts or in various myeloma lines (Blackwell et al., 1986; Engler and Storb, 1987). These results suggest that the recombinase is turned on only in certain specific lineages and is turned off after gene rearrangement has occurred. The control of immunoglobulin gene rearrangement is the topic of section III.

Very little is presently known about the enzymology of immunoglobulin gene rearrangement. The number of components involved, their cell type distribution, and their mode of action are all unknown. Presumably, gene rearrangement at its simplest would involve recognition of the RSS, endonucleolytic cleavage, and religation. It is conceivable that all these functions are carried out by a single enzyme, or it may be that multiple molecules participate.

Several groups have reported the presence of endonucleolytic activity in nuclear extracts of pre-B cells as well as in other sources (Kataoka et al., 1984; Desiderio and Baltimore, 1984; D'Agostaro et al., 1985; Hope et al., 1986). Double-strand cleavage has been shown to occur at sites near the RSSs, although in most cases substantial nonspecific cleavage was also seen. It appears that this nuclease is not restricted to lymphoid cells since HeLa- and L-cell extracts also have this activity. Thus, while the reported endonucleases do have certain of the properties expected of enzymes involved in immunoglobulin gene rearrangement, the lack of precise specificity and expression in a variety of cell types makes it unlikely that they are the sole component of the recombinase system. It is possible that in conjunction with other regulatory subunits they may be a component of a multiunit complex which carries out the recombination.

A protein has been described that is present only in pre-B cell lines and that binds specifically to a synthetic RSS oligonucleotide (Aguilera et al., 1987). This protein binds the RSS in a gel retardation assay and protects G residues from methylation in vitro primarily in the 7-mer region. It can be found in various pre-B and, to a lesser extent, some B cell lines but not in fibroblast, T, or mature myeloma cell lines. The binding protein is clearly distinct from the endonuclease, as they can be differentially eluted from a heparin-agarose column.

A mutation which apparently affects immunoglobulin and TCR gene rearrangement has been described (Bosma et al., 1983; Schuler et al., 1986). A mouse mutant (*scid*) with severe combined immune deficiency lacks functional B and T cells. Analysis of cell lines derived from the mice showed that rearrangements of H-chain and TCR β genes were defective: deletions of the entire J regions were common. In normal rearrangements deletions of a few nucleotides are common (section II.D.), but the extent of the deletions in the *scid* mice suggests that a component of the recombinase system in these mice is faulty.

C. Orientation of Recombining Segments

As summarized by Tonegawa et al. (1977), there are, in principle, four possible mechanisms for the somatic joining of immunoglobulin gene segments: (i) copy insertion, (ii) excision insertion, (iii) deletion, and (iv) inversion.

The copy insertion mechanism, in which one of the many V genes is duplicated and translocated to near the C gene segment, would result in the

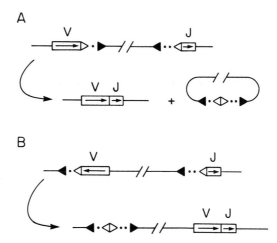

FIGURE 6. Schematic illustration of rearrangement by deletion (A) or by inversion (B). Straight arrows indicate transcriptional polarity of the gene segments. Open and filled triangles are the heptamer and nonamer sequences, and the dots indicate spacers (as in Fig. 4).

retention of all copies of V regions originally present. This model was put into doubt when it was shown that the restriction fragment carrying a particular V gene was present in embryonic DNA but not in the genome of a myeloma line expressing that V region (Tonegawa et al., 1977).

A prediction of the excision insertion model was that after translocation of a particular V gene to the C locus, all DNA between V and C would be retained. However, this was not found to be true for λ rearrangement (Sakano et al., 1979), for κ (Seidman et al., 1980), for V_H (Cory and Adams, 1980), or for D_H (Sakano et al., 1981). These results demonstrating loss of sequences strongly suggested that immunoglobulin gene rearrangement proceeded via an intramolecular deletion. The structure of the RSSs suggested a specific mechanism for this process: since the V and J RSSs are self-complementary, they could base pair to form a stem loop structure, thus bringing V and J into proximity. A cutting and ligation would complete the joining process. Although it has been substantially confirmed that intramolecular deletion is one possible rearrangement mechanism, the existence of the stem loop intermediate structure now seems unlikely.

That the situation was not so simple was hinted at by the finding, in certain immunoglobulin-producing cells, of sequences 3' of V segments and 5' of J segments—sequences that should have been eliminated if rearrangement was strictly due to intramolecular deletion. These rearranged sequences contain V_κ and J_κ RSSs fused in head-to-head fashion (i.e., 9-mer/spacer/7-mer/7-mer/spacer/9-mer; Fig. 6B). Significantly, these rearranged sequences all contain the $J_{\kappa 1}$ RSS fused to various V_κ RSSs; neither the V nor the J is the one used in the productive rearrangement found in the cell lines analyzed (Steinmetz et al., 1980; Hochtl et al., 1982; Selsing et al., 1984). It was suggested that this finding was best explained by an unequal sister chromatid exchange because the V-J joint and the reciprocal recombination product would segregate into different cells. However, these data are also consistent with V-J rearrangement via inversion followed by secondary rearrangements (see below).

Evidence for multiple rearrangement of endogenous κ gene segments together with the retention of the reciprocal products has been presented (Feddersen and Van Ness, 1985). The rearrangement in the plasmacytoma PC7043 consists of an aberrantly joined V-$J_{\kappa 2}$ which is not expressible because of a frameshift at the joint. Just downstream of the $J_{\kappa 2}$ are the 9-mer/7-mer of the $J_{\kappa 4}$ segment fused head-to-head with the 7-mer/9-mer from a second V gene. Most significantly, this second V is the one found productively rearranged to $J_{\kappa 4}$ in the same cell. These data show that multiple rearrangements can occur on the same allele and indicate that both reciprocal products of a V-J rearrangement can be found in the same cell, thus ruling out segregating mechanisms. The simplest interpretation is that two inversion events on the same chromosome are responsible. Clearly, a detailed understanding of the allowable joining events at the κ locus awaits a better understanding of the organization of the region—in particular the relative orientations of V regions and the J cluster as well as the orientation of V segments among themselves.

Some of this information is available for the TCR β locus (Malissen et al., 1986). One particular V segment ($V_{\beta 14}$) is found 10 kb 3' of a C region ($C_{\beta 2}$) and is in the opposite transcriptional orientation. Furthermore, a particular cell line is known that has productively joined these two segments. Therefore, a chromosomal inversion must have occurred to generate this structure.

In immunoglobulin gene rearrangement by inversion, both the coding joint and the reciprocal joint (head-to-head 7-mers) are retained, while in deletional rearrangement the intervening DNA is eliminated. However, in TCR α (Fujimoto and Yamagishi, 1987) and TCR β (Okazaki et al., 1987) gene rearrangement, the excision products of rearrangements by deletion have been isolated as circular DNA molecules. It has been known for some time that thymocytes contain up to 200 copies of circular DNA molecules per cell (Yamagishi et al., 1982); analysis of these circles has shown that some contain the reciprocal products of TCR gene rearrangement. The single reported J_{α}-containing circle (Fujimoto and Yamagishi, 1987) contained the RSS from a V_{α} fused head-to-head with the RSS from $J_{\alpha 2}$; presumably, the coding joint remained on the chromosome. Several D_{β}- or J_{β}-containing circles were described (Okazaki et al., 1987). One class consisted of a J_{β} RSS fused head-to-head with the RSS from a D_{β}. The other class consisted of the fusion of an upstream J_{β} to a downstream D_{β} (a pseudonormal joint [Alt and Baltimore, 1982]); in this case the head-to-head RSSs were presumably left on the chromosome. It is not known whether such circles exist in B cells.

Thus, it appears that whether rearrangement occurs via deletion or inversion (and this is determined solely by the relative orientation of the recombining segments), the same products result: fused coding regions plus fused RSSs. If the coding regions were originally in the same orientation, then the reciprocal product is deleted from the chromosome as a circle; if the coding regions were in opposite orientation, then the reciprocal product is retained on the chromosome. The fine structure of the coding joints and the reciprocal joints will be considered in the next section.

Further evidence that immunoglobulin gene rearrangement may occur by a deletion or an inversion mechanism has come from work involving transfer of recombination substrates into actively rearranging cell lines.

One recombination substrate contained a selectable marker flanked on one side by the $J_{\kappa1-5}$ cluster and on the other side by a V_κ gene segment in opposite orientation with respect to the J_κ cluster (Lewis et al., 1984, 1985). A rearrangement by inversion would flip the selectable marker into the proper orientation so that it could be transcribed from a long terminal repeat promoter contained in the retroviral vector. Analysis of drug-resistant cells obtained after infection of pre-B cells with this vector showed that rearrangement of V to either $J_{\kappa1}$ or $J_{\kappa4}$ had occurred by simple inversions. In addition to these simple inversions, more complex secondary rearrangements were also found. These rearrangements were the result of the V_κ RSS, after flipping during the first inversional rearrangement, now being oriented to rearrange by deletion. This sort of secondary rearrangement by deletion provides an explanation for the occurrence of fused head-to-head 7-mers not reciprocal to the coding joint often found in myeloma lines.

Other recombination substrates containing κ gene segments were designed for rearrangement by deletion (Engler and Storb, 1987; Akira et al., 1987) or by either deletion or inversion (Hesse et al., 1987). After transfer of these substrates into recombination-proficient pre-B cell lines, both types of rearrangement were frequently detected. When both deletion and inversion were allowed on the same episome (Hesse et al., 1987), inversions were somewhat less frequent than deletions. It is not clear whether this reflects an intrinsic preference of the recombination system or is an artifact resulting from the arrangement of the RSSs. Clearly, rearrangement is able to proceed by either inversion or deletion, and which one occurs is determined only by the orientation of the participating RSSs.

D. Structure of Recombinant Junctions

The joining of immunoglobulin gene segments is often imprecise; this imprecision can result in increased amino acid diversity at the antigen-binding region in the immunoglobulin molecule. Another consequence of this imprecision can be the generation of nonfunctional genes, caused by creation of nonsense codons or shifting of the reading frame. These faulty rearrangements clearly play an important role in allelic exclusion (unlike other autosomal genes, only a single H- or L-chain allele is expressed), although it is also clear that this is not the sole mechanism. Therefore, a detailed knowledge of the products of immunoglobulin gene rearrangement is important for understanding the functioning of the immune system as a whole as well as for determining the mechanism of the unusual enzyme system responsible for rearrangement.

A careful comparison of the amino acid sequences of 20 κ chains of the $V_{\kappa21}$ subgroup with the known DNA sequences of $V_{\kappa21}$ and the four functional J_κ segments led to the conclusion that the point of recombination is often imprecise (Weigert et al., 1980). The 95th amino acid is always specified by a V codon, and in many cases the 96th amino acid is encoded by the first codon of one of the Js. However, a number of amino acid sequences cannot be accounted for by a precise fusion of V to J; all of these sequences can be explained by recombination within, rather than between, codons.

The variability in the site of recombination does not always result in an in-frame joint. For example, the myeloma lines MOPC173 and MOPC21 have two V_κ-J_κ rearrangements (Max et al., 1980; Walfield et al., 1981). One of these rearrangements corresponds to the expressed κ chain; the other joint results in an out-of-phase reading frame and therefore in a prematurely terminated κ protein.

In contrast to the variability often seen at the κ V-J coding joint, the fused head-to-head 7-mers found at the κ locus have all been precise. As mentioned in the previous section, these all contain the 7-mer of $J_{\kappa 1}$ fused to the 7-mer of different (random?) V segments. Not one of the nine reported endogenous joints contains a deletion or insertion of even a single nucleotide.

The junctions of endogenous H gene segments also exhibit variability. In addition to deletions of variable length, the D_H-J_H joint often contains additions of one to several nucleotides that cannot be assigned to either the D or J coding region (Kurosawa and Tonegawa, 1982). It has been suggested that these regions (termed N for nucleotide) represent de novo additions of nucleotides at the point of recombination (Alt and Baltimore, 1982). The enzyme terminal deoxynucleo-tidyltransferase, which is able to catalyze the addition of nucleotides onto free 3′ ends without a template, may be responsible for N-region addition. In support of this hypothesis is the finding of N regions in four of six joints isolated from a terminal deoxynucleotidyltransferase-positive pre-B cell line, whereas no N regions were found in four joints formed in a cell line lacking this enzyme (Desiderio et al., 1984). However, N regions are present in TCR β-chain genes from the moth-eaten mutant mouse (Hagiya et al., 1986a), which is deficient in terminal deoxynucleotidyltransferase. Thus, the origin of these noncoded additions remains uncertain.

Since H-chain gene rearrangement is accompanied by the deletion of the RSSs and all intervening sequences, the reciprocal products of V_H-D_H or D_H-J_H are not available for analysis. There are, however, instances of aberrant inversional rearrangements of D_H and J_H which allow recovery of the reciprocal products of these joinings. In one rearrangement, involving a D to J_{H3} joint, perfectly fused 7-mers were found, just as in κ (Alt and Baltimore, 1982). But in another case also involving a D to J_{H3} inversion, four nucleotides separated the head-to-head 7-mers (Liu et al., 1987).

Other exceptions to the rule of precise reciprocal joints are found in the circles generated during TCR β gene rearrangement (Okazaki et al., 1987). In these cases it appears that the nucleotides between the head-to-head 7-mers are derived from the associated coding regions, suggesting that these sequences are the result of imprecise cleavage rather than N-region addition.

Analysis of the recombinant junctions of transfected substrates is very informative because, unlike the situation in myeloma lines, there is no selection for expression of a protein product and because multiple independent rearrangements of segments of known structure can be studied.

In the systems designed to allow V-J joining via inversion, both the coding joint and reciprocal joint are retained. In one study eight of nine reciprocal joints consisted of perfect head-to-head 7-mers; the other instance had a deletion of a single nucleotide at the junction (Lewis et al., 1985). Another system was designed so that the reciprocal joint was retained after deletion (in this case, because of the orientation of the recombining segments, the coding joint is deleted); inversions were also allowed, and in these events both the coding joints and reciprocal joints were retained (Hesse et al., 1987). Analysis of 99 reciprocal joints from 86 recombinant molecules showed that in every case the restriction site expected to be found by a perfect 7-mer–7-mer fusion was indeed present. Thus, the results from transfected substrates confirm the original observations that the joining of RSS to RSS is almost always precise. It is possible that the RSSs are protected from alteration by binding of a component of the recombinase; the coding joint may be more accessible to exonucleases.

The published sequences of the coding joints of transfected substrates are compiled in Fig. 7. Only 1 of the 29 junctions is the result of perfect fusion of the coding regions (although this would in fact be out of frame); all others have deletions and/or additions of variable length. Although the joining of the coding regions nearly always occurs in the same general area, the exact point of recombination appears to be essentially random. Most importantly, there is no significant bias toward one reading frame.

E. Class Switching

Immunoglobulin H-chain genes undergo another form of rearrangement, distinct from V-(D)-J joining, known as class switching. This recombination event results in the shifting of an already rearranged VDJ complex to a downstream C_H gene segment. Each C_H gene segment (with the exception of C_δ) has characteristic repeated sequences located ~2 kb 5' of the C coding region, and these appear to be involved in switch recombination.

Switch regions consist of randomly repeated short sequences GAGCT and GGGGT that are shared by all S (switch) regions as well as other repeated units that are more or less unique to the specific S region. The S_μ region consists of two distinct portions. The 5' half contains repeats of ${}^C_T$AGGTTG which are also found in other S regions, while the 3' half consists almost entirely of GAGCT repeats interspersed with an occasional GGGGT pentamer (Marcu et al., 1982). The S_ε and S_α regions are similar to each other and to S_μ, all having multiple repeats of the same pentameric sequences (Nikaido et al., 1982). The four S_γ regions are homologous, all being based on a 49-bp repeat, but are not closely related to the S_μ sequence (Szurek et al., 1985; Mowatt and Dunnick, 1986). It has been pointed out that homology with S_μ is greatest with $S_{\gamma 3}$ and decreases with downstream S regions (Shimizu et al., 1982).

The C_δ gene has no apparent S region, and in some cases IgD expression is the result of alternative splicing of a VDJ-C_μ-C_δ message (Moore et al., 1981; Maki et al., 1981). However, in certain IgD-producing cell lines it has been shown that the C_μ gene has been deleted at the DNA level (Moore et al., 1981; Maki et al., 1981), but no sequences related to S regions were found at the 3' deletion endpoint (Richards et al., 1983).

```
V_κ21      GGATCCTCCCACAGTG        Lewis, Gifford and Baltimore,
                                                        1984, 1985
           CACTGTGGTGGACGT     J_κ1
           CACTGTGATTCACGT     J_κ4
                                                   bp-    bp+
           GGATCC-----GGACGT   V -> J_1             5      0
           GGATCCT----GGACGT   V -> J_1             4      0
           GGATCCTC-GTGGACGT   V -> J_1             1      0
           GGATCCT---TGGACGT   V -> J_1             3      0
           GGATCCT-----ACGT    V -> J_4             6      0
           GGATCC-atATTCACGT   V -> J_4             3      2
           GGATCCTC-----ACGT   V -> J_4             5      0
           GGATCC-----GGACGT   V -> J_1             5      0
           GGATCC---GTGGACGT   V -> J_1             3      0


D_Q52      CAACTGGGACCACGGTG       Blackwell and Alt, 1984
                                   Blackwell et al., 1986

           TAGTGTGACTACTTTGACT   J_H2
           CAATGTGCCTGGTTTGCTT   J_H3
           TATTGTGATTACTATGCTA   J_H4
                                                   bp-    bp+
           CAACTGGG-----------CT    D -> J_2       12     0
           CAACTGG--------TTTGACT   D -> J_2        8     0
           CAAC-cc--GCCTGGTTTGCTT   D -> J_3        5     2
           CAACTG---------TTTGCTT   D -> J_3        9     0
           CAACTGGGAC---------TT    D -> J_3       10     0
           CAACTGGGA----GGTTTGCTT   D -> J_3        4     0
           CAACTGGGAC-----TATGCTA   D -> J_4        5     0
           CAACTGGG---tgag---GCTA   D -> J_4       10     4
           CAACTGGGA----ACTATGCTA   D -> J_4        4     0


D_β1.1     CAGGGGGCCCACGGTG       Yancopoulos et al., 1986

           CACTGTGCAAACACA     J_β1.1
           TGATGTGCAAACTCC     J_β1.2
                                                   bp-    bp+
           CAGG-atggg-ACACA    D -> J_1            7      5
           CAGGGG---ag---CA    D -> J_1            8      2
           CAGGG-----ACTCC     D -> J_2            5      0
           CAG-----cttcttgc    D -> ?             5      -


V_λ1       AACCATTTCCCACAATG      Blackwell et al., 1986

           CACAGTGCTGGGTGTT    J_λ1
                                                   bp-    bp+
           AACC-----CTGGGTGTT  V -> J              5      0


V_κM167    GTATCCTCCACAGTG        Engler and Storb, 1987

           CACTGTGGTGGACGT     J_κ1
                                                   bp-    bp+
           GTATCCTCGTGGACGT    V -> J              0      0
           GTATCCTC--GGACGT    V -> J              2      0
           GTATCCT-a-GGACGT    V -> J              3      1
           GTATC---GTGGACGT    V -> J              3      0
           GTA-CC--GTGGACGT    V -> J              3?     0
           GTATCCTCCGTGGACGT   V -> J              0      1?
```

FIGURE 7. Compilation of coding joints of recombined artificial substrates. The heptamers associated with the coding regions are underlined. Hyphens indicate bases lost during rearrangement; bases added are in lowercase type.

Class switching could proceed by several mechanisms: intrachromatid deletion, sister chromatid exchange, or recombination between homologs. To distinguish among these possibilities, Wabl et al. (1985b) examined a cell line that switches from C_μ to $C_{\gamma 2b}$ for retention of C_μ sequences. All the $\gamma 2b$-producing subclones had lost at least one copy of C_μ, while none contained three copies, suggesting that at least in this cell line, class switching is due to an intrachromatid deletion. The problem of class switching has recently been approached by using transfected substrates (Ott et al., 1987). A retroviral vector harboring a *tk* gene flanked on one side by S_μ sequences and on the other side by $S_{\gamma 2b}$ regions was introduced into fibroblasts and pre-B-cell lines. Deletion of *tk* in pre-B cells occurred at a rate of 1×10^{-5} to 5×10^{-5} cell^{-1} generation^{-1}, while *tk* activity was lost in fibroblasts about 10^3 times less frequently. The loss of activity in pre-B cells was the result of deletions involving the flanking S regions; other mechanisms were responsible for inactivation in fibroblasts. These results make it clear that there are cell-specific factors involved in S recombination and that it is not simply a matter of homologous recombination between adjacent repeated sequences. As is the case with V-(D)-J joining, this important event in B-cell development is under strict control.

III. CONTROL OF REARRANGEMENT

As we discussed in section I., only cells of the B lymphocyte lineage are capable of correctly rearranging and expressing immunoglobulin genes. In this section we consider how the ordered rearrangement of these genes is controlled during B-cell development.

A. Sequential Rearrangement of H and κ Genes

1. Turning on rearrangement

There is now good evidence that the same recombinase is responsible for the rearrangement of all types of immunoglobulin genes, as well as TCR genes (see also II.B.). Thus, the sequential rearrangement of endogenous H and L genes during B-cell development cannot be explained by a corresponding set of specific recombinases. Instead, it has been proposed that accessibility of the immunoglobulin genes to the recombinase determines whether rearrangement is turned on (Yancopoulos and Alt, 1985).

In support of this notion, in early pre-B cells, unrearranged V_H genes are transcribed at the same time that active rearrangement of H genes is ongoing (Yancopoulos and Alt, 1985). In more mature pre-B cells, B cells, and plasma cells, however, germ line V_H genes are not transcribed (Yancopoulos and Alt, 1985). Apparently, in the mature pre-B cells H gene rearrangement has ceased, and κ gene rearrangement has begun (Reth et al., 1985).

There is some evidence also which suggests that germ line V_κ genes may be in active chromatin before or at the start of κ gene rearrangement. A germ line V_κ

gene (V_κ MOPC167) has DNase I-hypersensitive sites and is partially under-methylated in the pre-B cell line 18-81 (U. Storb and B. Arp, unpublished data). The germ line V_κ167 and V_κ MPC-11 genes are not DNase I sensitive and are completely methylated in some, but not all, mature B-cell lines, myelomas, and hybridomas (Storb and Arp, 1983; Mather and Perry, 1983). Obviously, the question of regulation of κ gene rearrangement by the accessibility of germ line V_κ genes needs further analysis. It is clear, however, that germ line, unrearranged C_κ genes are transcribed in hybridomas from fetal liver pre-B cells (Perry et al., 1981b) and are DNase I sensitive in myelomas (Storb et al., 1981).

Further evidence that accessibility of the rearrangement substrate determines whether a gene will be rearranged comes from transfection experiments (Yanco-poulos et al., 1986). TCR genes are normally not rearranged in pre-B cells. However, a transfected TCR gene is rearranged and DNase I sensitive, while the endogenous TCR gene of the pre-B cell remains DNase I insensitive and unrearranged.

An unexplained phenomenon is the preferential rearrangement of the most J_H-proximal V_H genes in early pre-B-cell lines (Yancopoulos et al., 1984; Perlmutter et al., 1985). This may be related to accessibility or to chromosomal location, suggesting perhaps that the recombinase operates in a one-dimensional scanning mode (Yancopoulos et al., 1984).

The molecular composition of rearrangement-accessible DNA regions is unclear. Open chromatin or transcription, or both, may contribute. Transfection experiments with a rearrangement substrate containing a herpes simplex virus *tk* gene have supported a requirement for transcription (Blackwell et al., 1986). The *tk* gene was arranged with its own transcriptional control sequences about 1 kb away from the rearrangement substrate (D_H/J_H). In most of the transfected cells, *tk* was not expressed and the immunoglobulin sequences did not rearrange. However, by hypoxanthine-aminopterin-thymidine selection one in 10^4 to 10^5 cells could be isolated which transcribed *tk*. The majority of these also rearranged the immunoglobulin substrate. Only a few of the clones not selected in hypoxan-thine-aminopterin-thymidine selection showed rearrangement. In this experiment rearrangement appeared to be correlated with transcription and not open chromatin, since all of the transfected test genes seemed to be DNase I sensitive. This result is apparently in conflict with other observations. The chicken λ gene was rearranged in the thymus of transgenic mice despite lack of demonstrable transcription (Bucchini et al., 1987). Furthermore, deletion of the transcriptional promoter from a mouse κ rearrangement substrate had no effect on rearrangement frequency (unpublished data). A very low or transient level of transcription from host cell promoters at the integration site has not been ruled out in the latter experiments. However, in the experiment by Blackwell et al. (1986) it is unclear why the *tk* gene was not transcribed in the majority of cells. Perhaps the inactive *tk* gene may impose unusual constraints upon the neighboring rearrangement substrates. Obviously, the question of a transcriptional requirement for rearrange-ment needs further study using different test genes for transcription.

2. H genes rearrange first

The details of the molecular events which regulate the sequential rearrangement of H and κ genes are not known. A basic model which is compatible with most of the data is that H gene rearrangement is turned on by the production of the recombinase in pre-B cells and the simultaneous opening of the chromatin of H genes. H gene rearrangement is turned off when μ H chains are produced. In this model κ gene rearrangement is turned on by the presence of the μ protein and is turned off when κ chains have been produced which, together with H chains, provide a signal for complete shutoff of the recombinase. The following is a discussion of the experimental results relevant to this model.

The B-cell lineage develops late in ontogeny of the mouse. At about days 17 and 18 of gestation pre-B lymphocytes are present with relatively few of the later B-cell forms. At this time μ mRNA and μ chains are found in the fetal liver, but L-chain mRNA does not reach the level of μ chain until after birth (Levitt and Cooper, 1980; Siden et al., 1981). The conclusion that the sequential expression of H and L genes is due to sequential rearrangement is further supported by the analysis of A-murine leukemia virus (A-MuLV)-transformed lymphoid cell lines (Alt et al., 1981) and hybridomas of fetal liver cells (Maki et al., 1980; Perry et al., 1981b). Many of these have rearranged only the H genes, not the L genes. L gene rearrangement is only found in cells which also have rearrangement of H genes.

Within the H locus, rearrangement is apparently also ordered in a strict sequence (Alt et al., 1984). The rearrangement of J_H with D_H precedes the joining of D_H with V_H. No evidence has been obtained for V_H joining to an unrearranged D_H sequence (Alt et al., 1984; Aguilera et al., 1985). The first H gene joint, $D_H J_H$, results in the production of D_μ mRNA transcribed from promoters upstream of each D and translated into D_μ protein (Reth and Alt, 1984). It is unknown whether D_μ protein may play a regulatory role in immunoglobulin gene expression or B-cell differentiation. It has been pointed out that D_μ protein does not promote allelic exclusion (see below) and that, thus, V_H to DJ_H joining is responsible for allelic exclusion in H gene activation (Reth and Alt, 1984). The sequential DJ and then VD rearrangement may be controlled by open chromatin restricted to the D and J regions of immature pre-B cells, with only a later opening up of V_H, perhaps under the influence of D_μ protein. As yet, no data are available concerning this control. While, as expected from the one-turn/two-turn rule of rearrangement (see II.A.), no DD joints have been found (Alt et al., 1984), secondary DJ rearrangements have been described (Reth et al., 1986; Maeda et al., 1987). In an A-MuLV–transformed pre-B cell line one of the alleles went from a $D_{sp\ 2.8}$-J_{H3} joint to a $D_{FL16.1}$-J_{H4} joint; i.e., a D sequence upstream of the original DJ joint rearranged to a J_H located downstream of that joint.

As described above (section II.D.), there seems to be no mechanism to assure rearrangement in the correct reading frame. This is further supported by the analysis of A-MuLV cell lines in which random DJ and VD joints are found (Hagiya et al., 1986b). If the joining and N-region addition are random, approximately one-third of joining events would be in frame. Thus, $1/3^2$, or approximately one-tenth of the chromosomes, would have one correct VDJ_H joint. This is roughly the experimental result. About 90% of the alleles of a functional VDJ_H gene are out-of-frame DJ_H or VDJ_H rearrangements (Coleclough et al., 1981; Coleclough, 1983).

Cell lines obtained by A-MuLV transformation of pre-B cells can undergo immunoglobulin gene rearrangement in culture (Alt et al., 1981; Lewis et al., 1982; Whitlock et al., 1983). A particular early pre-B cell line has been observed to sequentially rearrange H and κ genes (Reth et al., 1985). The parent cell line has DJ rearrangements on both H alleles, one of which is in frame and encodes a D_μ protein. About 20 to 30% of sublines undergo one correct VDJ_H rearrangement and produce complete μ chains. All of the H-chain producers also rearrange κ genes, many of them productively. It was concluded from these data that κ gene rearrangement had been induced by the presence of the μ proteins. There is now some direct evidence that this may be the case: a nonproducing A-MuLV cell line which is transfected with a productive μ gene gives rise to some sublines which rearrange κ genes (M. Reth, personal communication).

H chains are apparently involved in the shutoff of H gene rearrangement once a productive VDJ_H gene has been assembled. Most of the evidence comes from transgenic mice with functional μ transgenes. In 40% of A-MuLV–transformed bone marrow pre-B cell lines from μ transgenic mice, at least one of the endogenous H-chain gene alleles is in germ line form, i.e., has not undergone a DJ_H rearrangement (Weaver et al., 1985). In A-MuLV cell lines from normal bone marrow all J_H alleles are generally rearranged. In more mature B cells from μ transgenic mice about 10% (Weaver et al., 1985; Rusconi and Köhler, 1985) or 18% (J. Manz, R. Brinster, and U. Storb, unpublished data) have germ line J_H genes. Furthermore, VD rearrangement seems also to be inhibited by the presence of a μ transgene in a fraction of B cells (Weaver et al., 1985; Rusconi and Köhler, 1985; Manz et al., unpublished data). It is not clear why the postulated feedback by the transgenic μ protein is so incomplete. Perhaps the level of transgenic μ expression at the time when the recombinase is turned on in pre-B cells is not sufficient. This question will have to be addressed further when more is known about the molecular mechanism of the feedback. Apparently, the membrane form of μ chains can produce the effect (Nussenzweig et al., 1987; Manz et al., unpublished data), while the secreted form of transgenic μ does not exert any feedback on endogenous H gene rearrangement (K. Denis, J. Manz, R. Brinster, and U. Storb, unpublished data). Furthermore, the human γ1 chain (Yamamura et al., 1986) or the mouse γ2b chain (K. Denis, H. Tsang, R. Brinster, and U. Storb, unpublished data) has no effect on H gene rearrangement. It appears, therefore, that feedback on H gene rearrangement can only be effected by μ chains with a transmembrane tail, not by secreted μ chains or membrane-bound or secreted γ chains.

3. Allelic Exclusion of κ Gene Rearrangement

Allelic exclusion of immunoglobulin gene rearrangement denotes the fact that a B cell produces H and L chains encoded by only one productive VDJ_H and VJ allele. As described above, H gene allelic exclusion appears to be due to the inactivation of germ line V_H genes by some unknown feedback mechanism which involves membrane-bound μ chains. It may be that κ gene rearrangement is turned on coordinately with the V_H gene inactivation. How is the observed allelic exclusion of κ genes controlled? Transgenic mice with a functional, rearranged κ transgene have given a partial answer (Brinster et al., 1983; Ritchie et al., 1984). It appears that this is again a feedback mechanism. However, κ chains alone do not signal the cessation of κ gene rearrangement. Only those B cells which produce functional H chains together with transgenic κ chains do not rearrange endogenous κ genes (Ritchie et al., 1984). (It is not clear how these findings can be compatible with the postulated requirement for H chains to turn on κ gene rearrangement [Reth et al., 1985], i.e., cells from κ-transgenic mice which have turned on endogenous κ gene rearrangement in the absence of H chains ought not to exist.) There is also some evidence which indicates that for the feedback on κ gene rearrangement membrane μ chains are required; there is no feedback in the case of transgenic mice with a κ transgene and a μ transgene which encodes only secreted μ chains (Manz et al., unpublished data). The feedback on κ gene rearrangement may require a balanced amount of μ and κ chains , because in κ transgenic mice, which produce 10 times less κ chains than μ chains, no feedback on endogenous κ gene rearrangement has been observed (Rusconi and Köhler, 1985). Furthermore, the requirements for the κ sequence may be quite specific. In transgenic mice with an unrearranged rabbit κ transgene, apparently κ allelic exclusion is absent or defective, i.e., both mouse and rabbit κ genes are often rearranged in the same B cell (Goodhardt et al., 1987).

The molecular basis for the feedback on κ gene rearrangement needs to be determined. The $\mu\kappa$ combination may operate by a direct intracellular mechanism or via insertion into the plasma membrane. B cells with surface immunoglobulin molecules may, for example, interact with T cells to receive a "maturation" signal which would lead to the cessation of rearrangement. Perhaps the feedback on κ gene rearrangement results in a complete shutoff of recombinase production. As discussed above (section II.B.), myeloma cells, which represent the final B-cell stage, do not have recombinase activity (Blackwell et al., 1986; Engler and Storb, 1987).

B. Rearrangement of λ Genes

Immunoglobulins secreted from plasma cells contain either κ or λ L chains, but not both. Thus, in these cells productive κ and λ genes are mutually exclusive. The phenomenon has been termed κ-λ isotypic exclusion. On the level of the respective genes, κ-producing cells have rearranged κ genes,but the λ genes are almost always in germ line configuration. However, in most λ-producing cells the κ genes are either nonproductively rearranged or deleted (Perry et al., 1981a; Hieter et al., 1981; Korsmeyer et al., 1981). These observations have led to a "sequential model" to explain κ-λ isotypic exclusion (Hieter et al., 1981). The model postulates that κ genes rearrange first, and only if both κ alleles are aberrantly rearranged or deleted do λ genes become accessible for rearrangement. The simple model was not concerned with the control of this sequence; however, a later observation has suggested a control mechanism. It was found that 3' of C_κ genes a sequence is located which is preceded by typical immunoglobulin gene rearrangement recognition units (7-mer, 23-bp spacer, and 9-mer) and which is rearranged in most λ-producing cells (Durdik et al., 1984; Moore et al., 1985; Siminovich et al., 1985). The rearrangement to a V_κ gene or to a 7-mer within the intron of J_κ-C_κ causes the deletion of the intervening DNA, including the J_κ genes and/or C_κ. The human κ deleting element has recently been mapped 24 kb downstream of C_κ (Klobeck and Zachau, 1986). It has been proposed that recombination of this element may activate a transacting factor that induces λ gene rearrangement (Siminovitch et al., 1985; Persiani et al., 1987). Transcripts from the κ deleting element or its murine homolog have, however, not been found consistently (E. Selsing, personal communication). Therefore, it has remained an open question whether rearrangement of these elements plays a regulatory role.

Recent findings in transgenic mice with a functionally rearranged κ transgene have challenged the sequential model of κ-λ gene expression (K. Gollahon, J. Hagman, R. L. Brinster, and U. Storb, *J. Immunol.*, in press). In keeping with the sequential model, one would expect that in κ-transgenic mice λ genes would not be rearranged in B cells which express the κ transgene. However, when λ-producing hybridomas were isolated after fusion of spleen cells from transgenic mice with a functional κ transgene, they were found to coexpress κ and λ chains. Most of the hybridomas also produced H chains which were associated with both the κ and λ chains. Thus, these cells violated the rule of isotypic exclusion. Furthermore, the λ-producing cells also did not obey κ gene allelic exclusion: despite the presence of the transgenic κ together with H chains, endogenous κ genes were often rearranged or deleted. Finally, some of the hybridomas continued to rearrange their immunoglobulin genes.

These results are incompatible with a strict sequential model in which κ genes must have exhausted the ability to be functional before λ genes can be turned on: the active κ transgene was always present, and in some of the λ hybridomas from κ transgenic mice the endogenous κ genes were unrearranged. Rather, it appears that λ expression is coupled with the absence of feedback inhibition of rearrangement at the pre-B-cell stage. This does not seem to be a phenomenon restricted to transgenic mice, since κλ double producing cells can be found in normal mice (J. Kearny, personal communication; Gollahon et al., in press). It is not clear whether the B cells which lack feedback are a separate cell lineage or whether a certain random fraction of the B-cell population does not acquire the feedback mechanism. In any case, we have postulated that in these cells κ and λ gene rearrangements continue until most sequences capable of rearrangement have been exhausted (Storb, 1987; Gollahon et al., in press). In the κ locus this may often lead to deletion of C_κ by recombination of the κ deleting element sequence (Durdik et al., 1984; Siminovitch et al., 1985; Moore et al., 1985). There is no evidence for such a recombining sequence in the λ locus (personal communications: E. Selsing for mouse, K. Siminovitch for the human λ locus). Therefore, λ genes will be retained, and if a correct λ gene rearrangement exists, after κ locus inactivation the B cell will have become a pure λ producer. The RSS was lacking from the κ transgene; therefore, a higher than normal proportion of κλ-coproducing cells was seen. This model does not explain λ allelic exclusion, i.e., the absence of two different λ chains in λ myelomas. This may be due to a low frequency of antigen triggering of B cells which expose scrambled antibodies at the cell surface. Furthermore, perhaps in λ-producing cells rearrangement stops eventually, for example, at the plasma cell stage. These problems need to be studied further. However, the strict sequential model which prohibits coexpression of κ and λ seems to be untenable.

C. Alteration of Already Rearranged Genes

1. Rearrangement of V_H to VDJ_H

In general, VDJ_H rearrangements are stable, because no unrearranged D regions remain on the same chromosome and because the rearrangement of an upstream V_H gene to a downstream J_H gene is prohibited by the one-turn/two-turn rule (see section II.A.). However, it has been found that most mouse and human V_H genes have a highly conserved internal 7-mer near the 3' end of the coding region (Reth et al., 1986; Kleinfield et al., 1986). Two cases have been described in which rearrangement of an upstream V_H gene via the internal 7-mer of an existing VDJ_H joint led to the expression of a new V_H sequence (Reth et al., 1986; Kleinfield et al., 1986). In one case, an out-of-frame VD joint was thus replaced by a joint which was translationally in frame; i.e., a null cell with respect to H-chain synthesis had become an H-chain producer (Reth et al., 1986). This mechanism thus creates the possibility of turning a nonfunctional B cell into a functional one. At the same time, it has the potential of destroying productive rearrangements. Presumably, this is a function of the normal recombinase recognizing the isolated 7-mer in the absence of a 9-mer and therefore not being constrained by the one-turn/two-turn spacer rule. It is not clear how frequently such a secondary rearrangement occurs and how its control is compatible with feedback inhibition of rearrangement. Perhaps only a subset of B lymphocytes undergo the secondary rearrangements, because they lack feedback control of immunoglobulin gene rearrangement at the pre-B-cell stage. Both cell lines in which the secondary rearrangements have been observed also rearrange λ genes in culture (Kleinfield et al., 1986; M. Reth, personal communication). As discussed previously (section III.B.), λ-producing B cells seem to lack the feedback response.

2. Somatic gene conversion

It has been generally accepted that gene conversion must have played a major role in the evolution of immunoglobulin genes (Baltimore, 1981). Recently, strong evidence has been obtained for the occurrence of somatic gene conversion in the creation of a diversified V_λ gene repertoire in the chicken (Reynaud et al., 1985, 1987; Thompson and Neiman, 1987). The chicken has a single functional V_λ gene which rearranges with JC_λ in pre-B cells early in ontogeny. Apparently, no VJ rearrangements occur after pre-B cells seed the bursa of Fabricius—the central organ for the generation of chicken B cells (Weill et al., 1986). However, the rearranged V_λ gene becomes highly diversified during the further development of B lymphocytes. This diversification seems to be due to gene conversion of the functional, rearranged V_λ gene with V_λ pseudogenes (Reynaud et al., 1985, 1987; Thompson and Neiman, 1987). At least 25 V_λ pseudogenes are located within 19 kb upstream of the functional V_λ gene and apparently serve as donor genes. The conversion tracts comprise from 10 to more than 120 bp, and a single V gene can receive segmental exchanges from up to six different donor genes (Reynaud et al., 1987). The conversions take place at a frequency of 0.05 to 0.1 per cell generation (Reynaud et al., 1987).

Thus, the described gene conversion events are the only way for diversification of the chicken V_λ gene (except for a few nucleotide changes, which may be due to somatic point mutation, because no donor sequences have been found [Reynaud et al., 1987]). It is not clear whether somatic gene conversion plays a role in the diversification of mammalian V genes. A single case has been described where a rearranged mouse V_H gene may have been converted by an upstream V_H gene (Krawinkel et al., 1983). However, a single intra- or transchromosomal crossover with an unknown V_H gene could not be ruled out in this case because of the large number of germ line V_H genes of the mouse.

3. Somatic hypermutability

It had long been suspected that immunoglobulin genes undergo a high rate of somatic mutation (see review by Baltimore, 1981). The existence of somatically introduced mutations was proved by comparing the DNA sequences of expressed immunoglobulin genes with their germ line counterparts (Selsing and Storb, 1981; Crews et al., 1981; Pech et al., 1981; Bothwell et al., 1981). The mutations seem to be restricted to the V regions of H and L chains. They occur over a stretch of about 1 kb comprising V, (D), and J sequences and a few hundred base pairs upstream and downstream (Kim et al., 1981; Gearhart and Bogenhagen, 1983). Because mutations were not seen in unrearranged genes (Nishioka and Leder, 1980; Selsing and Storb, 1981; Gorski et al., 1983), because of the way mutations were spaced around the joining sites, and because DJ joints had fewer mutations than VDJ joints (Sablitzky et al., 1985a), it had been proposed that the mutations may arise during the rearrangement event. Recent data with prerearranged κ transgenes have, however, shown that hypermutation of V regions occurs without ongoing rearrangement (O'Brien et al., 1987).

The mechanism for the hypermutability is unknown. Apparently, a specific mutator operates after antigenic stimulation, probably during the primary antibody response (Manser and Gefter, 1986; Siekevitz et al., 1987; Claflin et al., 1987). The mutations appear to accumulate during the immune response; i.e., a given immunoglobulin gene can probably undergo a mutation event more than once (Berek et al., 1985; Clarke et al., 1985; Sablitzky et al., 1985b). The mutation rate has been calculated to be approximately 10^{-3} bp per cell generation (Clarke et al., 1985). Mutations of immunoglobulin genes also occur in cultured cells, although at a 10 to 100 times lower rate (Wabl et al., 1985a). Perhaps in such cell lines a stage in B-cell development has been immortalized in which the putative mutator normally operates.

The mutations do not require *cis*-acting elements beyond about 15 kb of DNA which were present in the κ transgenes (O'Brien et al., 1987). In fact, mutations occurred at two different, random transgene integration sites. Most likely the V region and its close vicinity are the specific target, although mutations in the C region may not have been detected because of functionally greater stringency for C-region conservation. It had appeared that the V regions must be expressed in active antibodies or be in the nonproductively rearranged allele of a functional gene (Pech ct al., 1981). However, mutations were recently detected in the unrearranged V_λ genes of a cell which also had mutated rearranged V_λ genes (Weiss and Wu, 1987). This may suggest that any V regions which happen to be in active chromatin during operation of the postulated mutator can be a target for mutation.

Although gene conversion has been considered as a source of the mutated sequences in mammalian immunoglobulin genes (see also section III.C.2. above), there is so far no solid evidence that this is true; i.e., in most cases no obvious donor sequences have been found.

4. Control of switch recombination

As described in section II.E. above, B cells can switch from the expression of μ to other H-chain isotypes in the course of the immune response. The switch recombination between a μ and other C_H switch (S) sites occurs at frequencies of up to 10% per cell per generation in B lymphocytes which have been activated by a mitogen (Winter et al., 1987). This allows the progeny of a B cell to produce antibodies with the same variable region combined with different constant regions. Before the actual switch occurs, the switch regions of one or several C_H genes become undermethylated, suggesting opening up of chromatin (Stavnezer-Nordgren and Sirlin, 1986). Transcription from a promoter associated with the unrearranged downstream S region has also been observed (Stavnezer-Nordgren and Sirlin, 1986; Alt et al., 1986). It has been suggested that preceding the switch, recombination-activated B cells produce a long transcript comprising transcription from the VDJ-C_μ gene through all C_H genes located between C_μ and the final switch target C_H gene (Alt et al., 1982; Yaoita et al., 1982; Perlmutter and Gilbert, 1984). This possibility remains an open question, particularly since in a pre-B cell line with the potential to switch to γ2b no transcripts have been found from the $C_{\gamma3}$ gene, which is located between C_μ and $C_{\gamma2b}$ (Alt et al., 1986).

There seems to be general agreement that the final stage of the switch represents recombination of two S regions, generally S_μ with another S region, and deletion of the intervening DNA (Radbruch et al., 1986a, b). Some cases have also been described of S rearrangements downstream of an expressed H-chain gene (Hummel et al., 1987) and of successive switch events in which recombination between S_μ and $S_{\gamma3}$ was followed by a recombination with $S_{\gamma1}$ (Petrini et al., 1987).

All switch recombinations are probably carried out by the same switch recombinase (Petrini et al., 1987). This enzyme is presumed to be different from the recombinase responsible for V(D)J rearrangements. However, these different functions may be carried out by the same nuclease associated with different ancillary proteins which target the nuclease to specific sites. An interesting difference in switching ability has been observed between different cell lines (Ott et al., 1987). The 18-8 pre-B-cell line switches a switch-recombination test gene introduced by a retroviral vector as well as endogenous H genes, whereas two other pre-B-cell lines switch only the test gene, but not their endogenous H genes. They do not even show internal μ switch region recombinations, in contrast to the 18-8 cell line. This may reflect some unusual requirements for S-region accessibility of endogenous genes beyond what is required for transcription.

The switch target selection had originally been thought to be a random process, but currently the idea of a programmed switch recombination seems to be favored by the experimental evidence (Radbruch et al., 1986b). In the majority of switching B cells both alleles switch to the same S region (Radbruch et al., 1986a; Winter et al., 1987; Hummel et al., 1987). It is believed that lymphocytes may be involved in the switch targeting by releasing class-specific lymphokines (Teale and Abraham, 1987; Winter et al., 1987). Thus, whereas during B-cell development and before activation of mature B cells switching is extremely rare or nonexistent, in the course of an immune response extracellular factors apparently turn on a site-specific switching program in activated B cells.

IV. SUMMARY

In conclusion, it seems likely that recombination signal sequences consisting of a 7-mer/spacer/9-mer motif serve as recognition elements which direct a single recombinase system able to carry out rearrangement at any immunoglobulin or TCR locus. This recombinase does not have a strict preference or orientation of signal sequences; depending on the orientation, rearrangement can occur by inversion or deletion. Whether rearrangement proceeds by inversion or deletion, the signal sequences are generally precisely fused while the ends of the "coding joint" are subject to deletions or insertions, or both.

The mechanism of class switching is clearly different from V(D)J joining in that it involves homologous, rather than site-specific, recombination.

Turning on rearrangement is controlled by the presence of recombinase and accessibility of the rearranging sequences. After sequential rearrangement of first DJ_H and then VD_H sequences, the production of membrane-bound μ chains leads to the turning off of H gene rearrangement.

Turning off rearrangement of κ genes may be the result of cessation of recombinase synthesis. Apparently, it is controlled by feedback from a complete membrane-bound κμ immunoglobulin molecule. In contrast, expression of λ genes appears to occur in B cells which lack feedback inhibition.

Already rearranged immunoglobulin genes can be further altered by several mechanisms. A germ line V_H gene can rearrange into an internal heptamer of a rearranged V_H gene, and at least some expressed V genes can be altered by gene conversion and somatic hypermutation.

The molecular details of the immunoglobulin gene rearrangement process and of its control are only partially understood.

ACKNOWLEDGMENTS. We are grateful to H. Sakano and R. Aguilera for communication of unpublished results.

This work was supported by Public Health Service grants AI 24780, HD 23089, and GM 38649 from the National Institutes of Health. P.E. was supported by predoctoral training grants GM 07270 and CA 09537.

LITERATURE CITED

Aguilera, R. J., S. Akira, K. Okazaki, and H. Sakano. 1987. A pre-B cell nuclear protein that specifically interacts with the immunoglobulin V-J recombination sequences. *Cell* **51**:909–917.

Aguilera, R. J., T. J. Hope, and H. Sakano. 1985. Characterization of immunoglobulin enhancer deletions in murine plasmacytomas. *EMBO J.* **4**:3689–3693.

Akira, S., K. Okazaki, and H. Sakano. 1987. Two pairs of recombination signals are sufficient to cause immunoglobulin V-(D)-J joining. *Science* **238**:1134–1138.

Alt, F. W., and D. Baltimore. 1982. Joining of immunoglobulin heavy chain gene segments: implications from a chromosome with evidence of three D-J_H fusions. *Proc. Natl. Acad. Sci. USA* **79**:4118–4122.

Alt, F. W., T. K. Blackwell, R. A. DePinho, M. G. Reth, and G. D. Yancopoulos. 1986. Regulation of genome rearrangement events during lymphocyte differentiation. *Immunol. Rev.* **89**:5–30.

Alt, F. W., N. Rosenberg, R. J. Casanova, E. Thomas, and D. Baltimore. 1982. Immunoglobulin heavy-chain expression and class switching in a murine leukemia cell line. *Nature* (London) **296**:325–331.

Alt, F. W., N. Rosenberg, S. Lewis, E. Thomas, and D. Baltimore. 1981. Organization and reorganization of immunoglobulin genes in A-MuLV-transformed cells: rearrangement of heavy but not light chain genes. *Cell* **27**:381–390.

Alt, F. W., G. D. Yancopoulos, T. K. Blackwell, C. Wood, E. Thomas, M. Boss, R. Coffman, N. Rosenberg, S. Tonegawa, and D. Baltimore. 1984. Ordered rearrangement of immunoglobulin heavy chain variable region segments. *EMBO J.* **3**:1209–1219.

Baltimore, D. 1981. Somatic mutation gains its place among the generators of diversity. *Cell* **26**:295–296.

Berek, C., G. M. Griffiths, and C. Milstein. 1985. Molecular events during maturation of the immune response to oxazolone. *Nature* (London) **316**:412–418.

Blackwell, T. K., and F. W. Alt. 1984. Site-specific recombination between immunoglobulin D and J_H that were introduced into the genome of a murine preB cell line. *Cell* **37**:105–112.

Blackwell, T. K., M. W. Moore, G. D. Yancopoulos, H. Suh, S. Lutzker, E. Selsing, and F. W. Alt. 1986. Recombination between immunoglobulin variable region gene segments is enhanced by transcription. *Nature* (London) **324**:585–589.

Bosma, G. C., R. P. Custer, and M. J. Bosma. 1983. A severe combined immunodeficiency mutation in the mouse. *Nature* (London) **301**:527–530.

Bothwell, A. L. M., M. Paskind, M. Reth, T. Imanishi-Kari, K. Rajewsky, and D. Baltimore. 1981. Heavy chain variable region contribution to the NP b family of antibodies: somatic mutation evident in a gamma2a variable region. *Cell* **24**:625–637.

Brinster, R. L., K. A. Ritchie, R. E. Hammer, R. L. O'Brien, B. Arp, and U. Storb. 1983. Expression of a microinjected immunoglobulin gene in the spleen of transgenic mice. *Nature* (London) **306**:332–336.

Brodeur, P. H., and R. Riblet. 1984. The immunoglobulin heavy chain variable region (Igh-V) in the mouse. I. One hundred Igh-V genes comprise seven families of homologous genes. *Eur. J. Immunol.* **14**:922–930.

Bucchini, D., C.-A. Reynaud, M.-A. Ripoche, H. Grimal, J. Jami, and J.-C. Weill. 1987. Rearrangement of a chicken immunoglobulin gene occurs in the lymphoid lineage of transgenic mice. *Nature* (London) **326**:409–411.

Claflin, J. L., J. Berry, D. Flaherty, and W. Dunnick. 1987. Somatic evolution of diversity among anti-phosphocholine antibodies induced with *Proteus morganii*. *J. Immunol.* **138**:3060–3068.

Clarke, S. H., K. Huppi, D. Ruezinsky, L. Staudt, W. Gerhard, and M. Weigert. 1985. Inter- and intraclonal diversity in the immune response to the influenza hemagglutinin. *J. Exp. Med.* **161**:687–704.

Coleclough, C. 1983. Chance, necessity and antibody gene dynamics. *Nature* (London) **303**:23–26.

Coleclough, C., R. P. Perry, K. Karjalainen, and M. Weigert. 1981. Aberrant rearrangements contribute significantly to the allelic exclusion of immunoglobulin gene expression. *Nature* (London) **290**:372–378.

Cooper, M. D., A. Velardi, J. E. Calvert, W. E. Gathings, and H. Kubagawa. 1983. Generation of B-cell clones during ontogeny. *Prog. Immunol.* **5**:603–612.

Cory, S., and J. M. Adams. 1980. Deletions are associated with somatic rearrangement of immunoglobulin heavy chain genes. *Cell* **19**:37–51.

Cory, S., B. M. Tyler, and J. M. Adams. 1981. Sets of immunoglobulin V_k genes homologous to ten cloned V_k sequences: implications for the number of germline V_k genes. *J. Mol. Appl. Genet.* **1**:103–116.

Crews, S., J. Griffin, H. Huang, K. L. Calame, and L. Hood. 1981. A single V_H gene segment encodes the immune response to phosphorylcholine: somatic mutation is correlated with the class of the antibody. *Cell* **25**:59–66.

D'Agostaro, G., F. Hevia, G. E. Wu, and H. Murialdo. 1985. Site-directed cleavage of immunoglobulin segments by lymphoid cell extracts. *Can. J. Biochem. Cell Biol.* **63**:969–976.

Denny, C. T., G. F. Hollis, F. Hecht, R. Morgan, M. P. Link, S. D. Smith, and I. R. Kirsch. 1986. Common mechanism of chromosome inversion in B- and T-cell tumors: relevance to lymphoid development. *Science* **234**:197–200.

Desiderio, S., and D. Baltimore. 1984. Double-stranded cleavage by cell extracts near recombinational signal sequences of immunoglobulin genes. *Nature* (London) **308**:860–862.

Desiderio, S. V., G. D. Yancopoulos, M. Paskind, E. Thomas, M. A. Boss, N. Landau, F. W. Alt, and D. Baltimore. 1984. Insertion of N regions into heavy-chain genes is correlated with expression of terminal deoxynucleotidyltransferase in B cells. *Nature* (London) **311**:752–755.

Dreyer, W. J., and J. C. Bennett. 1965. The molecular basis of antibody formation: a paradox. *Proc. Natl. Acad. Sci. USA* **54**:864–869.

Durdik, J., M. W. Moore, and E. Selsing. 1984. Novel kappa light-chain gene rearrangements in mouse lambda light chain-producing B lymphocytes. *Nature* (London) **307**:749–752.

Early, P., H. Huang, M. Davis, K. Calame, and L. Hood. 1980. An immunoglobulin heavy chain variable region gene is generated from three segments of DNA: V_H, D and J_H. *Cell* **19**:981–992.

Engler, P., and U. Storb. 1987. High-frequency deletional rearrangement of immunoglobulin kappa gene segments introduced into a pre-B-cell line. *Proc. Natl. Acad. Sci. USA* **84**:4949–4953.

Feddersen, R. M., and B. G. Van Ness. 1985. Double recombination of a single immunoglobulin kappa-chain allele: implications for the mechanism of rearrangement. *Proc. Natl. Acad. Sci. USA* **82**:4793–4797.

Fujimoto, S., and H. Yamagishi. 1987. Isolation of an excision product of T-cell receptor alpha-chain gene rearrangements. *Nature* (London) **327**:242–243.

Gearhart, P. J., and D. F. Bogenhagen. 1983. Clusters of point mutations are found exclusively around antibody variable genes. *Proc. Natl. Acad. Sci. USA* **80**:3439–3443.

Goodhardt, M., P. Cavelier, M. A. Akimenko, G. Lutfalla, C. Babinet, and F. Rougeon. 1987. Rearrangement and expression of rabbit immunoglobulin kappa light chain gene in transgenic mice. *Proc. Natl. Acad. Sci. USA* **84**:4229–4233.

Gorski, J., P. Rollini, and B. Mach. 1983. Somatic mutations of immunoglobulin variable genes are restricted to the rearranged V gene. *Science* **220**:1179–1181.

Hagiya, M., D. D. Davis, L. D. Shultz, and H. Sakano. 1986a. Non-germ-line elements (NGE) are present in the T cell receptor beta-chain genes isolated from the mutant mouse, motheaten (me/me). *J. Immunol.* **136**:2697–2700.

Hagiya, M., D. D. Davis, T. Takahashi, K. Okuda, W. C. Raschke, and H. Sakano. 1986b. Two types of immunoglobulin-negative Abelson murine leukemia virus-transformed cells: implications for B-lymphocyte differentiation. *Proc. Natl. Acad. Sci. USA* **83**:145–159.

Hesse, J. E., M. R. Lieber, M. Gellert, and K. Mizuuchi. 1987. Extrachromosomal DNA substrates in pre-B cells undergo inversion or deletion at immunoglobulin V-(D)-J joining signals. *Cell* **49**:775–783.

Hieter, P. A., S. J. Korsmeyer, T. A. Waldman, and P. Leder. 1981. Human immunoglobulin kappa light-chain genes are deleted or rearranged in lambda-producing B cells. *Nature* (London) **290**:368–372.

Hochtl, J., C. R. Muller, and H. G. Zachau. 1982. Recombined flanks of the variable and joining segments of immunoglobulin genes. *Proc. Natl. Acad. Sci. USA* **79**:1383–1387.

Hope, T. J., R. J. Aguilera, M. E. Minie, and H. Sakano. 1986. Endonucleolytic activity that cleaves immunoglobulin recombination sequences. *Science* **231**:1141–1145.

Hummel, M., J. Kaminska-Berry, and W. Dunnick. 1987. Switch region content of hybridomas: the two spleen cell Igh loci tend to rearrange to the same isotype. *J. Immunol.* **138**:3539–3548.

Kataoka, T., S. Kondo, M. Nishi, M. Kodaira, and T. Honjo. 1984. Isolation and characterization of endonuclease J: a sequence-specific endonuclease cleaving immunoglobulin genes. *Nucleic Acids Res.* **12**:5995–6010.

Kim, S., M. Davis, E. Sinn, P. Patten, and L. Hood. 1981. Antibody diversity: somatic hypermutation of rearranged V_H genes. *Cell* **27**:573–581.

Kindt, T. J., and J. D. Capra. 1984. *The Antibody Enigma*. Plenum Publishing Corp., New York.

Kleinfield, R., R. R. Hardy, D. Carlinton, J. Dangl, L. A. Herzenberg, and M. Weigert. 1986. Recombination between and expressed immunoglobulin heavy-chain gene and a germline variable gene segment in a Ly 1+ B-cell lymphoma. *Nature* (London) **322**:836–846.

Klobeck, H.-G., and H. G. Zachau. 1986. The human C_K gene segment and the kappa deleting element are closely linked. *Nucleic Acids Res.* **14**:4591–4603.

Korsmeyer, S. J., P. A. Hieter, J. V. Ravetch, D. G. Poplack, T. A. Waldman, and P. Leder. 1981. Developmental hierarchy of immunoglobulin gene rearrangements in human leukemic pre-B-cells. *Proc. Natl. Acad. Sci. USA* **78**:7096–7100.

Krawinkel, U., G. Zoebelein, M. Bruggemann, A. Radbruch, and K. Rajewsky. 1983. Recombination between antibody heavy chain variable-region genes: evidence for gene conversion. *Proc. Natl. Acad. Sci. USA* **80**:4997–5001.

Kronenberg, M., G. Siu, L. E. Hood, and N. Shastri. 1986. The molecular genetics of the T-cell antigen receptor and T-cell antigen recognition. *Annu. Rev. Immunol.* **4**:529–591.

Kurosawa, Y., H. von Boehmer, W. Haas, H. Sakano, A. Trauneker, and S. Tonegawa. 1981. Identification of D segments of immunoglobulin heavy-chain genes and their rearrangement in T lymphocytes. *Nature* (London) **290**:565–570.

Kurosawa, Y., and S. Tonegawa. 1982. Organization, structure, and assembly of immunoglobulin heavy chain diversity DNA segments. *J. Exp. Med.* **155**:201–218.

Levitt, D., and M. D. Cooper. 1980. Mouse pre-B cells synthesize and secrete mu heavy chains but not light chains. *Cell* **19**:617–625.

Lewis, S., A. Gifford, and D. Baltimore. 1984. Joining of V_k to J_k gene segments in a retroviral vector introduced into lymphoid cells. *Nature* (London) **308**:425–428.

Lewis, S., A. Gifford, and D. Baltimore. 1985. DNA elements are asymmetrically joined during the site-specific recombination of kappa immunoglobulin genes. *Science* **228**:677–685.

Lewis, S., N. Rosenberg, F. W. Alt, and D. Baltimore. 1982. Continuing kappa-gene rearrangement in a cell line transformed by Abelson murine leukemia virus. *Cell* **30**:807–816.

Litman, G. W., L. Berger, K. Murphy, R. Litman, K. Hinds, C. L. Jahn, and B. W. Erickson. 1983. Complete nucleotide sequence of an immunoglobulin V_H gene homologue from *Caiman*, a phylogenetically ancient reptile. *Nature* (London) **303**:349–352.

Liu, Z., A. F. Wu, and T. T. Wu. 1987. Short gene inversion involving two adjacent heavy chain joining minigenes and one heavy chain diversity minigene in the nonsecretor Sp2/0-Ag14 myeloma cell line. *Nucleic Acids Res.* **15**:4688.

Maeda, T., H. Sugiyama, Y. Tani, S. Miyake, Y. Oka, H. Ogawa, Y. Komori, T. Soma, and S.

Kishimoto. 1987. Start of mu-chain production by the further two-step rearrangements of immunoglobulin heavy chain genes on one chromosome from a DJ_H/DJ_H configuration in an Abelson virus-transformed cell line: evidence of a secondary DJ_H complex formation. *J. Immunol.* **138:**2305–2310.

Maki, R., J. F. Kearney, C. Paige, and S. Tonegawa. 1980. Immunoglobulin gene rearrangement in immature B cells. *Science* **209:**1366–1369.

Maki, R., W. Roeder, A. Traunecker, C. Sidman, M. Wabl, W. Raschke, and S. Tonegawa. 1981. The role of DNA rearrangement and alternative RNA processing in the expression of immunoglobulin delta genes. *Cell* **24:**353–365.

Malissen, M., C. McCoy, D. Blanc, J. Trucy, C. Devaux, A.-M. Schmitt-Verhulst, F. Fitch, L. E. Hood, and B. Malissen. 1986. Direct evidence for chromosomal inversion during T-cell receptor beta-gene rearrangements. *Nature* (London) **319:**28–33.

Manser, T., and M. L. Gefter. 1986. The molecular evolution of the immune response: idiotope-specific suppression indicates that B cells express germ-line-encoded V genes prior to antigenic stimulation. *Eur. J. Immunol.* **16:**1439–1444.

Marcu, K. B., R. B. Lang, L. W. Stanton, and L. J. Harris. 1982. A model for the molecular requirements of immunoglobulin heavy chain class switching. *Nature* (London) **298:**87–89.

Mather, E. L., and R. P. Perry. 1983. Methylation status and DNase I sensitivity of immunoglobulin genes: changes associated with rearrangement. *Proc. Natl. Acad. Sci. USA* **80:**4689–4693.

Max, E. E., J. V. Maizel, Jr., and P. Leder. 1981. The nucleotide sequence of a 5.5-kilobase DNA segment containing the mouse kappa immunoglobulin J and C region genes. *J. Biol. Chem.* **256:**5116–5120.

Max, E. E., J. G. Seidman, and P. Leder. 1979. Sequences of five potential recombination sites encoded close to an immunoglobulin kappa constant region gene. *Proc. Natl. Acad. Sci. USA* **76:**3450–3454.

Max, E. E., J. G. Seidman, H. Miller, and P. Leder. 1980. Variation in the crossover point of kappa immunoglobulin gene V-J recombination: evidence from a cryptic gene. *Cell* **21:**793–799.

Moore, K. W., J. Rogers, T. Hunkapiller, P. Early, C. Nottenburg, I. Weissman, H. Bazin, R. Wall, and L. E. Hood. 1981. Expression of IgD may use both DNA rearrangement and RNA splicing mechanisms. *Proc. Natl. Acad. Sci. USA* **78:**1800–1804.

Moore, M. W., J. Durdik, D. M. Persiani, and E. Selsing. 1985. Deletions of kappa chain constant region genes in mouse lambda chain-producing B cells involve intrachromosomal DNA recombinations similar to V-J joining. *Proc. Natl. Acad. Sci. USA* **82:**6211–6215.

Mowatt, M. R., and W. Dunnick. 1986. DNA sequence of the murine gamma 1 switch segment reveals novel structural elements. *J. Immunol.* **136:**2674–2683.

Nikaido, T., Y. Yamawaki-Kataoka, and T. Honjo. 1982. Nucleotide sequence of switch regions of immunoglobulin epsilon and gamma and their comparison. *J. Biol. Chem.* **257:**7322–7329.

Nishioka, Y., and P. Leder. 1980. Organization and complete sequence of identical embryonic and plasmacytoma kappa V-region genes. *J. Biol. Chem.* **255:**3691–3694.

Nussenzweig, M. C., A. C. Shaw, E. Sinn, D. B. Danner, K. L. Holmes, H. C. Morse, and P. Leder. 1987. Allelic exclusion in transgenic mice that express the membrane form of immunoglobulin mu. *Science* **236:**816–819.

O'Brien, R., R. Brinster, and U. Storb. 1987. Somatic hypermutation of an immunoglobulin transgene in kappa transgenic mice. *Nature* (London) **326:**405–409.

Okazaki, K., D. D. Davis, and H. Sakano. 1987. T cell receptor beta gene sequences in the circular DNA of thymocyte nuclei: direct evidence for intramolecular DNA deletion in V-D-J joining. *Cell* **49:**477–485.

Ott, D. E., F. W. Alt, and K. B. Marcu. 1987. Immunoglobulin heavy chain switch region recombination within a retroviral vector in murine pre-B cells. *EMBO J.* **6:**577–584.

Pech, M., J. Hochtl, H. Schnell, and H. G. Zachau. 1981. Differences between germ-line and rearranged immunoglobulin V kappa coding sequences suggest a localized mutation mechanism. *Nature* (London) **291:**668–670.

Perlmutter, A. P., and W. Gilbert. 1984. Antibodies of the secondary response can be expressed without switch recombination in normal mouse B cells. *Proc. Natl. Acad. Sci. USA* **81:**7189–7193.

Perlmutter, R. M., J. F. Kearney, S. P. Chang, and L. E. Hood. 1985. Developmentally controlled expression of immunoglobulin V_H genes. *Science* 227:1597–1601.

Perry, R. P., C. Coleclough, and M. Weigert. 1981a. Reorganization and expression of immunoglobulin genes: status of allelic elements. *Cold Spring Harbor Symp. Quant. Biol.* 45:925–933.

Perry, R. P., D. E. Kelley, C. Coleclough, and J. F. Kearney. 1981b. Organization and expression of immunoglobulin genes in fetal liver hybridomas. *Proc. Natl. Acad. Sci. USA* 78:247–251.

Persiani, D. M., J. Durdik, and E. Selsing. 1987. Active lambda and kappa antibody gene rearrangement in Abelson murine leukemia virus-transformed pre-B cell lines. *J. Exp. Med.* 165:1655–1674.

Petrini, J., B. Shell, M. Hummel, and W. Dunnick. 1987. The immunoglobulin heavy chain switch: structural features of gamma 1 recombinant switch regions. *J. Immunol.* 138:1940–1946.

Radbruch, A., C. Burger, S. Klein, and W. Muller. 1986a. Control of immunoglobulin class switch recombination. *Immunol. Rev.* 89:69–83.

Radbruch, A., W. Muller, and K. Rajewsky. 1986b. Class switch recombination is IgG1 specific on active and inactive IgH loci of IgG1-secreting B-cell blasts. *Proc. Natl. Acad. Sci. USA* 83:3954–3957.

Reth, M., and F. W. Alt. 1984. Novel immunoglobulin heavy chains are produced from DJ_H gene segment rearrangements in lymphoid cells. *Nature* (London) 312:418–423.

Reth, M., P. Ammirati, S. Jackson, and F. W. Alt. 1985. Regulated progression of a cultured pre-B-cell line to the B-cell stage. *Nature* (London) 317:353–355.

Reth, M., P. Gehrmann, E. Petrac, and P. Wiese. 1986. A novel V_H to V_HDJH joining mechanism in heavy chain-negative (null) pre-B cells results in heavy-chain production. *Nature* (London) 322:840–842.

Reynaud, C.-A., V. Anquez, A. Dahan, and J.-C. Weill. 1985. A single rearrangement event generates most of the chicken immunoglobulin light chain diversity. *Cell* 40:283–291.

Reynaud, C.-A., V. Anquez, H. Grimal, and J.-C. Weill. 1987. A hyperconversion mechanism generates the chicken light chain preimmune repertoire. *Cell* 48:379–388.

Richards, J. E., A. C. Gilliam, A. Shen, P. W. Tucker, and F. R. Blattner. 1983. Unusual sequences in the murine immunoglobulin mu-delta heavy-chain region. *Nature* (London) 306:483–487.

Ritchie, K. A., R. L. Brinster, and U. Storb. 1984. Allelic exclusion and control of endogenous immunoglobulin gene rearrangement in kappa transgenic mice. *Nature* (London) 312:517–520.

Rusconi, S., and G. Köhler. 1985. Transmission and expression of a specific pair of rearranged immunoglobulin mu and kappa genes in a transgenic mouse line. *Nature* (London) 314:330–334.

Sablitzky, F., D. Weisbaum, and K. Rajewsky. 1985a. Sequence analysis of non-expressed immunoglobulin heavy chain loci in clonally related, somatically mutated hybridoma cells. *EMBO J.* 4:3435–3437.

Sablitzky, F., G. Wildner, and K. Rajewsky. 1985b. Somatic mutation and clonal expansion of B cells in an antigen-driven immune response. *EMBO J.* 4:345–350.

Sakano, H., K. Huppi, G. Heinrich, and S. Tonegawa. 1979. Sequences at the somatic recombination sites of immunoglobulin light-chain genes. *Nature* (London) 280:288–294.

Sakano, H., Y. Kurosawa, M. Weigert, and S. Tonegawa. 1981. Identification and nucleotide sequence of a diversity DNA segment (D) of immunoglobulin heavy-chain genes. *Nature* (London) 290:562–565.

Schuler, W., I. J. Weiler, A. Schuler, R. A. Phillips, N. Rosenberg, T. W. Mak, J. F. Kearney, R. Perry, and M. J. Bosma. 1986. Rearrangement of antigen receptor genes is defective in mice with severe combined immune deficiency. *Cell* 46:963–972.

Seidman, J. G., M. M. Nau, B. Norman, S.-P. Kwan, M. Scharff, and P. Leder. 1980. Immunoglobulin V/J recombination is accompanied by deletion of joining site and variable region segments. *Proc. Natl. Acad. Sci. USA* 77:6022–6026.

Selsing, E., J. Miller, R. Wilson, and U. Storb. 1982. Evolution of mouse immunoglobulin lambda genes. *Proc. Natl. Acad. Sci. USA* 79:4681–4685.

Selsing, E., and U. Storb. 1981. Somatic mutation of immunoglobulin light-chain variable-region genes. *Cell* 25:47–58.

Selsing, E., J. Voss, and U. Storb. 1984. Immunoglobulin gene 'remnant' DNA—implications for antibody gene recombination. *Nucleic Acids Res.* 12:4229–4246.

Shimuzu, A., N. Takahashi, Y. Yaoita, and T. Honjo. 1982. Organization of the constant-region gene family of the mouse immunoglobulin heavy chain. *Cell* **28:**499–506.

Siden, E., F. W. Alt, L. Shinefeld, V. Sato, and D. Baltimore. 1981. Synthesis of immunoglobulin mu chain gene products precedes synthesis of light chains during B-lymphocyte development. *Proc. Natl. Acad. Sci. USA* **78:**1823–1827.

Siekevitz, M., C. Kocks, K. Rajewsky, and R. Dildrop. 1987. Analysis of somatic mutation and class switching in naive and memory B cells generating adoptive primary and secondary responses. *Cell* **48:**757–770.

Siminovitch, K. A., A. Bakhshi, P. Goldman, and S. J. Korsmeyer. 1985. A uniform deleting element mediates the loss of kappa genes in human B cells. *Nature* (London) **316:**260–262.

Stavnezer-Nordgren, J., and S. Sirlin. 1986. Specificity of immunoglobulin heavy chain switch correlates with activity of germline heavy chain genes prior to switching. *EMBO J.* **5:**95–102.

Steinmetz, M., W. Altenburger, and H. G. Zachau. 1980. A rearranged DNA sequence possibly related to the translocation of immunoglobulin gene segments. *Nucleic Acids Res.* **8:**1709–1720.

Storb, U. 1987. Transgenic mice with immunoglobulin genes. *Annu. Rev. Immunol.* **5:**151–174.

Storb, U., and B. Arp. 1983. Methylation patterns of immunoglobulin genes in lymphoid cells: correlation of expression and differentiation with undermethylation. *Proc. Natl. Acad. Sci. USA* **80:** 6642–6646.

Storb, U., R. Wilson, E. Selsing, and A. Walfield. 1981. Rearranged and germline immunoglobulin kappa genes: different states of DNase I sensitivity of constant kappa genes in immunocompetent and nonimmune cells. *Biochemistry* **20:**990–996.

Szurek, P., J. Petrini, and W. Dunnick. 1985. Complete nucleotide sequence of the murine gamma3 switch region and analysis of switch recombination sites in two gamma3-expressing hybridomas. *J. Immunol.* **135:**620–626.

Teale, J. M., and K. M. Abraham. 1987. The regulation of antibody class expression. *Immunol. Today* **8:**122–126.

Thompson, C. B., and P. E. Neiman. 1987. Somatic diversification of the chicken immunoglobulin light chain gene is limited to the rearranged variable gene segment. *Cell* **48:**369–378.

Tonegawa, S., N. Hozumi, G. Matthyssens, and R. Schuller. 1977. Somatic changes in the content and context of immunoglobulin genes. *Cold Spring Harbor Symp. Quant. Biol.* **41:**877–889.

Wabl, M., P. D. Burrows, A. von Gabain, and C. Steinberg. 1985a. Hypermutation at the immunoglobulin heavy chain locus in a pre-B-cell line. *Proc. Natl. Acad. Sci. USA* **82:**479–482.

Wabl, M., J. Meyer, G. Beck-Engeser, M. Tenkhoff, and P. D. Burrows. 1985b. Critical test of a sister chromatid exchange model for the immunoglobulin heavy-chain class switch. *Nature* (London) **313:** 687–689.

Walfield, A., E. Selsing, B. Arp, and U. Storb. 1981. Misalignment of V and J gene segments resulting in a nonfunctional immunoglobulin gene. *Nucleic Acids Res.* **9:**1101–1109.

Weaver, D., and D. Baltimore. 1987. B lymphocyte-specific protein binding near an immunoglobulin kappa-chain gene J segment. *Proc. Natl. Acad. Sci. USA* **84:**1516–1520.

Weaver, D., F. Costantini, T. Imanishi-Kari, and D. Baltimore. 1985. A transgenic immunoglobulin mu gene prevents rearrangement of endogenous genes. *Cell* **42:**117–127.

Weigert, M., R. Perry, and D. Kelley. 1980. The joining of V and J gene segments creates antibody diversity. *Nature* (London) **283:**497–499.

Weill, J.-C., C.-A. Reynaud, O. Lassila, and J. R. L. Pink. 1986. Rearrangement of chicken immunoglobulin genes is not an ongoing process in the embryonic bursa of Fabricius. *Proc. Natl. Acad. Sci. USA* **83:**3336–3340.

Weiss, S., and G. E. Wu. 1987. Somatic point mutations in unrearranged immunoglobulin gene segments encoding the variable region of lambda light chains. *EMBO J.* **6:**927–932.

Whitlock, C. A., S. F. Ziegler, L. J. Treiman, J. I. Stafford, and O. N. Witte. 1983. Differentiation of cloned populations of immature B cells after transformation with Abelson murine leukemia virus. *Cell* **32:**903–911.

Winter, E., U. Krawinkel, and A. Radbruch. 1987. Directed Ig class switch recombination in activated murine B cells. *EMBO J.* **6:**1663–1671.

Yamagishi, H., T. Kunisada, and T. Tsuda. 1982. Small circular DNA complexes in eucaryotic cells. *Plasmid* **8:**299–306.

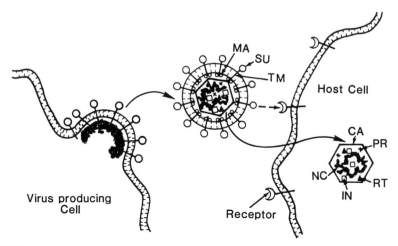

FIGURE 1. Diagrammatic representation of type C retrovirion formation from a virus-producing cell and infection of a susceptible cell. Virion envelope proteins accumulate in the cell membrane at the site of budding, and core precursor components aggregate underneath the membrane at the same sites. The free virion contains mature viral core proteins processed by the virion-encoded protease (PR) and viral envelope proteins processed by cellular proteases. The viral proteins include: the surface (SU) and transmembrane (TM) components of the envelope glycoprotein, an internal matrix protein (MA), a core consisting of the major capsid protein (CA), the major nucleocapsid protein (NC) which is bound to viral RNA, and the reverse transcriptase (RT) and integration-related protein (IN) which are contained within the virion core. Productive attachment to susceptible host cells requires interaction of the viral SU protein with specific cell surface receptors. The infecting virion core is presumed to be introduced into the cytoplasm of the infected cell after removal of the virus envelope via virus-cell membrane fusion.

the first "primers" for initiation of viral DNA synthesis. The core (*gag*) proteins and proteins encoded in the adjacent *pol* gene (which includes reverse transcriptase) are produced in the cytoplasm of infected cells as "immature" polyprotein precursors (Fig. 1). In the type C oncoviruses, processing of these precursors takes place at or shortly after the time when virions bud from the cell surface (Witte and Baltimore, 1978; Eisenman et al., 1980), by a virus-encoded protease which is part of the *gag* (avian viruses) or the *pol* (murine and some avian viruses) gene. It has been shown for the Moloney murine leukemia virus (MoMLV) that the *gag-pol* precursor lacks detectable reverse transcriptase activity (Panet and Baltimore, 1987; Witte and Baltimore, 1978), suggesting that this enzyme is activated by a proteolytic event. It seems likely that the functions of some of the other proteins which are included in the precursor also require maturation or processing.

The mature virus particle consists of a core surrounded by an envelope derived from the host cell membrane, but it includes viral glycoproteins as well. These viral glycoproteins, encoded in the *env* gene, are also processed from a precursor polypeptide, but in this case cellular proteases are involved (Wills et al., 1984; Leamnson and Halpern, 1976; Perez and Hunter, 1987; Schultz and Rein, 1985). This processing occurs as the precursor, synthesized on membrane-bound

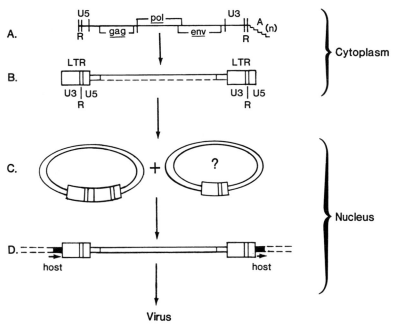

FIGURE 2. Replication of retroviral genomes. (A) The viral genome, a single-stranded RNA molecule of positive (mRNA) polarity with a polyA [A(n)] tail. The relative order of genes *gag* (core protein), *pol* (reverse transcriptase and endonuclease), and *env* (envelope proteins) is as noted. The locations of unique 5′ (U5), 3′ (U3), and short terminal repeat (R) noncoding regions are also indicated. (B) Linear double-stranded viral DNA synthesized in the cytoplasm of infected cells by reverse transcription. The gene order is colinear with the viral RNA genome. Genome length is approximately 7 kilobases for various type C oncoviruses; the LTRs vary in length from approximately 275 bp for the endogenous avian retroviruses to 1,400 bp for mouse mammary tumor virus. The origin of sequences in the LTRs is as indicated. (C) Circular viral DNAs (supercoils) found in the nuclei of infected cells. The form containing two LTRs is presumed to arise from the linear form (B) by blunt-end ligation. The origin of the form with one LTR (?) is unknown. (D) Integrated proviral DNA covalently linked to host DNA (dashed lines). The gene order is the same as that in the unintegrated linear molecule. The small black rectangles symbolize the short direct repeats in host DNA that are produced during integration.

polysomes, passes through the Golgi bodies and onto the surface of the host cell. In subsequent infection, the viral envelope proteins attach to specific receptors on susceptible cells. The envelope is later removed, presumably by fusion with cellular membranes, as the core enters the cytoplasm of infected cells. It is important to appreciate that the mature active viral proteins which are required for DNA synthesis and integration are part of the infecting virion core and probably function in a specific protein-nucleic acid complex within a particulate structure.

Reverse transcription of the single-stranded viral RNA genome(s) in the cytoplasm of infected cells results in the formation of a linear DNA molecule with two long terminal repeats (LTRs). Each LTR consists of a short repeat sequence (R) flanked by longer sequences that are derived from the 5′ (U5) and 3′ (U3) noncoding ends of the viral RNA (Fig. 2). The cDNA or minus strand of the linear molecules is synthesized in a continuous fashion from the tRNA primer, while the plus strand is synthesized in a discontinuous fashion using multiple initiation sites (Boone and Skalka, 1981; Kung et al., 1981; Hsu and Taylor, 1982a). Later, two forms of unintegrated covalently closed circular viral DNA molecules appear in the nucleus of the cell. One circular form contains two tandem copies of the LTR and, on the basis of sequence comparisons (Scott et al., 1981), could be formed by blunt-end ligation of the ends of a linear molecule in which the plus strand has been completed or repaired. The second circular form contains a single LTR. It has been proposed that this form arises via homologous recombination between two LTRs of a linear or circular molecule (Varmus and Swanstrom, 1982) or by premature repair and ligation of a circular replicative intermediate (Junghans et al., 1982). A significant fraction of the two-LTR circular molecules contain deletions or rearrangements at one or the other end of the juxtaposed LTR termini. These deletions or rearrangements are presumed to arise from ligation of incomplete linear molecules or aberrant integrative recombination reactions (Ju and Skalka, 1980; Shoemaker et al., 1981).

Integrated viral DNA, or proviral DNA as the integrated form is called, is observed somewhat later in infection. The proviral genetic map is colinear with the viral RNA genome and the unintegrated linear molecules. Transcription of the provirus by the host cell RNA polymerase II and subsequent RNA processing result in production of both progeny genomes and subgenomic viral mRNAs. Integration of retroviral DNA into the chromosome of its host cell is a normal, but not an obligatory, step in the retroviral life cycle (Panganiban and Temin, 1983; Harris et al., 1984); viral RNA can be transcribed and expressed from unintegrated viral DNA (Kopchick et al., 1981b).

III. RETROVIRAL DNA INTEGRATION

The recombination event that leads to integration of a provirus into the DNA of the host cell is specific for sites in the viral DNA and requires a virus-encoded function(s). Sequences near the ends of the LTRs in the viral DNA are joined to host DNA in the reaction. Many sites in the host DNA can serve as targets for integration. As with many other procaryotic and eucaryotic transposable elements, retroviral DNA integrations are marked by small duplications of host cell DNA at either end of the proviruses (see Fig. 2) (Syvanen, this volume). In addition, some viral DNA sequences, usually 2 base pairs (bp) from the end of each LTR, are lost during the integration reaction. *RNA Tumor Viruses* (Varmus and Swanstrom, 1982, 1985) contains a summary of our knowledge of the reaction as of 1985. In the following sections, I provide an update and focus on the questions that still remain, as well as some current approaches to their resolution.

A. Viral Donor Sites

Since the viral site appears to provide most of the specificity in the integration reaction, in this discussion it will be designated as the *donor* and the host cell site will be designated as the *target*. Because the ends of the LTRs are joined to host DNA in the provirus, the donor site is presumed to reside in sequences in the U3 and U5 termini, which are at the outer edges of the LTRs in linear DNA and at the LTR-LTR junction in the covalently closed, two-LTR circular form (Fig. 2). A comparison of the sequences at these U3 and U5 ends of several viral DNAs (Fig. 3) reveals a number of common features. When these sequences are juxtaposed, as in the two-LTR circle described above, they form inverted repeats. Some of the inverted repeats are quite long (e.g., the avian Rous-associated virus 2 and the murine MoMLV are 15 and 18 nucleotides long, respectively), but this feature is variable. At its minimum (e.g., in HTLV-I, a human T-cell lymphotrophic virus), the inverted repeat includes only the invariant AC . . . and . . . GT dinucleotides destined to become the ends of the LTRs in the integrated provirus. These same dinucleotides are found at the ends of a number of integrated transposable elements, some as phylogenetically distant as the procaryotic elements Tn9 and bacteriophage Mu (see Syvanen, this volume). The prevalence of the CA sequence at or near the site of a large variety of DNA rearrangement reactions is noteworthy (Rogers, 1983). This conservation suggests an evolutionary related-ness and perhaps a similarity in mechanism for such reactions. In many cases, the inverted repeats at the retroviral DNA termini include two nucleotides (TT and AA) which appear to be "lost" in the course of the integration reaction (Ju and Skalka, 1980; Majors and Varmus, 1981; Shoemaker et al., 1981). However, neither the sequence nor the number of lost nucleotides appears to be the same in all cases.

Two lines of evidence have helped to identify sequences that are required for integration and thus define the donor site. The first come from studies in which viral genomes were subjected to site-directed mutagenesis and the mutant viruses were then tested for their ability to integrate their DNA in vivo. Deletion of specific nucleotides near either end of the DNA of spleen necrosis virus (SNV) has established boundaries for the required region of within 12 bp of the U3 and 8 bp of the U5 terminus (Panganiban and Temin, 1983). The relevance of the region was supported by the observation that quantitative differences in the efficiency of integration occurred when certain nucleotides in the required U5 sequence were changed. The inverted repeat at the LTR termini is included in the required region and appears to be important for SNV integration: Panganiban and Luk (1987) reported that single point mutations in U5 that disrupt the inverted repeat and cause a decrease in integration can be compensated for by second site mutations in U3 which restore complementarity. This finding suggests either that the integration complex can "read" across viral DNA termini or that the termini may somehow interact by base pairing with each other during the reaction. Colicelli and Goff (1985) tested the infectivity of mutant MoMLV DNAs in which the U5 terminal deoxynucleotide was changed or in which different numbers of base pairs were deleted. They found that a change from T to A at the terminus had no affect on integration. Thus, for this virus, and presumably for others (Fig. 3), a perfect terminal inverted repeat is not required. These studies also showed that the loss of a single deoxynucleotide did not significantly reduce the efficiency of integration or alter the identity of the viral sequence which was joined to host

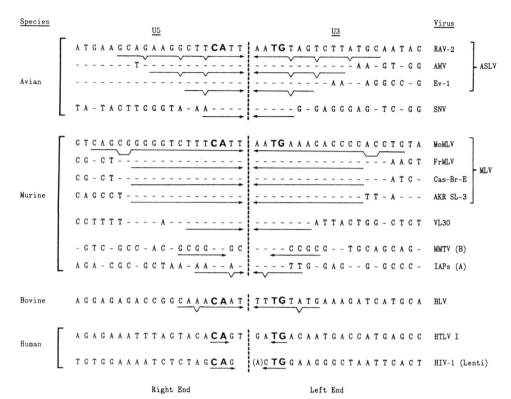

FIGURE 3. Comparison of known or predicted deoxynucleotide sequences at the termini of the DNA of several retroviruses. Only the plus strands are shown, 5' to 3'. The LTR region represented is indicated at the top and the relationship of the termini with respect to the viral genetic map (i.e., left end...*gag-pol-env*...right end) is indicated at the bottom. Dashes indicate identity of sequence. Horizontal arrows show inverted repeats; the breaks indicate the position of noncomplementary deoxynucleotides. Dinucleotides (CA and TG) that become the ends of the integrated provirus are shown in bold letters. For HIV-1, the A in parentheses indicates an ambiguity in prediction of the U3 terminus. Identification of the actual sequence awaits nucleotide sequencing of intact LTR-LTR junctions or ends of complete linear molecules. All of the viruses are of type C morphology except where indicated. RAV-2 (Rous-associated virus 2) is a member of a group of ASLVs that share significant sequence homology and can form viable recombinants with one another. The group includes AMV (avian myeloblastosis virus) and Ev-1 (chicken endogenous virus 1). SNV (spleen necrosis virus) is a member of a different group of avian viruses which have *pol* genes and proteins that are more closely related to those of the mammalian viruses. Unlike ASLV, SNV can replicate in some mammalian cells. A related group of MLVs includes MoMLV (Moloney murine leukemia virus), FrMLV (Friend murine leukemia virus), Cas-Br-E (Casitas brain ecotropic virus), and AKR SL-3, a pathogenic variant of AKV, the endogenous ecotropic virus of mice. VL30, from viruslike 30 RNA, is a murine endogenous viral genome that can be packaged, replicated and integrated with normal type C murine virus proteins. MMTV (mouse mammary tumor virus) has type B morphology, and IAPs (endogenous intracisternal A particles) have type A morphology. BLV is bovine leukemia virus. HTLV-I is a human T-cell lymphotrophic virus, and HIV-1 is a human lymphotropic virus belonging to the lentivirus subfamily. Details concerning the biology of these viruses and additional sequence data can be found in *RNA Tumor Viruses* (Weiss et al., 1982, 1985).

DNA in the integration reaction. However, larger deletions of 2 to 8 bp from the U5 end seriously impaired or abolished integration. These results, together with the comparisons shown in Fig. 3, suggest that deoxynucleotides at the very ends of the LTRs are not as important as internal sequences for recognition by the recombination complex.

The second line of evidence which bears on the identity of donor sites comes from biochemical studies of the *pol*-encoded endonuclease of the avian sarcoma-leukosis viruses (ASLV) (see Fig. 3). Genetic evidence that a function encoded in this domain of the *pol* gene is required for integration of ASLV and murine leukemia virus (MLV) is described below (section III.D.). In ASLV, the endonuclease domain is incorporated into two differentially processed major *pol* products (see Fig. 5). As part of the reverse transcriptase $\alpha\beta$ heterodimer, it exhibits an Mn^{2+}-dependent nicking activity which prefers DNA supercoils or single-stranded molecules over relaxed double-stranded DNA substrates (Leis et al., 1983). As part of the smaller pp32 protein, it has the same substrate preference, and although it is more active in the presence of Mn^{2+}, it also utilizes Mg^{2+} as a cofactor. The $\alpha\beta$ *pol* endonuclease cleaves preferentially between the conserved CA dinucleotide in substrates in which the viral LTRs are covalently joined (Duyk et al., 1985; Cobrinik et al., 1987). These sequences have been called "LTR-LTR junctions," or "circle junctions" to relate to their natural occurrence in the covalently closed circular forms of retroviral DNA that contain tandem LTRs (see Fig. 2). This $\alpha\beta$ *pol* endonuclease cleavage site is one deoxynucleotide upstream of the site predicted in the simplest model of breakage and joining proposed for integration (see Fig. 5). In the presence of Mg^{2+}, and limited digestion, the pp32 endonuclease cleaves preferentially after the CA dinucleotide, as expected for integration (Grandgenett et al., 1986). However, at higher concentrations of protein, with more extensive digestion, or in the presence of Mn^{2+} this enzyme cleaves between the CA dinucleotide, as does the $\alpha\beta$ form. Thus, these activities are clearly related, and on the basis of their preference for sequences at the LTR termini, either one or both seem likely to play a role in cutting viral, and perhaps host, DNA during integrative recombination.

Grandgenett et al. (1986) evaluated the ability of the ASLV pp32 endonuclease to cleave supercoiled DNA molecules that contained aberrant LTR-LTR junction sequences which were molecularly cloned from ASLV-infected cells. One clone, which contained an additional A-T base pair between the LTR-LTR junction, was cleaved like the wild type. This observation is consistent with genetic results with MoMLV (cited above) and supports the notion that there is some flexibility with respect to the exact number of nucleotides that are acceptable at the U3 and U5 termini. A second substrate that contained a substitution (deletion plus insertion) at the junction, which had replaced 12 bp from the U5 terminus and 22 bp from the U3 terminus with 35 bp of sequences from another region in the genome, was not cleaved near the junction.

In an extensive analysis of the Mn^{2+}-dependent $\alpha\beta$ *pol* endonuclease activity of ASLV, Duyk et al. (1985) and Cobrinik et al. (1987) found that sequences required for cleavage in the conserved CA dinucleotides at LTR-LTR junctions were slightly different in single-stranded and double-stranded supercoiled DNAs.

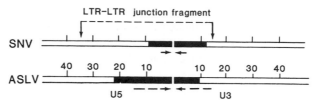

FIGURE 4. Proposed donor sites for integration of SNV and ASLV. Virus designations and DNA termini are as indicated in Fig. 3. Numbers of nucleotides from the termini are indicated in the scale above the map for ASLV. The solid areas indicate the extent of sequence information required for integration as measured by genetic (SNV; Panganiban and Temin, 1983) or biochemical analyses using double-stranded supercoiled DNA (ASLV; Cobrinik et al., 1987). LTR-LTR junction fragment indicates the SNV segment transposed to another genomic site where it functions in integrative recombination (Panganiban and Temin, 1984a).

However, in both cases a more extensive region was required for cleavage of the (−) strand than the (+) strand, and in both cases more of the U5 terminus was required than of the U3 terminus. The region required for cleavage of either strand in the double-stranded supercoils fell within 22 bp of the U5 terminus and within 8 bp of the U3 terminus. This U3 region coincides almost exactly with the section that is conserved among related strains of ASLV (see Fig. 3) in which genetic recombinants are known to be viable. Analysis of DNAs that contained a nested set of deletions near the U5 terminus showed that viral growth was not prevented when 22 bp of the U5 terminus remained intact but was abolished when only 15 bp remained. This coincides with the U5 sequence requirement for cleavage by $\alpha\beta$ *pol* endonuclease in the supercoiled substrate and suggests that double-stranded DNA may be the more biologically relevant substrate for integration. Insertion of an 8-bp palindromic sequence between the LTR-LTR junction or deletion of the 4 bp between the CA . . . TG dinucleotides abolished cleavage in such a substrate. Finally, the species specificity of the cleavage reaction was demonstrated by the fact that the ASLV endonuclease did not cleave preferentially near LTR-LTR junctions of either MoMLV or SNV DNAs. In analyses of molecular clones of circular ASLV DNA molecules, Olsen and Swanstrom (1985) found a relatively high representation of molecules in which one LTR contained large deletions that started at U5 and extended towards the viral *gag* gene. These authors suggest that such molecules may have arisen via aberrant cleavage by the viral *pol* endonuclease(s). This interpretation is consistent with the results described above (Duyk et al., 1985; Cobrinik et al., 1987) which showed that most of the sequences required for cutting at the U5 edge of an LTR lie within the U5 terminus.

Figure 4 shows a comparison of the proposed minimal virus donor sites required for integration of SNV and ASLV. The functionality of the SNV sequence has been verified by the demonstration that a 49-bp LTR-LTR junction fragment which encompasses this site can serve as an efficient donor for integrative recombination when it is inserted in another location in the viral genome (Panganiban and Temin, 1984a). Attempts to demonstrate integration from analogous fragments of MoMLV or ASLV LTR-LTR junctions in a similar way have so far proved unsuccessful (S. Goff, personal communication; Skalka, unpublished data). The reason for this difference is unclear; it may reflect some variability in mechanism (see section II.E.).

B. Cellular Target Sites

Restriction endonuclease analyses from many laboratories during the late 1970s showed that a number of different avian and murine retroviruses could integrate at many sites in the host DNA (summarized by Varmus and Swanstrom, 1982). Furthermore, there was no evidence for tandem integration or integration of a new provirus into a resident provirus as a normal occurrence. Thus, the number of possible host DNA target regions seemed to be large, and sequence homology did not appear to influence the reaction. More recent molecular cloning and DNA sequence analyses have so far failed to reveal any sequence consensus in the choice of retroviral integration target sites and have reinforced the notion that integration may occur at random in host DNA. These studies included sequence analyses of virus-cell junctions in seven independent clones of SNV integrated into chicken DNA (Shimotohno and Temin, 1980) and sequence analyses of six independent clones of unintegrated MoMLV DNA molecules that contained rearrangements which resemble intramolecular integrations (Shoe-maker et al., 1981). Since the sample number in each of these cases is relatively small, it is difficult to exclude some loose sequence specificity or recognition for sequences removed from the actual integration site, as is the case, for example, with class II restriction endonucleases or Chi-promoted general recombination (see Taylor, this volume; Mosig, this volume). A selected but strict specificity does seem to occur with ASLV integration (C. Shih and J. Coffin, personal communication). Analysis of molecular clones of randomly integrated ASLV proviruses has revealed a subclass of host target sites that are used for 1 in 2,000 to 5,000 integration events. When they occur at these preferred targets, independent integration reactions utilize exactly the same nucleotides in the host DNA. Shih and Coffin (personal communication) estimate that approximately 500 to 1,000 such preferred sites are present in avian cell DNA. So far, analysis of sequences 300 to 500 nucleotides to either side of one of the preferred target sites has revealed no notable features, with the exception of some AT-rich stretches.

It is worth noting that element 17.6, a retrotransposon of *Drosophila melanogaster*, shows site-specific insertion into the sequence, 5'-ATAT, which corresponds to the central portion of the TATA consensus box, TATA(A/T)A (A/T), in host promoter regions (Inouye et al., 1984). In addition, Gypsy, an unrelated *Drosophila* transposon, shows a preference for integration into host sequences 5'-TA(T/C)A(T/C)A (Parkhurst and Corces, 1986). A subset of these would include the 17.6 recognition site. Finally, in apparent exception to what has been observed for other retroviruses, infection of cultured human cells with the baboon endogenous retrovirus frequently leads to integration at a specific genetic locus on chromosome 6 (Lemons et al., 1978). Restriction endonuclease analysis of baboon endogenous retrovirus-infected human cells has revealed a common *Pst*I fragment believed to contain the downstream termini of proviral DNA and some adjacent host cell sequences (Cohen and Murphey-Corb, 1983). To explain this result, the authors proposed that baboon endogenous retrovirus may integrate preferentially adjacent to short, repeated host sequences which contain a *Pst*I restriction site. This interpretation has not yet been verified by additional independent results. In evaluating these observations, it is important to know whether baboon endogenous retrovirus truly represents an exception to the general rule for retroviral target sites. Alternatively, this phenomenon might provide a clue to details of a mechanism that is applicable to all retroviruses.

Early in vivo experiments using tissue culture cells were interpreted to imply that retroviruses can integrate only into newly synthesized DNA (Varmus et al., 1977). However, this cannot be a general rule for all retroviruses since at least one member of the lentivirus group (i.e., the slow virus, visna) does not appear to require host DNA replication for integration (Harris et al., 1984). More recent studies of in vivo targets of integration suggest that chromatin structure may influence the choice of site. Proviruses have been mapped within a few hundred base pairs of DNase I-hypersensitive sites in a number of cell types that have been selected for altered phenotypes: e.g., chicken bursal lymphomas induced by insertion of an avian leukosis provirus near the c-*myc* gene (Robinson and Gagnon, 1986; Schubach and Grodine, 1984), erythroblastosis induced by ALV insertion near the c-*erb* gene (Vijaya et al., 1986), and integration of MoMLV in the mouse alpha 1 collagen gene (Breindl et al., 1984). Although it may be supposed that such selection narrows the scope of events that may be scored, additional analyses of nonselected integrations have shown a similar correlation. For example, MoMLV proviruses in mouse or rat tissues or cultured cells also map close to DNase I-hypersensitive sites (Vijaya et al., 1986; Rohdewohld et al., 1987). Results from insertional mutagenesis studies of mouse cells infected with MLV (King et al., 1985) also suggest that not all regions of the respective genomes are equally receptive to retroviral integration. In *Saccharomyces cerevisiae*, the retrotransposon Ty appears to integrate preferentially into promoter regions of cellular genes in both selected and nonselected events (Eibel and Philippsen, 1984). This may reflect an analogy with retroviruses and *Drosophila* retrotransposons, since promoter regions which include TATA sequences are often also hypersensitive to DNAse I.

The recent development of cell-free systems which show correct integration of retroviral DNAs into exogenously added target DNA (Brown et al., 1987; Y. Lee and J. Coffin, personal communication; K. Mizuuchi, personal communication) should permit a more comprehensive evaluation of the question of target specificity. Analyses of extracts of virus-infected cells have shown that a subviral nucleoprotein complex present in the cytoplasm contains both DNA and protein components required for integration into exogenously added target DNA (B. Bowerman, personal communication). The activity was first detected by assaying for suppression of amber mutations in a bacteriophage lambda genome by integration of an MLV derivative that contained the *Escherichia coli supF* gene (Brown et al., 1987). The integrated proviruses bear the expected hallmarks (i.e., 4-bp flanking duplications of host DNA and loss of 2 bp from the ends of each LTR). So far, sequence analyses of seven independent MoMLV integrations into bacteriophage lambda DNA have revealed no consensus in the sequences extending 11 bp to either side of the proviral termini (Brown et al., 1987) or with these sequences and those in the MoMLV clones described above. Preliminary mapping of more than 30 proviruses in independent lambda clones showed that integration can take place in many locations (Brown et al., 1987). More detailed analyses should make it possible to further evaluate the influence of nucleotide sequence as well as other factors that may affect target selection. The cell-free system has provided the following additional important information: (i) exogenously added DNA is an acceptable target for retroviral integration; (ii) supercoiling is not required for the target; and (iii) the target DNA need not be transcriptionally active or newly replicated.

The simplest conclusions from results of the in vitro and in vivo studies to date are that DNA accessibility is an important factor in the choice of target site and that newly replicated sections, promoter regions, and DNase I-hypersensitive sites are preferred because they contain regions of exposed cellular DNA. An alternative possibility is that the integration complexes are active at such sites because they can or need to interact with one or more of the host proteins that accumulate there or that they act upon a product of these proteins (e.g., cleaved or structurally altered host DNA). Further work will be required to distinguish between these hypotheses.

C. Viral DNA Substrate and Products

Several lines of evidence, some already noted, suggest that the circular DNA molecule containing two tandem copies of the LTRs is the viral substrate for integrative recombination. (i) Genetic studies with SNV (Panganiban and Temin, 1984a) showed that integration can occur, with relatively high frequency, at LTR-LTR junction sequences that are introduced into an internal site in the viral genome and that the resulting integrated proviruses are in a circular permutation of the normal arrangement. (ii) Biochemical studies with the ASLV *pol* endonucleases show that supercoiled, covalently closed circular molecules are preferred substrates (Grandgenett et al., 1978; Leis et al., 1983) and that there is also a preference for binding (Misra et al., 1982) and cleavage (Duyk et al., 1983, 1985; Grandgenett and Vora, 1985; Grandgenett et al., 1986; Cobrinik et al., 1987) at LTR-LTR junctions. These facts and the apparent correlation between the minimal sequences required for cleavage at these junctions in supercoiled DNA substrates and sequences required for viral growth (Cobrinik et al., 1987) support the notion that circles containing LTR-LTR junctions are integrated. (iii) A number of experiments in which host cell functions were impaired by various drug treatments show an apparent correlation between the absence of closed circular (but not linear) DNA molecules and deficiency in integration and virus production (Chinsky and Soeiro, 1982; Hsu and Taylor, 1982b; Yang et al., 1980; Huleihel and Aboud, 1983). A similar correlation exists in visna virus infection of cultured cells, in which less than 0.1% of the newly synthesized viral DNA is in circular forms and very few infected cells contain integrated proviruses (Harris et al., 1981). However, virus production is not defective in this case, presumably because

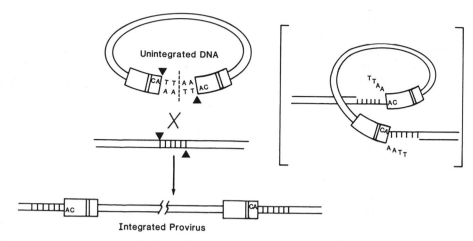

FIGURE 5. Integrative recombination. The diagram shows the dinucleotide (CA) believed to mark the site of donor cleavage and joining to host target DNA, based on DNA sequence analyses (for review, see Weiss et al., 1982, 1985) and biochemical studies (Leis et al., 1983; Duyk et al., 1985; Grandgenett et al., 1985, 1986; Cobrinik et al., 1987). A staggered cut of host DNA is suggested by the existence of short direct repeats (e.g., 6 bp for ASLV; indicated by vertical lines) flanking the integrated provirus. The drawing in brackets shows a hypothetical intermediate in the reaction in which single strands of host and viral DNAs are joined at the breaks. The 5'-terminal nucleotides of the virus, which are not complementary to the host DNA (four nucleotides, as indicated, for a viral substrate whose ends are ligated; two nucleotides for a substrate whose ends are not ligated), may be removed by DNA repair enzymes. The short flanking duplication of host sequences is produced by filling in the resulting short gaps.

unintegrated DNA can serve as a template for viral RNA synthesis (Harris et al., 1984).

Despite all of the evidence supporting the notion of a circular viral DNA substrate, some uncertainties remain. The possibility that linear molecules may also serve as integration substrates was not excluded in the SNV genetic studies cited in (i) above. In addition, it is not clear how far one can extrapolate from the ASLV endonuclease in vitro reaction (ii), since other proteins inside the cell could modulate its biochemical activity. Finally, the drug treatment experiments (iii) are difficult to interpret since defects in the synthesis of linear molecules could also account for such observations, and factors other than a limitation of appropriate substrates could affect visna virus integration (Blum et al., 1985). Thus, although the available evidence indicates that circular viral DNA can be a substrate in one case (SNV) and may also serve in other cases (ASLV), the possibility that molecules whose termini are not covalently joined can also function as substrates remains quite viable. Results with the MoMLV in vitro integration system (Brown et al., 1987) suggest that for this virus linear molecules may, in fact, be the sole or preferred substrate.

Comparisons of the nucleotide sequences of LTR termini in unintegrated and integrated DNA and of unoccupied and occupied cellular target sites have revealed two characteristic features of the integration reaction (Fig. 5). These are

(i) loss of deoxynucleotides, often 2 bp, from each LTR terminus and (ii) a short duplication of host cell sequences flanking the integrated provirus. A model which suggests how these features might originate is shown in Fig. 5. It differs only in detail from one proposed earlier by Shoemaker et al. (1980), and both are similar to those originally proposed for transposition of procaryotic elements (Grindley and Sherratt, 1979; Shapiro, 1979). A hypothetical intermediate is drawn in which a circular viral DNA molecule is the immediate substrate in a reaction in which the CA 3'-hydroxyl ends of viral DNA (as produced by the ASLV *pol* endonucleases [Leis et al., 1983]) are joined to protruding 5'-phosphate ends of host DNA. As was noted in the original proposal (Shoemaker et al., 1980), a similar intermediate could be drawn for a viral substrate whose ends are not ligated. In this case, two rather than four nucleotides would protrude from the 5' ends of the viral DNA. When the products of an MoMLV in vitro integration system are analyzed by gel electrophoresis, as much as 5 to 10% of the retroviral DNA appears to be joined to plasmid target DNA as judged by relative migration rates (Mizuuchi, personal communication). It seems likely that such joined molecules represent intermediates in the integrative recombination reaction (Brown et al., 1987; Mizuuchi, personal communication). The recent isolation and structural analysis of such intermediates have provided convincing evidence that the immediate precursor in integration of MoMLV DNA does not contain ligated ends (K. Mizuuchi, personal communication; P. Brown, personal communication).

The biochemical steps predicted for a complete integrative recombination reaction are (i) endonucleolytic cutting of virus and host DNA sequences, (ii) joining of one strand of viral DNA to one strand of host DNA at each of the DNA ends, (iii) DNA synthesis and repair to produce the host cell DNA duplications and to remove noncomplementary, protruding tails of viral DNA, and (iv) ligation of the repaired termini. The fact that viral sequences are "lost" in the process suggests that the integrative recombination reaction is not reversible. Proviral sequences could, however, be lost from chromosomal DNA via homologous recombination between LTRs in the same or different (Lazo and Tsichlis, 1988) proviruses. This is the supposed origin of "solo" LTRs identified in the DNA of several host species (for review, see Varmus and Swanstrom, 1985).

D. Enzymes/Proteins and Mechanisms

A role for viral proteins in the integration reaction is implied by the facts that the reaction is site-specific with respect to viral sequences and that the number of base pairs in the short duplication of host DNA which flanks the integrated provirus is characteristic of the virus and not the host cell in which it is grown (e.g., 6 bp for ASLV and 5 bp for SNV, both integrated in chicken cell DNA, and 4 bp for MoMLV and 6 bp for mouse mammary tumor virus, both integrated in mouse cell DNA [see Varmus and Swanstrom, 1982]). In addition, cloned retroviral DNA introduced into cells by transfection is not integrated at LTRs (Luciw et al., 1983). These facts, together with the observation that integration can occur in the absence of new viral protein synthesis (Varmus and Swanstrom, 1982), suggest that proteins brought in with the infecting particle are used for

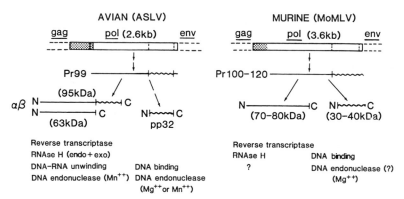

FIGURE 6. *pol* genes of avian and murine retroviruses, their protein products, and their known or presumed enzymatic activities. The stippled section in the genetic map at the top indicates the location of sequences encoding the viral protease which cleaves the viral precursor proteins into active polypeptides. Pr99 and Pr100-120 are precursor proteins of 99 and 100 to 120 kilodaltons (kDa), respectively. N and C indicate the amino and carboxyl termini of the processed proteins.

integration. More recent evidence has implicated functions encoded in the *pol* gene. This gene is the most highly conserved among retroviruses and other retrotransposons and is the only region in which extensive homologies in amino acid sequences exist between different retroviruses (Toh et al., 1983; Chiu et al., 1984).

Figure 6 summarizes some features of the *pol* genes of avian (ASLV) and murine (MLV) retroviruses and their products. Although nucleotide sequences of the genes and structural features of their proteins are different, certain similarities are apparent. In both cases, two domains important for DNA synthesis and integration can be identified. Both map immediately downstream of the sequence that encodes the viral acid protease which is responsible for processing the *gag* and *gag-pol* precursors. The first, and largest, of these domains encodes the reverse transcriptase activity, and the second, in ASLV at least, encodes a DNA endonuclease which is detected in virions. Earlier suggestions that the analogous domain in Rausher and Friend MLVs may encode the DNA endonuclease activity detected in murine virions (Kopchick et al., 1981a; Nissen-Meyer and Nes, 1980) have recently been brought into question. Genetic and biochemical analyses indicate that the major DNA endonuclease activities associated with MoMLV virions are not encoded in either *gag* or *pol* (Panet and Baltimore, 1987) and probably represent cellular "contaminants." Furthermore, biochemical studies with bacterially expressed fusion proteins which include this domain of the MoMLV genome have failed to reveal any DNA endonuclease activity (Goff, personal communication). It is, of course, possible that expression of the murine endonuclease activity requires an interaction between the protein that contains the endonuclease domain and other viral or cellular components. Alternatively, in MLV, the cleavage step and the initial ligation step which joins viral and host DNA during integration may be coupled. As noted by Brown et al. (1987), the absence of an energy requirement in the MoMLV cell-free integration system would seem to support this interpretation. For the purposes of this discussion, I will continue to refer to this region as the endonuclease domain, in analogy with ASLV.

The *pol* endonuclease activities of ASLV have been well characterized, and their viral origin has been verified by genetic analyses (Golomb et al., 1981). In addition, bacterially expressed, molecularly cloned sequences from this domain exhibit endonuclease activity that is very similar to that of the authentic viral protein (Terry et al., 1988). As mentioned earlier, biochemical studies on the ASLV viral endonuclease activities have revealed a preference for binding to the LTRs (Misra et al., 1982; Knaus et al., 1984), for cleaving at sites close to those joined to host DNA during integration (Duyk et al., 1983, 1985; Cobrinik et al., 1987; Grandgenett and Vora, 1985; Grandgenett et al., 1986; Skalka et al., 1984), and for sequence and species specificity consistent with a role in integrative recombination (Cobrinik et al., 1987). Some properties of the αβ form of the ASLV endonuclease suggest that it may be involved in cleavage of the host target site (Cobrinik et al., 1987). However, since the integration reaction requires not only breaking but also joining of viral and host DNA molecules, the endonuclease cleavage must be viewed as a partial reaction that may be modified by addition of other components of the replication complex.

Recently, Johnson et al. (1986) reported the results of computer-aided comparisons of the predicted amino acid sequences of proteins from the reverse transcriptase and endonuclease domains of five retroviruses with those of DNA polymerases and nucleases from various sources. In Fig. 7, the features which relate to the endonuclease domain are summarized in the context of ASLV. The computer studies uncovered a number of analogies as well as conserved amino acid sequences. These similarities were revealed by aligning similar sequences, allowing small "insertions" or "deletions" with respect to one another. In most instances these differences were minor. One significant exception was MoMLV, whose genome encodes two insertions with respect to the other viral proteins (see Fig. 7) which by virtue of location (in the N-terminal conserved region) or size (in the C-terminal region) may signify important differences in mechanisms of action. In an earlier computer comparison of proteins encoded in the "polymerase" domain (e.g., ASLV *pol* codons 21 to 235), which included the lentivirus visna as well as the five considered by the Johnson et al. (1986), Sonigo et al. (1985) concluded that the polymerase region of MoMLV diverged from the others at the earliest time in evolution. As noted by these authors, this seems consistent with the fact that the MoMLV reverse transcriptase prefers Mn^{2+} as a divalent cation whereas all of the others prefer Mg^{2+}. A comparison by the same authors of the N-terminal portion of the endonuclease domains (corresponding to ASLV *pol* codons 1 to 176) of nine viruses also suggested an early divergence for MoMLV (and for the SNV relative, reticuloendotheliosis virus).

The most notable features identified by Johnson et al. (1986) in their comparison of the endonuclease domains include a specific arrangement of two His and two Cys residues near the N terminus (see Fig. 7), which may be analogous to the Zn^{2+} "finger" DNA binding region which has been identified in a transcription factor of *Xenopus laevis* (TFIIIA [Miller et al., 1985]) and is

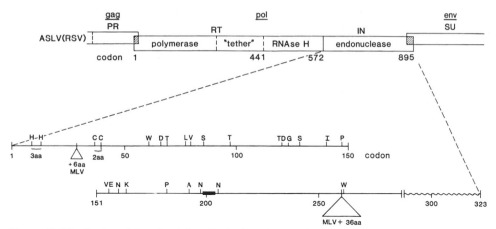

FIGURE 7. Distribution of functional domains in the *pol* gene of ASLV and conservation of sequences in the endonuclease protein. The figure is adapted from data of Johnson et al. (1986) in which predicted protein sequences from five retroviruses, HTLV-I, BLV (bovine leukemia virus), RSV (Rous sarcoma virus, a member of the ASLV family), HIV, and MoMLV, were compared with other polymerases and nucleases. Amino acids that are identical in analogous regions of all five viral proteins are indicated by the standard single-letter amino acid code on the ASLV map. The black rectangle shows the location of a four-amino-acid sequence which is found in ASLV but not in the other four viral proteins. Two regions where the MoMLV sequence is extended with respect to ASLV and the other viral proteins are also shown. Amino acids 9 to 40 in the ASLV endonuclease constitute the putative Zn finger. The break in the map followed by a wavy line shows the 4.1-kilodalton region that is removed from the end of the ASLV endonuclease during proteolytic processing (see also Fig. 6).

believed to be important for specific protein-DNA interactions in several other cases as well (see Berg, 1986; Brown and Argos, 1986). However, there is, as yet, no direct evidence that Zn^{2+} is required for the function of the retroviral endonuclease. The sequence contained within the finger of this domain is variable among the viruses compared, with the exception of an Arg-Ala sequence (at ASLV codons 18 and 19) and an Arg residue (at ASLV codon 30) which are found in all but the MoMLV sequence. This exception again suggests a functional difference for MLV. Other amino acids which are conserved in the proteins of all five viruses are indicated in Fig. 7. Among them are several Ser and Thr residues which are potentially modifiable by phosphorylation. Since the ASLV pp32 is phorphorylated in vivo (Schiff and Grandgenett, 1980), it will be of interest to determine whether one or more of these conserved residues is the natural substrate for phosphorylation and to evaluate its functional significance, if any. Because these and most of the other conserved amino acids are located in or near the N-terminal half of the protein, it seems possible that this end of the molecule encodes functions that are common to all the endonuclease proteins.

Maturation of the β and pp32 *pol* endonucleases of ASLV involves removal of the equivalent of 4.1 kilodaltons from the C terminus of the *pol* precursor, indicated as a wavy line in Fig. 7 and as the straight end in Fig. 6 (Grandgenett et al., 1985; Alexander et al., 1987). Deletion experiments in which this region was partially or precisely eliminated failed to reveal any altered phenotype with respect to viral growth (Katz and Skalka, 1988) or altered protein properties (Morris-Vasios et al., 1988; Terry et al., 1988). Thus, this region is either irrelevant to function or can be efficiently complemented by a host function. It is not yet known whether the analogous region is removed in other retroviral endonuclease domain proteins. Mutations that have been introduced into the protease processing site which separates the reverse transcriptase and endonuclease domain of both ASLV (M. Kotler, R. A. Katz, W. Danho, J. Leis, and A. M. Skalka, submitted for publication) and MoMLV (Goff, personal communication) are lethal when introduced into viral genomes. This result suggests that processing per se is required for the function of both *pol* proteins. However, since these mutations, of necessity, alter the C- and N-terminal amino acids of reverse transcriptase and endonuclease, it is not yet possible to rule out a direct effect on enzymatic activities.

Sequences within the endonuclease domain of *pol* are clearly essential for normal virus replication: numerous mutations introduced at various locations in this region produce replication-defective viruses. This is the case not only for ASLV, in which the domain is also included in reverse transcriptase (Hippenmeyer and Grandgenett, 1984), but also for SNV (Panganiban and Temin, 1984b) and MoMLV (Schwartzberg et al., 1984a; Goff, personal communication), in which the reverse transcriptase and endonuclease domains are separate on mature polypeptides (see Fig. 6). When such analyses have been possible, it has been shown that some endonuclease domain mutant virions contain normal reverse transcriptase and RNase H activities (Hippenmeyer and Grandgenett, 1984; Panganiban and Temin, 1984b; Schwartzberg et al., 1984b). Thus, this domain provides a distinct function for the virus.

Analysis of the properties of several endonuclease domain mutants with a partially defective phenotype has provided insight into the nature of this function and the significance of particular regions within the endonuclease domain. Hippenmeyer and Grandgenett (1984) produced two such mutants which contained single point mutations, in codon 7 or 8, that resulted in substitution of Asn for Asp and Phe for Leu, respectively. These changes, which lie immediately upstream of the first His residue in the proposed Zn^{2+} finger, result in virions that exhibit high reverse transcriptase activity but only 7.2 and 1.9% of the focus-forming activity of wild-type virions (Hippenmeyer and Grandgenett, 1985). Computerized structural analyses suggest that the changes occur at the border of two predicted α-helical regions which include the first 32 N-terminal amino acids of the endonuclease (Grandgenett et al., 1985). A small, in-frame deletion in this region (including amino acids 6 to 10) produced a completely defective virus. How each of these observations relates to function remains unclear. When infections with wild type and the partially defective mutants were normalized to equal reverse transcriptase units, the early replication events (i.e., virion entry, DNA synthesis) were indistinguishable, but it is not entirely clear whether integration was affected (D. Grandgenett, personal communication).

Mutant analyses of MoMLV by Goff and associates (personal communication) have also focused on residues included in the proposed Zn^{2+} finger region. They include a change of Ala to Val immediately upstream of the last Cys residue, substitution of Tyr for this Cys residue, and an insertion of four amino acids in front of the Ala residue that precedes this Cys. The first mutation, which is fairly conservative, did not seem to affect the ability of bacterially expressed protein to bind to DNA. The second two seemed to affect the ability of the proteins to refold after denaturation in vitro and thus to form a DNA-binding site. When introduced into viral genomes, the conservative change had no effect, but the second two mutations proved to be lethal. However, since the mutant protein could bind DNA if properly refolded, the relationship between DNA binding as measured in vitro and retroviral growth is still not clear.

Analyses of two additional partially defective endonuclease mutants of MoMLV (Schwartzberg et al., 1984a; Donehower and Varmus, 1984; Hagino-Yamagishi et al., 1987) and one of SNV (Panganiban and Temin, 1984a) have shown conclusively that the function impaired in each case is viral DNA integration. With these mutants, all late and most early functions appeared normal, but the viral DNA synthesized was shown to be not integrated (Panganiban and Temin, 1984b), integrated inefficiently (Hagino-Yamagishi et al., 1987), or not maintained in a productive fashion and presumed to be unintegrated (Schwartzberg et al., 1984a; Donehower and Varmus, 1984). With the SNV mutant, the function could be complemented by a second virus which was integration defective as a result of a cis-acting mutation in the virus donor site. As might be expected, under these conditions only the endonuclease mutant DNA was integrated (Panganiban and Temin, 1984b). The SNV mutation, a deletion that creates a frameshift, is presumably in a region encoding the N terminus of the endonuclease domain (Panganiban and Temin, 1984b), and from a comparison with the sequence of the related virus reticuloendotheliosis virus (Wilhelmsen et al., 1984), the mutation lies 72 codons upstream of the first His residue of the putative Zn^{2+} finger. One of the MoMLV mutants employed in analogous studies contained a frameshift-inducing deletion of 31 amino acids in the region corresponding to RSV residues 44 to 54, which starts 3 amino acids downstream of the proposed Zn^{2+} finger (Schwartzberg et al., 1984a). The second contained a substitution of Cys for a conserved Arg at a position analogous to Rous sarcoma virus residue 53 (Donehower and Varmus, 1984). (A third mutation which introduced a frameshift starting at the same residue gave the same general phenotype [Hagino-Yamagishi et al., 1987].) Analysis of the infrequent integrations which occur with the Arg for Cys substitution mutant of MoMLV appeared normal; they showed the same host sequence duplications and loss of viral sequence as the wild-type reaction. However, the frameshift mutant integrated aberrantly, as if its DNA was first oligomerized via replicative or general recombination mechanisms and then integrated randomly in the viral sequences much as any introduced plasmid DNA is integrated into eucaryotic DNA (see Roth and Wilson, this volume; Subramani and Seaton, this volume). Recently, T. Quinn and D. Grandgenett (personal communication) constructed an ASLV endonuclease mutant that is temperature sensitive for integration. The mutant (no. 107) contains a Pro to Ser change at codon 115 in the N-terminal coding region. Such conditional mutants should be especially useful in further efforts to elucidate the function and activities of this protein.

E. Future Directions

It is clear that despite the wealth of detail that has been accumulated concerning retroviral DNA integration over the past few years, we are still far from being able to describe specific molecular interactions, and several fairly fundamental questions remain to be resolved. For example, although it may be surmised to be important, it is still not clear how core particle architecture may influence the process. The active component in the MoMLV cell-free integration system behaves like a protein-DNA complex which includes at least the major core protein (Brown et al., 1987; B. Bowerman, personal communication). Earlier studies of host cell-specific restriction of MLV (encoded in the mouse Fv-1 locus) have also suggested a role for the major core protein of the virus in integration (Boone et al., 1983; DesGroseillers and Jolicoeur, 1983), and other proteins in the virus core may also affect both reverse transcription (Crawford and Goff, 1984; Schwartzberg et al., 1984b; Meric and Spahr, 1986) and integration.

Also to be resolved is the question of the structure of the immediate viral DNA precursor in the integration reaction. However, here the problem may be less real than apparent. For example, one may imagine that during the recombination reaction the two ends of the viral DNA which constitute the donor site must be brought together with the host target site, presumably by some integration protein complex. For such intermediates the stable protein-DNA complexes involved in the in vitro transposition of supercoiled phage Mu DNA would seem to provide an appropriate paradigm, since Mu transposes via a Shapiro (1979) or Grindley and Sherratt (1979) intermediate and several features of the biochemistry of this reaction appear to resemble retroviral integration (Mizuuchi, 1984; Craigie et al., 1984; Craigie and Mizuuchi, 1985a, b; Miller and Chaconas, 1986). In the case of Mu, a complex which includes the viral transposase protein does hold the interacting DNA sites together (Surette et al., 1987). Protein-protein interaction between such bound molecules in a retroviral integration complex could potentially juxtapose LTR termini. Unless there is some functional role for DNA ends per se, it may not matter whether the retroviral DNA termini in such a complex are (or become) ligated. As analyses of mutant donor sites have indicated, there is some flexibility in the nature and length of the terminal sequence. The apparent differences in viral DNA substrate preferences (i.e., ligated versus nonligated ends) may reflect species-specific protein differences such as those noted in section III.D.

Another fundamental problem in efforts to evaluate enzymatic mechanisms is the difficulty in separating reactions involved in synthesis of the substrate viral DNA from those required for integrative recombination. It is conceivable that the two processes may be inextricably linked by their mutual dependence on a complex that includes several viral proteins and perhaps host proteins as well. Functional interactions between reverse transcriptase and endonuclease domains are suggested by the fact that they are part of the same polypeptide in the β chain of ASLV reverse transcriptase and that the analogous domains are linked by disulfide bonds in MLV particles (Hu et al., 1986). As noted earlier, there is some suggestion that host cell functions may also be required for integration, if only for formation of appropriate viral substrates. Various cell-derived enzymes (DNA ligase, DNA endonucleases) are incorporated into virions (Varmus and Swanstrom, 1982; Panet and Baltimore, 1987); it is conceivable that these or other host nuclear enzymes may play a role as components of the recombination complex.

Several current experimental strategies seem likely to provide important new information in the near future. Continued application of cell-free integration systems should allow the identification of candidate proteins for the replication complex. This, in turn, may make it possible to reconstruct recombination systems with purified protein components. Genetic and mutational analyses should continue to provide important clues to viral gene function. For example, the ability to make specific changes in viral gene sequences, to test their effects with bacterially expressed proteins or plasmid substrates, and then to reinsert the changed sequences into viral genomes for biological analyses provides a particularly powerful approach. Finally, construction of systems in which selected viral proteins can be expressed in a variety of eucaryotic cellular backgrounds (e.g., Barr et al., 1987; Morris-Vasios et al., 1988) should allow mutant complementational analyses and other tests of protein functions. We can look forward to significant progress in our understanding of this very important reaction in the coming years.

ACKNOWLEDGMENTS. I am grateful to John Coffin, Steve Goff, Duane Grandgenett, B. Bowerman, and Kiyoshi Mizuuchi, who granted me permission to cite the details of their work prior to publication, and to many other colleagues who provided me with preprints and reprints. I also thank Jonathan Leis, John Coffin, John Taylor, Phil Tsichlis, and members of my laboratory, especially Richard Katz and Kathy Jones, for reading drafts of the manuscript and offering many helpful suggestions.

Our work is supported by Public Health Service grants CA-06927 and RR-05539 from the National Institutes of Health, by a Glenmede Award, and also by an appropriation from the Commonwealth of Pennsylvania.

LITERATURE CITED

Alexander, F., J. Leis, D. A. Soltis, R. M. Crowl, W. Danho, M. S. Poonian, Y.-C. E. Pan, and A. M. Skalka. 1987. Proteolytic processing of avian sarcoma and leukosis viruses *pol-endo* recombinant proteins reveals another *pol* gene domain. *J. Virol.* **61:**534–542.

Baltimore, D. 1985. Retrovirus and retrotransposons: the role of reverse transcription in shaping the eukaryotic genome. *Cell* **40:**481–482.

Barr, P. J., M. D. Power, C. T. Lee-Ng, H. L. Gibson, and P. Luciw. 1987. Expression of active human immunodeficiency virus reverse transcriptase in *Saccharomyces cerevisiae*. *Bio/Technology* **5:**486–489.

Berg, J. M. 1986. Potential metal-binding domains in nucleic acid binding proteins. *Nature* (London) **232:**485–487.

Blum, H. E., J. D. Harris, P. Ventura, D. Walker, K. Staskus, E. Retzel, and A. T. Haase. 1985. Synthesis in cell culture of the gapped linear duplex DNA of the slow virus Visna. *Virology* **142:**270–277.

Boone, L. R., and A. M. Skalka. 1981. Viral DNA synthesized in vitro by avian retrovirus particles permeabilized with melittin. II. Evidence for a strand displacement mechanism in plus-strand synthesis. *J. Virol.* **37:**117–126.

Boone, L. R., F. E. Myer, D. M. Young, C.-Y. Ou, C. K. Koh, L. E. Roberson, R. W. Tennant, and W. K. Yang. 1983. Reversal of *Fv*-1 host range by in vitro restriction endonuclease fragment exchange between molecular clones of N-tropic and B-tropic murine leukemia virus genomes. *J. Virol.* **48:**110–119.

Breindl, M., K. Harbers, and R. Jaenisch. 1984. Retrovirus induced lethal mutation in collagen I gene of mice is associated with an altered chromatin structure. *Cell* **38:**9–16.

Brown, P. O., B. Bowerman, H. E. Varmus, and J. M. Bishop. 1987. Correct integration of retroviral DNA *in vitro*. *Cell* **49:**347–356.

Brown, R. S., and P. Argos. 1986. Fingers and helices. *Nature* (London) **324:**215.

Chinsky, J., and R. Soeiro. 1982. Studies with aphidicolin on the *Fv*-1 host restriction of Friend murine leukemia virus. *J. Virol.* **43:**182–190.

Chiu, I.-M., R. Calahan, S. R. Tronick, J. Schlom, and S. A. Aaronson. 1984. Major *pol* gene progenitors in the evolution of oncoviruses. *Science* **223:**364–370.

Cobrinik, D., R. Katz, R. Terry, A. M. Skalka, and J. Leis. 1987. Avian sarcoma and leukosis virus *pol*-endonuclease recognition of the tandem long terminal repeat junction: minimum site required for cleavage is also required for viral growth. *J. Virol.* **61:**1999–2008.

Cohen, J. C., and M. Murphey-Corb. 1983. Targeted integration of baboon endogenous virus in the BEVI locus on human chromosome 6. *Nature* (London) **301:**129–132.

Colicelli, J., and S. P. Goff. 1985. Mutants and pseudorevertants of Moloney murine leukemia virus with alterations at the integration site. *Cell* **42:**573–580.

Craigie, R., and K. Mizuuchi. 1985a. Cloning of the A gene of bacteriophage Mu and purification of its product, the Mu transposase. *J. Biol. Chem.* **260:**1832–1836.

Craigie, R., and K. Mizuuchi. 1985b. Mechanism of transposition of bacteriophage Mu: structure of a transposition intermediate. *Cell* **41:**867–876.

Craigie, R., M. Mizuuchi, and K. Mizuuchi. 1984. Site-specific recognition of the bacteriophage Mu ends by the Mu A protein. *Cell* **39:**387–394.

Crawford, S., and S. P. Goff. 1984. Mutations in *gag* proteins p12 and p15 of Moloney murine leukemia virus block early stages of infection. *J. Virol.* **49:**909–917.

DesGroseillers, L., and P. Jolicocur. 1983. Physical mapping of the *Fv*-1 tropism host range determinant of BALB/c murine leukemia virus. *J. Virol.* **48:**685–696.

Donehower, L. A., and H. E. Varmus. 1984. A mutant murine leukemia virus with a single missense codon in *pol* is defective in a function affecting integration. *Proc. Natl. Acad. Sci. USA* **81:**6461–6465.

Duyk, G., J. Leis, M. Longiaru, and A. M. Skalka. 1983. Selective cleavage in the avian retroviral long terminal repeat sequence by the endonuclease associated with the $\alpha\beta$ form of avian reverse transcriptase. *Proc. Natl. Acad. Sci. USA* **80:**6745–6749.

Duyk, G., M. Longiaru, D. Cobrinik, R. Kowal, P. deHaseth, A. M. Skalka, and J. Leis. 1985. Circles with two tandem long terminal repeats are specifically cleaved by *pol* gene-associated endonuclease from avian sarcoma and leukosis viruses: nucleotide sequences required for site-specific cleavage. *J. Virol.* **56:**589–599.

Eibel, H., and P. Philippsen. 1984. Preferential integration of yeast transposable element Ty into a promoter region. *Nature* (London) **307:**386–388.

Eisenman, R. N., W. S. Mason, and M. Linial. 1980. Synthesis and processing of polymerase proteins of wild-type and mutant retroviruses. *J. Virol.* **36:**62–78.

Golomb, M., D. P. Grandgenett, and W. Mason. 1981. Virus-coded DNA endonuclease from avian retrovirus. *J. Virol.* **38:**548–555.

Grandgenett, D., T. Quinn, P. J. Hippenmeyer, and S. Oroszlan. 1985. Structural characterization of the avian retrovirus reverse transcriptase and endonuclease domains. *J. Biol. Chem.* **260:**8243–8249.

Grandgenett, D. P., and A. C. Vora. 1985. Site-specific nicking at the avian retrovirus LTR circle junction by the viral pp32 DNA endonuclease. *Nucleic Acids Res.* **13:**6205–6221.

Grandgenett, D. P., A. C. Vora, and R. D. Schiff. 1978. A 32,000 dalton nucleic acid binding protein from avian retrovirus cores possesses endonuclease activity. *Virology* **89:**119–132.

Grandgenett, D. P., A. C. Vora, R. Swanstrom, and J. C. Olsen. 1986. Nuclease mechanism of the avian retrovirus pp32 endonuclease. *J. Virol.* **58:**970–974.

Grindley, N. D., and D. Sherratt. 1979. Sequence analysis at IS *1* insertion sites: models for transposition. *Cold Spring Harbor Symp. Quant. Biol.* **43:**1257–1261.

Hagino-Yamagishi, K., L. A. Donehower, and H. E. Varmus. 1987. Retroviral DNA integrated during infection by an integration-deficient mutant of murine leukemia virus is oligomeric. *J. Virol.* **61:** 1964–1971.

Harris, J. D., H. Blum, J. Scott, B. Traynor, P. Ventura, and A. Haase. 1984. Slow virus visna: reproduction *in vitro* of virus from extrachromosomal DNA. *Proc. Natl. Acad. Sci. USA* **81:**7212–7215.

Harris, J. D., J. V. Scott, B. Traynor, M. Brahic, L. Stowring, P. Ventura, A. T. Haase, and R. Peluso. 1981. Visna virus DNA: discovery of a novel gapped structure. *Virology* **113:**573–583.

Hippenmeyer, P. J., and D. P. Grandgenett. 1984. Requirement of the avian retrovirus pp32 DNA binding protein domain for replication. *Virology* **137:**358–370.

Hippenmeyer, P. J., and D. P. Grandgenett. 1985. Mutants of the Rous sarcoma virus reverse transcriptase gene are nondefective in early replication events. *J. Biol. Chem.* **260:**8250–8256.

Hsu, T. W., and J. M. Taylor. 1982a. Single-stranded regions on unintegrated avian retrovirus DNA. *J. Virol.* **44:**47–53.

Hsu, T. W., and J. M. Taylor. 1982b. Effect of aphidicolin on avian sarcoma virus replication. *J. Virol.* **44:**493–498.

Hu, S. C., D. L. Court, M. Zweig, and J. G. Levin. 1986. Murine leukemia virus *pol* gene products: analysis with antisera generated against reverse transcriptase and endonuclease fusion proteins expressed in *Escherichia coli*. *J. Virol.* **60:**267–274.

Huleihel, M., and M. Aboud. 1983. Inhibition of retrovirus DNA supercoiling in interferon-treated cells. *J. Virol.* **48:**120–126.

Inouye, S., S. Yuki, and K. Saigo. 1984. Sequence-specific insertion of the Drosophila transposable genetic element 17.6. *Nature* (London) **310:**332–333.

Johnson, M. S., M. A. McClure, D.-F. Feng, J. Gray, and R. F. Doolittle. 1986. Computer analysis of retroviral *pol* genes: assignment of enzymatic functions to specific sequences and homologies with nonviral enzymes. *Proc. Natl. Acad. Sci. USA* **83:**7648–7652.

Ju, G., and A. M. Skalka. 1980. Nucleotide sequence analysis of the long terminal repeat (LTR) of avian retroviruses: structural similarities with transposable elements. *Cell* **223:**379–386.

Junghans, R. P., L. R. Boone, and A. M. Skalka. 1982. Products of reverse transcription in avian retroviruses analyzed by electron microscopy. *J. Virol.* **43:**544–554.

Katz, R. A., and A. M. Skalka. 1988. A C-terminal domain in the avian sarcoma-leukosis virus *pol* gene product is not essential for viral replication. *J. Virol.* **62:**528–533.

King, W., M. D. Patel, L. I. Lobel, S. P. Goff, and M. C. Nguyen-Huu. 1985. Insertion mutagenesis of embryonal carcinoma cells by retroviruses. *Science* **228:**554–558.

Knaus, R. J., P. J. Hippenmeyer, T. K. Misra, D. P. Grandgenett, U. R. Muller, and W. M. Fitch. 1984. Avian retrovirus pp32 DNA binding protein. Preferential binding to the promoter region of long terminal repeat DNA. *Biochemistry* **23:**350–359.

Kopchick, J. J., J. Harless, B. S. Geisser, R. Killam, R. R. Hewitt, and R. B. Arlinghaus. 1981a. Endodeoxyribonuclease activity associated with Rauscher murine leukemia virus. *J. Virol.* **37:**274–283.

Kopchick, J. J., G. Ju, A. M. Skalka, and D. W. Stacey. 1981b. Biological activity of cloned retroviral DNA in microinjected cells. *Proc. Natl. Acad. Sci. USA* **78:**4383–4387.

Kung, H.-J., Y. K. Fung, J. E. Majors, J. M. Bishop, and H. E. Varmus. 1981. Synthesis of plus strands of retroviral DNA in cells infected with avian sarcoma virus and mouse mammary tumor virus. *J. Virol.* **37:**127–138.

Lazo, P. A., and P. N. Tsichlis. 1988. Recombination between two integrated proviruses, one of which was inserted near c-*myc* in a retrovirus-induced rat thymoma: implications for tumor progression. *J. Virol.* **62:**788–794.

Leamnson, R. N., and M. S. Halpern. 1976. Subunit structure of the glycoprotein complex of avian tumor virus. *J. Virol.* **18:**956–968.

Leis, J., G. Duyk, S. Johnson, M. Longiaru, and A. M. Skalka. 1983. Mechanism of action of the endonuclease associated with the αβ and ββ forms of avian RNA tumor virus reverse transcriptase. *J. Virol.* **45**:727–739.

Lemons, R. S., W. G. Nash, S. J. O'Brien, and C. S. Sherr. 1978. A gene (*Bevi*) on human chromosome 6 is an integration site for baboon type C DNA provirus in human cells. *Cell* **14**:995–1005.

Luciw, P. A., J. M. Bishop, H. E. Varmus, and M. R. Capecchi. 1983. Location and function of retroviral and SV40 sequences that enhance biochemical transformation after microinjection of DNA. *Cell* **33**:705–716.

Majors, J. E., and H. E. Varmus. 1981. Learning about the replication of retroviruses from a single cloned provirus of mouse mammary tumor virus. *ICN-UCLA Symp. Mol. Cell. Biol.* **18**:241–253.

Meric, C., and P. F. Spahr. 1986. Rous sarcoma virus nucleic acid-binding protein p12 is necessary for viral 70S RNA dimer formation and packaging. *J. Virol.* **60**:450–459.

Miller, J., A. D. McLachlan, and A. Klug. 1985. Repetitive zinc-binding domains in the protein transcription factor III A from *Xenopus* oocytes. *EMBO J.* **4**:1609–1614.

Miller, J. L., and G. Chaconas. 1986. Electron microscope analysis of *in vitro* transposition intermediates of bacteriophage Mu DNA. *Gene* **48**:101–108.

Misra, T. K., D. P. Grandgenett, and J. T. Parsons. 1982. Avian retrovirus pp32 DNA-binding protein. I. Recognition of specific sequences on retrovirus DNA terminal repeats. *J. Virol.* **44**:330–343.

Mizuuchi, K. 1984. Mechanism of transposition of bacteriophage Mu: polarity of the strand transfer reaction at the initiation of transposition. *Cell* **39**:395–404.

Morris-Vasios, C., J. P. Kochan, and A. M. Skalka. 1988. Avian sarcoma-leukosis virus *pol*-endo proteins expressed independently in mammalian cells accumulate in the nuclease but can be directed to other cellular compartments. *J. Virol.* **62**:349–353.

Nissen-Meyer, J., and I. F. Nes. 1980. Purification and properties of DNA endonuclease associated with Friend leukemia virus. *Nucleic Acids Res.* **8**:5043–5055.

Olsen, J. C., and R. Swanstrom. 1985. A new pathway in the generation of defective retrovirus DNA. *J. Virol.* **56**:779–789.

Panet, A., and D. Baltimore. 1987. Characterization of endonuclease activities in Moloney murine leukemia virus and its replication-defective mutants. *J. Virol.* **61**:1756–1760.

Panganiban, A. T., and K.-C. Luk. 1987. Retroviral DNA integration and *pol* gene expression, p. 129–143. *In* W. Robinson, K. Koike, and H. Will (ed.), *Hepadna Viruses*. Alan R. Liss, Inc., New York.

Panganiban, A. T., and H. M. Temin. 1983. The terminal nucleotides of retrovirus DNA are required for integration but not virus production. *Nature* (London) **306**:155–160.

Panganiban, A. T., and H. M. Temin. 1984a. Circles with two tandem LTRs are precursors to integrated retrovirus DNA. *Cell* **36**:673–679.

Panganiban, A. T., and H. M. Temin. 1984b. The retrovirus *pol* gene encodes a product required for DNA integration: identification of a retrovirus *int* locus. *Proc. Natl. Acad. Sci. USA* **81**:7885–7889.

Parkhurst, S. M., and V. G. Corces. 1986. Retroviral elements and suppressor genes. *BioEssays* **5**:52–57.

Perez, L. G., and E. Hunter. 1987. Mutations within the proteolytic cleavage site of the Rous sarcoma virus glycoprotein that block processing to gp85 and gp37. *J. Virol.* **61**:1609–1614.

Robinson, H. L., and G. Gagnon. 1986. Patterns of proviral insertion and deletion in avian leukosis virus-induced lymphomas. *J. Virol.* **57**:28–36.

Rogers, J. 1983. CACA sequences—the ends and the means? *Nature* (London) **305**:101–102.

Rohdewohld, H., H. Weiher, W. Reik, R. Jaenisch, and M. Breindl. 1987. Retrovirus integration and chromatin structure: Moloney murine leukemia proviral integration sites map near DNase I-hypersensitive sites. *J. Virol.* **61**:336–343.

Schiff, R. D., and D. P. Grandgenett. 1980. Partial phosphorylation in vivo of the avian retrovirus pp32 DNA endonuclease. *J. Virol.* **36**:889–893.

Schubach, W., and M. Grodine. 1984. Alteration of c-*myc* chromatin structure by avian leukosis virus-induced lymphomas. *Nature* (London) **307**:702–708.

Schultz, A., and A. Rein. 1985. Maturation of murine leukemia virus *env* proteins in the absence of other viral proteins. *Virology* **145**:335–339.

Schwartzberg, P., J. Colicelli, and S. P. Goff. 1984a. Construction and analysis of deletion mutations in the *pol* gene of Moloney murine leukemia virus: a new viral function required for establishment of the integrated provirus. *Cell* **37**:1043–1052.

Schwartzberg, P. J., J. Colicelli, M. L. Gordon, and S. P. Goff. 1984b. Mutations on the *gag* gene of Moloney murine leukemia virus: effects on production of virions and reverse transcriptase. *J. Virol.* **49**:918–924.

Scott, M. L., K. McKereghan, H. S. Kaplan, and K. E. Fry. 1981. Molecular cloning and partial characterization of unintegrated linear DNA from gibbon ape leukemia virus. *Proc. Natl. Acad. Sci. USA* **78**:4213–4217.

Shapiro, J. 1979. Molecular model for the transposition and replication of bacteriophage Mu and other transposable elements. *Proc. Natl. Acad. Sci. USA* **76**:1933–1937.

Shimotohno, K., and H. M. Temin. 1980. No apparent nucleotide sequence specificity in cellular DNA juxtaposed to retrovirus proviruses. *Proc. Natl. Acad. Sci. USA* **77**:7357–7361.

Shoemaker, C., S. Goff, E. Gilboa, M. Paskind, S. W. Mitra, and D. Baltimore. 1980. Structure of a cloned circular Moloney murine leukemia virus DNA molecule containing an inverted segment: implications for retrovirus integration. *Proc. Natl. Acad. Sci. USA* **77**:3932–3936.

Shoemaker, C., J. Hoffman, S. Goff, and D. Baltimore. 1981. Intramolecular integration within Moloney murine leukemia virus DNA. *J. Virol.* **40**:164–172.

Skalka, A. M., G. Duyk, M. Longiaru, P. deHaseth, R. Terry, and J. Leis. 1984. Integrative recombination—a role for the retroviral reverse transcriptase. *Cold Spring Harbor Symp. Quant. Biol.* **49**:651–659.

Sonigo, P., M. Alizon, K. Staskus, D. Klatzmann, S. Cole, O. Danos, E. Retzel, P. Tiollais, A. Haase, and S. Wain-Hobson. 1985. Nucleotide sequence of the visna lentivirus: relationship to the AIDS virus. *Cell* **42**:369–382.

Surette, M. G., S. J. Buch, and G. Chaconas. 1987. Transpososomes: stable protein-DNA complexes involved in the *in vitro* transposition of bacteriophage Mu DNA. *Cell* **49**:253–262.

Temin, H. M. 1985. Reverse transcription in the eukaryotic genomes: retroviruses, pararetroviruses, retrotransposons and retrotranscripts. *Mol. Biol. Evol.* **2**:455–468.

Terry, R., D. A. Soltis, M. Katzman, D. Cobrinik, J. Leis, and A. M. Skalka. 1988. Properties of avian sarcoma-leukosis virus pp32-related *pol*-endonucleases produced in *Escherichia coli*. *J. Virol.* **62**:2358–2365.

Toh, H., H. Hayashida, and T. Miyata. 1983. Sequence homology between retroviral reverse transcriptase and putative polymerase of hepatitis B virus and cauliflower mosaic virus. *Nature* (London) **305**:827–829.

Varmus, H. E., T. Padgett, S. Heasley, G. Simon, and M. Bishop. 1977. Cellular functions are required for the synthesis and integration of avian sarcoma virus-specific DNA. *Cell* **11**:307–319.

Varmus, H. E., and R. Swanstrom. 1982. Replication of retroviruses, p. 359–512. *In* R. Weiss, N. Teich, H. Varmus, and J. Coffin (ed.), *RNA Tumor Viruses*. Cold Spring Harbor Laboratory, Cold Spring Harbor, N.Y.

Varmus, H. E., and R. Swanstrom. 1985. Replication of retroviruses, p. 75–134. *In* R. Weiss, N. Teich, H. Varmus, and J. Coffin (ed.), *RNA Tumor Viruses*. Cold Spring Harbor Laboratory, Cold Spring Harbor, N.Y.

Vijaya, S., D. L. Steffen, and H. L. Robinson. 1986. Acceptor sites for retroviral integrations map near DNase I-hypersensitive sites in chromatin. *J. Virol.* **60**:683–692.

Weiss, R., N. Teich, H. Varmus, and J. Coffin (ed.). 1982. *RNA Tumor Viruses*. Cold Spring Harbor Laboratory, Cold Spring Harbor, N.Y.

Weiss, R., N. Teich, H. Varmus, and J. Coffin (ed.). 1985. *RNA Tumor Viruses*. Cold Spring Harbor Laboratory, Cold Spring Harbor, N.Y.

Wilhelmsen, K. C., K. Eggleton, and H. M. Temin. 1984. Nucleic acid sequences of the oncogene v-*rel* in reticuloendotheliosis virus strain T and its cellular homolog, the proto-oncogene c-*rel*. *J. Virol.* **52**:172–182.

Wills, J. W., R. V. Shrinivas, and E. Hunter. 1984. Mutations of the Rous sarcoma virus *env* gene that affect the transport and subcellular location of the glycoprotein products. *J. Cell Biol.* **99**:2011–2023.

Witte, O. N., and D. Baltimore. 1978. Relationship of retrovirus polyprotein cleavages to virion maturation studied with temperature-sensitive murine leukemia virus mutants. *J. Virol.* **26**:750–761.

Yang, W. K., D. M. Yang, and J. O. Kiggans, Jr. 1980. Covalently closed circular DNAs of murine type C retroviruses: depressed formation in cells treated with cycloheximide early after infection. *J. Virol.* **36**:181–188.

Index